ADVANCES IN ATOMIC AND MOLECULAR PHYSICS AT THE INTERFACES WITH THE NATURAL SCIENCES, TECHNOLOGY AND MEDICINE

ADVANCES IN ATOMIC AND MOLECULAR PHYSICS AT THE INTERFACES WITH THE NATURAL SCIENCES, TECHNOLOGY AND MEDICINE

editor

Andrey V. Solov'yov

MBN Research Center, Frankfurt, Germany

World Scientific

'W JERSEY • LONDON • SINGAPORE • BEIJING • SHANGHAI • HONG KONG • TAIPEI • CHENNAI • TOKYO

Published by

World Scientific Publishing Europe Ltd.
57 Shelton Street, Covent Garden, London WC2H 9HE
Head office: 5 Toh Tuck Link, Singapore 596224
USA office: 27 Warren Street, Suite 401-402, Hackensack, NJ 07601

Library of Congress Cataloging-in-Publication Data
Names: Solov'yov, Andrey V. editor
Title: Advances in atomic and molecular physics at the interfaces with the natural sciences, technology and medicine / editor Andrey V. Solov'yov, MBN Research Center, Frankfurt, Germany.
Description: New Jersey : World Scientific, [2026] | Includes bibliographical references and index.
Identifiers: LCCN 2025032357 | ISBN 9781800618176 hardcover | ISBN 9781800618183 ebook for institutions | ISBN 9781800618190 ebook for individuals
Subjects: LCSH: Condensed matter | Molecules | Cosmochemistry | Multiscale modeling
Classification: LCC QC173.454 .A37 2026
LC record available at https://lccn.loc.gov/2025032357

British Library Cataloguing-in-Publication Data
A catalogue record for this book is available from the British Library.

For any available supplementary material, please visit
https://www.worldscientific.com/worldscibooks/10.1142/Q0534#t=suppl

Desk Editors: Murali Appadurai/Gabriel Rawlinson

Typeset by Stallion Press
Email: enquiries@stallionpress.com

https://doi.org/10.1142/9781800618183_fmatter

Preface

About the Book

This book is for anyone who has ever tried to answer the question: what are the frontiers of atomic, molecular, and optical physics and what lies beyond them?

Atomic and molecular physics is related to everything around us because everything in our immediate environment is made up of atoms and molecules. On the other hand, there are many more fascinating things in the world that cannot be thought of as merely a collection of atoms and molecules. How is it that, when matter grows from a single atom or molecule into a larger and larger system, a new reality emerges with all its specific features and properties? To what extent are the properties of larger systems determined by the properties of individual atoms and molecules, rather than by the properties of the whole system as it grows? At what scales do new properties appear in the system, and how do they influence other properties in the system? What are the laws of nature that govern this transition? Are they quantum or classical? And where and how do the quantum and classical worlds meet to form the reality in which we all live? This book does not provide the final answers to these questions, but rather tries to point the way to their elucidation and resolution by presenting the concrete solvable case studies from various fields that shed light on at least some of them.

These and many other such questions arise when we consider the transition of matter from a single atom to a larger system. The final system in this transition might be as large as a universe, but in this book, we limit our consideration to much smaller systems. Our focus is on systems of condensed matter of a size sufficient to manifest new properties of interest. Typically, this happens at system sizes of microns and above. However, there are many examples where new properties emerge in the system at the nanoscale and then determine system properties at all larger scales up to the macroscopic.

The atomic scale is not the smallest in the world. There is also the subatomic world, where nuclear interactions dominate. However, the discussion of the structure and properties of nuclei, nucleons, and quarks and the physics of elementary particles is beyond the scope of this book.

Atomic and molecular physics and its methods play an important role in each of the specific research areas discussed in this book. In many of the concrete case studies presented here, it is clear that an environment for each atom in the system under consideration plays an important role in determining its properties. This fact places atomic and molecular physics at the interface with many other natural sciences, such as chemistry, materials science, biology, and other life sciences, which study various types of matter, both inanimate and animate. The study of the properties of atoms and molecules in different environments is one of the central themes of this highly interdisciplinary field. This concerns not only the static properties of atoms and molecules, such as their electronic structure, but also their dynamic properties, such as the cross-sections of different collision processes or the rates of different chemical reactions.

The condensed matter systems discussed in this book are quite diverse in nature, composition, and complexity. They may be liquid or solid, organic, inorganic, or composite, inanimate or animate, and finite or infinite. Their number is vast – almost infinite. Despite their diversity, there are phenomena common to many, if not all, of them.

For example, the chemical transformations of condensed matter systems are determined by the properties of the interacting reactive atoms, their valences, and their interaction potentials, rather than by the properties of matter as a whole. The role of matter in this case is to provide the conditions under which chemical reactions take place, thus influencing their rates. Another example is the behaviour of condensed matter systems exposed to radiation. Despite the diversity of the condensed matter systems, the number of quantum processes that can be induced in such systems by radiation is very limited, and such processes occur in a similar way in very different systems. Again, the specificity of each particular system lies in the creation of specific conditions for the radiation-induced quantum processes to take place. Therefore, chemical and radiation-induced processes and phenomena are the main focus of this book due to their generality, universality, and importance. Furthermore, these processes are typically atomic and molecular in nature. They have been studied in the gas phase by the atomic, molecular, and optical physics communities for many decades. This provides a further link between atomic and molecular physics and other sciences to which certain condensed matter systems belong.

The study of various phenomena in biomolecular and biological systems is another focus of the book. This class of condensed matter systems and the atomic and molecular processes within them are very important because they form the link between atomic and molecular physics and the life sciences. Radiation-induced phenomena in biological systems, also known as radiobiological phenomena, are an excellent example of such an interface between the above-mentioned disciplines. These topics are discussed in several chapters of this book.

Many of the above problems are relevant to the emerging field of astrochemistry, which is concerned with the properties of atoms, molecules, and their condensates in space. Typical conditions in space are characterised by low temperatures, the presence of radiation, low densities of atomic and molecular species, and long time scales for the formation of condensates. Such conditions are difficult to reproduce in the laboratory but can be approximated to some extent. Due to the large amount of experimental data on molecular and condensed matter systems becoming available from satellite measurements in space, such related laboratory studies are becoming increasingly important. They aim to provide answers to many important fundamental questions, such as the origin of biological molecules in space that may have preceded the formation of living systems. This book provides an introduction to this area of modern research.

In recent years, methods originally developed to describe atomic and molecular systems and related phenomena have been applied to molecular and condensed matter systems of increasing size and complexity. This development has been driven by rapid advances in computational technologies and a significant increase in computing power. These developments are opening up new horizons for the application of traditional theoretical and computational methods to increasingly complex molecular systems and phenomena.

These advances have revealed the emergence of multiscale phenomena in complex molecular systems. A phenomenon is called 'multiscale' when it involves processes that occur at different temporal and spatial scales. The understanding of such phenomena should be based on the description of a cascade of processes occurring at all the temporal and spatial scales involved. Such a description is usually based on a multiscale approach that includes all the appropriate theoretical and computational methods relevant to each scale. The book provides an introduction to the techniques of multiscale modelling in modern research, such as reactive and irradiation-driven molecular dynamics and stochastic dynamics, and presents a number of concrete case studies and important advances based on their application.

It is also important to note that many of the research areas mentioned above are closely related to technological and medical applications. For example, studies of biological systems exposed to radiation and the multiscale phenomena of the radiation-induced biodamage are closely linked to the biomedical applications of radiation, such as radiotherapy. Radiation can also be used for the controlled fabrication of nanosystems with a resolution of only a few nanometres, or, in other words, for 3D nanoprinting. Self-assembly of nanosystems is considered to be the main process behind the bottom-up technological approach to the fabrication of nanosystems. All of these and many more connections from basic research to new and emerging technologies are discussed in the book.

Historical Milestones in the Development of the Research Area

The international interdisciplinary research community began to address the problems described above about 30 years ago. At that time, studies of atomic and molecular clusters and biological molecules in the gas phase were in their infancy. Clusters were seen as new physical objects with properties evolving from those of single atoms to those of the solid state. A similar vision existed for biological molecules, which could form larger molecular systems such as proteins and DNA from smaller molecular building blocks. Many of the problems mentioned above in this preface had not even been formulated. The study of collision processes with these new objects, along with the study of their structure and dynamics in such processes, was one of the main lines of research at that time. By the early 2000s, the research community had reached the necessary critical mass and the International Symposium on Atomic Cluster Collisions (ISACC) conference series was launched in 2003 in St. Petersburg, Russia. The conference photo from the ISACC 2023 is shown in Fig. 1.

To date, 11 ISACC conferences have been held[a] in Germany (twice), Russia (twice), the United States of America, China, Spain, Cuba, the United Kingdom, Italy, and – most recently–ISACC 2023 conference in Hveragerði, Iceland.[1–8] The next ISACC2025 will take place in Heidelberg, Germany in July 2025. Information about this event can be found on the web portal https://www.isacc-portal.org/.

Initially, ISACC was focused mainly on the dynamics of atomic clusters, especially in atomic cluster collisions.[1,2] Since then, its scope has been significantly broadened to include the dynamics of nanosystems, biomolecules, and macromolecules, with an emphasis on the similarity of clustering phenomena arising in different areas of physics, chemistry, and biology.[2,3]

Currently ISACC conferences provide a platform to discuss research, technological challenges, and related advances within the ISACC topic areas listed as follows:

- structure and dynamics of atomic clusters and nanoparticles (NPs);
- structure and dynamics of biomolecules;
- reactivity and nanocatalysis;
- clustering in systems of various dimensionality and degrees of complexity;
- photon, electron, and ion collisions with free atomic clusters and NPs;

[a]Information on all past and future ISACC conferences, including conference programmes and books of abstracts, can be found on a dedicated web portal of the ISACC Conference Series, https://www.isacc-portal.org/.

Fig. 1. Conference photo from the First International Symposium 'Atomic Cluster Collisions: fission, fusion, electron, ion and photon impact' (ISACC), 18–21 July 2003, St. Petersburg, Russia.

- radiation-induced phenomena with clusters and NPs deposited on surfaces and placed in a molecular environment;
- photon, electron, and ion collisions with biomolecules and molecular and biomolecular clusters;
- complex collision, radiative, and fragmentation processes;
- clusters and biomolecules in external fields, such as electric, magnetic, and laser fields;
- cluster and biomolecular research with free-electron lasers;
- related technological applications.

During the ISACC conferences, it became clear that there was a need for an interdisciplinary conference covering a wide range of topics related to the dynamics of nanoscale systems. It was therefore decided to start a new series of meetings in 2010 under the new title 'Dynamics of Systems on the Nanoscale', or DySoN, conference. This title better reflects the interdisciplinary nature of the topics that have emerged from the previous ISACC meetings and brings together all these topics of interest under a common theme.

The first international conference 'Dynamics of Systems on the Nanoscale' (DySoN 2010) was held in Rome, Italy, in November 2010. The conference photo from the DySoN 2010 is shown in Fig. 2. To date, eight DySoN conferences have been held[b] in Italy (twice), Russia, the United Kingdom, Germany (twice), the Czech Republic, and – most recently – the DySoN 2024 conference in Tbilisi, Georgia.[8–11]

Currently, DySoN conferences provide a platform to discuss research, technological challenges, and related advances within the DySoN topic areas listed as follows:

- structure and dynamics of molecules, clusters, and NPs;
- cluster and biomolecular ensembles and composite systems;
- clustering, self-organisation, and phase and morphological transitions on the nanoscale;
- nanostructured materials, surfaces, and interfaces;
- reactivity and nanocatalysis;

[b]Information on all past and future DySoN conferences, including conference programmes and books of abstracts, can be found on a dedicated web portal of the DySoN conference series, https://www.dyson-conference.org/.

Fig. 2. Conference photo from the First International Conference 'Dynamics of Systems on the Nanoscale' (DySoN 2010), 16–19 November 2010, Rome, Italy.

- electron and spin transport in molecular systems;
- collision and radiation processes, fusion, fission, and fragmentation;
- radiation-induced chemistry;
- irradiation-driven transformations, damage, and fabrication of MesoBioNano (MBN) systems;
- propagation of particles through media;
- biomedical and technological applications of radiation;
- related technologies, such as novel light sources, controlled nanofabrication, and functionalised materials.

The distinctive feature of the ISACC and DySoN conference series is that they present the most fundamental advances from the theoretical, computational, experimental, and technological perspectives in the research areas listed above.

In parallel with the ISACC and DySoN conference series, a number of other complementary conferences and workshops took place. Mention should be made of several COST Actions (RADAM, Nano-IBCT, and MultIChem) that have taken place since 2005 and are still ongoing. Each of these COST Actions has involved up to 350 researchers from 30–40 European countries, as well as a number of experts from around the world. They have organised conferences and several thematic workshops each year. These have all significantly contributed to the development of the research area described above. Several such events were organised in conjunction with the ISACC and DySoN conferences.

The COST Action entitled 'Radiation Damage in Biomolecular Systems' (RADAM) ran for four years from 2005 to 2008. It was dedicated to studies on the mechanism of radiation damage to biomolecules in the gas phase.

The subsequent COST Action 'Nanoscale Insights into Ion-Beam Cancer Therapy' (Nano-IBCT) ran from 2010 to 2014. It aimed to better understand the physical, chemical, and biological mechanisms at different time and space scales that lead to cell death under ion beam irradiation in order to realise the full potential of ion beam cancer therapy. This COST Action aimed to apply the multiscale approach proposed in Ref. 12 and exploit the unique experimental and theoretical expertise available in Europe to gain a better understanding of ion-beam-induced radiation damage at the nanoscopic and molecular levels underlying the ion beam cancer therapy.

Work in this area has been further supported through the FP7 project Initial Training Network (ITN) 'Advanced Radiotherapy, Generated by Exploiting Nanoprocesses and Technologies' (ARGENT, 2014–2018). The main objective of this intersectoral and multidisciplinary ITN was to create a new generation of researchers and experts capable of creating a platform on which the next-generation cancer therapy will be built. This work was conducted during 2019–2021 within the H2020-MSCA-IF project Radio-NP 'Computational Characterisation of Radiosensitising Nanoparticles and Their Properties' and several national projects across Europe. The Radio-NP project aimed at the atomistic computational analysis of (i) the structural properties of coated metal NPs in biological environments and (ii) the impact of these properties on the formation and transport of secondary electrons and radicals in the vicinity of NPs irradiated with ions.

The objective of the current COST Action 'Multiscale Irradiation and Chemistry Driven Processes and Related Technologies' (MultIChem, 2021–2025, https://www.mbnresearch.com/ca20129-multichem/main) is to establish a broad international, interdisciplinary, and cross-sectoral collaboration to advance the fundamental understanding of the multiscale irradiation-driven processes and related technologies, paving the way for major scientific and technological breakthroughs and socioeconomic impacts. The Action exploits irradiation-driven chemistry in selected applications and related technologies, such as methods for controlled nanofabrication, nanocatalysis, development of novel radiosensitising NPs, and radiotherapies, using advanced computational methods and modern experimental techniques.[13,14]

Work in the very broad, highly interdisciplinary, and extremely challenging area of research described above has also been supported by a large number of European projects under the European Commission's Framework Programmes FP6, FP7, Horizon 2020, and Horizon Europe. These have been focused on different problems within several research areas. Some of these projects have already been mentioned above. Others are briefly described in the following.

Materials science has been one of these areas. It is worth mentioning the Network of Excellence (NoE EXCELL, 2005–2010), which was set up under FP6 and aimed at overcoming the fragmentation of European research on multifunctional thin films. This network has been instrumental in establishing strong collaborations across Europe in this research area and has organised regular thematic meetings and conferences. A reflection of this collaborative work over the years is the fact that two teams from the former EXCELL NoE (the University of Cambridge team and the MBN Research Center team) have contributed several chapters to this book.

Another collaborative research project in materials science was VINAT. It was set up under FP7 and focused on the theoretical analysis, design, and virtual testing of the biocompatibility and mechanical properties of titanium-based materials. This project demonstrated the power of the computational modelling based on the molecular dynamics approach and the finite element method in materials science analysis and design, and it established strong collaborative links between groups in this field across Europe. The final conference of the VINAT project 'International Conference on Computational Modelling of Nanostructured Materials' was held in Frankfurt in September 2013. It was an important milestone in widening the use of advanced computational methods in materials science research.

The European Commission's Framework Programmes FP6, FP7, Horizon 2020, and Horizon Europe have supported a number of EU projects dedicated to the development and practical realisation of novel gamma-ray light sources operating at photon energies from 100 keV up to the GeV range. Such gamma-ray light sources can be constructed by exposing oriented crystals (linear, bent, and periodically bent) to the beams of ultra-relativistic charged particles. Here, we can only mention the titles and acronyms of these projects: FP6-NEST project 'Photon emission in crystalline undulators' (PECU), 2005–2009; FP7-PIRSES project 'Crystalline Undulator: Theory and Experiment' (CUTE), 2011–2015; H2020-MSCA-RISE 'Project Periodically Bent Crystals for Crystalline Undulators' (PEARL), 2016–2019; H2020-MSCA-RISE 'Project Novel Light Sources: Theory and Experiment' (N-LIGHT), 2020–2025; and the EIC-PATHFINDER project 'Emerging Technologies for Crystal-based Gamma-ray Light Sources' (TECHNO-CLS), 2022–2027. These projects are strongly focused on the development of the fundamental principles and the necessary technologies for the practical realisation of the novel gamma-ray light sources. It is important that these developments are successfully driven by theory and advanced multiscale computational modelling.

National projects in the field are not mentioned individually, but there many relevant ones during the past three decades that have contributed to the field.

The MesoBioNano Science Forum

The interdisciplinary research area presented above is called 'MesoBioNano science'.[1,2,13,15,16] It is the interdisciplinary research area that studies the structure formation and dynamics of animate and inanimate matter at the nano and meso scales up to the macro scales. Any form of condensed matter, including biological matter, is composed of many components linked by different interactions. Important efforts to deepen the molecular-level understanding of different forms of condensed matter and their dynamical behaviour concern the origin, nature, and evolution of different complex molecular systems, as well as the emergence of new features, properties, processes, and functions associated with the systems as they increase in size and complexity. Research in this area brings together many traditional topics and methods in theoretical physics and chemistry with problems in the life sciences and materials research under a common theme. The range of open and challenging scientific problems in this area is quite broad. The list of challenging topics in this area is growing rapidly, which also facilitates the development of relevant theoretical and computational methods.

The MBN science has many links to atomic and molecular physics. In particular, it involves theoretical and computational methods, such as many-body theory, density functional theory, and collision theory, as well as many algorithms that were originally developed for the simulation of atomic and molecular properties and later generalised to larger-scale simulations. Similarly, experimental methods, which have moved more and more from gas-phase experiments with single atoms and simple molecules to experiments on increasingly sophisticated molecular and condensed matter systems while maintaining atomic or at least nanoscopic resolution.

The theoretical and computational developments in MBN science are of great importance to the entire research community, leading to new ideas and predictions, revealing new phenomena, and opening up new research horizons. This can be seen from the development of the field over the past decades, where most of the projects and initiatives described above have been generated and coordinated by the team at the MBN Research Center, although most of the partner teams in these European projects were experimental.

The MBN Research Center (https://www.mbnresearch.com/, MBN RC) conducts an extensive programme of cutting-edge research. This is largely based on the use of the Centers's own state-of-the-art, universal, and powerful software packages, MBN Explorer[17] and MBN Studio.[18] MBN Explorer and MBN Studio are designed for advanced scientific research and computational modelling of the molecular structure and dynamics of complex MBN systems. These codes are under continuous development by the MBN RC team, who is committed to their wide dissemination and use. The MBN RC team, with a large number of partners worldwide, participates in collaborative research projects and organises thematic conferences.

Recently, the MBN RC team took the initiative of producing a roadmap paper entitled 'Condensed matter systems exposed to radiation: Multiscale theory, simulations, and experiment', together with a large group of co-authors representing the ongoing COST Action MultIChem described above. This was published in *Chemical Reviews* in 2024.[14] This roadmap reviewed the new, highly interdisciplinary field of research on the behaviour of condensed matter systems exposed to radiation. It highlighted recent advances in the field and provided a roadmap for the development of the field over the next decade.

The MBN RC team has long been the main driving force behind the organisation of the ISACC and DySoN conference series. It has become increasingly clear that the MBN science research area is becoming too broad even for the well-established conference series such as DySoN. Therefore, it has been decided that, from July 2026, the MBN Science Forum will cover the entire relevant research area described above. However, each forum event will focus on specific topics. For example, separate MBN Science Forums are planned for the ISACC, DySoN, RADAM and Nano-IBCT, MultIChem, TECHNO-CLS, and materials science research communities. This is not an exclusive list of topics. Other relevant research areas may be added to this list. Some forum events may be organised online, e.g., in the form of webinars, lectures, and practical tutorials.

These plans imply that this book is the first in a series of similar publications that will reflect developments in this fascinating area of research.

Overview of Book Chapters

The book is intended for master's and PhD students and specialists in the fields of physics, chemistry, biology, and medicine related to atomic and molecular physics, the physics and chemistry of condensed matter systems, multiscale phenomena in condensed matter systems, materials science, the interaction of radiation with condensed matter physics, astrochemistry, radiation nanophysics, biophysics, the physics of interaction of charged particles with matter and hadron or ion beam therapy, nanofabrication techniques, and multiscale computational modelling. The systematic introduction of many of the above-mentioned research areas within the interdisciplinary MBN research area is one of the important features and benefits of the book.

The chapter *Multiscale Computational Modelling in Modern Research* by A. V. Solov'yov provides an introduction to multiscale computational modelling in modern research. Multiscale computational modelling of complex condensed matter systems, sometimes referred to as MBN systems, is a recently emerging area of modern research. It has become one of the most popular research directions in recent years due to the rapid worldwide development of modern computer technology, computing power, and computational methods. It is now being applied in virtually all the traditional natural sciences, such as physics, chemistry, biology, and materials science, as well as in new interdisciplinary research areas at various interfaces between these fields. The focus of the chapter is on the behaviour of MBN systems exposed to radiation. Many of the key phenomena associated with the behaviour of irradiated systems are very similar and can be understood using the same theoretical principles and computational approaches. The multiscale nature of such phenomena requires the quantitative description of the radiation-induced effects occurring at different spatial and temporal scales, ranging from the atomic to the macroscopic. The practical computational implementation of multiscale modelling based on the advanced computational software MBN Explorer and MBN Studio is presented in this chapter, and examples of concrete case studies are briefly discussed.

The chapter *Interlinks between Atomic and Molecular Physics, Nanoscience, and Other Disciplines* by A. V. Verkhovtsev and A. V. Solov'yov discusses the application of well-established methods from atomic and molecular physics to the description of the electronic properties of nanosystems, such as atomic clusters and pristine and endohedral carbon fullerenes. Emphasis is placed on the description of the physical phenomena that manifest themselves in photon, electron, and ion collisions with these systems and involve the dynamics of electrons. Specific examples discussed in this chapter include (i) the diffraction and interference phenomena that occur during the interaction of nanosystems with electrons, (ii) the role of the multipole surface and volume plasmon excitations in the formation of electron emission spectra of metal clusters and fullerenes, (iii) the relaxation mechanisms of resonant electronic excitations in metal clusters and fullerenes, which are discussed in the context of evaluating the widths of the corresponding resonant processes, and (iv) the fundamental physical phenomena associated with the confinement of atoms and small molecules inside carbon fullerenes, which represent a nanometre-sized cavity.

The chapter *Magnetic Superatoms* by A. C. Reber and S. N. Khanna is focused on the progression of magnetic behaviours as the dimensions of the material are reduced to clusters containing a few atoms to a few hundred atoms and how novel classes of magnetic species could be created, offering potential pathways for designing cluster-assembled nanomaterials. The chapter describes the evolution of magnetic moments and their dynamical behaviour for itinerant ferromagnetic solids, such as Fe, Co, and Ni. The bulk of the chapter is devoted to a new area called magnetic superatoms. The quantum states in small compact metal clusters have orbital shapes resembling those in atoms, and the states are bunched into electronic shells, suggesting the classification of stable clusters as superatoms.

The chapter *An Introduction to Plasmonic Chemistry* by I. Bald provides an introduction to the physics and chemistry of plasmonic nanoparticles, such as gold and silver nanoparticles, which can efficiently interact with visible light via the excitation of the localised surface plasmon resonance (LSPR), which represents the collective oscillation of free electrons in the metal. This offers interesting possibilities for channelling the energy of light into chemical energy. This chapter introduces into the main concepts of the emerging field of plasmonic chemistry, starting with the optical properties of metals and then discussing LSPR, different interaction mechanisms with molecules, and selected examples of chemical systems studied in plasmonic chemistry.

The chapter *Multiscale Modelling and Characterisation of Nanosystems and Novel Materials* A. V. Verkhovtsev and A. V. Solov'yov presents some illustrative examples of computational atomistic and multiscale modelling of novel materials and nanosystems using the advanced software packages MBN Explorer and MBN Studio. This chapter presents two illustrative examples of the computational studies of such systems. One example concerns the characterisation of functionalised metal nanoparticles placed in molecular environments, which have been widely studied for possible applications in biotechnology. Another example concerns novel two-dimensional (2D) materials, which are currently being explored for their versatile electrical, optical, physical, chemical, and mechanical properties. An illustrative case study discussed in this chapter is a novel 2D material $NiCl_2$, which may be useful for catalysis and hydrogen storage applications. This chapter provides an overview of a computational characterisation of such systems using a multiscale modelling approach combining density functional theory calculations and classical molecular dynamics simulations.

The chapter *From Atomic Design to Cost-Efficient Solutions: Mastering Pt-Based Nanodendrite Formation for a Net-Zero Future* by S. Ming, N. Sammy, and A. E. H. Wheatley focuses on recent advances in the design of Pt-based bimetallic catalysts, where branched morphologies offer high surface-area-to-volume ratios and controllably intermixing Pt with a second metal offers enhanced activity, reduced expense, and minimisation of the kinetic drawbacks seen with monometallics. The ability of modelling to improve our understanding of the formation and behaviour of these systems is explored, as is the potential of predictable nanodendrite formation to shape the preparation of even more complex multimetallic catalysts.

The chapter *Reactive and Irradiation-Driven Molecular Dynamics with MBN Explorer* by A. V. Verkhovtsev and A. V. Solov'yov provides an overview of reactive molecular dynamics (RMD) and irradiation-driven molecular dynamics (IDMD) – advanced computational methods implemented in the MBN Explorer software package – which allow atomistic simulations of chemical and irradiation-driven transformations in a wide variety of MBN systems. Within the RMD and IDMD frameworks, quantum processes occurring in a system (e.g., bond cleavage and formation) are treated as random, fast, and local transformations that are stochastically incorporated into the classical MD framework, with probabilities elaborated on the basis of quantum mechanics. Major transformations of molecular systems (such as changes in molecular topology, redistribution of atomic partial charges, or changes in interatomic interactions) are simulated using the reactive CHARMM (rCHARMM) force field implemented in MBN Explorer. This chapter presents several exemplary case studies, illustrating the use of RMD and IDMD and their validation by comparison with experiments. Particular examples include simulations of the irradiation-induced fragmentation of organometallic and biomolecular systems and the formation of metal-containing nanostructures during focused electron beam induced deposition (FEBID).

The chapter *Molecular Mechanisms of Radiation Action on Peptides, Proteins, DNA Strands, and Antibiotics Studied in the Gas Phase* by T. Schlathölter and J.-C. Poully provides a review of experimental investigations that focus on three different classes of biomolecular systems, i.e., peptides and proteins, DNA strands, and antibiotics. Molecular processes such as ionisation, fragmentation, and H transfer are discussed. The chapter closes with a brief description of current experimental developments towards conformational control through ion mobility spectrometry, electron spectroscopy in crossed-beam experiments, and time-resolved measurements, e.g., using ultrashort pulses from free electron lasers.

The chapter *Radiation Damage of Biomolecular Systems: A Multiscale Approach* by A. V. Verkhovtsev and A. V. Solov'yov provides a comprehensive review of the multiscale approach (MSA) to the physics of radiation damage with ions and its application in ion beam cancer therapy. The assessment of radiation damage is based on a series of effects that occur on different time, space, and energy scales, starting with ion entering the tissue and ending with the analysis of the irreparable DNA damage events that lead to cell inactivation. The MSA allows the prediction of survival probabilities for cells irradiated with ions. These probabilities depend on the cell type, type of ions, the fluence of the beam, the depth in the tissue, the initial energy of ions, the concentration of oxygen in the tissue, and other external conditions. According to the MSA, the scenario of radiation damage with ions includes the ion-induced shock waves. Predictions of the calculated quantities are compared with different sets of experiments, including those involving different cell lines. The chapter shows applications of MSA to the analysis of the dependence of cell survival probability on depth in the case of spread-out Bragg peak and the dependence of the relative biological effectiveness on the linear energy transfer, particularly the overkill effect.

The chapter *Inorganic Nanoparticles and Ionising Radiation* by E. Brun and C. Sicard-Roselli gathers useful information on the subject from different scientific communities and highlights some specific points. First, it identifies the therapeutic contexts in which inorganic nanoparticles might interact with ionising radiation. It then presents the latest findings on the physicochemical mechanisms underlying nano-radiosensitisation. It also discusses the possible fate of the nanoparticles on their way from the chemist's bench to the patient and how nanoparticles can be modified by irradiation, as well as by their new cellular environment. Finally, some recent trends in the field are identified, including the latest discoveries on the biological responses that can be induced by nanomaterials in combination with radiation.

The chapter *Radiation-Induced DNA Damage and Repair in a Multiscale Chromatin Landscape: Past Insights, Current Understanding, and Future Complexities* by M. Falk, M. Schäfer, J. Weidner, I. Falk, J. Toufar, L. Dobesova, A. Bačíková, O. Kopečná, G. Hildenbrand, and M. Hausmann delves into the intricate relationship among ionising radiation, chromatin architecture, and the cellular response to DNA damage. With the advent of hadron radiotherapy and advancements in space research, understanding how the physical properties of ionising radiation, particularly high linear energy transfer (LET) heavy ions, affect biological systems has become increasingly critical. The chapter emphasises the role of chromatin architecture in the induction and repair of radiation-induced DNA damage, specifically double-strand breaks (DSBs). It explores how the chromatin environment at both local and global levels influences the damage response, which ultimately dictates the repair mechanism selected at each damage site. The use of advanced microscopy techniques, including super-resolution microscopy (SMLM), is highlighted as indispensable for studying these processes at the nanoscale. This chapter presents key findings that challenge existing paradigms in radiobiology and offers new insights into DNA repair mechanisms, chromosomal aberrations, and cellular radioresistance in the context of ionising radiation.

The chapter *Foundations of Astrochemistry I: Observations and Ion- and Electron-Induced Processes in Ices* by N. J. Mason, S. Ioppolo, H. M. Quitián-Lara, F. Fantuzzi, D. V. Mifsud, and J. Bocková discusses basics of astrochemistry – one of the newest but also most rapidly growing interdisciplinary fields of modern astronomy and space science. Built upon the foundations of atomic and molecular physics, astrochemistry relies on spectroscopy to observe the Universe and on collisional and radiation-induced processes to understand the evolution pathways of molecular synthesis, planet formation, and planetary atmospheres. This chapter provides an overview of the critical role atomic and molecular processes play in astrochemistry, emphasising the significance of both gas and condensed phases. The latter is particularly vital for the study of dust grains in the interstellar medium, where they frequently serve as sites for molecular synthesis. The chapter also discusses the essential contribution of experimental research and the considerable challenges involved in replicating space environments. As we enter a new era of astronomy and space science, a comprehensive understanding of chemical processes, alongside physical phenomena, is indispensable for addressing some of the most profound questions in modern science – questions that demand an even deeper insight into atomic and molecular physics.

The chapter *Foundations of Astrochemistry II: Photon- and Radical-Induced Processes in Ices, Computational Approaches, and Biomolecular Chirality* by N. J. Mason, S. Ioppolo, H. M. Quitián-Lara, F. Fantuzzi, D. V. Mifsud, and J. Bocková discusses key atomic and molecular processes shaping modern astrochemistry, with a focus on the role of spectroscopy in interpreting astronomical data. Photon-induced processes – absorption, ionisation, and dissociation – are explored for their impact on the chemical evolution of interstellar environments. The chapter discusses experimental investigations utilising synchrotron radiation facilities to study the chemistry of interstellar ices, dust grains, and molecular clouds.

Dedication to Professor Jean-Patrick Connerade

A special session in honour of the 80th birthday of Professor Jean-Patrick Connerade from Imperial College, London, was organised at the Seventh DySoN Conference, which took place at the Vila Lanna in Prague, the Czech Republic, on 24–26 April 2023. The conference photo of this event is shown in Fig. 3. The conference was followed by the Second Conference of the COST Action MultIChem held on 26–28 April 2023, as mentioned in the second section. The topics of the two conferences were closely linked to each other, so it was effectively a single scientific event spanning the whole week. The memorable week of these conferences in Prague in 2023 is directly linked to this book, as its content was thoroughly discussed and planned during that meeting. Much of the material described in the chapters of this book was presented at the DySoN 2023 and MiltIChem 2023 conferences.

The special session in honour of Professor J.-P. Connerade was organised in recognition of his prominent role in the establishment of both the ISACC and DySoN conference series. He was one of the pioneers in the field of atomic cluster collisions and sought to apply the methods of atomic and molecular physics to the study of the properties of atomic and molecular clusters.

The session was chaired by Professor Nigel J. Mason from the University of Kent, Canterbury, UK, and Professor Andrey V. Solov'yov of the MBN Research Center, Frankfurt am Main Germany, who began collaborating with J.-P. Connerade in the early stages of their research careers in the 1990s and met through this work. This collaboration has continued ever since, as evidenced by a number of joint publications cited in the previous sections.

After the announcement was made, there were a large number of people who wanted to speak at this session. It was not possible

Fig. 3. Conference photo from the Seventh International Conference 'Dynamics of Systems on the Nanoscale' (DySoN 2023), 24–26 April 2023, Vila Lanna, Prague, the Czech Republic.

to give everyone a chance. After much discussion, negotiation, and compromise, the following speakers and titles were selected for this session:

- **Mike Mansfield**, Physics Department, University College Cork, Ireland.
  Atomic Giant Resonances: Experiments in Frascati and in Bonn.
- **Aslam Baig**, National Centre for Physics, Quaid-i-Azam University, Islamabad, Pakistan.
  Work at Imperial College, Bonn University, and Islamabad.
- **George Philip**, India.
  The Laser Laboratory at the University of Kuwait.
- **Jon Marangos**, Physics Department, Imperial College London, United Kingdom.
  Dynamics of Molecular Photoionisation.
- **Jean-Claude Lehmann**, Kastler-Brossel Laboratory, Ecole Normale Supérieure, Paris, France,
  and
  Michel Broyer, Université Claude Bernard, Lyon, France.
  An 'English' physicist in France. Molecules, Superatoms, Sychrotron Radiation,... Friendship and Poetry.
- **Chris Mayhew**, Institut für Atomgasanalytik, Universität Innsbruck, Austria.
  J.-P. Connerade: 'The Atomic Physicist or There and Back Again' (with due credit to J. R. R. Tolkien).
- **Stephen Hogan**, Department of Physics and Astronomy, University College London, United Kingdom.
  Rydberg Atoms in Crossed Electric and Magnetic Fields: From Quantum Chaos to Geometric Phase.

The chairmen also spoke to Jean-Patrick. All other participants in the session had the opportunity to say a word to him at the end of the session during the open discussion session. All the speakers at this session shared their beautiful memories of the different periods of their lives, when they were involved in memorable endeavours with Jean-Patrick at different times of his career. These included, in particular, various aspects of his work, as well as many of the other shining moments of his life and of his personality. The following is a brief overview of some of the Jean-Patrick's most important contributions to research areas closely related to the topics discussed in several chapters of the present book.

Jean-Patrick contributed to the development of atomic physics using synchrotron radiation by introducing windowless furnace systems, which opened up the study of inner-shell excitation in metals, leading to the first observation of an atomic plasmon or 'Giant Resonance' in a free atom. This was significant because Tatiana Zimkina from St. Petersburg (Leningrad at that time) had observed such structures in metallic solids and resisted attributing them to solid-state effects. Indeed, she was the first to point out that they are completely atomic in origin and survive into different phases because of the geometry of the excitation. Her interpretation was confirmed experimentally by J.-P. Connerade and M. W. D. Mansfield (1974).[19] This result led to the broad subject of 'Giant Resonances' in atoms, solids, molecules,[20] and clusters,[21] studied both experimentally and theoretically as manifestations of collective phenomena of atomic origin using both many-body perturbation theory (developed in the US) and random phase approximation with exchange (pioneered in St. Petersburg). The connection between the 'atomic plasmon' and a classical Mie resonance was demonstrated and extended to clusters in a study by J.-P. Connerade and C. Bréchignac.[22]

Another field pursued by Jean-Patrick in the context of inner-shell excitations was orbital collapse in transition metals and lanthanides. This led to the discovery of instabilities of the atomic structure[23] and hence to the notion of a 'soft' chemistry in which, rather than molecular bonds being formed or broken being the driving mechanism, valence changes could be brought about by pressure effects, and phenomena such as lithiation in transition metals and rare-earths could arise merely by polaronic distortion in solids, thereby avoiding discontinuities in energy due to phase transitions. This provides an inherently reversible mechanism for the storage of lithium in the so-called 'rocking- chair' battery electrodes,[24] which have important applications.[25] Changes in atomic structure of rare-earths and transition metals in different phases[26] can also lead to sudden changes in valence. The phenomenon is related to orbital collapse or expansion in 'soft' effective potentials (inclusive of centrifugal barrier effects) being the root cause of all the anomalies in shell filling for elements in the long periods of the Periodic Table. J.-P. Connerade therefore suggested that orbital collapse might be controlled[27] by altering the mean field inside the atom concerned or by modifying its immediate environment, which led to the idea of a different sequence of filling of shells for the Periodic Table under pressure,[28] or atoms confined in cavities or shells.[29,30] It turned out that, since the atoms concerned lie among the 'heavy' elements of the Periodic Table, non-relativistic atomic theory is insufficient[31] to describe the sensitivities of atomic orbital collapse or make predictions about the distribution of electronic charge when it occurs. J.-P. Connerade therefore initiated studies of confinement for endohedrally trapped heavy atoms (initially Cs) using a fully relativistic model.[32] He also pointed out some shortcomings of relativistic theory in the context of atomic physics, particularly in regard to the nature of time.[33]

The endohedral confinement of atoms inside fullerene shells (the properties of metallofullerenes) became a widely studied theme of research.[34] Together with his colleagues V. K. Dolmatov and S. T. Manson, J.-P. Connerade pointed out the quantum origin of previously unknown confinement resonances due to the small size of the confining cavity.[35] These had previously been omitted from the theory, and their existence has since received preliminary experimental confirmation. Together with A. V. Solov'yov and other colleagues from the Ioffe Institute in St. Petersburg, J.-P. Connerade considered the case of an atom confined in distorted sphere or spheroidal cavities and showed that this could be modelled using some quasi-molecular parameters.[36] Again, together with A. V. Solov'yov and collaborators, mainly from Russia, J.-P. Connerade participated in several studies of the structure and properties of confined atoms and clusters (including the theory of multiphoton excitation for such systems[37]) and considered the interactions between resonances originating from different parts of such composite or hybrid systems. This included a study of the quantum analogue of the Faraday cage effect[38] for an atom confined in a conducting spherical shell and interacting with the plasmon of the shell. It was demonstrated that screening at the quantum level can become resonant at certain frequencies, leading to enhancement rather than attenuation in the range of the resulting 'Giant Resonance'. This result has implications for a proposal to construct a register for a quantum computer using an array of confined atoms. There were further themes of collaboration involving our different groups, which extended both to polarisational effects due to electrons and positrons[38–40] and to conference proceedings[1,2] and monographs.[41]

For many of his colleagues, J.-P. Connerade is more than just a colleague. His outgoing personality, his openness, his vision, his ability to think outside the box, and his friendly nature have attracted many people to him. From my own experience, I can say that our professional relationship with him has long since developed into a friendship. It is a pleasure to remember now the exciting and interesting times we spent with Jean-Patrick in Russia, in the UK, in Germany, and in many other places (Rome and Paris, Ann Arbor in the US, at the ISACC 2013 conference on the Yangtze River in China, etc.). Often, these meetings have taken place at the ISACC and DySoN conferences or in the context of joint research projects, but they have always been much more than mere joint research work. The topics discussed have been much broader, ranging from poetry,

Fig. 4. Chaunes (Jean-Patrick Connerade) receiving the Grand Prix de Poésie 2024 (Gold Medal) at the awards ceremony in Paris on 23 November 2024.

Fig. 5. Grand Prix de Poésie 2024: Gold Medal.

literature, art, history, culture, and ethics to politics, news, family, and everyday life.

Last but not least, another outstanding achievement of J.-P. Connerade deserves to be mentioned here. He proved that one can be both a scientist and a poet. J.-P. Connerade's pen name is Chaunes. His work in modern French poetry is widely recognised, and he has already won many prizes for it. He was recently awarded Grand Prix de Poésie 2024 by the Academy of French Poetry for his poetic work. We present photographs of J.-P. Connerade taken at the award ceremony held in his honour, as well as the gold medal he was awarded on that occasion (Figs. 4 and 5).

As the editor of this book, it is my pleasure to dedicate it to my dear colleague and friend, Professor J.-P. Connerade, for his outstanding and inspiring contributions to science and the humanities and to wish him continued new horizons in his endeavours and all the best in his life.

Andrey V. Solov'yov
MBN Research Center gGmbH
Frankfurt am Main, 31 December 2024

References

1. Solov'yov AV, Connerade J-P, editors. Latest advances in atomic cluster collisions: Fission, fusion, electron, ion and photon impact. London: Imperial College Press. 2004.

2. Connerade J-P, Solov'yov AV, editors. Latest advances in atomic cluster collisions: Structure and dynamics from the nuclear to the biological scale. London: Imperial College Press. 2008.
3. Solov'yov AV, Surdutovich E, editors. The fourth international symposium "Atomic cluster collisions: Structure and dynamics from the nuclear to the biological scale" (ISACC 2009). AIP Conf Proc. 2009;1197:1–227.
4. Korol AV, Solov'yov AV. Editorial: Atomic cluster collisions. Eur Phys J D. 2013;67:1.
5. Prosmiti R, *et al.* Editorial: Atomic cluster collisions: ISACC-2015 (7th International Symposium). Eur Phys J D. 2017;71:36.
6. Verkhovtsev A, *et al.* Editorial: Topical issue on atomic cluster collisions. Eur Phys J D. 2019;73:154.
7. Verkhovtsev A, *et al.* editors. Eur Phys J D. 2019.
8. Verkhovtsev AV, *et al.* Editorial: Topical issue "Dynamics of systems on the nanoscale (2021)". Eur Phys J D. 2023;77:83.
9. Solov'yov AV. Preface: International Conference on Dynamics of Systems on the Nanoscale (DySoN 2012). J Phys Conf Ser. 2013;438: 011001.
10. Korol AV, Solov'yov AV. Editorial: Dynamics of systems on the nanoscale. Eur Phys J D. 2017;71:339.
11. Bald I, *et al.* Editorial: Special issue: Dynamics of systems on the nanoscale (2018). Eur Phys J D. 2020;74:75.
12. Solov'yov AV, *et al.* Physics of ion beam cancer therapy: A multiscale approach. Phys Rev E. 2009;79:011909.
13. Solov'yov IA, Korol AV, Solov'yov AV. Multiscale modeling of complex molecular structure and dynamics with MBN Explorer. Cham: Springer International Publishing. 2017.
14. Solov'yov AV, *et al.* Condensed matter systems exposed to radiation: Multiscale theory, simulations, and experiment. Chem Rev. 2024;124:8014–8129.
15. Solov'yov AV, editor. Nanoscale insights into ion-beam cancer therapy. Cham: Springer International Publishing. 2017.
16. Solov'yov IA, *et al.* Dynamics of systems on the nanoscale. Cham: Springer Nature Switzerland. 2022.
17. Solov'yov IA, *et al.* MesoBioNano Explorer – A universal program for multiscale computer simulations of complex molecular structure and dynamics. J Comput Chem. 2012;33:2412–2439.
18. Sushko GB, Solov'yov IA, Solov'yov AV. Modeling MesoBioNano systems with MBN Studio made easy. J Mol Graph Model. 2019;88: 247–260.

19. Connerade J-P, Mansfield MWD. Observation of an 'atomic plasmon' resonance in the Ba I absorption spectrum between 10 and 200 Å. Proc Roy Soc A. 1974;341:267.
20. Connerade J-P, Esteva JM, Karnatak RC, editors. Giant resonances in atoms, molecules and solids. New York: Plenum Press. 1987.
21. Connerade J-P, Solov'yov AV. Giant resonances in photon emission spectra of metal clusters. J Phys B At Mol Opt Phys. 1996;29:3529–3547.
22. Bréchignac C, Connerade J-P. Giant resonances in free atoms and in clusters. J Phys B At Mol Opt Phys. 1994;27:3795.
23. Connerade J-P. Atomic spectra and instabilities of valence. J Alloys Compd. 1995;227:145.
24. Connerade J-P. Soft chemistry and reversible ion storage: The underlying atomic physics. J Alloys Compd. 1997;255:79–90.
25. Connerade J-P, Olivier-Fourcade J, Jumas J-C. Pressure-sensitive modelling of the reversible lithiation of vacant thiospinel $(In_{0.5}\check{Z}_{0.5})[In_{1.5}Sn_{0.5}]S_4$. J Solid State Chem. 2000;152:533.
26. Connerade J-P. The connection between intermediate valence and atomic wavefunction collapse. J Phys C Solid State Phys. 1982;15:L367.
27. Connerade J-P. Controlling the collapse of atomic wavefunctions. J Phys B At Mol Phys. 1978;11:L381.
28. Connerade J-P, *et al.* The filling of shells in compressed atoms. J Phys B At Mol Opt Phys. 2000;33:251–264.
29. Connerade J-P. Confining and compressing the atom. Eur Phys J D. 2020;74:211.
30. Dolmatov VK, *et al.* Structure and photoionization of confined atoms. Radiat Phys Chem. 2004;70:417–433.
31. Connerade J-P, Semaoune R. Atomic compressibility and reversible insertion of atoms into solids. J Phys B At Mol Opt Phys. 2000;33:3467–3484.
32. Connerade J-P, Semaoune R. Relativistic study of the electronic structure and 5d orbital of La confined inside a C_{60} fullerene cage. J Phys B At Mol Opt Phys. 2000;33:869–880.
33. Connerade J-P. The atom at the heart of physics. Atoms. 2023;11:32.
34. Connerade J-P, *et al.* A unique situation for an endohedral metallofullerene. J Phys B At Mol Opt Phys. 1999;32:L395–L403.
35. Connerade J-P, *et al.* On the nature and origin of confinement resonances. J Phys B At Mol Opt Phys. 2000;33:2279–2285.
36. Connerade J-P, *et al.* A simple atomic model for hydrogen confined inside a prolate-shaped C_{60} fullerene cage. J Phys B At Mol Opt Phys. 2001;34:2505.

37. Connerade J-P, Solov'yov AV. Formalism for multiphoton plasmon excitation in jellium clusters. Phys Rev A. 2002;66:013207.
38. Connerade J-P, Solov'yov AV. Dynamical screening of a confined atom by a fullerene. J Phys B At Mol Opt Phys. 2005;38:807.
39. Connerade J-P, *et al.* Polarization effects in electron attachment to metallic clusters. J Phys B At Mol Opt Phys. 1999;32:877–894.
40. Connerade J-P. Positronic lithium and the many-body problem. Nature. 1998;391:439.
41. Connerade J-P. Highly excited atoms. Cambridge: Cambridge University Press. 1998.

https://doi.org/10.1142/9781800618183_fmatter

About the Editor

A. V. Solov'yov graduated from Peter the Great St. Petersburg Polytechnic University and received his PhD in theoretical and mathematical physics from the A. F. Ioffe Physical-Technical Institute of the Russian Academy of Sciences in 1988. Since 2014, he has served as the scientific and executive director of the MBN Research Center in Frankfurt am Main, Germany. In 2016, he was awarded the academic title of Professor of Theoretical Physics by the Peter the Great St. Petersburg Polytechnic University. His primary research areas include theoretical atomic, molecular, chemical, and condensed matter physics, computational physics, MesoBioNano (MBN) science, and physics at the interface of the natural sciences, technology, and medicine. He has led numerous international and national research projects that have received funding from the European Commission's Framework Programmes, as well as from many other renowned science funding agencies in Germany, the UK, and Russia. He has published over 450 journal articles and book chapters, as well as 12 books, covering a wide range of topics within his field of research. He has also pioneered many interdisciplinary research directions. He has edited 6 books and 16 topical issues of journals devoted to various interdisciplinary topics.

https://doi.org/10.1142/9781800618183_fmatter

Contents

https://doi.org/10.1142/9781800618183_0001

Chapter 1

Multiscale Computational Modelling in Modern Research

Andrey V. Solov'yov
MBN Research Center,
Altenhöferallee 3, 60438 Frankfurt am Main, Germany
solovyov@mbnresearch.com

This chapter provides an introduction to multiscale computational modelling in modern research. Multiscale computational modelling of complex condensed matter systems, sometimes referred to as MesoBioNano (MBN) systems, is an emerging area of modern research. It has become one of the most popular research directions in recent years due to the rapid worldwide development of modern computer technology, computing power, and computational methods. It is now being applied in virtually all the traditional natural sciences, such as physics, chemistry, biology, and materials science, as well as in new interdisciplinary research areas at various interfaces among these fields. The focus of this chapter is on the behaviour of MBN systems exposed to radiation. These systems may be inorganic, organic, or biological; finite or infinite; composed of different molecular species or materials; exist in different phases; and operate under different thermodynamic conditions. Many of the key phenomena associated with the behaviour of irradiated systems are very similar and can be understood using the same theoretical principles and computational approaches. The multiscale nature of such phenomena requires the quantitative description of the radiation-induced effects occurring at different spatial and temporal scales, ranging from the atomic to the macroscopic. The multiscale nature of the effects and the similarity of their manifestation in systems of different origins necessarily bring all the above disciplines together, revealing numerous links and commonalities among them. This area of research is highly relevant to many new and emerging technologies and medical applications. Examples of concrete case studies will be presented and briefly discussed. The practical computational implementation of multiscale modelling based on the advanced computational software MBN Explorer and MBN Studio will also be presented, and a number of remaining challenges in the field will be highlighted.

1.1 Introduction

Any form of condensed matter, including biological matter, consists of many different components linked by many different interactions. The origin, nature, and evolution of different complex molecular systems, as well as the emergence of new features, properties, processes, and behaviours of these systems as they grow in size, are important aspects of our understanding of different forms of condensed matter and their dynamical behaviour. At the meso- and nanoscale, the physics and chemistry of biological and biomolecular systems, nanosystems, and materials are typically concerned with such behaviour. There are many examples of the emergence of qualitatively new features, such as the development of new collective properties in the transition from small molecules to large clusters and the aggregation of clusters on surfaces leading to the appearance of fractal morphologies. The fractal morphologies that emerge in nanoscale dynamical systems remain characteristic of many systems, including biological systems, at virtually all larger scales and are present in virtually all living systems.

There are many examples of complex molecular systems at the nano- and larger scales that exhibit unique features, properties, and functions. Figure 1.1 illustrates selected examples of these systems.

This field of research is sometimes referred to as MesoBioNano (MBN) science.[1] It studies the structure formation and dynamics of animate and inanimate matter at the nano- and mesoscale. In strict terms, the mesoscale is difficult to define because it is not an exact size range, as there is no 'mesometre'. Often mesoscopic systems are considered to be larger than atomic, molecular and nanosystems but smaller than a bulk and therefore of finite size. This research area brings together several traditional topics in theoretical physics under a common theme. The range

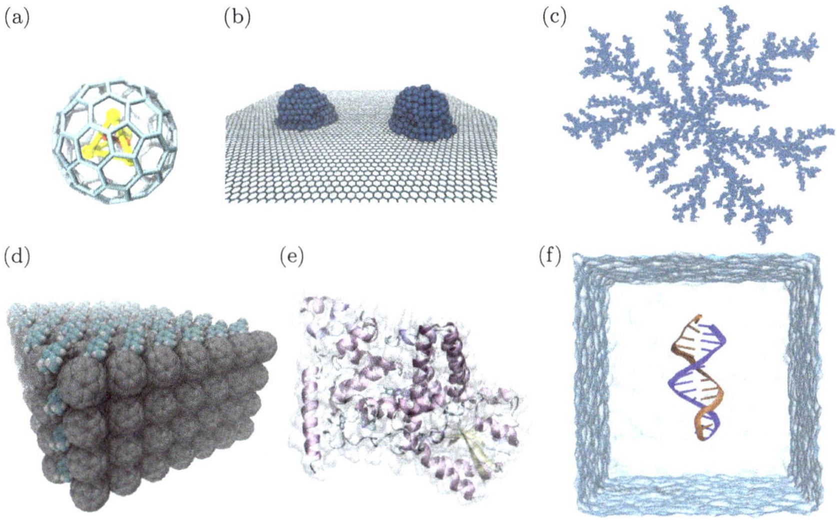

Fig. 1.1. Illustration of selected MBN systems: (a) encapsulated clusters, (b) deposited nanoparticles, (c) nanofractals, (d) composite nanowires, (e) proteins, (f) biomolecules, e.g., DNA, in solution.
Source: Reproduced from Ref. 65 with permission from Wiley-VCH Verlag.

of open, challenging scientific problems in this area is quite broad. They may include: structure and dynamics of clusters, nanoparticles (NPs), biomolecules, and many other nanoscopic and mesoscopic systems; clustering, self-organisation, growth and structure formation processes, and their multiscale nature; assemblies of clusters/NPs and bio-macromolecules, hybrid bio–nano systems, nanostructured materials; surface phenomena; nanoscale phase and morphological transitions; thermal, optical, and magnetic properties; collective or many-body phenomena; electron transport and molecular electronics; nuclear magnetic resonance; collision, fusion, fission, and fragmentation processes; channelling effects; radiation effects; and radiobiological effects. There are many important applications that are closely related to the field. The list of topics in the field is growing quite rapidly. This facilitates the development of the relevant theoretical and computational methods.

This chapter reviews recent efforts to develop a systematic approach to this area of research, which is based on the methods of theoretical and, in particular, computational physics of condensed matter systems. It is not surprising that this area of theoretical research has developed only recently, along with the development of powerful computers and advanced computational techniques. Indeed, the computational aspect of this research area provides the methodology for revealing novel features of the structure and dynamics of nanoscopic and mesoscopic molecular systems. It also supports the high level of interdisciplinarity of the research. Similar computational methods can easily be applied to molecular and condensed matter systems of very different nature and origin.

Despite the wide variety of systems, the methods used to describe and simulate them on computers are often very similar. It is impossible to describe all possible molecular and condensed matter systems in a single chapter. Therefore, only selected characteristic examples and case studies are presented in the following for illustrative purposes.

Condensed matter systems represent the states of matter typically in either solid or liquid phases. These states of matter are formed due to the interatomic forces acting within the system.[2–5] The nature of the forces can be different. Often, they have an electromagnetic origin, although due to the quantum motion of atoms and electrons therein, quantum phenomena in interatomic interactions play an essential role. The quantum nature of interatomic interactions under certain conditions can lead to the formation of quantum states of the entire condensed matter system, such as superconducting phases that can be observed in certain materials at low temperatures, the ferromagnetic and antiferromagnetic phases of spins on crystal lattices of atoms, and the Bose–Einstein condensate observed in ultracold atomic systems.

The nature of condensed matter systems can be very different, representing animate and inanimate matter, organic and inorganic, crystal and amorphous materials, glasses, etc. Biological condensed matter systems include large biomolecules, cells, biological media, and entire biological systems, including organisms. These large groups of condensed matter systems have different natures, areas of application, and properties but exhibit many similarities from a theoretical and computational point of view.[1,6] This makes unravelling the material properties of condensed matter systems a rather general and fundamental task.

Condensed matter systems can be finite and exist in the form of clusters, NPs, and droplets. They can also be macroscopically large; in such cases, they can be considered infinite. However, even for an infinite system, one should distinguish its bulk properties and system properties arising near the system surface. Interesting phenomena arise at the interfaces of different condensed matter systems in the same or different phases.

Condensed matter systems can have different geometries, being zero, one, two, or three dimensional. Concrete examples of such systems are atomic and molecular clusters, fullerenes, nanotubes, nanowires, graphene, monoatomic layers, and nanofilms. A condensed matter system often possesses some specific properties (thermomechanical, electromagnetic, optical, etc.) that make it functional and useful for applications.[1,6]

1.2 Multiscale Computational Modelling

In many disciplines and research areas, the key complex physical, chemical, and biological processes often involve different temporal and spatial scales. One has to deal with problems, mechanisms, and processes that occur in such systems at different scales, dimensions, spaces, etc., although they are connected to form a coherent scenario. Understanding such multiscale scenarios often holds the key to major breakthroughs.

The nano- and mesoscales define the research domain where the quantum world of single atoms, small atomic molecules, and clusters meets the bulk scale of classical physics. The mesoscale is a significant step up in the system size and complexity from the nanoscale. Many new physical and chemical properties of the system emerge already at the nanoscale, and this process continues at the mesoscale. For example, thermomechanical properties of matter, magnetism, and other collective phenomena cannot be understood at the atomic level. Although they already appear at the nanoscale, their evolution towards the bulk limit continues, with some new features appearing at the mesoscale. Superconductivity only appears in mesoscopically large systems.

It is more difficult to assign a specific scale to the emergence of the phenomenon of life, although it is obvious that that the building blocks of living systems, as well as their interactions and the principles of their dynamics, behaviour, and functioning, are determined at the nano- and mesoscales.

The operation of MBN systems and devices often involves different scales too. For example, batteries and other energy-related devices such as solar cells, fuel cells, and supercapacitors, which are considered representative of mesoscale challenges, depend on the ability of the battery to store electricity due to the nanoscale structure of individual components such as the anode, cathode, and electrolyte; however, the actual performance of the device in the real world depends on how all the components work together at the mesoscale.

From a fundamental point of view, every molecular and condensed matter system consists of two weakly bound subsystems: electronic and ionic.[2] The ground-state electronic structure of a system defines the interatomic interactions, atomic valences, types of interatomic bonds, and ultimately the atomic or molecular structure of the system. The motion of the ions in the system is often quasiclassical and slow with respect to the characteristic times of the electronic motion. This observation made by Born and Oppenheimer is of the fundamental importance. It means that the electronic subsystem becomes a source of the electric field acting on the ions in addition to their inter-ionic Coulomb interaction and that the motion of the ions can be described in terms of classical trajectories. The characteristic time scales of the ion motion are on the order of femtoseconds and above, while the essential electronic dynamics and quantum processes occur on the sub- or femtosecond time scales.

Such a situation opens up remarkable possibilities for the description of molecular and condensed matter systems by means of classical molecular dynamics (MD). The limits of classical MD can be extended by the application of various well-founded modelling approaches, which allow its coupling to purely quantum processes occurring under certain conditions in the system on time scales shorter than femtoseconds. The limits can also be extended by coupling the classical MD to purely probabilistic stochastic dynamics (SD) based on the Monte Carlo (MC) approach, which is highly suitable for analysing the structure and dynamics of large molecular and condensed matter systems on sufficiently long time scales. Such multiscale descriptions often play a key role in overcoming the complexity of molecular systems. Figure 1.2 illustrates the multiscale

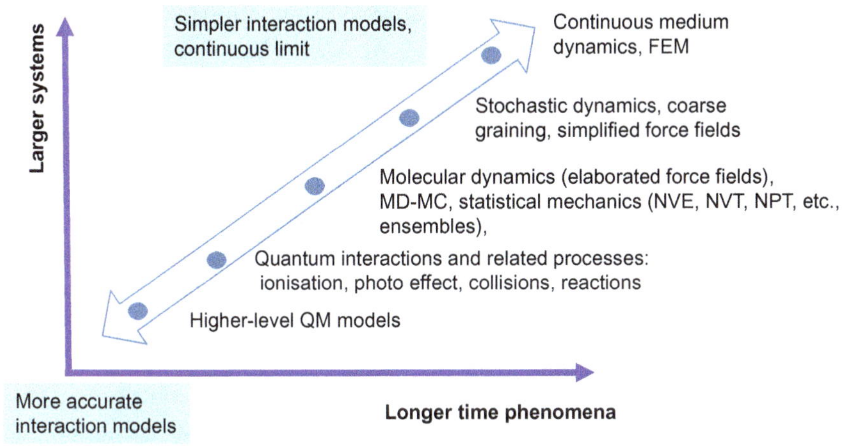

Fig. 1.2. Multiscale descriptions of MBN systems include higher-level quantum models (QMs), quantum interactions, and related processes, MD with elaborated force fields (such as those used for NVE, NVT, NPT, etc., statistical mechanics ensembles of particles (atoms and molecules)), coarse-grained approaches, simplified force fields, SD, continuous medium dynamics, and finite element method (FEM). Each of these methods operates at a particular characteristic scale of time and space.

Source: Adapted from Ref. 6 with permission from Springer Nature.

hierarchy of the main theoretical and computational approaches that are widely used in multiscale descriptions of MBN systems.

Figure 1.2 shows that multiscale modelling is widely used to describe complex phenomena (e.g., structure formation and evolution, materials under extreme conditions, and ageing) in many systems of different nature (e.g., bio- and nanostructured, composite materials), spanning a wide range of temporal and spatial scales. In their application to the description of multiscale phenomena, theoretical and computational methods specific to different spatial and temporal scales should be interlinked.

In many case studies, there is no alternative to multiscale computational modelling due to the complexity of the systems, processes, and phenomena under consideration and a lack of exact solutions for the corresponding equations. The methodology provides well-founded approximate numerical solutions based on validated simplifications verified by experiment or *ab initio* theories.

It has been recognised that computational multiscale modelling provides a powerful methodology by which the nano- and macroscale (quantum and classical worlds) can be intertwined.[1,7] It can be applied to processes and phenomena that occur in many condensed matter systems over different spatial and temporal scales, ranging from the atomic to the macroscopic. The potential of multiscale modelling in the study of various condensed matter systems has been successfully demonstrated in many case studies.[1,7] Advances in multiscale modelling offer the possibility of occasionally using it as an alternative to laboratory experiments or as a tool for predicting results that can be made prior to experimental measurements.

Technological applications of research based on multiscale modelling include such important tasks as advancement of 3D nanoprinting or controlled nanofabrication, the design and functionalisation of new materials, atomistic analysis of radiation damage and degradation of materials, optimisation of radiotherapies, developing technologies for radiation protection in space, and plasma and laser technologies. Thus, the implementation of multiscale modelling is expected to have significant socioeconomic impact in the coming decades, with multiscale modelling being adopted by the broad research community as a critical tool.[1,7] The EU research community is currently at the forefront of the development, commercialisation, and exploitation of multiscale modelling as a tool for modern research and technology.[1,7] It has been recognised as a rapidly emerging interdisciplinary research area developed through an ongoing COST Action CA20129 MultIChem, H2020 RISE projects RADON and N-LIGHT, the Horizon Europe Pathfinder project TECHNO-CLS, and many national projects.[7]

1.3 Multiscale Processes in Condensed Matter Systems: Clustering, Self-Organisation, and Growth

Clustering, self-assembly, structure formation, and growth processes are a group of key phenomena that occur in almost all types of condensed matter systems, including biological ones.[1,6]

The processes of aggregation of atoms and small molecules into clusters, NPs, and macromolecules, clustering (or coalescence) of NPs and biomolecules into nanostructures, nanostructured materials, biomolecular complexes, and hybrid systems with different morphologies allow the creation of a wide range of condensed matter systems.[8] Examples of self-organised systems may include aggregates of metal NPs,[9,10] nanofilms,[11,12] nanotubes,[13–15] nanowires,[16–18] functional nanocoatings,[12] nanofractals,[19–23] and many other ordered or disordered nanostructures with characteristic structure and properties. Some of these systems have only recently been synthesised and have become the subject of intense study because of their unique structural, optical, magnetic, thermomechanical, or thermoelectrical properties and their potential use in a wide range of important applications.[8] Clustering, self-organisation, and structure formation are general phenomena that manifest themselves at very different levels and scales of matter organisation or self-organisation and are therefore relevant to many dynamical systems studied by different natural sciences. They appear in many different fields of research, including physics, astrophysics, chemistry, biology, materials science, nanoscience, neuroscience, and even technology (clustering in the wireless, computer, windmill networks, etc). Although there are many examples of these processes, their mechanisms and driving forces are often not well understood.[1,6,24]

Apart from the fundamental value of understanding these processes, this knowledge is also highly relevant to the key problems of modern technology. An important example of a relevant technological process is the fabrication of nanostructures, nanosystems, and nanomaterials. This goal can be achieved by two conceptually different approaches.[25] The 'top-down' approach deals with various techniques that enable the production of smaller nanostructures by breaking down larger pieces of material. The 'bottom-up' approach typically relies on the self-assembly of atoms, molecules, and clusters into larger nanostructures, nanosystems, and nanomaterials.

The bottom-up approach is driven mainly by the diffusion, reaction, and self-organisation processes, which involve the dynamics of the system on quite different spatial and temporal scales. These processes arise from the diffusion of individual atoms and molecules within the system. The quantum interactions of atoms and molecules with their neighbours occur at the characteristic distances of a few angstroms and on sub-femtosecond to femtosecond time scales. However, the characteristic time scale for the diffusion processes of atoms and small molecules in a condensed matter system occurs on time scales orders of magnitude longer, depending on the phase, atomic composition, and temperature of the condensed matter system or interphase under consideration. The time scale for the self-assembly processes in a condensed matter system are much longer, reaching seconds, minutes, and longer. The spatial scale of such processes is comparable to the size of an entire system, which can be macroscopically large.

The bottom-up approach for the controllable, reproducible, and industrially viable fabrication of nanosystems with desired morphology and properties can be realised by various atomic, molecular, or cluster deposition processes with subsequent self-assembly of deposited species into the desired nanostructures.[26,27] Among the many different deposition techniques and characteristic patterns of deposited species on surfaces, the nanofractal shapes are the most studied ones[1,6,19,24,28] because the physics of fractals is of general nature, research interest, and technological importance. Here, we briefly discuss the formation, growth, and evolution of nanofractal structures that can be created on surfaces in the course of atomic, molecular, or cluster deposition processes as an exemplary case study to elucidate the multiscale nature of structure formation and dynamics of such systems.

As explained above and observed in numerous experiments,[20,29–31] nanofractal formation in the course of the surface deposition processes, growth, and evolution involves the dynamics of an enormous number of atoms on time scales which are far beyond the current limits of quantum mechanical and even classical MD simulations. However, the fractal dynamics can be understood through a multiscale approach. Such an approach invokes quantum mechanics to describe the parameters of the interatomic potentials used as inputs for MD simulations. The MD simulations provide a set of validated parameters for models based on SD.[32] SD can be used to simulate the structure formation and dynamics of fractal systems up to the macroscopically large spatial scales and the corresponding temporal scales defined by the performed experiments.[22,23,32,33]

Figure 1.3 depicts the key elementary processes that are crucial for the formation, evolution, and fragmentation of nanofractals. The described multiscale scenario and the corresponding multiscale approach can explain and quantify the most important experimental

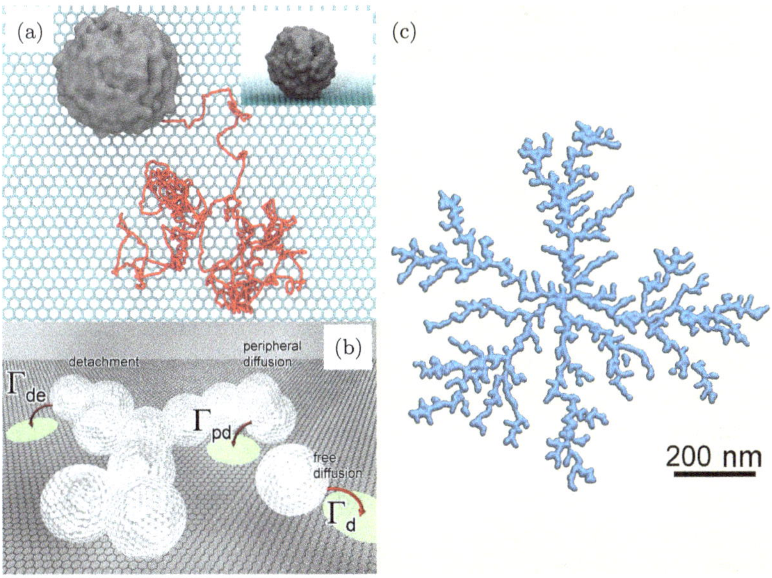

Fig. 1.3. Main elementary stochastic processes of atomic clusters deposited on a surface. Panel (a) shows the diffusion of a silver cluster Ag_{488} deposited on a graphite surface, with the trajectory of the cluster centre of mass derived from MD simulations. The figure illustrates that the deposited cluster undergoes a random, Brownian motion that can be parameterised by the corresponding cluster diffusion rate Γ_d, which is one of the input parameters for the stochastic dynamics simulation of such a process. Panel (b) explains that the long time scale stochastic motion of an ensemble of deposited clusters can be parameterised by the rates Γ_d, Γ_{pd}, and Γ_{de} of the three elementary stochastic processes describing (i) the diffusion of a cluster over a surface (Γ_d), (ii) the diffusion of a cluster along the periphery of an island on the surface (Γ_{pd}), and (iii) the detachment of a cluster from an island (Γ_{de}). Panel (c) shows that the random deposition of new particles on the surface, and the consideration of the aforementioned stochastic processes leads to the formation of the fractal structure shown.

Source: Reproduced from Ref. 33 with permission from Wiley-VCH Verlag.

observations on the formation and evolution of fractal systems and their morphology and properties.[21–23,32,33]

This illustrative example represents a large class of systems and dynamical processes that can be understood by using simulations based on the multiscale approach, which combines quantum and classical MD with SD.

1.4 Multiscale Computational Modelling of Condensed Matter Systems Exposed to Radiation

Let us now turn to the multiscale computational modelling of condensed matter systems exposed to radiation. It has already been pointed out that there are many different condensed matter systems in nature. Their number is practically infinite. The recent roadmap paper[7] has reviewed the state of the art in theoretical and computational methods that allow an accurate description of numerous condensed matter systems exposed to radiation and related experimental and technological developments. Many of the phenomena observed in this research area are multiscale in nature, i.e., they arise from the interrelated processes occurring at quite different temporal and spatial scales. Therefore, their understanding requires the application of appropriate multiscale theories and related computational tools, which are discussed as follows.

Radiation effects in condensed matter systems can be induced by the radiation of different modalities, which include photons, electrons, positrons, ions, neutrons, and other elementary particles. Typically, these particles are delivered in the form of a beam. The physical characteristics of the particle beams, such as energy, intensity, size, emittance, and divergence, as well as duration, bunching, fluence, duty cycle, etc., can vary significantly. This variety of particle beam options requires different approaches to study radiation-induced effects in different irradiation regimes.[34]

Irradiation of condensed matter systems can occur naturally, for example, by cosmic rays or the presence of a natural radiation background. In laboratories, it is usually provided by beams of particles of different energy, size, intensity, etc. At high radiation intensities, the condensed matter systems become highly excited, and extreme conditions of their state (high temperature, pressure, or ionisation state) can be created. When a system in such a state is brought into a gaseous phase, it becomes a plasma.

Despite the great diversity of condensed matter systems, their interaction with radiation and the associated radiation-induced phenomena share many common features. This is because the number of different radiation modalities is not so large, and the radiation-induced phenomena, defined by the atomic and molecular processes occurring in different condensed matter systems, including biological ones, operate in a similar way. In this context, there are similarities in the behaviour of the cross-sections of various elementary radiation-induced quantum processes in all condensed matter systems. Such similarities arise generally from a relatively weak dependence of the cross-sections on the environment in which the radiation-induced quantum processes occur. Typically, the cross-sections depend only on a limited number of relevant physical parameters, allowing us to establish the dependencies of the cross-sections on these parameters.[2] It should also be noted that the fundamental physical principles and the associated theories describing the radiation-induced processes are equally applicable to all condensed matter systems. The number of such theories is negligibly small compared to the number of different condensed matter systems.[35]

Let us now consider the main theoretical concepts utilised to study the various stages of the multiscale scenario of radiation-induced processes in condensed matter systems and related phenomena. The description of the entire multiscale scenario requires the utilisation of several methodologies, including those that enable interfacing of the methodologies operating in different spatial and temporal regimes.

1.4.1 *Main stages of the multiscale dynamical response of condensed matter systems to irradiation*

The dynamical response of condensed matter systems to irradiation typically involves a cascade of processes. The interaction of radiation with matter takes place on time scales determined by the characteristic interaction time of a projectile particle with an atom or a molecule.[2,36] This can vary considerably, from attoseconds to femtoseconds, depending on the velocity of the projectile and the size of an atom or a molecule. The interaction of projectile particles with atoms and molecules induces their quantum transformations, such as excitation, ionisation, dissociation, and electron attachment. These transformations occur via corresponding radiation-induced quantum processes, the duration of which is relatively short compared to the total duration of the radiation-induced processes in the condensed matter systems. However, the initial radiation-induced quantum transformations are followed by further processes that may extend over much larger spatial and temporal scales than those typical for the initial quantum processes. For example, irradiation of biological systems can lead to persistent radiobiological effects, such as cell damage, DNA repair processes, mutations, and cell apoptosis. Irradiation of organometallic molecules deposited on a surface with electrons or ions leads to their degradation and the formation of metal clusters, crystals, or nanostructures.

Initial quantum interactions of the radiation with the system can lead to chemical transformations of molecules and their reactions, thermomechanical transformations of the medium and its dynamics, various many-body/collective effects, material ageing, radiobiological effects, etc. The temporal evolution of such cascades of processes involves variations of different characteristics of the system, such as particle distributions, energy, and thermodynamic variables (temperature, pressure, etc.) within a system. This evolution can consist of many different stages before the system reaches equilibrium. On even larger scales, it is meaningful to consider opening the system to larger-scale environments and study the related larger-scale processes. These stages and the corresponding states of the system can be studied experimentally, and such experiments are typically stage-specific.

It is quite common that irradiation of condensed matter systems to radiation triggers a sequence of interrelated processes (physical, chemical, and biological) that manifest themselves on different temporal and spatial scales, creating a multiscale scenario for the final observables induced by the initial irradiation. This multiscale feature of radiation-induced processes in condensed matter systems has only recently been understood and elucidated in a number of concrete case studies.[32,33,37–43] This feature is general and common to different types of condensed matter systems and radiation modalities. Understanding the radiation effects of radiation in the condensed matter systems therefore requires an understanding of both the immediate effects of the interaction of radiation with matter and the behaviour of the matter (in particular, its dynamics) over long periods after its irradiation. This second and essential part of the problem is far from being understood and is currently the subject of intensive research and development.[7]

The fundamental understanding of all these complex phenomena is only possible through the comprehensive multiscale approaches that include initial-, intermediate-, and final-stage processes involved.

Currently, only a few advanced multiscale scenarios have been developed in different research fields.[1,6,33,37–40,42,44] Further development directions for the theory and related computational methods for the simulation of processes at all levels of the multiscale scenarios and their interfaces are also a focus of the current developments.[7] Concrete examples of such developments are given in the following sections.

One can find many similarities in the multiscale scenarios for different condensed matter systems and radiation modalities. Indeed, there are characteristic stages that appear in most of them. **Stage 1: *radiation-induced quantum processes*** is characterised by initial quantum interactions of radiation with atoms and molecules and radiation-induced quantum processes within the system. This is followed by **stage 2: *particle propagation***, in which the transport of the primary radiation through a condensed matter system occurs with the generation and transport of secondary (and also tertiary, quaternary, etc.) particles created in the system. The transport of the primary radiation and the produced (secondary, etc.) particles through a system and their interaction with the surrounding molecules results in energy transfer into the system and relaxation processes in the medium, leading to the next **stage 3: *irradiated medium dynamics*** of the multiscale scenario. This stage is characterised by specific dynamic and thermodynamic effects in the medium, leading to its molecular and chemical transformations. After some period of time, the irradiated system may reach **stage 4: *chemical and thermodynamic equilibrium***. However, the multiscale scenario may continue even further, involving larger-scale environments and manifesting larger temporal and spatial scale phenomena, such as biological (e.g., radiation damage repair, cell apoptosis, and mutations), structural formation and evolution, material ageing, and morphological transitions. Typically, each of these processes represents a separate case study or even a focused area of research, but together, they can be assigned to **stage 5: *large-scale processes*** of the multiscale scenario.

The whole cascade of processes is essentially multiscale and requires a multiscale theory to understand and describe it, as shown in Fig. 1.4. The diagram shows the aforementioned main stages of a cascade of processes occurring in a condensed matter system after irradiation. The stages correspond to different regions of the

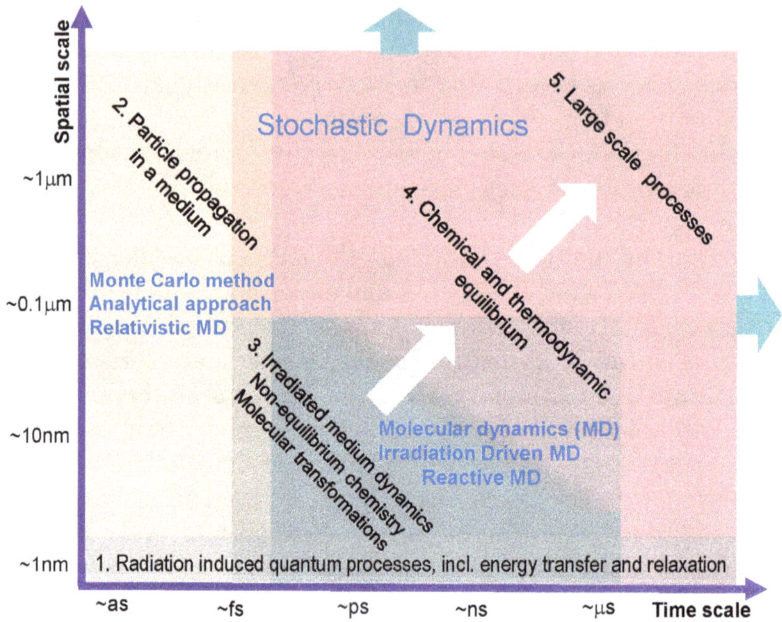

Fig. 1.4. A schematic space-time representation of the main stages of the multiscale scenario of radiation-induced processes in condensed matter systems, with the corresponding methods for their description. The colours of the different areas on the diagram indicate the ranges for the manifestation of the corresponding phenomena and the application limits of the associated methods.
Source: Reproduced from Ref. 7 published under an open access Creative Commons CC BY license.

temporal and spatial scales characteristic of their manifestation. In Fig. 1.4, they are marked with black letters and numbered according to their temporal sequence. The coloured areas in Fig. 1.4 introduce the basic theoretical approaches and methods (labelled with blue letters) used to describe the corresponding stages of the radiation-induced cascades. The boundaries of the coloured areas indicate the limits of applicability of the corresponding theoretical and computational approaches and methods. The origin in Fig. 1.4 is placed at the beginning of the multiscale cascade of the considered radiation-induced processes and is typically associated with the primary/initial radiation event. Note here that Fig. 1.4 may correspond to a scenario arising from a single initial radiation event, e.g., the passage of an ion through a condensed matter system; however, in the case of multiple events, their statistical analysis and overlap should be considered and analysed. This may result in the dispersion of the interfaces between different regions shown in Fig. 1.4.

Let us now discuss each element in Fig. 1.4. The rectangular areas indicated by different colours correspond to the ranges of applicability of different key theoretical and computational methods used for the multiscale simulations of radiation-induced processes. The range for each coloured area indicate the physical limit for the associated processes or may correspond to the computational limit of the indicated method with today's most powerful computational tools.

1.4.2 *Radiation-induced quantum processes*

The interaction of radiation with a condensed matter system occurs via the radiation-induced elementary quantum processes; see the horizontal greyish rectangular area at the bottom of Fig. 1.4. The most characteristic example of such a process is atomic or molecular ionisation. It can be induced by the radiation of any modality, although the ionisation cross-section depends on the type of ionised atoms, the type of projectile particles, and their energy. In the case of molecular ionisation, the ionised molecule may undergo fragmentation or bond cleavage during or after the process. Electronic excitation of atoms and molecules is another quantum process that can be induced by radiation of any modality in any irradiated system. Electronic excitation in a molecular target can also lead to molecular fragmentation or a bond cleavage. Both excitation and ionisation of atoms and molecules can involve different electronic shells. Atomic and molecular excitations can exhibit different energy relaxation mechanisms, e.g., the Auger process, radiative de-excitation, and electron–phonon coupling.

Other atomic and molecular processes may play an important role in the interaction of radiation with condensed matter systems, such as electron capture and dissociative electron attachment (DEA). Only some of these are discussed here. However, it is important to emphasise that the temporal and spatial scales for most (if not all) radiation-induced atomic and molecular elementary quantum processes vary within the ranges shown in Fig. 1.4. The lower temporal and spatial limits are determined by the characteristic time and distance in a single particle collision with an atom. The upper spatial limit is determined by the collision kinematics of energetic particles with atoms. The upper time limit is related to the characteristic relaxation times of different excitation processes in the system (electronic or nuclear). These can occur via the emission of photons, electrons, or other particles (atoms, ions, and elementary particles), tunnelling effects, and energy transfer to the vibrational degrees of freedom of the system via electron–phonon coupling of the electronic and ionic subsystems.

It should also be noted that irradiation of condensed matter systems with gamma rays or particles of sufficiently high energy can induce nuclear reactions and associated transformations of the irradiated condensed matter systems. When these nuclear processes do occur, they take place on much smaller spatial scales, although they can last for periods that are orders of magnitude longer than those shown in Fig. 1.4. Although they are of a different nature from atomic and molecular processes, we do not distinguish them from other quantum processes shown in Fig. 1.4. The formal separation of nuclear processes from atomic and molecular processes would not add any additional aspects to the following considerations.

The quantum processes are characterised by the corresponding probabilities, which can be derived from quantum mechanics.[2] In the case of collision processes, the probability of a process is determined by the product of its cross-section and the flux density of the projectile particles. The cross-sections of collision processes can be calculated using collision theory.[45,46] Depending on the problem, the flux density of the projectile particles is determined by the primary radiation characteristics or the primary radiation, together with the radiation of backscattered and secondary particles. The latter can be derived from the particle transport theory.

The important characteristics of radiation-induced quantum processes are the energy and momentum transferred from the projectile particles into the system. These characteristics can be derived from quantum mechanics and collision theory. The mechanisms of the energy and momentum transfer, as well as their analysis and quantification, are essential for understanding radiation-induced phenomena in condensed matter systems and their efficient computational simulation and quantitative description. Elementary quantum processes involving nuclei, atoms, and molecules are usually treated by theoretical and computational methods of quantum mechanics and collision theory, including many-body theory, density functional theory (DFT), and time-dependent density functional theory (TDDFT). These theoretical methods and their computational realisations and limitations are not shown in Fig. 1.4 due to space constraints. However, all these aspects are discussed in the roadmap paper,[7] together with all other theoretical and computational methods relevant to the multiscale approach illustrated in Fig. 1.4.

1.4.3 *Particle propagation through a medium*

Consider the vertically oriented, rectangular, brown area in Fig. 1.4, which corresponds to particle transport theories: *MC method*, *analytical approach*, and *relativistic MD*. These methods describe two important phenomena related to particle transport through a condensed matter system, namely (i) the primary particle propagation and energy loss, and (ii) the creation and propagation of the secondary and subsequent generations of particles. All these phenomena occur in the multiscale scenario of radiation-induced processes shown in Fig. 1.4.

The first phenomenon is the transport of primary particles/radiation through a condensed matter system. The primary particles undergo multiple collisions with atoms and molecules as they move through the system. During these interactions, the particles change their direction of motion and lose energy and momentum, which is transferred to the medium. The dynamics of such particles is usually considered in a static medium of a given density. Knowing the density of the medium and the cross-sections of collision processes of incident particles with atoms in the medium, one can simulate the propagation of incident primary particles through the medium using the *MC method*. Alternatively, the propagation of particles through a medium can be simulated using the *Relativisic MD approach*[47] or various *analytical approaches* to describe the propagation of particles through a condensed matter. These methods are discussed in more detail in the roadmap paper.[7] The distances to which ultrarelativistic particles can propagate through the medium can be macroscopically large. Therefore, the upper spatial limit for the brown rectangular area in Fig. 1.4 is indicated as open, extending towards the larger scales. The time scale of primary particle propagation through a medium is determined by the time required for a particle to pass through a medium of a given size. Primary particle propagation is a fast process due to the fast, if not relativistic or ultrarelativistic, nature of their motion. For system sizes below $\sim$1 μm and relativistic particle velocities ($v \sim c$, where c is the speed of light), the particle propagation times become shorter than $\sim$10 fs.

The second important phenomenon associated with the transport of particles through a condensed matter system is the process of propagation of the secondary and subsequent generations of particles. The secondary generation of particles is produced by the primary particles/radiation. These secondary particles typically carry less energy than the primary particles. In condensed matter systems, ionised electrons play an important role as secondary particles. However, if the energy of the primary radiation is sufficiently high, the secondary particles can be produced by the nuclear reactions induced by the primary particles/radiation. Secondary particles can, in turn, produce tertiary, quaternary, etc., generations of particles before particle propagation stops. The energy of the secondary particles and their ranges of their propagation are smaller than those of the primary particles and depend on the energy of the primary particles. The energy of the primary particles decreases with their propagation into a medium.

The production of secondary particles and their subsequent processes induced by them in the medium can be simulated using the MC method, similarly to the propagation of primary particles. An alternative to the above-mentioned MC approach for the simulation of the propagation of radiation-induced secondary and subsequent generations of particles is based on continuous transport theories, e.g., diffusion equation, diffusion-reaction equation, and kinetic equation. When justified, these methods can provide faster and more reliable solutions for the cumulative dynamics of propagating secondary and higher generations of particles. Thus, in the vicinity of the Bragg peak, the propagation of secondary and all subsequent generations of particles in the medium stops on time scales $\sim 10^1$–10^2 fs.[38,48] The spatial scales characteristic of the propagation of secondary particles are much smaller than those of the primary particles and depend on the energies at which they are emitted. For example, δ-electrons emitted after the ionisation of a medium by relativistic heavy ions can propagate on the micrometre scale, while electrons emitted near the Bragg peak for an ion can only move a few nanometres away from the ion track.[38,48] Using these arguments, the temporal limit of the particle transport region in Fig. 1.4 was established.[48]

These considerations explain the main features of the particle transport domain shown in Fig. 1.4. Further details of the main methods, their computational realisations and limitations, and the interfaces between them are given in the roadmap paper.[7]

1.4.4 *Irradiated medium dynamics and related phenomena*

The next stage of the multiscale scenario involves *irradiated medium dynamics*, *non-equilibrium chemistry*, and *molecular transformations*, which occur in most of the case studies considered. This stage is introduced in the central part of Fig. 1.4. It can be seen that the phenomena mentioned above can be simulated using different types of *MD*, including standard *classical MD*, *reactive MD*,[49] and *irradiation-driven MD*.[40] The temporal and spatial ranges accessible for these methods are introduced in Fig. 1.4 by the blueish rectangular area in the middle part of the figure. These methods and their limitations are discussed in detail in Ref. 7. It is worth noting that the ranges of the blueish rectangular area correspond to the maximum temporal and spatial scales that can be accessed in simulations of condensed matter systems with the aforementioned computational methods using the most powerful supercomputers. The transparency of a part of the blueish rectangular area is related to the feasibility of the simulations; the larger system sizes can be simulated on much shorter time scales, and the longer temporal behaviour can only be explored for relatively small systems.

On the left, this blueish area overlaps with stage 2, corresponding to *particle transport*, as it is seen from a different colour of the area where the overlap occurs. The lower part of the blueish area overlaps with the area of *radiation-induced quantum processes*.

An alternative to MD-based simulations within the limits of their feasibility is based on the *SD* approach;[32] see the reddish area in Fig. 1.4. The SD methodology extends well beyond the capabilities of the MD approach; see the roadmap paper for more details.[7]

Stages 1 and 2 of the multiscale scenario described in Sections 1.4.2 and 1.4.3 represent numerous radiation-induced quantum processes in the medium, leading to its transformations such as the breaking of molecular bonds and the creation of molecular fragments, the formation of defects, ionised centres, and free solvated electrons and holes. Such an excited medium is created in a state far from equilibrium and evolves towards equilibrium through a cascade of subsequent processes. At the end of the particle transport stage, a significant fraction of the energy of the primary radiation is transferred to the electronic degrees of freedom of the system. Therefore, stage 3 of the multiscale scenario continues with the energy transfer processes from the electronic to the ionic degrees of freedom in a condensed matter system. This involves various quantum mechanisms, including electron–phonon coupling, leading to chemical transformations in the system, such as the formation and subsequent closure of dangling molecular bonds, the creation of new chemical species as products of the irradiated molecules in the system, or their subsequent chemical reactions. At the end of this intermediate state, most of the energy delivered to the medium by the primary radiation is transferred to the ionic degrees of freedom in the system, but the system is still not thermalised. These processes typically occur on a picosecond time scale.

The dynamic response of the medium upon its irradiation does not end with the transfer of energy to the ionic degrees of freedom in the system because, at this stage, the energy is distributed inhomogeneously throughout the system. Therefore, the follow-up process in the multiscale scenario describes the redistribution of the energy transferred to the medium in the vicinity of the particle tracks over the entire volume of the system and among all its degrees of freedom. The details of this process depend on the irradiation conditions and the amount of energy transferred. For sufficiently large LET values, it initiates strong medium dynamics. Under certain conditions, it can lead to severe distortion of the medium due to a significant increase in temperature and pressure in relatively small volumes where the energy deposition has taken place. Such conditions have been shown to lead to the formation of nanoscopic shock waves. The strength of these shock waves may be sufficient to cause irreversible changes in the medium, such as breaking of molecular bonds, defects, formation of craters on surfaces, and lethal damage to cells; see Section 1.6 for further details.

The dynamic response of the medium begins on the picosecond time scale after the redistribution of the energy transfer to the ionic degrees of freedom. The dynamic response of the excited medium leads to its relaxation on the timescale of tens to hundreds of picoseconds, resulting in a more homogeneous distribution of the transferred energy among the ionic degrees of freedom of the system.

The complete thermodynamic equilibration of the system can take up to the nanosecond time scale or even longer, depending on the size of the system and the amount of energy transferred to it. During this time, the distributions of various quantities, such as particle velocities and vibrational excitations, evolve and reach their equilibrated forms, which are consistent with those that follow from statistical mechanics for a system in thermal equilibrium. During this equilibration process, various physico-chemical properties also evolve towards their equilibrated values. This affects, for example, the diffusion coefficients of atoms, molecules, and other molecular species present in the system after its irradiation. The diffusion coefficients of atoms and molecules determine the resulting chemical transformations of the irradiated medium.

It is important to emphasise that chemical transformations in the system continue throughout this entire stage, and their outcomes are strongly influenced by the dynamics of the medium described above. Therefore, such phenomena within the multiscale scenario are referred to as 'non-equilibrium chemistry'. The spatial scales that characterise the domain of non-equilibrium chemistry are determined by the spatial distances at which a complete set of possible chemical reactions within the system can take place and reach chemical equilibrium during the period of the medium dynamics and the subsequent thermal relaxation.

This non-equilibrium chemistry can be studied using RMD and IDMD, introduced above and discussed further in Section 1.5, within the ranges indicated for these methods in Fig. 1.4. For the larger system sizes, MD simulations of the dynamic response of the medium to irradiation and induced non-equilibrium chemical processes become increasingly difficult or even impossible. Nevertheless, the dynamic behaviour of such systems can be studied using the SD approach introduced above. The discussion of this approach and its interfaces with other methods mentioned above and the corresponding stages of the multiscale scenario shown in Fig. 1.4 are given in the roadmap paper.[7]

1.4.5 *Post-irradiation chemical and thermodynamic equilibration*

The characteristic size of an irradiated system at which it can be chemically equilibrated is determined by the chemical species present in the system at the lowest concentration. Its specific value depends on the atomic and molecular composition of the system, the irradiation modality used, the radiation dose, the LET, the temperature, etc. It can vary from tens of nanometres to micrometres and more.

The chemical and thermodynamic equilibrium stage is shown in Fig. 1.4 as the fourth stage of the multiscale scenario, following the irradiated medium dynamics stage, non-equilibrium chemistry and molecular transformations. This stage describes the system in a state where all possible chemical transformations are in equilibrium, and thus all the chemical products exist at specific equilibrated concentrations. Once achieved, the chemical equilibrium can, in principle, last indefinitely, provided that the state of the system is not affected by any further external factors.

As can be seen from Fig. 1.4, simulations of a system in its chemical equilibrium using RMD or IDMD can be challenging due to the relatively large system sizes involved. However, these simulations can be performed efficiently using SD due to the significant computational advantages gained using the MC approach as the basis for SD.

SD describes dynamical processes in almost all complex systems, including condensed matter systems, that are probabilistic in nature. Such processes can occur over a wide range of temporal and spatial scales. Therefore, the reddish area, representing the SD methodology and related processes, occupies the largest area in the Fig. 1.4. The concept of SD and its implementation in the popular software package MBN Explorer are discussed in Section 1.7 and the roadmap paper.[7] It is worth noting that SD allows the simulation of physical, chemical, and biological processes. It is relevant for modelling the multiscale phenomena presented in this paper and most other case studies in the field.[7]

1.4.6 *Large-scale post-irradiation processes*

The last (fifth) stage of the multiscale scenario shown in Fig. 1.4 corresponds to the large-scale processes that occur in an irradiated system after it has reached a chemical equilibrium. This stage of the multiscale scenario typically involves macroscopic observables that characterise irradiated systems, resulting from the radiation-induced processes or transformations that have occurred in the system in all previous stages of the multiscale scenario. The macroscopic observables typically involve larger spatial scales and/or longer time scales. These large-scale processes arise because the radiation-induced phenomena occurring in the system and passing through the multiscale scenario described above are usually embedded in a larger-scale environment. This may be a bulk material or a biological medium that is part of a cell, tissue, organ, or whole organism; it may be a gas or a plasma volume surrounded by a wall that is part of a device, etc.

There are many examples of large-scale processes being triggered in systems by their irradiation. For biological systems, this could include most of the radiobiological phenomena. At the cellular level, this could be, for example, repair mechanisms of the DNA damage caused by irradiation of a cell nucleus, cell apoptosis following irradiation, or changes in the chromatin properties following irradiation. At the intercellular level, this could be, for instance, a bystander effect showing the response of one cell to the irradiation of a neighbouring cell. Reactions of the immune system to irradiation of a part of the organism could be a representative example of a large-scale process at the level of the organism. There are also many examples of large-scale processes in non-biological systems. For example, irradiation of materials leads to the changes in their various bulk properties, such as elasticity, hardness, and recyclability, due to the formation of defects in the materials. Radiation-induced defects in electronic chips can lead to errors in the operation of electronic devices, etc.

In order to understand and quantitatively assess the large-scale effects, it is necessary to establish a relationship between the radiation-induced phenomena or transformations resulting from the post-irradiation equilibrium state of the system and the macroscopic observables. This goal has been achieved in the exemplary case studies presented in Section 1.6. For example, a relationship has been established between the complexity of the DNA damage at which it becomes irreparable and the probability of cell survival. The formation of such complex damage can be quantified within the multiscale approach, and an important macroscopic observable/characteristic, the probability of cell survival, can be calculated on this basis. It is also worth noting that for many systems, the links between microscopic radiation effects with macroscopic observables and the large-scale processes are not yet established and are currently the subject of intensive investigations in many research areas.

It is obvious that the lower limits for the temporal and spatial scales of large-scale processes should correspond to the scales at which such processes occur. These are often the scales at which the irradiated parts of the system reach thermodynamic and/or chemical equilibrium. As mentioned above, these scales will vary from case to case. The relevant upper spatial and temporal limits for the large-scale processes often become macroscopic. Therefore, the ranges of the SD, which is the most appropriate approach for simulating the large-scale processes, are indicated in Fig. 1.4 as extendable, with the large arrows pointing parallel to the temporal and spatial axes.

Large-scale processes can often be modelled using SD because they occur probabilistically. The probabilistic nature of large-scale processes is related to the fact that they are not fully controlled, and their key characteristics may, for example, depend on the environment in which the processes occur or involve some other known or even as yet unknown phenomena. Establishing probabilities for large-scale processes is often a difficult task. Therefore, there is no general recipe for their determination other than to experimentally measure the probabilities of relevant events occurring in a system during its SD dynamics. It is also possible to derive such probabilities theoretically or computationally, but this is usually done on a case-by-case basis.

1.5 A Multiscale Scenario for the Radiation-Driven Fabrication of Nanosystems

Let us now consider an illustrative example of a multiscale scenario involving radiation-driven processes. Radiation and radiation-driven processes can be used as a tool to create condensed matter systems with desired structure and properties at nanoscale resolution.

For the sake of simplicity, such systems are referred to as nanosystems. The importance of such a goal for nanotechnology was briefly discussed in Sections 1.1 and 1.3. Such technologies are relevant to a wide range of materials and can involve any of the radiation modalities.

The fabrication of nanometre-sized devices is one of the main goals of the nanotechnology industry due to the unique electronic, magnetic, superconducting, mechanical, and optical properties that emerge at the nanoscale.[50–52] Thus, the fabrication of ever smaller structures has been the goal of the electronics industry for several decades, since the smaller the fabricated systems, the higher the performance of the nanodevices fabricated on their basis. Until recently, the well-known Moore's Law has been adhered to in the semiconductor industry, enabling the production of ever smaller devices (mainly through the 'top-down' approach) and thus improving the operational performance within a fixed-size device. However, as the structure size decreases below 10 nm, traditional manufacturing methods (e.g., ultraviolet lithography, plasma etching, and plasma-enhanced chemical vapor deposition) struggle to meet Moore's Law. They cannot control the material properties and produce structures of the desired size, shape, and chemical composition. Therefore, there is an urgent need to develop new nanofabrication methods (which should be based on a 'bottom-up' approach), among which focused electron beam induced deposition (FEBID) is one of the most promising, allowing the controlled creation of nanostructures with nanometer resolution.[53] FEBID[54,55] exploits the irradiation of nanosystems with collimated electron beams. These can be used to create specific structural motifs of metal nanosystems for catalytic and nanoelectrochemical applications[56,57] to fabricate metal nanostructures for sensors, nanoantennas, magnetic devices, and surface coatings and to prepare thin films with tailored properties for electronic devices and other applications.

FEBID is an example of the 'bottom-up' approach to nanofabrication introduced in Section 1.3, where clustering, self-assembly, structure formation, and growth processes in condensed matter systems are discussed. In the case of FEBID, however, the structure formation of a fabricated nanosystem is additionally controlled by its irradiation with a focused electron beam. Other radiation modalities (ions and photons) can also be used for such purposes.[58,59] They have many similarities to the example considered in the following and will therefore not be discussed in any detail here.

The key element of the FEBID multiscale scenario is the irradiation of precursor molecules (usually organometallic, hereafter referred to as precursors)[60] with keV electron beams during their deposition on a substrate. Electron-induced fragmentation of the irradiated precursors releases metal atoms, forming a metal-rich deposit on the surface with a size similar to that of the incident electron beam (a few nanometres).[59] This phenomenon is multiscale and quite complex. It involves precursor deposition, diffusion, aggregation, clustering, and self-organisation processes, as already discussed in Section 1.3. In addition, the FEBID multiscale scenario must take into account adsorption and desorption processes; ionisation, excitation, and fragmentation of precursors induced by the primary as well as backscattered and secondary electrons emitted from the irradiated surface; electron transport in the substrate; processes of system relaxation after its electronic excitation, including energy transfer from electronic to vibrational degrees of freedom and the resulting thermomechanical effects; and chemical reactions between different reactive species produced upon system irradiation.[40,41,61,62]

In addition, the formation of 2D and 3D FEBID structures on larger spatial scales involves the movement of the electron beam (patterning) and multiple irradiations of the already formed structures. The decomposition of precursors is primarily induced by low-energy secondary electrons produced as the primary beam impinges on the surface of substrate. The electron beam can be controlled in both position and pulse duration with sub-nanometer and sub-microsecond precision such that complex structures can be fabricated in a single step.[63] This allows the controlled direct writing fabrication of complex, free-standing 3D structures with feature sizes already produced down to ~10 nm.[63,64] Naturally, such processes involve larger time scales. The quantitative description and characterisation of the entire FEBID scenario can only be achieved within a multiscale framework, taking into account a complex interplay of the above-mentioned phenomena occurring on many different temporal and spatial scales.[1] Although the first essential steps in this direction have been taken recently,[1,40,41] the full development and exploitation of the methodology remains an important task for the field. The importance of this task is clear. The multiscale approach to FEBID can guide technology in the selection of optimal nanofabrication regimes of the FEBID systems with desired properties, e.g., metal content, mechanical, electrical, and magnetic properties.

Until recently, the theoretical analysis of FEBID was based on the MC approach and numerical solutions of the diffusion-reaction equation. These methods provide the average characteristics of the FEBID structures but cannot provide details at the molecular-level. In Ref. 40, an atomistic approach to FEBID simulations has been developed. This approach is capable of providing the atomistic insights into created FEBID nanostructures. This approach is based on reactive molecular dynamics (RMD)[49] and irradiation-driven molecular dynamics (IDMD),[40] which take into account the quantum and chemical transformations within the absorbed molecular system. The case study performed in Ref. 40 considered the FEBID process of tungsten hexacarbonyl $W(CO)_6$ on a hydroxylated SiO_2 surface. It was carried out using MBN Explorer,[65] the software package in which RMD and IDMD are implemented. The simulation results were verified by their comparison with experimental data.[66]

Figure 1.5 shows a snapshot of the MD simulation[40] of the deposition of $W(CO)_6$ precursors on the SiO_2 substrate, depicting the initial stages of the electron beam irradiation process (represented by a green semi-transparent cylinder). It can be seen that most of the precursors located within the irradiated area undergo fragmentation. The rate of this process has been evaluated from the experimental data.[66]

The atomistic approach for FEBID, RMD, and IDMD involves the recently developed rCHARMM force field,[49] which requires the specification of several parameters, such as the equilibrium bond lengths, bond stiffness, and dissociation energies. In addition, the dissociative chemistry of the precursors should be defined, including in particular the definition of the molecular fragments and atomic valences.

The understanding of the irradiation-driven chemistry in the FEBID process can be advanced by a computational multiscale modelling (MM) approach that describes the full set of FEBID-related processes occurring at different time and spatial scales and by establishing procedures for experimental validation of the MM results. The MM approach should link different spatial and temporal scales and the corresponding processes and phenomena shown in Fig. 1.4. Such an approach should combine the following: (i) *ab initio* and

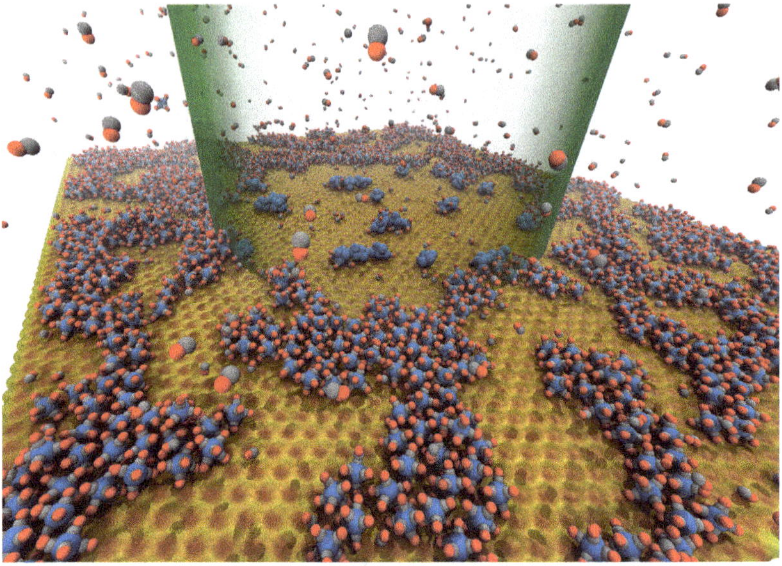

Fig. 1.5. An example of MD simulation[40] of the deposition of $W(CO)_6$ precursors atop the SiO_2 substrate, showing the initial stages of the electron beam irradiation process (a green semi-transparent cylinder). The interaction of the adsorbed precursors with the primary and secondary electrons emitted from the substrate leads to precursor fragmentation and the formation of clusters of tungsten atoms, shown as blue spheres.

Source: Reproduced from Ref. 40 with permission from Springer-Verlag.

DFT methods to evaluate parameters of irradiation- and chemically induced quantum transformations occurring in the systems at the molecular level; (ii) classical and reactive MD to study the fragmentation of precursor molecules[67,68] and their interaction with the substrates on time and spatial scales accessible in classical MD; (iii) evaluation of cross-sections of relevant collision-induced quantum processes (e.g., electronic excitation, ionisation, dissociative ionisation, and DEA) by *ab initio* calculations and analytical estimates or using data available in atomic and molecular databases; (iv) track-structure MC simulations to study the fluxes and fluences of primary and secondary electrons in order to access the probabilities of the above quantum processes; (v) IDMD to model random interactions of the electron beam and secondary electrons with the growing nanostructure taking into account possible chemical transformations therein; and (vi) the SD method to model the evolution of many-body systems on the time scales far beyond those accessible in MD simulations. References for all the above methods can be found in the roadmap paper.[7]

In MBN Explorer[65] different parts of the MM methodology have been successfully interlinked to study the mechanisms of formation and growth of metal-containing nanostructures under the action of focused electron beams.[40,41,61,62,69,70]

Figure 1.6 illustrates the ability of the MM methodology to characterise grown nanostructures at the atomistic level, predicting their morphology, growth rate, and geometrical (e.g., lateral size, height, and volume) and chemical (metal content) characteristics. The morphology of deposits is an important feature that determines many physical properties, such as electrical and thermal conductivity and magnetic properties.[71,72] Figure 1.6(a) shows snapshots of the IDMD simulations[62] of the FEBID process for $Fe(CO)_5$, one of the most common FEBID precursors used to fabricate magnetic nanostructures. Varying the electron current during the FEBID process significantly changes the morphology of the deposit (Figure 1.6(a)) and the elemental composition (Figure 1.6(b)). At a low beam current corresponding to a low degree of precursor fragmentation, the deposits consist of isolated small size iron clusters surrounded by organic ligands; see the top panel in Fig. 1.6(a). In this irradiation regime, metal clusters formed as a result of electron irradiation-induced fragmentation of precursor molecules do not agglomerate with increasing electron fluence, resulting in relatively low metal content in the deposit. Higher beam current facilitates the precursor fragmentation, and the metal clusters coalesce into dendrite-like structures of the size corresponding to the PE beam; see the lower panel of Fig. 1.6(a). In this regime, the metal content of the deposit increases twofold compared to the low current case, as shown in Fig. 1.6(b).

The results of atomistic simulations of the FEBID process using the IDMD approach can be used to construct stochastic models of FEBID using the SD methodology. An illustration of the application of SD for modelling the FEBID process has recently been presented.[70] It has been shown that the FEBID process can be described by transformations occurring with particles of different types, representing intact and fragmented precursor molecules, ligands, isolated metal atoms, and the substrate. The probabilities of the underlying

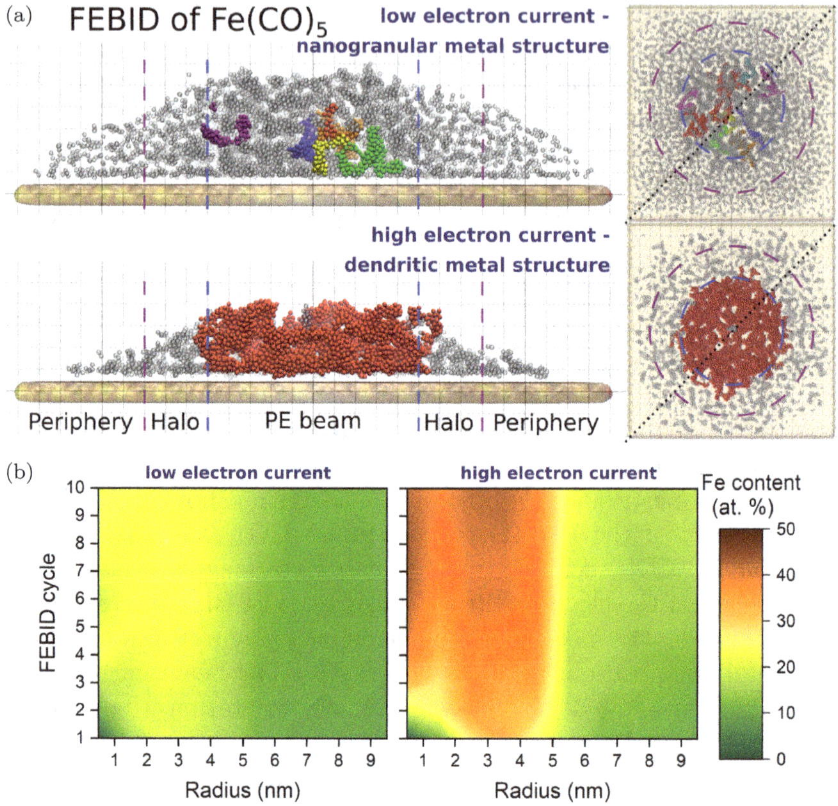

Fig. 1.6. Results of the IDMD simulations[62] of the FEBID process for $Fe(CO)_5$. Panel (a) shows the snapshots of the simulated iron-containing nanostructures: side view on diagonal cross-sections indicated by dotted black lines (left) and top view (right). The top and bottom snapshots in panel (a) correspond to the electron current of 1 and 4 nA, respectively. For clarity, only iron atoms are shown. Topologically unconnected metal clusters containing more than 100 iron atoms are shown in different colours (red, blue, green, yellow, orange, and purple) in the central part of the nanostructures. Smaller clusters with less than 100 iron atoms are shown as pale grey spheres. The boundaries of the primary electron beam, halo and edge regions are indicated by dashed blue and purple lines in the left column and by circles in the right column. The grid line spacing is equal to 1 nm in all dimensions. Panel (b) shows the atomic Fe content of the grown iron-containing structures as a function of the number of simulated irradiation-replenishment FEBID cycles for electron currents of 1 nA (left) and 4 nA (right). *Source*: Reproduced from Ref. 62 with permission from Royal Society of Chemistry.

processes occurring in the system can be determined through atomistic MD simulations and track-structure MC calculations.

The advancement of irradiation-driven 3D nanofabrication methods should be based on a deeper understanding of the molecular interactions and the key dynamical phenomena in irradiated nanosystems. This goal can be achieved by using modern computational MM tools in combination with experimental studies to validate such simulations. Future directions to achieve this goal require the construction and validation of the models based on SD that allow the simulation of the growth of nanostructures and the characterisation of their properties on the time and spatial scales much larger than those accessible in purely atomistic MD simulations. The parameters of the SD models should be thoroughly validated by 'lower-scale' modelling (e.g., by quantum-mechanical calculations or IDMD simulations) and experimental data.

Typically, FEBID is processed through successive cycles of precursor replenishment and irradiation stages. A popular class of FEBID precursors is represented by metal carbonyls $Me_m(CO)_n$,[53,73] which consist of one or more metal atoms (Me) chemically bonded to multiple carbonyl ligands. Metal carbonyls have been extensively studied experimentally, and much information on their thermal and electron-induced fragmentation has been collected.[74–84] Particular attention has been paid to the structure and properties of these precursors and their peculiar structure with strong C–O bonds and relatively weak Me–C bonds. While the former are relatively difficult to cleave, the latter are readily cleaved, typically by sequential loss of CO groups, once sufficient energy has been transferred to the molecule.

The further development and wider use of a MM approach for molecular-level studies of the physicochemical processes behind FEBID would enable the prediction of the elemental composition (especially the metal content) and the nano- and microstructure of the growing nanostructures, e.g., their granularity properties and morphological transitions therein. Therefore, the MM approach to describe FEBID 3D nanofabrication can provide the necessary insights into the fundamental knowledge of the radiation chemistry required to optimise the FEBID regimes and thus establish a routine methodology for FEBID 3D nanofabrication of new nanostructured systems with desired architecture and properties, e.g., electrical, magnetic, superconducting, plasmonic, mechanical, or thermomechanical ones.

It is worth noting that the ability to navigate the charged particle beam in a well-defined manner is used in the related technique of focused ion beam induced deposition (FIBID),[54,58,85] where the adsorbed precursor molecules are decomposed by irradiation with a focused ion beam (typically Ga^+ but also lighter ions, such as He^+).[86–89] Since FEBID and FIBID are based on the similar principle of charged-particle beam-induced deposition, they can complement each other, e.g., in terms of the growth dynamics for 3D fabrication, achieving different material properties and structural resolution.[63]

The reliable transition from initial 3D design to the routine delivery of real nanoarchitectures remains a significant challenge. The current major obstacle is the lack of molecular-level understanding of the irradiation-driven chemistry that governs nanostructure formation and growth during the FEBID and FIBID processes.[40,90] Of particular concern is the incorporation of unwanted chemical elements (such as carbon) into the fabricated metal nanostructures,[91] which can reduce or even mask the intended material properties. Most of the available post-processing purification methods developed to mitigate this drawback are not readily applicable to 3D structures due to the severe structural deformation resulting from the high volume loss during impurity removal.[92,93]

FEBID is considered one of the most promising technologies for the controlled direct-write fabrication of complex nanostructures with nanometer resolution.[63] Nanostructures created using this method can be used in electronic devices and other applications, including catalysis and nanoelectrochemistry[56,57] and as sensors, nanoantennas, magnetic devices, surface coatings, and thin films with tailored properties.

An advantage of FEBID 3D nanofabrication is that it can be performed using a conventional scanning electron microscope (SEM) with a mounted gas-injecting system to inject gaseous precursor molecules inside the SEM vacuum chamber. Considering the large number of SEMs installed at universities and research centers worldwide, the development of a reliable and easy-to-use methodology for 3D nanofabrication using SEMs would open up great opportunities for fundamental and applied research.

Novel and more efficient methods of 3D nanofabrication will allow for the miniaturisation of the created electronic nanodevices and their cost-effective production. A better understanding of the mechanisms of radiation-induced formation, growth, and modification of nanostructures will enable the effective optimisation of existing nanofabrication technologies, allowing more precise or better-controlled fabrication and targeting specific compositions and morphologies of the fabricated nanostructures with tailored properties.

Finally, multiscale models similar to those applied for FEBID simulations can be utilised for simulations of other nanofabrication techniques using focused beams of charged particles, such as FIBID, or for other nano-processing methods exploiting ion beams, e.g., focused ion beam-induced etching.[59,94] Further technological applications of the multiscale models include simulations of non-irradiative chemical synthesis[95] for the formation of nanofractals and other complex nanoobjects.

1.6 Multiscale Scenario for Radiation Damage to Biological Systems

The next illustrative example of a multiscale scenario concerns radiation damage (RADAM) caused by irradiation of condensed matter systems with ions. Special attention is given to the radiation-induced phenomena in biological systems. This example highlights an important issue of how the physical radiation-induced processes are coupled to biological systems and can lead to large-scale biological effects, such as cell deactivation, cell mutation, and the bystander effect.[37] This issue is closely related to biomedical applications of radiation, such as different types of radiotherapy, which aim to irradiate tumour cells for their subsequent destruction using different radiation modalities. It should be noted that the basic ideas discussed in this section are applicable to different irradiation modalities and different types of irradiated materials.

The elucidation of the fundamental mechanisms of ion-induced RADAM of biomolecular and biological systems has attracted great interest in recent decades.[37,96] It was motivated by the development of ion beam radiotherapy[38,97,98] and other applications of ions

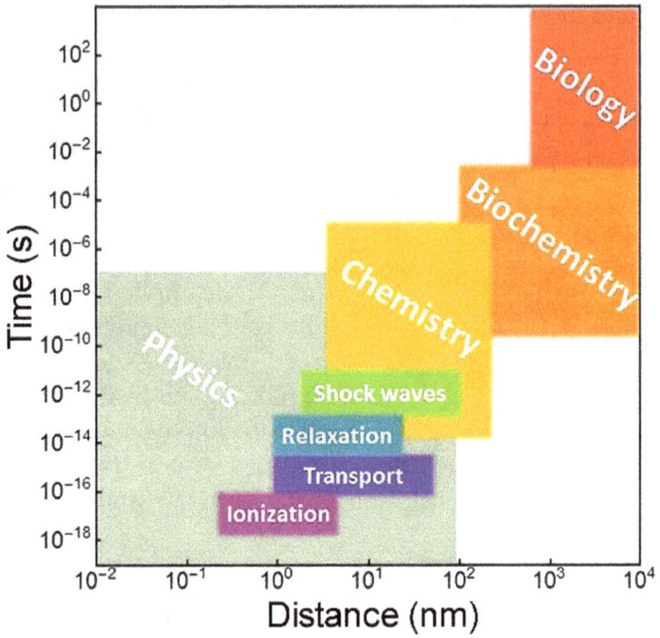

Fig. 1.7. A space-time diagram of features, processes, and disciplines associated with hadron-/ion beam therapy, showing the approximate scales of the key physical phenomena.

Source: Adapted from Ref. 38 with permission from Springer-Verlag.

interacting with biological targets, such as radiation protection in space.[99,100]

Understanding of the cascade of processes induced by irradiation of biomolecular and biological targets with ions and other radiation modalities requires an MM approach[37,38] that could bridge the (sub)-nanoscale atomic and molecular physics with the macroscale biophysics, biochemistry, and biology; see Fig. 1.7.

The multiscale approach (MSA) to ion beam cancer therapy (IBCT) was proposed in Ref. 101, where the first steps were taken to develop this approach. The cited work initiated the development of a phenomenon-based approach to the assessment of RADAM with ions that is fundamentally different from any other method used in the field. The primary objective of the MSA was to understand and quantitatively describe the multiscale scenario of RADAM with ions in the language of physical, chemical, and biological effects at the molecular level. The aim was to relate the initial physical effects of energy loss by projectiles to the biological effects that define cell inactivation. It should be emphasised that the MSA is a non-dosimetric approach, i.e., in the MSA, no damage is defined solely by the locally deposited dose.

The multiscale scenario of the RADAM by ions[6,37,38,101] and its key processes are shown schematically in Fig. 1.8, where panel (a) shows a schematic representation of the scenario[101] and panel (b) shows a qualitative and artistic view.[38] The figure shows several pathways leading from energy loss of a propagating ion to cell apoptosis.

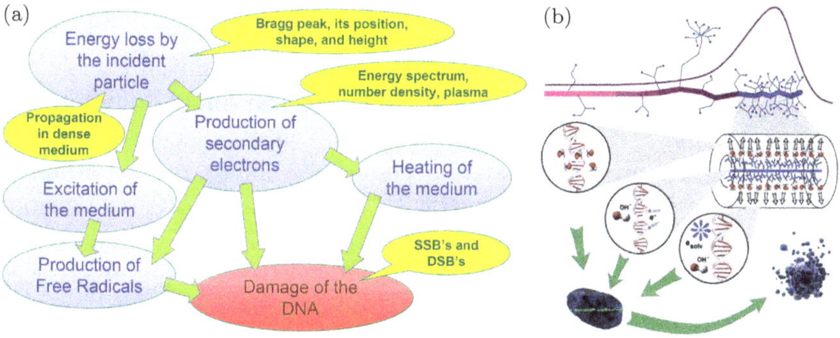

Fig. 1.8. The scenario of biological damage with ions: (a) a schematic representation[101] and (b) an artistic view.[38] Ion propagation ends with a Bragg peak, shown in the top-right corner of panel (b). Panel (b) also shows an ion track segment at the Bragg peak in more detail. Secondary electrons and radicals propagate radially from the ion's path, damaging biomolecules (central circle). These reactive species transfer the energy to the medium within the hot cylinder,[102] which causes a rapid temperature and pressure increase inside this cylinder. The emerging shock wave (shown in the expanding cylinder) due to this local pressure increase may damage biomolecules by stress (left circle).[103–105] Moreover, the shock wave also effectively propagates reactive species (radicals and solvated electrons) to larger distances (right circle).[38,102,105] The bottom-left corner of panel (b) shows an image of a cell nucleus, which is crossed by an ion track visualised through foci (visible in the stained cells). The foci arise where DNA lesions are created and then repaired by enzymes carrying luminescent markers. Unsuccessful repair efforts lead to eminent cell death; an apoptotic cell is shown in the lower-right corner of panel (b).

Source: Panel (a) is reproduced from Ref. 101 with permission from American Physical Society. Panel (b) is reproduced from Ref. 38 with permission from Springer-Verlag.

Figures 1.7 and 1.8 illustrate that the multiscale scenario involves several temporal and spatial scales. It has been shown that the ion energy, and hence the energy of the secondary electrons, is essential for defining the multiscale scenario. The MSA treats the relevant physical, chemical, and biological effects within a single comprehensive framework. The scenario starts with the propagation of ions through the tissue, which in most studies is replaced by liquid water since water makes up about 75% of the tissue mass. The dominant process accompanying the propagation of ions is the ionisation of the molecules of the medium. It is characterised by the presence of a prominent feature known as the Bragg peak in the depth-dose curve. The position of the Bragg peak depends on the initial energy of the ions. In radiotherapy applications, the initial energy of the ions can be varied to place the Bragg peak in the tumour. The position and profile of the Bragg peak as a function of initial energy can be obtained analytically[37,38,106,107] based on the singly differentiated cross-sections (SDCSs) of the ionisation of water molecules with ions. This analytical approach was successfully validated by comparing the calculated depth-dose curve with the results of MC simulations and experiments.[108] The developed methodology is practical for the rapid evaluation of the depth-dose curve analysis and can be used for treatment planning.

An important physical effect predicted by the MSA concerns the significant RADAM arising from the nanoscale shock waves (SWs)[102] generated by ions in a dense dielectric medium (such as liquid water or a biological system) at high LET. This phenomenon occurs because ions can deposit a significant amount of energy on the nanometer scale, leading to strong heating of the medium in the vicinity of ion tracks. The energy lost by the ion is deposited in the medium due to (i) the production, transport, and stopping of secondary electrons and (ii) the relaxation of the electronic excitation energy of the medium into its vibrational degrees of freedom via the electron–phonon coupling mechanism.[102]

It has been shown that the ion-induced SWs play an essential role in the multiscale RADAM scenario; see the reviews in Refs. 6, 37–39,109 and references therein. Two possible mechanisms of DNA damage by the ion-induced SWs have been proposed.[38,103] First, the SWs may cause damage through thermomechanical stress and induce covalent bond breakage in the DNA molecule.[103–105,110–113]

As the strength of ion-induced SWs increases with LET, the SWs contribute significantly to radiation biodamage around the Bragg peak region of ion trajectories.[103] A recent study[105] showed that the thermomechanical stress of the DNA molecule caused by the ion-induced SWs is the dominant mechanism of complex DNA damage for high LET (e.g., iron) ion irradiation, leading to cell inactivation. In addition, the radial collective motion of the medium induced by the SWs contributes to the propagation of highly reactive molecular species, such as OH radicals and solvated electrons, to distances of up to tens of nanometres from the ion track, thus preventing their recombination.[37,38]

Let us now discuss the characteristic energy, time, and spatial scales at which the various processes mentioned above occur. The key phenomena and processes treated by the MSA are (i) the propagation of charged particles in biological media, (ii) radiation-induced fast quantum processes within biomolecular environments, (iii) the time and spatial evolution of track structures and localised energy deposition into biological media, (iv) slower nanoscale post-irradiation relaxation, chemical, and thermalisation processes occurring in the irradiated biological media, and (v) the evaluation of ion-induced biodamage and its link to larger-scale radiobiological phenomena, such as cell survival probabilities and RBE; see Figs. 1.7 and 1.8, as well as Refs. 37–39,101.

The analysis of the differential electron impact ionisation and excitation cross-sections of the water molecules[107,114] has revealed essential features in the energy spectrum of the ejected secondary electrons on the time scales of 10^{-18}–10^{-17} s after the passage of the ion. It has been shown that the dominant fraction of ionised secondary electrons emitted from molecules into the medium in collisions with ions have energies below ~50 eV. More energetic δ-electrons are kinematically suppressed in the Bragg peak region. They are less likely to be emitted in the plateau region preceding the peak. At energies of ~50 eV, the electron transport through the medium is ballistic. At such energies, the cross-sections of electronic collisions with molecules of the medium are nearly isotropic.[115] These facts justified the use of the random walk approximation (i.e., the diffusion mechanism) to describe the secondary electron transport. The approach based on the diffusion equation was successfully developed in Refs. 38,48,101,116,117.

There are several features of secondary electron transport that are relevant to the multiscale scenario of ion-induced RADAM:

- In the vicinity of the Bragg peak, the secondary electrons lose most of their energy within 1–1.5 nm of the ion's path. This process ends within 50 fs of the ion's track.[48] The biologically relevant radiation-induced damage, such as the single- and double-strand breaks (SSBs and DSBs) in the nuclear DNA can be caused by inelastic collisions of secondary electrons with DNA. Low-energy electrons can also create these lesions via their dissociative attachment to DNA. All these processes take place within 3–5 nm of the ion's track.
- The average energy of secondary electrons in the vicinity of the Bragg peak is nearly independent of the projectile's energy and does not depend on the LET of projectiles. Most of these electrons can ionise a few molecules of the medium.[106] Therefore, the number of secondary electrons in the vicinity of the Bragg peak is roughly proportional to the LET.
- Due to the energy loss by secondary electrons within 50 fs and within 1–1.5 nm of the ion's track, the so-called 'hot' cylinder is created. There are no means of immediate or fast transport of this energy away from the cylinder because heat conductivity and diffusion occur slowly on the picosecond timescale. Therefore, the pressure rises within the hot cylinder during the 50–1000 fs period, being the characteristic duration of the electron–phonon coupling processes responsible for the energy transfer from electronic degrees of freedom in the system to its vibrational degrees of freedom. The maximum value of pressure is proportional to the LET. By the end of this period, a significant collective flow associated with an induced shock wave starts, given a sufficiently large LET. Ion-induced shock waves were predicted by the MSA and have been thoroughly studied in a series of works, both analytically and computationally.[38,102–105,110–112,118,119]
- A manifold of reactive species is formed from the molecules ionised by primary projectiles and secondary electrons. The effect of these reactive species on DNA is deemed to be more significant than the direct effect of secondary electrons. Therefore, understanding their production and transport is vital for the assessment of RADAM. The initial fraction of reactive species is formed within

1–2 ps of the ion's passage, i.e., during the transport of secondary electrons followed by the relaxation of molecular excitations and energy transfer to the vibrational degrees of freedom of the system. The number density of such reactive species may be large and, in the first approximation, grows linear with the LET. However, their recombination rates are proportional to the square of their number density. At large LET values, the recombination of reactive species may dominate the transport by diffusion, resulting in the suppression of the number of species that diffuse out of ion tracks. On the other hand, a strong collective flow due to an ion-induced shock wave initiated during 1–2 ps after the ion's passage can propagate the reactive species away from the tracks more effectively than diffusion. This process reduces the recombination rate of the initially created reactive species, thus affecting the initial conditions for the chemical phase.[38,48,111]

- For larger LETs, corresponding to carbon or heavier ions, the nanoscopic cylindrical shock waves created in the vicinity of ion tracks become sufficiently strong to create breakages of molecular covalent bonds, including those in the DNA strands.[103] This process provides an essential – and, at very high LET, the dominating – contribution to the nuclear DNA damage in cells irradiated with ions.[105]

The elements of the MSA described above are related to its physical part. The analytical methods developed for analysing the physical part of the MSA also provide an efficient methodology for assessing chemical effects. On this basis, a biological model for cell inactivation, which involves the concept of a lethal DNA lesion, has been developed. Introducing the cell lethality criterion, the number of such lesions per unit length of the ion's path can be calculated, and the cell survival probability can be obtained. Two hypotheses underlie the concept of lethal damage within the MSA: (i) the inactivation of cells irradiated with ions occurs mainly due to nuclear DNA damage, and (ii) a DNA lesion of a certain complexity is lethal. The second hypothesis originates from a series of papers,[120–124] spanning over three decades. These papers analysed simple DNA lesions (such as SSBs or base damage), DSBs, and complex lesions consisting of a DSB and several simple lesions. After a series of investigations, it was postulated that complex lesions consisting of a DSB and at least two more simple lesions within a length of two DNA twists are lethal, at least for a normal cell.[38,125] This criterion for cell lethality implicitly includes the probability of enzymatic DNA repair. This criterion may be modified for different cancerous cells and some special cell lines.[125] Even more important is that, within the MSA, in contrast to other approaches, each lesion has been associated with an action of an agent, such as a primary particle, secondary electron, or a reactive species, or can be caused by high pressure gradients arising in the medium at the fronts of the nanoscopic cylindrical shock waves induced by propagating ions. An 'action' here means that a lesion is caused by one of the aforementioned processes treated within the MSA that are not necessarily determined by a particular energy deposition but also by many other factors. This feature significantly distinguishes the MSA from the nano- and micro-dosimetric approaches used previously and in most treatment planning models.

With the defined criterion of lethality, the fluence of agents on a specific DNA segment (located at a given distance from an ion's path) is calculated in accordance with the transport mechanism, taking into account collective flows due to ion-induced shock waves. The fluences are weighted with probabilities of chemical processes leading to lesions. The number of SSBs caused by the direct action of ion-induced shock waves at a given LET is derived from reactive MD simulations. Then, the probability of SSBs caused by each particular mechanism leading to such events and a cumulative probability of SSBs are derived based on the Poisson statistics. The yield of lethal lesions per unit length of an ion's path is also subsequently calculated using the Poisson statistics.[38,125] It turns out that this quantity depends on three physical characteristics, namely ion fluence, the LET, and the dose deposited in the cell nucleus, as well as on the biological characteristics — the genome size in an irradiated cell. The average length of all tracks through the cell nucleus can be calculated if two of these physical characteristics are treated independently, e.g., the LET and the dose. The yield of lethal lesions per cell could then be calculated as the product of this length and the yield of lethal lesions per unit length of the ion's path. This analysis enables us to explain the 'overkill' effect, manifesting itself as a decrease in the biological effectiveness of ionising radiation at high LET values. The explanation of this effect is that at high LET values, the energy is deposited into a target cell nucleus by a small number of ions, and this energy is larger than that needed for cell inactivation. As a result, high-LET irradiation produces higher DNA damage than actually required, which leads to a reduction in biological effectiveness.[126] The analysis carried out using MSA also demonstrates that the yield of lethal lesions depends on the dose, the LET, and the oxygen concentration in the medium. The relative biological effectiveness (RBE) can also be derived from the calculated cell survival curves. In Ref. 125, the theoretically established survival curves were successfully compared with those experimentally obtained for several cell lines.

The MSA is a fairly unique framework. There is no other approach that addresses the entire scenario described above. For example, the transport of secondary electrons and radicals and the RADAM induced by these particles have often been studied computationally using track structure MC simulations.[37,96] However, such simulations consider the transport of particles in a static medium at equilibrium. Therefore, they cannot include the full set of physical, chemical, and biological processes discussed above and shown in Figs. 1.4, 1.7, and 1.8. Therefore, these simulations are usually only relevant to the very early stages of the MSA scenario.

Further development of the current understanding of the ion-induced RADAM processes, including the thermomechanical mechanism caused by the ion-induced SWs, will require further elaboration and extension of the following interlinks:

- *An interlink between the track-structure MC methods and classical MD through the IDMD approach.*[40] This will permit the efficient simulations of irradiation-driven chemistry processes in complex molecular systems exposed to radiation. It will also allow IDMD simulations combined with track-structure MC simulations and analytical particle-transport models, enabling the study of the transport of secondary electrons and the dynamics of the irradiated medium in the vicinity of the track of an ion propagating through biological media, such as water or bio molecules/systems (e.g., DNA, proteins, and membranes), in which the track is created and the energy is deposited into the medium. This will enable simulations of the evolution of ion track dynamics on the sub-picosecond and picosecond time scales.

- *An interlink between quantum-chemical methods and MD through the development of reactive force fields for RMD simulations.*[49] On this basis, one can investigate RADAM phenomena caused in the biological medium by the ion-induced SW effect using the RMD approach and establishing the role of the thermomechanical mechanism in different scenarios of RADAM for different ions and biological targets, at different LET values, etc.[103,105,110]
- *An interlink between the outcomes of MD, RMD, and IDMD and the inputs for SD, MC, or any other theoretical approach for modelling radiation-induced biological effects.* This will allow us to simulate various large-scale dynamical processes on a probabilistic level.[49] This development will enable the analysis of the post-SW relaxation dynamics of biological media on time scales ranging from nanoseconds up to seconds, covering the chemical stage of RADAM; see Fig. 1.4.
- *An interlink between the outcomes of RMD and SD with the statistical methods for the evaluation of probabilities of complex RADAM events.* This will allow us to quantify the formation of SSBs and DSBs in DNA, complex DNA damages, etc.[37,38,103,105,110]

A more detailed description of all these and other relevant methodologies can be found in the roadmap paper[7] and references therein.

1.7 MBN Explorer and MBN Studio: Advanced Computational Tools for Multiscale Modelling

The practical computational realisation of the MSA is based on interlinking of the methods used to simulate processes at all the stages of the multiscale scenario shown in Fig. 1.4. The roadmap paper[7] gives an overview of the available theoretical and computational methods with their limits, as well as the associated well-established computer codes. It is important to emphasise that, in the near future, it will not be possible to explore all the scale ranges within the multiscale scenarios, as those discussed above, by simply extending one computational method to the other domain. Therefore, the MM approach is concerned with the methodologies that establish theoretical and computational interlinks between the key methods of different scales, allowing the description and numerical quantification of multiscale phenomena at different stages of the multiscale scenarios. By interlinking different methods, it is possible to overcome the limitations of each individual method and pave the way for a coherent MSA that unifies the description of radiation-induced processes.

Let us summarise the current achievements in the practical computational realisation of the MSA through interlinking different theoretical and computational methods used at different stages of the multiscale scenario of radiation-induced processes shown in Fig. 1.4. Each stage of the multiscale scenario can be treated with theoretical methods describing the dynamics of systems at the corresponding temporal and spatial scales. The following interfaces between the five main stages of the multiscale scenario and related methods can be seen in Fig. 1.4:

- *Interface between stages 1 and 2* deals with the coupling of methods for the description of radiation-induced quantum processes with particle transport theories and associated codes.
- *Interface between stages 1 and 3* deals with the coupling of the QM-based methods with classical MD to describe the dynamics of the irradiated medium dynamics and related phenomena.
- *Interface between stages 2 and 3* deals with the coupling of the methods describing particle propagation in a medium (MC method, analytic approach, and relativistic MD) with the IDMD method. Through this interlink, one can develop and validate the SD-based models interlinking the aforementioned particle propagation techniques with the SD-based description of the irradiated medium dynamics and related phenomena.
- *Interface of stage 3 with stages 4 and 5* deals with the coupling of stages 2 and 3, described by different types of non-relativistic or relativistic MD, with stages 4 and 5 simulated by means of SD. It allows the simulation and analysis of post-irradiation processes in the medium on large temporal and spatial scales.
- *Interface of stages 3 and 4 with subsequent larger-scale stages.* This interface deals with the coupling of the outputs of MD and SD with the inputs of thermodynamics and other macroscopic theories on a larger scale.

Significant efforts have been made in recent years to develop these interfaces and related methodological interlinks.[1,6,37] As a result of this effort, a unique computational platform that allows for the treatment of all of them has been developed on the basis of the MBN Explorer[65] and MBN Studio[127,128] software packages. A review of the practical realisation of all these interfaces and interlinks on the basis of MBN Explorer and MBN Studio, as well as their validation, has been made in the recent roadmap paper.[7]

MBN Explorer[65] is a software package for advanced multiscale simulations of the structure and dynamics of complex molecular MBN systems. It has many unique features and a wide range of applications in physics, chemistry, biology, materials science, industry, and medicine.;[1,6,7] see Fig. 1.9 It is suitable for classical non-relativistic and relativistic MD, Euler dynamics, RMD and IDMD simulations, and SD or MC simulations of various randomly moving MBN systems or processes. These algorithms are applicable to a wide range of molecular systems of different types, such as nano- and biological systems, composite/hybrid systems, nanostructured materials, gases, plasmas, liquids, solids, and their interfaces, with sizes ranging from atomic to mesoscopic. An important feature of MBN Explorer is its ability to study the thermomechanical, chemical, and radiation-induced transformations in all of the above systems.

Such knowledge is needed for a wide range of industrial applications, such as functionalised surface coatings and nanostructured materials, stronger and lighter materials for aircraft and automobiles (performance analysis under extreme conditions), superhard nanostructured materials for mechanical engineering, nanostructured implants for medical applications, superplasticisers for the cement industry (needed for the production of concrete with higher compressive strength), highly efficient batteries and catalysts, drugs, etc. It is also used in a large number of applications of radiation in technology, biological and chemical engineering, and medicine. In most of these applications, it is necessary to identify and/or design specific properties of the system, determined by its molecular structure at the nanoscale, and to ensure their transfer to the macroscopic scale in order to make them functional and usable. Such a transition implies multiscale modelling, which is supported in MBN Explorer by a combination of different simulation techniques, e.g., MD and SD.

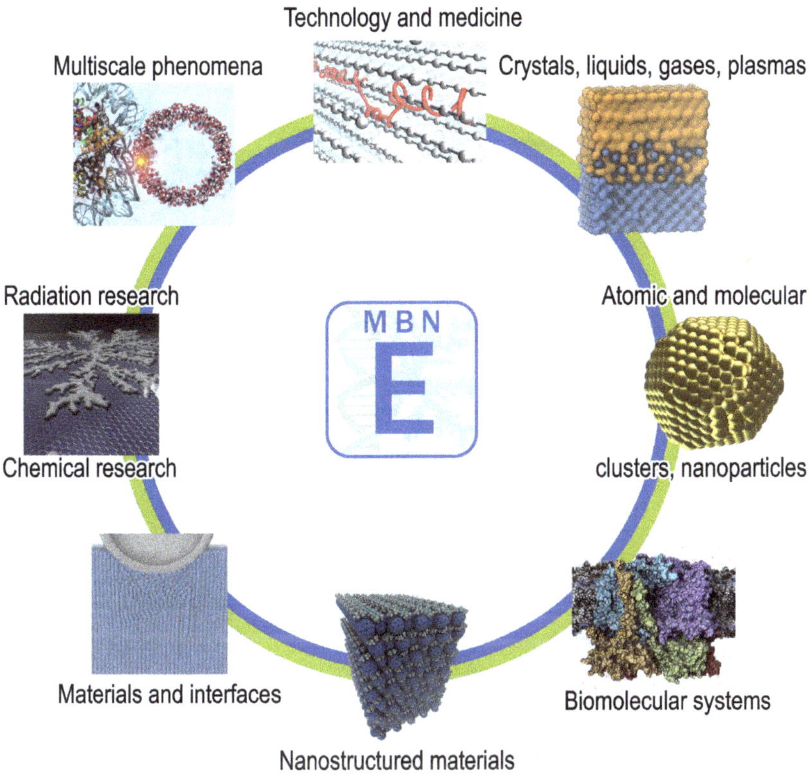

Fig. 1.9. Illustration of the different areas of application of MBN Explorer. *Source*: Image courtesy of MBN Research Center gGmbH,[129] adapted from Ref. 6 with permission from Springer Nature.

To facilitate practical work with MBN Explorer, a special multi-task software toolkit called MBN Studio has been developed.[127,128] It simplifies the modelling of MBN systems, setting up and starting calculations with MBN Explorer, monitoring their progress, and examining the calculation results. MBN Studio can be used for all types of calculations supported by MBN Explorer. The main features of MBN Studio are shown in Fig. 1.10.

MBN Studio allows setup for both standard and application specific projects. Application-specific projects are usually designed for specific systems or processes relevant to new or emerging technologies, and they often involve additional specific parameters and

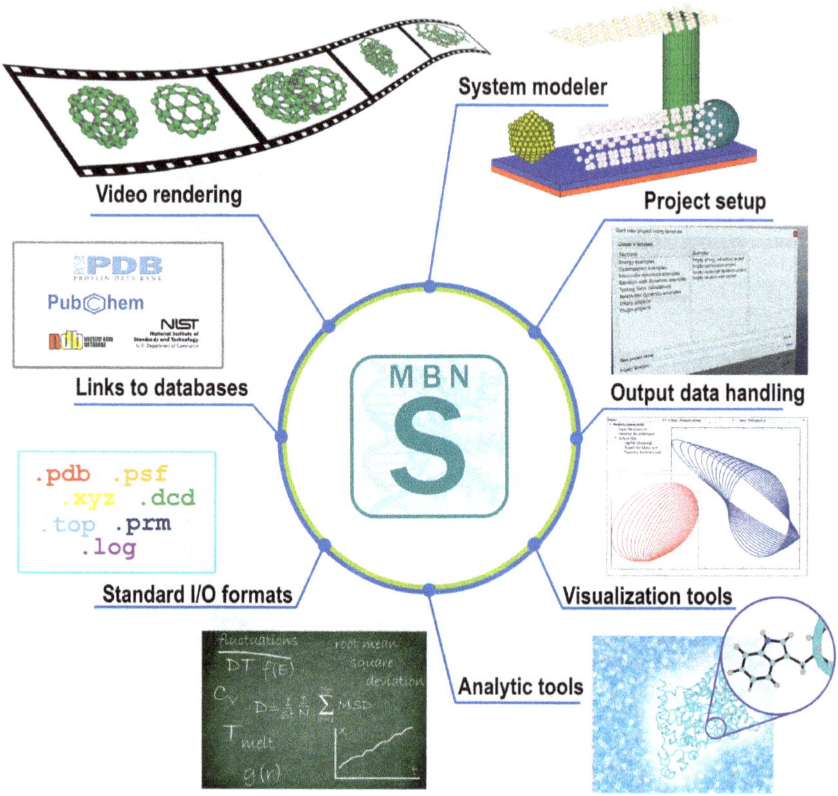

Fig. 1.10. The main features of MBN Studio.
Source: Reproduced from Ref. 128 with permission from Elsevier.

special algorithms. A special modelling plug-in allows one to prepare application-specific projects for simulation quickly and efficiently.

MBN Studio has an advanced built-in tool for MBN system design, which is called MBN system modeller. This plug-in makes it easy to assemble molecular systems of different geometry and composition for further simulation with MBN Explorer.

MBN Studio supports several standard input/output data formats and links to numerous online databases and libraries with coordinates and geometries for atomic clusters, nanoparticles, biomolecules, crystals, and other molecular systems that can be simulated with MBN Explorer.

MBN Studio is equipped with the output data handling, visualisation, and analytic tools, which allow the calculation and analysis of specific properties determined by the output of simulations, such as diffusion coefficients, heat capacities, melting temperatures for solids, and atomic radial distribution functions.

MBN Studio also allows video rendering of all types of dynamics of MBN systems simulated with MBN Explorer.

A detailed description of MBN Explorer and MBN Studio can be found in the books dedicated to their different aspects. For example, all the technical details about the MBN Explorer and MBN Studio software packages and how to use them to simulate the various MBN systems and their features are described in the *Users' Guide*[130] and in *Tutorials*.[129,131] The systematic description of numerous case studies carried out with MBN Explorer and MBN Studio are collected in the dedicated monographs on multiscale modelling.[1,6,44]

1.8 Conclusions and Outlook

This chapter has provided a brief introduction to a recently emerging area of modern research referred to as multiscale computational modelling. This research area has been discussed for complex condensed matter systems, also known as MBN systems, and their behaviour under irradiation. This research area has become one of the most popular research directions in recent years due to the rapid worldwide development of modern computer technology, computing power, and computational methods. Multiscale modelling methods are now being applied to problems in virtually all the traditional natural sciences, such as physics, chemistry, biology, and materials science, as well as in new interdisciplinary research areas at various interfaces between these fields.

The concept of multiscale computational modelling and its practical implementation in the MBN Explorer and MBN Studio software discussed in this chapter is applicable to a wide range of MBN systems, which may be inorganic, organic, or biological, finite or infinite, composed of different molecular species or materials, existing in different phases, and operating under different thermodynamic conditions. Many of the key phenomena associated with the behaviour of irradiated systems are very similar and can be understood using the same theoretical principles and computational approaches. The multiscale nature of such phenomena requires the quantitative description of the radiation-induced effects occurring at different spatial and temporal scales, ranging from the atomic to the macroscopic. The multiscale nature of the effects and the similarity of their manifestation in systems of different origin necessarily bring together all the above disciplines, revealing numerous links and commonalities between them.

The chapter discusses the realistic multiscale scenarios that can be developed to quantitatively describe the multiscale radiation-induced processes and gives examples of concrete case studies that have already been successfully investigated in recent years. The presented examples of recent progress show that the research field is open and requires further investigations of condensed matter systems exposed to radiation of different modalities. The number of new case studies in this research area and related technologies is growing rapidly. The recent roadmap paper[7] provides an overview of the state-of-the-art achievements in this field, the main direction of its development, paving the way for numerous novel challenging multiscale case studies and their connections to important applications in technology and medicine. Therefore, for the discussion of all these issues, refer to this comprehensive work.

This research area is highly relevant to many new and emerging technologies and medical applications. Many lines of research in this area aim to elucidate different phenomena in the context of their potential technological or medical applications. Technological applications within this research area include such important tasks as the optimisation of ion beam cancer therapy and other radiotherapies, the advancement of 3D nanoprinting or controlled nanofabrication, space technologies for radiation protection, atomistic analysis of RADAM and material degradation, the design of new materials, the elucidation of the nature of radiation-induced biological effects, and plasma technologies.

The chapter has also introduced the advanced computational software MBN Explorer and MBN Studio, which allow the practical computational implementation of multiscale modelling. It has highlighted a number of current challenges, particularly in the research areas related to the case studies considered. It has been shown that a theoretical description of the production and propagation of reactive species in biologically relevant media should be extended beyond traditional MC approaches by considering the coupling of the reactive products with the thermally driven mobile biological medium in which they are produced. This can be done by elaborating the links between the theoretical and computational methods describing successive stages of the multiscale scenario for RADAM. Triggered by multiscale simulations, new radiation-induced biodamage mechanisms, such as SW-induced damage, should also be explored experimentally. The first experimental studies of the effects associated with the complex ultrafast dynamics of the medium in the vicinity of ion tracks have recently been performed by exploring the time-resolved picosecond dynamics of liquid water irradiated with laser-accelerated protons.[132,133]

The computational MM approach to study ion-induced RADAM in biomolecular and biological targets will provide further nanoscopic insights into the key phenomena induced by ion irradiation, particularly the mechanisms of biodamage by the ion-induced SWs and the role of thermomechanical mechanisms in the overall scenario of biodamage.[37,38,105] This knowledge can be integrated into current clinical radiotherapy planning protocols, which are based on the concept of 'macrodosimetry', i.e., macroscale calculations of the radiation energy deposited per unit mass of patient. Advances in this direction may lead to the optimisation of the currently used radiotherapy treatment models by taking into account the molecular-level phenomena of biodamage.

An improved understanding of the fundamental molecular mechanisms underlying the effects of ionising radiation on DNA, achieved through MM, will have direct implications for biotechnological and biomedical applications of ion beam irradiation, particularly hadron therapy. Currently, there are fundamental gaps in our understanding of therapeutically relevant issues, such as the operating principles of radiosensitising NPs or the FLASH mechanism, which uses ion beam irradiation at ultrahigh dose rates[134] to spare healthy cells. Advanced atomistic and nanoscale understanding of the above phenomena will open new horizons for their efficient exploitation in biomedical applications.

These and other relevant research areas where multiscale modelling methods have been applied, or have great potential for application, are discussed in the following chapters of this book. However, there are many more relevant topics to be considered in the future.

References

1. Solov'yov IA, Korol AV, Solov'yov AV. Multiscale modeling of complex molecular structure and dynamics with MBN Explorer. Cham: Springer International Publishing. 2017.
2. Landau LD, Lifshitz EM. Quantum mechanics: Non-relativistic theory. 3rd ed. Oxford: Butterworth-Heinemann. 1981.
3. Landau LD, Lifshitz EM. Statistical physics. 3rd ed. Oxford: Pergamon Press. 1981.
4. Landau LD, Lifshitz EM. Statistical physics: Theory of the condensed state. Oxford: Butterworth-Heinemann. 1980.
5. Lifshitz EM, Pitaevskii LP. Physical kinetics. Oxford: Butterworth-Heinemann. 1981.
6. Solov'yov IA, Verkhovtsev AV, Korol AV, Solov'yov AV. Dynamics of systems on the nanoscale. Cham: Springer Nature Switzerland. 2022.
7. Solov'yov AV, *et al.* Condensed matter systems exposed to radiation: Multiscale theory, simulations, and experiment. Chem Rev. 2024;124:8014–8029.
8. Connerade J-P, Solov'yov AV, Greiner W. The science of clusters: An emerging field. Europhys News. 2002;33:200–202.
9. Kim B, Tripp SL, Wei A. Self-organization of large gold nanoparticle arrays. J Am Chem Soc. 2001;123:7955–7956.
10. Eichhorn SH, Yu JK. Directed assembly and self-organization of metal nanoparticles in two and three dimensions. In: Li Q, editor. Anisotropic nanomaterials. Cham: Springer. 2015. p. 289–336.
11. Kim J-Y, *et al.* Crystallization growth and micropatterning on self-assembled conductive polymer nanofilms. J Phys Chem C. 2007;111:11252–11258.
12. Zhang S, Xing M, Li B. Biomimetic layer-by-layer self-assembly of nanofilms, nanocoatings, and 3D scaffolds for tissue engineering. Int J Mol Sci. 2018;19:1641.
13. Mae K, *et al.* Self-organized micro-spiral of single-walled carbon nanotubes. Sci Rep. 2017;7:5267.
14. Macak JM, *et al.* Tio_2 nanotubes: Self-organized electrochemical formation, properties and applications. Curr Opin Solid State Mater Sci. 2007;11:3–18.
15. Shimizu T, *et al.* Self-organized nanotube materials and their application in bioengineering. Polym J. 2014;46:831–858.
16. Fan HJ, Werner P, Zacharias M. Semiconductor nanowires: From self-organization to patterned growth. Small. 2006;2:700–717.

17. Yi D, *et al.* Self-organization and tunable characteristic lengths of two-dimensional hexagonal superlattices of nanowires directly grown on substrates. Small. 2020;16:1606–1613.
18. Zhang X, *et al.* Self-organized growth of nanowires on a graphene film. Cryst Growth Des. 2023;23:3813–3819.
19. Jensen P. Growth of nanostructures by cluster deposition: experiments and simple models. Rev Mod Phys. 1999;71:1695–1735.
20. Lando A, *et al.* Coarsening and pearling instabilities in silver nanofractal aggregates. Phys Rev Lett. 2006;97:133402.
21. Dick VV, Solov'yov IA, Solov'yov AV. Nanoparticles dynamics on a surface: fractal pattern formation and fragmentation. J Phys Conf Ser. 2010;248:012025.
22. Dick VV, Solov'yov IA, Solov'yov AV. Fragmentation pathways of nanofractal structures on surface. Phys Rev B. 2011;84:115408.
23. Panshenskov M, Solov'yov IA, Solov'yov AV. Efficient 3D kinetic Monte Carlo method for modeling of molecular structure and dynamics. J Comput Chem. 2014;35:1317–1329.
24. Bréchignac C, Houdy P, Lahmani M. Nanomaterials and nanochemistry. Cambridge: Cambridge University Press. 2007.
25. Yu H-D, *et al.* Chemical routes to top-down nanofabrication. Chem Soc Rev. 2013;42:6006–6018.
26. Biswas A, *et al.* Advances in top-down and bottom-up surface nanofabrication: Techniques, applications & future prospects. Adv Colloid Interface Sci. 2012;170:2–27.
27. Shimomura M, Sawadaishi T. Bottom-up strategy of materials fabrication: A new trend in nanotechnology of soft materials. Curr Opin Colloid Interface Sci. 2001;6:11–16.
28. Meiwes-Broer K-H. Metal clusters at surfaces: Structure, quantum properties, physical chemistry. Berlin Heidelberg: Springer-Verlag. 2000.
29. Lando A, *et al.* Chemically induced morphology change in cluster-based nanostructures. Eur Phys J D. 2007;43:151–154.
30. Bréchignac C, *et al.* Thermal and chemical nanofractal relaxation. Eur Phys J D. 2003;24:265–268.
31. Liu H, Reinke P. C_{60} thin film growth on graphite: Coexistence of spherical and fractal-dendritic islands. J Chem Phys. 2006;124:164707.
32. Solov'yov IA, *et al.* Multiscale modeling of stochastic dynamics processes with MBN Explorer. J Comput Chem. 2022;43:1442–1458.
33. Solov'yov IA, *et al.* Thermally induced morphological transition of silver fractals. Phys Stat Sol B. 2014;70:251.

34. Workman RL, *et al.* (Particle Data Group). Review of particle physics. Prog Theor Exp Phys. 2022;2022:083C01.
35. Landau LD, Lifshitz EM, Pitaevskii LP. Electrodynamics of continuous media. 2nd ed. Oxford: Elsevier Butterworth-Heinemann. 1984.
36. Berestetskii VB, Lifshitz EM, Pitaevskii LP. Quantum electrodynamics. 2nd ed. Oxford: Elsevier Butterworth-Heinemann. 1982.
37. Solov'yov AV, editor. Nanoscale insights into ion-beam cancer therapy. Cham: Springer International Publishing. 2017.
38. Surdutovich E, Solov'yov AV. Multiscale approach to the physics of radiation damage with ions. Eur Phys J D. 2014;68:353.
39. Surdutovich E, Solov'yov AV. Multiscale modeling for cancer radiotherapies. Cancer Nanotechnol. 2019;10:6.
40. Sushko GB, Solov'yov IA, Solov'yov AV. Molecular dynamics for irradiation driven chemistry: Application to the FEBID process. Eur Phys J D. 2016;70:217.
41. de Vera P, *et al.* Multiscale simulation of the focused electron beam induced deposition process. Sci Rep. 2020;10:20827.
42. Korol AV, Solov'yov AV. Novel lights sources beyond free electron lasers. Cham: Springer Nature Switzerland. 2022.
43. Korol AV, Solov'yov AV. Crystal-based intensive gamma-ray light sources. Eur Phys J D. 2020;74:201.
44. Korol AV, Solov'yov AV, Greiner W. Channeling and radiation in periodically bent crystals. 2nd ed. Heidelberg: Springer-Verlag. 2014.
45. Taylor JR. Scattering theory: The quantum theory of nonrelativistic collisions. Mineola: Dover Publications Inc. 2006.
46. Newton RG. Scattering theory of waves and particles. 2nd ed. Berlin: Springer. 1982.
47. Sushko GB, *et al.* Simulation of ultra-relativistic electrons and positrons channeling in crystals with MBN Explorer. J Comput Phys. 2013;252:404–418.
48. Surdutovich E, Solov'yov AV. Transport of secondary electrons and reactive species in ion tracks. Eur Phys J D. 2015;69:193.
49. Sushko GB, *et al.* Studying chemical reactions in biological systems with MBN Explorer: Implementation of molecular mechanics with dynamical topology. Eur Phys J D. 2016;70:12.
50. Huth M, Porrati F, Dobrovolskiy OV. Focused electron beam induced deposition meets materials science. Microelectron Eng. 2018; 185–186:9–28.
51. Plant SR, Cao L, Palmer RE. Atomic structure control of size-selected gold nanoclusters during formation. J Am Chem Soc. 2014;136:7559–7562.

52. De Teresa JM, *et al.* Review of magnetic nanostructures grown by focused electron beam induced deposition (FEBID). J Phys D Appl Phys. 2016;49:243003.
53. Huth M, *et al.* Focused electron beam induced deposition: A perspective. Beilstein J Nanotechnol. 2012;3:597–619.
54. Utke I, Moshkalev S, Russell P. Nanofabrication using focused ion and electron beams. New York: Oxford University Press. 2012.
55. Cui Z. Nanofabrication: Principles, capabilities and limits. Cham: Springer International Publishing. 2017.
56. Xu W, *et al.* Single-molecule nanocatalysis reveals heterogeneous reaction pathways and catalytic dynamics. Nat Mater. 2008;7: 992–996.
57. Murray RW. Nanoelectrochemistry: Metal nanoparticles, nanoelectrodes, and nanopores. Chem Rev. 2008;108:2688–2720.
58. De Teresa JM, editor. Nanofabrication: Nanolithography techniques and their applications. Bristol: IOP Publishing Ltd. 2020.
59. Utke I, Hoffmann P, Melngailis J. Gas-assisted focused electron beam and ion beam processing and fabrication. J Vac Sci Technol B. 2008;26:1197–1276.
60. Barth S, Huth M, Jungwirth F. Precursors for direct-write nanofabrication with electrons. J Mater Chem C. 2020;8:15884–15919.
61. Prosvetov A, *et al.* Irradiation-driven molecular dynamics simulation of the FEBID process for $Pt(PF_3)_4$. Beilstein J Nanotechnol. 2021;12:1151–1172.
62. Prosvetov A, *et al.* Atomistic simulation of the FEBID-driven growth of iron-based nanostructures. Phys Chem Chem Phys. 2022;24:10807–10819.
63. Winkler R, *et al.* 3D nanoprinting via focused electron beams. J Appl Phys. 2019;125:210901.
64. Huth M, Porrati F, Barth S. Living up to its potential — direct-write nanofabrication with focused electron beams. J Appl Phys. 2021;130:170901.
65. Solov'yov IA, *et al.* MesoBioNano Explorer — A universal program for multiscale computer simulations of complex molecular structure and dynamics. J Comput Chem. 2012;33:2412–2439.
66. Fowlkes JD, Rack PD. Fundamental electron-precursor-solid interactions derived from time-dependent electron-beam-induced deposition simulations and experiments. ACS Nano. 2010;4:1619–1629.
67. de Vera P, *et al.* Reactive molecular dynamics simulations of organometallic compound $W(CO)_6$ fragmentation. Eur Phys J D. 2019;73:215.

68. Andreides B, *et al.* Role of the molecular environment in quenching the irradiation-driven fragmentation of $Fe(CO)_5$: A reactive molecular dynamics study. J Phys Chem A. 2023;127:3757–3767.
69. Prosvetov A, *et al.* Atomistic modeling of thermal effects in focused electron beam-induced deposition of Me_2Au(tfac). Eur Phys J D. 2023;77:15.
70. Solov'yov IA, *et al.* Stochastic dynamics simulation of the focused electron beam induced deposition process. In: The Second Conference "Multiscale Irradiation and Chemistry Driven Processes and Related Technologies"; 2023 Apr 26–28; Prague, Czech Republic. Book of Abstracts. 2023. p. 26.
71. Huth M, *et al.* Conductance regimes of W-based granular metals prepared by electron beam induced deposition. New J Phys. 2009;11:033032.
72. Huth M, *et al.* Temperature-dependent growth characteristics of Nb- and CoFe-based nanostructures by direct-write using focused electron beam-induced deposition. Micromachines. 2020;11:28.
73. Ragesh Kumar TP, *et al.* Electron interactions with the heteronuclear carbonyl precursor $H_2FeRu_3(CO)_{13}$ and comparison with $HFeCo_3(CO)_{12}$: From fundamental gas phase and surface science studies to focused electron beam induced deposition. Beilstein J Nanotechnol. 2018;9:555–579.
74. Wysocki VH, Kenttämaa HI, Cooks RG. Internal energy distributions of isolated ions after activation by various methods. Int J Mass Spectrom Ion Proc. 1987;75:181–208.
75. Beranova S, Wesdemiotis C. Internal energy distributions of tungsten hexacarbonyl ions after neutralization-reionization. J Am Soc Mass Spectrom. 1994;5:1093–1101.
76. Cooks RG, *et al.* Internal energy distributions deposited in doubly and singly charged tungsten hexacarbonyl ions generated by charge stripping, electron impact, and charge exchange. J Am Soc Mass Spectrom. 1990;1:16–27.
77. Wnorowski K, *et al.* Electron ionization of $W(CO)_6$: Appearance energies. Int J Mass Spectrom. 2012;314:42–48.
78. Wnorowski K, *et al.* Low-energy electron interactions with tungsten hexacarbonyl — $W(CO)_6$. Rapid Commun Mass Spectrom. 2012;26:2093–2098.
79. Neustetter M, *et al.* Fragmentation pathways of tungsten hexacarbonyl clusters upon electron ionization. J Chem Phys. 2016;145:054301.
80. Lacko M, *et al.* Electron-induced ionization and dissociative ionization of iron pentacarbonyl molecules. Eur Phys J D. 2015;69:84.

81. Lengyel J, Fedor J, Fárnik MF. Ligand stabilization and charge transfer in dissociative ionization of $Fe(CO)_5$ aggregates. J Phys Chem C. 2016;120:17810–17816.
82. Lengyel J, *et al.* Water-assisted electron-induced chemistry of the nanofabrication precursor iron pentacarbonyl. J Phys Chem A. 2021;125:1919–1926.
83. Massey S, Bass AD, Sanche L. Role of low-energy electrons (<35 eV) in the degradation of $Fe(CO)_5$ for focused electron beam induced deposition applications: Study by electron stimulated desorption of negative and positive ions. J Phys Chem C. 2015;119:12708–12719.
84. Bilgilisoy E, *et al.* Low energy electron- and ion-induced surface reactions of $Fe(CO)_5$ thin films. J Phys Chem C. 2021;125: 17749–17760.
85. Reyntjens S, Puers R. Focused ion beam induced deposition: Fabrication of three-dimensional microstructures and Young's modulus of the deposited material. J Micromech Microeng. 2000;10:181–188.
86. Kometani R, Ishihara S. Nanoelectromechanical device fabrications by 3-D nanotechnology using focused-ion beams. Sci Technol Adv Mater. 2009;10:034501.
87. Cui A, *et al.* Freestanding nanostructures for three-dimensional superconducting nanodevices. Appl Phys Lett. 2012;100:143106.
88. Nanda G, *et al.* Helium ion beam induced growth of hammerhead AFM probes. J Vac Sci Technol B. 2015;33:06F503.
89. Córdoba R, *et al.* Vertical growth of superconducting crystalline hollow nanowires by He^+ focused ion beam induced deposition. Nano Lett. 2018;18:1379–1386.
90. Swiderek P, Marbach H, Hagen CW. Chemistry for electron-induced nanofabrication. Beilstein J Nanotechnol. 2018;9:1317–1320.
91. Botman A, Mulders JJL, Hagen CW. Creating pure nanostructures from electron-beam-induced deposition using purification techniques: A technology perspective. Nanotechnology. 2009;20:372001.
92. Geier B, *et al.* Rapid and highly compact purification for focused electron beam induced deposits: A low temperature approach using electron stimulated H_2O reactions. J Phys Chem C. 2014;25:14009–14016.
93. Fowlkes JD, *et al.* Electron nanoprobe induced oxidation: A simulation of direct-write purification. Phys Chem Chem Phys. 2015;17:18294–18304.
94. Stanford MG, *et al.* Laser-assisted focused He^+ ion beam induced etching with and without XeF_2 gas assist. ACS Appl Mater Interfaces. 2016;16:29155–29162.

95. Ming S, Wheatley AEH. Manipulating morphology and composition in colloidal heterometallic nanopods and nanodendrites. Nanoscale. 2023;15:8814–8824.
96. Garcia Gómez-Tejedor G, Fuss MC, editors. Radiation damage in biomolecular systems. Dordrecht: Springer. 2012.
97. Amaldi U, Kraft G. Radiotherapy with beams of carbon ions. Rep Prog Phys. 2005;68:1861–1882.
98. Schardt D, Elsässer T, Schulz-Ertner D. Heavy-ion tumor therapy: Physical and radiobiological benefits. Rev Mod Phys. 2010;82:383–425.
99. Durante M, Cucinotta FA. Physical basis of radiation protection in space travel. Rev Mod Phys. 2011;83:1245–1281.
100. Kronenberg A, Cucinotta FA. Space radiation protection issues. Health Phys. 2012;103:556–567.
101. Solov'yov AV, *et al.* Physics of ion beam cancer therapy: A multiscale approach. Phys Rev E. 2009;79:011909.
102. Surdutovich E, Solov'yov AV. Shock wave initiated by an ion passing through liquid water. Phys Rev E. 2010;82:051915.
103. Surdutovich E, Yakubovich AV, Solov'yov AV. Biodamage via shock waves initiated by irradiation with ions. Sci Rep. 2013;3:1289.
104. Yakubovich AV, Surdutovich E, Solov'yov AV. Thermomechanical damage of nucleosome by the shock wave initiated by ion passing through liquid water. Nucl Instrum Meth B. 2012;279:135–139.
105. Friis I, *et al.* Lethal DNA damage caused by ion-induced shock waves in cells. Phys Rev E. 2021;104:054408.
106. Surdutovich E, *et al.* Ion-induced electron production in tissue-like media and DNA damage mechanisms. Eur Phys J D. 2009;51:63–71.
107. Scifoni E, Surdutovich E, Solov'yov AV. Spectra of secondary electrons generated in water by energetic ions. Phys Rev E. 2010;81: 021903.
108. Pshenichnov I, Mishustin I, Greiner W. Comparative study of depth-dose distributions for beams of light and heavy nuclei in tissue-like media. Nucl Instrum Meth B. 2008;266:1094–1098.
109. de Vera P, Surdutovich E, Solov'yov AV. The role of shock waves on the biodamage induced by ion beam radiation. Cancer Nanotechnol. 2019;10:5.
110. de Vera P, *et al.* Molecular dynamics study of accelerated ion-induced shock waves in biological media. Eur Phys J D. 2016;70:183.
111. de Vera P, *et al.* Simulation of the ion-induced shock waves effects on the transport of chemically reactive species in ion tracks. Eur Phys J D. 2018;72:147.

112. Friis I, *et al.* Modeling the effect of ion-induced shock waves and DNA breakage with the reactive CHARMM force field. J Comput Chem. 2020;41:2429–2439.
113. Bottländer D, Mücksch C, Urbassek HM. Effect of swift-ion irradiation on DNA molecules: A molecular dynamics study using the REAX force field. Nucl Instrum Meth B. 2015;365:622–625.
114. de Vera P, *et al.* Semiempirical model for the ion impact ionization of complex biological media. Phys Rev Lett. 2013;110:148104.
115. Nikjoo H, *et al.* Track-structure codes in radiation research. Radiat Meas. 2006;41:1052–1074.
116. Surdutovich E, Solov'yov AV. Double strand breaks in DNA resulting from double ionization events. Eur Phys J D. 2012;66:206.
117. Bug MU, *et al.* Nanoscale characterization of ion tracks: MC simulations versus analytical approach. Eur Phys J D. 2012;66:291.
118. Yakubovich AV, Surdutovich E, Solov'yov AV. Atomic and molecular data needs for radiation damage modeling: Multiscale approach. AIP Conf Proc. 2011;1344:230–238.
119. de Vera P, *et al.* Radial doses around energetic ion tracks and the onset of shock waves on the nanoscale. Eur Phys J D. 2017;71:281.
120. Ward JF. DNA damage produced by ionizing radiation in mammalian cells: Identities, mechanisms of formation and repairability. Prog Nucleic Acid Res Mol Biol. 1988;35:95–125.
121. Ward JF. Radiation mutagenesis: The initial DNA lesions responsible. Radiat Res. 1995;142:362–368.
122. Malyarchuk S, Castore R, Harrison L. DNA repair of clustered lesions in mammalian cells: Involvement of non-homologous end-joining. Radiat Res. 2008;36:4872–4882.
123. Malyarchuk S, Castore R, Harrison L. Apex1 can cleave complex clustered DNA lesions in cells. DNA Repair. 2009;8:1343–1354.
124. Sage E, Harrison L. Clustered DNA lesion repair in eukaryotes: Relevance to mutagenesis and cell survival. Mutat Res. 2011;711: 123–133.
125. Verkhovtsev A, Surdutovich E, Solov'yov AV. Multiscale approach predictions for biological outcomes in ion-beam cancer therapy. Sci Rep. 2016;6:27654.
126. Linz U. Ion beam therapy: Fundamentals, technology, clinical applications. Berlin-Heidelberg: Springer-Verlag. 2012.
127. Sushko GB, Solov'yov IA, Solov'yov AV. Meso-Bio-Nano (MBN) Studio. 2015. Available from: http://mbnresearch.com/mbn-studio.
128. Sushko GB, Solov'yov IA, Solov'yov AV. Modeling MesoBioNano systems with MBN Studio made easy. J Mol Graph Model. 2019;88: 247–260.

129. Verkhovtsev AV, *et al.* MBN Explorer and MBN Studio Tutorials. Version 5.0. Frankfurt am Main: MesoBioNano Science Publishing. 2024. ISBN: 979-8328861403.
130. Solov'yov IA, Sushko GB, Solov'yov AV. MBN Explorer users's guide. Version 3.0. Frankfurt am Main: MesoBioNano Science Publishing. 2017. ISBN-13: 978-1975639044.
131. Solov'yov IA, *et al.* MBN Explorer and MBN Studio Tutorials. Version 3.0. Frankfurt am Main: MesoBioNano Science Publishing. 2017. ISBN-13: 978-1976460920.
132. Senje L, *et al.* Experimental investigation of picosecond dynamics following interactions between laser accelerated protons and water. Appl Phys Lett. 2017;110:104102.
133. Prasselsperger A, *et al.* Real-time electron solvation induced by bursts of laser-accelerated protons in liquid water. Phys Rev Lett. 2021;127:186001.
134. Favaudon V, *et al.* Ultrahigh dose-rate FLASH irradiation increases the differential response between normal and tumor tissue in mice. Sci Transl Med. 2014;6:245ra93.

https://doi.org/10.1142/9781800618183_0002

Chapter 2

Interlinks between Atomic and Molecular Physics, Nanoscience, and Other Disciplines

Alexey V. Verkhovtsev* **and Andrey V. Solov'yov†**

MBN Research Center

Altenhöferallee 3, 60438 Frankfurt am Main, Germany

**verkhovtsev@mbnresearch.com*

†*solovyov@mbnresearch.com*

This chapter discusses the application of well-established methods from atomic and molecular physics to the description of the electronic properties of nanosystems, such as atomic clusters and pristine and endohedral carbon fullerenes. Emphasis is placed on the description of the physical phenomena that manifest themselves in photon, electron, and ion collisions with these systems and involve the dynamics of electrons. Specific examples discussed in this chapter include (i) the diffraction and interference phenomena that occur during the interaction of nanosystems with electrons, (ii) the role of the multipole surface and volume plasmon excitations in the formation of electron emission spectra of metal clusters and fullerenes, (iii) the relaxation mechanisms of resonant electronic excitations in metal clusters and fullerenes, which are discussed in the context of evaluating the widths of the corresponding resonant processes, and (iv) the fundamental physical phenomena associated with the confinement of atoms and small molecules inside carbon fullerenes, which represent a nanometre-sized cavity.

2.1 Introduction

This chapter provides an overview of the existing theoretical and computational methodologies, originating from the atomic and molecular physics, for the study of radiation-induced processes in various many-electron systems.[1]

The interaction of radiation with molecular systems takes place via radiation-induced elementary quantum processes, such as ionisation, electronic excitation, electron attachment, charge transfer, and energy relaxation. The quantitative description of these processes is achieved using theoretical and computational methods based on quantum mechanics (*ab initio* methods), such as the Hartree–Fock (HF) method[2] and density functional theory (DFT).[3] These methods have been widely used for decades to calculate the electronic properties of many-body systems, such as atoms and molecules.[2,4] Since the 1980s, these methods have also been increasingly and successfully applied to larger systems, including atomic and molecular clusters[5–7] and biomolecular systems.[8] Due to the increased size and complexity of such systems, their electronic properties have often been studied within simplified model approaches that are based on fundamental physics principles; see Ref. 1 and references therein.

There are numerous books and reviews[2,4,9–12] (with the most recent ones being Refs. 13,14) devoted to methods based on quantum mechanics, especially HF and DFT methods. Various theoretical and computational methods for the description of molecular and condensed matter systems exposed to radiation have recently been reviewed in great detail in Ref. 1. Therefore, the aim of this chapter is not to describe in detail the formulation of the existing methods but to outline the key ideas behind these methods.

A significant part of this chapter is devoted to the discussion of one of the most prominent phenomena related to the electronic subsystem: collective electronic excitations, also known as plasmons (see also earlier reviews on this topic in Refs. 15,16). Therefore, a major focus of this chapter is on the collective phenomena related to electron dynamics in nanosystems, particularly metal clusters and fullerenes (both pristine and endohedral).

2.2 Quantum-Mechanical Description of the Electronic Subsystem

2.2.1 *Many-body theory*

The HF method[2,12] is a computational physics and chemistry tool for solving the time-independent Schrödinger equation for a many-body electronic system. Since there are no known analytical solutions for many-electron systems, the problem is solved numerically. The basic idea of the HF method is that the total N-body electronic wave function of a many-electron system is approximated as an antisymmetrised product of N one-electron wave functions characterised by a set of quantum numbers. By invoking the variational method, one can derive a set of N-coupled one-electron equations, the solution of which yields an HF wave function and the energy of a many-electron system.

The *Dirac–Hartree–Fock (DHF) method* is a well-established method[17,18] for atomic, molecular, cluster, and condensed matter systems containing heavy elements that require accurate treatment of relativistic effects,[19] such as core orbital contraction, spin–orbit coupling, and spin–spin interactions. In this method, a many-electron wave function, the solution of the relativistic Dirac equation, is constructed as an antisymmetrised product of molecular spinors.

The HF method does not take into account the dynamical electron correlation because it uses the closed-shell Slater determinant as the ground-state electronic configuration, where the electrons are forced to be confined to particular orbitals. Three different classes of *ab initio* methods[20] have been developed to calculate the dynamical electron correlation energy, allowing the study of excitation energies, transition states, spectroscopic properties, etc. These methods use the HF electronic configuration as the starting point for the calculations.

In *coupled cluster (CC)*[21–23] and *configuration interaction (CI)*[24] methods, excited determinants are generated from the HF determinant and added to the Slater determinant with appropriate coefficients to improve the wave function while keeping the Hamiltonian fixed. In the CC approach, the correlated wave function of a many-body system is defined as an exponential of the cluster operator T acting on a 'reference' HF state. In CI, the electronic ground-state wave function is constructed as a linear combination of configuration-state functions. Full CI[25] provides a numerically exact solution (within the complete basis set of configuration-state functions, which includes all Slater determinants obtained by exciting all possible electrons to all possible virtual orbitals) to the time-independent, non-relativistic Schrödinger equation. However, due to the complexity of the full CI method, it has only been applied to systems with a few electrons.[26] Full CC is equivalent to full CI; however, due to the high computational costs, it is only applicable to small-size systems containing up to a few tens of atoms.[27]

The most common approximation is the truncation of the CC operator or CI space expansion according to the excitation level relative to the 'reference' HF state. Commonly used truncated CC methods include (i) the CC singles and doubles (CCSD) approach,[28,29] where the cluster operator T is truncated at the two-body component T_2, and (ii) the quasi-perturbative correction to CCSD due to the three-body component T_3, defining the widely used CCSD(T) approximation.[30] For the truncated CI method, the widely employed CI singles and doubles (CISD) wave function includes only those N-electron basis functions representing single or double excitations relative to the reference state.

The other class of post-HF methods is based on the *many-body perturbation theory (MBPT)*, also known as the *Møller–Plesset (MP) perturbation theory.*[31,32] It improves on the HF method by adding electron correlation effects of the second (MP2), third (MP3), fourth (MP4), or higher order. In this approach, the unperturbed Hamiltonian is replaced by a perturbed Hamiltonian with the constraint that the perturbation must be small. The perturbed wave function and perturbed energy of a many-body system are expressed as a power series of a parameter that controls the size of the perturbation. MP methods are computationally less demanding than the CC- and CI-based methods, with MP2 being the least expensive and most widely used *ab initio* method for correcting HF results for correlation effects. MP2, MP3, and MP4 are standard computational methods in many quantum chemistry codes. Higher-level MP calculations, mainly using MP5, can be performed with some codes (e.g., Gaussian), but these methods are rarely used due to their high computational cost.

The HF and DHF methods have been widely used to calculate the electronic structure of atoms, various molecules, atomic clusters, and periodic systems.[33,34] The HF method is also often used as a starting point for more sophisticated *ab initio* methods, such as the MBPT or the random phase approximation method.[35] The ground- and excited-state electronic wave functions for atoms, molecules, and atomic clusters, calculated using the HF and DHF methods, have been commonly used to calculate matrix elements for photo- and electron-impact ionisation and other processes and their corresponding cross-sections (CSs)[15,36] (see also Section 2.2.4).

2.2.2 *Density functional theory*

Density functional theory (DFT) is one of the basic tools for describing the ground-state electronic properties of finite and periodic many-body systems and nanomaterials. The modern version in use today is the Kohn–Sham (KS) DFT.[3] It defines self-consistent equations that are solved for a set of electronic orbitals whose density, $\rho(r)$, is defined to be exactly that of the real system. In DFT, the energy of a many-electron system includes the so-called exchange-correlation (XC) energy, which is defined in terms of $\rho(r)$.

The DFT approach requires the definition of the XC potential to solve the single-electron KS equations, although the form of this potential in the general case is unknown.[37] Many different approximations have therefore been introduced to solve many-electron problems and to describe the physical properties of many-electron systems.[9] The simplest XC approximation is the local density approximation (LDA),[3] which became the popular standard for calculating the properties of solids in the 1970s and 1980s.[4] In the late 1980s, generalised gradient approximations (GGA)[10,38] reached a sufficient level of accuracy for chemical calculations. In the early 1990s, hybrid XC functionals were introduced, where a fraction of the GGA exchange interaction was replaced by the HF exchange contribution, leading to the ubiquitous B3LYP functional. Later, more advanced long-range XC functionals, such as CAM-B3LYP,[39] LC-ωPBE,[40] and ωB97X-D,[41] were developed, allowing for a more accurate description of the long-range electron–electron exchange interactions, which are essential for the correct description of charge transfer and nonlinear optical properties of molecular and condensed matter systems.[42,43]

Conventional local and semi-local XC functionals do not describe long-range dispersion (van der Waals) interactions, which play an important role in the formation, stability, and functioning of molecules and materials. To address this problem, dispersion corrections to the standard KS DFT have been developed; see, e.g., the reviews in Refs. 44,45.

DFT has found many applications in chemistry and materials science by calculating the electronic ground-state properties of various systems.[11,46–49] In solid-state calculations, the LDA is still commonly used together with plane-wave basis sets, as an electron-gas approach is more appropriate for electrons delocalised through an infinite solid. However, in calculations involving finite molecular systems, more accurate long-range corrected XC functionals are typically required for a quantitative description of charge transfer processes, e.g., in photochemistry and quantum biology studies.

2.2.3 *Time-dependent density functional theory*

The ground-state DFT cannot describe electronic excited states; therefore, many important irradiation-driven physical properties, such as optical absorption and emission, response to time-dependent fields, and the dynamical dielectric function. *Time-dependent density functional theory (TDDFT)* extends the basic ideas of ground-state DFT to allow the treatment of electronic excitations or, more generally, time-dependent phenomena.[50]

TDDFT is an approximate method for solving the time-dependent Schrödinger equation, which allows one to study the properties of many-electron systems as a function of time. In this approach, the time-dependent Schrödinger equation is substituted with a set of time-dependent single-particle KS equations.[51] In analogy to the KS method,[3] which is the basic principle of the ground-state DFT, TDDFT considers an external time-dependent potential $v_{\text{ext}}(\mathbf{r}, t)$ to describe the motion of a system of non-interacting particles and the time-dependent electron density $\rho(\mathbf{r}, t)$ of a real many-electron system.

TDDFT is usually realised either in the linear-response regime, where the electronic density of a system is considered in the frequency domain as a first-order response to an external perturbation potential, or directly in the time domain (the so-called real-time

TDDFT) by evolving the KS wave functions in time.[52] In the time domain, the time evolution of an induced dipole moment is studied by integrating time-dependent KS equations, while in the frequency domain the Fourier transform of time-dependent quantities is performed. Depending on the method used to represent single-particle KS orbitals in TDDFT calculations, either a three-dimensional real-space grid or a plane-wave basis set is used. The main advantage of such an approach is that the convergence of the numerical solution can be controlled in each case by a single parameter, namely the grid spacing in real space or the kinetic energy cutoff in a plane-wave approach.[53]

TDDFT has been used extensively to calculate excitation energies and optical absorption spectra of atoms, molecules, atomic clusters (containing up to $\sim 10^2$ atoms), and solids.[54–56] It has been widely used for photophysical and photochemical applications to study the excited-state dynamics (photoluminescence, electron transfer, etc.) in organic dye molecules and biomolecules.[57–60] It has also been used to calculate electron energy loss spectra for different solids at different values of the transferred momentum.[61,62] With the advent of attosecond laser pulses, there is a great demand for efficient methods to study the time-dependent evolution of systems beyond the linear response. As such, the real-time TDDFT is particularly useful for studying the ultrafast dynamics of atoms, molecules, atomic clusters, and solids in strong laser fields.[63,64]

2.2.4 *Ab initio and model approaches for calculating scattering cross-sections*

The scattering CS is an important quantity for understanding the physical interactions between radiation and a target. The quantum-based theory of scattering developed over the past century has been described in numerous textbooks on quantum mechanics, many-body quantum theory, and quantum collision theory.[65–69] These books describe in great detail the theoretical methods (such as the many-body scattering theory, Green's function, S-matrix formalism, partial-wave expansion, perturbation theory, the Born approximation, eikonal approximation, and quasiclassical approximation) used to calculate the scattering wave function and related physical quantities, such as scattering amplitudes, scattering phase shifts, and scattering CSs. In addition, various theoretical methods have been developed to describe scattering processes involving relativistic particles.[70]

From a computational point of view, the scattering processes involving single atoms and simple few-atom molecules have been extensively described by various *ab initio* methods, including *many-body scattering theory*,[71,72] *configuration interaction*,[73] and *R-matrix methods*.[74–76] Due to the computational complexity of the *ab initio* methods, they have mainly been used to calculate the CSs of low-energy electrons scattering from atoms and simple few-atom molecular targets, while they are impractical (or even unfeasible) for larger molecular systems and CSs above the ionisation potential of most molecular targets.

The need for photon-, electron-, and ion-impact molecular ionisation CSs as input data for various radiation/particle transport modelling tools has stimulated the use of simpler (empirical and semi-empirical) theoretical models for the evaluation of atomic and molecular ionisation CSs. The simplest approaches are based on the additivity rule,[77] where the molecular ionisation CS is derived by adding the ionisation CSs, corresponding to the atomic constituents of a molecular system. In addition to these relatively simple approaches, several semi-empirical approaches have been developed. The most commonly used methods for calculating ionisation CSs are the *Deutsch–Märk formalism*[78] and the more rigorous *binary-encounter dipole (BED)* and *binary-encounter Bethe (BEB) models*,[79] which combine the additivity rule concept with molecular structure information calculated using quantum mechanics.

Both the Deutsch–Märk formalism and the BED/BEB formalism use an additivity concept, so that the ionisation CS of a molecule is added up from the contributions resulting from the ejection of an electron from the different molecular orbitals. A detailed review of the BED, BEB, and related models can be found in Ref. 80. The Deutsch–Märk formalism was reviewed in Ref. 78. In brief, the BED model requires, as input,the differential dipole oscillator strength (DOS) values of the target, which can be derived from theoretical or experimental photoionisation CSs. The BEB model is a simplification of the BED model, in which the DOS term is approximated by a simple function of the secondary electron energy.[79] In the BEB framework, molecular ionisation CSs are evaluated using an analytical formula that requires only the incident particle energy and the binding and kinetic energies of molecular orbitals of the target. These energies can be obtained from electronic structure calculations, e.g., using the HF method (see Section 2.2.1). The BEB method calculates the electron-impact molecular ionisation CS over the incident electron energies ranging from the ionisation threshold up to a few keV or even up to thousands of keV using the relativistic version of BEB.[81]

Other analytical approaches exist for calculating the CSs of inelastic scattering (in terms of the energy transferred from the projectile to the medium) or ionisation (in particular, as a function of the kinetic energy of the emitted secondary electrons) and the related characteristics, such as the stopping power, defined as the average energy loss of the projectile per its unit path length. The ionisation CS for different atomic and molecular targets irradiated with protons and heavier ions is commonly calculated using the semi-empirical Rudd's model,[82] which is based on a combination of experimental data and calculations within the plane-wave Born approximation. An alternative theoretical method[83] for the calculation of ionisation CSs is based on the calculation of the energy-loss function of the target medium, $\mathrm{Im}(-1/\varepsilon(E,q))$, where $\varepsilon(E,q)$ is the complex dielectric function and $\hbar q$ and E are the momentum and energy transferred in the electronic excitation, respectively. This formalism allows us to obtain the charged-particle-impact ionisation CSs for various condensed media, including liquid water and other biologically relevant media containing sugars, amino acids, etc., which are of great importance for the study of radiation damage phenomena in biological systems.[84,85]

2.3 Jellium Model for Metal Clusters and Fullerenes

Metallic clusters are characterised by the complete delocalisation of their valence electrons. To some extent, this applies to some non-metal systems, such as carbon fullerenes, where the delocalisation of electrons takes place in the vicinity of the fullerene cage. The delocalised valence electrons often play an important role in the formation of the CSs of various collision processes, such as photoabsorption and photoionisation, as well as elastic and inelastic scattering of charged particles. These phenomena can be studied using various quantum-mechanical approaches described in Section 2.2.

Solutions to many problems in atomic cluster physics are based on the so-called jellium model,[5,6,15,86–88] which assumes that a many-electron system is the sum of two interacting subsystems: a valence-electron subsystem and a positively charged ionic core. The valence electrons move in the field created by all the cluster ions. In the jellium model, the detailed ionic structure of the system is smeared out and substituted by the uniform (often spherically symmetric) distribution of the positive charge:

$$n(r) = \frac{N_{\rm at}}{V}\,\Theta(R-r). \tag{2.1}$$

Here, $N_{\rm at}$ is the number of atoms in the cluster; $V = \frac{4}{3}\pi R^3$ is the cluster volume; R is the core radius, which depends on the number of atoms in the cluster as $R = r_s\, N_{\rm at}^{1/3}$; r_s is the so-called Wigner–Seitz radius, which describes the average distance between atoms in the bulk material; and $\Theta(R-r)$ is the Heaviside step function.

The potential generated by the spherically symmetric ionic core of the radius R reads as

$$U_{\rm core}(r) = \begin{cases} -\dfrac{Z}{2R}\left[3 - \left(\dfrac{r}{R}\right)^2\right], & r \le R \\ -\dfrac{Z}{r}, & r > R, \end{cases} \tag{2.2}$$

where Z is the charge of the ionic core. For a neutral cluster, Z is equal to the number of delocalised valence electrons N, which is equal to the number of atoms $N_{\rm at}$ for clusters of monovalent elements (e.g., sodium or potassium). For singly charged cluster anions, the relation is $Z = N - 1$.

This model describes the electron density in terms of single-electron wave functions that extend over the entire cluster. In a spherically symmetric field, a single-electron wave function is written as a product of radial, angular, and spin wave functions:

$$\varphi_{nlm\sigma} = \frac{1}{r}P_{nl}(r)Y_{lm}(\theta,\phi)X_{\sigma}, \tag{2.3}$$

where n, l, m, and σ are the principal, orbital, magnetic, and spin quantum numbers, respectively. The self-consistent solution of the system of N single-particle equations yields the single-electron wave functions φ_i ($i \equiv nlm\sigma$) and the corresponding energies ε_i.

The electronic shell structure of neutral and charged metal clusters corresponds to that in a short-range spherical potential. If the spin–orbit interaction is neglected, the shell with quantum numbers $\{n, l\}$ accommodates $2(2l+1)$ electrons. For example, the ground-state electronic configurations of the selected neutral sodium clusters are defined as follows:

$$1s^2 2p^6 \text{ for } \mathrm{Na}_8,$$
$$1s^2 2p^6 3d^{10} 2s^2 \text{ for } \mathrm{Na}_{20},$$
$$1s^2 2p^6 3d^{10} 2s^2 4f^{14} 3p^6 5g^{18} \text{ for } \mathrm{Na}_{58}.$$

This approach yields the electronic shell structure of free alkali metal clusters, which was described theoretically and observed experimentally in Refs. 89,90.

Shortly after the discovery of the electronic shell structure in free alkali metal clusters,[89,90] it was realised that the detailed dependence of ionisation potentials and other properties of small metal clusters on the number of cluster atoms can be understood as a consequence of non-spherical cluster shapes[91] by analogy with the nuclear shell model. Direct evidence for cluster deformation was found experimentally.[92–94]

Kohn–Sham calculations for spherical[90,95] metal clusters have been generalised to spheroidal[96,97] and more general axial shapes of light clusters.[98–101] In Ref. 102, the shapes of deformed axially symmetric metal clusters were studied in the framework of the HF jellium model.

In Refs. 102–104, the dynamic jellium model for metal clusters was proposed. This model treats simultaneously the vibrational modes of the ionic jellium background, the quantised electron motion, and the interaction between the electronic and ionic subsystems. Quantised electron motion was described on the basis of the HF approach. This model enabled the description of the widths and relaxation times of electron excitations in metal clusters beyond the adiabatic approximation.

Many-electron correlations for deformed metal clusters were analysed in Ref. 105. In the cited paper, the binding energies per atom, ionisation potentials, deformation parameters, and optimised values of the Wigner–Seitz radii were calculated using the HF and LDA methods for neutral and singly charged sodium clusters with the number of atoms as $N \leq 20$. The characteristics calculated within the framework of the deformed jellium model were compared with the results obtained from *ab initio* simulations of the electronic and ionic structures of the clusters based on DFT and systematic post-HF many-body perturbation theory, taking into account all the electrons in the system. The comparison revealed a significant role of the cluster shape deformations in the formation of cluster properties and a reasonable level of applicability of the deformed jellium model.

The jellium model has also been widely used to study the electronic structure of hollow cluster systems, particularly carbon fullerenes.[106–111] The molecule C_{60} is probably the most famous example of a hollow, highly symmetric, almost spherical icosahedral structure. The valence $2s^22p^2$ electrons in each carbon atom form a cloud of delocalised electrons, while the inner-shell $1s^2$ electrons are treated as frozen and not taken into account. Therefore, 240 delocalised electrons in C_{60} are considered. The valence electrons move in a spherically symmetric central field, so that one can construct the electronic configuration described by the unique set of quantum numbers $\{n, l\}$, where n and l are the principal and orbital quantum numbers, respectively.

2.4 Plasmon Resonance Approximation

As described in the previous section, the delocalised valence electrons often play an important role in the formation of the CSs of various collision processes involving metal clusters and nanoparticles (NPs), fullerenes, and other nanoscale systems, such as polycyclic aromatic hydrocarbons. Plasmon excitations, i.e., collective excitations of delocalised valence electrons, appear as prominent resonances in the CSs, and the position of the resonance peak depends strongly on the type of the system.[112–115] The investigation of electron collisions with metal clusters and fullerenes in the gas phase has led to the manifestation of the interference and diffraction phenomena,[116–120] the role of surface and volume plasmon excitations in the formation of electron energy loss spectra,[116–119,121–123] and the total inelastic scattering,[116,117] photoabsorption[124] and photoionisation CSs.[107,125]

An effective method for evaluating the contribution of plasmons to the excitation/ionisation spectra of the aforementioned systems is the so-called plasmon resonance approximation (PRA).[15,124,126] This approach postulates that the dominant contribution to the CS in the vicinity of the plasmon resonance comes from collective electron excitations, while single-particle effects give much smaller contributions.[117,121] Over the past decades, the PRA approach has provided a clear physical explanation for the resonant-like structures in photoionisation spectra[124,125] and differential inelastic scattering CSs[116,118,119,121–123] of metal clusters and carbon fullerenes under photon, electron, and ion impact.

The PRA relies on a few parameters, including the oscillator strength of the plasmon excitation, as well as the position of the plasmon resonance peak and its width. The choice of these parameters can be justified by comparing the model-based spectra either with experimental data or with the results of more accurate *ab initio* calculations. The main aspects of the PRA are outlined in this section.

Let us consider a system with a homogeneous, spherically symmetric distribution of the negative charge over the system's volume. The equilibrium electron density distribution, $\rho_0(r)$, is expressed via the number N of delocalised electrons and the system's volume V:

$$\rho_0 = \begin{cases} N/V, & R_1 \le r \le R_2 \\ 0, & r < R_1,\ r > R_2. \end{cases} \tag{2.4}$$

In the general case, the volume of the system can be expressed as

$$V = \frac{4\pi}{3}(R_2^3 - R_1^3) = \frac{4\pi}{3}R_2^3(1 - \xi^3), \tag{2.5}$$

where R_1 and R_2 are the inner and outer radii of the system, respectively, and $\xi = R_1/R_2 \le 1$ is the ratio of the two radii.

The spherical-shell model defined by Eqs. (2.4) and (2.5) is applicable for any spherically symmetric system with an arbitrary value of the ratio ξ. The limiting case of $\xi \to 0$ describes a metallic cluster or NP with an electron density distribution over the full sphere of a radius R:

$$\rho_0(r) = \frac{N}{V}\,\Theta(R - r). \tag{2.6}$$

In this limit, $R_1 \to 0$ and $R \equiv R_2$ so that the cluster volume is $V = 4\pi R^3/3$. The limit $\xi = 1$ or $R_1 \to R_2 \equiv R$ represents a system with an electron density distribution over an infinitely thin sphere R. This model was used previously to study the photoionisation[107] and the electron scattering[116,118,119] processes of carbon fullerenes.

The total electron density of the system at point r and time t is introduced as $\rho(\mathbf{r}, t) = \rho_0(\mathbf{r}) + \delta\rho(\mathbf{r}, t)$, where $\rho_0(\mathbf{r})$ denotes the stationary distribution of the negative charge at point r and $\delta\rho(\mathbf{r}, t)$ is the density variation caused by the interaction with an external electric field. The collective motion of the electron density is described by the Euler and continuity equations.[124]

The Euler equation couples the acceleration, $\mathrm{d}\vec{v}(\mathbf{r}, t)/\mathrm{d}t$, of the electron density to the total electric field $\mathbf{E}$ acting on the system at the point $(\mathbf{r}, t)$:

$$\frac{\mathrm{d}\vec{v}(\mathbf{r}, t)}{\mathrm{d}t} = \mathbf{E}(\mathbf{r}, t). \tag{2.7}$$

The electric field $\mathbf{E}$ includes both the external field acting on the system and the polarisation contribution due to the electron density variation $\delta\rho(\mathbf{r}, t)$:

$$\mathbf{E}(\mathbf{r}, t) = -\vec{\nabla}\phi(\mathbf{r}, t) - \vec{\nabla}\int \frac{\delta\rho(\mathbf{r}', t)}{|\mathbf{r} - \mathbf{r}'|}\mathrm{d}\mathbf{r}', \tag{2.8}$$

where $\phi(\mathbf{r}, t)$ is the scalar potential of the external field.

The motion of the electron density in the system obeys the continuity equation, which reads

$$\frac{\partial\rho(\mathbf{r}, t)}{\partial t} + \vec{\nabla}\cdot\left(\rho(\mathbf{r}, t)\,\vec{v}(\mathbf{r}, t)\right) = 0. \tag{2.9}$$

In the case of a spherically symmetric density distribution, $\rho_0(\mathbf{r}) = \rho_0(r)$, and one can exclude angular variables from the above equations by expanding the functions $\phi(\mathbf{r})$, $\delta\rho(\mathbf{r})$, and $|\mathbf{r} - \mathbf{r}'|^{-1}$ into spherical harmonics and integrating over spherical angles of the vector $\mathbf{r}$. By performing some algebraic transformations,[126] one obtains a general equation for the variation of the electron density in an arbitrary spherically symmetric system:

$$(\omega^2 - 4\pi\rho_0(r))\delta\rho_l(r) + 4\pi\frac{\rho_0'(r)}{\Pi_l^2}\int_0^\infty g_l(r,r')\,\delta\rho_l(r')\,\mathrm{d}r'$$
$$= q^2\rho_0(r)\phi_l(r) - \rho_0'(r)\phi_l'(r), \tag{2.10}$$

where $\phi_l(r)$ is the scalar potential, ω is the frequency of the external field, and the notation $\Pi_l = \sqrt{2l+1}$ is introduced. The function $g_l(r, r')$ is defined as

$$g_l(r,r') = l\left(\frac{r}{r'}\right)^{l-1}\Theta(r'-r) - (l+1)\left(\frac{r'}{r}\right)^{l+2}\Theta(r-r'),$$

where $\Theta(x)$ is the Heaviside step function.

Consider a system with a homogeneous charge distribution over a spherical shell of a finite width, $\Delta R = R_2 - R_1$ (see the left panel of Fig. 2.1):

$$\rho_0(r) = \rho_0\,\Theta(r - R_1)\Theta(R_2 - r), \tag{2.11}$$

where ρ_0 is given by Eq. (2.4). The derivative of $\rho_0(r)$ is given by

$$\rho_0'(r) = \rho_0\,(\delta(r - R_1) - \delta(r - R_2)), \tag{2.12}$$

where $\delta(x)$ is the delta function. The solution to Eq. (2.10) for such a system is sought in the following form:

$$\delta\rho_l(r) = \delta\varrho_l(r)\Theta(r - R_1)\Theta(R_2 - r) + \sigma_l^{(1)}\delta(r - R_1) + \sigma_l^{(2)}\delta(r - R_2), \tag{2.13}$$

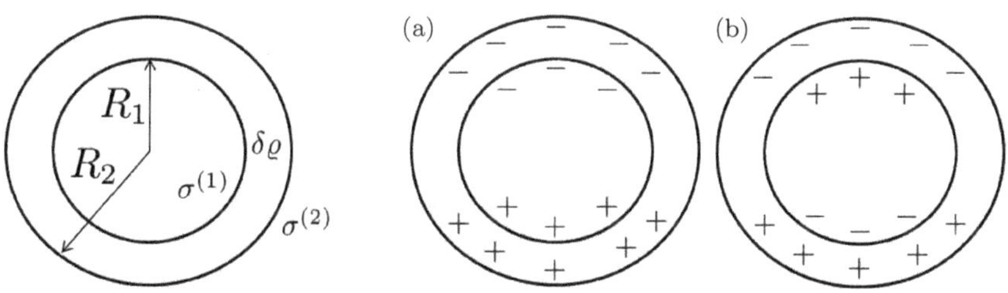

Fig. 2.1. Left panel: exposure of a spherical shell of a finite width, $\Delta R = R_2 - R_1$, to an external electric field causes a variation in the surface charge densities, $\sigma^{(1,2)}$, and the volume charge density, $\delta\varrho$. Right panel: representation of the symmetric (a) and antisymmetric (b) modes of the surface plasmon. *Source*: Reproduced from Ref. 126 with permission from Springer Nature.

where $\delta\varrho_l(r)$ describes the volume density variation inside the finite-width spherical shell and $\sigma_l^{(1,2)}$ are variations of the surface charge densities at the inner and outer surfaces of the shell, respectively (see Fig. 2.1). The volume density variation leads to the formation of the volume plasmon, while the surface density variations lead to the formation of the surface plasmon, which has two normal modes: the symmetric and the antisymmetric.

2.4.1 *Interaction with a uniform external field*

The case of a uniform external field describes the interaction of a system with an electromagnetic field. It is assumed that the wavelength of the electromagnetic radiation is much larger than the typical size of the system, i.e., the condition $\omega R \ll 1$ is fulfilled. This condition implies the validity of the dipole approximation. In this limit, the wave vector $q = 0$ and Eq. (2.10) for the multipole variation of the electron density in a spherically symmetric system becomes the following:

$$(\omega^2 - 4\pi\rho_0(r))\delta\rho_l(r) + 4\pi\frac{\rho_0'(r)}{\Pi_l^2}\int_0^\infty g_l(r,r')\delta\rho_l(r')\mathrm{d}r' = -\rho_0'(r)\phi_l'(r). \tag{2.14}$$

Substituting Eqs. (2.11)–(2.13) into (2.14) and taking into account that, in the dipole approximation, the field intensity ϕ_l' does not depend on the spatial coordinate, $\phi_l'(R_1) = \phi_l'(R_2) \equiv \phi_l'$, one derives the following equation:

$$\begin{aligned}
&(w-1)\delta\varrho_l(r)\Theta(r-R_1)\Theta(R_2-r)\\
&\quad + \left(w\sigma_l^{(1)} + I_l^{(1)} - \frac{l+1}{\Pi_l^2}\sigma_l^{(1)} + \frac{l}{\Pi_l^2}\sigma_l^{(2)}\,\xi^{l-1}\right)\delta(r-R_1)\\
&\quad + \left(w\sigma_l^{(2)} + I_l^{(2)} + \frac{l+1}{\Pi_l^2}\sigma_l^{(1)}\xi^{l+2} - \frac{l}{\Pi_l^2}\sigma_l^{(2)}\right)\delta(r-R_2)\\
&= -\frac{\phi_l'}{4\pi}\delta(r-R_1) + \frac{\phi_l'}{4\pi}\delta(r-R_2),
\end{aligned} \tag{2.15}$$

where

$$w = \omega^2/\omega_p^2, \tag{2.16}$$

ω_p is the volume plasmon frequency associated with the density ρ_0,

$$\omega_p = \sqrt{4\pi\rho_0} = \sqrt{\frac{3N}{R_2^3 - R_1^3}}, \tag{2.17}$$

and

$$I_l^{(1)} = \frac{l}{\Pi_l^2} R_1^{l-1} \int_{R_1}^{R_2} \frac{\delta\varrho_l(x)}{x^{l-1}}\,\mathrm{d}x, \quad I_l^{(2)} = \frac{l+1}{\Pi_l^2}\frac{1}{R_2^{l+2}} \int_{R_1}^{R_2} x^{l+2}\delta\varrho_l(x)\mathrm{d}x. \tag{2.18}$$

Matching the terms of different types on the right- and left-hand sides of Eq. (2.15) gives expressions which define the variations of the volume and surface charge densities. The solution corresponding to the variation of the volume density is as follows:

$$(w-1)\delta\varrho_l(r)\Theta(r-R_1)\Theta(R_2-r) = 0. \tag{2.19}$$

This means that no volume plasmon can be formed in the system due to interaction with a uniform external field. The volume plasmon only manifests itself when the system interacts with a non-uniform external field, for example, in collisions with charged particles. The total variation of the electron density is therefore described only by the surface density contributions:

$$\delta\rho_l(r) = \sigma_l^{(1)}\delta(r-R_1) + \sigma_l^{(2)}\delta(r-R_2), \tag{2.20}$$

which can be defined from the following system of coupled equations:

$$\begin{cases} \left(w - \dfrac{l+1}{\Pi_l^2}\right)\sigma_l^{(1)} + \dfrac{l}{\Pi_l^2}\xi^{l-1}\sigma_l^{(2)} = -\dfrac{\phi_l'}{4\pi} \\ \dfrac{l}{\Pi_l^2}\xi^{l+2}\sigma_l^{(1)} + \left(w - \dfrac{l}{\Pi_l^2}\right)\sigma_l^{(2)} = \dfrac{\phi_l'}{4\pi}. \end{cases} \tag{2.21}$$

Variation in the surface charge densities, $\sigma_l^{(1,2)}$, results in the formation of two coupled modes of surface plasmon oscillations.[126–129] Frequencies of the symmetric, $\omega_l^{(s)}$, and antisymmetric, $\omega_l^{(a)}$, surface plasmons of multipolarity l are given by the following expression:

$$\omega_l^{(s/a)} = \left(1 \mp \frac{1}{2l+1}\sqrt{1 + 4l(l+1)\xi^{2l+1}}\right)^{1/2} \frac{\omega_p}{\sqrt{2}}, \tag{2.22}$$

where '−' and '+' stand for symmetric and antisymmetric modes, respectively. In the symmetric mode, the charge densities of the two surfaces oscillate in phase, while in the antisymmetric mode, they are out of phase (see Fig. 2.1). Since only dipole excitations may arise in the system due to interaction with the external electromagnetic field, the case of interest is $l = 1$. The expression for the resonant frequencies of the surface plasmons is then reduced to

$$\omega^{(s/a)} = \left[\frac{N^{(s/a)}}{2(R_2^3 - R_1^3)} (3 \mp p) \right]^{1/2} , \tag{2.23}$$

where $p = \sqrt{1 + 8\xi^3}$, with $\xi = R_1/R_2$ being the ratio of the inner to the outer radius. The values

$$N^{(s)} = N \frac{p+1}{2p}, \quad N^{(a)} = N \frac{p-1}{2p} \tag{2.24}$$

are the number of delocalised electrons which are involved in each plasmon mode. They obey the sum rule $N^{(s)} + N^{(a)} = N$, where N is the total number of delocalised electrons participating in the collective excitation.

The CS of photoionisation by a single photon is given by the general expression

$$\sigma_\gamma(\omega) = \frac{4\pi\omega}{c} \operatorname{Im} \alpha(\omega), \tag{2.25}$$

where $\alpha(\omega)$ is the dynamic polarisability related to the external electric field $E(\omega)$ and the induced dipole moment.

The dipole polarisability $\alpha(\omega)$ can be expressed in the following form:[126]

$$\alpha(\omega) \propto N \left(\frac{1}{\omega^2 - \omega^{(s)2}} + \frac{1}{\omega^2 - \omega^{(a)2}} \right). \tag{2.26}$$

This expression clearly shows that the photoionisation CS is defined by the two surface plasmons with the frequencies $\omega^{(s)}$ and $\omega^{(a)}$.

The final expression for the photoionisation CS within the PRA is obtained by taking into account the damping of the plasmon oscillations due to the decay of the collective mode to the incoherent sum of single-electron excitations (see Section 2.5.3). This is done by introducing the finite widths, $\Gamma^{(s)}$ and $\Gamma^{(a)}$, of the plasmon resonances and making the following substitutions on the right-hand side of Eq. (2.26). The final formula for the photoionisation CS within the PRA is written in the following form:[126,130]

$$\sigma(\omega) = \frac{4\pi\omega^2}{c}\left(\frac{N^{(s)}\,\Gamma^{(s)}}{\left(\omega^2 - \omega^{(s)2}\right)^2 + \omega^2\Gamma^{(s)2}} + \frac{N^{(a)}\,\Gamma^{(a)}}{\left(\omega^2 - \omega^{(a)2}\right)^2 + \omega^2\Gamma^{(a)2}}\right). \tag{2.27}$$

2.4.2 *Interaction with a non-uniform external field*

In the case of an interaction with a non-uniform electric field, e.g., in collisions with charged particles, the multipole variation of the electron density, Eq. (2.13), is defined as the solution of a general Eq. (2.10).

Carrying out the transformations, similar to those described above for the case of a uniform external field, one derives the following equation:

$$\begin{aligned}
&(w-1)\delta\varrho_l(r)\Theta(r-R_1)\Theta(R_2-r)\\
&\quad+\left(w\sigma_l^{(1)} + I_l^{(1)} - \frac{l+1}{\Pi_l^2}\sigma_l^{(1)} + \frac{l}{\Pi_l^2}\sigma_l^{(2)}\,\xi^{l-1}\right)\delta(r-R_1)\\
&\quad+\left(w\sigma_l^{(2)} + I_l^{(2)} + \frac{l+1}{\Pi_l^2}\sigma_l^{(1)}\xi^{l+2} - \frac{l}{\Pi_l^2}\sigma_l^{(2)}\right)\delta(r-R_2)\\
&= q^2\frac{\phi_l(r)}{4\pi}\Theta(r-R_1)\Theta(R_2-r)\\
&\quad-\frac{1}{4\pi}\left(\phi_l'(R_1)\delta(r-R_1) - \phi_l'(R_2)\delta(r-R_2)\right).
\end{aligned} \tag{2.28}$$

Matching the terms of different types on the left- and right-hand sides of Eq. (2.28) gives three equations: one for the variation of the volume density and the other two for the variation of the surface charge densities. The solution of Eq. (2.28) corresponding to the volume density variation reads as:

$$\delta\varrho_l(r) = \frac{q^2}{w-1}\frac{\phi_l(r)}{4\pi}, \tag{2.29}$$

and the density variation due to the surface plasmon modes is

$$\sigma_l(r) = \sigma_l^{(1)}\delta(r - R_1) + \sigma_l^{(2)}\delta(r - R_2). \tag{2.30}$$

The quantities $\sigma_l^{(1)}$ and $\sigma_l^{(2)}$ satisfy the following system of coupled equations:

$$\begin{cases} \left(w - \dfrac{l+1}{\Pi_l^2}\right)\sigma_l^{(1)} + \dfrac{l}{\Pi_l^2}\xi^{l-1}\sigma_l^{(2)} = F_1, \\ \dfrac{l}{\Pi_l^2}\xi^{l+2}\sigma_l^{(1)} + \left(w - \dfrac{l}{\Pi_l^2}\right)\sigma_l^{(2)} = F_2, \end{cases} \tag{2.31}$$

where

$$F_{1,2} = \mp\frac{\phi_l'(R_{1,2})}{4\pi} - I_l^{(1,2)} \tag{2.32}$$

and the functions $I_l^{(1,2)}$ were defined above in (2.18).

Solutions of the system (2.31) are given by the following expression:

$$\begin{aligned} \sigma_l^{(1)} &= \frac{1}{\Delta}\left[F_1 w - \frac{l}{\Pi_l^2}\left(F_1 + \xi^{l-1}F_2\right)\right] \equiv \frac{\mathcal{F}_1}{\Delta}, \\ \sigma_l^{(2)} &= \frac{1}{\Delta}\left[F_2 w - \frac{l+1}{\Pi_l^2}\left(F_2 + \xi^{l+2}F_1\right)\right] \equiv \frac{\mathcal{F}_2}{\Delta}, \end{aligned} \tag{2.33}$$

where Δ is the determinant of the system.

Using Eqs. (2.29) and (2.33) in Eq. (2.13), one obtains the expression for the multipole variation of electron density in a spherically symmetric hollow system under the action of the multipole component $\phi_l(r)$ of the external field:

$$\begin{aligned} \delta\rho_l(r) = &\frac{q^2}{w-1}\frac{\phi_l(r)}{4\pi}\Theta(r - R_1)\Theta(R_2 - r) \\ &+ \frac{\mathcal{F}_1}{\Delta}\delta(r - R_1) + \frac{\mathcal{F}_2}{\Delta}\delta(r - R_2). \end{aligned} \tag{2.34}$$

The first term leads to the formation of the volume plasmon, while the two other terms are responsible for the formation of two coupled modes of the surface plasmon.

2.5 Illustrative Case Studies

2.5.1 *Application of the jellium model to the study of photodetachment from metal clusters*

Significant progress in the experimental study of photoabsorption processes by metal clusters has been achieved by using new sources for free neutral[131] and charged[132] size-selected clusters. These measurements were followed by new theoretical calculations of the photoionisation CS[133–135] and the photoelectron angular distribution for neutral and charged metal clusters.[136,137]

From a theoretical point of view, the valence electron-shell structure and optical properties of various atomic clusters were successfully described using the jellium model (see Section 2.3). This section highlights some of the achievements in the application of the jellium model to the study of the photodetachment process from metal cluster anions.[137]

The first measurements of the angle-resolved photoelectron spectra of negatively charged sodium clusters were reported in Refs. 138, 139. The experiments were carried out in a wide range of cluster sizes, $3 \leq Z \leq 147$, and allowed the angular momenta of the single-electron orbitals to be probed. It was also shown that simple models based on a single-electron treatment of the photoionisation process fail to describe the angular anisotropy of photoelectrons. The first calculations using the jellium model for the ionic core and the random phase approximation with exchange (RPAE) for the valence electrons showed a crucial role of many-electron correlations in describing the correct behaviour of the photoelectron angular distribution.[137] The essential contribution of many-electron correlations to the photoionisation process is not surprising for neutral metal clusters. The correlations are even more pronounced for negative ions due to the weaker binding of the valence electrons. As a result, the role of the continuous excitation spectrum becomes even more important in describing the photoionisation of negative ions, which is very interesting from a theoretical point of view.

In Ref. 137 two magic sodium clusters, namely Na_{19}^- and Na_{57}^-, have been studied. These clusters can be treated as spherically symmetric objects due to their completely filled electronic shells. The ground-state electronic configurations of the Na_{19}^- and Na_{57}^- clusters are defined as follows:

$$1s^2 2p^6 3d^{10} 2s^2 \text{ for } \mathrm{Na}_{19}^- , \quad 1s^2 2p^6 3d^{10} 2s^2 4f^{14} 3p^6 5g^{18} \text{ for } \mathrm{Na}_{57}^- .$$

The wave functions of a photoelectron were obtained from the HF equations as the solution corresponding to the energy $\varepsilon = \hbar\omega - I_p$ and satisfying certain asymptotic behaviour (see, e.g., Ref. 140), where I_p is the ionisation potential and ω is the photon energy. The wave functions of the excited states were calculated either in the field of the 'frozen' core with the created vacancy or in the field of the rearranged residual electron structure of the cluster. The obtained HF radial wave functions $P_{nl}(r)$ and $P_{\varepsilon l}(r)$ are used to calculate the dipole-photon amplitudes, as well as the Coulomb matrix elements, to account for many-electron correlations.

The partial photodetachment CS $\sigma_{nl}(\omega)$ of the nl shell is given by

$$\sigma_{nl}(\omega) = \frac{4\pi^2 \alpha N_{nl}}{3(2l+1)\omega} \left(|d_{l+1}|^2 + |d_{l-1}|^2 \right) , \tag{2.35}$$

where α is the fine-structure constant, ω is the photon energy, and N_{nl} is the number of electrons in the nl shell. The reduced HF dipole matrix elements $d_{l\pm1}$ (in the length form) are defined as follows:

$$d_{l\pm1} \equiv \langle \varepsilon l \pm 1 | \, d \, | nl \rangle = (-1)^l \sqrt{l_>} \int_0^\infty P_{nl}(r) P_{\varepsilon l\pm1}(r) \, r \, dr, \tag{2.36}$$

where $l_> = l+1$ for the $l \to l+1$ transition and $l_> = l$ for the $l \to l-1$ transition. The total photodetachment CS is obtained by the sum over all partial CSs.

Many-electron correlations were taken into account using the RPAE scheme,[140] which describes the dynamic collective response of an electron system to an external electromagnetic field. Within the RPAE, the dipole-photon amplitudes $\langle \varepsilon l \pm 1 | \, D(\omega) | \, nl \rangle$ are obtained by solving the following integral equation:[140]

$$\langle \varepsilon l \pm 1 | \, D(\omega) | \, nl \rangle = \langle \varepsilon l \pm 1 | \, d \, | \, nl \rangle + \left(\sum_{\substack{\nu_2 > F \\ \nu_1 \le F}} - \sum_{\substack{\nu_1 > F \\ \nu_2 \le F}} \right) \times \frac{\langle \nu_2 | \, D(\omega) | \, \nu_1 \rangle \langle \nu_1, \varepsilon l \pm 1 | \, U \, | \, \nu_2, nl \rangle}{\omega - \varepsilon_2 + \varepsilon_1 + \imath\delta} . \tag{2.37}$$

Here, the indices ν_1 and ν_2 denote the quantum numbers $\{n(\varepsilon), l\}$ of the virtual electron-hole states, F is the Fermi energy, and the matrix element $\langle \ldots | U | \ldots \rangle$ stands for the sum of the direct and exchange Coulomb matrix elements.[140]

The angular distribution of photoelectrons detached from the nl shell by an unpolarised photon is determined by the differential CS of the electron emission in the solid angle $\mathrm{d}\Omega = \sin\theta \mathrm{d}\theta \mathrm{d}\phi$:

$$\frac{\mathrm{d}\sigma_{nl}(\omega)}{\mathrm{d}\Omega} = \frac{\sigma_{nl}(\omega)}{4\pi}\left[1 - \frac{1}{2}\beta_{nl}(\varepsilon)P_2(\cos\theta)\right], \tag{2.38}$$

where $P_2(\cos\theta)$ is the Legendre polynomial, $\beta_{nl}(\varepsilon)$ is the angular distribution anisotropy parameter, and ε is the photoelectron energy.

The main feature of the total photoabsorption CS in metal clusters is the giant plasmon resonance at about 2–3 eV, which appears due to the collective response of the systems.[15] For neutral and positively charged metal clusters, the plasmon resonance lies in the discrete spectrum of electronic excitations. The simple HF scheme produces the set of single-electron excitations below the ionisation potential and the corresponding distribution of the oscillator strengths. When the correlations are taken into account, most of the oscillator strength is concentrated at the energy close to 2–3 eV, correctly describing the plasmon excitation.

The peculiarity of negatively charged clusters is that they have either no or only one or two discrete dipole excitations. Therefore, most of the oscillator strength of the partial transitions is distributed in the continuous spectrum of the photodetachment CS. This feature is manifested in both the HF approximation and the RPAE. Thus, the plasmon resonance lies in the continuum, and the consideration of many-electron correlations leads to a significant change in the resonance position, its maximum value, and the shape of the resonance curve.[135]

The typical photodetachment CS behaviour for metal cluster anions (considering Na_{19}^- as a case study) is shown in Fig. 2.2(a) and reveals the powerful maxima due to the collective electron excitation. Six dipole transitions were taken into account in the RPAE calculations for Na_{19}^-. The dipole transitions from the outer $3d$ and $2s$ shells give the main contribution to the total CS in the vicinity of the plasmon resonance, which is located at the energy $\hbar\omega \approx 2.4$ eV. It should be noted, however, that it is not sufficient to consider only for the transitions from the outer shells in order to obtain the final profile of the plasmon excitation. Instead, the interactions between all the valence electrons should be included. For larger cluster anions, the number of transitions to be accounted for increases significantly.

The total photodetachment CS is not a very sensitive indicator of the applicability of the jellium model to metal clusters. The photoelectron angular distribution is a much better test for the conventional jellium model since the behaviour of the anisotropy parameter $\beta(\varepsilon)$ can show whether the shell structure is real and the orbital momentum l is a good quantum number in the cluster system.

The comparison of the HF, the RPAE, and the experimental data[139] for the outer $3d$ shell in the Na_{19}^- anion is shown in Fig. 2.2(b). The experimental points labelled A, B, C, and D correspond to the photoionisation from the sub-levels of the $3d$ orbital which is split by the crystalline field of the cluster in the range 1.75–2.15 eV. It is interesting to note that the general behaviour of the photoelectron angular distribution is approximately the same for all sub-levels.

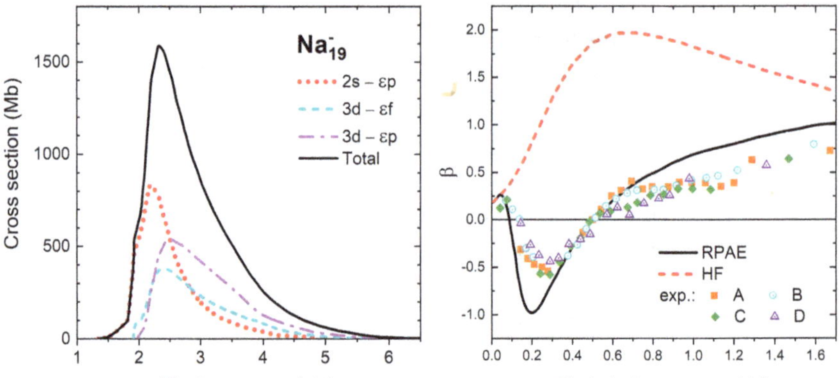

(a) (b)

Fig. 2.2. Panel (a): the RPAE partial (dashed, dash-dotted, and dotted lines) and total (solid line) photodetachment CSs for the Na_{19}^- anion. Panel (b): the angular anisotropy parameter $\beta(\varepsilon)$ for the $3d$ shell of the Na_{19}^- anion as a function of photoelectron energy. Comparison of the HF (dashed line), RPAE (solid line), and the experiment.[139] (symbols) Experimental data marked as A, B, C, and D correspond to the photoionisation from the sub-levels of the $3d$ orbital split by the crystalline field.

Source: Redrawn from data presented in Refs. 137,141. Reproduced from Ref. 16 with permission from Springer International Publishing.

This indicates that all electrons in the sub-levels can be characterised by the same orbital quantum number l. The $3d$ binding energy, equal to 1.92 eV in the HF approximation, is very close to the centre of the multiplet. The calculations in the HF framework (dashed line in Fig. 2.2(b)) do not explain the dependence of the angular anisotropy parameter on the photoelectron energy. On the contrary, the RPAE results (solid line in Fig. 2.2(b)) are in a good agreement with the experimental data. However, the HF and RPAE schemes based on the jellium model ignore the splitting and, thus, provide the average dependence of the angular anisotropy parameter.

As for larger clusters, the calculations of the angular anisotropy parameter $\beta(\varepsilon)$ were performed for several outer shells of the Na_{57}^- anion. The experimental measurements of the photoelectron spectra were not performed for the closed-shell Na_{57}^- anion, but for Na_{55}^-, which has the unfilled $5g^{16}$ shell.[139] Note that despite the different cluster anions considered theoretically and experimentally, as well as the differences in ionisation potentials, the general behaviour of the angular anisotropy parameter for inner shells (e.g., for the $4f$ shell, see Fig. 2.3(a)) is well reproduced by the many-body calculations performed in Ref. 137. However, the calculated dependence $\beta(\varepsilon)$ of Na_{57}^- for the outer $5g$ shell only agrees qualitatively with the experimental data for Na_{55}^-, as shown in Fig. 2.3(b). The quantitative discrepancy can be explained by several factors. First, the calculations[137] neglected the polarisation potential, which is rather large for neutral metal clusters and thus modifies the dynamics of the outgoing photoelectron. Second, the electrons in the open $5g^{16}$ shell in Na_{55}^- are much more affected by the crystalline field than the electrons in the filled $5g^{18}$ shell of Na_{57}^-.

The comparison of the existing experimental and theoretical data indicates that further investigation of the photodetachment process is needed. The theoretical consideration of the many-electron system was carried out in Ref. 137 using the simple jellium model for the ionic core. A more detailed analysis of the problem should go beyond this model and take into account the realistic geometrical structure of the core. However, the agreement between the experimental data on the angular distribution and the results of the current theory indicates that the definite shell structure of the valence electrons is reproduced quite well even if the interaction with the real crystalline field is neglected.

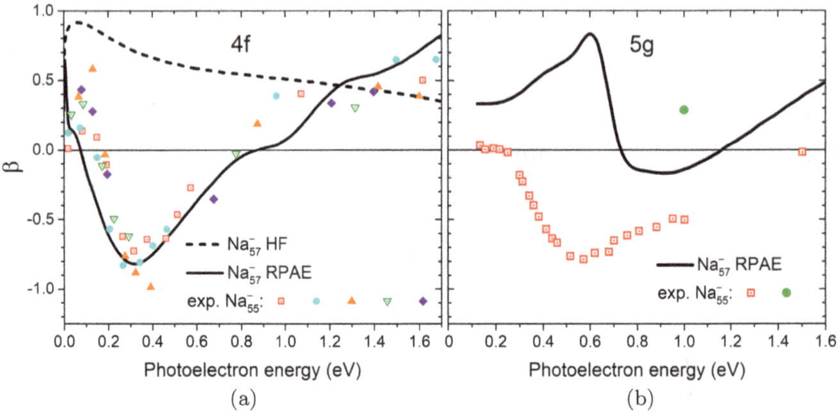

Fig. 2.3. The angular anisotropy parameter for the $4f$ (panel (a)) and $5g$ (panel (b)) shells of the Na_{57}^- anion as a function of the photoelectron energy. The results of the HF (dashed line) and RPAE (solid line) calculations are compared with the experimental data for Na_{55}^- (symbols) taken from Ref. 139. Different experimental points correspond to the photoionisation from the sub-levels of the orbital split by the crystalline field of the cluster.

Source: Redrawn from data presented in Refs. 137,141. Reproduced from Ref. 16 with permission from Springer International Publishing.

2.5.2 *Plasmon excitations in photo- and electron impact ionisation of atomic clusters and fullerenes*

Photoionisation of fullerenes and other nanoscale systems is a complex phenomenon with a number of features that can be studied using different theoretical methods. Being a quantum phenomenon by nature, the photoionisation process can be described within the *ab initio* framework, e.g., based on TDDFT;[50] see Section 2.2.3. However, it is well established that the photoionisation of nanoscale carbon systems, particularly fullerenes, as well as various metallic clusters and NPs, takes place via collective plasmon excitations. The plasmon excitations correspond to oscillations of the electron density with respect to the positively charged ions and are described in the classical physics terms[37,124] (see Section 2.4).

When a fullerene is ionised either by a photon or by a charged projectile, different types of collective excitations are formed in the system, characterised by prominent resonant-like structures in the ionisation spectra. The most prominent structure, located in the excitation energy range of 20–30 eV, is formed by collective oscillations of both σ and π delocalised electrons of a system, while a smaller narrow peak in the low-energy region of the spectrum (below 10 eV) is attributed to the collective excitation of only π-electrons. The σ- and π-electrons occupy the σ- and π-orbitals of a fullerene, respectively, which are formed by the sp^2-hybridisation of the electronic orbitals of carbon atoms. The resonance peaks in the ionisation spectra are described by some characteristic widths Γ, which have a quantum origin and appear due to the decay of the collective excitation modes into the incoherent sum of single-electron excitations (see Section 2.5.3).

An effective tool for evaluating the contribution of plasmon excitations to the ionisation spectra is based on the plasmon resonance approximation (PRA); see Section 2.4. The advantage of this approach is that it provides a clear physical explanation of the resonant-like structures in the photoionisation[124,125] and inelastic scattering CSs[116–118,122,123] on the basis of excitation of plasmons by the photon or electron impact.

2.5.2.1 *Photoionisation of fullerenes*

The contributions of various classical and quantum physics phenomena occurring in the photoionisation process of C_{60} fullerene were studied in Ref. 125. By comparing the photoabsorption spectra of C_{60} calculated by TDDFT and the PRA, the well-resolved features of the photoabsorption spectrum of C_{60} were assigned to different types of single-particle and collective electron excitations of different physical nature. It was shown that the peculiarities arising in the spectrum atop the dominant plasmon excitations are of quantum origin. A series of individual peaks in the continuous part of the excitation spectrum were assigned to the particular single-electron transitions caused by the ionisation of the inner molecular orbitals of the fullerene.

TDDFT calculations were performed in the linear regime within the dipole approximation. The dynamical polarisability tensor $\alpha_{ij}(\omega)$, which describes the linear response of the dipole to the external electric field, was calculated, and the photoabsorption CS $\sigma(\omega)$ was related to the imaginary part of $\alpha_{ij}(\omega)$.

Within the PRA, the dynamical polarisability $\alpha(\omega)$ has a resonance behaviour in the range of frequencies where collective electron modes in a fullerene can be excited; see Eq. (2.26). In Ref. 125, both π- and $(\sigma+\pi)$-plasmons involving only π or both $\sigma+\pi$ delocalised electrons of the system were studied. The photoionisation CS, $\sigma_{\rm pl}(\omega) \propto {\rm Im}\,\alpha(\omega)$, is then defined as a sum of the two plasmons, $\sigma_{\rm pl}(\omega) = \sigma^{\pi}(\omega) + \sigma^{\sigma+\pi}(\omega)$, and the contribution of each plasmon is governed by the symmetric and antisymmetric modes, according to Eq. (2.27).

Figure 2.4(a) shows the photoabsorption spectrum of C_{60} calculated within the *ab initio* and classical approaches in the photon energy region up to 100 eV. The thin solid (black) line represents the results of TDDFT calculations within the LDA approach, and the thick solid (green) line shows the contribution from the plasmon excitations. The main resonant structure presented shown in Fig. 2.4(a) is formed by collective oscillations of both σ- and π-electrons of the system, while a prominent peak in the low-energy region of the spectrum (shown in the inset) is attributed to the collective excitation of only π-electrons. The dashed (red) and dash-dotted (blue) lines

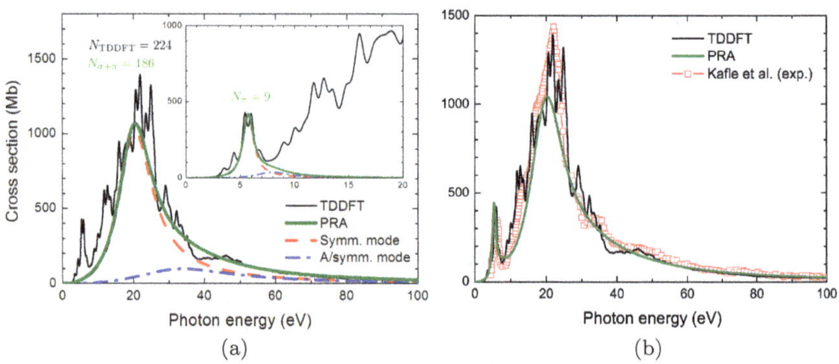

Fig. 2.4. The photoabsorption CS of C_{60} calculated within the TDDFT method (thin black line) and the plasmon resonance approximation (thick green line).[125] The curves obtained within the classical approach describe the dominating plasmon resonance, which is formed due to collective oscillations of $(\sigma+\pi)$ delocalised electrons of the system, and a narrow low-energy peak below 10 eV which is attributed to the collective excitation of only π-electrons. Panel (a) shows the contributions of the symmetric (dashed red lines) and antisymmetric (dash-dotted blue lines) modes of the two plasmons. In panel (b), the theoretical curves are compared with the experimental data.[144]

Source: Reproduced from Ref. 125 with permission from the American Physical Society.

show the contributions of the symmetric and antisymmetric plasmon modes to the CS, respectively. The width $\Gamma_s^{\sigma+\pi} = 11.4$ eV of the symmetric $(\sigma + \pi)$-plasmon mode corresponds to the experimental values obtained from the photoionisation and energy loss experiments on neutral C_{60}.[114,119] For the antisymmetric mode, the value $\Gamma_a^{\sigma+\pi} = 33.2$ eV was used, which corresponds to the width of the second plasmon resonance obtained in the study of the photoionisation of C_{60}^{q+} $(q = 1 - 3)$ ions.[142]

The PRA describes the main features of the spectrum, such as the height, width, and position of the plasmon resonance peaks. The spectrum calculated using the TDDFT approach shows a more detailed structure formed on top of the plasmon resonances, representing a series of individual peaks. The oscillator strengths, calculated using TDDFT and the PRA in the photon energy range up to 100 eV, are equal to 224 and 195, respectively. Analysis of the plasmon contribution to the CS shows that about 9 π-electrons are involved in the low-energy collective excitation below 10 eV. This value corresponds to the experimentally evaluated sum rule of the oscillator strength up to the ionisation threshold of C_{60}, $I_p \approx 7.6$ eV, which gives the value of 7.8.[143]

In Fig. 2.4(b), the theoretical curves are compared with the results of experimental measurements of the photoabsorption of C_{60}.[144] The oscillator strength calculated using TDDFT is very close to the experimentally measured value of 230.5.[144] The PRA also provides an adequate description of the experimental results. The detailed structure of the spectrum, which is described by the TDDFT approach, is not seen in the experimental curve due to a high operating temperature of 500–700°C.[143] In the experiments, the linewidths of single-electron excitations are broadened in the vicinity of the main plasmon resonance due to the coupling of the electron excitations with the vibrational modes of the ionic background.[104]

2.5.2.2 *Inelastic scattering of electrons from atomic clusters and fullerenes*

By studying the photoionisation process, only the dipole plasmon excitation mode can be analysed.[15,126] Electron collective modes with higher angular momenta $(l > 1)$ can be studied in the fast electron–cluster collisions if the scattering angle of the electron is large enough.[116–119] Dipole plasmon resonances of the same physical nature as in the case of the photoabsorption or photoionisation dominate the electron energy loss spectrum (EELS) if the scattering angle of the electron, and thus its transferred momentum, is sufficiently small. As the scattering angle increases, plasmon excitations with higher angular momenta become more probable. The actual number of multipoles involved depends on the cluster size.

In the process of inelastic scattering, the projectile electron undergoes a transition from the initial state $(\varepsilon_1, \mathbf{p}_1)$ to the final state $(\varepsilon_2, \mathbf{p}_2)$, which is accompanied by the ionisation (or excitation) of a target from the initial state i with the energy ε_i to the final state f with ε_f. The matrix element M, which defines the amplitude of the inelastic scattering, is given by

$$
\begin{aligned}
M &= \left\langle f, 2 \left| \sum_a \frac{1}{|\mathbf{r} - \mathbf{r}_a|} \right| 1, i \right\rangle \\
&= \sum_a \int \psi_2^{(-)*}(\mathbf{r}) \psi_f^*(\{\mathbf{r}_a\}) \frac{1}{|\mathbf{r} - \mathbf{r}_a|} \psi_i(\{\mathbf{r}_a\}) \psi_1^{(+)}(\mathbf{r}) \{\mathrm{d}\mathbf{r}_a\} \mathrm{d}\mathbf{r},
\end{aligned}
\tag{2.39}
$$

where $\{\mathbf{r}_a\} = \mathbf{r}_1 \ldots \mathbf{r}_N$ are the position vectors of the delocalized electrons in the target, $\mathbf{r}$ is the position vector of the projectile, and $\psi_1^{(+)}(\mathbf{r})$ and $\psi_2^{(-)}(\mathbf{r})$ stand for the initial- and final-state wave functions of the projectile, respectively. Superscripts $(+)$ and $(-)$ indicate that asymptotic behaviour of the wave functions is 'plane wave + outgoing spherical wave' and 'plane wave + incoming wave', respectively.

The matrix element (2.39) can be written as follows:

$$
M = \int \frac{4\pi}{q^2} \frac{\mathrm{d}\mathbf{q}}{(2\pi)^3} \langle 2 | \mathrm{e}^{-\mathrm{i}\mathbf{q}\cdot\mathbf{r}} | 1 \rangle \left\langle f \left| \sum_a \mathrm{e}^{\mathrm{i}\mathbf{q}\cdot\mathbf{r}_a} \right| i \right\rangle, \tag{2.40}
$$

where $\mathbf{q} = \mathbf{p}_1 - \mathbf{p}_2$ is the transferred momentum.

If the velocity of a projectile is high and significantly exceeds the characteristic velocities of the delocalised electrons in the target, the first Born approximation is applicable.[116] Within this approximation, the initial and final states of the incident electron can be described by plane waves:

$$\psi_1^{(+)}(\mathbf{r}) = \mathrm{e}^{\mathrm{i}\mathbf{p}_1\cdot\mathbf{r}}, \quad \psi_2^{(-)}(\mathbf{r}) = \mathrm{e}^{\mathrm{i}\mathbf{p}_2\cdot\mathbf{r}}. \tag{2.41}$$

Then, the amplitude of the process reduces to

$$M = \frac{4\pi}{q^2}\left\langle f \left| \sum_a \mathrm{e}^{\mathrm{i}\mathbf{q}\cdot\mathbf{r}_a} \right| i \right\rangle_{\mathbf{q}=\mathbf{p}_1-\mathbf{p}_2}. \tag{2.42}$$

The magnitude q^2 is related to the scattering angle as $q^2 \approx p_1^2\theta^2$ under the assumption that the energy loss $\omega = \varepsilon_1 - \varepsilon_2$ is small, $\omega \ll \varepsilon_1$ (which implies that $p_1 \approx p_2$) and the scattering angle is small, $\theta \ll 1$ rad.

Performing the multipole expansion of the exponential factors in Eq. (2.42) (see, e.g., Ref. 145), one obtains

$$M = 4\pi \sum_{lm} \mathrm{i}^l\, Y_{lm}^*(\mathbf{q}) \left\langle f \left| \sum_a \phi_l(r_a) Y_{lm}(\mathbf{r}_a) \right| i \right\rangle, \tag{2.43}$$

where the notation

$$\phi_l(r) = 4\pi \frac{j_l(qr)}{q^2} \tag{2.44}$$

is introduced and j_l is a spherical Bessel function of the order l.

Let us consider a general expression for the CS of the scattering process:

$$\mathrm{d}\sigma = \frac{2\pi}{p_1}\,\delta(\omega_{fi} - \omega) \sum_{\mathrm{pol}_f} \overline{\sum_{\mathrm{pol}_i}} |M|^2 \frac{\mathrm{d}\mathbf{p}_2}{(2\pi)^3}\, \mathrm{d}\rho_f. \tag{2.45}$$

Here, $\omega_{fi} = \varepsilon_f - \varepsilon_i$, the sign $\sum_{\mathrm{pol}_f}$ denotes the summation over the projection of the final-state f orbital momentum, whereas $\overline{\sum}_{\mathrm{pol}_i}$ denotes the averaging over the projections of the initial-state orbital momentum, and $\mathrm{d}\rho_f$ is the density of final states of the target.

Substituting the scattering amplitude (2.43) into Eq. (2.45), one derives the doubly differential CS (DDCS):

$$\frac{d^2\sigma}{d\varepsilon_2 d\Omega_{\mathbf{p}_2}} = \frac{1}{\pi}\frac{p_2}{p_1}\sum_{lm}\int\left|\left\langle f\left|\sum_a V_{lm}(\mathbf{r}_a)\right|i\right\rangle\right|^2 \delta(\omega_{fi}-\omega)\,d\rho_f, \tag{2.46}$$

where

$$V_{lm}(\mathbf{r}) = \phi_l(r)Y_{lm}(\mathbf{r}) \tag{2.47}$$

is the multipolar potential of the fast projectile, $d\Omega_{\mathbf{p}_2}$ denotes the differentiation over the solid angle of the scattered electron, and the sign $\int d\rho_f$ means the summation over the final states (which includes the summation over the discrete spectrum and the integration over the continuous spectrum).

According to Kubo's linear response theory,[116,146] the integral in Eq. (2.46) can be related to the variation of the electron density caused by an external electric field, and the following substitution can be made:

$$\int\left|\left\langle f\left|\sum_a V_{lm}(\mathbf{r}_a)\right|i\right\rangle\right|^2 \delta(\omega_{fi}-\omega)d\rho_f$$
$$\to \frac{1}{\pi}\,\mathrm{Im}\int V_{lm}^*(\mathbf{r})\delta\rho_l(\omega,q;\mathbf{r})d\mathbf{r}. \tag{2.48}$$

Here, $\delta\rho_l(\omega,q;\mathbf{r})$ is the density variation due to the exposure of the system to the multipolar potential $V_{lm}(\mathbf{r})$. In a general case, this variation depends on the transferred energy ω, transferred momentum q, and the position vector $\mathbf{r}$.

As described in Section 2.4, atomic clusters and fullerenes can be considered, in the general case, as spherically symmetric systems where the charge is homogeneously distributed between two concentric spheres; see Fig. 2.1. For such a geometry, the multipole variation of the electron density is given by the sum of the volume density variation occurring inside the shell and the variations of the surface charge densities at the inner and outer surfaces of the shell, respectively; see Eqs. (2.13) and (2.34). The volume density variation causes the formation of the volume plasmon, while the surface density variations correspond to the symmetric and antisymmetric modes of the surface plasmon (see Fig. 2.1).

Presenting the multipole variation of the electron density, $\delta\rho_l(\omega, q; \mathbf{r})$, as a sum of three contributions (see Eq. (2.13)), using the explicit expression for the multipolar potential $V_{lm}(r)$ (see Eqs. (2.44) and (2.47)), and performing some algebraic transformations, one obtains the formula for the differential inelastic scattering CS:[126]

$$\frac{\mathrm{d}^2\sigma}{\mathrm{d}\varepsilon_2 \mathrm{d}\Omega_{\mathbf{p}_2}} = \frac{\mathrm{d}^2\sigma^{(\mathrm{v})}}{\mathrm{d}\varepsilon_2 \mathrm{d}\Omega_{\mathbf{p}_2}} + \frac{\mathrm{d}^2\sigma^{(\mathrm{s}_1)}}{\mathrm{d}\varepsilon_2 \mathrm{d}\Omega_{\mathbf{p}_2}} + \frac{\mathrm{d}^2\sigma^{(\mathrm{s}_2)}}{\mathrm{d}\varepsilon_2 \mathrm{d}\Omega_{\mathbf{p}_2}}, \tag{2.49}$$

where

$$\begin{aligned}
\frac{\mathrm{d}^2\sigma^{(\mathrm{v})}}{\mathrm{d}\varepsilon_2 \mathrm{d}\Omega_{\mathbf{p}_2}} &= \frac{2R_2 p_2}{\pi q^4 p_1}\,\omega \sum_l \frac{\omega_{\mathrm{p}}^2\, \Gamma_l^{(\mathrm{v})}\, V_l(q)}{\left(\omega^2 - \omega_{\mathrm{p}}^2\right)^2 + \omega^2 \left(\Gamma_l^{(\mathrm{v})}\right)^2}, \\
\frac{\mathrm{d}^2\sigma^{(\mathrm{s}_1)}}{\mathrm{d}\varepsilon_2 \mathrm{d}\Omega_{\mathbf{p}_2}} &= \frac{2R_2 p_2}{\pi q^4 p_1}\,\omega \sum_l \frac{\omega_{1l}^2\, \Gamma_{1l}^{(\mathrm{s})}\, S_{1l}(q)}{\left(\omega^2 - \omega_{1l}^2\right)^2 + \omega^2 \left(\Gamma_{1l}^{(\mathrm{s})}\right)^2}, \\
\frac{\mathrm{d}^2\sigma^{(\mathrm{s}_2)}}{\mathrm{d}\varepsilon_2 \mathrm{d}\Omega_{\mathbf{p}_2}} &= \frac{2R_2 p_2}{\pi q^4 p_1}\,\omega \sum_l \frac{\omega_{2l}^2\, \Gamma_{2l}^{(\mathrm{s})}\, S_{2l}(q)}{\left(\omega^2 - \omega_{2l}^2\right)^2 + \omega^2 \left(\Gamma_{2l}^{(\mathrm{s})}\right)^2}.
\end{aligned} \tag{2.50}$$

In this expression, ω_p is the volume plasmon frequency, Eq. (2.17), which is independent of l. ω_{1l} and ω_{2l} are the frequencies of the symmetric and antisymmetric surface plasmon modes of multipolarity l, respectively, defined by Eq. (2.22). The functions $V_l(q)$, $S_{1l}(q)$ and $S_{2l}(q)$ are the diffraction factors depending on the transferred momentum q. They determine the relative significance of the multipole plasmon modes in different ranges of the projectile's scattering angles. The dominant contribution of different multipole modes results in a significant angular dependence for the differential EELS.[123] Explicit expressions for these functions are given in Ref. 126.

The DDCS (2.49) can also be written in terms of the energy loss $\omega = \varepsilon_1 - \varepsilon_2 \equiv \Delta\varepsilon$ of the incident projectile of energy ε_1. Integration of $\mathrm{d}^2\sigma/\mathrm{d}\Delta\varepsilon\, \mathrm{d}\Omega_{\mathbf{p}_2}$ over the solid angle leads to the single differential CS:

$$\frac{\mathrm{d}\sigma}{\mathrm{d}\Delta\varepsilon} = \int \mathrm{d}\Omega_{\mathbf{p}_2} \frac{\mathrm{d}^2\sigma}{\mathrm{d}\Delta\varepsilon\, \mathrm{d}\Omega_{\mathbf{p}_2}} = \frac{2\pi}{p_1 p_2} \int_{q_{\min}}^{q_{\max}} q\, \mathrm{d}q \frac{\mathrm{d}^2\sigma}{\mathrm{d}\Delta\varepsilon\, \mathrm{d}\Omega_{\mathbf{p}_2}}. \tag{2.51}$$

It was shown[116] that the excitations with large angular momenta l have a single-particle rather than a collective nature. This follows from the fact that with increasing l, the wavelength of the plasmon excitation becomes smaller than the characteristic wavelength of the delocalised electrons in the target. In the case of C_{60}, the estimates show[116] that the excitations with $l > 3$ are formed by single-electron transitions rather than by the collective excitations. Therefore, only terms corresponding to the dipole ($l = 1$), quadrupole ($l = 2$), and octupole ($l = 3$) plasmon excitations should be included in the sum over l in Eqs. (2.49) and (2.50).

The theory presented relies on the number of multipole terms to be considered and the widths of the plasmon resonances, $\Gamma_{jl}^{(s)}$ and $\Gamma_l^{(v)}$, which are not fitting parameters of the model but the real physical quantities. A precise calculation of the widths can be made by analysing the decay of the collective excitation mode into the incoherent sum of single-electron excitations. This process should be considered in the quantum-mechanical framework[121] and cannot be treated in the classical physics framework as the PRA does. Section 2.5.3 describes the application of such an analysis to evaluate the surface and volume plasmon widths for a Na_{40} cluster.[121]

The widths of the plasmon excitations in the fullerene can be estimated using the relation similar to the Landau damping of plasmon oscillations. Such an estimate gives $\Gamma_{jl}^{(s)} \sim l_{v_F}/\Delta R$, where v_F is the velocity of the fullerene electrons at the Fermi surface. A similar estimate was successfully applied to the study of collective excitations in metal clusters.[121]

The formation of the volume plasmon in the electron impact ionisation of metal clusters and carbon fullerenes was demonstrated in Refs. 121–123. The model accounting for the contribution of different plasmon modes was successfully used to describe the experimentally observed variation of the EELS of C_{60} in collisions with fast electrons.[122,123]

Figure 2.5 shows the DDCS $\mathrm{d}^2\sigma/\mathrm{d}\Delta\varepsilon\,\mathrm{d}\Omega_{\mathbf{p}_2}$ as a function of the transferred energy $\Delta\varepsilon$ calculated for the electron-Na_{40} collision.[118] Panel (a) shows that at the scattering angle $\theta = 1°$, the dipole plasmon excitation ($l = 1$) dominates in the electron energy loss spectrum while the quadrupole excitation ($l = 2$) makes a relatively small contribution. The contributions of the monopole ($l = 0$) and

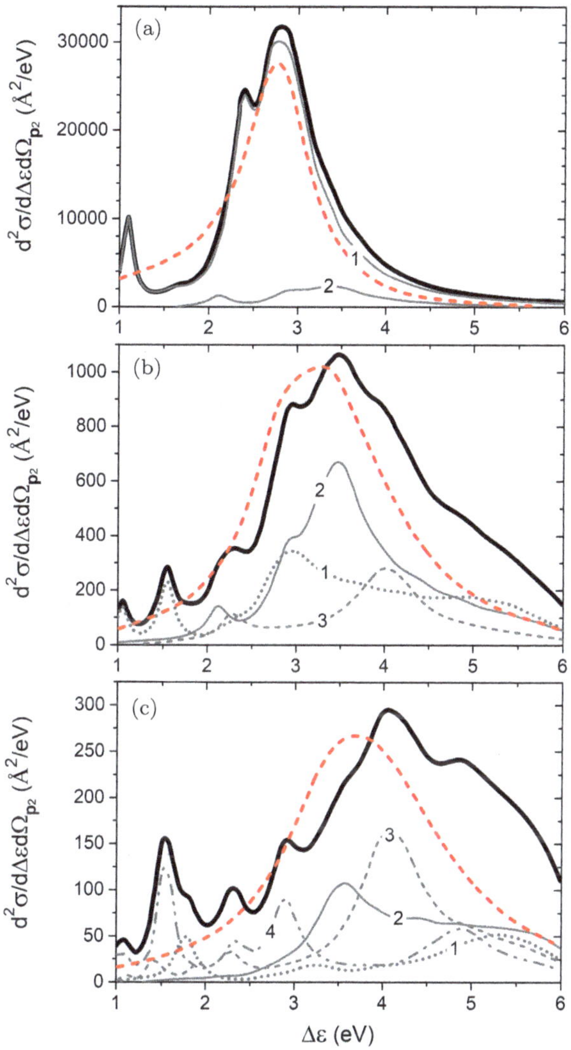

Fig. 2.5. The doubly differential CS $d^2\sigma/d\Delta\varepsilon\, d\Omega_{\mathbf{p}_2}$ as a function of the transferred energy $\Delta\varepsilon$ calculated for the electron–Na_{40} collision.[118] The impact electron energy is $\varepsilon = 50$ eV. Panels (a), (b), and (c) correspond to the electron scattering angles of $\theta = 1°$, $6°$, and $9°$, respectively. Solid lines represent the results of the RPAE calculation with the HF jellium model basis wave functions. Thick solid (black) line is the total EELS. Thin solid (grey) lines marked with the angular momentum number represent various multipole contributions to the EELS. Dashed line shows the EELS calculated within the PRA.

Source: Redrawn from data presented in Ref. 118.

all higher multipole excitations (not shown in Fig. 2.5(a)) are almost negligible at this scattering angle. Figure 2.5(b) shows that at $\theta = 6°$, the quadrupole excitation becomes the leading excitation in the electron energy loss spectrum, shifting the maximum of the spectrum towards higher energies and also changing the profile of the resonance. The dipole and the octupole ($l = 3$) excitations also make significant contributions over a wide range of transferred energies, broadening the spectrum. The dominance of the quadrupole excitation is not as large as for the dipole excitation at $\theta = 1°$. At $\theta = 9°$ (see Fig. 2.5(c)), the picture becomes more complex. In this case, the octupole excitations make the dominant contribution to the spectrum in the vicinity of the maximum of the EELS at $\Delta\varepsilon \approx 4$ eV. Beyond this region, the dipole, quadrupole, and even excitations with angular momentum $l = 4$ give comparable contributions to the EELS and form a rather broad structure. The monopole excitation and the excitations with angular momentum $l = 5$ and higher (not shown in Fig. 2.5(c)) are almost negligible. The excitations with $l = 4$ become more important with increasing scattering angle. However, the corresponding spectrum does not show any resonance behaviour, as it is mainly formed by single-electron transitions.

A comparison of the results obtained from the RPAE calculations (see solid lines in Fig. 2.5) with those obtained from the PRA (dashed lines) shows that, despite its simplicity, the PRA gives a good agreement with the consistent many-body quantum calculation. The main discrepancy between the two approaches arises from the single-particle transitions that are omitted in the PRA but included in the RPAE calculation. These transitions add some structure to the final EELS on top of the smooth resonance behaviour reproduced by the PRA (see also the case of photoionisation of C_{60} described in Section 2.5.2). At larger scattering angles, plasmons with larger angular momenta can be excited. However, excitations with sufficiently large angular momenta occur due to single-particle transitions rather than due to collective excitations. Therefore, the agreement between the classical PRA and the quantum-based RPAE is better at small angles.

Figure 2.6 shows the EELS of C_{60} calculated in the PRA (see Eqs. (2.49) and (2.50)). The calculations were performed for the scattering angles $\theta = 3°, 5°, 7°$, and $9°$. The calculated curves

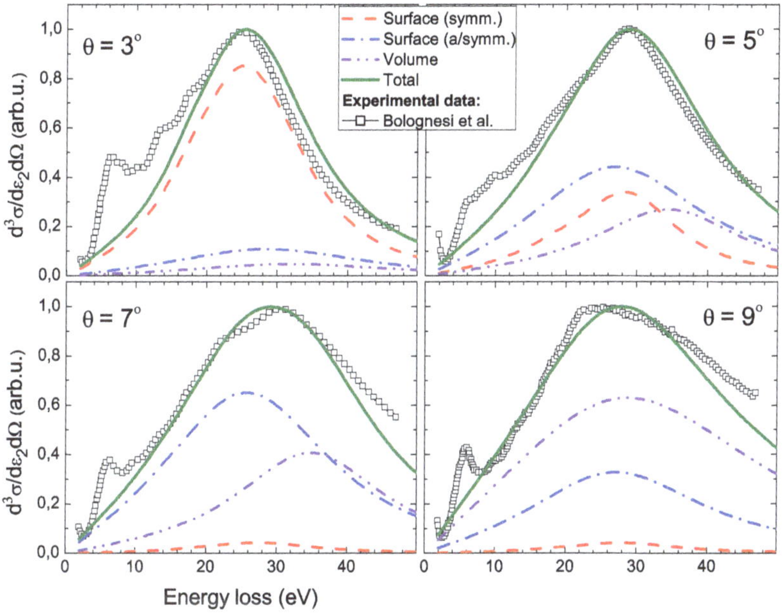

Fig. 2.6. Comparison of the EELS, calculated by the PRA, with the experimental spectra for the incident energy range 1002 – 1050 eV and for the scattering angles $\theta = 3^\circ - 9^\circ$. The symmetric and antisymmetric modes of the surface plasmon are shown by the dashed red and dash-dotted blue lines, respectively. The volume plasmon contribution is shown by the double-dotted purple line. The total CS is represented by the thick green line. Open squares represent the experimental data.[122,123] For the sake of convenience, both the experimental and theoretical curves are normalised to 1 at the maximum point. The energy scale is the same for all panels.

Source: Reproduced from Ref. 147 published under an open access Creative Commons CC BY license.

are compared with the results of experimental measurements of the inelastic scattering of fast (1 keV) electrons on C_{60}.[122,123]

At the small scattering angle, $\theta = 3^\circ$, the symmetric mode of the surface plasmon dominates the CS. A similar behaviour is observed in the photoionisation process (see Fig. 2.4). In fact, in the case of a uniform external field ($q \to 0$), there is no volume plasmon excitation in the system (see Section 2.4), and the symmetric plasmon mode significantly exceeds the antisymmetric mode. The non-uniformity of the external field leads to the formation of the volume plasmon, whose contribution to the CS is insignificant at small scattering angles. As the scattering angle increases ($\theta = 5°$ and $7°$), the symmetric mode of the surface plasmon becomes less relevant, while the antisymmetric mode becomes more prominent. At the larger angle ($\theta = 9°$), the symmetric surface plasmon almost does not contribute to the CS, while the volume plasmon becomes dominant. The contribution of the antisymmetric surface and the volume plasmons can explain the origin of the two peaks in the energy loss range from 20 to 30 eV at the scattering angle $\theta = 7°$.

The peak position of each plasmon resonance (2.50) and the resulting CS $d^3\sigma/d\varepsilon_2 d\Omega_{\mathbf{p}_2}$ is influenced by the manifestation of the diffraction effects. In Ref. 119, it was shown that plasmon modes with different angular momenta make dominant contributions to the differential CS at different scattering angles, leading to the significant angular dependence of the energy loss spectrum. This phenomenon was described in terms of the electron diffraction at the fullerene edge.[119] As can be seen from Eq. (2.50), the resonance peak of each plasmon is defined not only by the plasmon frequencies ω_p, ω_{1l}, and ω_{2l}, but also by the multipolar diffraction factors $V_l(q)$, $S_{1l}(q)$, and $S_{2l}(q)$, which depend on the transferred momentum q. In the limiting case of an infinitely thin layer, this dependence is described by the spherical Bessel functions $j_l^2(qR)$, which oscillate with q and, thus, lead to suppression and enhancement of the partial plasmon modes at certain angles.[119] The incident energy of the projectile does not affect this behaviour and only defines the absolute value of the CS.

2.5.3 *Electron energy loss spectra of metal clusters: Contribution of the surface and volume plasmon excitations*

The damping of the plasmon oscillations is related to the decay of the collective electron excitations to the single-particle ones similar to the mechanism of Landau damping in an infinite electron gas. The frequencies of the surface plasmon excitations in neutral metal clusters lie in the vicinity of the ionisation threshold. In small sodium clusters, for example, they are below the ionisation potential, and single-particle excitations in the vicinity of the surface plasmon resonance therefore have a discrete spectrum. In this case, the width of a surface plasmon excitation caused by the Landau damping should be treated as the width of the distribution of the oscillator strengths in the vicinity of the resonance. The problem of the formation of the surface plasmon resonance widths in clusters was studied in Refs. 121,148–150.

The resonance frequencies of volume plasmon excitations in metal clusters are typically located above the ionisation threshold. This means that the volume plasmon excitations are quasi-stable and have the real channel of the Landau damping, leading to the ionisation of the cluster.[121] Thus, the process of inelastic scattering for transferred energies above the ionisation threshold can be described as follows. The projectile particle induces the oscillations of the electron density in the cluster; these in turn cause oscillations of the electric field, leading to the ionisation of the cluster. A similar scenario occurs with the damping of the surface plasmon resonances in fullerenes,[151] which also decay via the autoionisation channel.

The DDCS of the electron inelastic scattering on metal clusters obtained in the PRA while accounting for both surface and volume excitations is given by[121]

$$
\begin{aligned}
\frac{\mathrm{d}^2\sigma}{\mathrm{d}\varepsilon_2 \mathrm{d}\Omega_{\mathbf{p}_2}} &= \frac{4Rp_2}{\pi q^4 p_1} \sum_l (2l+1)^2 j_l^2(qR) \frac{\omega_l^2\,\omega\,\Gamma_l^{(\mathrm{s})}}{(\omega^2-\omega_l^2)^2+\omega^2\left(\Gamma_l^{(\mathrm{s})}\right)^2} \\
&\quad + \frac{2R^3p_2}{\pi q^2 p_1} \sum_l (2l+1) \frac{\omega_{\mathrm{p}}^2\,\omega\,\Gamma_l^{(\mathrm{v})}}{(\omega^2-\ \omega_{\mathrm{p}}{}^2)^2+\omega^2\left(\Gamma_l^{(\mathrm{v})}\right)^2} \\
&\quad \times \left[j_l^2(qR) - j_{l+1}(qR) j_{l-1}(qR) - \frac{2}{qR} j_{l+1}(qR) j_l(qR) \right], \qquad (2.52)
\end{aligned}
$$

where $\omega_{\mathrm{p}} = \sqrt{3N_{\mathrm{e}}/R^3}$ is the volume plasmon resonance frequency, $\omega_l = \sqrt{l/(2l+1)}\omega_{\mathrm{p}}$ is the frequency of a surface plasmon excitation with the angular momentum l, and $\Gamma_l^{(\mathrm{v})}$ and $\Gamma_l^{(\mathrm{s})}$ are the corresponding widths. The CS (2.52) is similar to the expression obtained in Ref. 152 for electron scattering on small metal particles by means of classical electrodynamics.

According to Ref. 121, the width of the surface plasmon resonance in the PRA is equal to

$$
\Gamma_l^{(\mathrm{s})} = \frac{4\pi\,\omega_l}{(2l+1)R} \sum_{\nu,\mu} \left| \left\langle \psi_\mu \left| \varphi_l^{(\mathrm{s})}(\mathbf{r}) \right| \psi_\nu \right\rangle \right|^2 \delta(\omega_l - \varepsilon_\mu + \varepsilon_\nu), \qquad (2.53)
$$

where $\varphi_l^{(s)}(\mathbf{r}) = \sum_m \left[(r/R)^l \theta(R-r) + (R/r)^{(l+1)}\theta(r-R)\right] Y_{lm}(\mathbf{n})$. Note that the same expression was obtained in earlier studies[148,149] using other methods. Evaluation of the expression (2.53) for sufficiently large clusters leads to the well-known result for the Landau damping of the surface plasmon oscillations: $\Gamma_l^{(s)} = 3l\, v_F/R$, where v_F is the velocity of the cluster electrons on the Fermi surface.[149]

The PRA one can used to determine the autoionisation width of the volume plasmon resonance,[121] which is equal to

$$\Gamma_l^{(v)} = \frac{8\pi^2\omega_p}{q^2R^3} \frac{\sum_\nu \int |\langle\psi_\mu|\varphi_l^{(v)}(\mathbf{r})|\psi_\nu\rangle|^2\, \delta(\omega_p - \varepsilon_\mu + \varepsilon_\nu)\mathrm{d}\mu}{j_l^2(qR) - j_{l+1}(qR)j_{l-1}(qR) - \frac{2}{qR}j_{l+1}(qR)j_l(qR)}, \tag{2.54}$$

where $\varphi_l^{(v)}(\mathbf{r}) = \sum_m [j_l(qr) - j_l(qR)(r/R)^l]\theta(R-r)Y_{lm}(\mathbf{n})$. The summation is performed over the occupied single-electron states ν, and the integration is performed over the electronic states μ of the continuous spectrum.

The projectile particle simultaneously excites many modes of the volume plasmon. The sum of the potentials of all the modes gives the resulting potential $\varphi_l^{(v)}(\mathbf{r})$. It is essential that all normal modes of the volume plasmon have the same resonance frequency ω_p, but the excitation probability for these modes depends on the collision kinematics. This leads to the dependence of the volume plasmon potential $\varphi_l^{(v)}(\mathbf{r})$ on the transferred momentum. The oscillations of the volume plasmon potential lead to the ionisation of the cluster; its probability and the volume plasmon resonance width depend on the transferred momentum q. However, numerical analysis[121] has showed that the dependence of $\Gamma_l^{(v)}$ on q is rather weak in the region of $q \ll 1$, where collective electron oscillations mainly take place. Therefore, the volume plasmon resonance width with the given l can be approximated by a limiting value, which follows from Eq. (2.54) at $q = 0$:

$$\Gamma_l^{(v)} = (2l+5)\frac{\pi^2\omega_p}{R} \sum_\nu \int |\langle\psi_\mu|\varphi_l^{(v)}|_{q=0}(\mathbf{r})|\psi_\nu\rangle|^2\, \delta(\omega_p - \varepsilon_\mu + \varepsilon_\nu)\mathrm{d}\mu. \tag{2.55}$$

Figure 2.7 shows the dependence of the autoionisation width $\Gamma_l^{(v)}$ on the transferred momentum q for the volume plasmon modes,

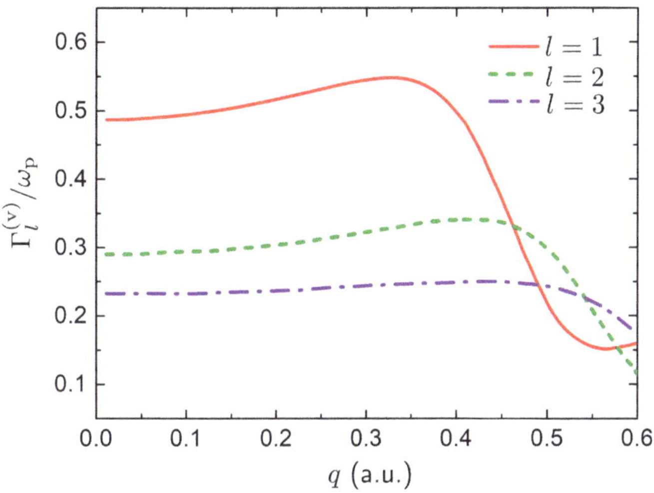

Fig. 2.7. Autoionisation width $\Gamma_l^{(v)}$ of the dipole ($l = 1$), quadrupole ($l = 2$), and octupole ($l = 3$) volume plasmon excitations as a function of transferred momentum q.

Source: Redrawn from data presented in Ref. 121.

which make a significant contribution to the EELS. The widths of the dipole, quadrupole, and octupole volume plasmon resonances have been calculated according to Eq. (2.54). The transferred momentum q plays the role of the wave vector for the volume plasmon excitations. All three plasmon modes have a similar dependence of $\Gamma_l^{(v)}$ on q. The width increases slowly for small q and decreases rapidly for larger q. In the latter region, the probability of volume plasmon excitation by the incoming electron is correspondingly reduced. Note that the wavelength of a collective electron oscillation should be larger than the inter-electron distance in the cluster, i.e., the plasmon wave vector should be smaller than the Fermi momentum of the cluster electrons. In the region $q < 0.5$, where the latter condition is fulfilled, the dependence of $\Gamma_l^{(v)}$ on q is rather weak, and the resonance width can be approximated by the following values: $\Gamma_1^{(v)} \simeq 0.5\omega_p$, $\Gamma_2^{(v)} \simeq 0.3\omega_p$, and $\Gamma_3^{(v)} \simeq 0.23\omega_p$. In contrast to surface plasmons, the autoionisation width of a volume plasmon decreases with increasing angular momentum.

Figure 2.8 shows the EELS of a sodium Na_{40} cluster in collision with a 50 eV electron at the scattering angle $\theta = 9°$. The figure

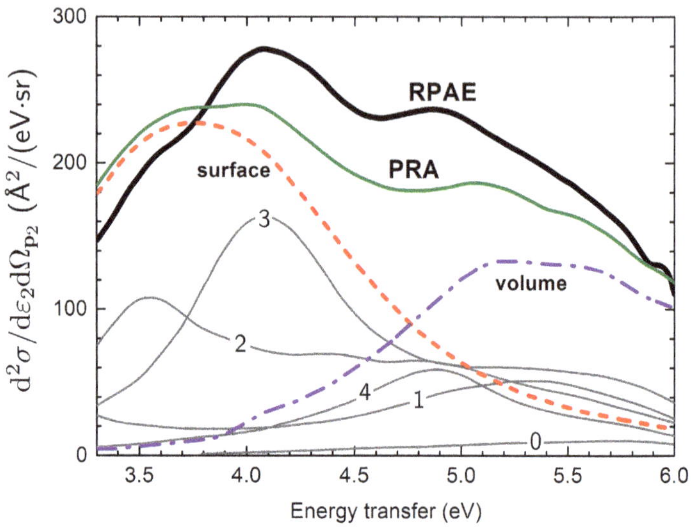

Fig. 2.8. Differential CS $d^2\sigma/d\varepsilon_2 d\Omega_{\mathbf{p}_2}$ as a function of the transferred energy ω calculated for the collision of a 50 eV electron with a Na_{40} cluster for the scattering angle $\theta = 9°$.[121] The thick solid black line labeled 'PRAE' and the thin solid grey lines numbered from 0 to 5 represent the RPAE results (see the text for further details). Contributions of the surface and volume plasmons calculated in the PRA, Eq. (2.52), are shown by dashed (red) and dash-dotted (purple) lines, respectively. The solid green line labeled 'PRA' represents the sum of these two contributions.

Source: Redrawn from data presented in Ref. 121.

illustrates the region of transferred energy above the ionisation potential, $\omega > 3.3$ eV, where the volume plasmon modes become significant. The solid curves illustrate the spectrum calculated using the random phase approximation with exchange (RPAE) method[15,117] – an *ab initio* method that takes into account many-electron correlations in a many-body system. The thick curve in Fig. 2.8 corresponds to the total EELS calculated using RPAE, while the thin curves labelled with numbers show different partial contributions corresponding to angular momenta l = 0–4. The partial contributions to the EELS with $l < 3$ have the broad maximum in the vicinity of $\omega \simeq 5.1$ eV. Comparison of the EELS calculated within the *ab initio* RPAE approach and the PRA (dash-dotted curve) confirms the idea that the peculiarity in the EELS around $\omega \sim 5$ eV is related to the volume plasmon excitation. Figure 2.8 shows that collective excitations make a dominant contribution to the total EELS and determine its pattern.

2.5.4 *Relaxation of electronic excitations in metal clusters*

As described above, the plasmon resonances in metal clusters may lie below the ionisation thresholds, i.e., in the region of the discrete spectrum of electron excitations. This fact raises an interesting physical problem about the eigenwidths of these electronic excitations, which have large oscillator strengths and form the plasmon resonances. Knowledge of these widths is necessary for a complete description of the EELS, electron attachment, polarisation *bremsstrahlung*, and photoabsorption CSs in the vicinity of the plasmon resonances and the description of their dependence on the cluster temperature. The temperature dependence of the plasmon resonance photoabsorption patterns of metal clusters was studied experimentally in Ref. 153.

In metal clusters, the origin of the electron excitation widths is mainly related to the dynamics of the ionic cluster core.[103,104,154–158] Let us focus on the influence of the dynamics of ions on the motion of delocalised electrons in metal clusters and discuss it on the basis of the dynamic jellium model proposed in Ref. 103 and further developed in Refs. 102,104. This model generalises the static jellium model,[95,159,160] which treats the ionic background of an atomic cluster as frozen by taking into account vibrations of the ionic background near the equilibrium point. The dynamic jellium model treats simultaneously the vibrational modes of the ionic jellium background, the quantised electron motion, and the interaction between the electronic and ionic subsystems. In Ref. 103, the dynamic jellium model was applied to provide a consistent description of physical phenomena arising from the vibrational dynamics of ions.

An important example of the effect resulting from the interaction of the ionic vibrations with delocalised electrons is the broadening of the electron excitation lines. Interest in the problem of electron excitation line broadening in metal clusters was stimulated by numerous experimental data on photoabsorption spectra, most of which were addressed in the region of dipole plasmon resonances.[86,92,161]

The dynamic jellium model[103] allows one to calculate the widths of the electron excitations in metal clusters caused by the dynamics of ions and their temperature dependence, taking into account the two mechanisms of the electron excitation line broadening, namely, *adiabatic* and *non-adiabatic* (or dynamic).

The *adiabatic* mechanism is related to the averaging of the electron excitation spectrum over the temperature fluctuations of the ionic background in a cluster. This phenomenon has also been studied in a number of papers.[103,104,154–158] The adiabatic linewidth is equal to

$$\Gamma = \sqrt{\frac{4\ln 2}{m\Omega}\,\mathrm{cth}\left(\frac{\Omega}{2k_{\mathrm{B}}T}\right)}\,\left|\hat{V}_{\mathrm{nn}}\right|. \tag{2.56}$$

Here, m and Ω are, respectively, the mass and frequency corresponding to the generalised oscillatory mode considered, T is the cluster temperature, k_{B} is the Boltzmann constant, and $\left|\hat{V}_{\mathrm{nn}}\right|$ is the matrix element of the electron–phonon coupling, calculated for surface and volume cluster vibrational modes.[104]

The mechanism of *non-adiabatic* electron excitation line broadening has been considered for the first time in Refs. 103,104. This mechanism originates from the real multiphonon transitions between the excited electron energy levels. Therefore, the dynamic linewidths characterise the real lifetimes of the electronic excitations in a cluster.

According to Ref. 103, the probability of a multiphonon transition from an excited cluster state with electronic and phononic quantum numbers n and N to all possible states (n', N') is equal to

$$\Gamma = \frac{2\pi}{\Omega}\,|A|^2 = \Omega\left|\frac{H^2_{\mathrm{n'n}}}{v(q_0)(V_{\mathrm{n'n'}} - V_{\mathrm{nn}})}\right| e^{2(\varphi_{\mathrm{n'}}-\varphi_{\mathrm{n}})}. \tag{2.57}$$

Here, $H_{\mathrm{n'n}}$ is the half-distance between the electron energy levels $\varepsilon_{\mathrm{n}}(q)$ and $\varepsilon_{\mathrm{n'}}(q)$ in the tangent point;

$$v(q_0) = \sqrt{\Omega\left[l^2 - 2S(2N - l - 1) + S^2\right]/2S} \tag{2.58}$$

is the ion velocity in the tangent point, which is expressed via the number of emitted phonons $l = N' - N$, where N' and N are the phonon numbers; and the parameter $S = |\hat{V}_{\mathrm{nn}} - \hat{V}_{\mathrm{n'n'}}|^2/2m\Omega^3$; $\varphi_{\mathrm{n}}, \varphi_{\mathrm{n'}}$ are the phases of ionic motion, arising from the distance between the turning points and the tangent point, being equal to

$$\varphi_{\mathrm{n}} = \frac{Z_{\mathrm{n}}\sqrt{Z_{\mathrm{n}}^2 - 2N - 1}}{2} + \frac{2N+1}{4}\ln\left(\frac{Z_{\mathrm{n}} + \sqrt{Z_{\mathrm{n}}^2 - 2N - 1}}{2N+1}\right), \tag{2.59}$$

where $Z_{\mathrm{n}} = (l - S)/\sqrt{2S}$. The expression for $\varphi_{\mathrm{n'}}$ is the same, but the parameter $Z_{\mathrm{n'}}$ is equal to $Z_{\mathrm{n'}} = (l + S)/\sqrt{2S}$.

The adiabatic broadening mechanism explains the temperature dependence of the photoabsorption spectra in the vicinity of the plasmon resonance via the coupling of the dipole excitations in a cluster with the quadrupole deformation of the cluster surface. The photoabsorption spectra were calculated within the framework of the deformed jellium model using either the plasmon pole approximation[154,155] or the local density approximation.[156–158,162,163] In Ref. 158, the octupole deformation of the cluster surface was taken into account. It was shown that the octupole deformation increases the Landau damping as a result of breaking the selection rule, leading to a mixture of dipole and quadrupole electronic excitations. Through this mechanism, the octupole deformations of the cluster surface provide the dominant contribution to the thermal broadening of the electron excitation lines in small metal clusters.

Both the adiabatic and non-adiabatic linewidths of electron excitations in the vicinity of the plasmon resonance caused by the coupling of electrons with different ionic vibrational modes were calculated in Refs. 103,104. The behaviour of the adiabatic and non-adiabatic linewidths as a function of temperature is shown in Fig. 2.9.

The non-adiabatic linewidths characterise the real lifetimes of cluster electron excitations. Naturally, the non-adiabatic widths are much smaller than the adiabatic ones due to the slower motion of ions in the cluster. However, the adiabatic linewidths do not completely mask the non-adiabatic ones because the two types of linewidths manifest themselves differently. The adiabatic broadening determines the pattern of the photoabsorption spectrum in the linear regime. The non-adiabatic linewidths are important for the processes where the real lifetime of the electron excitations and the electron–ion energy transfer are essential. The information about the non-adiabatic electron–phonon interactions in clusters is necessary for the description of the inelastic electron scattering on clusters,[116–118,121]

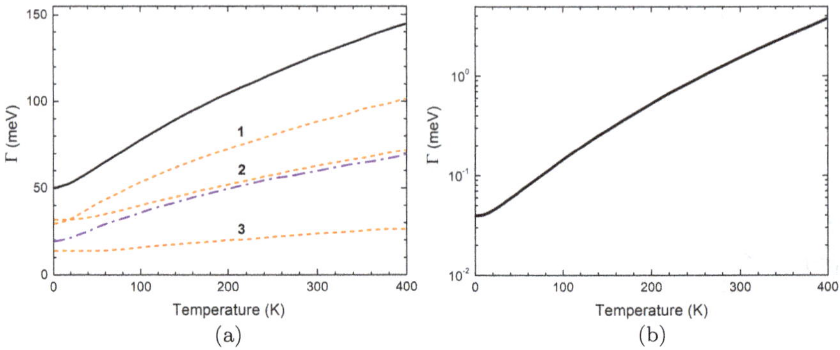

Fig. 2.9. Panel (a): temperature dependence of the adiabatic linewidth Γ, calculated according to Eq. (2.56) for the dipole electron excitation with the energy $\omega_n = 3.013$ eV in the Na_{40} cluster. The dashed curves labelled 1, 2, and 3 show the adiabatic widths corresponding to the electron coupling with the first three volume vibrational modes. The dashed-dotted curve shows the adiabatic width resulting from the electron coupling with the surface vibrations of the cluster. The solid curve shows the total adiabatic linewidth. Panel (b): temperature dependence of the non-adiabatic width calculated according to Eq. (2.57) for the dipole excitation with the energy $\omega_n = 3.013$ eV in the Na_{40} cluster.

Source: Redrawn from data presented in Refs. 103,104.

including the processes of electron attachment,[164–166] the nonlinear photoabsorption and bremsstrahlung,[167–170] the problem of cluster stability, and fission. The non-adiabatic linewidths, determined by the probability of multiphonon transitions, are also essential for the treatment of the relaxation of electronic excitations in clusters and the energy transfer from the excited electrons to the ions, which occurs after the impact- or photoexcitation of the cluster.

The role of the volume and surface vibrations of the ionic cluster core in the formation of the electron excitation linewidths was studied in Refs. 103,104. It was shown that the volume and surface vibrations make comparable contributions to the adiabatic linewidths, but the surface vibrations are much more essential for the non-adiabatic multiphonon transitions than the volume vibrations.

2.5.5 *Diffraction of fast electrons on atomic clusters and fullerenes*

The phenomenon of elastic scattering of fast electrons on metal clusters and fullerenes arises because the ionic density distribution in a cluster is typically characterised by a rigid boundary. The presence of a surface in a cluster leads to the specific oscillatory behaviour of the electron elastic scattering CSs, which can be interpreted in terms of electron diffraction at the cluster surface.[116,119] The detailed theoretical treatment of the diffraction phenomena arising in electron scattering on metal clusters and fullerenes was given in Refs. 116–118. Experimentally, diffraction in electron elastic scattering CSs on gas-phase fullerenes was first reported in Ref. 119.

The CS of elastic scattering of a fast electron on a cluster in the first Born approximation reads as[65]

$$\frac{\mathrm{d}\sigma}{\mathrm{d}\Omega_{\mathbf{p}_2}} = \frac{4}{q^4}\,F(q)^2. \tag{2.60}$$

Here, $F(q)$ is the form factor of the cluster, $q = |\mathbf{p}_2 - \mathbf{p}_1|$ is the momentum transfer, with $\mathbf{p}_1$, $\mathbf{p}_2$ being the momenta of the electron in the initial and final states, respectively, and $\mathrm{d}\Omega_{\mathbf{p}_2}$ denotes the differentiation over the solid angle of the scattered electron. The magnitude of q^2 is related to the scattering angle $\theta = \widehat{\mathbf{p}_1\,\mathbf{p}_2} \ll 1$ rad via

$$q^2 = p_1^2 + p_2^2 - 2p_1p_2\cos\theta = 2p_1^2(1-\cos\theta) \approx p_1^2\theta^2. \tag{2.61}$$

The form factor of the target, $F(q)$, can be expressed as the product of the form factor of the atomic concentration, $n(q)$, and the form factor of a single atom, $F_{\mathrm{A}}(q)$:

$$F(q) = F_{\mathrm{A}}(q)\sum_j e^{\mathrm{i}\mathbf{q}\mathbf{r}_j} = F_{\mathrm{A}}(q)\,n(q), \tag{2.62}$$

where the summation is performed over the coordinates $\mathbf{r}_j$ of all the atoms in the cluster. The applicability of this approximation was examined in Ref. 119 for metal clusters and fullerenes.

The form factor of the atomic concentration, $n(q)$, depends on the geometry of the cluster. In the case of the *metal cluster*, considering the cluster as a spherically symmetric system of the radius R and assuming a homogeneous distribution of atoms in the volume of the cluster, one derives

$$n(q) = 3N\left[\frac{\sin(qR)}{(qR)^3} - \frac{\cos(qR)}{(qR)^2}\right] = 3N\,\frac{j_1(qR)}{qR}, \tag{2.63}$$

where $j_1(qR)$ is the spherical Bessel function of the first order and N is the number of atoms in the cluster.[116] The simplest approximation for describing *a fullerene* is to assume that the carbon atoms are uniformly distributed on the surface of a sphere of the radius R. In this case,

$$n(q) = N\,\frac{\sin(qR)}{qR}. \tag{2.64}$$

This form factor oscillates with a period of $q = 2\pi/R \ll 1$. These oscillations form the diffraction pattern of the differential CS, Eq. (2.60), which possesses a series of diffraction maxima and minima whose positions are mainly determined by the radius of the target.

Figure 2.10 shows the dependence of the CS $\mathrm{d}\sigma/\mathrm{d}\Omega_{\mathbf{p}_2}$ on the scattering angle θ for an elastic collision of a 809 eV electron with the C_{60} fullerene.[119] The figure shows that the CS has a number of diffraction maxima and minima. The experimental data points obtained in Ref. 119 in the two sets of measurements are represented by open and

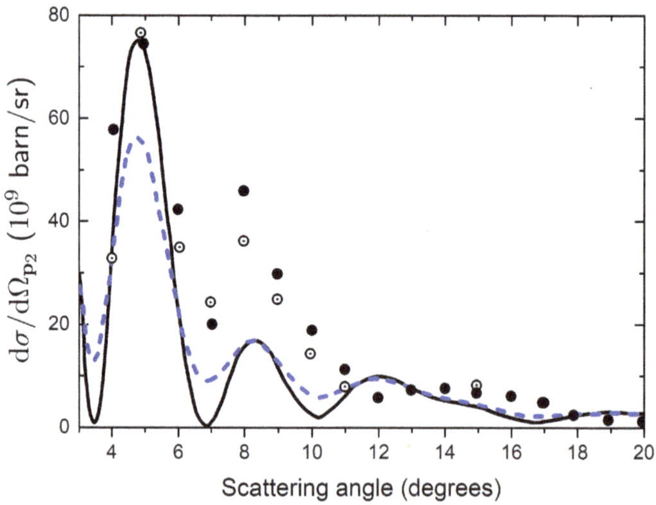

Fig. 2.10. Experimental (symbols) and theoretical (solid curve) angular dependence of the differential elastic scattering CS in the collision of an 809 eV electron with the C_{60} molecule.[119] The symbols correspond to the two independent sets of measurements. The dashed curve is the CS for the mixture containing 60% of C_{60} and 40% of equivalent isolated carbon atoms.

Source: Redrawn from data presented in Ref. 119.

closed circles. The theoretically obtained CS dependence is shown by a solid curve. The experimental data have been normalised to the theoretical CS at the second diffraction maximum ($\theta = 5°$). Figure 2.10 shows a good agreement between the experimental and theoretical results in the position of the first and second maxima. The entire pattern of the differential CS obtained theoretically is very similar to that measured experimentally.

In the vicinity of the diffraction maxima at $\theta < 10°$, the CS greatly exceeds the elastic scattering CS on the equivalent number of isolated atoms due to the coherent interaction of the projectile electron with the fullerene sphere. In the region $\theta > 10°$, where $q > 1$, the projectile electron scatters on individual carbon atoms of the fullerene rather than on the entire molecule. Therefore, the diffraction features of the CS in the region $\theta < 10°$ are much more pronounced than in the region $\theta > 10°$.

In the region $\theta < 10°$, where $q < 1$, the theoretical CS has zeros while the experimental one does not. The presence of zeros at $q \approx \pi k/R < 1$, where k is an integer, in the theoretical curve is the consequence of the coherent scattering of the electron on the fullerene sphere.[a] In the experiment, however, zeros in the CS may disappear due to the presence of carbon atoms or some other impurities in the gas cell. Figure 2.10 also shows the differential CS for the mixture containing 60% of C_{60} and 40% of isolated carbon atoms (the dashed curve). The differential electron elastic scattering CS on individual carbon atoms has no diffraction oscillations and, therefore, forms a smooth background, which removes the zeros in the angular dependence of the CS.

2.5.6 *Dynamical screening effects in endohedral fullerenes*

Another area of interest for the atomic cluster community in recent decades has been related to endohedral fullerenes – systems

[a]In the region $\theta > 10°$, where $q > 1$, the process of elastic scattering at the fullerene shell with the subsequent excitation of surface multipole plasmons becomes dominant. This process is described by the formulas of the second Born approximation, which was used[119] to correct the calculated CS at large values of the transferred momentum.

consisting of an atom or a small molecule encapsulated in a fullerene cage (see, e.g., reviews in Refs. 171–173 and references therein). Great scientific interest has been driven by the numerous potential applications[172] of atoms encapsulated in spherical or near-spherical fullerenes and the fundamental aspects of an atom confined in a nanometer-sized cavity.[174]

The confinement of atoms in fullerene cages has been addressed by numerous theoretical approaches (see, e.g., Refs. 129,175–180), but experimental studies have been relatively scarce,[181–185] mainly due to the technical difficulties in producing such targets in significant quantities and with high purity. The most definitive experimental measurement to directly access the physics of confined atoms is the observation of their response to photon irradiation. Several experimental studies[181,184] focused on the photoionisation of endohedral fullerene ions, considering the mutual influence of the fullerene cage and the encapsulated species on the photoionisation process. A recent review of these studies can be found in Ref. 186.

The fundamental interest in the study of endohedral fullerenes has been driven by the possibility of obtaining spectra of a novel type in which atomic excitations interact with cavity resonances,[187] and by the fact that the confined atom is screened by a fullerene shell, whose properties are similar to those of a hollow metal sphere.[176] In essence, the fullerene shell behaves like a 'Faraday cage' of very small dimensions in which, in the static limit, external fields cannot penetrate and the entire enclosed volume is therefore at the same potential.

Numerous theoretical studies (e.g., Refs. 175,178 to name a few) have predicted the existence of so-called 'confinement resonances' in the photoionisation spectra of endohedral fullerenes. Experimentally, such resonances were detected for the first time in the photoionisation spectra of $Xe@C_{60}$.[181,182] Confinement resonances appear as a result of interference between a direct wave of the photoelectron escaping the atom and the waves due to scattering from the atoms of the fullerene cage. Depending on the photoelectron momentum, the interference can be constructive or destructive. As a result, the spectrum of the confined atom acquires additional oscillations compared to that of the free atom.

The role of dynamical screening in the photoionisation or photoabsorption process of an atom confined within a fullerene shell was elucidated in Ref. 176. When an endohedral fullerene is exposed to an external electromagnetic field, the fullerene shell dynamically screens the confined atom. The atom experiences a field that is enhanced or suppressed depending on the frequency of the light, ω. The result is that, for the same external field, the photoabsorption rate of the confined atom differs from that of the free atom. A dynamical screening factor $\mathcal{F}(\omega)$ can be defined to relate the photoabsorption CSs of these two atoms:

$$\mathcal{F}(\omega) = \frac{\sigma_{\text{conf}}(\omega)}{\sigma_{\text{free}}(\omega)}. \tag{2.65}$$

The screening is strongly dependent on the frequency of the light (hence the term dynamical screening) and leads to an enhancement of the amplitude of the electromagnetic wave within a specific range of frequencies. This range is determined by the characteristics of the fullerene plasmon (or, more generally, by the plasmon of the confining system).

The problem of dynamical screening of an atom confined in the centre of a fullerene cage of finite thickness (see Fig. 2.11(a)) was studied theoretically in Ref. 129. In a later study,[188] the method was generalised for an arbitrary position of the confined atom inside the fullerene (see Fig. 2.11(b)). Similar to other case studies presented in this chapter, a fullerene was treated as a spherical shell of thickness ΔR with a dielectric function $\epsilon \equiv \epsilon(\omega)$. Following the Drude model, the dielectric function ϵ is expressed as:

$$\epsilon(\omega) = 1 - \frac{\omega_{\text{p}}^2}{\omega^2}, \tag{2.66}$$

where the plasma frequency ω_{p} is described by the total number of delocalised electrons N_e (four valence electrons from each carbon atom) and the volume of the spherical shell.

The point-like atom located at $r = 0$ is characterised by the polarisability $\alpha_{\text{a}}(\omega)$. This system is exposed to a monochromatic electromagnetic wave. Using the dipole approximation and neglecting the magnetic part, this is treated as an external uniform electric field $\mathbf{E}_{\text{ext}}(t) = \mathbf{E}_0 e^{i\omega t}$.

The photoabsorption rate of an object is proportional to its photoabsorption CS. Therefore the dynamical screening factor $\mathcal{F}(\omega)$ can be calculated by comparing the absorption rate of the confined atom

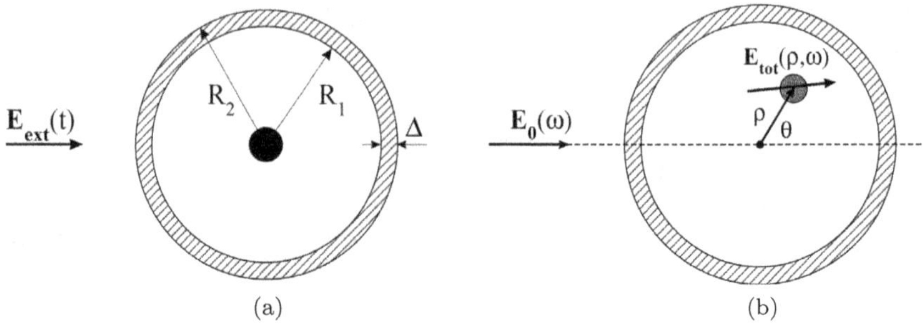

Fig. 2.11. Panel (a): The atom, indicated by the solid circle, is confined at the centre of a fullerene of average radius R and thickness ΔR. The inner and the outer radii of the fullerene are given by R_1 and R_2 respectively. The system is exposed to an external field $\mathbf{E}_{\rm ext}$. Panel (b): The confined atom is located at distance ρ from the centre of the fullerene. The field at the position of the atom is given by $\mathbf{E}_{\rm tot}(\rho,\omega)$.

Source: Reproduced from Ref. 129 with permission from IOP Publishing and Ref. 188 with permission from the American Physical Society.

with that of the free atom. In the dipole limit, an object exposed to an electromagnetic wave will absorb its energy at a rate Q given by:[189]

$$Q = \frac{\omega}{2}\,{\rm Im}\int \mathbf{E}^*(\mathbf{r})\,{\rm d}\mathbf{D}(\mathbf{r}). \tag{2.67}$$

Here the integration is performed over the volume of the considered system. $\mathbf{E}(\mathbf{r})$ is the electric field strength at the position $\mathbf{r}$ and ${\rm d}\mathbf{D}(\mathbf{r})$ is the dipole moment of the integration element at that position. For a point-like atom located at $\mathbf{r}$, Eq. (2.67) reads as

$$Q_{\rm a} = \omega\,{\rm Im}\left(\mathbf{E}^*(\mathbf{r})\mathbf{d}(\mathbf{r})\right)/2 = \omega\,{\rm E}_0^2\,{\rm Im}\left(\alpha_{\rm a}(\omega)\right)/2, \tag{2.68}$$

where $\mathbf{d}(\mathbf{r})$ is the dipole moment of the atom. Thus, the dynamical screening factor becomes

$$\mathcal{F}(\omega) = \frac{{\rm Im}\left(\mathbf{E}^*_{\rm tot}(\mathbf{r})\,\mathbf{d}_{\rm tot}(\mathbf{r})\right)}{E_0^2\,{\rm Im}(\alpha_{\rm a}(\omega))}, \tag{2.69}$$

where $\mathbf{E}_{\rm tot}(\mathbf{r})$ and $\mathbf{d}_{\rm tot}(\mathbf{r})$ are the total electric field and total dipole moment at $\mathbf{r}$, the position of the encapsulated atom.

As demonstrated in Ref. 129, the dynamical screening factor $\mathcal{F}(\omega)$ of an atom confined *at the centre of the fullerene* of finite width (see

Fig. 2.11a) is given by

$$\mathcal{F}(\omega) = \left| \frac{w}{1 - \alpha_a(\omega)\tilde{w}} \right|^2, \tag{2.70}$$

where

$$w = 1 + \frac{2(N_1 - N_2)}{R_2^3} \left(\frac{1}{\omega^2 - \omega_1^2 + i\Gamma_1\omega} - \frac{1}{\omega^2 - \omega_2^2 + i\Gamma_2\omega} \right), \tag{2.71}$$

$$\tilde{w} = -\frac{2}{R_1^3 R_2^3} \left(\frac{N_2}{\omega^2 - \omega_1^2 + i\Gamma_1\omega} + \frac{N_1}{\omega^2 - \omega_2^2 + i\Gamma_2\omega} \right), \tag{2.72}$$

where

$$\omega_1^2 = \frac{\omega_p^2}{6}(3 - p), \quad \omega_2^2 = \frac{\omega_p^2}{6}(3 + p), \quad p = \sqrt{1 + 8\xi^3}. \tag{2.73}$$

The frequencies ω_1 and ω_2 are the eigenfrequencies of the symmetric and antisymmetric surface plasmon modes of the fullerene, respectively. N_1 and N_2 are the number of delocalized electrons involved in each mode:

$$N_1 = N\frac{p+1}{2p} \quad N_2 = N\frac{p-1}{2p}. \tag{2.74}$$

Γ_1 and Γ_2 are the widths of the two surface plasmon modes. These describe the decay rate from the collective excitation to the incoherent sum of single-electron excitations. The case study described in the following corresponds to the C_{60} fullerene with a mean radius of $R = 3.5$ Å and a thickness of $\Delta R = 1.5$ Å.

Figure 2.12 shows the dynamical screening factor calculated for the Ar@C_{60} system, where the Ar atom is located in the centre of the C_{60} cage (see dashed green line).[188] The dynamical screening factor has been calculated for the photon energies above the ionisation threshold of the argon atom, $I = 16$ eV. For photon energies $\omega < I$, the calculation of the screening factor is meaningless according to Eq. (2.65) since the atomic photoionisation CS σ_{free} is equal to zero. The curve is compared with the limiting case where the interaction between the atom and the fullerene is neglected (thin solid green line). There is a notable difference between the two cases, which

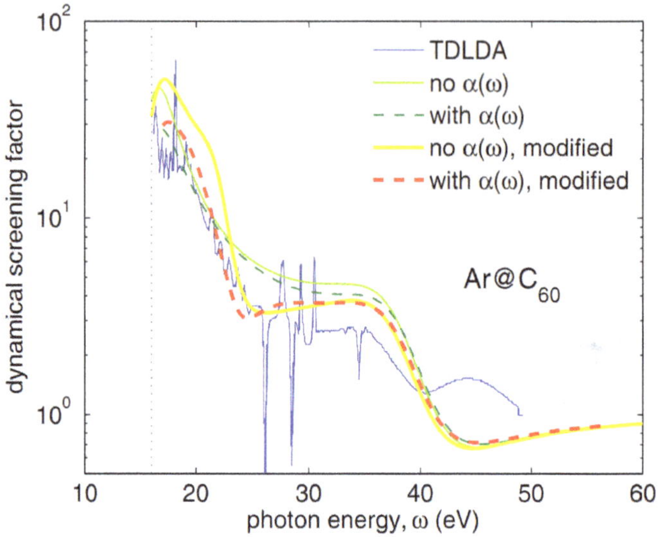

Fig. 2.12. The dynamical screening factor for the Ar atom located at the centre of the C_{60} cage. The dynamical screening factor was calculated for two cases: (i) neglecting the atomic feedback, labelled 'no $\alpha(\omega)$', and (ii) with accounting for this feedback, labelled 'with $\alpha(\omega)$', represented by the solid and dashed green lines, respectively. These are compared with the TDLDA calculations (thin solid blue line) from Ref. 177. The ionisation potential of the endohedral atom is indicated by the vertical dotted line. The thick solid (yellow) and dashed (red) lines represent the dynamical screening factor calculated using the modified model, which includes the contributions of the σ and π plasmons.

Source: Reproduced from Ref. 188 with permission from the American Physical Society.

can be attributed to the large dipole polarisability of argon in the energy range considered. The results of the model calculations are compared with the results of the TDLDA calculations from Ref. 177 (thin blue line).

Figure 2.12 shows a good overall agreement between the model results and the TDLDA calculation. The inclusion of the interaction between the polarised atom and the fullerene in the dynamical screening factor improves the agreement. Even better agreement with the results of the TDLDA calculations was achieved by taking into account the σ and π plasmons (which involve σ and π delocalised electrons of the fullerene, respectively), each split into a symmetric and an antisymmetric mode.

The modified dynamical screening factors are shown in Fig. 2.12 by thick solid (yellow) and dashed (red) curves. The solid line is the dynamical screening factor for the limiting case where the interaction between the atom and the fullerene is neglected, and the dashed line takes into account the interaction between the polarised atom and the polarised fullerene. It is evident that the modification of the screening factor by taking into account σ and the π plasmons leads to a better agreement with the TDLDA calculations.[177]

In reality, the endohedral atom is not fixed at the centre of the fullerene. Thermal vibrations, van der Waals interaction with the fullerene, and other effects, such as electron transfer, can cause the atom to be preferentially displaced from the centre. Figure 2.13 illustrates the dynamical screening factor for Ar@C_{60} where the Ar atom is arbitrarily positioned inside the fullerene. The four panels in

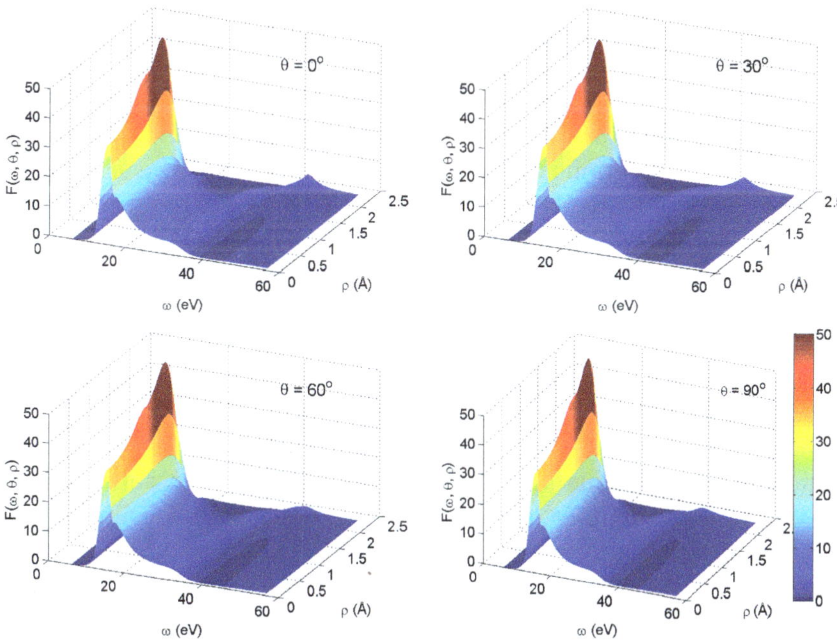

Fig. 2.13. The dynamical screening factor for Ar@C_{60} as a function of photon energy ω and the radial distance of the atom from the fullerene's centre, ρ, calculated for different values of the angle θ between ρ and $\mathbf{E_0}$.

Source: Reproduced from Ref. 188 with permission from the American Physical Society.

Fig. 2.13 show the screening factor as a function of photon energy ω and radial distance from the centre, ρ; see Fig. 2.11(b). Each panel corresponds to the indicated value of the angle θ between ρ and $\mathbf{E_0}$. Figure 2.13 shows that the dynamical screening factor depends rather weakly on θ. The radial distance from the centre, ρ, has a much larger influence on the dynamical screening factor, especially at large distances. As the atom is moved away from the fullerene centre, the screening factor increases in the vicinity of the dipole-symmetric plasmon mode. This increase becomes more dramatic as the atom approaches the fullerene shell.

The role of the off-centre position of an encapsulated atom in the photoionisation spectra of endohedral fullerenes (specifically, the presence of confinement resonances in the spectra) was theoretically elucidated in Ref. 179. The results of the experiments[181] which contradicted the earlier theoretical predictions about the existence of confinement resonances in specific endohedral systems were explained.

The position of the endohedral atom can be characterised by the displacement vector $\mathbf{a}$ from the centre of the fullerene cage. The equilibrium position of an endohedral atom and the amplitude of the thermal vibrations depend on the atom type, the cage size, and the temperature. Due to the thermal motion, the atom is not fixed in any particular position inside the fullerene. For meaningful observable results, it is therefore necessary to average over the values $a = |\mathbf{a}|$. As shown in Ref. 179, the structure of confinement resonances in the photoionisation CS of an endohedral atom is highly sensitive to the mean displacement $\langle a \rangle$ of the atom from the cage centre, and in many cases averaging over the values a destroys the confinement resonance structure. Qualitatively, this can be explained as follows. The confinement resonances arise from the interference of the two waves of the photoelectron, the direct and scattered ones, both originating from the confined atom. The atom at the centre can be treated as a point-like source. For $a \neq 0$; the source acquires a size of $D \approx 2\langle a \rangle$, where $\langle a \rangle$ is the mean distance from the centre. The finite size of the source affects the interference pattern. When D exceeds half the wavelength, the angular averaging can destroy the oscillatory structure.

On this basis, a criterion was formulated which allows one to estimate, for a given endohedral system, the interval of photon energies in which the resonances can appear. It was shown[179] that the resonance structure is suppressed for the electron momenta $p > p_{\min} \approx \pi/D$. The corresponding range of photon energies is $\omega > p_{\min}^2/2 + I_0$, where I_0 is the ionisation potential of the atomic shell.

References

1. Solov'yov AV, *et al.* Condensed matter systems exposed to radiation: Multiscale theory, simulations, and experiment. Chem Rev. 2024;124:8014–8129.
2. Froese Fischer C. The Hartree-Fock method for atoms: A numerical approach. Hoboken: John Wiley & Sons Inc. 1977.
3. Kohn W, Sham LJ. Self-consistent equations including exchange and correlation effects. Phys Rev. 1965;140:A1133–A1138.
4. Parr RG, Yang W. Density-functional theory of atoms and molecules. New York: Oxford University Press. 1989.
5. Ekardt W, editor. Metal clusters. Chichester: Wiley. 1999.
6. Connerade JP, Solov'yov AV, editors. Latest advances in atomic clusters collision: Fission, fusion, electron, ion and photon impact. London: Imperial College Press. 2004.
7. Mariscal MM, Oviedo OA, Leiva EPM, editors. Metal clusters and nanoalloys: From modeling to applications. New York: Springer Science+Business Media. 2013.
8. Jalkanen KJ, Suhai S, Bohr H. Quantum molecular biological methods using density functional theory. In: Bohr HG, editor. Handbook of molecular biophysics. Methods and applications. Weinheim: Wiley-VCH Verlag. 2009. p. 7–66.
9. Becke AD. Fifty years of density-functional theory in chemical physics. J Chem Phys. 2014;140:18A301.
10. Scuseria GE, Staroverov VN. Development of approximate exchange-correlation functionals. In: Dykstra C, Frenking G, Kim K, Scuseria G, editors. Theory and applications of computational chemistry: The first forty years. Amsterdam: Elsevier. 2005. p. 669–724.
11. Jones RO. Density functional theory: Its origins, rise to prominence, and future. Rev Mod Phys. 2015;87:897–923.
12. Leszczynski J, *et al.*, editors. Handbook of computational chemistry. 2nd ed. Cham: Springer International Publishing. 2017.
13. Kaplan AD, Levy M, Perdew JP. The predictive power of exact constraints and appropriate norms in density functional theory. Annu Rev Phys Chem. 2023;74:193–218.

14. Cancès E, Friesecke G, editors. Density functional theory: Modeling, mathematical analysis, computational methods, and applications. Cham: Springer Nature Switzerland AG. 2023.
15. Solov'yov AV. Plasmon excitations in metal clusters and fullerenes. Int J Mod Phys B. 2005;19:4143–4184.
16. Solov'yov IA, *et al.*, editors. Dynamics of systems on the nanoscale. Cham: Springer International Publishing. 2022.
17. Grant IP. Relativistic calculation of atomic structures. Adv Phys. 1970;19:747–811.
18. Mohanty A, Clementi E. Dirac-Fock self-consistent field method for closed-shell molecules with kinetic balance and finite nuclear size. Int J Quantum Chem. 1991;39:487–517.
19. Pyykkö P. Relativistic effects in structural chemistry. Chem Rev. 1988;88:563–594.
20. Shavitt I, Bartlett RJ. Many-body methods in chemistry and physics: MBPT and coupled-cluster theory. Cambridge: Cambridge University Press. 2009.
21. Cizek J, Paldus J. Coupled cluster approach. Phys Scr. 1980;21: 251–254.
22. Bartlett RJ, Musiał M. Coupled-cluster theory in quantum chemistry. Rev Mod Phys. 2007;79:291–352.
23. Zhang IY, Grüneis A. Coupled cluster theory in materials science. Front Mater. 2019;6:123.
24. Szalay PG, *et al.* Multiconfiguration self-consistent field and multireference configuration interaction methods and applications. Chem Rev. 2012;112:108–181.
25. Knowles PJ, Handy NC. A determinant based full configuration interaction program. Comput Phys Commun. 1989;54:75–83.
26. Rontani M, *et al.* Full configuration interaction approach to the few-electron problem in artificial atoms. J Chem Phys. 2006;124: 124102.
27. Xu E, Uejima M, Ten-no SL. Full coupled-cluster reduction for accurate description of strong electron correlation. Phys Rev Lett. 2018;121:113001.
28. Purvis GD III, Bartlett RJ. A full coupled-cluster singles and doubles model: The inclusion of disconnected triples. J Chem Phys. 1982;76:1910–1918.
29. Cullen JM, Zerner MC. The linked singles and doubles model: An approximate theory of electron correlation based on the coupled-cluster ansatz. J Chem Phys. 1982;77:4088–4109.
30. Raghavachari K, *et al.* A fifth-order perturbation comparison of electron correlation theories. Chem Phys Lett. 1989;157:479–483.

31. Møller C, Plesset MS. Note on an approximation treatment for many-electron systems. Phys Rev. 1934;46:618–622.
32. Cremer D. Møller–Plesset perturbation theory: From small molecule methods to methods for thousands of atoms. WIREs Comput Mol Sci. 2011;1:509–530.
33. March NH. Electronic structure of atoms, molecules, and clusters: Hartree-Fock and beyond. In: Maruani J, Lefebvre R, Brändas EJ, editors. Advanced topics in theoretical chemical physics. Dordrecht: Springer. 2003. pp. 53–70.
34. Evarestov RA. Quantum chemistry of solids: LCAO treatment of crystals and nanostructures. 2nd ed. Berlin Heidelberg: Springer-Verlag. 2012.
35. Ren X, *et al.* Random-phase approximation and its applications in computational chemistry and materials science. J Mater Sci. 2012;47:7447–7471.
36. Starace A. Photoionization of atoms. In: Drake G, editor. Springer handbook of atomic, molecular, and optical physics. 2nd ed. New York: Springer. 2006. pp. 379–390.
37. Lundqvist S, March NH. Theory of the inhomogeneous electron gas. New York: Plenum Press. 1983.
38. Perdew JP, Burke K, Ernzerhof M. Generalized gradient approximation made simple. Phys Rev Lett. 1996;77:3865–3868.
39. Yanai T, Tew DP, Handy NC. A new hybrid exchange–correlation functional using the Coulomb-attenuating method (CAM-B3LYP). Chem Phys Lett. 2004;393:51–57.
40. Vydrov OA, *et al.* Importance of short-range versus long-range Hartree-Fock exchange for the performance of hybrid density functionals. J Chem Phys. 2006;125:074106.
41. Chai J-D, Head-Gordon M. Long-range corrected hybrid density functionals with damped atom-atom dispersion corrections. Phys Chem Chem Phys. 2008;10:6615–6620.
42. Tsuneda T, Hirao K. Long-range correction for density functional theory. WIREs Comput Mol Sci. 2014;4:375–390.
43. Li M, *et al.* Accurate prediction of the properties of materials using the CAM-B3LYP density functional. J Comput Chem. 2021;42: 1486–1497.
44. Grimme S. Density functional theory with London dispersion corrections. WIREs Comput Mol Sci. 2011;1:211–228.
45. Grimme S, *et al.* Dispersion-corrected mean-field electronic structure methods. Chem Rev. 2016;116:5105–5154.
46. Makkar P, Ghosh NN. A review on the use of DFT for the prediction of the properties of nanomaterials. RSC Adv. 2021;11:27897–27924.

47. Thoben N, *et al.* Density functional theory calculations for multiple conformers explaining the regio- and stereoselectivity of Ti-catalyzed hydroaminoalkylation reactions. ChemPhysChem. 2023;24:e202300370.
48. Mohrhusen L, *et al.* Multidentate interaction of methylamine with rutile TiO_2 (110). J Phys Chem C. 2021;125:11975–11986.
49. Gerhards L, *et al.* Peculiar differences between two copper complexes containing similar redox-active ligands: Density functional and multiconfigurational calculations. Inorg Chem. 2024;63:961–975.
50. Runge E, Gross EKU. Density-functional theory for time-dependent systems. Phys Rev Lett. 1984;52:997–1000.
51. Ullrich CA. Time-dependent density-functional theory: Concepts and applications. Oxford: Oxford University Press. 2012.
52. Maitra NT. Fundamental aspects of time-dependent density functional theory. J Chem Phys. 2016;144:220901.
53. Walker B, Gebauer R. Ultrasoft pseudopotentials in time-dependent density-functional theory. J Chem Phys. 2007;127:164106.
54. Marques MAL, *et al.*, editors. Time-dependent density functional theory. Berlin Heidelberg: Springer. 2006.
55. Adamo C, Jacquemin D. The calculations of excited-state properties with time-dependent density functional theory. Chem Soc Rev. 2013;42:845–856.
56. Lian C, *et al.* Photoexcitation in solids: First-principles quantum simulations by real-time TDDFT. Adv Theory Simul. 2018;1:1800055.
57. Malcioğlu OB, *et al.* Dielectric and thermal effects on the optical properties of natural dyes: A case study on solvated cyanin. J Am Chem Soc. 2011;133:15425–15433.
58. Sjulstok E, Olsen JMH, Solov'yov IA. Quantifying electron transfer reactions in biological systems: What interactions play the major role? Sci Rep. 2015;5:18446.
59. Shao Y, *et al.* Benchmarking the performance of time-dependent density functional theory methods on biochromophores. J Chem Theory Comput. 2020;16:587–600.
60. Timmer D, *et al.* Structural flexibility slows down charge transfers in diaminoterephthalate-C_{60} dyads. J Phys Chem C. 2024;128:2380–2391.
61. Timrov I, *et al.* Electron energy loss and inelastic X-ray scattering cross sections from time-dependent density-functional perturbation theory. Phys Rev B. 2013;88:064301.
62. Nicholls RJ. Advances in modelling electron energy loss spectra from first principles. J Phys Mater. 2021;4:024008.

63. Moitra T, *et al.* Accurate relativistic real-time time-dependent density functional theory for valence and core attosecond transient absorption spectroscopy. J Phys Chem Lett. 2023;14:1714–1724.
64. Ghosal A, Roy AK. A real-time TDDFT scheme for strong-field interaction in Cartesian coordinate grid. Chem Phys Lett. 2022;796:139562.
65. Landau LD, Lifshitz EM. Quantum mechanics: Non-relativistic theory. 3rd ed. Oxford: Butterworth-Heinemann. 1981.
66. Taylor JR. Scattering theory: The quantum theory of nonrelativistic collisions. Mineola: Dover Publications Inc. 2006.
67. Newton RG. Scattering theory of waves and particles. 2nd ed. Berlin Heidelberg: Springer. 1982.
68. Friedrich H. Scattering theory. Berlin Heidelberg: Springer. 2013.
69. Dreizler RM, Kirchner T, Lüdde C. Quantum collision theory of nonrelativistic particles. Berlin: Springer-Verlag GmbH Germany. 2022.
70. Wachter A. Relativistic quantum mechanics. Dordrecht: Springer. 2010.
71. Hofierka J, *et al.* Many-body theory of positron binding to polyatomic molecules. Nature. 2022;606:688–693.
72. Rawlins CM, *et al.* Many-body theory calculations of positron scattering and annihilation in H_2, N_2, and CH_4. Phys Rev Lett. 2023;130:263001.
73. Samanta K, *et al.* Electron–atom and electron–molecule resonances: Some theoretical approaches using complex scaled multiconfigurational methods. Adv Quantum Chem. 2018;77:317–390.
74. Schneider B. R-matrix theory for electron-atom and electron-molecule collisions using analytic basis set expansions. Chem Phys Lett. 1975;31:237–241.
75. Burke PG, Mackey I, Shimamura I. R-matrix theory of electron–molecule scattering. J Phys B At Mol Phys. 1977;10:2497–2512.
76. Tennyson J. Electron–molecule collision calculations using the R-matrix method. Phys Rep. 2010;491:29–76.
77. Otvos JW, Stevenson DP. Cross-sections of molecules for ionization by electrons. J Am Chem Soc. 1956;78:546–551.
78. Deutsch H, *et al.* Theoretical determination of absolute electron-impact ionization cross sections of molecules. Int J Mass Spectrom. 2000;197:37–69.
79. Kim Y-K, Rudd ME. Binary-encounter-dipole model for electron-impact ionization. Phys Rev A. 1994;50:3954–3967.
80. Tanaka H, *et al.* Scaled plane-wave Born cross sections for atoms and molecules. Rev Mod Phys. 2016;88:025004.

81. Kim Y-K, Santos JP, Parente F. Extension of the binary-encounter-dipole model to relativistic incident electrons. Phys Rev A. 2000;62: 052710.
82. Rudd ME, *et al.* Electron production in proton collisions with atoms and molecules: Energy distributions. Rev Mod Phys. 1992;64:441–490.
83. de Vera P, *et al.* Semiempirical model for the ion impact ionization of complex biological media. Phys Rev Lett. 2013;110:148104.
84. Solov'yov AV, editor. Nanoscale insights into ion-beam cancer therapy. Cham: Springer International Publishing. 2017.
85. Surdutovich E, Solov'yov AV. Multiscale approach to the physics of radiation damage with ions. Eur Phys J D. 2014;68:353.
86. Haberland H, editor. Clusters of atoms and molecules, theory, experiment and clusters of atoms. Berlin: Springer. 1994.
87. Ekardt W. Dynamical polarizability of small metal particles: Self-consistent spherical jellium background mode. Phys Rev Lett. 1984;52:1925–1928.
88. Connerade JP, Solov'yov AV, editors. Latest advances in atomic clusters collision: Structure and dynamics from the nuclear to the biological scale. London: Imperial College Press. 2008.
89. Knight WD, *et al.* Electronic shell structure and abundances of sodium clusters. Phys Rev Lett. 1984;52:2141–2143.
90. Ekardt W. Work function of small metal particles: Self-consistent spherical jellium-background model. Phys Rev B. 1984;29:1558–1564.
91. Clemenger K. Ellipsoidal shell structure in free-electron metal clusters. Phys Rev B. 1985;32:1359–1362.
92. de Heer WA. The physics of simple metal clusters: Experimental aspects and simple models. Rev Mod Phys. 1993;65:611–676.
93. Brack M. The physics of simple metal clusters: Self-consistent jellium model and semiclassical approaches. Rev Mod Phys. 1993;65: 677–732.
94. Bréchignac C, Connerade JP. Giant resonances in free atoms and in clusters. J Phys B At Mol Opt Phys. 1994;27:3795–3828.
95. Ekardt W. Collective multipole excitations in small metal particles: Critical angular momentum lcr for the existence of collective surface modes. Phys Rev B. 1985;32:1961–1970.
96. Ekardt W, Penzar Z. Self-consistent Sommerfeld droplet as a simple model for an accurate prediction of the electronic properties of small metal particles. Phys Rev B. 1988;38:4273–4276.
97. Ekardt W, Penzar Z. Collective excitations in open-shell metal clusters: The time-dependent local-density approximation applied to the self-consistent spheroidal jellium particle. Phys Rev B. 1991;43: 1322–1330.

98. Montag B, *et al.* Shape isomerism in sodium clusters with $10 \leq Z \leq 44$: Jellium model with quadrupole, octupole, and hexadecapole deformations. Phys Rev B. 1995;52:4775–4778.
99. Koskinen M, Lipas PO, Manninen M. Unrestricted shapes of jellium clusters. EPL. 1995;30:519–524.
100. Reimann SM, *et al.* Magic triangular and tetrahedral clusters. Phys Rev B. 1997;56:12147–12150.
101. Rytkönen A, Häkkinen H, Manninen M. Melting and octupole deformation of Na_{40}. Phys Rev Lett. 1998;80:3940–3943.
102. Lyalin AG, *et al.* Hartree-Fock deformed jellium model for metal clusters. J Phys B At Mol Opt Phys. 2000;33:3653–3664.
103. Gerchikov LG, Solov'yov AV, Greiner W. Dynamical jellium model for metallic clusters. Int J Mod Phys E. 1999;8:289–298.
104. Gerchikov LG, *et al.* Nonadiabatic electron-ion coupling in dynamic jellium model for metal clusters. J Phys B At Mol Opt Phys. 2000;33:4905–4926.
105. Matveentsev A, *et al.* On the applicability of jellium model to the description of alkali clusters. Int J Mod Phys E. 2003;12:81–107.
106. Yabana K, Bertsch GF. Electronic structure of C_{60} in a spherical basis. Phys Scr. 1993;48:633–637.
107. Ivanov VK, *et al.* Photoionization cross sections of the fullerenes C_{20} and C_{60} calculated in a simple spherical model. J Phys B At Mol Opt Phys. 2001;34:L669–L677.
108. Polozkov RG, Ivanov VK, Solov'yov AV. Photoionization of the fullerene ion C_{60}^{+}. J Phys B At Mol Opt Phys. 2005;38:4341–4348.
109. Madjet ME, *et al.* Photoionization of C_{60}: A model study. J Phys B At Mol Opt Phys. 2008;41:105101.
110. Polozkov RG, *et al.* Stability of metallic hollow cluster systems: Jellium model approach. Phys Rev A. 2009;79:063203.
111. Verkhovtsev AV, *et al.* Hybridization-related correction to the jellium model for fullerenes. J Phys B At Mol Opt Phys. 2012;45:215101.
112. Bréchignac C, *et al.* Collective excitation in closed-shell potassium cluster ions. Chem Phys Lett. 1989;164:433–437.
113. Selby K, *et al.* Surface plasma resonances in free metal clusters. Phys Rev B. 1989;40:5417–5427.
114. Hertel IV, *et al.* Giant plasmon excitation in free C_{60} and C_{70} molecules studied by photoionization. Phys Rev Lett. 1992;68: 784–787.
115. Ling Y, Lifshitz C. Plasmon excitation in polycyclic aromatic hydrocarbons studied by photoionization. Chem Phys Lett. 1996;257: 587–591.

116. Gerchikov LG, *et al.* Scattering of electrons on metal clusters and fullerenes. J Phys B At Mol Opt Phys. 1997;30:4133–4161.
117. Gerchikov LG, Ipatov AN, Solov'yov AV. Many-body treatment of electron inelastic scattering on metal clusters. J Phys B At Mol Opt Phys. 1997;30:5939–5959.
118. Gerchikov LG, *et al.* Excitation of multipole plasmon resonances in clusters by fast electron impact. J Phys B At Mol Opt Phys. 1998;31:3065–3077.
119. Gerchikov LG, *et al.* Diffraction of fast electrons on the fullerene C_{60} molecule. Phys Rev Lett. 1998;81:2707–2710.
120. Liebsch T, *et al.* Angle-resolved photoelectron spectroscopy of C_{60}. Phys Rev A. 1995;52:457–464.
121. Gerchikov LG, *et al.* Surface and volume excitations in electron inelastic scattering on metal clusters. Phys Rev A. 2000;62:043201.
122. Verkhovtsev AV, *et al.* Interplay of the volume and surface plasmons in the electron energy loss spectra of C_{60}. J Phys B At Mol Opt Phys. 2012;45:141002.
123. Bolognesi P, *et al.* Collective excitations in the electron energy loss spectra of C_{60}. Eur Phys J D. 2012;66:254.
124. Connerade JP, Solov'yov AV. Formalism for multiphoton plasmon excitation in jellium clusters. Phys Rev A. 2002;66:013207.
125. Verkhovtsev AV, Korol AV, Solov'yov AV. Quantum and classical features of the photoionization spectrum of C_{60}. Phys Rev A. 2013;88:043201.
126. Verkhovtsev AV, Korol AV, Solov'yov AV. Formalism of collective electron excitations in fullerenes. Eur Phys J D. 2012;66:253.
127. Lambin P, Lucas AA, Vigneron JP. Polarization waves and van der Waals cohesion of C_{60} fullerite. Phys Rev B. 1992;46:1794–1803.
128. Östling D, Apell P, Rosen A. Theory for collective resonances of the C_{60} molecule. Europhys Lett. 1993;21:539–544.
129. Lo S, Korol AV, Solov'yov AV. Dynamical screening of an atom confined within a finite-width fullerene. J Phys B At Mol Opt Phys. 2007;40:3973–3982.
130. Korol AV, Solov'yov AV. Comment on 'Photoexcitation of a volume plasmon in C_{60} ions'. Phys Rev Lett. 2007;98:179601.
131. Huttula M, *et al.* Size-varied photoelectron spectroscopy of metal clusters using the exchange-metal cluster source. J Electron Spectros Relat Phenom. 2010;181:145–149.
132. von Issendorff B, Cheshnovsky O. Metal to insulator transitions in clusters. Annu Rev Phys Chem. 2005;56:549–580.

133. Jänkälä K, Mikkelä MH, Huttula M. Valence photoionization of free, neutral, and size-varied alkali metal clusters. J Phys B At Mol Opt Phys. 2011;44:105101.
134. Jänkälä K, *et al.* Photon energy dependent valence band response of metallic nanoparticles. Phys Rev Lett. 2011;107:183401.
135. Polozkov RG, *et al.* Photodetachment of metal cluster negative ions within many-body theory. Eur Phys J D. 2012;66:287.
136. Wopperer P, *et al.* Angular distributions of photoelectrons from free Na clusters. Phys Rev A. 2010;82:063416.
137. Solov'yov AV, Polozkov RG, Ivanov VK. Angle-resolved photoelectron spectra of metal cluster anions with a many-body-theory approach. Phys Rev A. 2010;81:021202.
138. Kostko O, *et al.* Structure determination of medium-sized sodium clusters. Phys Rev Lett. 2007;98:043401.
139. Bartels C, *et al.* Probing the angular momentum character of the valence orbitals of free sodium nanoclusters. Science. 2009;323:1323–1327.
140. Amusia MYa. Atomic photoeffect. New York: Plenum Press. 1990.
141. Polozkov RG, *et al.* New applications of the jellium model for the study of atomic clusters. J Phys Conf Ser. 2013;438:012009.
142. Scully SWJ, *et al.* Photoexcitation of a volume plasmon in C_{60} ions. Phys Rev Lett. 2005;94:065503.
143. Berkowitz J. Sum rules and the photoabsorption cross sections of C_{60}. J Chem Phys. 1999;111:1446–1453.
144. Kafle BP, *et al.* Absolute total photoionization cross section of C_{60} in the range of 25–120 eV: Revisited. J Phys Soc Jpn. 2008;77:014302.
145. Varshalovich DA, Moskalev AN, Khersonskii VK. Quantum theory of angular momentum. Singapore: World Scientific Publishing. 1988.
146. Kubo R. Electronic properties of metallic fine particles. I. J Phys Soc Jpn. 1962;17:975–986.
147. Verkhovtsev AV, Korol AV, Solov'yov AV. Plasmon excitations in photo- and electron impact ionization of fullerenes. J Phys Conf Ser. 2013;438:012011.
148. Lushnikov AA, Simonov AJ. Surface plasmons in small metal particles. Z Phys. 1974;270:17–24.
149. Yannouleas C, Broglia RA. Landau damping and wall dissipation in large metal clusters. Ann Phys. 1992;217:105–141.
150. Yannouleas C. Microscopic description of the surface dipole plasmon in large Na_n clusters ($950 \leq n \leq 12050$). Phys Rev B. 1998;58: 6748–6751.

151. Bertsch GF, *et al.* Collective plasmon excitations in C_{60} clusters. Phys Rev Lett. 1991;67:2690–2693.
152. Lushnikov AA, Simonov AJ. Excitation of surface plasmons in metal particles by fast electrons and X rays. Z Phys B. 1975;21:357–362.
153. Ellert C, *et al.* Temperature dependence of the optical response of small, open shell sodium clusters. Phys Rev Lett. 1995;75:1731–1734.
154. Pacheco JM, Broglia RA. Effect of surface fluctuations in the line shape of plasma resonances in small metal clusters. Phys Rev Lett. 1989;62:1400–1402.
155. Bertsch GF, Tomanek D. Thermal line broadening in small metal clusters. Phys Rev B. 1989;40:2749–2751.
156. Penzar Z, Ekardt W, Rubio A. Temperature effects on the optical absorption of jellium clusters. Phys Rev B. 1990;42:5040–5045.
157. Montag B, Reinhard PG, Meyer J. The structure-averaged jellium model for metal clusters. Z Phys D. 1994;32:125–136.
158. Montag B, Reinhard PG. Width of the plasmon resonance in metal clusters. Phys Rev B. 1995;51:14686–14692.
159. Guet C, Johnson WR. Dipole excitations of closed-shell alkali-metal clusters. Phys Rev B. 1992;45:11283–11287.
160. Kharchenko VA, *et al.* Size dependence of electronic structure and adiabatic type of collective vibration in small metal clusters. Phys Rev A. 1994;50:1459–1464.
161. Kreibig U, Vollmer M. Optical properties of metal clusters. Berlin-Heidelberg: Springer-Verlag. 1995.
162. Wang Y, *et al.* Collective electronic excitations and their damping in small alkali clusters. Chem Phys Lett. 1993;205:521–528.
163. Pacheco JM, Schöne WD. Shape phase transitions in the absorption spectra of atomic clusters. Phys Rev Lett. 1997;79:4986–4989.
164. Connerade JP, Solov'yov AV. Radiative electron capture by metallic clusters. J Phys B At Mol Opt Phys. 1996;29:365–375.
165. Ipatov A, *et al.* Electron attachment to metallic clusters. J Phys B At Mol Opt Phys. 1998;31:L27–L34.
166. Connerade JP, *et al.* Polarization effects in electron attachment to metallic clusters. J Phys B At Mol Opt Phys. 1999;32:877–894.
167. Connerade JP, Solov'yov AV. Giant resonances in photon emission spectra of metal clusters. J Phys B At Mol Opt Phys. 1996;29:3529–3547.
168. Gerchikov LG, Solov'yov AV. Photon emission in electron–cluster collision in the vicinity of plasmon resonance. Z Phys D. 1997;42:279–287.

169. Korol AV, Solov'yov AV. Polarizational bremsstrahlung of electrons in collisions with atoms and clusters. J Phys B At Mol Opt Phys. 1997;30:1105–1150.
170. Gerchikov LG, Ipatov AN, Solov'yov AV. Many-body treatment of the photon emission process in electron-clusters collisions. J Phys B At Mol Opt Phys. 1998;31:2331–2341.
171. Popov AA, Yang S, Dunsch L. Endohedral fullerenes. Chem Rev. 2013;113:5989–6113.
172. Yang S, Wang CR, editors. Endohedral fullerenes: From fundamentals to applications. Singapore: World Scientific. 2014.
173. Jalife S, *et al.* Noble gas endohedral fullerenes. Chem Sci. 2020;11:6642–6652.
174. Connerade JP. Confining and compressing the atom. Eur Phys J D. 2020;74:211.
175. Puska MJ, Nieminen RM. Photoabsorption of atoms inside C_{60}. Phys Rev A. 1993;47:1181–1186.
176. Connerade J-P, Solov'yov AV. Dynamical screening of a confined atom by a fullerene. J Phys B At Mol Opt Phys. 2005;38:807–813.
177. Madjet ME, Chakraborty HS, Manson ST. Giant enhancement in low energy photoemission of Ar confined in C_{60}. Phys Rev Lett. 2007;99:243003.
178. Dolmatov VK. Photoionization of atoms encaged in spherical fullerenes. In: Adv Quantum Theory. New York: Academic. 2009;58:13–68.
179. Korol AV, Solov'yov AV. Confinement resonances in the photoionization of endohedral atoms: myth or reality? J Phys B At Mol Opt Phys. 2010;43:201004.
180. Korol AV, Solov'yov AV. Vacancy decay in endohedral atoms: the role of an atom's non-central position. J Phys B At Mol Opt Phys. 2011;44:085001.
181. Kilcoyne ALD, *et al.* Confinement resonances in photoionization of $Xe@C_{60}^{+}$. Phys Rev Lett. 2010;105:213001.
182. Phaneuf RA, *et al.* Probing confinement resonances by photoionizing Xe inside a C_{60}^{+} molecular cage. Phys Rev A. 2013;88:053402.
183. Stefanou M, *et al.* Angle-resolved photoelectron spectroscopy and scanning tunnelling spectroscopy studies of the endohedral fullerene $Li@C_{60}$. Nanoscale. 2019;11:2668–2678.
184. Müller A, *et al.* Photoionization and photofragmentation of singly charged positive and negative $Sc_3N@C_{80}$ endohedral fullerene ions. Phys Rev A. 2019;99:063401.

185. Obaid R, *et al.* Intermolecular coulombic decay in endohedral fullerene at the 4d → 4f resonance. Phys Rev Lett. 2020;124:113002.
186. Müller A, *et al.* Experimental studies on photoabsorption by endohedral fullerene ions with a focus on Xe@C_{60}^{+} confinement resonances. Phys Scr. 2021;96:064004.
187. Connerade JP, *et al.* Controlled strong non-dipole effects in photoionization of confined atoms. J Phys B At Mol Opt Phys. 2000;33: L275–L282.
188. Lo S, Korol AV, Solov'yov AV. Dynamical screening of an endohedral atom. Phys Rev A. 2009;79:063201.
189. Landau LD, Lifshitz EM. Electrodynamics of continuous media. Oxford: Butterworth-Heinemann. 1984.

https://doi.org/10.1142/9781800618183_0003

Chapter 3

Magnetic Superatoms

A. C. Reber* and S. N. Khanna
Department of Physics, Virginia Commonwealth University, Richmond VA, 23220, USA
**acreber@vcu.edu*

From motors and refrigerators to magnetic memory storage, junctions, sensing devices, and drug delivery, magnetic materials find widespread applications. As the pace of technology carries us to smaller devices, the study of magnetic behaviour at reduced sizes has become an important area. The current chapter is focused on the progression of magnetic behaviours as the dimensions of the material are reduced to clusters containing few to a few hundred atoms and how novel classes of magnetic species could be created, offering potential for designer cluster-assembled nanomaterials.

The chapter first describes the evolution of magnetic moments and their dynamical behaviour for itinerant ferromagnetic solids such as Fe, Co, and Ni. The reduction in size reduces the anisotropy energy, and small clusters behave like giant moments that can fluctuate in direction under thermal fluctuations, making them undergo superparamagnetic relaxation. Clusters of non-magnetic elements, such as Rh, become magnetic, while those of antiferromagnetic solids, such as Mn, also tend to become ferromagnetic.

The bulk of the chapter is devoted to a new area called magnetic superatoms. The quantum states in small compact metal clusters have orbital shapes resembling those in atoms, and the states are bunched into electronic shells, suggesting the classification of stable clusters as superatoms. The filling of superatomic orbitals, however, does not generally follow Hund's rule, as the clusters can undergo Jahn–Teller structural distortion spanning over multiple atoms, which enhances stability via filled shells with large gaps between occupied and unoccupied states. The chapter outlines how stable magnetic species could be created by impregnating metal clusters with atoms having localised moments or by enhancing exchange splitting in supershells. The chapter presents evidence for the existence of magnetic superatoms and describes a technique for measuring their spin magnetic moment by combining first-principles theory with negative-ion photodetachment experiments. The chapter ends with an entirely novel approach, where a novel class of magnetic clusters can be stabilised by filling subshells belonging to different magic numbers.

3.1 Introduction

Extraordinary things are found to happen when the dimension of materials is reduced to a nanometre or sub-nanometre length scale[1–5] These materials include atomic clusters, nanostructured materials, and ultrathin films. Their electronic, magnetic, and optical properties are very different from those of the bulk and are a unique function of their size, shape, and composition[2,5] These findings have given rise to the hope that a new class of materials with desirable properties can be synthesised[6] These materials are expected to revolutionise materials science as we enter the new millennium. Most appealing among these classes of materials are clusters that display interesting behaviours, whose composition can be selectively chosen and whose individual characteristics might be retained when assembled into an extended material[7–10] As we have recently demonstrated in a series of papers, many of these stable clusters mimic the chemical behaviour of elements in the periodic table and hence can be regarded as ‘superatoms’. The conceptual basis of superatoms has provided a unifying framework for not only the grouping of clusters but also understanding the periodic trends observed in the properties of clusters[11,12] Quantum confinement effects often govern the behaviour of matter in this size regime, and one of the areas with potential applications relates to magnetic properties.

Magnetism, despite its long history, continues to be a research topic of great current interest. This is due to the important role that magnetic materials play in a wide range of technologies. As the pace of technology carries magnetic devices to smaller and smaller length scales, considerable attention is being focused on new effects such as novel magnetic order in clusters,[13–15] oscillatory exchange coupling in magnetic multilayers[16,17] and its modification by doping with hydrogen,[18] giant magnetoresistance (GMR) in multilayers,[19] GMR in granular alloys composed of magnetic grains or clusters,[20] quantum tunnelling in molecular magnets,[21,22] and optically or chemically induced magnetic transitions.[23,24] These developments have a strong economic impact on non-volatile digital magnetic storage, sensing devices, and hard permanent magnets. Thus, a fundamental understanding of various factors that affect magnetic order in nanoscale materials may lead to a new generation of atomically engineered materials with tailored magnetic properties.

The purpose of this chapter is to focus on the stability of magnetic moments in superatomic species. As we will show, the quantum effects lead to the grouping of electronic states into small clusters, and systems with filled electronic shells are more stable. On the other hand, the existence of magnetic moments requires an imbalance between majority and minority spin manifolds, which is most easily realised in clusters with unfilled shells. This dichotomy raises the question as to how one can design stable magnetic clusters. In this chapter, we demonstrate how such a situation can be attained.

We first briefly outline the conceptual basis of the superatomic framework and then briefly review the developments in magnetic properties of clusters. We then introduce the conceptual basis of magnetic superatoms and how stable magnetic species can be formed in a variety of ways.

3.2 Superatomic Framework in a Confined Nearly Free Electron Gas and Its Extension to Covalent Systems

Extensive synergistic studies combining the state-of-the-art experiments and theoretical studies on a variety of metal clusters now confirm that a confined nearly free electron (CNFE) gas (a simplified example is the Jellium model) can provide a reasonably simplistic and fairly accurate description of their observed properties, including stability, redox properties, and reactivity profiles.[4,25–33] The quantum states in such a spherical, CNFE gas are bunched into shells that can be labelled by their principal and angular momentum quantum numbers. Unlike atoms, where the potential has a $1/r$ dependence, the relation between principal and angular momentum quantum numbers in a CNFE gas are different. The states are ordered as 1S, 1P, 1D, 2S, 1F, 2P, and so on. The model accounted for the observed magic numbers in alkali clusters, as well as reaction profiles in aluminium clusters. More importantly, the model provided insights into many of the observed periodicities in properties and offered an organisational principle to classify clusters into groups depending on the occupancy and nature of their electronic states. While the initial studies focused on the electronic properties, studies on larger clusters showed the importance of geometric structures on the stability and their chemical behaviours.[34] For example, abundance spectra in larger alkali and alkaline earth systems showed enhanced stability of clusters with complete geometric shells. The geometric structures are also found to affect the chemical behaviours.[35,36] For example, from studies on reactivity with polar etchants such as water or methanol, certain aluminium cluster anions that have closed electronic shells were found to be highly reactive.[37–40] Clusters with open geometric shells exhibit defect-like structures that result in an uneven charge distribution across the surface of the cluster, which results in active sites and enhanced reactivity. When the cluster forms a more compact and symmetric structure, the charge distribution becomes more even, quenching the active sites and making all sites on the cluster equally reactive. The second organising principle for understanding the reactivity and stability of clusters is that clusters with closed geometric shells exhibit reduced reactivity, while those with geometries analogous to defects display enhanced reactivity.[3]

The above discussion shows that both electronic and geometric shells are important for understanding the stability, redox properties, reactivity, and other properties of clusters. Symmetric compact clusters have electronic states grouped into electronic shells, and the occupancy of these electronic shells determines the stability, redox properties, and chemical reactivities of clusters. Like atoms, clusters with filled shells are found to be more stable than those with unfilled shells, which tend to exhibit chemical valence as atoms try to lose or gain electrons to acquire a filled shell status. These features that were originally realised for the CNFE gas are not specific to metal clusters. So, while the nature of electronic interactions leading to filled electronic shells may vary depending on the cluster type, numerous stable clusters do exhibit well-defined valence.

These observations led to the proposition that well-defined clusters can be regarded as superatoms forming a third dimension to the periodic table. Earlier studies by Khanna and Castleman and their co-workers had identified superatoms resembling alkali, alkaline earth, and multiple valence atoms.[25–27,41] In a later development, Khanna and co-workers extended these concepts to magnetic superatoms, a development that has attracted considerable attention and will form the basis of the current chapter.[42,43] The superatom analogy led Reber and Khanna to propose a periodic table of cluster orbitals much like the periodic table for atomic orbitals (Fig. 3.1).[3]

While these initial developments focused on clusters of metallic elements, it is important to point out that the superatom concept has recently been extended to numerous other classes of clusters. An important class of such clusters that have drawn considerable attention in recent years are ligated metal–chalcogenide clusters.[6,44–48] These clusters have been shown to form solids when combined with counterions. A few examples of these reported clusters

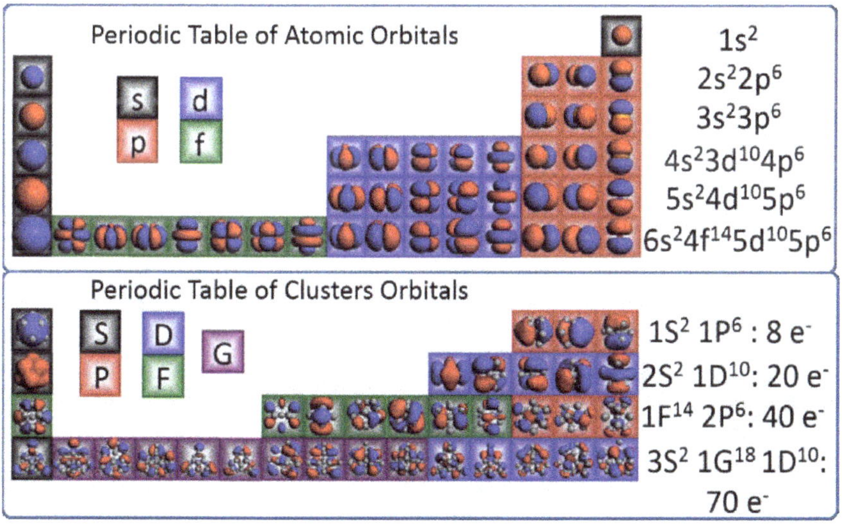

Fig. 3.1. The periodic patterns in the orbitals of the atoms of the periodic table and aluminium clusters. Each column represents a new orbital, not an additional electron, so each column represents the addition of two electrons. The orbitals are described by the shell structure shown on the right.

Source: Reprinted with permission from Reber and Khanna (2018).[3] Copyright 2018 American Chemical Society.

are $Co_6Se_8(PEt_3)_6$, $Cr_6Te_8(PEt_3)_6$, $Ni_9Te_6(PEt_3)_8$, $Ni_6Se_5(PR_3)_8$, $Co_6Te_8(PR_3)_6$, $Ni_9Te_6(CO)_8$, $Co_{11}Te_7(CO)_{10}$, and $Co_9Te_6(CO)_8$, which include donor (PEt_3, PR_3) and acceptor (CO) ligands. The ligated clusters consist of a central core of transition metals and chalcogens that are bound by covalent bonds and are stabilised by ligands, such as phosphines, CO, or other ligands attached to the metal sites. This interesting class of superatoms are found to be highly stable and can be prepared in solution in large quantities. Further, studies in our group have shown that, in addition to protecting the clusters, the ligands offer an additional pathway for tuning the redox properties of a cluster, either by stabilising a closed electronic shell, inducing active sites, or by controlling the donor–acceptor characteristics.[44,49–51] Further, depending on the ligand, they can have charge donor/acceptor characteristics that allow them to form a variety of solids when combined with complementary ions, including C_{60}, $FeCl_4$, and other clusters with high electron affinity, maintaining the identity of the metal–chalcogenide core. In Fig. 3.2, we show some of the solids that have been formed using these clusters.

Unlike metal clusters, the bonding in these clusters is a mixture of covalent bonds and charge-transfer complexes, and a CNFE gas approximation is not applicable. Further, the clusters exhibit a variety of atomic arrangements and require a different approach to be

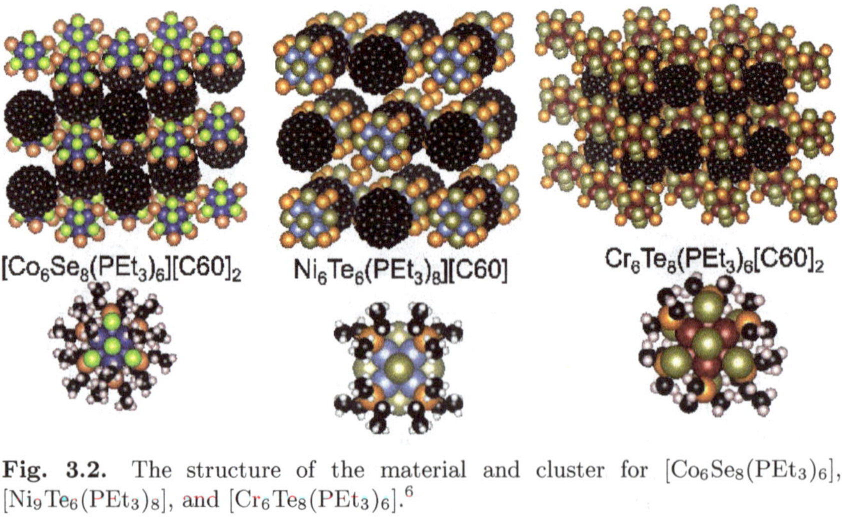

Fig. 3.2. The structure of the material and cluster for [$Co_6Se_8(PEt_3)_6$], [$Ni_9Te_6(PEt_3)_8$], and [$Cr_6Te_8(PEt_3)_6$].[6]

included in the superatom family. As previously stated, superatoms are clusters whose chemical and electronic properties are dominated by their closeness to a zero-valent state, similar to atoms in the periodic table. For determining the valence, one needs to find valence electron counts that correspond to enhanced stability through a closed electronic shell. In clusters with CNFE gas, we have already discussed how the electronic structure leads to electron counts resulting in enhanced stability.

We recently undertook a study of the metal–chalcogenide clusters to determine the electronic counts leading to closed electronic shells.[12,52] An important class of clusters are the octahedral $TM_6E_8L_6$ (see structures in Figs. 3.3 and 3.4), where TM denotes a transition metal, E denotes a chalcogen, and L denotes a ligand. In our studies, we primarily focused on clusters that exhibit high symmetry, which results in higher electronic stability at certain electron counts.

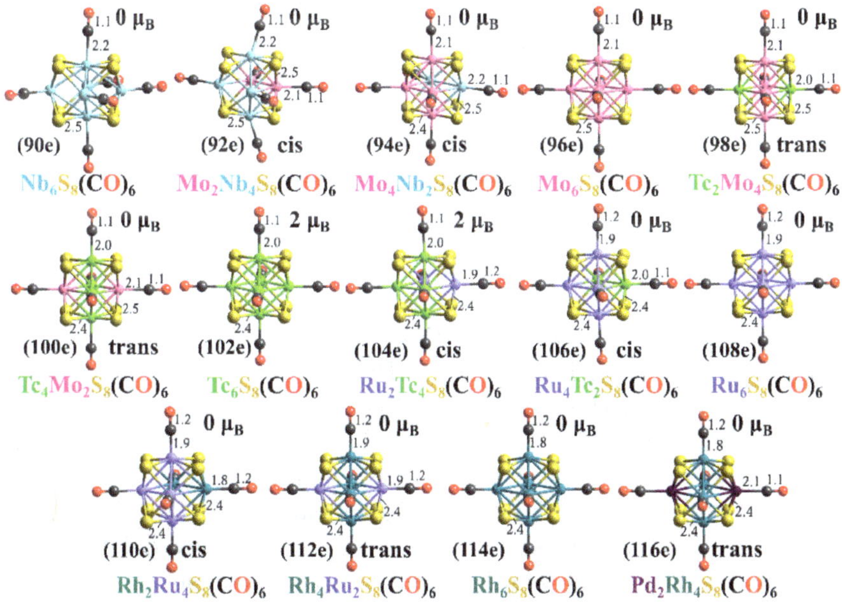

Fig. 3.3. Optimised ground state structures of 6 CO-ligated TM_6S_8 (TM = 4d metals) clusters. The number of total valence electrons, magnetic moment (in μ_B), and the type of the conformer (*cis* or *trans*) are mentioned. The bond distances are in Å.

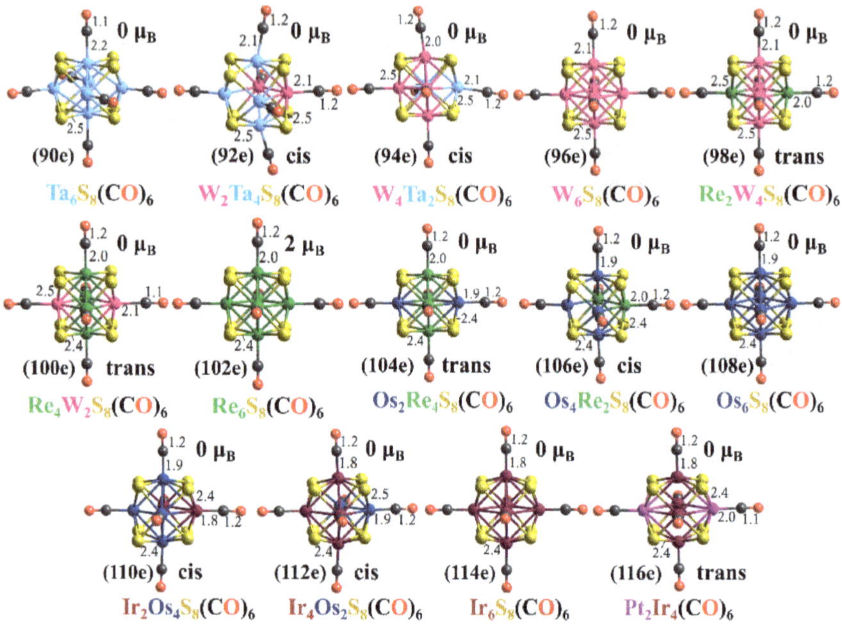

Fig. 3.4. Optimised ground state structures of 6 CO-ligated TM_6S_8 (TM = 5d metals) clusters. The number of valence electrons, magnetic moment (in μ_B), and the type of the conformer (*cis* or *trans*) are mentioned. The bond distances are in Å.

Our studies were mainly confined to $TM_6S_8(CO)_6$ clusters, where the TM included transition metals from the 4d and 5d series. The case of 3d metals that have localised 3d states leading to stable spin moments is currently being investigated. To determine their valence, we explored the electron counts that enhance stability as marked by a filled electronic shell and a large HOMO–LUMO gap. Our additional pointers for the attainment of a zero-valent superatomic state examined a high ionisation energy and a low electron affinity. Here, we focused on octahedral metal chalcogenide clusters of the form $M_6S_8L_6$, with S as the chalcogen and the 4d and 5d transition metals, Nb-Pd, and Ta-Pt. The valence electron counts were determined by including the outer s and d electrons of the transition metal, six valence electrons from each S, and two electrons per CO and PH_3 ligand.

In order to examine the effect of electron counts on stability, pairs of transition metal atoms were replaced by neighbouring elements to maintain symmetry. The electron count varied from 90 to 116 valence electrons, in steps of two. The clusters included in these investigations consisted of $Nb_6S_8(CO)_6$, $Mo_2Nb_4S_8(CO)_6$, $Mo_4Nb_2S_8(CO)_6$, $Mo_6S_8(CO)_6$, $Tc_2Mo_4S_8(CO)_6$, $Tc_4Mo_2S_8(CO)_6$, $Tc_6S_8(CO)_6$, $Ru_2Tc_4(CO)_6$, $Ru_4Tc_2(CO)_6$, $Ru_6S_8(CO)_6$, $Rh_2Ru_4S_8(CO)_6$, $Rh_4Ru_2S_8(CO)_6$, $Rh_6S_8(CO)_6$, and $Pd_2Rh_4(CO)_6$, all containing CO ligands and elements from the 4d series, and $Ta_6S_8(CO)_6$, $W_2Ta_4S_8(CO)_6$, $W_4Ta_2S_8(CO)_6$, $W_6S_8(CO)_6$, $Re_2W_4S_8(CO)_6$, $Re_4W_2S_8(CO)_6$, $Re_6S_8(CO)_6$, $Os_2Re_4(CO)_6$, $Os_4Re_2(CO)_6$, $Os_6S_8(CO)_6$, $Ir_2Os_4S_8(CO)_6$, $Ir_4Os_2S_8(CO)_6$, $Ir_6S_8(CO)_6$, and $Pt_2Ir_4(CO)_6$ from the 5d series.

Figures 3.3 and 3.4 show the ground-state structures of the clusters, the metal–S, CO, and metal–CO bond lengths in Å units, and the number of valence electrons. For mixed species, the location of heteroatoms leads to *cis* or *trans* structures, as shown in the figures. The studies showed that the metal–carbon bond length generally decreases from 2.2 Å to 1.8 Å with increasing valence count due to the decrease in the metal–sulphur bond caused by a decrease in charge transfer to the sulphur atom. The other interesting trend is that clusters with some electron counts show more geometrical distortions than others. These distortions are reminiscent of Jahn–Teller distortions in simple metal clusters that generally occur in clusters with partially filled shells.

Figures 3.5(a)–(c) show the HOMO–LUMO gap, locations of HOMO and LUMO, ionisation energy (IE), and electron affinity (EA) for both 4d and 5d series. It is particularly interesting to identify clusters with large HOMO–LUMO gaps. As discussed previously, in the case of a CNFE gas, these occur at the electron counts of 2, 8, 18, 20, 34, 40, and so on. Figure 3.5(a)–(c) show that the HOMO–LUMO gaps in the present case also show large variations with peaks at the electron counts of 96, 100, and 114 valence electrons. The HOMO–LUMO gap in the 4d series also presents a maximum at 108 electrons. However, this is not supported by IE or EA, which do not present any maxima or minima, and it is also not shared by the 5d series. As in metallic clusters, clusters with filled shells exhibit local peaks in the ionisation energy and lower electron affinity. These clusters, therefore, are resistant to donating or accepting electrons, which is indicative of inert behaviour. Since the original discovery of magic species resulted from the observed abundance spectra in

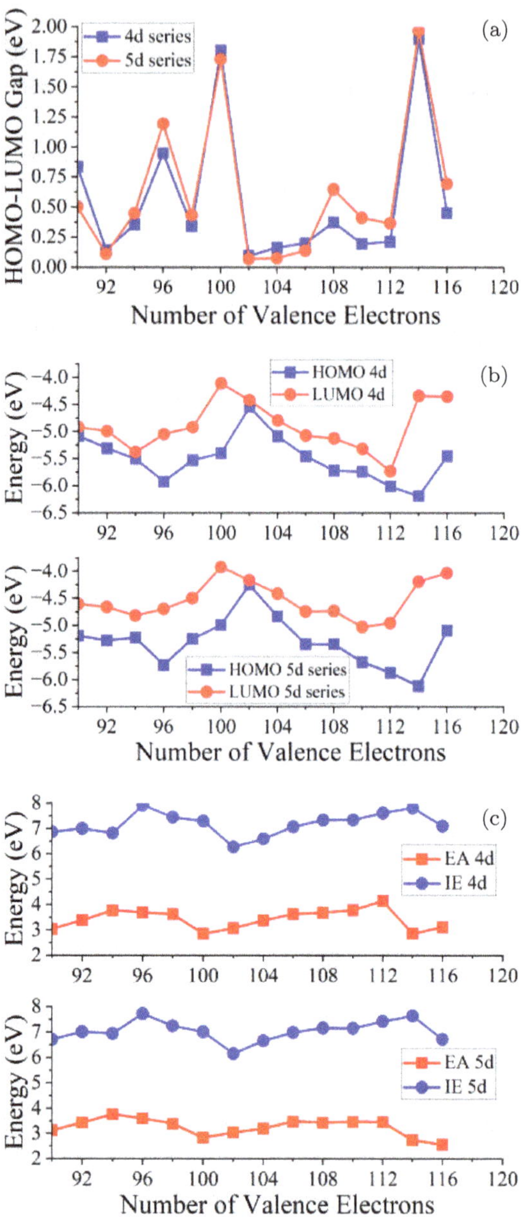

Fig. 3.5. (a) HOMO–LUMO gap, (b) energies of HOMO and LUMO, and (c) the IE and EA (adiabatic) of 4d and 5d series $TM_6S_8(CO)_6$ clusters.[52]

beams, it is reasonable to ask whether the closed shell is also consistent with energetic stability. Our studies did indicate peaks in the binding energies at these electron counts.

The previous discussions show that in simple and metal–chalcogenide species, we can identify the electron counts, leading to closed electronic shells and enhanced stability. The formation of magnetic species requires spin-up and spin-down manifolds with different electron counts to enable finite spin moments. How does one then form stable magnetic species that do not sacrifice stability and yet lead to magnetic superatoms. We now discuss a conceptual framework that leads to magnetic superatoms. But before we address this framework, we first present a brief review of magnetic properties in small clusters.

3.3 Brief Review of Magnetic Properties in Small Clusters

Some of the early work on the magnetic properties of clusters focused on the evolution of magnetic moments as the size is reduced from the bulk value. In fact, numerous experimental and theoretical studies on the magnetism of low-dimensional materials (surfaces, overlayers, sandwich layers)[53,54] have shown that not only magnetic moments of ferromagnetic elements can be significantly enhanced over their bulk values but also non-magnetic elements can become magnetic.[55] These findings suggest that studies of the magnetic properties of clusters may lead to many novel results, as the structure, symmetry, inter-atomic separation, and coordination number are size-specific. De Heer *et al.* carried out systematic experiments on the magnetism of Fe_n clusters using a Stern–Gerlach field.[56] The results were surprising. Unlike the case of atoms, where the beam splits into various angular momentum states, all the clusters deviated in the same direction. The magnetic moment per atom was also much below the bulk value, and it increased with size — contrary to the theoretical expectation that the clusters should be more magnetic than the bulk. The studies on CO_n clusters, as reported by Bloomfield and co-workers, not only led to similar findings but also showed that the magnetisation per atom increased linearly with size.[57]

The apparent disagreement between the experiment and the existing theory was almost immediately resolved by Khanna and Linderoth.[58] They suggested that, since the cluster sizes are much smaller than the typical size of bulk magnetic domains that can contain thousands of atoms, the atomic spins in the cluster are coupled via exchange interactions. This causes the cluster to behave like a giant magnet, with its moment being the sum of the individual atomic moments. They argued, however, that the reduction in size leads to a reduction in anisotropy energy, and for Fe, Co, or Ni clusters containing 10 to a few hundred atoms, the anisotropy energy is much smaller than the thermal energy at the cluster temperatures. Except under special circumstances, the combined moment can change direction under thermal fluctuations, making the cluster undergo superparamagnetic (SP) relaxation. In a magnetic field B, the total magnetic moment of the uni-domain cluster containing N atoms is $N\mu$, where μ is the intrinsic moment per atom in the cluster. The moment tends to align in the direction of the magnetic field. Thermal fluctuations, however, tend to counter this alignment. The key thing to note is that the typical SP relaxation time is 10^{-10}–10^{-12} s whereas the passage time through the magnet is in the range of 1–1000 μs. Hence, in actual experiments, one measures the average magnetisation, M_{ave}, which is related to the intrinsic moment via the paramagnetic relation

$$M_{\text{ave}} = N\mu[\coth(N\mu B/k_B T) - k_B T/N\mu B], \tag{3.1}$$

where T is the temperature of the cluster. Note that since each cluster spans the entire phase space during its passage through the magnet, they all exhibit the same average magnetisation, and hence all the clusters are deviated in the same direction. The uniform deflection does, however, raise an interesting dilemma as to how the clusters are different from single atoms. For single atoms, the Stern–Gerlach profile corresponds to $2J+1$ multiplets, where J is the total angular momentum quantum number. Why is it then that the cluster beam does not exhibit the $2J+1$ lines? The reason is the conservation of total angular momentum. In both cases, the presence of the external field causes $2J+1$ fold equally spaced energy values. However, in the case of atoms, the conservation of angular momentum prevents any relaxation of the states since there is no angular momentum associated with the rotation of the atom, and the beam splits into $2J+1$

directions. In the case of clusters, in addition to the electronic angular momentum, the cluster has an angular momentum associated with its own rotation. It can be shown that a coupling between the electronic momentum and the massive rotational angular momentum is present due to an effective spin–orbit-like interaction. This interaction can be demonstrated by transforming to a reference frame that is attached to an electron (which is needed to separate spin and charge degrees of freedom) where the ions rotate around the axis. It is this coupling that allows for the relaxation of the magnetic moment in clusters, provided that the thermal energy is greater than the anisotropy energy. For cluster sizes of interest, $N\mu B$ is less than k_BT, and Equation (3.1) reduces to

$$M_{\text{ave}} = N^2\mu^2 B/3\ k_BT,$$

which corresponds to an observed effective moment, μ_{eff}, per atom of

$$\mu_{\text{eff}} = N\mu^2 B/3\ k_BT.$$

This was found to be in agreement with the experiments that indeed showed that the observed moment increased linearly with the size of the cluster and the field. In fact, the above equation allows one to extract the true cluster moment per atom, μ, from the measured μ_{eff}, knowing the cluster temperature and the cluster size. This model has been routinely used by experimentalists to analyse the experimental data, and it is found that the bulk magnetic moment is enhanced by almost 35% in small clusters, as suggested by theoretical investigations.

One of the interesting questions in cluster science was if clusters of otherwise non-magnetic elements could become magnetic at reduced sizes. Theoretical investigations carried out by Khanna *et al.* on Rh_n clusters predicted that, while bulk Rh is paramagnetic, a Rh_{13} cluster is ferromagnetic.[55] Following this prediction, Bloomfield and coworkers carried out Stern–Gerlach experiments and indeed found that Rh_n clusters up to $n \leq 80$ are magnetic.[57] Similarly, Khanna *et al.* predicted that small clusters of antiferromagnetic elements, such as Mn[59] and Cr,[60] and those of MnO[61] would not only be ferromagnetic but also exhibit isomers that can have ferrimagnetic order. Experimental verification of this possibility was indeed confirmed by experiments on Cr_n clusters.[62]

Khanna *et al.* proposed the relationship between magnetism and the topology of clusters quite some time ago based on studies of Li_4 and Na_4 clusters.[63] They were the first to point out that the preferred spin multiplicity (and hence the number of unpaired spins) of alkali metal clusters can be linked to their topology. In particular, an alkali metal tetramer confined to a two-dimensional (rhombus) structure has zero magnetic moment, whereas in a three-dimensional configuration (tetrahedral), it has a spin moment of 2 μ_B. Since then, several studies by our group and others, as well as experiments, have substantiated this finding.

An exciting phenomenon in magnetism is the possibility of an oscillatory exchange coupling between localised moments. A well-known example in bulk solids is the RKKY interaction, where the magnetic interaction between localised magnetic moments is mediated by conduction electrons and oscillates as a function of distance with a period determined by the Fermi wave number of the electron gas. Khanna and co-workers proposed a new oscillatory exchange interaction where the magnetic coupling between two localised moments oscillates due to changes in chemical bonding. In fact, they showed that the coupling between two Cr atoms in a Cr_2O_n cluster oscillates as the number of O atoms is varied.[23] A detailed investigation of the electronic structure revealed that the oscillations are a direct consequence of the changes in the electronic states between Cr and O atoms in the vicinity of the Fermi energy.

The possibilities that clusters of ferromagnetic elements can have magnetic moments enhanced over their bulk value, paramagnetic elements can become ferromagnetic, anti-ferromagnetic elements can exhibit ferro- or even ferrimagnetic order, and the magnetic coupling can be controlled by changing the chemical composition provided initial impetus to the potential development of new magnetic materials with novel features with clusters as building blocks.

3.4 Magnetic Superatoms

3.4.1 *Magnetic moments due to localised states on an atom*

The initial idea of superatoms, as applied to free or ligated clusters, hinged on acquiring stability through closed electronic shells of paired electrons. Consequently, it was restricted to non-magnetic species because magnetism requires unpaired species. It was proposed that magnetic superatoms could be realised if electronic shells of the valence pool had localised as well as diffuse subspaces. One way to accomplish this is to consider composite systems where some atoms have relatively localised atomic orbitals (e.g., d or f) with a large separation between their spin-up and spin-down levels, while others have s and p electrons that become delocalised upon assembly and form shells. Such systems would acquire magnetic moments via orbitals that are localised at the atomic sites and those that do not overlap significantly as atoms are assembled, while the nearly free electron SP shell orbitals could impart stability via filled electronic shells.

Khanna and co-workers demonstrated this intriguing possibility by considering a free cluster composed of a vanadium (V) atom surrounded by alkali atoms.[42,64] Their initial choice of V was motivated by the studies of V impurities in alkali hosts. V atoms have a valence electronic configuration of $3d^3 4s^2$ and their behaviour in bulk alkali hosts provides a unique opportunity for the stabilisation of magnetic moments in a sea of conduction electrons. Studies on the anomalous Hall effect measurements show a large atomic moment of 6.6–7.0 μ_B per V atom for Na and K and $4\,\mu_B$ for Cs films.[65,66] Insight into observed moments was provided by theoretical studies of V impurities in alkali hosts, which indicated that large magnetic moments resulted from the filled majority d-states in a V atom, whereas the minority d-states were above the Fermi energy. Work by Khanna and co-workers extends these observations to the role of magnetic impurities in a CNFE gas.[42] They studied VA_n (where A = Na, Cs) clusters containing up to 12 atoms. Their studies examined gains in energy as alkali atoms were added to the cluster. The first observation was that specific clusters, such as VA_8, were particularly stable, as seen through gains in energy as atoms were added. Figure 3.6 shows the gain in energy as a Na or Cs atom is added to a pure Na_n cluster and a VNa_n cluster.

The results showed that clusters containing eight alkali atoms exhibited pronounced stability. The electronic structure of a V atom is d^3s^2 with three unpaired electrons in the majority shell. The addition of eight Na atoms creates a valence pool of ten sp electrons. For a CNFE gas, the magic size corresponds to eight and not ten

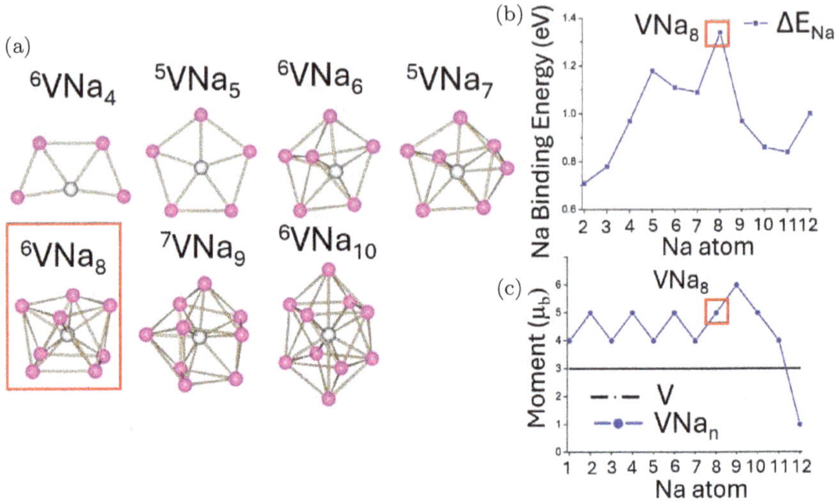

Fig. 3.6. (a) The structure of VNa_n clusters, $n = 4$–10, (b) the size-dependent Na binding energy, and (c) the magnetic moment of the VNa_n clusters.

valence electrons. The reason for the maxima with ten SP electrons is attributed to the fact that a V atom has a partially filled majority d-subshell with three electrons. As successive alkali atoms are added, the initial electrons are transferred to fill the d-subshell on V, increasing the local spin moment from 3.0 μ_B to 5.0 μ_B. Once the majority d-subshell is filled, the remaining electrons occupy SP states of the delocalised electron gas, leading to oscillations of the total moment between 4 and 5 μ_B depending on whether the cluster has an odd or even number of SP electrons. A VNa_8 cluster, therefore, has a half-filled d-subshell with a moment of 5 μ_B while the remaining electrons form a filled shell of $1S^2$ $1P^6$ electrons, leading to a stable magnetic species. These distributions are clearly seen by examining the one-electron energy levels shown in Fig. 3.7. In fact, Fig. 3.7(a) shows the spin density (red colour) and the valence charge density (grey), whereas Fig. 3.7(b) shows the various electronic states. In these studies, the authors identified the symmetry of the orbitals by examining the distribution of the charge density. The studies also looked at the overall charge distribution, which should be nearly spherical in a cluster with a filled shell.

As discussed earlier, there is considerable interest in ligated clusters; in particular, Au_n-based ligated clusters have attracted

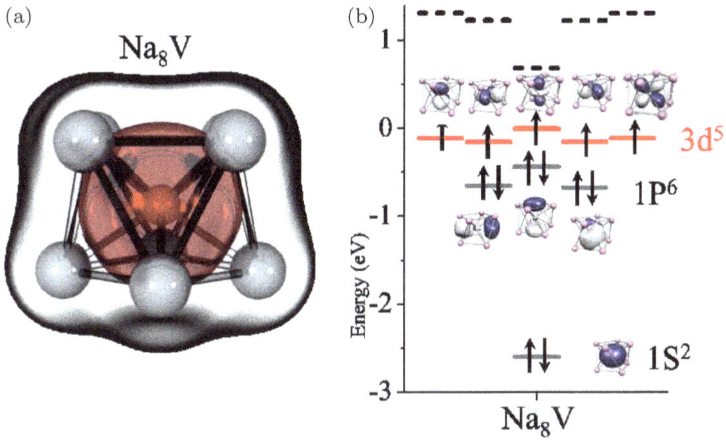

Fig. 3.7. (a) The net spin charge density (red central isosurface) and the total electron charge density (outer grey isosurface) in a VNa_8 cluster. (b) One-electron energy levels in a VNa_8 cluster. Uppercase letters refer to superatom orbitals, while the lowercase 3d corresponds to the orbitals that have dominant atomic 3d character.

tremendous interest as atomically precise clusters. It was demonstrated quite early on that the superatomic framework could be extended to ligated species. Khanna and co-workers examined a $Au_{25}(SCH_3)_{18}^-$ cluster, where the central core of gold atoms has been shown to be highly stable due to shell closure with eight SP electrons.[42] In fact, the 18 ligands each remove one state from the nearly free electron pool to localised states, thus leaving a metallic core with eight nearly free electrons that fill 1S and 1P shells. Khanna and co-workers showed that replacement of a central Au atom with a Mn atom having a $3d^5 4s^2$ electronic configuration leads to a stable neutral $MnAu_{24}(SH)_{18}$ species where the Mn site provides two SP electrons, with the remaining free electrons coming from six Au atoms.[42] These allocations were confirmed by the density of states, which showed a large HOMO–LUMO gap. This is clearly seen in Fig. 3.8, which shows the total valence charge density, the spin density, and the one-electron levels in the cluster confirming S^2P^6 allocation.

While the above theoretical considerations outlined a framework for the existence of magnetic superatoms, Khanna *et al.* collaborated with Bowen *et al.*'s experimental group to actually provide

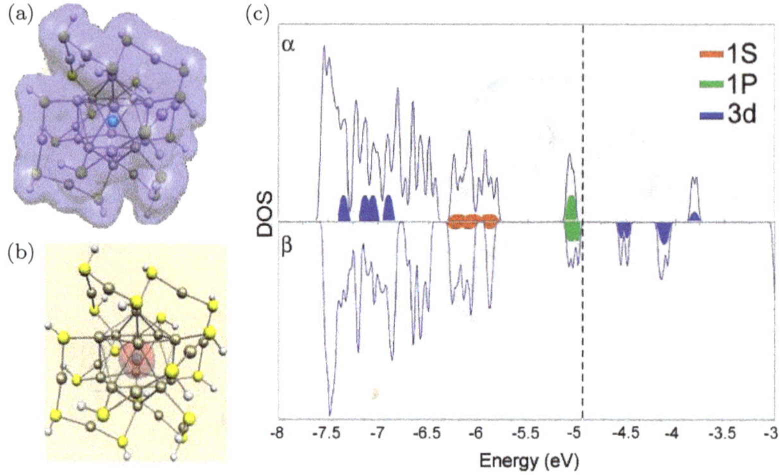

Fig. 3.8. (a) The electronic charge density, (b) the spin density, and (c) the density of states of the $MnAu_{24}(SH)_{18}$ cluster. The shells fill as 3d, 1S, and 1P on the spin-up side, and order as 1S and 1P, and then the 3d shell is unfilled.

evidence for the existence of such stable clusters, as well as a characterisation of their magnetic state.[67] As outlined in Section 3.3, the measurement of the magnetic moment at small cluster sizes is a difficult task due to experimental limitations. This is because such measurements for small clusters are usually carried out using the Stern–Gerlach setup: a beam of clusters passes through a gradient magnet that orients the cluster magnetic moments in the direction of the field, while the gradient magnetic field deflects the beam. As pointed out earlier, the reduction in size reduces the magnetic anisotropy energy, causing the small clusters to undergo superparamagnetic relaxation. Consequently, the net magnetisation of such paramagnetic species that result from a competition between the magnetic alignment energy and thermal fluctuations is small. For clusters having moments of only a few Bohr magneton, the deflections can be small and at the limit of the accuracy of the experiments. In this joint effort, an alternative approach that combines negative-ion photodetachment spectroscopy with first-principles calculations was used. In the experiments, mass-selected cluster anions are subjected to laser irradiation, where the photon detaches the excess electron. The detachment involves a transition from the ground state of the anion to the ground and excited states of the neutral. Consequently, a measurement of the energy of the resulting electron allows the photoelectron spectra to provide a fingerprint of the electronic states of the corresponding neutral clusters. These measurements can be compared with theoretical calculations via the calculated transition energies. A match between the calculated transitions and the experiments provides an indirect determination of the multiplicity and the spin magnetic moment of the cluster. Previous studies on numerous clusters have indicated that such a procedure is quite accurate and can even distinguish between isomers with closely lying energies.

In these experiments, $V_{1,2}Na_n^-$ clusters were generated using a pulsed arc cluster ionisation source.[67] The details of the experimental approach are described elsewhere. Anions generated by this method were then mass-selected for photoelectron spectroscopic studies. Anion photoelectron spectroscopy was carried out by detaching the electrons via a laser beam and measuring the energy of the photodetached electrons. Using the relation, $h\nu = \text{EBE} + \text{EKE}$, where $h\nu$, EBE, and EKE are the photon energy, electron binding energy (transition energy), and the electron kinetic energy, respectively, one can determine the transition energy from an anion to the neutral species. The details of the work have been described in a joint paper. The other part of the approach is to compare the transition energies with theoretical predictions. To accomplish this, Khanna and coworkers carried out first-principles electronic structure calculations on neutral and anionic V_xNa_y clusters within a generalised gradient density functional theory (DFT) formalism, and for comparison with the experiments, time-dependent functional theory (TDDFT) calculations were performed for the neutral clusters at the anion's optimised geometries.[67] Starting from an anionic cluster of multiplicity m, they calculated the total energies and excitation energies of the corresponding neutral clusters at $m + 1$ and $m - 1$ multiplicities, as the photodetached electron could originate either from the majority or the minority levels. Then, they added the excitation energies to obtain the total energies of the neutral excited states. Next, they calculated the vertical detachment energies (VDEs) as the energy difference between the anionic cluster and all the calculated neutral states to position the excitation spectra. The energy differences were broadened by Gaussians of width 0.02 eV, and the two spectra of the $m+1$ and $m-1$ multiplicities were added to generate the theoretical spectra. Figure 3.9 shows the experimental and

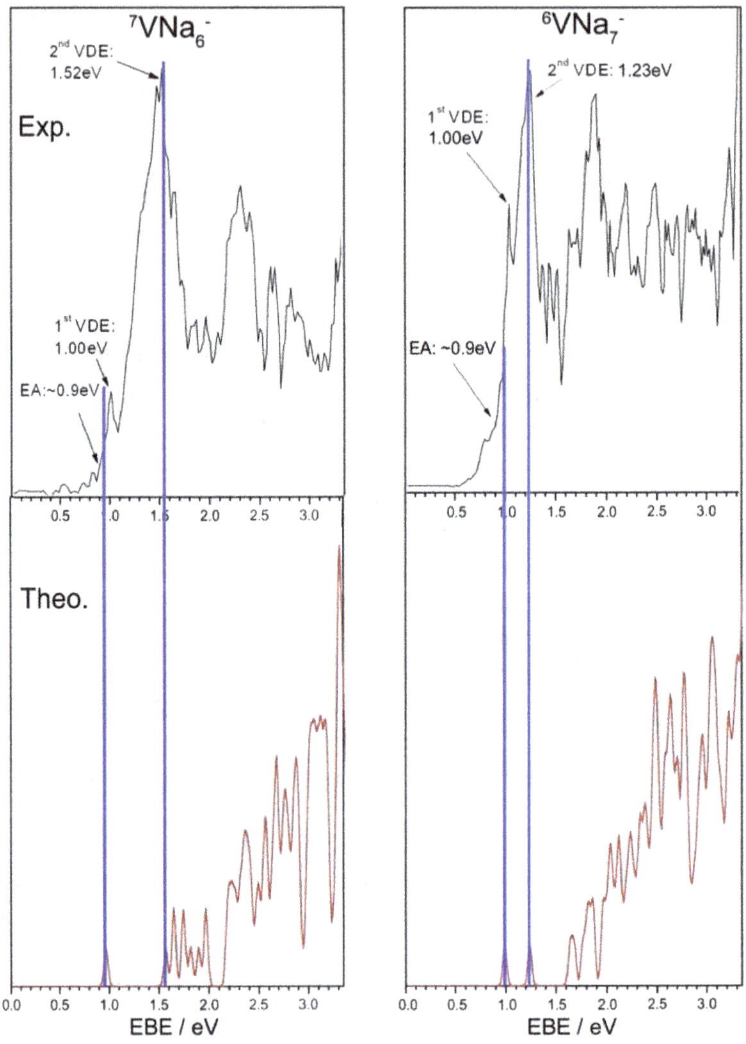

(a) (b)

Fig. 3.9. The photodetachment spectra for (a) VNa_6^- and (b) VNa_7^-. The vertical blue lines correspond to the theoretical values, and the red lines in the lower spectra correspond to the experimental spectra.

theoretical spectra, which also show the peak positions corresponding to vertical transitions from the anion to the neutral cluster. The close agreement between the theory and experiment spectra confirmed that VNa_8^- has a multiplicity of 7, while neutral VNa_8 has a ground-state multiplicity of 6 or a spin magnetic moment of 5 μ_B, as predicted earlier. As previously shown in Fig. 3.6, the one-electron energy levels and the molecular orbital charge density isosurfaces in VNa_8 confirm filled 3d orbitals localised largely on the V site and 8 SP diffuse electrons, confirming the existence of a magnetic superatom.

3.4.2 *Magnetic superatoms with intrinsic moments from superatomic shells*

In the previous section, we have outlined how magnetic superatoms can be realised through localised d states derived from a heteroatom while the superatomic shells bring about the stability via the filled SP shells. As pointed out earlier, the electronic states in superatoms are delocalised and are spread over multiple sites as opposed to localised atomic states. This delocalisation affects the order in which the electrons occupy the superatomic states. In single atoms, the electronic states are filled according to Hund's rule, where the electronic levels are exchange split, and electrons first fill all the majority states before going to minority states. In clusters, while Hund's rule favours high-spin states in open-shell systems stabilised by exchange coupling, and while high-spin multiplicities have been observed in some cases, the delocalisation of electronic orbitals leads to another competing possibility. For clusters with unfilled shells, small clusters can undergo structural distortion to break the degeneracy. This effect, known as the Jahn–Teller effect, can counteract the degeneracy favouring higher spins. In most cases, the energy gain through Jahn–Teller distortion dominates, and unless structural distortions could be constrained by forcing symmetric geometry, the filling of electronic orbitals does not follow Hund's rule of maximising the spin, resulting in superatoms with pairs of electrons favouring non-magnetic species.

One possibility for stabilising the magnetic state originating from the filling of superatomic orbitals is to somehow enhance the exchange splitting of the delocalised states. Khanna and co-workers demonstrated this intriguing possibility by considering a composite cluster with a central site occupied by a transition metal atom. Their studies first considered $FeMg_n$ ($n \leq 12$) clusters, as the valence configuration of an Fe atom is $3d^6 4s^2$ with a spin multiplicity of 5 and theoretical studies indicate that the up and down spin manifolds in the Fe atom are separated by 0.8 eV.[43] Figure 3.10 shows the ground-state geometries of the $FeMg_n$ clusters. Starting from $FeMg_7$, where Mg atoms form a pentagonal bi-pyramid, the Fe atom occupies an interior site. Their studies further found that structures where the Fe atom was forced to occupy an exterior site resulted in ground states that were 0.27 eV higher in energy. A primary focus of the study was the stability of clusters. Figure 3.10 shows the gain in energy as successive Mg atoms were added.

The structure of $FeMg_8$ was quite intriguing. The Mg sites formed a square anti-prism with an interior Fe atom. $FeMg_9$ has a Mg atom added to the square face of the bi-prism $FeMg_8$. Finally, $FeMg_{12}$ was a hexagonal cage of Mg sites with an interior Fe atom. To examine stability, the studies considered the gain in energy as successive Mg atoms were added. The results showed a distinct maximum at $FeMg_8$. More importantly, the results showed (Fig. 3.10) a sharp drop in the gain in energy immediately following $FeMg_8$, a sign of the enhanced stability and the associated magic character in metallic clusters.

Another focus of the studies was on the evolution of the magnetic moment. Figure 3.10 shows the spin magnetic moment of clusters with the successive addition of Mg atoms. The studies also examined the spin moment of the nearest higher isomer to ascertain its stability. Figure 3.10 shows that there is a sharp drop in spin moment after $FeMg_9$. Therefore, $FeMg_8$ has higher energetic stability, along with a higher moment. Note that the electronic structure of a Fe atom is $3d^6 4s^2$. Thus, the addition of eight Mg atoms creates a valence pool of 24 electrons, and ordinarily, that does not correspond to the electron count for a shell filling within a CNFE gas. The origin of the magic character of $FeMg_8$, therefore, is quite interesting. The spin magnetic moment of $FeMg_8$ is 4.0 μ_B, which is the same as that of a free Fe atom. Since the Fe atom maintains its spin magnetic moment, one could think that a Fe atom contributes two sp electrons to the valence pool. Combined with 16 sp electrons from eight Mg atoms, one could think of $FeMg_8$ as a stable cluster with a shell filling at 18 electrons. This allotment is supported by the one-electron levels and the associated electronic orbitals shown in Fig. 3.11. The lowest state is spread out over multiple atoms and clearly has a 1S shape. The next two states have nodal 1P distribution. These are followed by four D orbitals. A Mulliken analysis indicates that these D orbitals

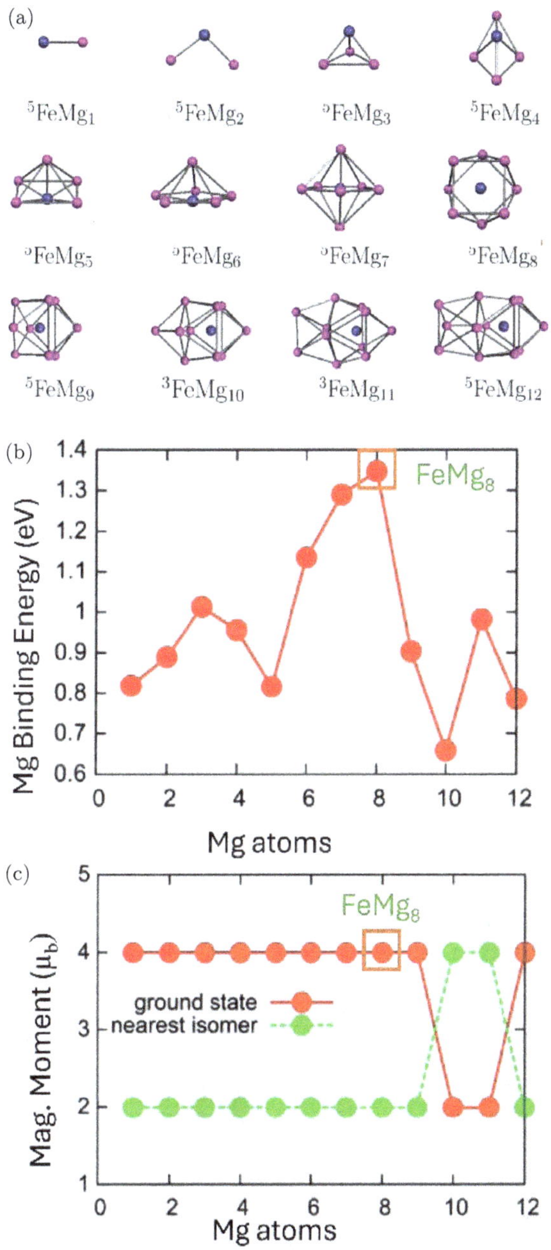

Fig. 3.10. (a) The structure of $FeMg_n$ clusters, $n = 1 - 12$, (b) the size-dependent Mg binding energy, and (c) the magnetic moment of the $FeMg_n$ clusters.

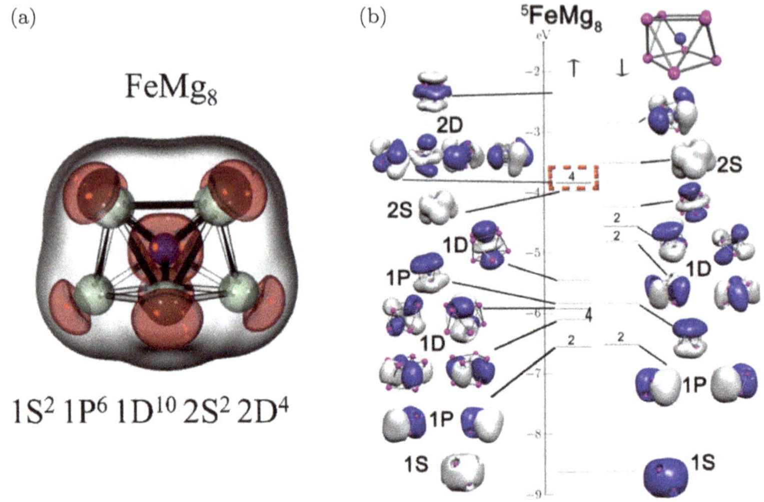

Fig. 3.11. (a) The structure of $FeMg_8$, with the spin density shown in the central red isosurface, and the charge density shown in the outer grey isosurface. (b) The molecular orbitals of $FeMg_8$ with the spin-up channel.

have 20% contribution from the Fe site, and the remainder comes from the Mg atoms. The set of D orbitals is followed by a P_z orbital and a D orbital corresponding to D_z^2 symmetry. These states are followed by the 2S state. The $1S^2$, $1P^6$, $1D^{10}$, and $2S^2$ states of both spins are filled with 20 valence electrons coming from Fe and Mg sites, leaving one with four of the total of 24 valence electrons. The splitting of the 2D orbitals occurs due to the shape of the cluster, as the Mg cores in $FeMg_8$ have a square antiprism arrangement, as seen in Fig. 3.10. In the potential of the ionic core, the D states are split into a group of four (D_{xy}, $D_{x^2-y^2}$, D_{xz}, and D_{yz}) and a D_{z^2} state, leading to a stable species with four unpaired electrons, as shown in a red dotted square in Fig. 3.10. The capacity of clusters to gain stability through sub-orbital splitting via the internal ionic rearrangements is quite unique and, in this case, results in magnetic species with a spin magnetic moment of 4.0 μ_B. The 2D states have a node in the centre and are composed of atomic d states localised on the Fe atom that mix with superatomic D states. In fact, a Mulliken population analysis confirmed that the 2D states have almost 85% contribution from the Fe d states. To further demonstrate the role of Hund's rule in the filling of D shells in cluster superatoms, the authors

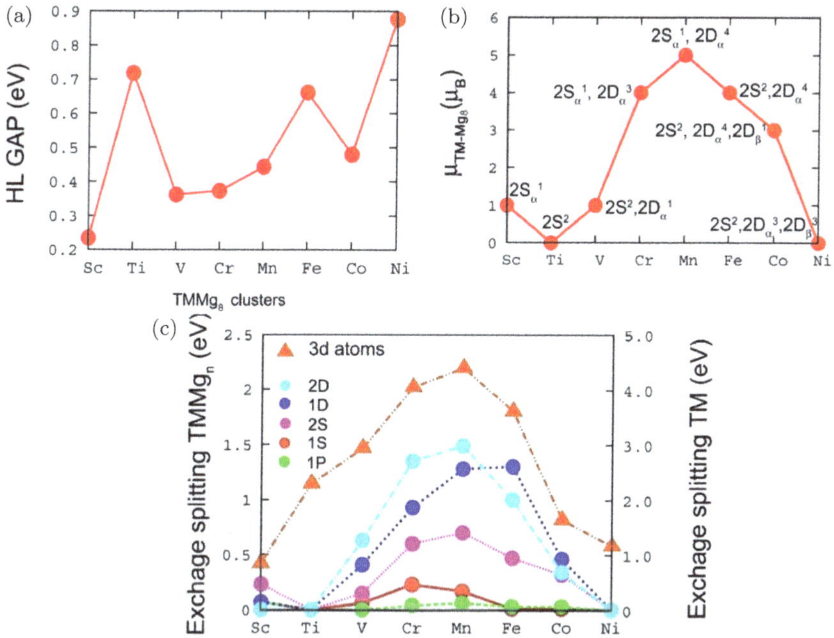

Fig. 3.12. (a) The HOMO–LUMO gap energy, (b) the spin magnetic moment, and (c) the exchange splitting of the TMMg$_8$ clusters, with the exchange splitting of the atoms for comparison. The ordering from low to high exchange splitting is generally 1P, 1S, 2S, 1D, 2D, and 3d atoms.

carried out a study of the spin magnetic moment of TMMg$_8$ clusters with the central site occupied by 3d transition elements. atom.

The results are shown in Fig. 3.12. While ScMg$_8$ has a configuration of 1S^2 1P^6 1D^{10} 2S^1, and TiMg$_8$ is the first cluster with a 20-electron closed-shell configuration 1S^2 1P^6 1D^{10} 2S^2. Further addition of d electrons results in the filling of the 2D subshell, leading to a configuration of 2S^1 2D^4 in the Mn, as shown on the left, and a spin-down configuration, as shown on the right. Four nearly degenerate 2D orbitals are circled in red.

Another interesting aspect is the exchange splitting for various transition metals as the transition metal d-states hybridise with the superatomic–orbital D-states, resulting in the splitting between the majority and minority sets of the nearly free electron gas D supershells. Figure 3.12 also shows the energy difference between the lowest majority and minority states of the 1S, 1D, 2S, and 2D supershells (exchange splitting) in the $TMMg_8$ (TM = Sc, Ti, V, Cr, Mn, Fe, Co, and Ni) clusters and the corresponding exchange splitting in the 3d TM atoms. It can be seen that the splitting in the TM atoms is the principal driving force for the splitting of the supershells. This demonstrates Hund's rule in the $TMMg_8$ clusters and further demonstrates how the splitting can be controlled by the nature of the central atom.

3.4.3 *Magnetic superatoms with filled subshells*

The conventional mechanism that leads to stable magnetic moments is based on Hund's rule in atomic physics and the Stoner model in solid-state physics. In both cases, the valence electrons avoid pairing by partially occupying subshells. The partial filling increases the exchange energy and lowers the Coulomb repulsion between electrons, leading to stable magnetic moments in partially filled shells. In a recent effort combining theoretical efforts at VCU and experimental work at Columbia University, the researchers proposed a new mechanism for the formation of magnetic moments but with filled shells.[68] This mechanism can lead to clusters that combine energetic stability while offering stable moments. The proposed scenario occurs when the quantum confinement of electrons in clusters splits their electronic structures into distinct subshells corresponding to different magic numbers. The filling of subshells belonging to different magic numbers to reach a closed-shell electronic configuration stabilises the cluster but creates differential spin occupations, resulting in a net magnetic moment. The proposed mechanism bears a striking resemblance to a similar process in magnetic semiconductors, but it uniquely operates at small sizes (clusters or nanostructures). As discussed in earlier sections, the Jahn–Teller distortions typically prevent the formation of high-spin states in clusters due to structural distortions that split a degenerate state. The dual subshell filling mechanism, however, results in both high electronic stability and a large magnetic moment that is robust against structural distortions and defects.

To describe this new approach, we begin by considering magic numbers in metal–chalcogenide clusters. As discussed in Section 3.2, the ligated transition metal chalcogenide clusters exhibit high symmetry, which leads to significant degeneracy in their electronic states, resulting in high electronic stability for certain electron counts. In particular, studies on various combinations of $TM_6S_8(CO)_6$ clusters showed that clusters with electronic counts of 96, 100, and 114 are particularly stable. The authors considered $Co_6E_8L_6$ and $[Re_6E_8L_6]^{2+}$, two of the most ubiquitous clusters in the literature: $Co_6E_8L_6$ has 114 electrons distributed in 57 orbitals, and $[Re_6E_8L_6]^{2+}$ has 100 electrons distributed in 50 orbitals. While both electron counts lead to remarkably stable cores due to their closed shells, the clusters are diamagnetic due to filled shells.

The idea to stabilise a magnetic moment (net spin moment) involves combining the orbital subspaces from two different closed-shell clusters using the subshell spin-up channel of one magic size and the spin-down spin channel of the other magic size to create a hybrid electronic structure. Figure 3.13 illustrates how such a situation may be realised by selecting a cluster based on Fe. Consider the prototypical diamagnetic closed-shell clusters $Co_6S_8(CO)_6$ and $[Re_6S_8(CO)_6]^{2+}$. $Co_6S_8(CO)_6$ has an electron count of 114 valence electrons (6×9 electrons derived from Co, 8×6 electrons from S, and 6×2 electrons in the Co–CO bonds) and a large HOMO–LUMO gap of 1.71 eV. Note that this valence electron count does not include the electrons within the ligand itself. $[Re_6E_8(CO)_6]^{2+}$, on the other hand, has an electron count of 100 and a HOMO–LUMO gap of 1.88 eV. When compared to $[Re_6E_8(CO)_6]^{2+}$, the additional 14 electrons of $Co_6S_8(CO)_6$ occupy A_{2g}, T_{1g}, and T_{2u} orbitals; these close-lying sets of high-symmetry orbitals are the reason why certain electron counts are favoured.

To design an electronically stable dual subshell magnetic cluster, consider a cluster $[Fe_6S_8(CN)_6]_{5-}$ that has an ideal electron count of 107 electrons (see Fig. 3.13). In $[Fe_6S_8(CN)_6]^{5-}$, the majority-filled subshell consists of orbitals corresponding to the 114 electron configuration (spin-up channel), with 57 electrons before a large energy gap. The minority subshell configuration (spin-down channel) consists of orbitals derived from the 100-electron configuration, with 50 electrons below its energy gap. The result is a cluster with a sizeable HOMO–LUMO gap and a large spin magnetic moment resulting from seven excess (unpaired) electrons in the majority spin channel (shown in red in Fig. 3.13).

Can such a cluster be synthesised in actual experiments? The experimental group at Columbia University synthesised the novel

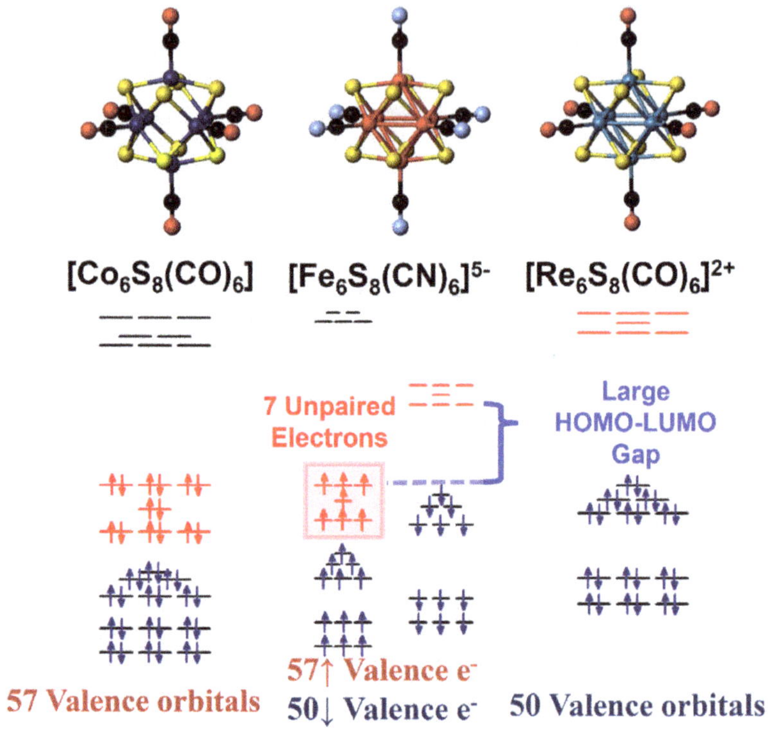

Fig. 3.13. The electronic structure of $Co_6S_8(CO)_6$, $[Fe_6S_8(CN)_6]^{5-}$, and $[Re_6S_8(CO)_6]^{2+}$. Valence electrons for the cluster core are assigned by counting all valence electrons from the transition metals and sulphur and two valence electrons per CO or CN molecule.

cluster $[Fe_6S_8(CN)_6]^{5-}$ using electron-accepting cyanide ligands to stabilise the high charge state. The reason for focusing on this novel cluster, instead of another previously reported $[Fe_6S_8(PEt_3)_6]^+$ cluster with the same electron count, was that cyanides can be used as bridging ligands capable of mediating strong magnetic exchange, for example in Prussian Blue-type materials. Consistent with theoretical predictions, the experiment studies found that the cluster $[Fe_6S_8(CN)_6]^{5-}$ is stable and has an isolated magnetic spin state of $S = 7/2$.

The experimental group also synthesised a cluster-assembled solid with ligated clusters as the building blocks (Fig. 3.14).[68] In the solid state, the magnetic moment was found to be smaller than expected,

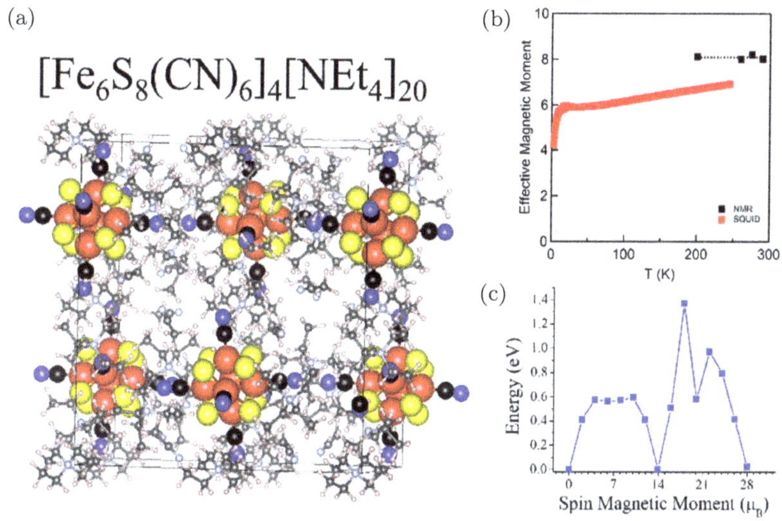

Fig. 3.14. (a) The crystal structure of $[Fe_6S_8(CN)_6]_4[NEt_4]_{20}$, (b) the SQUID values for the effective magnetic moment of the material, and (c) the calculated spin magnetic moment of the solid.

suggesting the presence of antiferromagnetic intercluster interactions. This finding was further supported by theoretical calculations of the magnetic configuration of the solid-state structure, indicating that the antiferromagnetic arrangement is lowest in energy. On the whole, the dual subshell filling of $[Fe_6S_8(CN)_6]^{5-}$ produces a unique magnetic superatom, combining a well-defined valence, a large HOMO–LUMO gap, and a high spin.

3.5 Conclusions

The current chapter focused on the magnetic behaviour at small sizes and how novel magnetic units described as magnetic superatoms can be designed or synthesised in a variety of ways. The first approach consists of breeding the magnetic character by introducing magnetic atoms with localised states in a nearly free electron gas. In this approach, the delocalised filled states from the nearly free electron gas offer the stability, while the localised d states provide magnetic moments. An alternative strategy to design magnetic motifs in the superatom family consists of inducing exchange splitting through hybridisation of the atomic d states of the transition metal atom with the superatomic D states derived from the free electron gas. Taking the example of an Fe atom surrounded by Mg atoms, the hybridisation leads to a large exchange splitting of the D superatom orbitals. These systems can be further stabilised by using the electrostatic field of the Mg ions as the crystal field that splits the D state into groups of four and one, leading to highly stable Mg_8Fe species with an even larger HOMO–LUMO gap. Finally, a novel class of magnetic superatoms can be designed by having a cluster where the majority electrons fill a subshell belonging to one magic number, while the minority electrons fill the subshell belonging to another magic number. Here, as shown by the example of the ligated metal chalcogenide cluster $[Fe_6S_8(CN)_6]^{5-}$, the cluster has a net spin magnetic moment even though the subshells are filled. This is a unique example of magnetism with filled shells.

While one can design magnetic clusters, can such units be assembled into molecules and higher units that maintain their identity and what novel features are offered by such units? Work by Khanna and co-workers showed that in a $(Mg_8Fe)_2$ supermolecule, the individual clusters did retain their shapes, and the molecule was marked by a small energy difference between the ferromagnetic and antiferromagnetic states. Previous work by the authors has shown that such units have the potential for spin-polarised currents through the application of small fields.

The current chapter has largely focused on the formation of spin magnetic moments. The other important quantity is the magnetic anisotropy energy, which determines the energy barrier for the moments to reorient under thermal or other fields. This is critical for forming, for example, permanent magnets. One of the main contributors to the anisotropy energy is the spin–orbit coupling. Some of these issues are discussed in previous works by the current authors, and these discussions will form the basis for further chapters.

Acknowledgements

This work was supported by a grant from US Department of Energy (DOE) under the award DE-SC0006420.

References

1. Khanna SN, Jena P. Atomic clusters: Building blocks for a class of solids. Phys Rev B. 1995;51(19):13705–13716.
2. Castleman AW, Khanna SN. Clusters, superatoms, and building blocks of new materials. J Phys Chem C. 2009;113(7):2664–2675.
3. Reber AC, Khanna SN. Superatoms: Electronic and geometric effects on reactivity. Acc Chem Res. 2017;50(2):255–263.
4. Knight WD, *et al.* Electronic shell structure and abundances of sodium clusters. Phys Rev Lett. 1984;52(24):2141–2143.
5. El-Sayed MA. Small is different: Shape-, size-, and composition-dependent properties of some colloidal semiconductor nanocrystals. Acc Chem Res. 2004;37(5):326–333.
6. Roy X, *et al.* Nanoscale atoms in solid-state chemistry. Science. 2013;341(6142):157–160.
7. Castleman AW, *et al.* From designer clusters to synthetic crystalline nanoassemblies. Nano Lett. 2007;7(9):2734–2741.
8. Claridge SA, *et al.* Cluster-assembled materials. ACS Nano. 2009;3(2):244–255.
9. Chaki NK, *et al.* Controlling band gap energies in cluster-assembled ionic solids through internal electric fields. ACS Nano. 2010;4(9): 5813–5818.
10. Qian M, *et al.* Cluster-assembled materials: Toward nanomaterials with precise control over properties. ACS Nano. 2009;4(1):235–240.
11. Khanna SN, Reber AC, Bista D, Sengupta T, Lambert R. The superatomic state beyond conventional magic numbers: Ligated metal chalcogenide superatoms. J Chem Phys. 2021;155(12):120901.
12. Ward AFM, Reber AC, Khanna SN. Periodic trends in the infrared and optical absorption spectra of metal chalcogenide clusters. J Phys Chem A. 2023;127(1):38–45.
13. Islam MF, Khanna SN. Stable magnetic order and charge induced rotation of magnetization in nano-clusters. Appl Phys Lett. 2014; 105(15):152409.
14. Sutradhar P, Chauhan V, Khanna SN, Atulasimha J. Magnetic behavior of superatomic-fullerene assemblies. Phys Chem Chem Phys. 2017;19(2):996–1002.
15. Sutradhar P, Khanna SN, Atulasimha J. Magnetic behaviour of assemblies of interacting cobalt-carbide nanoparticles. J Magn Magn Mater. 2019;469:128–132.
16. Xiao JQ, Jiang JS, Chien CL. Giant magnetoresistance in nonmultilayer magnetic systems. Phys Rev Lett. 1992;68(24):3749–3752.

17. Bloemen PJH, van Kesteren HW, Swagten HJW, de Jong WJM. Oscillatory interlayer exchange coupling in Co/Ru multilayers and bilayers. Phys Rev B. 1994;50(19):13505–13514.
18. Klose F, Rehm C, Nagengast D, Maletta H, Weidinger A. Continuous and reversible change of the magnetic coupling in an Fe/Nb multilayer induced by hydrogen charging. Phys Rev Lett. 1997;78(7):1150–1153.
19. Baibich MN, *et al.* Giant magnetoresistance of (001)Fe/(001)Cr magnetic superlattices. Phys Rev Lett. 1988;61(21):2472–2475.
20. Parent F, *et al.* Giant magnetoresistance in Co-Ag granular films prepared by low-energy cluster beam deposition. Phys Rev B. 1997; 55(6):3683–3687.
21. Mannini M, Pineider F, Sainctavit P, *et al.* Magnetic memory of a single-molecule quantum magnet wired to a gold surface. Nat Mater. 2009;8(3):194–197.
22. Akin ST, Liu J, Xu S, *et al.* Size-dependent ligand quenching of ferromagnetism in $Co_3(benzene)_n^+$ clusters studied with X-ray magnetic circular dichroism spectroscopy. J Phys Chem Lett. 2016;7(20): 4568–4575.
23. Reddy BV, Khanna SN. Chemically induced oscillatory exchange coupling in chromium oxide clusters. Phys Rev Lett. 1999;83(15):3170–3173.
24. Roy DR, Robles R, Khanna SN. Magnetic moment and local moment alignment in anionic and/or oxidized Fen clusters. J Chem Phys. 2010;132(19):194305.
25. Bergeron DE, Castleman AW Jr, Morisato T, Khanna SN. Formation of $Al_{13}I^-$: Evidence for the superhalogen character of Al_{13}. Science. 2004;304(5667):84–87.
26. Bergeron DE, Roach PJ, Castleman AW, Jones NO, Khanna SN. Al cluster superatoms as halogens in polyhalides and as alkaline earths in iodide salts. Science. 2005;307(5707):231–235.
27. Reveles JU, Khanna SN, Roach PJ, Castleman AW. Multiple valence superatoms. Proc Natl Acad Sci USA. 2006;103(49):18405–18410.
28. Jones CE, *et al.* Al_nBi clusters: Transitions between aromatic and jellium stability. J Phys Chem A. 2008;112(56):13316–13325.
29. Roach PJ, Woodward WH, Reber AC, Khanna SN, Castleman AW. Crystal field effects on the reactivity of aluminum-copper cluster anions. Phys Rev B. 2010;81(19):195404.
30. Reber AC, Khanna SN, Roach PJ, Woodward WH, Castleman AW. Spin accommodation and reactivity of aluminum-based clusters with O_2. J Am Chem Soc. 2007;129(50):16098–16101.
31. Reber AC, Gamboa GU, Khanna SN. The oblate structure and unexpected resistance in reactivity of Ag_{15}^+ with O_2. J Phys Conf Ser. 2013;438(1):012002.

32. Luo Z, Grover CJ, Reber AC, Khanna SN, Castleman AW. Probing the magic numbers of aluminum–magnesium cluster anions and their reactivity toward oxygen. J Am Chem Soc. 2013;135(11):4307–4313.
33. Armstrong A, Reber AC, Khanna SN. Multiple-valence aluminum and the electronic and geometric structure of Al_nO_m clusters. J Phys Chem A. 2019;123(22):5114–5121.
34. Reber AC, Roach PJ, Woodward WH, Khanna SN, Castleman AW. Edge-induced active sites enhance the reactivity of large aluminum cluster anions with alcohols. J Phys Chem A. 2012;116(33):8085–8091.
35. Martin TP. Shells of atoms. Phys Rep. 1996;273(4):199–241.
36. Krishnadas KR, Baksi A, Ghosh A, Natarajan G, Pradeep T. Manifestation of geometric and electronic shell structures of metal clusters in intercluster reactions. ACS Nano. 2017;11(6):6015–6023.
37. Roach PJ, Woodward WH, Castleman AW, Reber AC, Khanna SN. Complementary active sites cause size-selective reactivity of aluminum cluster anions with water. Science. 2009;323(5916):492–495.
38. Reber AC, Khanna SN, Roach PJ, Woodward WH, Castleman AW. Reactivity of aluminum cluster anions with water: Origins of reactivity and mechanisms for H2 release. J Phys Chem A. 2010;114(22):6071–6081.
39. Abreu MB, Powell C, Reber AC, Khanna SN. Ligand-induced active sites: Reactivity of iodine-protected aluminum superatoms with methanol. J Am Chem Soc. 2012;134(51):20507–20512.
40. Luo Z, Reber AC, *et al.* What determines if a ligand activates or passivates a superatom cluster? Chem Sci. 2016;7(5):3067–3074.
41. Reber AC, Khanna SN, Castleman AW. Superatom compounds, clusters, and assemblies: Ultra alkali motifs and architectures. J Am Chem Soc. 2007;129(31):10189–10194.
42. Reveles JU, Dissanayake S, Roach PJ, *et al.* Designer magnetic superatoms. Nat Chem. 2009;1(5):310–315.
43. Medel VM, *et al.* Hund's rule in superatoms with transition metal impurities. Proc Natl Acad Sci USA. 2011;108(24):10062–10066.
44. Chauhan V, Reber AC, Khanna SN. Metal chalcogenide clusters with closed electronic shells and the electronic properties of alkalis and halogens. J Am Chem Soc. 2017;139(5):1871–1877.
45. Chen Z. Multicentered bonding and quasi-aromaticity in metal-chalcogenide cluster chemistry. J Clust Sci. 1995;6(3):357–377.
46. Duan X, Wang C, Pan A, Yu R, Duan X. Two-dimensional transition metal dichalcogenides as atomically thin semiconductors: Opportunities and challenges. Chem Soc Rev. 2015;44(9):8859–8876.
47. Sarkar D, Xie X, Zhao W, *et al.* Functionalization of transition metal dichalcogenides with metallic nanoparticles: Implications for doping and gas-sensing. Nano Lett. 2015;15(5):2852–2862.

48. Reber AC, Khanna SN. $Co_6Se_8(PEt_3)_6$ superatoms as tunable chemical dopants for two-dimensional semiconductors. Npj Comput Mater. 2018;4(1):33.
49. Chauhan V, Reber AC, Khanna SN. Transforming Ni_9Te_6 from electron donor to acceptor via ligand exchange. J Phys Chem A. 2016;120(31):6644–6649.
50. Reber AC, Bista D, Chauhan V, Khanna SN. Transforming redox properties of clusters using phosphine ligands. J Phys Chem C. 2019;123(16):8983–8989.
51. Reber AC, Chauhan V, Bista D, Khanna SN. Superatomic molecules with internal electric fields for light harvesting. Nanoscale. 2020;12(9):4736–4742.
52. Reber AC, Sengupta T, Bista D, Khanna SN. Magic numbers in octahedral ligated metal-chalcogenide superatoms. Inorg Chem. 2022;61(40):16003–16008.
53. Liu F, Press MR, Khanna SN, Jena P. Magnetism and local order: Ab initio tight-binding theory. Phys Rev B. 1989;39(10):6914–6924.
54. Robles R, Khanna SN. Magnetism in assembled and supported silicon endohedral cages: First-principles electronic structure calculations. Phys Rev B. 2009;80(11):115414.
55. Reddy BV, Khanna SN, Dunlap BI. Giant magnetic moments in 4d clusters. Phys Rev Lett. 1993;70(22):3323–3326.
56. Billas IML, Châtelain A, de Heer WA. Magnetism from the atom to the bulk in iron, cobalt, and nickel clusters. Science. 1994;265(5179): 1682–1684.
57. Cox AJ, Louderback JG, Apsel SE, Bloomfield LA. Magnetism in 4d-transition metal clusters. Phys Rev B. 1994;49(17):12295–12298.
58. Khanna SN, Linderoth S. Magnetic behavior of clusters of ferromagnetic transition metals. Phys Rev Lett. 1991;67(6):742–745.
59. Pederson MR, Reuse F, Khanna SN. Magnetic transition in Mnn (n=2–8) clusters. Phys Rev B. 1998;58(9):5632–5636.
60. Reddy BV, Khanna SN, Jena P. Structure and magnetic ordering in Cr8 and Cr13 clusters. Phys Rev B. 1999;60(22):15597–15600.
61. Pederson MR, Khanna SN. Magnetic anisotropy barrier for spin tunneling in $Mn_{12}O_{12}$ molecules. Phys Rev B. 1999;60(14):9566–9572.
62. Payne FW, Jiang W, Bloomfield LA. Magnetism and magnetic isomers in free chromium clusters. Phys Rev Lett. 2006;97(19):193401.
63. Rao BK, Khanna SN, Jena P. Evolution of the electronic and structural properties of microclusters. Phys Rev B. 1987;36(2):953–960.
64. Pradhan K, Sen P, Reveles JU, Khanna SN. First principles study of Sc, Ti, and V doped Nan (n=4,5,6) clusters: Enhanced magnetic moments. Phys Rev B. 2008;77(4):045408.

65. Song F, Bergmann G. Strongly enhanced magnetic moments of vanadium impurities in thin films of sodium and potassium. Phys Rev Lett. 2002;88(16):167202.
66. Bergmann G, Song F. Electronic transition of vanadium impurities in different alkali hosts. J Magn Magn Mater. 2004;272–276: E863–E864.
67. Zhang X, *et al.* On the existence of designer magnetic superatoms. J Am Chem Soc. 2013;135(13):4856–4861.
68. Bista D, *et al.* High-spin superatom stabilized by dual subshell filling. J Am Chem Soc. 2022;144(12):5172–5179.

https://doi.org/10.1142/9781800618183_0004

Chapter 4

An Introduction to Plasmonic Chemistry

Ilko Bald

University of Potsdam, Institute of Chemistry,
Karl-Liebknecht-Str. 24-25, 14476 Potsdam, Germany
bald@uni-potsdam.de

Plasmonic nanoparticles, such as gold and silver nanoparticles, efficiently interact with visible light through excitation of their localised surface plasmon resonance (LSPR), which represents the collective oscillation of free electrons in the metal. This provides interesting opportunities to channel the energy of light into chemical energy. There are different ways in which molecules adsorbed to plasmonic nanoparticles can interact with the LSPR, with the most important ones being enhanced electromagnetic fields, charge transfer, and heat. This chapter introduces the main concepts of the emerging field of plasmonic chemistry, starting from the optical properties of metals to LSPR, different mechanisms of interaction with molecules, and selected examples of chemical systems studied in plasmonic chemistry.

4.1 Introduction

Plasmon-induced chemistry is an emerging field with potential applications in the fabrication of solar fuels, CO_2 reduction, polymers, new nanomaterials, and organic molecules.[1–3] Plasmonic nanostructures allow the conversion of the energy of visible light into chemical energy, which represents an attractive advantage over traditional photochemistry. The field of plasmonic chemistry brings together different communities and requires an understanding of various scientific disciplines, ranging from solid-state physics to interface chemistry, nanoscience, and physical chemistry. This book chapter aims to provide a tutorial introduction into the field.

4.2 Optical Properties of Metals and Metal Nanoparticles

The core of plasmonics is the effective interaction of light, ranging from the UV region to the near-infrared part of the electromagnetic spectrum with plasmonic nanostructures. Most plasmonic nanomaterials are made from precious metals such as Au, Ag, and Cu, which show intense extinction bands in the visible part of the electromagnetic spectrum. Before we can discuss the origin of these extinction bands, we have to briefly discuss the optical properties of metals in general.

The electric field component of light $\vec{E}$ is able to polarise materials – i.e., to shift the electrons in the material and to induce a dipole moment. In the case of a linear, isotropic, and homogeneous dielectric, the electronic polarisability $\vec{P}$ is directly proportional to $\vec{E}$, with the permittivity of vacuum ε_0 and the electric susceptibility χ_e being the proportionality constants:[4]

$$\vec{P} = \chi_e \varepsilon_0 \vec{E} = (\epsilon_r - 1)\varepsilon_0 \vec{E}, \tag{4.1}$$

where $\vec{P}$ describes the dipole moment in a material per volume (in other words, it corresponds to a dipole moment density); it has the SI unit Cm^{-2}. The relation $\chi_e = (\epsilon_r - 1)$ follows from the definition of the electric displacement field $\vec{D}$:

$$\vec{D} = \epsilon_r \varepsilon_0 \vec{E} = \varepsilon_0 \vec{E} + \vec{P}. \tag{4.2}$$

Chemists are familiar with molar polarisation, which is given by the Clausius–Mossotti equation and has the unit of a molar volume ($\mathrm{m}^3 \cdot \mathrm{mol}^{-1}$):

$$P_m = \frac{\varepsilon_r - 1}{\varepsilon_r + 2} \cdot \frac{M}{\rho} = \frac{N_A}{3\epsilon_0}\alpha_m, \tag{4.3}$$

where α_m is the molecular polarisability, ρ is the density of the material, M is the molar mass, and N_A is Avogadro's constant. For our purposes, the Clausius–Mossotti factor is useful, which is also valid for homogeneous solids and liquids and where N is the number density of the medium:

$$\frac{\varepsilon_r - 1}{\varepsilon_r + 2} = \frac{N}{3\epsilon_0}\alpha_m. \tag{4.4}$$

The Clausius–Mossotti factor can be converted into the Lorentz–Lorenz equation using the relation $\varepsilon_r = n^2$, with n being the refractive index of the material:

$$\frac{n^2 - 1}{n^2 + 2} = \frac{N}{3\epsilon_0}\alpha_m. \tag{4.5}$$

The atomic or molecular polarisability α_m is a microscopic quantity, while the electric susceptibility χ_e describes the macroscopic response of a materials to light or, generally, to an external electric field. Both are interconnected through the Clausius–Mossotti equation.

4.2.1 *The optical properties of metals as described by the Drude model*

Metals possess a large number of electrons that can move relatively freely through the metal. In a very simple model, it is assumed that the valence electrons move freely, like an electron gas, around the lattice of positively charged nuclei, which is known as the Drude model.[4] These free electrons also determine the optical properties of the metal, as they can interact with electromagnetic radiation, whose electric field component is described by

$$\vec{E}(t) = \vec{E}_0 e^{-i\omega t}, \tag{4.6}$$

where $\vec{E}_0$ is the amplitude of the oscillating wave and ω is its frequency. The oscillating electric field leads to oscillating polarisation of electrons, with the polarisation of the electron gas $\vec{P}$ (which is equivalent to the electronic polarisability introduced above) due to the local dipole moments $\vec{p}$ and the number density of electrons n_e:

$$\vec{P} = n_e\vec{p}. \tag{4.7}$$

If we consider only one-dimensional motion in the x direction and solve the Newtonian equations, we obtain

$$\vec{P} = -n_e ex = \frac{n_e e^2}{m(\omega^2 + i\gamma\omega)}\vec{E}_0 e^{-i\omega t} = \chi_e \varepsilon_0 \vec{E}, \tag{4.8}$$

where γ describes the damping of the oscillating electrons and can be understood as the collisional frequency of the electrons, which is around 10^{14} Hz.

Thus, the electric susceptibility can be expressed as

$$\chi_e = -\frac{n_e e^2}{\varepsilon_0 m(\omega^2 + i\gamma\omega)} = -\frac{\omega_p^2}{\omega^2 + i\gamma\omega}. \tag{4.9}$$

Here, we introduced the plasma frequency of a Drude metal:

$$\omega_p = \sqrt{\frac{n_e e^2}{m\varepsilon_0}}. \tag{4.10}$$

Considering the definition of the dielectric susceptibility introduced above, we can now write the dielectric function of a Drude metal:

$$\varepsilon(\omega) = 1 + \chi_e = 1 - \frac{\omega_p^2}{\omega^2 + i\gamma\omega}. \tag{4.11}$$

As the damping γ is typically significantly smaller than ω (the frequency of visible light is of the order of 10^{15} Hz), the dielectric function can also be written in a simplified form:

$$\varepsilon(\omega) \approx 1 - \frac{\omega_p^2}{\omega^2}. \tag{4.12}$$

The plasma frequency can explain the reflectivity and thus the optical appearance of a metal, i.e., for $\omega < \omega_p$, light is reflected from the metal, while for $\omega > \omega_p$, the metal is transparent. Typically, ω_p lies in the UV range, which leads to the well-known metallic shining appearance of metals. Some metals exhibit significant absorption in the visible range, which leads to the characteristic colours of e.g., gold and copper. The absorption is described in the imaginary part of the dielectric function:

$$\varepsilon = \varepsilon' + i\varepsilon''. \tag{4.13}$$

4.2.2 *Metal nanoparticles*

In metal nanoparticles, the motion of free electrons is limited by the boundary of the particle, which gives rise to *localised surface plasmon resonances* (LSPRs). LSPRs are the collective motion of free electrons in the metal. Here, we consider a spherical nanoparticle and assume that its electronic polarisation is exactly in phase with the excitation field, which is known as the *quasistatic approximation.*[4] The excitation field $\vec{E}_0$ induces an electric dipole moment in the metal sphere:

$$\vec{p} = \alpha \vec{E}_0. \tag{4.14}$$

The polarisability α can be described by

$$\alpha = 4\pi\varepsilon_0 a^3 \frac{\varepsilon - \varepsilon_{\text{diel}}}{\varepsilon + 2\varepsilon_{\text{diel}}}. \tag{4.15}$$

This relationship already shows a very important property of metal nanoparticles: The polarisability will become very large when the denominator $(\varepsilon + 2\varepsilon_{\text{diel}})$ becomes small, and in fact a maximum is reached when the following condition is met, which is known as the *Fröhlich condition*, or simply the *resonance condition*:

$$\varepsilon \approx -2\varepsilon_{\text{diel}}. \tag{4.16}$$

In this case, the polarisability will become very large, and only the fact that the permittivity of the environment is a complex number prevents α from becoming infinity. By taking the dielectric function from Eq. (4.12) and using the resonance condition equation (4.16), one can estimate the frequency of the LSPR to be

$$\omega_{\text{LSPR}} = \frac{\omega_p}{\sqrt{1 + 2\varepsilon_{\text{diel}}}} \approx \frac{\omega_p}{\sqrt{3}}. \tag{4.17}$$

In the latter step, it has been assumed that the medium is air, for which $\varepsilon_{\text{diel}} \approx 1$.

The polarisability is connected to other important quantities, such as the scattering and absorption cross-sections and the electric field.

The scattering cross-section is given by

$$\sigma_{\text{scat}} = \frac{k^4}{6\pi\epsilon_0^2}|\alpha|^2 = \frac{8\pi}{3}k^4a^6\left|\frac{\varepsilon-\varepsilon_{\text{diel}}}{\varepsilon+2\varepsilon_{\text{diel}}}\right|^2, \quad (4.18)$$

and the absorption cross section is given by

$$\sigma_{\text{abs}} = \frac{k}{\varepsilon_0}\text{Im}[\alpha] = 4\pi ka^3\text{Im}\left[\frac{\varepsilon-\varepsilon_{\text{diel}}}{\varepsilon+2\varepsilon_{\text{diel}}}\right]. \quad (4.19)$$

These relations indicate that, at the resonance condition, both scattering and absorption peak, which gives rise to the well-known phenomenon of LSPR, which is shown in Fig. 4.1. For gold spheres, this

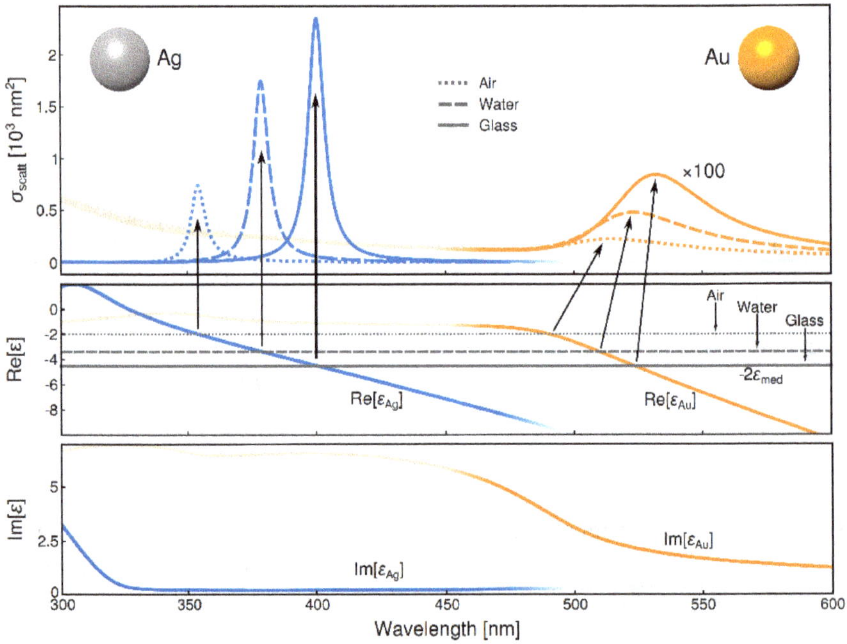

Fig. 4.1. Scattering cross-sections of silver (blue lines on the left) and gold (orange lines on the right) nanospheres having a diameter of 25 nm in air (dotted lines, $\epsilon_{\text{diel}} = 1$), water (dashed lines, $\epsilon_{\text{diel}} = 1.7$) and glass (solid lines, $\epsilon_{\text{diel}} = 2.3$). The central plot illustrates the Fröhlich condition: The crossing points of the real parts of the permittivities of silver and gold, respectively, with $-2\epsilon_{\text{diel}}$ indicating the spectral position where the Fröhlich condition is met. The imaginary parts of ϵ for silver and gold are shown in the bottom plot, which are important to understand the amplitudes of the scattering cross-sections.

Source: Reprinted with permission from Ref. 5 by Dr. Felix Stete; the data for the permittivities of gold and silver are based on Ref. 6.

condition is met around 520 nm, while for silver spheres, it is met at around 400 nm, depending on the specific value of $\varepsilon_{\text{diel}}$.

The electric field outside a nanoparticle decays according to

$$\vec{E}_{\text{ext}} - \vec{E}_0 \sim \frac{1}{r^3}. \tag{4.20}$$

This relation is important for surface-enhanced spectroscopy and electric-field-induced chemical reactions, which are discussed in the following. Equation (4.20) describes the optical near field.

4.2.3 *Which materials are plasmonic?*

Plasmonic materials need a high concentration of mobile charge carriers to be able to support LSPRs. Gold and silver meet this requirement very well, but both are rare and costly elements, and especially for larger-scale applications, it is interesting to consider other, more abundant, and hence sustainable materials. This leads to the question of what specifically characterises a plasmonic material.

The Fröhlich condition (Eq. (4.16)) already states that the permittivity of the material must be negative in order to support resonance. As the permittivity is a complex number, we can be more specific by stating that the real part of the permittivity must be negative, while its imaginary part should be small. As the imaginary part of the permittivity reflects the absorption of the material, the absorption should be as small as possible in order to prevent significant losses.[4] Ag exhibits very small absorption at the LSPR position, while Au shows significant interband transitions, which lead to a damping of the LSPR. Figure 4.2 compares the real and imaginary parts of different plasmonic metals. Apart from Au and Ag, Cu also shows promising plasmonic properties in the visible part of the spectrum (i.e., between 1.6 and 3.0 eV), but this metal is highly prone to oxidation, which compromises its plasmonic properties.[7,8] Pt, Al, and Ni also support the formation of plasmons, but their LSPRs are strongly damped by significant absorption (the imaginary part of the permittivity) in the visible range. Only Al is considered a promising material primarily for the UV part of the electromagnetic spectrum owing to its abundance.

Apart from pure metals, doped semiconductors and some transition metal oxides, chalcogenides, and nitrides can have a sufficiently

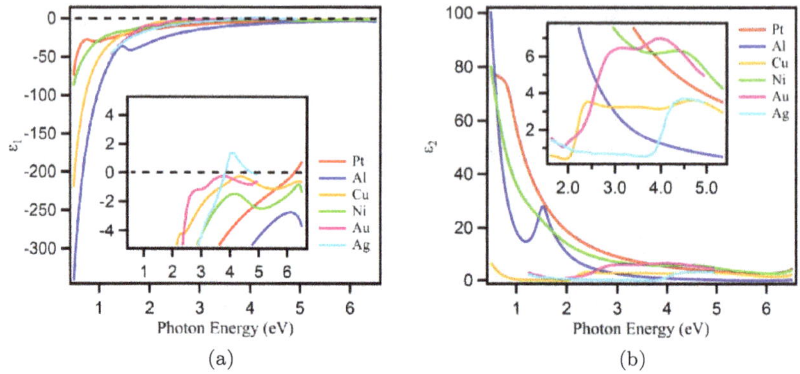

Fig. 4.2. (a) Real part (ϵ_1) and (b) imaginary part (ϵ_2) of the permittivities of Pt, Al, Cu, Ni, Au, and Ag, which demonstrate the spectral range of plasmonic activity. Al shows the lowest ϵ_1 of 1 eV at low photon energies, followed by Cu, Ni, and Pt. ϵ_2 is highest for Pt, lying between 1 and 3 eV, followed by Ni and Al, while Cu shows the lowest ϵ_2 values, lying below 2 eV. Plasmonic materials should have a negative real part of the permittivity and a small imaginary part of the permittivity.

Source: Reprinted with permission from Ref. 8.

high charge carrier density, which should be higher than 10^{21} cm^{-3}.[7,9] For example, copper chalcogenides and metal nitrides, such as $Cu_{2-x}S$, $Cu_{2-x}Se$, $CuFeS_2$ TiN, and HfN, represent interesting candidates for plasmonic materials; however, the band structure of semiconductors can lead to significant absorption, which in turn dampens the LSPR.[7,10]

Another important aspect for practical applications is chemical stability. Au is a very popular material due to its high chemical inertness, while Ag is easily oxidised. Cu and Al are even more prone to oxidation, making practical applications as plasmonic materials difficult, although they possess interesting optical properties and are much more abundant than Au and Ag.

4.3 Plasmon-Induced Chemical Reactions

The excitation of the LSPR triggers various processes that can be exploited in different ways in physics and chemistry. First of all, LSPR excitation leads to an enhanced electromagnetic field around the metal nanoparticles. Through electron-surface scattering, the

LSPR can decay to form single hot electron–hole pairs, which then thermalise through electron–electron scattering. Finally, electron–phonon coupling transfers energy into the lattice vibrations, resulting in an increase in the temperature of the nanoparticle, as well as that of the surrounding. Molecules located at the nanoparticles can now be influenced by each step of these processes, which are discussed in more detail as follows.

4.3.1 *Enhanced electromagnetic fields*

Depending on the material of the nanoparticles and their shape and arrangement, the electric field can be enhanced by up to a few hundredfold compared to the electric field of the incoming light. This is exploited in, e.g., surface-enhanced spectroscopy, particularly surface-enhanced Raman scattering (SERS), which is an important and widely used tool, including to study plasmon-induced chemical processes.[11,12]

Upon excitation of the LSPR, the nanoparticles act like a lens, in which the electromagnetic field is focused onto a small volume in close vicinity to the nanoparticles. When the excitation energy matches an electronic excitation in the close-by molecules, the photon absorption and possible photochemical processes can be enhanced (Fig. 4.3). This can lead to interesting applications, such as photo polymerisation.[13] The reactions can be triggered with high local control, i.e., only in close vicinity to the nanoparticles. This locality is particularly pronounced in two-photon photopolymerisation, where high intensities are required to trigger a reaction. Another interesting aspect is that the electronic structure of molecules can change close to a nanoparticle surface; for example, homo-LUMO gaps can be decreased, enabling photochemical processes at longer wavelengths in the presence of plasmonic nanoparticles than for molecules without nanoparticles.

4.3.2 *Charge transfer*

Charge transfer is the most interesting method to induce chemical modifications because it enables not only an acceleration of chemical processes in the sense of catalysis but also opens up new chemical pathways.

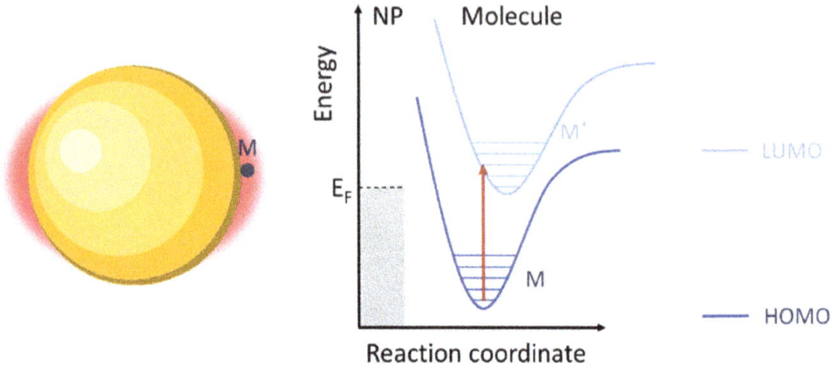

Fig. 4.3. Schematic diagram of a molecule adsorbed to a nanoparticle and located in the enhanced electric field due to excitation of the LSPR. On the right, a potential energy diagram is shown, indicating electronic transition within a molecule that is enhanced by the enhanced electric field at the surface of the nanoparticle (NP). E_F refers to the Fermi energy, and the grey area below E_{F} indicates the occupied states in the NP.

The collective oscillation of the free electrons in the metal upon excitation of the LSPR can decay via scattering of the electrons at the surface of the nanoparticles. This can lead to the generation of a single electron–hole pair, a process referred to as *Landau damping*. The electrons and holes generated in this way are typically termed as *hot electrons* or *hot holes*, respectively. The term 'hot electron' is not well defined, and different authors might use it with slightly different definitions. Here, a hot electron refers to an electron that is not part of a thermal distribution, i.e., a non-thermal electron. The surface scattering, and thus the generation of hot electrons, is more effective for small nanoparticles because their surface-to-volume ratio is higher. This is quantitatively described by *Kreibig's term*, which describes the rate of electron-surface scattering:

$$\gamma_{e-\mathrm{surf}} = A\frac{v_F}{a}, \tag{4.21}$$

where A is a numerical constant, v_F is the Fermi velocity, and a is the particle diameter. The theory of hot electron generation, in the meantime, has been further refined,[14] but Kreibig's term remains a good approximation.

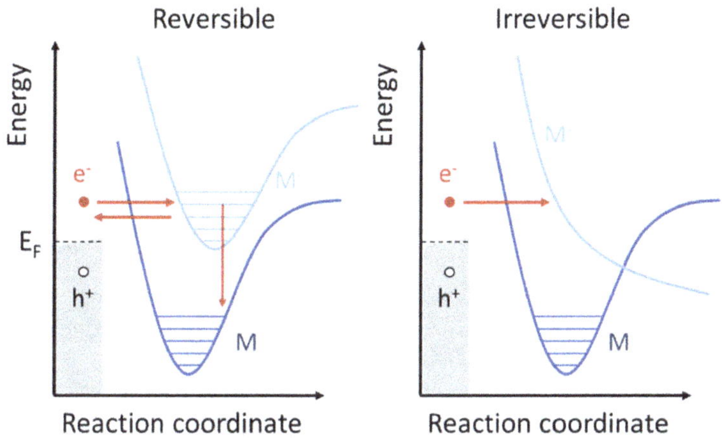

Fig. 4.4. Schematic potential energy diagram depicting a charge transfer between a nanoparticle and an adsorbed molecule (horizontal red arrows) after the generation of a hot electron–hole pair due to excitation of the LSPR. As a result, a molecular ion M^- is formed. The charge transfer can be reversible, resulting in a vibrationally excited neutral molecule (a vertical red arrow indicates the transfer of the M^- anion to a neutral, vibrationally excited molecule M), or irreversible, resulting in an ion that might dissociate.

Once a hot electron–hole pair is formed, the hot charges can be transferred between the nanoparticle and a molecule. When an electron is transferred to a molecule M, a transient negative ion (TNI) M^- is formed (Fig. 4.4). Depending on the properties of this TNI, different subsequent processes are possible. The excess electron can be transferred back into the metal, leaving the neutral molecule behind, which might now be vibrationally strongly excited. This is similar to the well-studied process of desorption induced by electron transfer (DIET), and this mechanism was suggested, e.g., for the plasmon-induced oxidation of ethylene, in which O_2 is activated by a reversible electron transfer.[15] Alternatively, a TNI can be formed in a strongly repulsive state, i.e., the TNI is strongly unstable with respect to dissociation. Then, the dissociation along a specific coordinated bond is faster than the back transfer of an electron, and this process corresponds to the dissociative electron attachment (DEA) mechanism. As an example, this mechanism has been proposed for the plasmon-induced decomposition of 8-bromoadenine (BrA).[16,17]

The decay of the LSPR through Landau damping and subsequent electron transfer to adsorbed molecules is sometimes argued to be rather ineffective. Instead of first forming an electron–hole pair, the LSPR could also decay by a direct transfer of an electron from the valence band of the metal to a vacant orbital of an adsorbed molecule. As this process depends significantly on the availability of empty molecular states, it is referred to as *chemical interface damping* (CID). The existence of CID was demonstrated using dark-field scattering spectroscopy, in which elastically scattered light is collected from single nanoparticles and spectrally analysed. In this way, the LSPR band of single nanostructures can be measured, and it was shown that the signal broadens (i.e., its lifetime decreases) in the presence of specific molecules.[18]

The rates of hot-electron-induced reactions can be effectively monitored using SERS, as the vibrational bands specific for reactants and products can be recorded over time (in the seconds-to-minutes range), and the time dependence of the signal intensities gives rise to rate constants. This is shown in Fig. 4.5 for the example of the debromination of BrA, resulting in adenine (A), which is triggered by hot electrons:

$$\mathrm{BrA} + e^- \rightarrow \mathrm{A} + \mathrm{Br}^-.$$

Mechanistic understanding can be gained by proposing a reaction mechanism, then deriving a rate law, and finally comparing it with the time-dependent SERS signals.

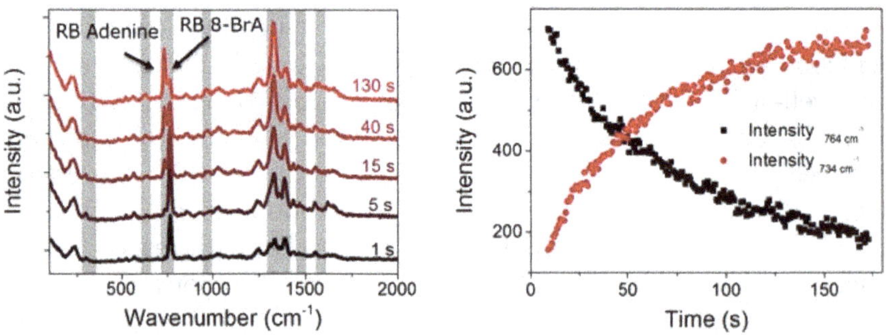

(a) (b)

Fig. 4.5. (a) SERS spectra of BrA on AgNPs dried on a Si-substrate for different illumination times. (b) Intensities of the ring-breathing mode of BrA at 767 cm^{-1} and that of A at 735 cm^{-1} over time.

Source: Reprinted with permission from Ref. 16.

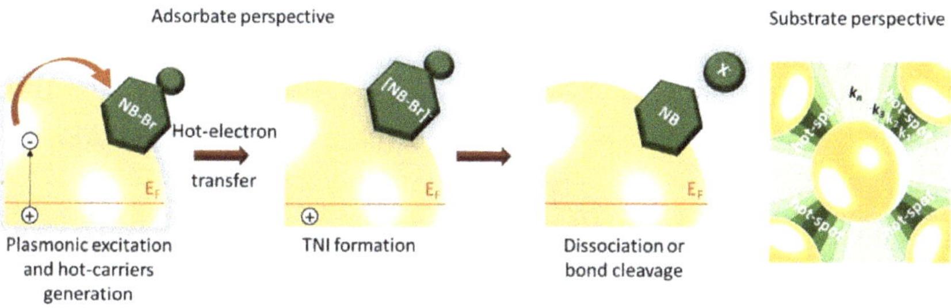

Fig. 4.6. Illustration of the individual reaction steps: from LSPR excitation and formation of an electron–hole pair to hot-electron transfer and dissociation of the adsorbed molecule. On the extreme right, the inhomogeneous field enhancement and the formation of hot spots are depicted, which result in an inhomogeneous distribution of reaction rates at the nanoparticle surface.

Source: Reprinted with permission from Ref. 17.

The total reaction can be decomposed into several elementary steps, as depicted in Fig. 4.6. Assuming that the electron attachment to the molecule is the rate-limiting step, the rate law for the overall reaction is of pseudo-first order:

$$\frac{d[\mathrm{BrA}]}{dt} = -k_{\mathrm{obs}}[\mathrm{BrA}]. \tag{4.22}$$

This consideration assumes that the reaction occurs with the same reaction rate at all locations on the surface of the nanoparticles. However, this is certainly not the case when the reaction is induced by hot electrons because hot electrons are generated with a higher probability at locations with high field enhancement, i.e., either at the nanoscale gaps between two nanoparticles or at the sharp edges of non-spherical particles. These locations act as *hot spots* for the SERS signals, but they can also be understood as *chemical hot spots*.[19] The consequence is that the hot-electron-induced reaction takes place with different reaction rates at the different sites on the nanoparticle surface (see the substrate perspective in Fig. 4.6). However, in the SERS experiments, only a rate constant averaged over many nanoparticle sites can be determined, and at the same time, the SERS hot spots are over-represented in the SERS intensity. This will lead to a time-dependent rate constant, which can be described by fractal kinetics:[12]

$$k = k_f \cdot t^{-h}. \tag{4.23}$$

Here, t is the time, h is the fractal dimension ($0 \leq h \leq h$), and k_f is the time-independent fractal rate constant. k_{obs} is replaced by k, and in SERS measurements, the molecular concentrations are replaced by the SERS intensities I, which leads to the following expression:

$$\ln \frac{I_0}{I} = \frac{k}{1-h} t^{1-h}. \tag{4.24}$$

Experimentally, a simple curve fitting is employed:

$$\ln \frac{I_0}{I} = a t^b. \tag{4.25}$$

This yields the rate constant from the fitting parameters a and b:

$$k = a \cdot b. \tag{4.26}$$

In this context, it should be noted that the term 'plasmonic catalysis' refers to situations in which the plasmonic nanoparticles actually serve as catalysts, i.e., they increase the rate of a chemical process by reducing the activation barrier according to the Arrhenius equation:

$$k = k_0 e^{-\frac{E_A}{\mathrm{RT}}}. \tag{4.27}$$

This is indeed possible,[20] but in fact the chemical processes taking place upon excitation of the LSPR are typically more complex, as explained above, and they often show characteristics that are distinctly different from classical heterogeneous catalysis. For example, in classical catalysis, the reaction takes place without the catalyst as well, and the presence of the catalyst accelerates the reaction. This is only possible when the Gibbs free energy of the reaction is negative. In contrast, plasmon-induced chemical reactions are often not possible without nanoparticles because the light-induced plasmonic excitation is a necessary precondition for the production of hot carriers. Consequently, processes with positive Gibbs free energy become possible, and in fact light becomes a necessary ingredient in these reactions and should be regarded as a reactant rather than a simple means of adding energy into the system. Therefore, in this chapter, the term 'plasmonic catalysis' is only used for processes that already have a negative Gibbs free energy.

SERS is a highly versatile method for determining the rates of plasmon-induced reactions because basically any molecule adsorbed on the plasmonic nanoparticles can be recorded, and the excitation laser can be used for both triggering the reaction and monitoring the SERS signal. Apart from SERS, photometry can also be used for reactions involving reactants and/or products that have characteristic UV–Vis absorption bands.[21] For gas-phase reactions, mass spectrometry is a well-established technique.[15,22]

In order to compare the performance of different plasmonic materials, the yield of a hot-electron-induced reaction has been decomposed into different contributions:[23]

$$\text{yield} \sim \sigma_{\text{abs}} \cdot \varphi_{\text{hote}^-} \cdot \gamma_{\text{transfer}}, \tag{4.28}$$

where σ_{abs} is the absorption cross-section and describes the efficiency of photon absorption by the nanoparticle. For gold spheres, the absorption cross-section is of the order of 10^4 nm^2 (10^{-10} cm^2) at the peak of the LSPR, which is orders of magnitude higher than typical absorption cross-sections of molecules (which typically range between 10^{-21} and 10^{-18} cm^2). This is one of the major reasons why plasmonic nanoparticles are such an attractive option to harvest visible light and transform its energy into chemical energy. σ_{abs} depends strongly on the material, size, and shape of the nanoparticles. φ_{hote^-} is the generation efficiency of hot electrons and is described, to a first approximation, by Kreibig's term (Eq. (4.21)). γ_{transfer} describes the transfer rate of hot electrons to the adsorbed molecules.

Quite often, it has been observed through SERS that silver nanoparticles give rise to higher rate constants than gold nanoparticles in hot-electron-induced reactions. One explanation is that silver nanoparticles provide more efficient hole deactivation reactions compared to Au (see the following paragraph for a short discussion on hole-induced processes). Secondly, it has been recently concluded that this observation can be attributed to a larger hot carrier generation efficiency in silver compared to gold (Fig. 4.7).

So far, the discussion centred mainly around hot electrons inducing reduction reactions. However, plasmon decay also generates hot holes, which can induce oxidation reactions. This has been demonstrated, e.g., for the oxidation of aminothiophenol to dimercaptoazobenzene (DMAB) and the decarboxylation of mercaptobenzoic acid. Another important aspect is that the efficiency of hot-electron-induced reactions can be significantly reduced when the remaining holes are not deactivated in a counter reaction. Consequently, it can

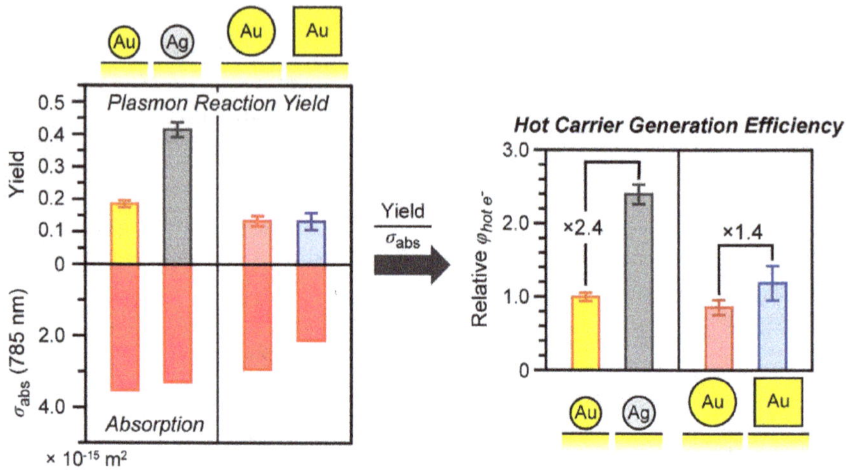

Fig. 4.7. Left: Comparison of the experimentally measured reaction yields (upper panel) and the absorption cross-sections at 785 nm, σ_{abs} (785 nm), for each nanoparticle-on-a-mirror system (lower panel). Right: Normalised hot carrier generation efficiency ϕ_{hote^-} obtained by dividing the measured reaction yields by the calculated absorption cross-sections.

Source: Reprinted with permission from Ref. 23. Copyright 2022 American Chemical Society.

be useful to add sacrificial agents into the reaction system that are oxidised by the generated holes. Examples of such sacrificial agents are alcohols.

4.3.3 *Elevated temperatures*

The energy stored in single electron–hole pairs can be distributed among many electrons through electron–electron interactions within 10–100 fs, and eventually, electron–phonon scattering results in heating of the nanoparticle lattice and its environment, which occurs beyond 10 ps. One can distinguish between the temperature of the electron gas, which can reach thousands of Kelvin, and the lattice temperature of the nanoparticle, which is considerably lower (Fig. 4.8). This is referred to as the *two-temperature model*. The power absorbed by the nanoparticle Q is directly related to the light intensity I (in W/cm^2, also referred to as irradiance) and

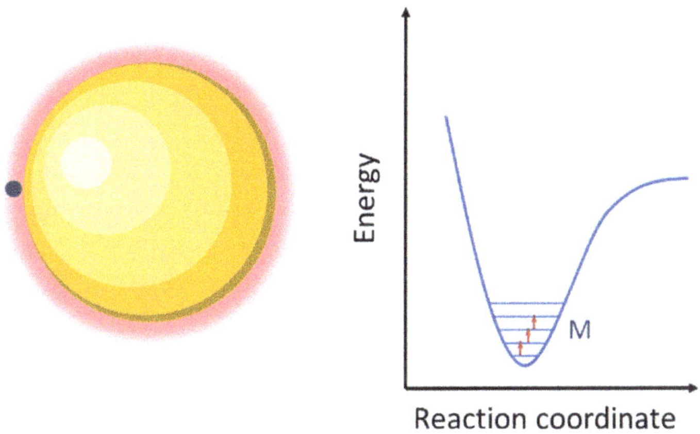

Fig. 4.8. Schematic diagram of a molecule adsorbed to a nanoparticle and subject to a temperature increase due to LSPR excitation. On the right, a potential energy diagram is depicted, showing the vibrational excitation of the molecule due to heating.

the nanoparticle's absorption cross-section σ_{abs}:

$$Q = \sigma_{\text{abs}} \cdot I. \tag{4.29}$$

Assuming that all the absorbed power is transferred as heat to the surrounding, one can use Fourier's law (with the thermal conductivity of the environment κ and the particle radius R) to derive the temperature increase of the environment ΔT:

$$\Delta T = \frac{I\sigma_{\text{abs}}}{4\pi\kappa R}. \tag{4.30}$$

The temperature increase typically ranges between a few Kelvin up to several tens of Kelvin when using continuous-wave lasers and affects the nanoparticle in a homogeneous manner. The temperature increase can then lead to a significant increase in the reaction rates of chemical processes, according to the Arrhenius equation (Eq. (4.27)), and can even trigger processes such as polymerisations. The heating of nanoparticles is also the basis of photothermal therapy, in which gold shells or rods are injected into tumour tissue and then irradiated with near-IR lasers in order to kill the tumour tissue.

4.3.4 *How to distinguish between the different contributions of heat and charges*

The field of plasmonic chemistry was and continues to be strongly dominated by a controversial debate about the most important mechanisms driving the processes observed in different experiments.[24–26] The main contributions that have been discussed are charge transfer and heat. While many light-induced chemical reactions have been explained by the transfer of hot charges, other authors explain the findings by purely thermal processes. This debate led to two important conclusions:

(i) Whether a plasmon-induced chemical reaction is mainly driven by heat or hot charges needs to be studied through careful experiments.[27] For example, SERS experiments can be performed at different laser powers. For a purely charge-induced process, the reaction rate should increase linearly with the laser power, while for a temperature-induced process, the reaction rate should increase exponentially with the laser power. This is due to the fact that the number of hot electrons should be directly proportional to the laser power, while the rate is influenced by the temperature of the system according to the Arrhenius equation.

(ii) Different processes can occur at the same time. For example, higher temperatures can favour electron transfer, which in turn might be required to induce a specific reduction reaction. It should also be noted that it is difficult to avoid heating of the nanoparticles completely. By properly choosing the experimental conditions (laser power, continuous wave versus pulsed laser excitation, wavelength, nanoparticle species and arrangement, etc.), heating can be kept to a minimum, but it can't be avoided completely. On the other hand, heating might be desired to drive temperature-sensitive processes.

4.4 Examples of Specific Chemical Processes

In the meantime, plasmon-induced chemical reactions have been demonstrated for many different chemical systems. They can be roughly divided into the four following groups: reactions involving small aromatic organic molecules, reactions involving gas-phase molecules, polymerisation reactions, and nanoparticle growth.

4.4.1 *Reactions involving small aromatic organic molecules*

The first observations of plasmon-induced chemical reactions have been made by SERS using the molecules para-aminothiophenol (ATP) and para-nitrothiophenol (NTP).[28] Initially, unexpected bands have been reported that were not associated with the original molecules until the now established interpretation has been provided that ATP and NTP dimerise through a plasmon-induced process to dimercaptoazobenzene (DMAB). Since then, both reactions have been studied in great detail through SERS.[28–30] It is interesting to note that the formation of DMAB from ATP represents an oxidation reaction that is driven by hot holes in the presence of oxygen, while the formation of DMAB from NTP represents a reduction reaction driven by hot electrons:

$$2\text{ATP} + 4h^+ \xrightarrow{\text{O}_2,\text{H}_2\text{O}} \text{DMAB}, \tag{4.31}$$

$$2\text{NTP} + 6e^- \rightarrow \text{DMAB}. \tag{4.32}$$

Both reactions occur on gold nanoparticles, as well as on silver nanoparticles. Small aromatic molecules are easy to observe using SERS due to their large Raman cross-section and well-defined SERS bands; consequently, other reactions have also been reported in the meantime, such as the dehalogenation of bromonucleobases discussed already above,[17] dehalogenation of halogenthiophenols,[31] and decarboxylation of mercaptobenzoic acid.[32]

An interesting example that has been reported recently involved the transformation of a secondary aromatic amide (N-methyl-4-sulfanylbenzamide; NMSB), which represents a model system for a peptide bond (Fig. 4.9).[33] Previously, it has been shown by tip-enhanced Raman scattering (TERS) that this molecule is transformed into a primary amide via a hot-electron-induced reaction, and it was proposed that at elevated temperatures, the reaction pathway changes to form an alternative product.[34] However, this could not be confirmed by SERS.[33] Instead, it was shown that the reaction

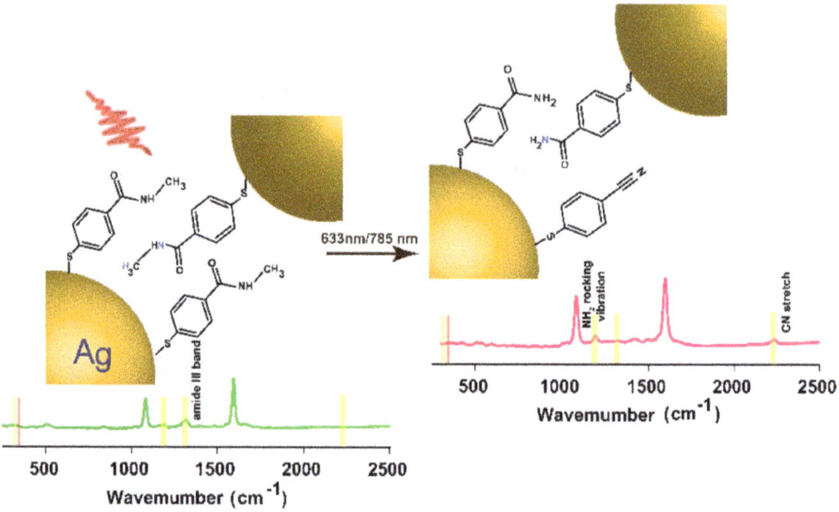

Fig. 4.9. Schematic diagram showing the transformation of a secondary amide (indicated by a decrease in the intensity of the Amide III band at 1320 cm^{-1}) to a primary amide (rising bands at 1189 cm^{-1} and 1412 cm^{-1}) followed by the formation of an aromatic nitrile (CN stretching vibration at 2230 cm^{-1}). The reactants and products have been detected and monitored using SERS.

Source: Reprinted with permission from Ref. 33.

is rather independent of the temperature when tested under external heating and that another subsequent reaction takes place: the formation of an aromatic nitrile.

The reactants in most reactions mentioned above are anchored to the nanoparticle surfaces by thiols, which represents a good method to provide well-defined self-assembled monolayers on metal surfaces and thus serve as well-defined conditions for the study of fundamental characteristics of plasmon-induced reactions. However, it needs to be noted that such systems do not represent classic examples of catalysis because the reactants and products stick to the nanoparticle surface, and the nanoparticles are not used as catalysts for further chemical transformations. Nevertheless, such systems provide useful conditions for studying fundamental aspects of plasmon-induced chemical reactions.

The transformation of small aromatic organic molecules is also very important for the development of plasmon-induced coupling reactions as a tool for organic synthesis. The formation of DMAB from ATP or NTP represents an N–N coupling that is triggered by excitation of Au or Ag nanoparticles. Another example is the C–N coupling of benzylamine, which is also induced by plasmon excitation of Au nanoparticles.[35] In organic synthesis, C–C couplings are quite often realised by Pd-catalysed reactions, such as Suzuki–Miyaura and Heck couplings, and it has already been demonstrated that the yield of such couplings can be improved in the presence of plasmonic nanoparticles and light irradiation, and the reactions can be performed at lower temperatures.[36,37] Current research efforts are focusing on the question of whether plasmon-induced molecular activation can not only improve the yield of Pd-catalysed cross couplings but also enable new reaction pathways and consequently novel products and synthesis schemes.

4.4.2 *Reactions involving gas-phase molecules*

Gas-phase molecules can be easily monitored by mass spectrometry, and plasmon-induced chemical reactions have been demonstrated for various small molecules, e.g., the dissociation of H_2. Also, many industrially relevant reactions have been shown to be improved by the irradiation of plasmonic nanoparticles with visible light, e.g., the oxidation of ethylene[15] and H_2 generation from NH_3.[2] In general, there is great potential to use plasmonic nanoparticles for hydrogen production, CO_2 reduction, and solar fuels.[1]

4.4.3 *Polymerisation reactions*

Plasmonic nanoparticles can be used to trigger polymerisation reactions using all three mechanisms discussed above.

The transfer of an electron from a plasmonic nanoparticle to a molecule generates a radical, which can be used for radical polymerisations. Thus, the excitation of plasmonic nanoparticles serves as a radical starter, and this was demonstrated, e.g., for the radical polymerisation of divinyl benzene.[38] Also, the elevated temperature around the nanoparticles can be used to trigger a polymerisation reaction, as demonstrated with the formation of polydimethylsiloxane (PDMS) at Au nanostructures.[39] As a third way to exploit plasmonic excitation, the enhanced near field around plasmonic nanoparticles has been used for photopolymerisation, e.g., with triacrylate as a monomer and a photoinitiator. The excitation conditions are chosen such that the threshold for photopolymerisation is only reached in the near field around the nanoparticles.[40]

Apart from fundamental proof-of-concept studies, plasmon-induced photopolymerisation has been used for imaging near fields around nanoparticles through electron microscopy and atomic force microscopy[41] and for positioning nanoemitters in plasmonic nanocavities.[3]

4.4.4 *Nanoparticle growth*

The growth of nanoparticles can also be triggered by plasmonic excitation, and in fact this represents one of the simplest chemical reactions to be studied in this context. This is because the typical nanocharacterisation tools, such as electron microscopy and dark-field scattering, can be used to monitor the nanoparticles before, during, and after irradiation with light. In this context, the growth of a silver shell around gold nanorods has been monitored using dark-field scattering spectroscopy, and it was concluded that the light-induced growth is driven by interband transitions in gold; that is, electrons excited from the 5d band of gold are transferred to silver ions, which then get reduced. In particular, the deactivation of the 5d holes is the rate-determining step.

Similar reactions can be used to control the formation of nanoparticle shells by light, thus enabling the creation of more sophisticated nanoarchitectures such as chiral plasmonic nanostructures using circularly polarised light.[42]

4.5 Conclusions

Plasmonic chemistry is an emerging field that exploits the light-harvesting properties of plasmonic nanoparticles, promising to collect light from the UV to near-infrared spectral region and convert it into chemical energy. Different processes contribute to plasmon-mediated chemical reactions, which provide the possibility of driving chemical processes that are fundamentally different from those in thermal chemistry. Plasmonic nanostructures are primarily composed of Au and Ag, but novel alternatives are also currently being developed using more abundant materials. Also, hybrid materials, consisting of a plasmonic material and, e.g., a semiconductor, promise to enhance the efficiency of photochemical processes. Specific chemical processes range from nanoparticle growth, hydrogen evolution, and bond breaking as well as bond forming reactions in organic molecules up to polymerisation reactions. With such possibilities, plasmonic chemistry provides ample potential for attractive applications.

References

1. Ezendam S, *et al.* Hybrid plasmonic nanomaterials for hydrogen generation and carbon dioxide reduction. ACS Energy Lett. 2022;7:778–815.
2. Yuan Y, *et al.* Earth-abundant photocatalyst for H_2 generation from NH_3 with light-emitting diode illumination. Science. 2022;378:889–893.
3. Chen M, *et al.* Approaches for positioning the active medium in hybrid nanoplasmonics. Focus on plasmon-assisted photopolymerization. ACS Photonics. 2024;11:3933–39532.
4. Pluchery O, Bryche J-F. An introduction to plasmonics. London: World Scientific; 2024.
5. Stete F. Gold at the nanoscale. Universität Potsdam; 2021.
6. McPeak KM, *et al.* Plasmonic films can easily be better: Rules and recipes. ACS Photonics. 2015;2:326–333.
7. West PR, *et al.* Searching for better plasmonic materials. Laser Photonics Rev. 2010;4:795–808.
8. Satrya CD, Darma Y. Dielectric-function analysis of metals for plasmonic-device application. J Phys Conf Ser. 2017;877:12040.
9. Agrawal A, *et al.* Localized surface plasmon resonance in semiconductor nanocrystals. Chem Rev. 2018;118:3121–3207.
10. Bhattacharyya B, *et al.* Copper iron chalcogenide semiconductor nanocrystals in energy and optoelectronics applications—State of the art, challenges, and future potential. Adv Optical Mater. 2023;11:2202411.
11. Stefancu A, *et al.* Impact of surface enhanced Raman spectroscopy in catalysis. ACS Nano. 2024;18:29337–29379.

12. Schürmann R, *et al.* Microscopic understanding of reaction rates observed in plasmon chemistry of nanoparticle–ligand systems. J Phys Chem C. 2022;126:5333–5342.
13. Kameche F, *et al.* Plasmon-triggered living photopolymerization for elaboration of hybrid polymer/metal nanoparticles. Mater Today. 2020;40:38–47.
14. Besteiro LV, Kong X-T, Wang Z, Hartland G, Govorov AO. Understanding hot-electron generation and plasmon relaxation in metal nanocrystals: quantum and classical mechanisms. ACS Photonics. 2017;4:2759–2781.
15. Christopher P, Xin H, Linic S. Visible-light-enhanced catalytic oxidation reactions on plasmonic silver nanostructures. Nat Chem. 2011;3:467–472.
16. Schürmann R, Bald I. Real-time monitoring of plasmon induced dissociative electron transfer to the potential DNA radiosensitizer 8-bromoadenine. Nanoscale. 2017;9:1951–1955.
17. Dutta A, *et al.* Kinetics and mechanism of plasmon-driven dehalogenation reaction of brominated purine nucleobases on Ag and Au. ACS Catal. 2021;11:8370–8381.
18. Foerster B, Spata VA, Carter EA, Sönnichsen C, Link S. Plasmon damping depends on the chemical nature of the nanoparticle interface. Sci Adv. 2019;5:eaav0704.
19. Gargiulo J, Berté R, Li Y, Maier SA, Cortés E. From optical to chemical hot spots in plasmonics. Acc Chem Res. 2019;52:2525–2535.
20. Kim Y, Dumett Torres D, Jain PK. Activation energies of plasmonic catalysts. Nano Lett. 2016;16:3399–3407.
21. Kim Y, Smith JG, Jain PK. Harvesting multiple electron-hole pairs generated through plasmonic excitation of Au nanoparticles. Nat Chem. 2018;10:763–769.
22. Mukherjee S, *et al.* Hot electrons do the impossible: Plasmon-induced dissociation of H_2 on Au. Nano Lett. 2013;13:240–247.
23. Huynh LTM, Kim S, Yoon S. Effect of material and shape of nanoparticles on hot carrier generation. ACS Photonics. 2022;9:3260–3267.
24. Zhou L, *et al.* Quantifying hot carrier and thermal contributions in plasmonic photocatalysis. Science. 2018;362:69–72.
25. Sivan Y, Baraban J, Un IW, Dubi Y. Comment on "quantifying hot carrier and thermal contributions in plasmonic photocatalysis". Science. 2019;364:eaaw9367.
26. Jain PK. Taking the heat off of plasmonic chemistry. J Phys Chem C. 2019;123:24347–24351.

27. Baffou G, Bordacchini I, Baldi A, Quidant R. Simple experimental procedures to distinguish photothermal from hot-carrier processes in plasmonics. Light Sci Appl. 2020;9:108.
28. Chen X-J, Cabello G, Wu D-Y, Tian Z-Q. Surface-enhanced Raman spectroscopy toward application in plasmonic photocatalysis on metal nanostructures. J Photochem Photobiol C Photochem Rev. 2014;21:54–80.
29. Schürmann R, Ebel K, Nicolas C, Milosavljević AR, Bald I. Role of valence band states and plasmonic enhancement in electron-transfer-induced transformation of Nitrothiophenol. J Phys Chem Lett. 2019;10:3153–3158.
30. Koopman W, *et al.* The role of structural flexibility in plasmon-driven coupling reactions: kinetic limitations in the dimerization of Nitrobenzenes. Adv Mater Interfaces. 2021;8:2101344.
31. Schürmann R, *et al.* Plasmonic reactivity of halogen thiophenols on gold nanoparticles studied by SERS and XPS. J Chem Phys. 2022;157:84708.
32. Huh H, Trinh HD, Lee D, Yoon S. How does a plasmon-induced hot charge carrier break a C-C bond? ACS Appl Mater Interfaces. 2019;11:24715–24724.
33. Dutta A, *et al.* Plasmon-driven chemical transformation of a secondary amide probed by surface enhanced Raman scattering. Commun Chem. 2024;7:188.
34. Szczerbiński J, Metternich JB, Goubert G, Zenobi R. How peptides dissociate in plasmonic hot spots. Small. 2020;16:1905197.
35. Swaminathan S, Rao VG, Bera JK, Chandra M. The pivotal role of hot carriers in plasmonic catalysis of C-N bond forming reaction of Amines. Angew Chem Int Ed. 2021;60:12532–12538.
36. Sarina S, *et al.* Enhancing catalytic performance of palladium in gold and palladium alloy nanoparticles for organic synthesis reactions through visible light irradiation at ambient temperatures. J Am Chem Soc. 2013;135:5793–5801.
37. Verkaaik M, *et al.* Suzuki-Miyaura cross-coupling using plasmonic Pd-decorated Au nanorods as catalyst: A study on the contribution of laser illumination. ChemCatChem. 2019;11:4974–4980.
38. Ding T, Mertens J, Lombardi A, Scherman OA, Baumberg JJ. Light-directed tuning of plasmon resonances via plasmon-induced polymerization using hot electrons. ACS Photonics. 2017;4:1453–1458.
39. Fedoruk M, Meixner M, Carretero-Palacios S, Lohmüller T, Feldmann J. Nanolithography by plasmonic heating and optical manipulation of gold nanoparticles. ACS Nano. 2013;7:7648–7653.

40. Kameche F, *et al.* Probing plasmon-induced chemical mechanisms by free-radical nanophotopolymerization. J Phys Chem C. 2021;125:8719–8731.
41. Horrer A, *et al.* Local optical chirality induced by near-field mode interference in achiral plasmonic metamolecules. Nano Lett. 2020;20:509–516.
42. Saito K, Nemoto Y, Ishikawa Y. Circularly polarized light-induced chiral growth of achiral plasmonic nanoparticles dispersed in a solution. Nano Lett. 2024;24:12840–12848.

https://doi.org/10.1142/9781800618183_0005

Chapter 5

Multiscale Modelling and Characterisation of Nanosystems and Novel Materials

Alexey V. Verkhovtsev* **and Andrey V. Solov'yov**†
MBN Research Center,
Altenhöferallee 3, 60438 Frankfurt am Main, Germany
**verkhovtsev@mbnresearch.com*
†*solovyov@mbnresearch.com*

This chapter presents some illustrative examples of computational atomistic and multiscale modelling of novel materials and nanosystems using the advanced software packages MBN Explorer and MBN Studio. These software tools allow the computational design and characterisation of a wide variety of systems of different dimensionality, materials, and material interfaces, as well as the study of various physicochemical processes with such systems. The chapter presents two illustrative examples of the computational studies of such systems. One example concerns the characterisation of functionalised metal nanoparticles placed in molecular environments, which have been widely studied for possible applications in biotechnology. Another example concerns novel two-dimensional (2D) materials, which are currently being explored for their versatile electrical, optical, physical, chemical, and mechanical properties. An illustrative case study discussed here is a novel 2D material, $NiCl_2$, which may be useful for catalysis and hydrogen storage applications. The chapter provides an overview of the computational characterisation of such systems using a multiscale modelling approach combining density functional theory calculations and classical molecular dynamics simulations. Many other computational projects in this research area can be pursued using the algorithms and tools available in MBN Explorer and MBN Studio.

5.1 Introduction

Over the past decades, computational modelling of structural, mechanical, thermal, and transport properties of materials has been pursued through various methods of computational condensed matter physics and materials science.[1–4] Recent advances in this research area have often been achieved through computational multiscale modelling (MM) and high-performance computing.[1,5–7]

The most widely used approaches for the computational modelling of material properties are: (i) methods based on quantum mechanics, such as density functional theory (DFT),[8] (ii) the molecular dynamics (MD) approach,[1,2,9] which is widely used to study material properties and related physical phenomena on the nanometer spatial scale and on time scales from sub-nanoseconds to sub-microseconds, (iii) the kinetic Monte Carlo (kMC) and stochastic dynamics (SD) methods,[1,2,10,11] which provide a computationally efficient way to study the structural evolution and dynamics of systems on larger spatial and longer temporal scales through a probabilistic description of elementary dynamical processes, and (iv) continuum medium computational models, e.g., employing the finite element method (FEM).[12,13]

Computational MM plays a prominent role in developing the theoretical understanding of the connections between the atomic-level structure and the effective (macroscopic) properties of materials. For example, the MM approach can be used to test the mechanical models from continuum-based elasticity theory by comparing them with the results of MD simulations. In this way, the concepts from continuum theories, such as stress analysis using the stress tensor, elastic constants, and elastic moduli, can be re-examined at the atomistic level. Despite extensive studies in this field on different materials and their properties, many questions remain unanswered. This is reflected in the ongoing search for new materials and materials with improved or tailored properties by laboratories and research groups around the world. It is important that this work is carried out both experimentally and theoretically.

In recent decades, atomistic modelling has been at the forefront of computational studies of nanostructured and composite materials. Nanostructured materials are characterised by the characteristic sizes of their structural elements on the order of several hundred nanometres or less in at least one dimension. Composite materials are materials made from two or more constituent materials whose physical or chemical properties differ significantly from those of their constituent materials. The description of composite systems and materials is usually more complicated than that of homogeneous systems because composite systems involve different types of interactions within a single system, as well as interfaces between different materials or parts of the system. Atomistic modelling has revealed a wealth of information about the structure and properties of individual structural elements of such materials (e.g., various nanolayers, nanoparticles (NPs), nanofibres, nanowires, and nanotubes), as well as the characteristics of the interfacial regions and the modification of the material properties at the nanoscale.

This chapter presents some illustrative examples of computational atomistic and multiscale modelling and solutions to the above problems based on the use of the advanced software packages MBN Explorer[14] and MBN Studio.[15] These software tools allow the computational design and characterisation of a wide variety of materials and material interfaces,[1,2,16] as well as the study of different physical processes with such systems (e.g., nanoindentation, elastic and plastic deformations, and structural and phase transitions).[1,2,17–23]

5.2 Characterisation of Functionalised Metal Nanoparticles

Functionalised metal NPs have become a subject of intense research due to their potential applications in electronics, optics, chemistry, biosensing, imaging, drug delivery, and microfabrication.[24–26] Currently, there is a strong scientific interest in studying the interaction of nanoscale systems, such as atomic clusters and NPs, with biological media because of the large number of possible applications in nanomedicine.[27–29] One of the most promising ideas is the use of metal-based NPs or metal clusters in cancer treatments with ionising radiation.[30–33]

It has been suggested that such NPs have the ability to enhance the biological damage induced by energetic photon and ion-beam irradiation, i.e., to act as radiosensitisers.[30,34–36] When accumulated in the tumour region, these NPs can locally enhance the radiation damage of tumour cells compared to normal tissue, thereby increasing the efficiency of radiotherapy.

The radiosensitising effect of metal NPs is generally attributed to the strong radiation-induced emission of secondary electrons.[36–38] Secondary electrons induce radiolysis of the surrounding molecular medium (which consists largely of water) and thus increase the yield of hydroxyl ($OH^{\bullet}$) radicals and other reactive oxygen species that can produce damage to tumour cells.[39,40] Exploration of the nanoscale mechanisms underlying radiosensitisation by NPs is the subject of current efforts by numerous research teams worldwide.[41–44]

Metal NPs for radiotherapy and other biomedical applications are typically synthesised with an organic coating in order to reduce toxicity, improve stability under physiological conditions, and target specific biological sites.[45–47] A large number of possible coating molecules, including acids, polymers, sugars, and proteins, form a vast landscape of core–coating combinations.

A vast number of parameters can be varied to optimise the radiosensitising properties of NPs (such as the size, shape, and composition of a metal core; the thickness, composition, and density of a coating; and the composition and density of a molecular medium), making it a formidable task to systematically explore each of them experimentally. It is therefore desirable to develop a computational and theoretical framework for a quantitative atomistic description of the structural properties of coated metal NPs in biologically relevant environments, irradiation-induced effects (such as electron emission and production of reactive species), and the kinetics of irradiation-induced chemical reactions in the vicinity of NPs. Such descriptions based on the computational modelling approach may reduce the experimental costs for optimisation, control, and functionalisation of NPs in biomedical applications, thus facilitating significant progress in the field.[5]

In Ref. 48, a detailed atomistic approach for the computational modelling and structural characterisation of coated metal NPs in molecular media was presented using all-atom MD simulations with MBN Explorer[14] and MBN Studio.[15] As an illustrative and experimentally relevant case study, nanometre-sized gold NPs functionalised with thiol-poly(ethylene glycol)-amine ($S\text{-}(CH_2)_2\text{-}PEG_n\text{-}NH_2$), one of the coating materials frequently used in radiobiological experiments with metal NPs,[49–51] were considered.

In Ref. 48, the analysis was performed for 1.6 nm gold NPs coated with thiol-PEG$_n$-amine ($n = 2-11$) molecules and solvated in water at ambient conditions. The chosen size of the metal core was close to the size of the NPs studied experimentally (1.9 ± 0.8) nm.[49] In a follow-up study,[52] this analysis was extended to a larger, 3.7 nm, gold NP coated with thiol-PEG$_n$-amine ($n = 2 - 14$) ligands. These systems were chosen based on experimental observations that gold NPs smaller than 10 nm are typically coated with PEG chains containing less than 10 monomers.[45,46,53,54] The number of ligands attached to the gold core was varied from 8 to 60 for the 1.6 nm gold NP and from 44 to 300 for the 3.7 nm NP, corresponding to the range of ligand surface densities, $\sigma = 1.0 - 7.0$ molecules per nm^2. This range of ligand densities has been considered in the vast majority of experimental studies with gold NPs.[55]

In the MD simulations performed in Refs. 48,52, coated metal NPs were created by simulating a non-site-specific attachment of thiol-PEGn-amine molecules to the NP surface. The interaction between all the atoms of the metal core (including those located on the NP surface and in its interior) was modelled using the many-body Gupta potential. The covalent interaction between the gold atoms located on the metal core surface and sulphur atoms of the ligands was modelled using the Morse potential. Parameters of this interaction and of the non-covalent interactions between gold atoms and atoms of the ligands were validated through DFT calculations, as described in Ref. 48. All interatomic interactions within the thiol-PEG-amine ligands and their interactions with water molecules after solvation were described using the CHARMM molecular mechanics force field.

After the ligand attachment, the coated metal NPs were annealed and then introduced to an explicit water solvent. Then, the solvated systems were first optimised and then equilibrated at 300 K using the Langevin thermostat for 5 ns. Further details on the utilised computational procedure can be found in Refs. 48,52.

5.2.1 *Structural characterisation of coated nanoparticles*

As a first step of the analysis performed in Ref. 48, the structure of the thiol-PEG-amine coatings of different lengths, both annealed in vacuum and solvated in water, was analysed. Particular attention was

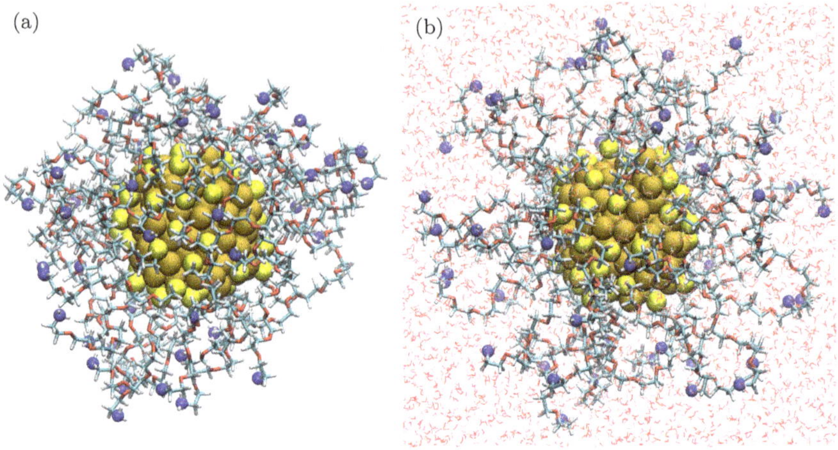

Fig. 5.1. Structure of a 1.6 nm gold NP coated with 48 thiol-PEG$_5$-amine molecules.[48] Panel (a): final structure obtained after annealing of the NP in a vacuum Panel (b): final structure obtained after solvation of the NP in water and equilibration of the whole system for 5 ns. Glossy yellow spheres in the central part of each panel represent sulphur atoms. Smaller blue spheres at the periphery of each panel represent nitrogen atoms.

Source: Reproduced from Ref. 48 with permission from the American Chemical Society.

paid to the analysis of the radial distribution of different atoms in the coatings. It was found that the structure of solvated systems differs significantly from the structure of free NPs annealed in vacuum; see Fig. 5.1.

Figure 5.2(a) shows the number density of all atoms in the coating made of 48 thiol-PEG$_8$-amine molecules for a solvated 1.6 nm NP and the same NP annealed in vacuum. The number densities are plotted as a function of radial distance from the centre of mass (CoM) of the gold core. In this analysis, the whole space with the origin at the CoM was divided into spherical bins of 0.1 nm width. The number of atoms of each type located in each bin was then calculated and divided by the volume of each bin.

Figure 5.2(a) shows a significant difference between the radial distributions of atoms in the dense thiol-PEG$_8$-amine coatings annealed in a vacuum and equilibrated in water. The atomic number density distribution for the free NP in a vacuum (the solid line) is characterised by two peaks located at 1.1 and 1.5 nm from the CoM. Each

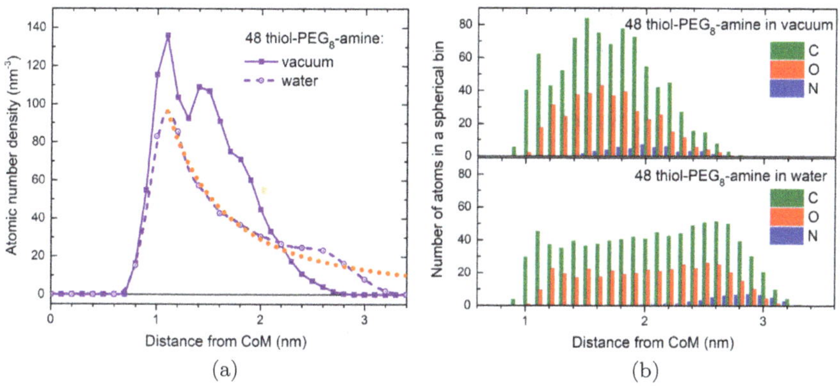

Fig. 5.2. Panel (a): number density $n(r)$ of all atoms in the coating made of 48 thiol-PEG8-amine molecules for NPs annealed in vacuum and solvated in water.[48] The number densities are plotted as a function of radial distance r from the centre of mass (CoM) of the gold core. The dotted orange line shows the $n(r) \propto r^{-2}$ dependence. Panel (b): the number of C and O atoms of the PEG backbone of thiol-PEG8-amine and the number of N atoms in the terminal NH_2 group located in spherical bins of 0.1 nm thickness. Carbon atoms (highest bars in green) and oxygen atoms (second-highest bars in red) are located at distances ranging from ~1 up to 3 nm from the CoM. Nitrogen atoms (shortest bars in blue) are located at distances of ~1.5–2.5 nm for the NP annealed in vacuum and at distances of 2.4–3 nm for the solvated NP.
Source: Reproduced from Ref. 48 with permission from the American Chemical Society.

of these peaks is attributed to the formation of a dense layer of ligand molecules. The first peak, located closer to the CoM, was observed for short and low-density coatings, whereas the second peak emerges with increasing coating density or ligand length.[48] The atomic number density distribution for the coating of the solvated NP (dashed line) spans over larger distances from the CoM. The maximum value of the atomic number density for the coating of the solvated NP is about 70% of the maximum density for the NP in a vacuum. This indicates that the thiol-PEG-amine ligands in water are loosely packed compared to the free NP.

Figure 5.2(b) shows the number of carbon and oxygen atoms in the PEG backbone and the number of nitrogen atoms in the terminal NH_2 group located in spherical bins of 0.1 nm thickness. In the solvated NP (see the lower panel of Fig. 5.2(b)), the number of carbon and oxygen atoms located in each spherical bin varies only slightly over the range of distances from ~1.1 to 2.5 nm from the CoM. The relatively uniform distribution of the PEG backbone atoms in the coating layer can explain the decrease in atomic number density for the solvated NP, which is approximately proportional to r^{-2}, where r is the distance from CoM (see the dotted orange line in Fig. 5.2(a)). This dependence arises because the surface area of a spherical bin increases quadratically with respect to r, while the thickness of each bin and the number of atoms located in it remain nearly constant. A shoulder in the number density distribution located at distances from 2.4 to 3 nm corresponds to the position of the terminal NH_2 group, as indicated by the distribution of nitrogen atoms in the lower panel of Fig. 5.2(b).

In polymer physics, conformational states describe the structures of polymer coatings attached to surfaces. There are two main conformational states, namely 'mushroom' and 'brush' conformations.[56–58] The mushroom conformation occurs when the ligands are loosely packed together and do not interact with each other. In this case, the ligands are strongly bent. Mushroom conformations typically occur at lower ligand surface densities. The brush conformation occurs at higher ligand densities when the ligands begin to interact with and repel each other. This interaction causes the ligands to straighten.[56,57,59,60] As elaborated in more detail in the following, the appearance of the shoulder seen in Fig. 5.3 is associated with a change between the conformation states of the coating molecules from mushroom-like to brush-like structures.[48]

The results of the atomistic simulations performed in Ref. 48 allow us to elucidate some common structural features of the thiol-PEG-amine coatings in an aqueous environment. Figure 5.3 shows the number density of all atoms in the selected thiol-PEG-amine coatings in water. Figure 5.3(a) shows how the atomic number density changes when the number of ligands attached to the metal core is varied; molecules containing eight ethylene glycol monomers (PEG_8) are considered as an example. Figure 5.3(b) shows the atomic number density for the coatings made of 32 molecules (corresponding to a ligand surface density of 4 nm^{-2}) of different lengths.

Figure 5.3(a) shows that the shape of each number density curve remains roughly the same for a given ligand length, regardless of the number of ligands. All number density curves peak at about 1.1 nm from the CoM, which is about 0.3 nm away from the surface

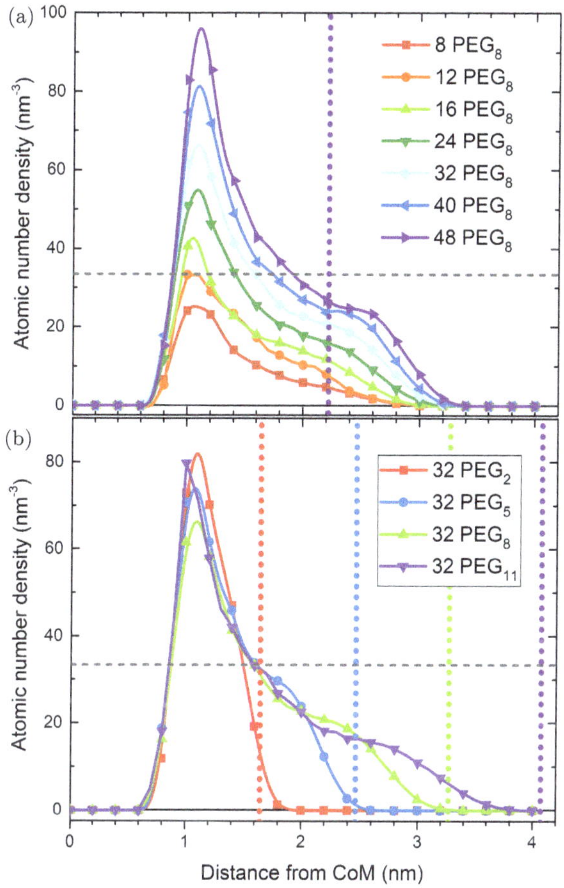

Fig. 5.3. Atomic number density for the thiol-PEG-amine coatings in water as a function of distance from the CoM of the gold core.[48] Panel (a): number density distribution for thiol-PEG$_8$-amine for different number of ligands attached. Panel (b): number density distribution for the coatings made of 32 ligand molecules with different number of ethylene glycol monomers. Horizontal dashed lines show the number density of water molecules in ambient water, $n_{\text{wat}} = 33.4\ \text{nm}^{-3}$.

Source: Reproduced from Ref. 48 with permission from the American Chemical Society.

of the metal core. This peak is broader for small numbers of ligands attached, while it becomes narrower and sharper as the coating density increases. A broader shoulder centred at about 2.5 nm distance from the CoM appears for denser coatings, as shown for the case of

48 thiol-PEG$_8$-amine molecules. The formation of such a shoulder indicates that molecules in dense coatings of solvated NPs tend to unfold into elongated brush-like structures; see Fig. 5.1. This effect has a geometrical explanation related to the dense packing of the ligands in the coating. The vertical dotted line in Fig. 5.3(a) shows the thickness of a spherical shell surrounding the gold core with a volume equal to the volume occupied by a single thiol-PEG$_8$-amine molecule in a vacuum, multiplied by 48. Due to electrostatic and van der Waals interactions between ligands and water molecules, such a dense packing of ligands is not possible, and the ligands are stretched further away from the metal core.

5.2.2 *Mushroom to brush transition with an increase in ligand surface density*

Figure 5.4 shows the radial distribution of nitrogen atoms in thiol-PEG$_n$-amine coatings with different lengths and surface densities of ligands. The number of nitrogen atoms located in spherical bins of 0.1 nm thickness was calculated and divided by the number of ligand molecules in a given coating. The overall probability of finding a nitrogen atom, integrated over the distance from the CoM, is thus equal to one. The results presented were averaged over 40 independent MD simulation frames. At ligand surface densities below 3 molecules per nm^2 (corresponding to up to 24 attached molecules), the coating layer represents a mixture of different conformational states. Nitrogen atoms in the low-density coatings are scattered over a wide range of distances from the CoM of the core; this effect is particularly evident for thiol-PEG$_2$-amine coatings (see Fig. 5.4(a)). The plot for the low-density coatings containing PEG$_5$ molecules (Fig. 5.4(b)) also shows a bimodal distribution for the position of the nitrogen atoms, whereas this effect becomes less pronounced for the coatings made of PEG$_8$ and PEG$_{11}$ molecules (Fig. 5.4(c) and Fig. 5.4(d), respectively). Nitrogen atoms are preferentially distributed at larger distances from the CoM, while small islands corresponding to a higher probability of finding a nitrogen atom are also clearly visible at about 1 nm distance from the CoM. The positions of the nitrogen atoms are close to the calculated average position of the sulphur atoms, $\sim$0.85 nm from the CoM. This suggests that low-density coatings made of shorter ligands have several different

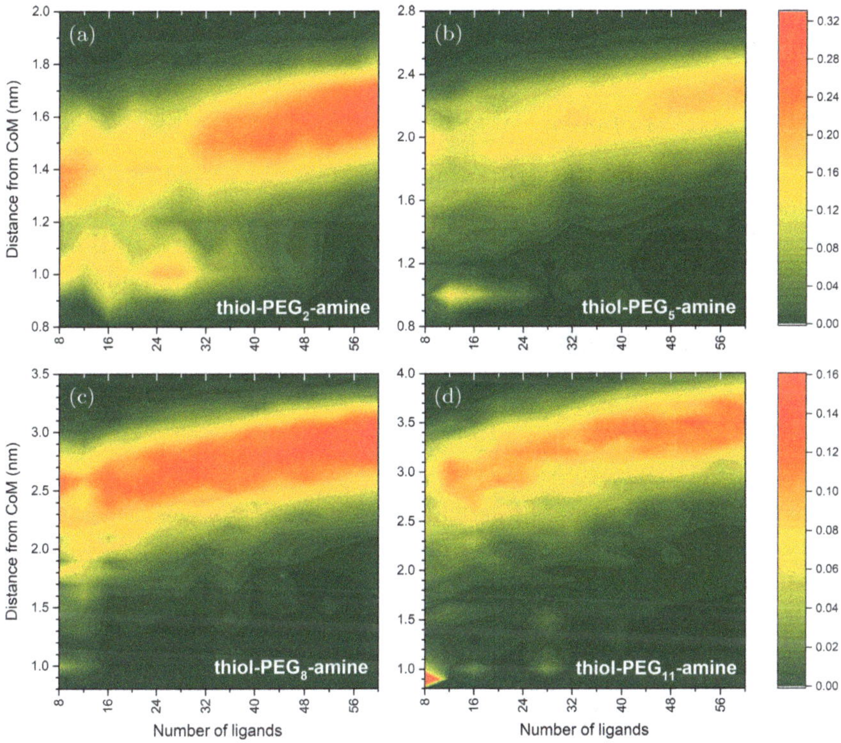

Fig. 5.4. Probability distribution of nitrogen atoms in thiol-PEG$_n$-amine ($n = 2, 5, 8, 11$) coatings made of 8–60 ligand molecules.[48] Several distinct conformational states of ligand molecules are revealed at low-density coatings containing up to 24 ligands; this effect is especially pronounced for the coatings made of shorter PEG$_2$ and PEG$_5$ molecules (panels (a) and (b), respectively). The nitrogen atoms in the low-density coatings are scattered over a wide range of distances from the CoM of the core, as shown by the yellow and orange islands in the probability distributions. As both the number of ligands in the coatings and the ligand length increase, the molecules acquire the elongated brush-like shape. In this case, the nitrogen atoms are preferentially distributed at larger distances from the CoM. The colour scale shows the probability of finding a nitrogen atom at a given distance from the CoM.

Source: Reproduced from Ref. 48 with permission from the American Chemical Society.

conformational states, namely molecules tightly curled around the core as well as elongated brush-like structures. As both the surface density and the length of the ligands increase, the molecules predominantly acquire the brush-like shape. This behaviour is evidenced by a gradual increase in the characteristic distance from the nitrogen atoms to the CoM with increasing ligand density.

5.2.2.1 *Radius of gyration*

Further insight into the structural properties and conformation of the ligands in thiol-PEG-amine coatings can be obtained by analysing the radius of gyration of the ligands.[52] The radius of gyration, R_g, is commonly used to characterise the size of a coiled polymer chain; it is defined as the root-mean-square distance of all atoms in the molecule from its CoM:[61]

$$R_g = \left[\frac{1}{M} \sum_i m_i \left(\mathbf{r}_i - \mathbf{r}_{\mathrm{CoM}} \right)^2 \right]^{1/2} . \tag{5.1}$$

Here, M is the mass of the ligand molecule and $\mathbf{r}_{\mathrm{CoM}}$ is the position vector of its CoM; m_i and $\mathbf{r}_i$ are the mass and position vector of a constituent atom i, respectively.

In Ref. 52, the radius of gyration for thiol-PEG$_n$-amine ($n = 2 - 14$) molecules was determined through complementary MD simulations of single thiol-PEG$_n$-amine molecules solvated in water and thermalised at 300 K. For each molecule, the radius of gyration was calculated according to Eq. (5.1) by averaging over 50 frames of the simulated trajectory. The calculated values of R_g for thiol-PEG$_n$-amine molecules of different sizes are shown in Fig. 5.5 by red squares.

The calculated values of R_g for thiol-PEG$_n$-amine ($n = 2 - 14$) molecules have been compared to the radii of gyration for PEG$_n$ molecules (also known as polyethylene oxide, PEO) reported in the literature. Figure 5.5 presents the comparison of the data[52] with those obtained from other MD simulations[62–65] and experimental data obtained from static light scattering measurements.[66] In the latter case, the scaling law of $R_g = 0.0202 M_{\mathrm{w}}^{0.55}$ between the radius of gyration (given in nanometres) and the molecular weight M_{w} of a polymer molecule was proposed.[66] The green crosses in Figure 5.5 describe the values of R_g calculated using the indicated scaling law for the exact molecular weights of thiol-PEG$_n$-amine ligands studied in Refs. 48,52.

Open blue symbols in Fig. 5.5 show the results of the study,[65] where the R_g values were determined for different PEO$_n$ ($n = 12, 20, 36, 45$) molecules on the basis of MD simulations,

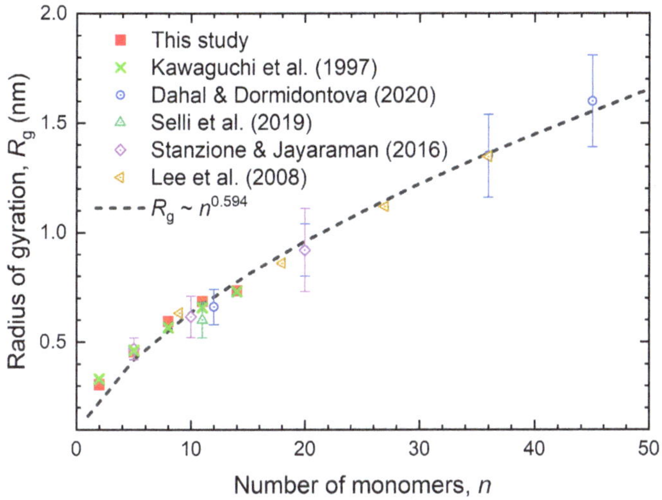

Fig. 5.5. The radius of gyration R_g of thiol-PEG$_n$-amine ligands calculated in Ref. 52 (red squares) and the radii of gyration for PEG$_n$ molecules (also known as polyethylene oxide, PEO) obtained from other MD simulations[62–65] (open symbols) and experimental data obtained from static light scattering measurements[66] (green crosses). The dashed line shows a power-law fit, $R_g \propto n^{\nu}$ with $\nu \approx 0.594$, for the results from Ref. 52 for $n = 2-14$ monomers and those reported in Ref. 65 for $n = 12$–45.
Source: Reproduced from Ref. 52 published under an open access Creative Commons CC BY license.

The figure shows the dependencies of R_g on the number of monomers n, obtained in Ref. 52 in the range of $n = 2 - 14$ and in Ref. 65 for $n = 12 - 45$, follow a similar trend. Fitting the combination of the calculated R_g values and those from Ref. 65 yields a power-law dependence $R_g \propto n^{\nu}$, where $\nu \approx 0.594$ (see the dashed curve in Fig. 5.5). This result is in agreement with the power-law dependence of $R_g \propto n^{0.549}$, determined through MD simulations for polyacrylic acid (PAA) polymers.[67] The exponent in these equations is known as the Flory exponent ν and defines the type of polymer chain described. In the case of a real polymer chain, $\nu = 3/5$,[68] which is close to the value $\nu = 0.594$ determined in Ref. 52.

5.2.2.2 *Ligand density for the mushroom to brush transition*

Using the values of R_g determined above, one can evaluate the ligand surface density of thiol-PEG$_n$-amine molecules at which the transition from mushroom- to brush-like structures occurs for the ligands of different sizes. According to the polymer theory,[60] at low ligand surface densities, $\sigma \lesssim 1/R_g^2$, the radius of gyration of the ligands does not exceed the spacing between the neighbouring ligands attached to a surface. In this case, individual ligands do not interact with adjacent ligands, thus acquiring mushroom-like shapes. Consequently, the coating thickness in this regime is equal to $L \approx 2R_g$.[56,60] When the mushroom-like structures begin to overlap (which happens at surface densities $\sigma \gtrsim 1/R_g^2$), a mushroom-to-brush transition occurs, and the ligands acquire brush-like shapes.

The transition ligand density between the mushroom and brush regimes has been estimated from the results of the performed MD simulations[48,52] by analysing the probability distributions of the terminal nitrogen atoms within the NP coatings, similar to the results plotted in Fig. 5.4. In this analysis, the probabilities of finding nitrogen atoms at distances $r < 2R_g$ and $r > 2R_g$ from the NP metal surface were summed up for various surface densities of the ligands.

In the mushroom regime, the height of the polymer coating is approximately equal to twice the radius of gyration, $h \approx 2R_g$. Hence, by taking the sum of the nitrogen atom probability in the range of radial distances $R_c \leq r \leq R_c + 2R_g$ (where R_c is the radius of the metal NP core), the transition ligand density has been evaluated for each particular case. For an estimate, it was assumed that the transition density corresponds to the density at which less than 10% of nitrogen atoms are located at distances $r < 2R_g$ and more than 90% of nitrogen atoms are located at $r > 2R_g$ from the metal core surface. Figure 5.6(a) illustrates this method for the case of the 1.6 nm gold NP coated with thiol-PEG$_8$-amine at different surface densities. The transition ligand density calculated this way for 1.6 and 3.7 nm NPs coated with thiol-PEG$_n$-amine ligands of different lengths is plotted in Fig. 5.6(b). The discrepancy between the data points of the estimated transition surface density and the $1/R_g^2$ dependence (see the dashed curve) can be attributed to the nature of this transition. As shown in Fig. 5.4, at lower surface densities, the coating structure is not exclusively mushroom but rather a mixture of both mushroom- and brush-like conformations.

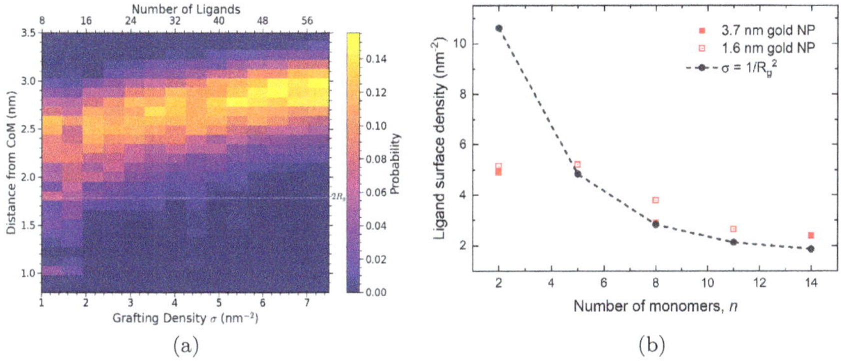

Fig. 5.6. Panel (a): plot of the probability distribution of nitrogen atoms located at a given distance from the NP CoM for various ligand surface densities/number of thiol-PEG$_8$-amine ligands. Twice the gyration radius is indicated by a dashed line to show the approximate mushroom-like structure height. Panel (b): estimated transition ligand density for 1.6 and 3.7 nm NPs coated with thiol-PEG$_n$-amine ligands of different lengths (closed and open red symbols). The dashed curve with black symbols shows the $\sigma = 1/R_g^2$ dependence, where the radius of gyration R_g for the ligands containing n monomer units has been calculated, as plotted in Fig. 5.5.

Source: Reproduced from Ref. 52 published under an open access Creative Commons CC BY license.

5.2.3 *Distribution and structural properties of water around coated NPs*

The computational framework described above provides the detailed information necessary for a quantitative atomistic description of the structural properties of coated metal NPs in biological environments. It can thus provide a basis for a more detailed atomistic analysis of the irradiation-induced transformations and chemical reactions in the vicinity of NPs.

The atomistic MD simulations performed in Refs. 48,52 provided valuable insight into the water content and spatial distribution of water molecules in the vicinity of coated metal NPs. An increase in water content near the surface of the NP has been observed at low densities of the coating, as shown in Fig. 5.7. When the metal core is covered by a small number of ligands, a significant part of the metal surface is open to the surrounding water. Due to strong van der

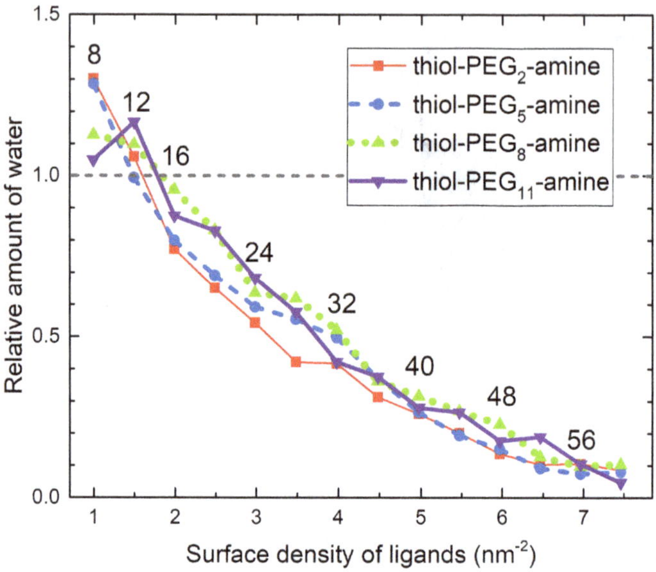

Fig. 5.7. Relative change in the content of water around the 1.6 nm gold NP as a function of ligand length and the surface density of ligands.[48] The corresponding numbers of attached ligands are also indicated. The calculated number densities are normalised to the ambient number density of water (shown by the horizontal dashed line). An increase in the water content by 10–30% is seen at low-density ligands made of up to 12 molecules.

Source: Reproduced from Ref. 48 with permission from the American Chemical Society.

Waals interactions between gold atoms and oxygen atoms of water molecules, the latter become attracted to the core and form a dense monomolecular water layer near the NP surface.[48] The maximum number density of water molecules in the thiol-PEG$_2$-amine coating is $\sim$43 nm^{-3}, which is about 30% higher than the ambient water density of $\sim$33.4 nm^{-3}.

An increase in water content in close proximity to the metal core may have an important effect on the production of hydroxyl radicals around the NP. As shown in Refs. 36,38, irradiation of nanometer-sized metallic NPs with ions at Bragg peak energies induces the formation of collective electronic (plasmon) excitations in the NPs. Decay of these collective electron excitations results in a strong emission of low-energy electrons with kinetic energy of a few electronvolts. Electrons of such energies can travel short distances (several Ångstroms) from the surface of the NP.[69] An increase in water density near the surface may thus increase the probability of electron interactions with water molecules (e.g., dissociative electron attachment) and the subsequent formation of hydroxyl radicals.

Figure 5.7 shows the number density of water molecules in a 0.1 nm thick spherical shell in proximity to the metal core for different coating densities and ligand lengths. The calculated number densities are normalised to the ambient number density of water, 33.4 nm^{-3}. As shown in the figure, an increase in the amount of water has been observed for the low-density coatings of different length (with the density up to 1.5 molecules per nm^2). The opposite effect has been observed for denser coatings where the amount of water near the NP surface is significantly reduced. When the ligand density is above 6 molecules per nm^2, the number of water molecules in the coating region around the metal surface is 5–10 times lower than the number of water molecules in ambient water. Interestingly, this result is almost independent of the length of ligands for the molecules considered in Ref. 48. A reduced amount of water in close proximity to the metal core will suppress the production of OH radicals by low-energy electrons emitted from the metal NP. High-density coatings (even made of short molecules like thiol-PEG_2-amine) may thus impair the radiosensitising properties of metal NPs with regard to this molecular mechanism of radiosensitisation.

The computational framework employed in Refs. 48,52 provides a basis for future studies of the chemical effects arising from metal NPs, including chemical modifications of the ligands due to low-energy electrons emitted from the metal core and the kinetics of irradiation-induced chemical reactions in the vicinity of NPs. The first step in this direction was made in Ref. 69, which presented a novel theoretical and computational approach to analyse the formation and transport of secondary electrons and hydroxyl radicals after irradiation of coated metal NPs with ions. It was the first atomistic study that accounted for the detailed atomistic structure of the coating material around the metal core[48,70] and also accounted for the relevant low-energy and many-body phenomena[36,38,71] in the formation and transport of secondary electrons and free radicals after irradiation of coated metal NPs.

The presented computational approach is general and not restricted to this particular system, thus providing a convenient framework for studying any core–coating combination. A range of metal cores of different composition and size (including highly symmetric ligand-protected gold clusters,[72–74]) coatings (e.g., PEG molecules conjugated with other functional groups,[75] or other ligands with hydrophilic or hydrophobic properties) and environmental conditions could be explored more systematically and compared with experimental observations.

5.3 Characterisation of Novel 2D Materials

Another illustrative example of advanced computational modelling and characterisation of novel materials using MBN Explorer and MBN Studio concerns the study of novel two-dimensional (2D) materials.

Currently, the need to develop novel, sustainable, and cost-effective materials with specific properties has increased significantly. For example, the successful discovery and isolation of graphene,[76] as well as the subsequent synthesis of layered semiconductors[77,78] and heterostructures beyond graphene[79–81] have led to a surge in the study of 2D materials.[82] These materials have attracted attention for their potential use in various technological applications, including optoelectronics, spintronics, sensors, and superconductors, as well as thermoelectric, photoelectric, and energy storage devices.

Among the many different 2D materials, those containing transition metals have been widely studied due to their versatile and unique electrical, optical, physical, chemical, and mechanical properties.[83,84] A particular subset of such systems are nickel-containing 2D materials,[85–88] which exploit the remarkable properties of nickel as a catalyst.[89] For instance, Ni-doped graphene materials provide excellent sensing properties[90–92] and can effectively catalyse technologically important reactions, such as hydrogen production from water[93] and ethanol vapour.[94] Another important application of Ni-doped graphene is hydrogen storage.[95–97] Due to the $3d$ electrons and the spillover effect, nickel can hold hydrogen molecules. At the same time, graphene is not the only 2D material that can be doped with nickel.

Other 2D materials, such as silicene, germanene, and MoS_2, have also been used as substrates for nickel atoms and nanoparticles.[98–100]

The aforementioned properties of Ni-doped 2D materials are based on a combination of the mechanical characteristics of a 2D sheet with the adsorption and catalytic properties of nickel. A further development of this idea is that nickel can be not only an alloying impurity but also the basis for a 2D material. Pure nickel does not exist as a 2D allotrope. However, an extensive computational search for layered crystals and related 2D materials[101] has predicted the existence of several 2D nickel-based materials of the composition NiX_2 (where X = O, Cl, Br, I, S, or Se). Some of these materials have already been synthesised. Technologies developed for the widespread 2D MoS_2 and $MoSe_2$ materials[102] and their analogues based on niobium,[103] titanium,[104] and tungsten[105] have been proved useful for the synthesis of NiX_2 monolayers. Large-area NiO_2 monolayers separated by lanthanum atoms have been observed using scanning tunnelling microscopy.[106] Recent experiments demonstrated the feasibility of chemical vapour deposition synthesis of nanometre-sized layers of $NiCl_2$,[107] NiS_2,[108] and $NiSe_2$.[109] In addition, $NiCl_2$ films with up to four layers were synthesised recently.[107] In all these materials, nickel atoms are organised in a regular 2D lattice, giving the materials enhanced adsorption and catalytic properties.

At present, there is limited experimental information on the properties of such 2D materials, as only a few laboratories have synthesised them in limited quantities. Atomistic computer simulations can serve as a useful complementary approach to characterise the structure of such novel materials and explore or predict their properties.

This chapter gives a brief overview of a recent computational characterisation of a novel 2D material, $NiCl_2$, using a MM approach.[23] combining DFT calculations and classical MD simulations DFT calculations provide reference data on the geometrical characteristics of the 2D $NiCl_2$ material. On this basis, a new classical interatomic potential has been developed and benchmarked against the results of quantum-mechanical calculations. The validated potential has been used for the atomistic modelling of the mechanical and thermal stability of $NiCl_2$ using classical MD using the MBN Explorer[14] and MBN Studio[15] software packages.

Through an illustrative case study of $NiCl_2$, a general methodology is presented that can be used for the computational characterisation of other novel materials, including the recently synthesised 2D NiO_2,[106] NiS_2,[108] and $NiSe_2$.[109] Apart from Ref. 23, 2D $NiCl_2$ has only been studied computationally in a few studies.[110,111] Ref.[110] investigated the mechanical, electronic, and magnetic properties of monolayer and bilayer NiX_2 (X = Cl, Br, or I) 2D structures using DFT. The recent study in Ref.[111] investigated structural defects in $NiCl_2$ and similar materials and confirmed their stability in the environment, also using DFT calculations.

5.3.1 *Computational methodology*

The combination of quantum and classical approaches into a unified multiscale methodology allows a comprehensive investigation of the structure, mechanical properties, and thermal stability of materials, in particular the 2D $NiCl_2$ considered here as a case study.

5.3.1.1 *DFT calculations*

DFT calculations of the structural properties of a $NiCl_2$ sheet have been performed in Ref. 23 using the QUANTUM Espresso 6.5 software package.[112,113] The plane-wave basis set for valence electron states, generalized gradient approximation (GGA) in the Perdew–Burke–Ernzerhof (PBE) functional form for the exchange-correlation energy[114] and the projector-augmented-wave (PAW) pseudopotentials[115,116] for core–electron interactions were used to perform the calculations. The kinetic energy cutoff for wave functions of 100 Ry (1360 eV) and that for charge density and potential of 600 Ry (8160 eV) were employed. In addition, the van der Waals interactions were considered through the D3 Grimme (DFT-D3) dispersion corrections.[117] The DFT-D3 approach possesses improved accuracy due to the use of environment-dependent dispersion coefficients and the inclusion of a three-body component in the dispersion correction energy term.

The distance between separate $NiCl_2$ sheets was set equal to 30 Å, which provides sufficient space separation to avoid non-physical interactions. Thus, optimisation of the lattice parameter along the axis perpendicular to the $NiCl_2$ sheet plane was unnecessary. The atomic equilibrium positions were obtained by the total energy minimisation

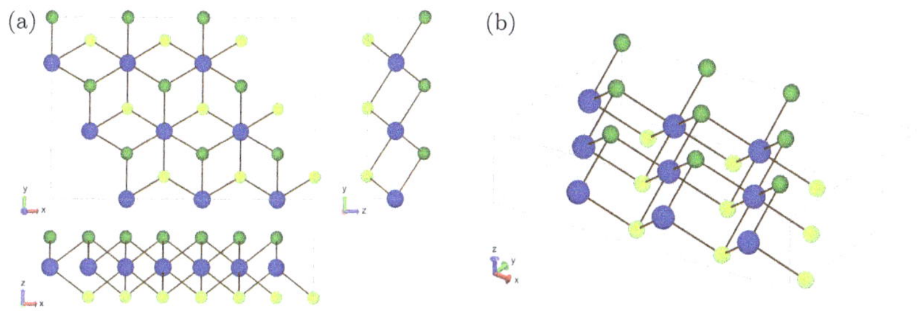

Fig. 5.8. Structure of a $NiCl_2$ supercell used in the DFT calculations.[23] The supercell contains 9 nickel atoms (shown as the larger blue spheres) and 18 chlorine atoms (shown as the smaller light- and dark-green spheres). Chlorine atoms in the top and bottom atomic layers are shown by dark-green and light-green colours, respectively. Panel (a) shows three orthographic projections of the considered supercell: top and side views (top row) and a front view (bottom row). Panel (b) shows a three-dimensional projection of the supercell.

Source: Reproduced from Ref. 23 with permission from Elsevier.

of the supercell using the calculated forces and stress on the atoms. The convergence criterion for self-consistent calculations for ionic relaxations was set to 10^{-10} eV between two consecutive steps. The geometry optimisation of the $NiCl_2$ sheet was carried out without symmetry constraints until the Hellman–Feynman forces acting on the atoms became smaller than 10^{-4} hartree/bohr. Such criteria ensure that the absolute value of stress is less than 0.01 kbar. The parameters of the supercell were also optimised. The first Brillouin zone integrations were performed using the Monkhorst–Pack k-point sampling scheme[118] with a $6 \times 6 \times 1$ mesh grid and the Methfessel–Paxton smearing[119] with a smearing width of 0.02 Ry.

The $NiCl_2$ sheet is represented by a hexagonal Ni_9Cl_{18} periodic cell containing 3×3 elementary $NiCl_2$ cells; see Fig. 5.8. The hexagonal symmetry corresponds to the results obtained earlier for this monolayer.[101] The lattice constant determined from the geometry optimisation calculation is $a = 3a_0 = 10.328$ Å, where a_0 is the parameter of the $NiCl_2$ primitive unit cell. The equilibrium Ni–Cl bond length obtained from the DFT calculations is equal to 2.38 Å.

5.3.1.2 *Classical MD simulations*

Classical geometry optimisation calculations and MD simulations have been performed in Ref. 23 by means of the MBN Explorer software package.[14] MBN Explorer permits the simulation of a wide range of MesoBioNano (MBN) systems,[1,2] including nanosystems, nanostructured materials, and composite and hybrid materials with sizes ranging from the atomic to the mesoscopic[18,120–124] The dedicated MBN Studio toolkit[15] has been used to create the systems, prepare all necessary input files, and also to analyse and visualise simulation outputs.

5.3.2 *Determination of the classical force field for $NiCl_2$*

The 2D $NiCl_2$ material has not been studied earlier using classical MD. Therefore, the first part of the study[23] was devoted to the determination of the parameters of a classical force field for $NiCl_2$. The geometry of the $NiCl_2$ supercell obtained after DFT-based optimisation (see Fig. 5.8) was taken as the initial geometry for the calculations using the classical force field. The $NiCl_2$ supercell was translated in space along the translation vectors, and the generated crystal was cut to fill in a rectangular simulation box, which was replicated in space using periodic boundary conditions. The resulting structure of a 2D $NiCl_2$ system is shown in Fig. 5.9.

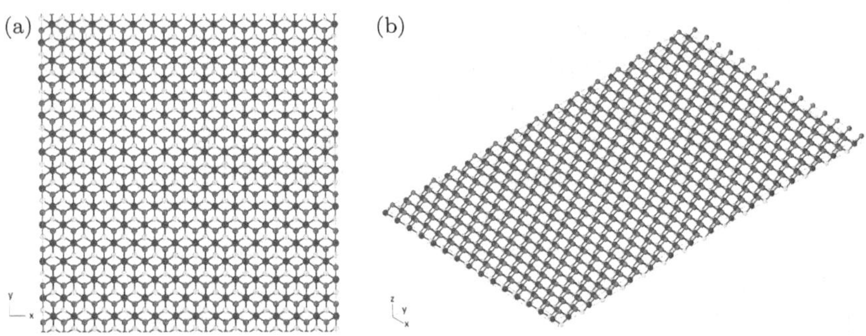

Fig. 5.9. Structure of a $NiCl_2$ sheet used in classical MD simulations performed in Ref. 23. Panel (a) shows the top view of a fragment of a $NiCl_2$ sheet. Panel (b) shows a three-dimensional projection of the entire system containing 1,350 atoms. Ni and Cl atoms are shown in blue and green colours, respectively. Cl atoms in the top and bottom atomic layers are shown by dark-green and light-green colours, respectively.

Source: Reproduced from Ref. 23 with permission from Elsevier.

The simulation box used for the classical simulations has a size of $52.08 \times 90.30 \times 100.00$ Å^3 and contains 1,350 atoms.

The force field considered in the simulations[23] was constructed as the sum of the short-term exponential potential, the long-term power potential, and the Coulomb potential. Such force fields have been widely used in MD simulations of metal halides and oxides, as well as various ionic crystals and other ionocovalent systems (e.g., glasses).[125–128]

$NiCl_2$ is an ionic crystal in which polarisation effects due to the ions of opposite charge may play an important role. Several computational approaches to describe the ion-induced polarisation interaction in various molecular systems within the classical MD framework have been discussed in the literature, including the addition of a r^{-4} term to a r^{-6} long-range power potential[129–131] or the consideration of a r^{-4} term as the only contribution to the long-range interaction.[132,133] The latter approach was used in Ref. 23 to minimise the number of force field parameters to be determined. The resulting total interaction potential between atoms i and j of the system is given by the following expression:

$$U(r_{ij}) = A_{ij}\, e^{-\alpha_{ij} r_{ij}} - \frac{C_{ij}}{r_{ij}^4} + \frac{q_i q_j}{4\pi\varepsilon_0 r_{ij}} \,. \tag{5.2}$$

Parameters of the force field for the 2D $NiCl_2$ system were determined using the Nelder–Mead optimisation algorithm.[134] This algorithm is simplex-based, meaning that in a n-dimensional space, it maintains a set of $n + 1$ points called simplex. After calculating the value of the function at each point of the simplex, the algorithm extrapolates the function's behaviour to find a new test point and replaces the test point with the worst result with a new one. This process continues until the result converges to the predefined tolerance value or reaches the maximum number of iterations. The Nelder–Mead algorithm was implemented in Python via the scipy.optimize.minimize library.[135] This algorithm was used to fit the geometrical parameters of $NiCl_2$ calculated using MBN Explorer[14] to the corresponding parameters obtained through the DFT-based structure optimisation. Results of this analysis are summarised in Section 5.3.3.1.

Six parameters of the force field, Eq. (5.2), have been optimised in Ref. 23: $A_{\rm Ni-Cl}$, $\alpha_{\rm Ni-Cl}$, $A_{\rm Cl-Cl}$, $\alpha_{\rm Cl-Cl}$, $C_{\rm Cl-Cl}$ and $q_{\rm Cl}$. Partial charges on Ni and Cl atoms were set to ensure that the system is electrically neutral, so that $|q_{\mathrm{Ni}}| = 2|q_{\mathrm{Cl}}|$. Following the approach adopted in several earlier studies[136–140] for a number of ionic crystals and ionocovalent systems, the non-Coulombic contributions between the positively charged (in our case, nickel) ions were not considered, i.e., $A_{\mathrm{Ni-Ni}} = 0$ and $C_{\mathrm{Ni-Ni}} = 0$. Following the considerations made in Ref. 137, the large partial atomic charge on the cations ensures that they are always well separated from each other, and their repulsive and van der Waals interactions can be neglected. The parameter $C_{\mathrm{Ni-Cl}}$ for the long-range interaction between the ions of different signs was also set equal to zero (e.g., following Refs. 140–142) on the grounds that unlike (i.e., Ni and Cl) nearest neighbours are close enough for the van der Waals interaction to be quenched.[136,137] A 7 Å cutoff was applied for the exponential and power terms of the potential. The Coulomb potential was calculated using the Ewald summation method[143,144] with a cutoff of 12 Å.

The following geometrical parameters of $NiCl_2$ were compared with the results of the DFT calculations: (i) Ni–Cl bond length ($d_{\mathrm{Ni-Cl}}$), (ii) Ni–Ni bond length ($d_{\mathrm{Ni-Ni}}$), (iii) the in-plane Cl–Cl bond length ($d_{\mathrm{Cl-Cl}}$), and (iv) the interplanar Cl–Ni–Cl angle ($\varphi_{\mathrm{Cl-Ni-Cl}}$), hereafter referred to as φ for brevity. In addition, the energy of the system was analysed as a function of the Ni–Cl interplanar distance l. The distance between the Ni and Cl atomic planes was varied stepwise, and single-point energy calculations were performed using MBN Explorer for different values of l. The resulting dependence of the system's energy on the Ni–Cl interplanar distance was fitted by a quadratic function. The value of l corresponding to the system's minimum energy was determined from classical force field calculations and compared with the results of the DFT calculations. The standard deviation

$$\sigma = \sqrt{\sum_i \left(\frac{d_i - d_i^{\mathrm{DFT}}}{d_i^{\mathrm{DFT}}}\right)^2 + \left(\frac{\varphi - \varphi^{\mathrm{DFT}}}{\varphi^{\mathrm{DFT}}}\right)^2 + \left(\frac{l - l^{\mathrm{DFT}}}{l^{\mathrm{DFT}}}\right)^2} \tag{5.3}$$

was then calculated and minimised using the Nelder–Mead algorithm.[134] The summation was performed over all Ni–Ni, Ni–Cl, and Cl–Cl covalent bonds of the system.

The procedure for determining the force field parameters for 2D $NiCl_2$ consisted of two consecutive steps. First, a series of classical

MD simulations with a simulation time of 30 ps and a time step of 1 fs, a Langevin thermostat temperature of 300 K, and a damping time of 0.1 ps were performed for each generated parameter set. Each round of force field parameter optimisation consisted of 30 iterations, and the parameters resulting from each optimisation round were used as the starting parameters for the following round. In total, 400 consecutive optimisation rounds were performed.

Due to relatively long simulation times, the MD-based parameter optimisation procedure is inefficient for scanning a multidimensional parameter surface over a wide range of parameters. Therefore, the second step of force field parameter optimisation using the Nelder–Mead algorithm[134] was performed through a series of classical structure optimisation calculations. The parameters generated in the previous step through MD simulations were used as a starting point. The classical structure optimisation calculations were performed with MBN Explorer using the velocity quenching algorithm for 5,000 optimisation steps with a time step of 0.1 fs.

The much shorter computational time of a structure optimisation calculation compared to an MD simulation allows a wider range of force field parameter values to be explored. However, according to the observations made in Ref. 23, the optimisation of all force field parameters, including partial charges, leads to unrealistic non-physical values of partial charges (much larger than the formal charges of +2 and −1 on Ni and Cl atoms, respectively) due to the decrease in the total potential energy of the system with an increase in the attractive electrostatic interaction between positively charged Ni atoms and negatively charged Cl atoms. In order to avoid this non-physical behaviour, the partial charges on Ni and Cl were considered in Ref. 23 as fixed parameters, ensuring that the total charge of the whole system is zero. Ten independent rounds of force field parameter optimisation were performed with constant values of atomic partial charges q_{Cl} in the range from $-1.0|e|$ to $-0.5|e|$. The force field parameters obtained after each round of structure optimisation calculations were tested by analysing the stability and planarity of the $NiCl_2$ system over a one-nanosecond-long MD simulation at 300 K. The final choice of the atomic partial charges was made to obtain the closest agreement with the DFT-based results on the geometrical and mechanical properties of the 2D $NiCl_2$ system, as described in Sections 5.3.3.1 and 5.3.3.2.

Table 5.1. Parameters of the force field, Eq. (5.2), derived in Ref. 23 for the description of a 2D $NiCl_2$ material.

	A_{ij} (eV)	α_{ij} (Å^{-1})	C (eV Å^4)		q, $\|e\|$
Ni–Cl	10715.83	5.034	–	Ni	1.34
Cl–Cl	1439.08	3.494	0.779	Cl	−0.67

It should be noted that the question of the choice of atomic partial charges for the description of ionic crystals and other ionocovalent systems (such as silicate glasses) has been discussed in the literature; see, e.g., Ref. 145 and references therein. MD simulations of SiO_2 systems have shown that the use of formal charges (i.e., +4 for Si and −2 for O atoms) requires the use of additional angular three-body energy terms to properly describe the geometry of the system.[146] Therefore, a commonly adopted approach for the simulation of ionocovalent systems is to use partial (rather than formal) atomic charges, which are treated as an additional fitting parameter[138,139,147] in order to reproduce the correct geometry of the system without introducing additional energy terms in the empirical force field. This approach was used in Ref. 23 to determine the partial charges on nickel and chlorine atoms.

The final values of the force field parameters and atomic partial charges for $NiCl_2$, derived from the comparison of the $NiCl_2$ geometrical parameters with the DFT results, are listed in Table 5.1. These values have been used in all the classical (geometry optimisation and MD) simulations described in Section 5.3.3.

5.3.3 *Results and discussion*

5.3.3.1 *Evaluation of the accuracy of the constructed force field*

$NiCl_2$ is a novel 2D material for which little experimental and computational information has been obtained. Thus, it is important to evaluate the accuracy of the developed classical force field to make computational predictions of the structural and thermomechanical properties of $NiCl_2$.

Table 5.2. Comparison of geometrical parameters of a $NiCl_2$ 2D sheet after geometry optimisation using the classical force field, Eq. (5.2) (column labelled 'FF'), and DFT using the PBE exchange-correlation functional. The results obtained in Ref. 23 are also compared with the results of recent DFT calculations.[110,111,148]

	FF		DFT		
Optimisation method	Ref. 23		Ref. 111	Ref. 110	Ref. 148
$d_{\text{Ni-Cl}}$, Å	2.24	2.38	2.38	2.42	2.42
$d_{\text{Ni-Ni}}$, Å	3.57	3.44	–	–	3.56
$d_{\text{Cl-Cl}}$ (in-plane), Å	3.47	3.44	–	–	–
$l_{\text{Cl-Cl}}$ (interplanar), Å	2.68	3.28	–	2.65	–
$\varphi_{\text{Cl-Ni-Cl}}$, deg	73.5	87.2	–	–	–

Table 5.2 lists the values of the Ni–Cl, Ni–Ni, and Cl–Cl bond lengths ($d_{\text{Ni-Cl}}$, $d_{\text{Ni-Ni}}$, and $d_{\text{Cl-Cl}}$), as well as the Cl–Cl interplanar distance ($l_{\text{Cl-Cl}}$) and the Cl–Ni–Cl angle ($\varphi_{\text{Cl-Ni-Cl}}$) obtained in Ref. 23 by means of classical force field optimisation (column labelled 'FF') and DFT optimisation calculations using the PBE exchange-correlation functional. The results obtained are compared with several recent DFT calculations for a free 2D $NiCl_2$ monolayer[110,111] and that encapsulated inside a metal–organic framework.[148]

The results listed in Table 5.2 show a good overall agreement (with the relative discrepancy of ∼5%) between the results of the classical structure optimisation calculations and the DFT calculations performed in Ref. 23 and reported in the literature.[110,111,148] The vertical interlayer distance between two chlorine planes, $l_{\text{Cl-Cl}}$, obtained from the classical calculations is about 20% shorter than that obtained by DFT, while it is in good agreement with the results of the recent DFT-based study.[110] A smaller Cl–Cl interplanar distance results in smaller values of the angles $\varphi_{\text{Cl-Ni-Cl}}$ compared to the results of the DFT calculations.

Overall, the classical force field, Eq. (5.2), with the parameters listed in Table 5.1, describes the geometrical parameters of a 2D $NiCl_2$ sheet reasonably accurately. The range of discrepancies between the results of classical force field-based and DFT-based geometry optimisation calculations is similar to that found in previous studies of transition-metal clusters and bulk materials.[18,149]

5.3.3.2 *Mechanical properties of $NiCl_2$*

The mechanical properties of a $NiCl_2$ sheet were analysed in Ref. 23 using both DFT and classical force field calculations. The DFT-based calculations used as input the optimised $NiCl_2$ supercell, which has been described above in Section 5.3.1.1.

The uniform stretching of the $NiCl_2$ sheet was simulated by simultaneously increasing the lattice parameters of the supercell along the x- and y-directions (parallel to the $NiCl_2$ plane) and further optimising the atomic positions in the supercell. Classical geometry optimisation calculations were performed using MBN Explorer[14] with the velocity quenching algorithm for 20,000 optimisation steps and a time step of 1 fs.

During the biaxial stretching, the lattice constant a increases as

$$a = a_0(1 + \varepsilon) \ , \tag{5.4}$$

where a_0 is the equilibrium lattice constant and ε is the relative deformation (strain). The deformation energy ΔE was calculated as follows:

$$\Delta E = E - E_0 \ , \tag{5.5}$$

where E is the total potential energy of the deformed system and E_0 is the potential energy of the unperturbed system. As the simulation boxes considered in the DFT-based and classical force field optimisation calculations contain different numbers of atoms, the calculated deformation energy was divided by the number of $NiCl_2$ units in the respective simulation box.

Figure 5.10(a) shows the dependence of the $NiCl_2$ sheet deformation energy on the strain ε. In the region of small deformations (for strain values $\varepsilon \lesssim 0.05$), the dependence of deformation energy follows a quadratic dependence on ε, indicating elastic deformation. For larger deformations, the calculated dependence begins to deviate from the quadratic dependence. The calculations using the classical force field, Eq. (5.2), predict a deformation threshold at $\varepsilon \sim 0.11$, when the 2D sheet breaks into smaller 'islands' (see Fig. 5.10(b)), causing a sharp drop in the potential energy of the system and, hence, in its deformation energy.

The mechanical properties of 2D materials are commonly characterised by the 2D Young's modulus E_{2D}, which, unlike the

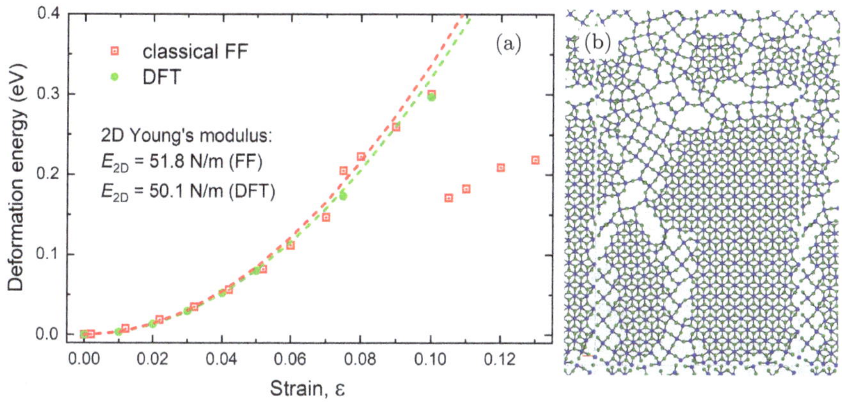

Fig. 5.10. Panel (a): Deformation energy ΔE of a biaxially stretched 2D $NiCl_2$ sheet as a function of relative deformation ε. Classical geometry optimisation calculations (red symbols) have been performed using the force field, Eq. (5.2), and the parameters listed in Table 5.1. Panel (b): Structure of $NiCl_2$ at strain $\varepsilon = 0.11$, which corresponds to the mechanical stability threshold for this material as predicted by the classical force field calculations. Ni and Cl atoms are shown in blue and green colours, respectively.

Source: Reproduced from Ref. 23 with permission from Elsevier.

'conventional' 3D Young's modulus, is independent of the sheet thickness.[150,151] The 2D Young's modulus is calculated as

$$E_{2D} = \frac{1}{2S_0} (\Delta E)'' , \tag{5.6}$$

where S_0 is the surface area of the relaxed 2D system and $(\Delta E)''$ is the second derivative of the deformation energy with respect to the biaxial deformation ε. A quadratic fit to the dependence of the total deformation energy of the system on ε, calculated by means of DFT, gives the value $E_{2D} = 50.1$ N/m. The classical force field, Eq. (5.2), predicts a close value of 51.8 N/m. These values are in agreement with the value $E_{2D} = 54$ N/m, obtained for $NiCl_2$ through DFT calculations in Ref. 110. Similar values of the 2D Young's modulus were reported in the cited study[110] for related Ni-based 2D materials, $NiBr_2$ (50 N/m) and NiI_2 (45 N/m). These results indicate that 2D $NiCl_2$ and other nickel dihalides are less rigid materials compared to other well-studied 2D sheets such as graphene or MoS_2, which have 2D Young's modulus of ~340 N/m and 170 N/m, respectively.[152,153]

5.3.3.3 *Thermal stability, melting, and solidification of $NiCl_2$*

The interatomic force field described in Section 5.3.2 was also used in Ref. 23 to study the thermal properties of a 2D $NiCl_2$ sheet and to evaluate its melting temperature using classical MD. The following computational procedure was used. The optimised structure of $NiCl_2$ was gradually heated by 50 degrees (in the temperature range from 0 to 650 K) or by 100 degrees (in the range from 800 to 3200 K) over 200 ps and then thermalised over 1 ns at each temperature. The simulations were performed for the NVT statistical ensemble, with the temperature controlled by the Langevin thermostat. The x and y dimensions of the 2D sheet were fixed according to the initial size of the simulation box defined during the geometry optimisation. To avoid artefacts related to the incomplete thermalisation of the system, the first 400 ps of each trajectory simulated at a given temperature were excluded from the analysis. The average value of the total energy of the system at a given temperature was determined over the last 600-ps part of each trajectory. A smaller temperature step of 10 K was considered in the temperature range from 650 to 800 K, where the melting phase transition occurs (see the following). In this case, simulations with a duration of 10 ns were performed at each temperature, and the total energy of the system was analysed over the last 6 ns of the simulated trajectories. The resulting caloric curve, i.e., the dependence of the time-averaged total energy of the system on temperature, $\langle E_{\mathrm{tot}} \rangle (T)$, is indicated in Fig. 5.11(a) by symbols. Snapshots of the $NiCl_2$ system at different temperatures and time instants are shown in Fig. 5.12.

In the early stages of $NiCl_2$ heating, the total energy of the system grows linearly with temperature, as shown in the inset of Fig. 5.11(a). At temperatures below $\sim$680 K, the crystalline structure of the $NiCl_2$ sheet is strongly deformed, but the system retains its integrity, as shown in Fig. 5.12(b). A jump in the total energy of the system was observed at temperatures between 680 and 700 K. This rapid energy increase is attributed to the formation of defects (holes) in the $NiCl_2$ sheet; see Fig. 5.12(c).

Figure 5.11(b) shows the temperature dependence of the heat capacity at constant volume C_{V}, defined as the derivative of the internal energy of the system with respect to temperature:

$$C_{\mathrm{V}} = \left(\frac{\partial E}{\partial T} \right)_{\mathrm{V}} . \tag{5.7}$$

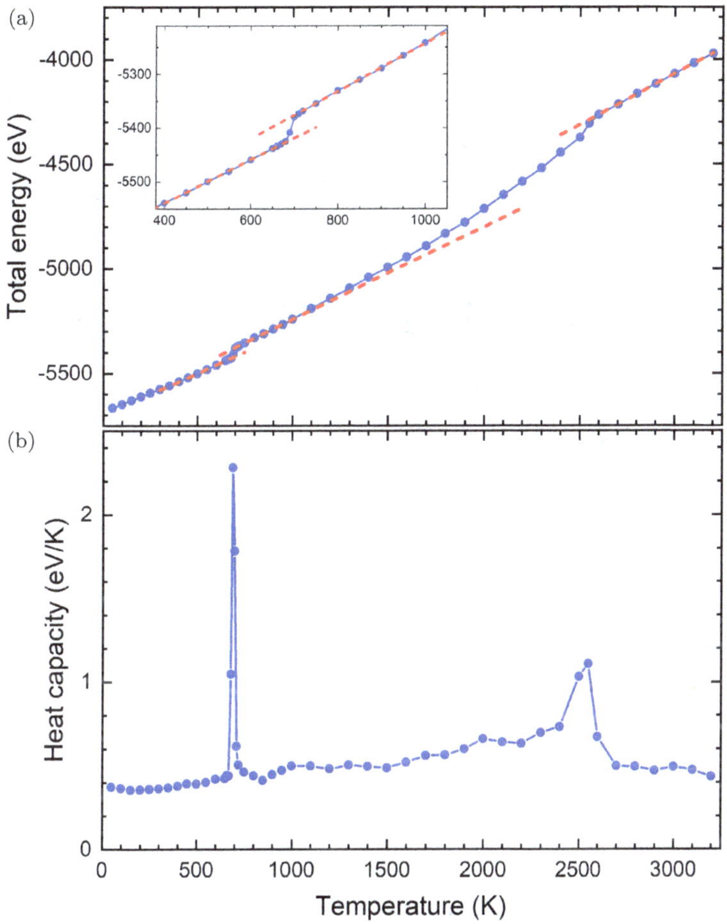

Fig. 5.11. Caloric curve for a 2D $NiCl_2$ material (panel (a)) and the corresponding heat capacity C_v as a function of temperature (panel (b)) obtained using classical MD simulations.[23] The peak in the $C_v(T)$ dependence at $T \approx 695$ K indicates the melting temperature of $NiCl_2$. A broader peak centred at $T \sim 2550$ K indicates sublimation into a gas of $NiCl_2$ molecules and small atomic and molecular fragments.

Source: Reproduced from Ref. 23 with permission from Elsevier.

The melting process in a macroscopic system occurs at a specific temperature and is revealed as a first-order phase transition by a peak in the heat capacity of the system at the transition temperature.

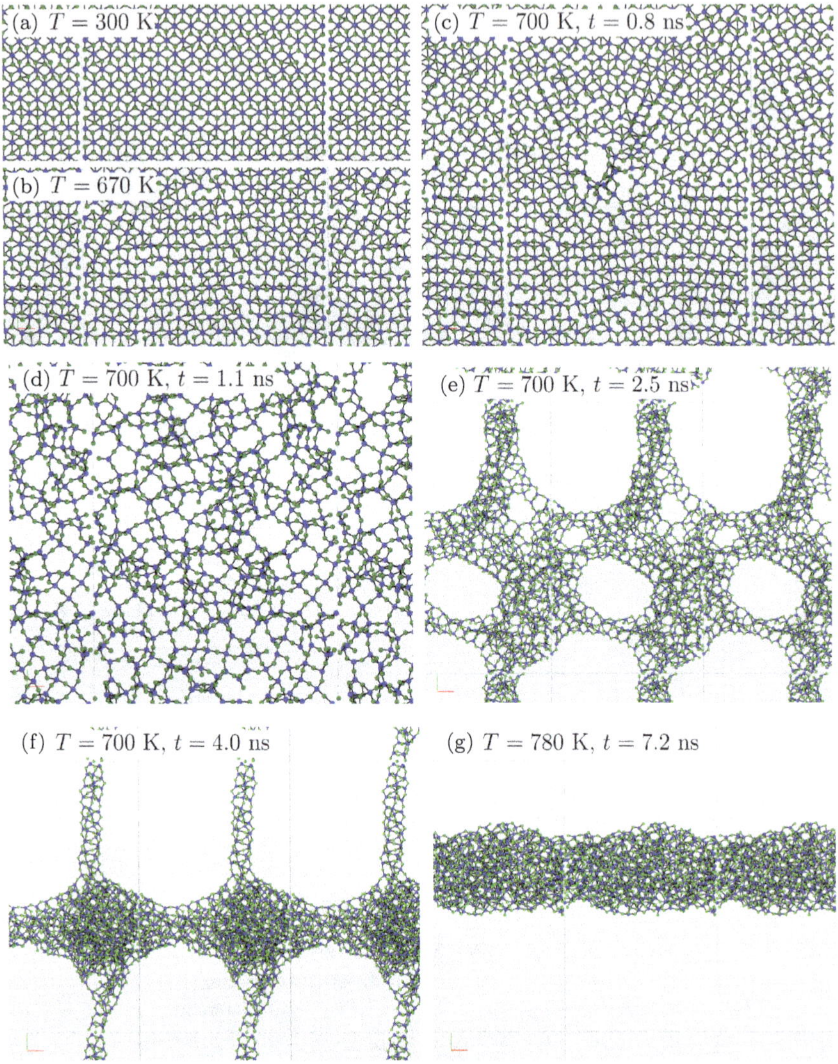

Fig. 5.12. Snapshots of the $NiCl_2$ system at different temperatures T: (a) 300 K, (b) 670 K, (c–f) 700 K, and (g) 780 K.[23] Nickel and chlorine atoms are shown in blue and green colours, respectively. Panels (c)–(g) illustrate the system's structure at different time instants corresponding to labels (1)–(5) in Fig. 5.13.

Source: Reproduced from Ref. 23 with permission from Elsevier.

The calculated heat capacity curve has a sharp maximum at $T \approx$ 695 K, indicating the melting temperature of $NiCl_2$.

The melting temperature of the 2D $NiCl_2$ sheet obtained from the simulations, $T \approx 695$ K, is lower than the melting temperature of bulk nickel, $T_m(Ni) = 1728$ K, and of bulk $NiCl_2$, $T_m(NiCl_2^{bulk}) = 1274$ K.[154] At the same time, the evaluated melting temperature of 2D $NiCl_2$ is within the temperature range ($T \sim 500 - 1000$ K) where other 2D metal halides, such as NaCl, become thermally unstable and lose their structural integrity, as determined computationally from *ab initio* MD simulations.[155]

Of particular interest is to explore in more detail the phase transition from the crystalline $NiCl_2$ to its molten state. Figure 5.13 shows

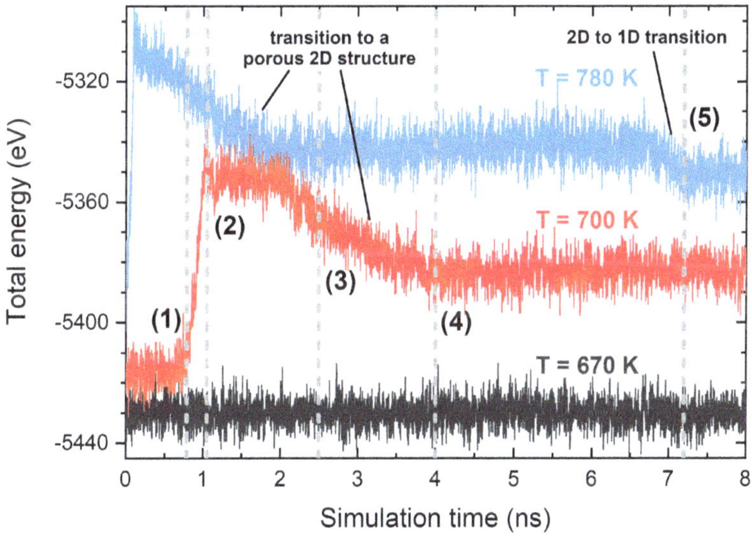

Fig. 5.13. Variation of the total energy of the system, E_{tot}, as a function of the simulation time t. Coloured curves show the $E_{tot}(t)$ dependence for different temperatures, namely $T = 670$ K (below the phase transition temperature), $T = 700$ K (in the phase transition region), and $T = 780$ K (above the phase transition temperature). Labels (1)–(4) correspond to different time instants (dashed lines) for the curve at $T = 700$ K, while label (5) indicates the time instant for the curve at $T = 780$ K. The instants (1)–(5) correspond to the system's snapshots shown in panels (c)–(g) of Fig. 5.12.

Source: Reproduced from Ref. 23 with permission from Elsevier.

the variation of the total energy of the system, E_{tot}, as a function of the simulation time t; the first 8 ns of the simulations (out of 10 ns) are shown for better clarity. Coloured curves show the $E_{\text{tot}}(t)$ dependence for three different temperatures, namely $T = 670$ K (below the phase transition temperature), $T = 700$ K (in the phase transition region), and $T = 780$ K (above the phase transition temperature).

At $T = 670$ K (black curve in Fig. 5.13), the total energy of $NiCl_2$ fluctuates around the average value of about -5430 eV, indicating that the system is thermally stable at this temperature over the considered simulation time. At $T = 700$ K (red curve in Fig. 5.13), the energy of the system remains almost constant within the first $\sim$700 ps of the simulated trajectory, followed by a rapid increase in $E_{\text{tot}}(t)$ within the next $\sim$300 ps. This jump in the total energy is attributed to the formation of defects (holes) in the $NiCl_2$ sheet, as shown in Fig. 5.12(c). The period of about 1 ns long when the total energy of the system remains nearly constant (see the period from $\sim$1 to $\sim$2 ns in the red curve shown in Fig. 5.13) corresponds to a transformation from the deformed $NiCl_2$ sheet into a disordered 2D system in which Ni and Cl atoms form a net-like structure of 1-atom-thick chains; see Fig. 5.12(d). After that, the total energy of the system decreases as the system relaxes into a porous 2D structure, as shown in Fig. 5.12(e). This structure is metastable and evolves into another 2D structure in which denser regions are connected by thin links made of nickel and chlorine atoms; see Fig. 5.12(f).

The structure shown in Fig. 5.12(f) is stable at $T = 700$ K over a simulation of 10 ns duration. However, the 'dissolution' of the thin $NiCl_2$ links seen in Fig. 5.12(f) might occur on a longer time scale, leading to the formation of a 1D structure. Upon analysing the dynamics of the $NiCl_2$ system at higher temperatures, a 2D-to-1D structural transformation can be observed within the considered simulation times. Indeed, at $T = 780$ K, the $NiCl_2$ system transforms into a 1D nanowire (see Fig. 5.12(g)), which is thermally stable at temperatures below $\sim$1500 K within the considered simulation times. This transformation is characterised by a decrease in the total energy of the system; see the blue curve in Fig. 5.13.

At temperatures above the melting point, the total energy of the system, $\langle E_{\text{tot}} \rangle$, continues to grow linearly with T (see the inset in Fig. 5.11(a)) until the onset of another structural transformation at

$T \sim 1500$ K. This transformation takes place over a wide temperature range, from about 1500 to 2700 K, as indicated by the deviation of the $\langle E_{\rm tot}\rangle(T)$ dependence from a linear function and a peak in the $C_{\rm v}(T)$ dependence in the temperature range $\sim$ 2400–2700 K. At temperatures of $\sim$1500 K, $NiCl_2$ monomers begin to evaporate from the 1D $NiCl_2$ nanowire. As the temperature increases, more $NiCl_2$ units detach from the nanowire, resulting in the sublimation of the nanowire into a gas of $NiCl_2$ molecules and smaller fragments such as NiCl dimers and isolated Ni and Cl atoms. This phenomenon is characterised by a change in the $\langle E_{\rm tot}\rangle(T)$ slope shown in Fig. 5.11(a) and a peak in the heat capacity curve centred at $T \approx 2550$ K.

5.4 Conclusions

This chapter has illustrated the use of the advanced software packages MBN Explorer[14] and MBN Studio[15] for the computational modelling and characterisation of composite nanosystems and novel materials, considering functionalised coated gold NPs and a 2D $NiCl_2$ material as two illustrative case studies.

MBN Explorer and MBN Studio allow the creation and characterisation of various nanosystems, including NPs of varying size, shape, and composition of a metal core; thickness, composition, and density of a coating; and composition and density of a molecular medium. The computational and theoretical framework briefly described in this chapter enables a quantitative atomistic description of the structural properties of coated metal NPs in biologically relevant environments. The information obtained can be used to study irradiation-induced effects (such as electron emission and production of reactive species) as well as the kinetics of irradiation-induced chemical reactions in the vicinity of such NPs, which is important for the application of such systems in biotechnological and biomedical applications.

This chapter has overviewed the results of a computer simulation of the recently proposed novel 2D material $NiCl_2$ by means of DFT calculations and classical MD simulations. In order to characterise the geometrical and thermomechanical properties of the 2D $NiCl_2$ system, a classical interatomic force field was developed and verified through a comparison with the results of DFT calculations.

The developed classical force field describes reasonably accurately the geometrical parameters of the 2D $NiCl_2$ material, determined on the basis of the DFT calculations. Regarding the analysis of the mechanical properties, it has been found that the $NiCl_2$ sheet is mechanically stable upon biaxial stretching at the relative deformation up to 11%. 2D Young's modulus for $NiCl_2$ obtained by DFT calculations and calculations employing the classical force field is ~50 N/m, in agreement with the results of recent DFT calculations.[110] The performed MD simulations of 2D $NiCl_2$ melting indicate that the $NiCl_2$ sheet is thermally stable up to the melting point of ~695 K. The MD simulations of the cooling process show that the molten $NiCl_2$ system solidifies into an amorphous porous 2D structure at $T \sim 450$ K. The resulting structure has lower cohesive energy (i.e., a smaller absolute value of the total energy per atom) with respect to the initial 2D sheet. At temperatures above the melting point, structural degradation of the $NiCl_2$ material has been observed, which involves several subsequent structural transformations into a porous 2D sheet and a 1D nanowire, followed by sublimation into a gas of $NiCl_2$ monomers and one- and two-atom fragments.

The methodology presented through the illustrative case study of $NiCl_2$ can be utilised for the computational characterisation of other novel (including 2D) materials.

References

1. Solov'yov IA, Korol AV, Solov'yov AV. Multiscale modeling of complex molecular structure and dynamics with MBN Explorer. Cham: Springer International Publishing. 2017.
2. Solov'yov IA, *et al.*, editors. Dynamics of systems on the nanoscale. Cham: Springer International Publishing. 2022.
3. Bozzolo G, *et al.*, editors. Applied computational materials modeling: Theory, simulation and experiment. New York: Springer. 2007.
4. Bromley ST, Zwijnenburg MA, editors. Computational modeling of inorganic nanomaterials. Boca Raton: CRC Press. 2016.
5. Solov'yov AV, *et al.* Condensed matter systems exposed to radiation: Multiscale theory, simulations, and experiment. Chem Rev. 2024;124:8014–8129.

6. Tadmor EB, Miller RE. Modeling materials: Continuum, atomistic and multiscale techniques. Cambridge: Cambridge University Press. 2011.
7. Vaz M Jr, *et al.*, editors. Advanced computational materials modeling: From classical to multi-scale techniques. Weinheim: Wiley-VCH. 2010.
8. Giustino F. Materials modelling using density functional theory. Oxford: Oxford University Press. 2014.
9. Pal S, Reddy KV. Molecular dynamics for materials modeling. Boca Raton: CRC Press. 2024.
10. Andersen M, Panosetti C, Reuter K. A practical guide to surface kinetic Monte Carlo simulations. Front Chem. 2019;7:202.
11. Solov'yov IA, *et al.* Multiscale modeling of stochastic dynamics processes with MBN Explorer. J Comput Chem. 2022;43:1442.
12. Liu GR, Quek SS. The finite element method: A practical course. 2nd ed. Oxford: Butterworth-Heinemann. 2014.
13. Whiteley J. Finite element methods: A practical guide. Cham: Springer International Publishing. 2017.
14. Solov'yov IA, *et al.* MesoBioNano Explorer - A universal program for multiscale computer simulations of complex molecular structure and dynamics. J Comput Chem. 2012;33:2412.
15. Sushko GB, Solov'yov IA, Solov'yov AV. Modeling MesoBioNano systems with MBN Studio made easy. J Mol Graph Model. 2019;88:247.
16. Verkhovtsev A, *et al.* MBN Explorer and MBN Studio Tutorials: Version 5.0. Frankfurt am Main: MesoBioNano Science Publishing. 2024.
17. Verkhovtsev AV, *et al.* Molecular dynamics simulations of nanoindentation process of titanium crystal. Comput Mater Sci. 2013;76:20.
18. Sushko GB, Verkhovtsev AV, Solov'yov AV. Validation of classical force fields for the description of thermo-mechanical properties of transition metal materials. J Phys Chem A. 2014;118:8426.
19. Kexel C, Schramm S, Solov'yov AV. Atomistic simulation of martensite-austenite phase transition in nanoscale nickel-titanium crystals. Eur Phys J B. 2015;88:221.
20. Yakubovich AV, *et al.* Kinetics of liquid-solid phase transition in large nickel clusters. Phys Rev B. 2013;88:035438.
21. Sushko GB, *et al.* Reconciling simulated melting and ground-state properties of metals with a modified embedded-atom method potential. J Phys Condens Matter. 2016;28:145201.
22. Kexel C, *et al.* Toward the exploration of the NiTi phase diagram with a classical force field. J Phys Chem C. 2016;120:25043.

23. Kalika EB, *et al.* Computational characterization of novel nanostructured materials: A case study of $NiCl_2$. Comput Mater Sci. 2024;239:112975.
24. Katz E, Shipway AN, Willner I. Chemically functionalized metal nanoparticles: Synthesis, properties and applications. In: Liz-Marzán LM, Kamat PV, editors. Nanoscale materials. New York: Springer Science+Business Media. 2003. p. 5–78.
25. Wang S, *et al.* Surface functionalization of metal and metal oxide nanoparticles for dispersion and tribological applications – A review. J Mol Liquids. 2023;389:122821.
26. Sharma S, Sharma H, Sharma R. A review on functionalization and potential application spectrum of magnetic nanoparticles (MNPs) based systems. Chem Inorg Mater. 2024;2:100035.
27. Salata OV. Applications of nanoparticles in biology and medicine. J Nanobiotechnol. 2004;2:3.
28. Murthy SK. Nanoparticles in modern medicine: State of the art and future challenges. Int J Nanomed. 2007;2:129.
29. Jain K, Jain NK, editors. Multifunctional and targeted theranostic nanomedicines. Singapore: Springer. 2023.
30. Porcel E, *et al.* Platinum nanoparticles: A promising material for future cancer therapy? Nanotechnology. 2010;21:085103.
31. McMahon S, *et al.* Biological consequences of nanoscale energy deposition near irradiated heavy atom nanoparticles. Sci Rep. 2011;1:18.
32. Zhang XD, *et al.* Ultrasmall glutathione-protected gold nanoclusters as next generation radiotherapy sensitizers with high tumor uptake and high renal clearance. Sci Rep. 2015;5:8669.
33. McQuaid HN, *et al.* Imaging and radiation effects of gold nanoparticles in tumour cell. Sci Rep. 2016;6:19442.
34. Butterworth KT, *et al.* Physical basis and biological mechanisms of gold nanoparticle radiosensitization. Nanoscale. 2012;4:4830.
35. Kuncic Z, Lacombe S. Nanoparticle radio-enhancement: Principles, progress and application to cancer treatment. Phys Med Biol. 2018;63:02TR01.
36. Verkhovtsev AV, Korol AV, Solov'yov AV. Revealing the mechanisms of the low-energy electron yield enhancement from sensitizing nanoparticles. Phys Rev Lett. 2015;114:063401.
37. Xiao F, *et al.* On the role of low-energy electrons in the radiosensitization of DNA by gold nanoparticles. Nanotechnology. 2011;22:465101.
38. Verkhovtsev AV, Korol AV, Solov'yov AV. Electron production by sensitizing gold nanoparticles irradiated by fast ions. J Phys Chem C. 2015;119:11000.

39. Sicard-Roselli C, *et al.* A new mechanism for hydroxyl radical production in irradiated nanoparticle solutions. Small. 2014;10:3338.
40. Gilles M, Brun E, Sicard-Roselli C. Gold nanoparticles functionalization notably decreases radiosensitization through hydroxyl radical production under ionizing radiation. Colloids Surf B Biointerfaces. 2014;123:770.
41. Solov'yov AV, editor. Nanoscale insights into ion-beam cancer therapy. Cham: Springer International Publishing. 2017.
42. Schuemann J, *et al.* Roadmap for metal nanoparticles in radiation therapy: Current status, translational challenges, and future directions. Phys Med Biol. 2020;65:21RM02.
43. Kempson I. Nanoparticle-based radiosensitization. Int J Mol Sci. 2020;21:2879.
44. Sajo E, Zygmanski P, editors. Nanoparticle enhanced radiation therapy: Principles, methods and applications. Bristol: IOP Publishing. 2020.
45. Ghosh P, *et al.* Gold nanoparticles in delivery applications. Adv Drug Delivery Rev. 2008;60:1307.
46. Kim ST, *et al.* The role of surface functionality in determining nanoparticle cytotoxicity. Acc Chem Res. 2013;46:681.
47. Azharuddin M, *et al.* A repertoire of biomedical applications of noble metal nanoparticles. Chem Commun. 2019;55:6964.
48. Verkhovtsev AV, *et al.* Molecular dynamics characterization of radiosensitizing coated gold nanoparticles in aqueous environment. J Phys Chem A. 2022;126:2170.
49. Grellet S, *et al.* Cancer-selective, single agent chemoradiosensitising gold nanoparticles. PLoS One. 2017;12:e0181103.
50. Zhu C-D, *et al.* Synthesis of novel galactose functionalized gold nanoparticles and its radiosensitizing mechanism. J Nanobiotechnol. 2015;13:67.
51. Tzelepi K, *et al.* Galactose: PEGamine coated gold nanoparticles adhere to filopodia and cause extrinsic apoptosis. Nanoscale Adv. 2019;1:807.
52. Dickers MD, *et al.* Atomistic modelling and structural characterisation of coated gold nanoparticles for biomedical applications. Eur Phys J D. 2023;77:141.
53. Ghosh P, *et al.* Intracellular delivery of a membrane-impermeable enzyme in active form using functionalized gold nanoparticles. J Am Chem Soc. 2010;132:2642.
54. Lin J, *et al.* PEGylation on mixed monolayer gold nanoparticles: Effect of grafting density, chain length, and surface curvature. J Colloid Interface Sci. 2017;504:325.

55. Smith AM, *et al.* Ligand density quantification on colloidal inorganic nanoparticles. Analyst. 2017;142:11.
56. de Gennes PG. Conformations of polymers attached to an interface. Macromolecules. 1980;13:1069.
57. Brittain WJ, Minko S. A structural definition of polymer brushes. J Polym Sci A Polym Chem. 2007;45:3505.
58. Binder K, Milchev A. Polymer brushes on flat and curved surfaces: How computer simulations can help to test theories and to interpret experiments. J Polym Sci B Polym Phys. 2012;50:1515–1555.
59. Wijmans CM, Zhulina EB. Polymer brushes at curved surfaces. Macromolecules. 1993;26:7214.
60. Dukes D, *et al.* Conformational transitions of spherical polymer brushes: Synthesis, characterization, and theory. Macromolecules. 2010;43:1564.
61. Rudin A, Choi P. The elements of polymer science & engineering. 3rd ed. Waltham: Academic Press. 2013.
62. Selli D, Motta S, Di Valentin C. Impact of surface curvature, grafting density and solvent type on the PEGylation of titanium dioxide nanoparticles. J Colloid Interface Sci. 2019;555:519.
63. Stanzione F, Jayaraman A. Hybrid atomistic and coarse-grained molecular dynamics simulations of polyethylene glycol (PEG) in explicit water. J Phys Chem B. 2016;120:4160.
64. Lee H, *et al.* Molecular dynamics studies of polyethylene oxide and polyethylene glycol: Hydrodynamic radius and shape anisotropy. Biophys J. 2008;95:1590.
65. Dahal U, Dormidontova EE. Chain conformation and hydration of polyethylene oxide grafted to gold nanoparticles: Curvature and chain length effect. Macromolecules. 2020;53:8160.
66. Kawaguchi S, *et al.* Aqueous solution properties of oligo- and poly(ethylene oxide) by static light scattering and intrinsic viscosity. Polymer. 1997;38:2885.
67. Kawagoe Y, *et al.* Molecular dynamics study on thermal energy transfer in bulk polyacrylic acid. AIP Advances. 2019;9:025302.
68. Teraoka I. Polymer solutions: An introduction to physical properties. New York: Wiley-Interscience. 2002.
69. Haume K, *et al.* Transport of secondary electrons through coatings of ion-irradiated metallic nanoparticles. Eur Phys J D. 2018;72:116.
70. Haume K, Mason NJ, Solov'yov AV. Modeling of nanoparticle coatings for medical applications. Eur Phys J D. 2016;70:181.
71. Solov'yov AV. Plasmon excitations in metal clusters and fullerenes. Int J Mod Phys B. 2005;19:4143–4184.

72. Pohjolainen E, *et al.* A unified AMBER-compatible molecular mechanics force field for thiolate-protected gold nanoclusters. J Chem Theory Comput. 2016;12:1342.
73. Jin R. Quantum sized, thiolate-protected gold nanoclusters. Nanoscale. 2010;2:343.
74. Wang Y, Bürgi TB. Ligand exchange reactions on thiolate-protected gold nanoclusters. Nanoscale Adv. 2021;3:2710.
75. Shi L, *et al.* Effects of polyethylene glycol on the surface of nanoparticles for targeted drug delivery. Nanoscale. 2021;13:10748.
76. Novoselov KS, *et al.* Electric field effect in atomically thin carbon films. Science. 2004;306:666.
77. Song X, Hub J, Zeng H. Two-dimensional semiconductors: Recent progress and future perspectives. J Mater Chem C. 2013;1:2952.
78. Kim KS, *et al.* The future of two-dimensional semiconductors beyond Moore's law. Nat Nanotechnol. 2024;19:895.
79. Geim AK, Grigorieva IV. Van der Waals heterostructures. Nature. 2013;499:419.
80. Wang H, *et al.* Two-dimensional heterostructures: Fabrication, characterization, and application. Nanoscale. 2014;6:12250.
81. Novoselov KS, *et al.* 2D materials and van der Waals heterostructures. Science. 2016;353:aac9439.
82. Lin Y-C, *et al.* Recent advances in 2D material theory, synthesis, properties, and applications. ACS Nano. 2023;17:9694.
83. Manzeli S, *et al.* 2D transition metal dichalcogenides. Nat Rev Mater. 2017;2:17033.
84. Joseph S, *et al.* A review of the synthesis, properties, and applications of 2D transition metal dichalcogenides and their heterostructures. Mater Chem Phys. 2023;297:127332.
85. Luan C, *et al.* Structure effects of 2D materials on α-nickel hydroxide for oxygen evolution reaction. ACS Nano. 2018;12:3875.
86. Wang Y, *et al.* Environmentally-friendly exfoliate and active site self-assembly: Thin 2D/2D heterostructure amorphous nickel -iron alloy on 2D materials for efficient oxygen evolution reaction. Small. 2019;15:1805435.
87. Liang Q, *et al.* Interfacing epitaxial dinickel phosphide to 2D nickel thiophosphate nanosheets for boosting electrocatalytic water splitting. ACS Nano. 2019;13:7975.
88. Shu Y, *et al.* Facile synthesis of ultrathin nickel -cobalt phosphate 2D nanosheets with enhanced electrocatalytic activity for glucose oxidation. ACS Appl Mater Interfaces. 2018;10:2360.
89. Tasker SZ, Standley EA, Jamison TF. Recent advances in nickel catalysis. Nature. 2014;509:299.

90. Phan D-T, Chung G-S. Reliability of hydrogen sensing based on bimetallic Ni–Pd/graphene composites. Int J Hydrogen Energy. 2014;39:20294.
91. Ren Z, *et al.* Glucose sensor based on porous Ni by using a graphene bottom layer combined with a Ni middle layer. Carbon. 2019;149:609.
92. Deokar G, *et al.* Wafer-scale few-layer graphene growth on Cu/Ni films for gas sensing applications. Sens Actuators B Chem. 2020;305:127458.
93. Zhang W, Li Y, Peng S. Template-free synthesis of hollow Ni/reduced graphene oxide composite for efficient H_2 evolution. J Mater Chem A. 2017;5:13072.
94. Chen D, Wang W, Liu C. Ni-encapsulated graphene chainmail catalyst for ethanol steam reforming. Int J Hydrogen Energy. 2019;44:6560.
95. Zhou C, Szpunar JA, Cui X. Synthesis of Ni/graphene nanocomposite for hydrogen storage. ACS Appl Mater Interfaces. 2016;8:15232.
96. Zhou D, *et al.* Enhanced hydrogen-storage properties of MgH_2 by Fe–Ni catalyst modified three-dimensional graphene. Int J Hydrogen Energy. 2021;46:34369.
97. Safina LR, *et al.* Crumpled graphene-storage media for hydrogen and metal nanoclusters. Materials. 2021;14:2098.
98. Manjanath A, Kumar V, Singh AK. Mechanical and electronic properties of pristine and Ni-doped Si, Ge, and Sn sheets. Phys Chem Chem Phys. 2014;16:1667.
99. Li S, *et al.* Enhanced stability and induced magnetic moments of silicene by substitutional doping of nickel. Chem Phys Lett. 2018;706:202.
100. Li Y, *et al.* Adsorption behavior of COF_2 and CF_2 gas on the MoS_2 monolayer doped with Ni: A first-principles study. Appl Surf Sci. 2018;443:274.
101. Mounet N, *et al.* Two-dimensional materials from high-throughput computational exfoliation of experimentally known compounds. Nat Nanotechnol. 2018;13:246.
102. McCreary KM, *et al.* Large-area synthesis of continuous and uniform MoS_2 monolayer films on graphene. Adv Funct Mater. 2014;24:6449.
103. Wang H, *et al.* High-quality monolayer superconductor $NbSe_2$ grown by chemical vapour deposition. Nat Commun. 2017;8:394.
104. Sherrell PC, *et al.* Thickness-dependent characterization of chemically exfoliated TiS_2 nanosheets. ACS Omega. 2018;3:8655.
105. Cong C, *et al.* Synthesis and optical properties of large-area single-crystalline 2D semiconductor WS_2 monolayer from chemical vapor deposition. Adv Opt Mater. 2013;2:131.

106. Ikeda A, *et al.* Direct observation of infinite NiO_2 planes in $LaNiO_2$ films. Appl Phys Express. 2016;9:061101.
107. Luo Z, *et al.* Novel $NiCl_2$ nanosheets synthesized via chemical vapor deposition with high specific energy for thermal battery. ACS Appl Mater Interfaces. 2020;12:34755.
108. Dai C, *et al.* Controllable synthesis of NiS and NiS_2 nanoplates by chemical vapor deposition. Nano Res. 2020;13:2506.
109. Liu S, *et al.* Two-dimensional $NiSe_2$ N-rich carbon nanocomposites derived from Ni-hexamine frameworks for superb Na-ion storage. ACS Appl Mater Interfaces. 2018;10:34193.
110. Lu M, *et al.* Mechanical, electronic, and magnetic properties of NiX_2 (X = Cl, Br, I) layers. ACS Omega. 2019;4:5714.
111. Kistanov AA, *et al.* Family of two-dimensional transition metal dichlorides: Fundamental properties, structural defects, and environmental stability. J Phys Chem Lett. 2022;13:2165.
112. Giannozzi P, *et al.* QUANTUM ESPRESSO: A modular and open-source software project for quantum simulations of materials. J Phys Condens Matter. 2009;21:395502.
113. Giannozzi P, *et al.* Advanced capabilities for materials modelling with Quantum ESPRESSO. J Phys Condens Matter. 2017;29:465901.
114. Perdew JP, Burke K, Ernzerhof M. Generalized gradient approximation made simple. Phys Rev Lett. 1996;77:3865.
115. Blöchl PE. Projector augmented-wave method. Phys Rev B. 1994; 50:17953.
116. Kresse G, Joubert D. From ultrasoft pseudopotentials to the projector augmented-wave method. Phys Rev B. 1999;59:1758.
117. Grimme S, *et al.* A consistent and accurate ab initio parametrization of density functional dispersion correction (DFT-D) for the 94 elements H-Pu. J Chem Phys. 2010;132:154104.
118. Monkhorst HJ, Pack JD. Special points for Brillouin-zone integrations. Phys Rev B. 1976;13:5188.
119. Methfessel M, Paxton AT. High-precision sampling for Brillouin-zone integration in metals. Phys Rev B. 1989;40:3616.
120. Verkhovtsev AV, Schramm S, Solov'yov AV. Molecular dynamics study of the stability of a carbon nanotube atop a catalytic nanoparticle. Eur Phys J D. 2014;68:246.
121. Sushko GB, Solov'yov IA, Solov'yov AV. Molecular dynamics for irradiation driven chemistry: Application to the FEBID process. Eur Phys J D. 2016;70:217.
122. de Vera P, *et al.* Multiscale simulation of the focused electron beam induced deposition process. Sci Rep. 2020;10:20827.

123. Geng J, *et al.* Uncovering a solvent-controlled preferential growth of buckminsterfullerene (C_{60}) nanowires. J Phys Chem C. 2009;113: 6390.
124. Moskovkin P, *et al.* Simulation of nanowire fragmentation by means of kinetic Monte Carlo approach: 2D case. Phys Stat Sol B. 2014; 251:1456.
125. Matsui M, Akaogi M. Molecular dynamics simulation of the structural and physical properties of the four polymorphs of TiO_2. Mol Simul. 1991;6:239.
126. Mattoni A, *et al.* Methylammonium rotational dynamics in lead halide perovskite by classical molecular dynamics: The role of temperature. J Phys Chem C. 2015;119:17421.
127. Morgan LM, *et al.* Protecting ceria nanocatalysts – the role of sacrificial barriers. ACS Appl Mater Interfaces. 2018;10:32510.
128. Pedone A, *et al.* Interatomic potentials for oxide glasses: Past, present, and future. J Non-Cryst Solids: X. 2022;15:100115.
129. Li P, Song LF, Merz KM Jr. Parameterization of highly charged metal ions using the 12-6-4 LJ-type nonbonded model in explicit water. J Phys Chem B. 2015;119:883.
130. Turupcu A, Tirado-Rives J, Jorgensen WL. Explicit representation of cation–π interactions in force fields with $1/r^4$ nonbonded terms. J Chem Theory Comput. 2020;16:7184.
131. Li P, Merz KM Jr. Metal ion modeling using classical mechanics. Chem Rev. 2017;117:1564–1686.
132. Mason EA, O'Hara H, Smith FJ. Mobilities of polyatomic ions in gases: Core model. J Phys B At Mol Phys. 1972;5:169.
133. Nelson D, *et al.* Basic data of polyatomic ion-molecule systems for flue gas discharge modelling. J Phys D Appl Phys. 2001;34:3247.
134. Nelder JA, Mead R. A simplex method for function minimization. Comput J. 1965;7:308.
135. Gao F, Han L. Implementing the Nelder-Mead simplex algorithm with adaptive parameters. Comput Optim Appl. 2012;51:259–277.
136. Catlow CRA, Dixon M, Mackrodt WC. Interionic potentials in ionic solids. In: Catlow CRA, Mackrodt WC, editors. Computer simulation of solids. Berlin: Springer. 1982. pp. 130–161.
137. Gillan MJ. Collective dynamics in superionic CaF_2. I. Simulation compared with neutron-scattering experiment. J Phys C Solid State Phys. 1986;19:3391.

138. van Beest BWH, Kramer GJ, van Santen RA. Force fields for silicas and aluminophosphates based on ab initio calculations. Phys Rev Lett. 1990;64:1955.
139. Sundararaman S, *et al.* New optimization scheme to obtain interaction potentials for oxide glasses. J Chem Phys. 2018;148:194504.
140. Atkinson K, Grimes R, Owens S. Simulation of fluorine migration in NiF_2. Solid State Ionics. 2002;150:443.
141. Wilson M, Hutchinson F, Madden PA. Simulation of pressure-driven phase transitions from tetrahedral crystal structures. Phys Rev B. 2002;65:094109.
142. Hutchinson F, *et al.* The "Ionic" to "Molecular" transitions in $AlCl_2$ and $FeCl_2$ as predicted by an ionic interaction model. J Chem Phys. 1999;110:5821.
143. Ewald P. Die Berechnung optischer und elektrostatischer Gitterpotentiale. Ann Phys. 1921;369:253–287.
144. Solov'yov IA, Sushko G, Solov'yov AV. MBN Explorer User's Guide (Version 3.0). Frankfurt am Main: MesoBioNano Science Publishing. 2017.
145. Liu H, *et al.* Exploring the landscape of Buckingham potentials for silica by machine learning: Soft vs hard interatomic forcefields. J Chem Phys. 2020;152:051101.
146. Woodcock LV, Angell CA, Cheeseman P. Molecular dynamics studies of the vitreous state: Simple ionic systems and silica. J Chem Phys. 1976;65:1565.
147. Kramer GJ, *et al.* Interatomic force fields for silicas, aluminophosphates, and zeolites: Derivation based on ab initio calculations. Phys Rev B. 1991;43:5068.
148. Alnemrat S, Tomlinson WW. Magnetic ground state of $NiCl_2/Br_2$ monolayers encapsulated in UiO-67 MOF, effect of partially filled pores: Ab-initio and Monte Carlo theoretical study. J Magn Magn Mater. 2023;566:170317.
149. Verkhovtsev AV, *et al.* Characterization of small pure and Ni-doped titanium clusters: Ab initio versus classical approaches. Comput Mater Sci. 2013;76:80.
150. Liu K, Wu J. Mechanical properties of two-dimensional materials and heterostructures. J Mater Res. 2016;31:832.
151. Zhang B, *et al.* 2D Young's modulus of black phosphorene with different layers. J Phys Chem C. 2022;126:1094.

152. Lee C, *et al.* Measurement of the elastic properties and intrinsic strength of monolayer graphene. Science. 2008;321:385.
153. Liu K, *et al.* Elastic properties of chemical-vapor-deposited monolayer MoS_2, WS_2, and their bilayer heterostructures. Nano Lett. 2014;14:5097.
154. Lide DS, editor. CRC handbook of chemistry and physics. 84th ed. Boca Raton: CRC Press. 2003.
155. Luo B, *et al.* Graphene-like monolayer monoxides and monochlorides. Proc Natl Acad Sci USA. 2019;116:17213.

https://doi.org/10.1142/9781800618183_0006

Chapter 6

From Atomic Design to Cost-Efficient Solutions: Mastering Pt-Based Nanodendrite Formation for a Net-Zero Future

Siyi Ming, Nicholas Sammy, and Andrew E. H. Wheatley*

University of Cambridge, Yusuf Hamied Department of Chemistry, Lensfield Road, Cambridge CB2 1EW, UK

**aehw2@cam.ac.uk*

Limiting anthropogenic greenhouse gas emissions is fundamental to achieving the 'net-zero emissions' target that underpins efforts to mitigate the global effects of climate change. One strategy targeting reduced emissions is the development of a green hydrogen economy. Central to this is the improvement of Pt-based nanocatalysts that offer environmentally friendly solutions in the transport sector. This chapter focuses on recent advances in the design of Pt-based bimetallic catalysts, where branched morphologies offer high surface-area-to-volume ratios and controllably intermixing Pt with a second metal offers enhanced activity, reduced costs, and minimisation of the kinetic drawbacks seen with monometallics. In this context, the need for non-empirical routes to individually bimetallic, uniformly intermixed nanodendrites is discussed. The relationship between isotropic and anisotropic growth in Pt-based bimetallics is summarised. The recently established relationship between nanopod concentration and nanodendrite formation is then explored and its implications for the facile, non-empirical synthesis of high-surface-area nanocatalysts with controllable composition are considered.

The ability of modelling to improve our understanding of the formation and behaviour of these systems is explored, as is the potential of predictable nanodendrite formation to shape the preparation of even more complex multimetallic catalysts.

6.1 The Importance of Designing New Pt-Based Catalysts for Net-Zero

6.1.1 *The need for net-zero emissions*

The expression 'net-zero emissions' refers to achieving an equilibrium between anthropogenic greenhouse gases being emitted into the atmosphere and those removed from it.[1] Reaching net-zero is crucial for mitigating climate change and avoiding its most catastrophic impacts.[2] Empirical evidence presented at recent COP summits[3] indicates that global average temperatures have increased by approximately 1.2°C since the pre-industrial era.[4] This upward trend, primarily driven by human activities, underscores the urgent need for international cooperation and robust policy interventions to mitigate climate change effects. Moreover, while 1.2°C might seem modest, its already significant consequences are being seen in erratic weather patterns, including heatwaves, floods, and increasingly severe storms, coupled with the loss of polar ice, ocean acidification, and rising sea levels.[5] Despite the global commitment to limit temperature rise to well below 2°C, and preferably to 1.5°C, above pre-industrial levels, as outlined in the Paris Agreement, current projections based on existing policies suggest that temperatures could escalate by as much as 2.7°C by 2100. This trajectory highlights the significant gap between current efforts and the more aggressive targets needed to avert catastrophic climate scenarios.[6]

To limit global warming to the 1.5°C target set by the 2015 Paris Agreement, the world must achieve net-zero emissions by around 2050.[7] This is essential for stabilising global temperatures and minimising severe impacts on ecosystems, economies, and human health. Achieving this requires rapid and comprehensive transformations, including those in the energy sector, transportation, agriculture, and land use.[8] Nearly all electricity will need to be supplied by zero-carbon energy sources by that date, and significant reductions in emissions from transportation and other sectors will be required.[9] Moreover, improving the efficiency of food production, changing dietary choices, restoring degraded land, and reducing food loss and waste will be critical contributors. Coordinated global action and innovation across all sectors are imperative to meet these challenges.[10]

6.1.2 *The role of Pt-based catalysts in a net-zero future*

Achieving the 2050 climate goals, as outlined in Section 6.1.1, necessitates substantial and aggressive decarbonisation across diverse economic sectors. One pivotal strategy for reaching net-zero emissions by 2050 will be shifting from a fossil fuel to a hydrogen economy. Hydrogen is a clean energy carrier by virtue of emitting no CO_2 when used, making it a promising alternative to fossil fuels. When hydrogen is utilised in fuel cells or combustion, the primary byproduct is water vapour, making it highly desirable for applications in the transportation and energy generation sectors.[11] In 2020, global hydrogen consumption was approximately 80 million tonnes, out of the 120 million tonnes produced annually.[12] The discrepancy suggests that about 40 million tonnes of produced hydrogen were not utilised directly in that year, and this was attributable to a combination of storage, losses, or inefficiencies in distribution and usage processes. The International Energy Agency (IEA) projects that hydrogen consumption will increase to 530 million tonnes by 2050.[11] However, most current hydrogen production is not carbon-free, as it predominantly serves ammonia/methanol production and petroleum refining in the chemical industry.[13] These production methods emit significant amounts of CO_2.[14,15] Hydrogen also plays a role in manufacturing chemicals and fuels that, despite their clean production processes, result in carbon emissions during their subsequent use.[16] This creates a complex dynamic, where the benefits of hydrogen can be offset by the carbon-intensive processes it supports.

For hydrogen to contribute effectively to reductions in anthropogenic CO_2 emissions and achieving zero-carbon goals, it must transition from being grey hydrogen, which is derived from the carbon-intensive processes described above, to being green hydrogen. This is produced by electrolysing water using renewable energy sources[17] – a method that stands in contrast to steam methane reforming and coal gasification, which are responsible for the vast majority of hydrogen production globally and result in substantial CO_2 emissions.[18] Indeed, in 2021, steam methane reforming alone accounted for about 50% of all hydrogen produced worldwide. In the industrial sector, green hydrogen can facilitate decarbonisation by replacing carbon-intensive reducing agents, such as coal in steel production, and substituting fossil fuels in the generation of heat, steam, or electricity.[19] Meanwhile, in the transportation sector, it can power fuel cells for electric vehicles and enable the electrification of systems in spacecraft and aircraft.[20] Although the transition to green hydrogen faces initial challenges, such as its current higher production costs compared to grey hydrogen, the long-term potential is substantial. Estimates suggest that with adequate support, green hydrogen could account for up to 22% of the world's energy demand by 2050.[21]

Platinum is an indispensable metal in a range of high-technology applications,[22] primarily due to its excellent catalytic properties,[23] resistance to corrosion,[24] and high electrical conductivity.[25] While Pt is one of the most effective among the platinum group metals (PGMs) for applications, it is particularly valued for its role in green hydrogen production.[26] In this sector, Pt catalysts are essential in proton exchange membrane (PEM) electrolysers. These produce hydrogen and oxygen by splitting water using renewable electricity.[27] In these systems, Pt catalyses water oxidation at the anode by lowering the activation energy required for the water-splitting reaction. This enhances the efficiency of hydrogen production by reducing the overall energy consumption of the process.[28] The rest of the PEM cell is designed to support this reaction efficiently. The PEM itself plays a critical role: it conducts protons from the anode to the cathode while blocking electrons. This selective permeability is crucial, as it ensures the separation of the hydrogen and oxygen generated at the different electrodes.

Based on the above, Pt-based catalysts are crucial for achieving net-zero emissions in the transportation sector. PEM fuel cells are already used to power hydrogen fuel cell vehicles (FCVs), including automobiles, buses, and trucks, providing a zero-emission alternative to conventional engines.[29] FCVs offer advantages such as longer ranges and faster refuelling compared to battery-electric vehicles. The latter operate by storing electrical energy in rechargeable lithium-ion or other types of batteries, which then power an electric motor.[30] While such vehicles produce zero emissions during operation, they typically require longer charging times and currently have shorter range capabilities compared to their FCV counterparts.[31] Currently, research is underway to utilise Pt catalysts for hydrogen-powered ships and aircraft to reduce emissions in these highly polluting sectors.[32,33] The adoption of hydrogen technology in aviation and maritime transport has been slower due to several factors. For aircraft, the challenges include the need for hydrogen storage systems that can meet safety and weight requirements.[34] Meanwhile for ships, the transition involves redesigning or retrofitting large and complex propulsion systems to accommodate hydrogen fuel,[35] a process that requires significant money and time. However, as technology advances and the urgency with which global emissions must be reduced increases, hydrogen is gaining traction as a feasible alternative fuel for these modes of transport.

Pt catalysts can also contribute to carbon removal strategies, a critical component of achieving net-zero emission targets. For instance, in automotive applications, they are utilised in catalytic converters to reduce harmful emissions, such as nitrogen oxides, carbon monoxide, and volatile organic compounds, from internal combustion engines, thus improving air quality.[36] In industrial settings, Pt catalysts are employed in processes that limit emissions of greenhouse gases and other pollutants, supporting cleaner operations by enhancing the efficiency and selectivity of chemical reactions.[37] These cleaner operations are often critical in industries such as chemical manufacturing and petroleum refining, where emissions control is a regulatory and environmental necessity. In the realm of carbon capture and utilisation (CCU), Pt catalysts enable the catalytic conversion of captured CO_2 into valuable chemicals and fuels.[38] These processes often involve the hydrogenation of CO_2 to produce methanol, a versatile chemical that can be used as a fuel, solvent, or feedstock for other chemical syntheses.[39] By facilitating these transformations, Pt catalysts help to close the carbon loop, effectively reducing net emissions by converting greenhouse gases into commercially valuable products. This dual role of reducing emissions at source and converting emissions into useful substances underscores the multifaceted potential of Pt catalysts in sustainability and environmental management efforts.[27]

Apart from Pt itself, Pt-based catalysts, which include Pt-based alloys, are fundamentally important across multiple sectors involved in achieving net-zero emissions. For example, though iridium presents its own sustainability issues, PtIr alloys, known for their exceptional durability and resistance to chemical attack, are currently valuable as electrodes used in water electrolysis systems, which are essential for producing green hydrogen.[40] The robustness of PtIr ensures long-term operation even under the harsh, acidic conditions often found in these systems. On the other hand, PtCo alloys are prominently used in the automotive sector, specifically in the cathodes of PEM fuel cells.[41] Adding Co to Pt not only reduces the overall cost by decreasing the amount of Pt required but also enhances the catalytic activity essential for improving the efficiency of hydrogen fuel cells. In addition, Pt-based catalysts have an important role in environmental remediation, including catalysing the breakdown of pollutants and the more eco-friendly synthesis of chemicals, which is crucial for technologies aimed at targeting the treatment of air and water pollutants.[42,43] The integration of Pt-based catalysts into various sectors, from automotive to industrial manufacturing, thus presents significant possibilities for working towards meeting global climate goals and ensuring a sustainable transition away from fossil fuels.

This chapter aims to summarise recent developments (typically, though not exclusively, since 2016) in the field of Pt-based (electro)catalysts. This derives from the desirability of reducing Pt dependence. A major focus will, moreover, be on efforts to raise activity by forming systems with high surface-area-to-volume ratios, which offer a greater number of exposed active sites. As such, discussion will revolve around a subset of anisotropic nanomaterials.[44] The aim will be to bring up-to-date work on the systematic synthesis of highly branched nanodendritic structures in which platinum is combined with a second metal,[45,46] necessitating simultaneous morphological and compositional control. Doing this will clarify how it has now emerged that systematising nanodendrite preparation involves understanding the corresponding chemistry of the simpler anisotropic particles – nanopods.[47] With the emphasis placed on the synthesis of these materials, applications will be alluded to, though the reader is directed elsewhere for more thorough treatments of these in monometallic Pt.[48–50] Meanwhile, electrocatalytic Pt-based systems have received coverage in summary form,[24,50] with more focused articles on morphologically isotropic systems[51] and control of surface composition[52] appearing as well. Recent coverage has also expanded to the search for improved durability in electrocatalytic applications.[53,54]

6.1.3 *Challenges with current Pt-based catalysts*

Having established the importance of Pt(-based) catalysts and their coverage, it is important to consider at greater length the effects of the fact that Pt itself is not abundantly available. Beyond its intrinsic scarcity, the majority of the global supply of Pt is concentrated in only a few countries. Specifically, South Africa, Russia, and Zimbabwe together account for over 90% of the world's Pt production, with respective shares of 67.6%, 13.9%, and 9.1% as of 2020.[55] Such a geographical concentration poses significant risks to global energy security, as geopolitical tensions or supply disruptions involving these areas can lead to substantial fluctuations in availability and pricing. The traditional approach of recycling spent Pt catalysts, although beneficial,[56] involves complex, costly, and energy-intensive chemical processes.[57] These challenges, along with economic inefficiency and high energy consumption in recycling, escalate costs and compound environmental impact by increasing greenhouse gas emissions and generating potentially hazardous waste.

Given these constraints, there is an imperative to innovate within the realm of Pt-based catalysts. Developing systems that maintain or enhance performance while reducing the quantity of Pt used is crucial. This can be achieved through the use of multimetallics, which incorporate less expensive metals to reduce costs and dependency on Pt. The introduction of these additional metals is frequently achieved on the nanoscale and can not only lower materials costs[58] but also enhance intrinsic activity through synergistic effects[59] and the creation of unique active sites, particularly at metal–metal interfaces[52] in heterodimeric[60] or core–shell nanoparticles.[61] These processes can improve reaction kinetics. For example, transformations that are typically sluggish using Pt alone, such as the oxygen reduction reaction (ORR) in fuel cells, can benefit significantly.[62] Alternatively, engineering the morphology of Pt-based alloys in nanoparticulate catalysts involves precisely controlling particle size, shape, or distribution. These nanoparticles are the primary functional units in many catalysts, with their physical dimensions and structural characteristics critically influencing performance.[63] Nanoparticle size affects the surface-area-to-volume ratio; smaller nanoparticles typically expose more surface area relative to their volume, which can increase the number of catalytically active sites available for chemical reactions.[64] Meanwhile, the shape of nanoparticles also plays a crucial role in catalysis.[65] Different shapes expose different crystal facets, with some facets being more catalytically active than others due to variations in atomic arrangements and bonding.[66] For example, cubic nanoparticles expose {100} facets, while octahedral nanoparticles expose {111} facets, either of which may be more or less catalytically effective depending on the reaction.[67] Lastly, the size distribution of these nanoparticles within a catalyst system affects how consistently reactants are processed. Uniform and narrow distributions encourage all reactants to have similar probabilities of encountering active sites, which benefits both efficiency and selectivity. By altering one or more of these aspects – nanoparticle size, shape, and size distribution – it becomes possible to enhance surface area and so active site availability, thereby improving activity.[68] Further, durability and stability can also be extended by engineering morphology, thus extending the life and efficacy of these materials.[24,69] Evidently, pursuing these advances is essential not only for short-term economic viability but also for meeting the demands of emerging energy technologies necessary for a carbon-zero future.

6.2 Strategies to Reduce Pt Dependence

6.2.1 *Combining Pt with transition metals*

Reducing reliance on Pt while simultaneously lowering costs and enhancing catalytic performance can be achieved by combining it with transition metals.[70–72] There are two primary methods to create such bimetallic catalysts: heterostructuring and alloying. Heterostructures involve assembling different metals in distinct layers or clusters, which can lead to varied electronic and structural characteristics at the interfaces, beneficial for specific catalytic reactions.[73] However, heterostructured materials will not be a major focus of this chapter. Instead, compositionally uniform systems will be emphasised. As such, alloys consist of a homogeneous mixture of two or more metals that are either uniformly or randomly mixed. They often exhibit enhanced catalytic properties due to the consistent distribution of electronic properties across the catalyst. This uniformity can lead to more reliable catalytic activity and a greater resistance to deactivation.[74] Earth-abundant first-row transition metals, such as iron,[75] cobalt,[76] and nickel,[61] represent attractive options for intermixing with Pt due to their high availability, low cost, and minimal toxicity, as well as their own unique catalytic properties. In alloys, these metals modify the electronic structure and geometric arrangement of Pt-based active sites, which can enhance the adsorption and activation of reactants, leading to improved efficiency and selectivity.[77] While discussion herein will focus on Fe, Co, and Ni, on account of their recent role in developing predictable routes to morphologically complex nanoparticles, metals such as copper[78] and manganese[79] have also been used in conjunction with Pt in a range of catalytic applications.[68] Compared to heterostructures, Pt alloys tend to exhibit greater stability under operational conditions due to the uniform distribution of metal atoms within the alloy.[50] This uniformity provides consistent catalytic activity and enhances resistance to deactivation under the demanding environments frequently encountered in applications like fuel cells and electrolysis. For example, in PEM fuel cells, the operating environment is highly acidic, typically involving strong acids such as sulfonated tetrafluoroethylene-based fluoropolymer-copolymer electrolytes. These conditions typically accelerate performance degradation through the oxidation and dissolution of carbon supports commonly used with Pt-based catalysts. This both reduces the physical stability of the catalyst, causing a loss of active material, and diminishes catalytic activity. Additionally, metal leaching, in which acids cause the leaching of metal out of active sites, and Ostwald ripening, which involves the migration and coalescence of nanoparticles, diminish catalyst activity by reducing the surface area available for reactions. Conversely, in alkaline electrolysis systems that use solutions such as potassium hydroxide for water splitting, challenges include sintering, which reduces the active surface area, poisoning of active sites by impurities like carbonates, and structural modification by hydroxide formation at non-noble metal catalyst sites.[28]

The durability of the catalysts (and their supports) is therefore crucial, as Pt catalysts are often rendered more expensive by their poor longevity in such harsh conditions. Additionally, transition metals also contribute to the structural enhancement of the catalyst. They increase the number of accessible active sites and improve mass and charge transfer characteristics by creating structural features such as morphological anisotropy and a greater surface area. These allow for better performance by facilitating more effective interaction between the catalyst and the reactants.[59]

To improve the performance of Pt catalysts, the development of alloys incorporating at least one earth-abundant transition metal offers enormous possibilities. Two primary strategies to achieve this can be pursued: (i) enhancing the intrinsic activity of individual active sites,[80] and (ii) increasing the number of accessible active sites.[81] Employing either strategy individually offers certain benefits, but using them in tandem promises the most significant benefits. Such an integrated approach effectively addresses issues associated with high catalyst loadings, including problems of mass transfer, as well as the obvious cost and sustainability drawbacks. For instance, typical Pt loadings in automotive catalysts might range from 10 to 100 g/ft^3. By implementing strategies that enhance the intrinsic activity of the catalyst, it might be possible to reduce these loadings to 5–50 g/ft^3.[82] Simultaneously, strategies that increase the number of accessible active sites on the catalyst can ensure more efficient utilisation of (lower amounts of) Pt while maintaining effective coverage of the electrode surface. This dual approach not only reduces the quantity of precious metal required but also maintains, or even enhances, overall performance.[59,82] Returning to the issue of using increased amounts of catalyst, there is a practical limit to how much can be loaded onto a surface (e.g., an electrode) before charge and mass transfer limitations develop. These arise because the flow of reactants and products to and from active sites becomes restricted. Related to this, catalyst design must also consider the geometric configuration of the catalyst. This can take the form of creating nanoporous structures or altering the shape to maximise surface area and ensure more effective mass transfer.[53] It is an approach that facilitates a higher number of reactions per unit of surface area without compromising the catalyst's efficiency due to excessive loading.[83]

Mass transfer limitation due to excessive catalyst loading is illustrated in Fig. 6.1, which shows a levelling-off in performance improvement at higher catalyst loadings due to limitations in mass and charge transfer. The figure graphically demonstrates that beyond a certain point, simply adding more catalyst does not proportionally enhance the reaction rate, but rather leads to inefficiencies such as reduced accessibility to active sites and diminished conversion rates. It is crucial to understand that while the effectiveness of catalysts can vary widely, with intrinsic activity differences between higher- and lower-performing catalysts potentially exceeding ten orders of magnitude, variations due to adjustments in catalyst loading typically range within one to three orders of magnitude.[83] This underscores the substantial impact that enhancing intrinsic activity, rather than merely increasing the quantity of catalyst, can have. The strategies to optimise catalyst performance, as delineated in Fig. 6.1, include ideas such as nanostructuring,[84] which can enhance surface area and the number of available active sites, the use of supports to improve dispersion and stability, and also the manipulation of particle shape to expose more reactive facets. Additionally, selecting different polymorphs and incorporating adsorbates can alter the surface chemistry

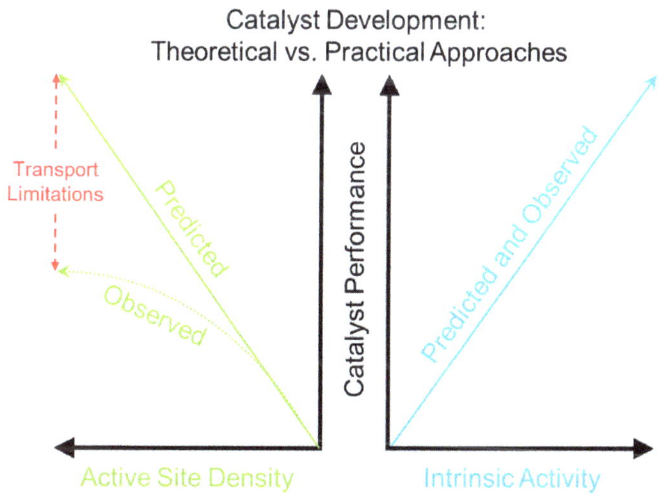

Fig. 6.1. A range of catalyst development strategies are available for enhancing activity. Approaches can increase the number of available active sites expressed (left), the intrinsic activity of each site (right), or both.

to favour specific reactions.[85] Core–shell structures and intercalation techniques can modify the electronic environment of the catalysts, while confinement within certain geometries can enhance reactivity by increasing reactant concentrations at catalytic sites.[86] Lastly, alloying with other metals can create beneficial synergistic effects that significantly boost both activity and lifetime.[87] These strategies collectively aim to circumvent the drawbacks of high catalyst loadings by fundamentally enhancing the catalyst's intrinsic activity and making each active site more effective.[59] The following sections explore these ideas in more depth, examining how each approach contributes to the overarching goal of optimising catalyst performance for a variety of applications, particularly in contexts where catalytic efficiency and sustainability are paramount.[55]

6.2.2 *Raising intrinsic activity*

Optimising the intrinsic catalytic activity of Pt-based catalysts is crucial for improving both the cost and sustainability of catalysts across a range of applications and involves adjustments to a catalyst's electronic characteristics[68,71,88] Enhancing performance requires the modulation of these characteristics, and key to this are the Pt-support interaction, facet, and ligand effects. Each profoundly impacts the electronic structure of Pt atoms and consequently their activity, as outlined in the following.

6.2.2.1 *Pt–support interaction effect*

This effect refers to the influence that the support material on which Pt catalysts reside has on activity and selectivity. The interaction can significantly alter the electronic environment of Pt, affecting its ability to adsorb and activate reactants and desorb products. It is a critical factor in developing applications such as hydrogenation reactions and environmental catalysis,[89] where precise control over reaction pathways is essential for achieving desired outcomes with the greatest efficiency. A recent comprehensive review of these interactions provides deeper insights into the mechanisms at play.[90] However, to consider a specific example, Zhou *et al.* enhanced the effectiveness of Pt monolayer electrocatalysts for the ORR by fine-tuning the interaction between a Pt monolayer and its support.[91]

The team utilised an annealed $Pd_3Fe(111)$ single-crystal alloy support with a segregated Pd layer that forms on the surface following high-temperature annealing. This Pd layer, structurally similar to Pd(111), showed considerably higher ORR activity than traditional Pd(111), with kinetics comparable to Pt(111). The increased catalytic activity was attributed to the modification of the surface electronic properties of Pd by the underlying Fe. However, the ORR activity of the $Pd_3Fe(111)$ support dramatically decreased when the subsurface Fe was depleted through potential cycling in acidic conditions. Despite this, the $Pd_3Fe(111)$ surface proved to be an excellent substrate for a Pt monolayer ORR catalyst, as demonstrated by enhanced kinetics on a Pt monolayer deposited over the $Pd/Pd_3Fe(111)$ structure. Density functional theory (DFT) studies suggested that the improved ORR activity mainly originated from the destabilisation of OH binding on the Pt layer and reduced Pt–OH coverage on the $Pt/Pd/Pd_3Fe(111)$ surface. That is, the activity of the Pt monolayer on Pd(111) and pure Pt(111) was controlled by OH removal, whereas for the Pt monolayer on $Pd/Pd_3Fe(111)$, it was controlled by O–O bond cleavage. In another work, Chen *et al.* developed a three-dimensional N-doped graphene material with Pt_3Ni nanoparticles uniformly dispersed on the substrate. The study referred to 'nitrogen defects' to mean specific sites within the graphene structure where nitrogen atoms replaced carbon atoms. These substitutions not only created unique electronic properties that prevented the agglomeration of Pt_3Ni nanoparticles, thus stabilising the catalyst, but also enhanced its effectiveness. The strong metal–support interaction, induced by nitrogen doping, created tensile strain at the Pt_3Ni surface, enhancing hydrogen evolution reaction (HER) activity.[92] This was rationalised by considering the optimised adsorption energies of hydrogen, as indicated by changes in the enthalpy of adsorption due to nitrogen doping. These changes led to a more favourable interaction that accelerated the HER. The result was an impressive performance of the system, achieving a current density of 100 mA/cm^2 at an overpotential of just 86 mV.

Such interactions are not unique to nitrogen-doped graphene; they are also commonly found with metal oxide supports.[93–95] For example, $Pt/Ti_{0.7}Mo_{0.3}O_2$ was shown to exhibit an ORR activity approximately sevenfold that of commercial Pt/C catalysts. Similar enhancements have also been seen with Pt/CeO_2 and Pt/FeO_x, where the metal oxides not only provided physical stability but also chemically interacted with Pt to modify its electronic properties.[96] Experimental evidence from X-ray photoelectron spectroscopy (XPS) and extended X-ray absorption fine structure (EXAFS) analyses supported these findings, showing changes in the electronic structure and oxidation state of Pt when it was located on either oxide, with corresponding commensurate improvements in catalytic performance. These examples illustrate how both nitrogen doping and the choice of support material are critical in tuning the electronic properties of Pt-based catalysts to optimise activity and stability.[97–99]

6.2.2.2 *Facet effect*

This refers to how the geometric arrangement of atoms at the surface of a material can influence its catalytic properties. The effect is particularly prominent in materials like Pt, where the density of states (DOS) for d-orbitals is highly sensitive to surface geometry.[100] Each crystallographic plane presents a unique coordination number, leading to a distinct d-band width which, in turn, influences the material's electronic properties and catalytic activity. For example, in Pt, the coordination numbers of atoms on the (111), (100), and (110) surfaces are 9, 8, and 7, respectively. These variations correlate directly with d-band centre energies of −2.76 eV, −2.90 eV, and −2.54 eV.[66] Such differences significantly impact the activity sequence for reactions such as the ORR. In experimental settings, such as in perchloric acid, Pt(110) exhibits the highest ORR activity, followed by Pt(111) and then Pt(100).[101] This variation is partly due to how the d-band position influences electron donation to adsorbed oxygen, affecting its activation and subsequent reactions. However, the electrochemical environment also plays a crucial role.[102] For instance, different studies have explored how variations in the type of acid used as the electrolyte, such as sulfuric or phosphoric acids, can influence outcomes. These acids may interact differently with the catalyst surfaces due to variations in ionic strength, pH, or specific ion effects, thereby affecting the overall ORR performance[103,104] Surfaces with high-index planes, characterised by their concave sites and crystal steps, typically exhibit reduced binding energies for adsorbed OH and thus enhanced ORR activity when compared to systems expressing more densely packed Pt(111) planes.[105,106] Of course, real catalytic surfaces often feature these irregular, high-index planes, which may significantly differ from the idealised, smooth planes studied in controlled experiments.[107,108] Illustrating this, Stamenkovic *et al.* demonstrated that $Pt_3Ni(111)$ exhibited a higher ORR performance compared to $Pt_3Ni(110)$ and $Pt_3Ni(100)$, indicating that the optimal ORR performance was influenced by both the catalyst's composition and its crystalline facets.[66]

6.2.2.3 *Ligand effect*

This refers to changes in the electronic properties and chemical reactivity of a metal centre due to the influence of surrounding ligands. The latter, which can be atoms, ions, or molecules, affect the metal by altering its electron density and the spatial distribution of its d electrons. This modulation of electronic properties can significantly impact the catalytic behaviour of the metal, influencing its ability to adsorb and activate reactants. In a study by Xie *et al.*,[109] the ability of nitrogen to enhance orbital interactions was explored to address the challenge of sluggish water dissociation kinetics at PtNi nanowires in the HER context in alkaline environments. Researchers pinpointed unfavourable orbital orientations at the Ni sites as the root cause of the inefficiency, which hampered effective water molecule activation. To counter this, nitrogen was introduced into the nanowires, a modification that profoundly altered the electronic environment around the nickel atoms. This alteration was focused on the d_{z^2} orbitals, leading to the creation of new, empty d_{z^2} orbitals that were better oriented for water adsorption and activation. This facilitated the cleavage of O–H bonds in water, with doping inducing a remarkable reduction in overpotential, achieving an ultralow performance of 13 mV at 10 mA/cm^2 – a new benchmark for alkaline HER. These empirical findings were supported by structural characterisation, which confirmed N integration, and DFT analysis, which confirmed changes in electronic density and orbital orientation. In another work, Wu *et al.* demonstrated significant advances in ORR performance through the innovative design of Pt-based core–shell nanocrystals.[110] In this study, Pb was alloyed with Pd cores and Pt shells to create $Pd_3Pb@Pt_mPb$

(m = 3, 4) nanocubes, resulting in catalysts that exhibited superior activity and stability compared to commercial Pt/C catalysts. Specifically, $Pd_3Pb@Pt_3Pb$ nanocubes achieved mass and specific activities in alkaline media that were approximately 40 and 25 times higher, respectively, than those of commercial Pt/C. Additionally, these nanocubes demonstrated remarkable durability, with only a 9.3% loss in mass activity after 10,000 cycles. This was in stark contrast to the 59.9% decrease observed in commercial Pt/C. Theoretical analyses, including geometric phase analysis (GPA) and DFT calculations, revealed that strain contributed only minimally (~4%) to the enhancement in ORR activity. Instead, the ligand effect played a dominant role, driven by the negligible lattice mismatch between Pd_3Pb and Pt_mPb. This minimal mismatch was argued to facilitate effective electronic coupling between Pt and Pd, while the strong p-d orbital hybridisation between Pt and Pb crucially downshifted the d-band centre of Pt. This shift dramatically boosted ORR activity, showcasing how altering the electronic structure through ligand effects could lead to significantly enhanced catalytic performance. In a similar vein, Pan *et al.* investigated the modification of the atomic and electronic structure of Pt-based alloys to improve their activity and resistance to CO poisoning in hydrogen oxidation reactions (HORs).[111] A key aspect of their study was the introduction of single-atom copper (Cu) modifiers, which dynamically influenced the electronic properties of the Pt-based alloy. By leveraging the ligand effect, the single-atom Cu was found to modulate the d-band centre in PtCuPd, significantly enhancing HOR performance. Utilising *in situ* X-ray absorption spectroscopy, a potential-driven structural rearrangement into high-coordination Cu–Pt/Pd configurations emerged. This rearrangement intensified the ligand effect within the alloy, optimising H_2 and OH binding energies and simultaneously reducing CO adsorption energies. These modifications significantly boosted the kinetics of the HOR and enhanced the alloy's tolerance to CO, making the $PtPdCu_1/C$ catalyst remarkably resistant to CO poisoning, even at 1,000 ppm impurity levels.

Beyond manipulating the three effects listed above, elevating intrinsic catalytic activity also relies on achieving precise compositional control. While adjusting reaction stoichiometry and process conditions might suggest straightforward methods, the reality is often complex, involving advanced synthesis techniques such as atomic layer deposition or controlled chemical vapour deposition.[52] These approaches are required to fine-tune the catalyst surface with high precision, and the ease of this control can vary significantly based on the system and its scale of production. In some established systems, such as those involving Pt-based catalysts, methodologies might be well-developed. But, in newer or less understood compositions, achieving precise control is far from straightforward. The need for such control over composition is crucial for optimising the adsorption strength of reactants and intermediates.[52] If adsorption is too strong, intermediates will accumulate at the nanoparticle surface, thereby blocking access to active sites and limiting catalytic performance.[52,60] Conversely, too weak an adsorption will fail to effectively enable activation of bonds in the adsorbed species, retarding reaction.[59] Ideally, composition control should aim to achieve consistent and minimal activation energies throughout the catalytic process. Bringing activation energies closer to the theoretical minimum enhances the catalyst's effectiveness, allowing it to operate under milder conditions and have an extended operational lifespan. Moreover, consistent activation energies across the reaction pathway ensure that no intermediate steps become rate-limiting, thereby enhancing the overall reaction rate and selectivity. The manipulation of the catalyst's composition and structure, such as through doping or creating bimetallic interfaces, is essential to fine-tune these energy barriers, shifting the catalyst's performance towards theoretical ideals.[58] However, significant challenges arise when attempting to modify a catalyst's composition in a reduction-based synthesis without compromising its morphology. Changes in composition typically require adjustments in the reduction rates of the different metal ions present, affecting the nucleation and growth rates of the resulting particles.[58] This influence on nucleation and growth rates is critical because it dictates the ultimate size and shape of the catalyst particles.[112] Particle shape, in particular, depends on the relative growth rates of exposed crystal facets, which in turn are influenced by factors such as relative rates of adatom deposition and surface energy variations between facets.[113] As these growth dynamics change with modifications in composition, they may correspondingly alter both particle size and shape. For example, increasing the reduction rate of one metal component might promote faster nucleation, leading to smaller particles, while a slower growth rate might result in elongated or irregular shapes.[60] Such changes can have profound effects on catalytic activity and selectivity, potentially undermining the benefits derived from precise morphological control.[52] Therefore, maintaining a balance between composition and morphology during catalyst synthesis is essential for optimising performance and achieving desired catalytic properties.

6.2.3 *Increasing the number of surface active sites*

Developing nanostructured Pt–transition metal alloys serves two crucial purposes in catalyst design. Firstly, it enhances the number of accessible active sites on catalysts, and secondly, it modifies the electronic properties of Pt, providing new pathways for chemical interactions. Techniques such as nanosizing, which involves reducing particle size to the nanometre scale, and creating high-surface-area structures are pivotal to increasing the efficiency of catalysts by exposing a greater number of active sites.[81] Nanosizing not only increases the surface area relative to volume but also often leads to enhanced catalytic activity due to quantum size effects and increased surface reactivity. Moving on to nanostructuring, continued research into the modification of alloys is vital for two main reasons. It reduces dependence on scarce materials such as Pt by effectively utilising smaller amounts in a dispersed form, and it improves the economic feasibility of catalytic technologies. Nanostructuring can include anisotropic growth, leading to structures with specific shapes such as rods[114] or sheets,[115] or it can involve the creation of single-atom catalysts,[116] where individual metal atoms are dispersed on a support. Both approaches aim to maximise the catalytic surface area and modulate the electronic environment of the active sites. For example, single-atom metal catalysts have shown exceptional promise in enhancing catalytic efficiency. In one recently reported hybrid electrocatalyst composed of atomically dispersed Pt and Fe atoms alongside PtFe alloy nanoparticles, this was achieved through the simple combination of Zn^{2+}, Fe^{2+} and 2-methylimidazole, followed by calcination at 1000°C. This contrasted with the more usual formation of single-atom sites using techniques such as atomic layer deposition or chemical vapour deposition. In the current case, the simply prepared catalysts were characterised, for example, using scanning transmission electron microscopy (STEM) and XPS, confirming the atomic dispersion of Pt and Fe. This catalyst demonstrated an ORR mass activity in PEM fuel cells that was 3.7 times greater than that of commercial Pt/C.[116] Transitioning from atomic-scale modifications to nanostructured alloys, ultrathin, crystalline Pt_3Ni nanocages represent a significant recent advance. These structures have been created through controlled reduction and dealloying processes that selectively removed less noble metals to form cages with ORR-favoured facets and high electrochemically active surface areas (ECSAs). An evaluation of ECSAs, defined as surface areas of catalysts that are available and reactive for electrochemical processes, is crucial for understanding catalyst efficiency. The ultrathin nature of the Pt_3Ni nanocages prepared in the current case allowed for a dramatically increased ECSA, delivering mass and specific activities up to 4.71 A/mg_{Pt} and 5.14 mA/cm^2, respectively – surpassing those of commercial Pt/C by factors of 26 and 20.[61] Furthermore, distinctive mesoporous Pt@Pt-skin Pt_3Ni core–shell framework nanowires have been developed to enhance surface site availability even further. These nanowires were synthesised through a multistep process involving the deposition of Pt_3Ni, followed by a controlled dealloying to create a Pt-rich skin. This resulted in a highly anisotropic, mesoporous structure that maximised the exposure of active sites. Its remarkable mass and specific activities of 6.69 A/mg_{Pt} and 8.42 mA/cm^2 at 0.9 V versus reversible hydrogen electrode (RHE), which are 29 and 26 times higher than those of commercial Pt/C, respectively, illustrate the effectiveness of this structural approach.[62]

While catalysts with complex morphologies such as ultrathin nanocages[117] and core–shell nanowires[118] have clearly yielded noteworthy activities in ORR, their precisely engineered properties often come at a cost. They can require intricate, multistep, and costly synthetic methods that limit their widespread application.[119] Additionally, these highly structured catalysts are not readily amenable to composition manipulation; changes in the population of one component likely alter the intricate structures that are crucial for their high performance, thereby diminishing their effectiveness. Consequently, there is growing interest in developing highly branched nanostructures,[120,121] such as nanopods and nanodendrites, which offer a simpler synthesis without necessarily depending on highly specific interatomic ratios.[60] These structures are valued for their high surface-area-to-volume ratios, providing extensive active sites relatively easily compared to more complex geometries.[121] Nanodendrites, in particular, feature abundant edge and corner atoms, as well as high porosities.[122,123] The latter not only increases the number of accessible active sites due to the larger surface area but also enhances the reactivity of the catalyst.[124] Meanwhile, the plentiful edge and corner atoms provide unsaturated sites that are more chemically active. The presence of stepped edges further increases reactivity by offering additional unique active sites that facilitate bond breaking and formation during chemical reactions.[125] The hierarchical structures of nanodendrites consist of three-dimensional interconnected networks of ultrathin branches (with widths approximately in the nanometre range) that facilitate electron transfer during chemical reactions.[126,127] Moreover, these branches, with their multiple sharp tips, not only promote electrodeposition but also generate high electric fields, accelerating reactant transport.[122,128] However, while their desirability is clear, the exact mechanisms governing the growth of nanodendritic structures[129] remain poorly understood, and their synthesis has tended to be largely based on empirical approaches.

6.3 Self-Organisation and Nanodendrite Formation

6.3.1 *Challenges of mastering composition control*

The synthesis of nanodendrites has traditionally hinged on empirical methods, largely because the growth mechanisms of these complex nanostructures have not been fully understood. This is particularly true of multielement systems, as outlined in the following. Briefly, at the core of nanodendrite formation is a kinetic growth process characterised by diffusion-controlled aggregation.[58] In this process, monomers (e.g., individual metal atoms) are initially produced by the chemical reduction or thermal degradation of metal-based reagents.[113] The collision-based agglomeration of these then gives seeds, which may further grow or dissipate on account of Ostwald ripening.[130] Thereafter, seeds with at least a critical size, moving randomly due to Brownian motion, aggregate to form dendritic structures on account of collision-based attachment mechanisms.[131] These may incur oriented or non-oriented development, depending on collision rates.[132] This characteristic dendritic growth profile contrasts with that of simpler yet anisotropic nanopods. These form on account of the selective overgrowth of specific high-energy facets expressed by seed surfaces.[112,133] Nanodendrites instead require that nanoparticles freely diffuse and collide, leading to spontaneous organisation into intricate patterns.[134]

Achieving precise control over both the morphology and composition of multielement NPs is critical for enhancing intrinsic catalytic activity and increasing the density of accessible active sites.[52] In the case of Pt-based nanodendrites, generating bimetallic systems presents significant technical challenges that impact both the feasibility and effectiveness of these systems. That said, the imperative for addressing the issues is clear. As already noted, Pt-based catalysts are effective in catalytic converters[135] and hold huge promise in fuel cell technologies, specifically in enhancing the electrochemical reduction of oxygen in PEM fuel cells.[136] Moreover, their application in organic synthesis, especially in the hydrogenation of fine chemicals, demonstrates improved selectivity and activity due to the unique electronic properties of the bimetallic interfaces.[137] In the case of high-surface-area catalysts, integrating multiple metals into a single dendritic structure without compromising its morphological and functional properties is complex due to the differing kinetic behaviour exhibited during metal deposition.[138] This has its origin in the different metals displaying varying reduction potentials and solubility.[139] This variability can result in non-uniform reduction, nucleation, and adatom deposition rates, leading to inconsistent growth,[140] heterostructure formation,[141] or the isolation of multiple products.[61] These disparities can distort the morphology of the dendrites, leading to structures that are not only irregular but also potentially weakened or fragmented. This distortion and potential compromise in structural integrity are critical because they can impact the effectiveness of dendrites in applications where uniformity in structure and composition is crucial. Additionally, structural weaknesses may render the dendrites less durable under operational conditions, thereby affecting their longevity. However, effectively manipulating the relationship between growth kinetics and compositional accuracy presents a formidable challenge. The synthesis of nanoparticles often reveals that directing kinetic growth pathways and achieving precise composition control are interdependent.[58] This interconnectedness is particularly pronounced in nanodendritic morphologies, which, due to their complex branching patterns, demand meticulous control over nucleation and growth to maintain their structures. These morphologies nevertheless offer significant advantages, facilitating the expression of a greater number of surface catalytic sites.[60] In comparison, other nanoparticle morphologies, such as pseudospheres, offer less impressive surface-area-to-volume ratios yet still suffer from challenges whereby particle size and distribution are easily affected if individual particle composition is varied.[58] These comments highlight a general issue in nanoparticle synthesis: altering one parameter (e.g., dopant inclusion) can unpredictably affect others (e.g., size or shape). Self-assembled structures (such as nanodendrites) in particular exhibit unique challenges, as they rely heavily on the balance of intermolecular forces and precise molecular configurations to form desired arrangements. Slight deviations in composition can disrupt these forces, leading to significant variations in the final structure. Synthetic control is therefore essential, as detailed in the literature, where studies have shown that even minor inconsistencies in nanoparticle synthesis can lead to substantial declines in performance due to changes in surface properties and active-site availability.[142,143]

A notable challenge in controlling morphology and composition in nanoparticles when seeking to incorporate metals such as Pt is the relatively positive reduction potential of this metal. This frequently leads to its preferential reduction when compared to that of other metal ions present.[113,144] This can lead to the overrepresentation of Pt in emerging nanoparticles, overshadowing other metals with more negative reduction potentials.[113] Addressing this synthetic challenge necessitates the adjustment of the reduction rates for the various metal ions present to steer them towards a more controlled thermodynamic pathway. On the face of it, this is straightforward, requiring only strongly reducing conditions. However, in practice, adjustments must be finely tuned to balance the nucleation and growth rates across different metals, with thermodynamic reduction followed by kinetic conditions that do not favour the rapid formation of nanoparticles composed of one metal over others. This balance is crucial because any shift in kinetic conditions could significantly alter the growth rates, potentially leading to deviations from the desired nanodendrite morphology. For instance, modifying conditions to slow down the reduction of Pt might inadvertently enhance the growth kinetics of other metals such as Ni or Co, meaning that not only composition but also growth patterns (morphology) are altered.[60] Successfully navigating these complexities requires an understanding of both the chemical kinetics involved in nanoparticle formation and the thermodynamic aspects influencing material composition.

6.3.2 *Current approaches to nanodendrite formation*

Previous explorations of nanodendrite formation have illuminated several key mechanistic points based on the use of different synthetic methods. In all cases, however, the fabrication of these intricate structures has been through empirical means. Nonetheless, each method has highlighted a different aspect of the growth process, and collectively, they have provided insights into effective control strategies. First, in 2014, Lv *et al.* introduced an innovative approach to synthesise porous PtPd nanodendrites through an environmentally benign, one-pot, wet chemical method.[145] This employed poly(vinylpyrrolidone) (PVP) and urea as both co-stabilising and co-structure-directing agents. The synthesis began with the addition of hydrazine hydrate, which rapidly generated a surge of Pt and Pd atoms. These metal atoms quickly reached a hypersaturated state in the solution, prompting the formation of dense nuclei through self-aggregation, designed to minimise total surface free energy – a common driving force in nanoparticle formation.[146] Having formed, these nuclei began to organise into complex dendritic structures. This was facilitated by mechanisms such as twinning and imperfectly oriented attachment, processes that are not only characteristic of dendritic growth but also inherently unpredictable and sensitive to synthesis conditions. In time-dependent morphology evolution tests, it was observed that increasing reaction duration allowed nanoparticles formed later in the reaction to deposit onto initially formed ones, evolving into mature nanodendrites. The presence of both PVP and urea played a critical role in directing the self-organisation of the initially formed seeds, with neither species working in isolation and high temperatures disfavouring nanodendrites by encouraging PVP de-coordination. While this work was of enormous interest, its empirical nature was problematic, as it required nuanced control over the conditions that were necessary to achieve a dendritic morphology. The synthesis depended significantly on the concentration of PVP, the amount of urea, the temperature of the reaction, and the precise timing of reagent addition. This made the process somewhat unpredictable if parameters should be varied and highly dependent on the experimental setup, illustrating the challenges of replicating and scaling such synthetic strategies.

Building on the above report, Nosheen *et al.* further explored nanostructure fabrication.[147] They demonstrated a route to create uniform and hierarchical PtCu with highly detailed tetragonal superstructures. The process introduced some significant aspects that were not seen in Lv's work. The new synthesis still utilised PVP but introduced additional reagents, such as NaI. The mixture was subjected to hydrothermal conditions at 200°C for 1 h, though duration had little effect on morphology. In this approach, both tryptophan and iodide proved pivotal. The former was added as a complexing agent, potentially influencing the reduction rates and nucleation processes of the metal ions, and the exact amount added proved to significantly influence the outcome. More systematically, an elevated concentration of NaI generated more I_2 *in situ*. This underpinned product formation by etching the initially formed CuI nanocrystals, while the oxidation of iodide simultaneously promoted the reduction of Cu^+, whereupon the galvanic replacement reaction between Cu and Pt(IV) produced alloyed superstructures. This complex process underscored the critical importance of adjusting precursor concentrations and carefully managing the reaction conditions to steer nanodendrite growth.

The connection between these two studies lay in their use of similar stabilising agents (PVP chains). Both used protocols that required the fine-tuning of chemical environments and the addition of multiple capping agents and/or other additives. They clearly illustrated the nuanced challenges of directing the kinetic growth of nanoparticles, and both showed how slight variations in conditions could lead to significantly different outcomes. This highlights the experimental nature of nanomaterials research, where empirical methods have typically been essential for uncovering the precise conditions needed to tailor complex architectures. Notably, these studies showed different proclivities for reaction time to influence outcomes. Kuang *et al.* contributed further to this field in 2015 by detailing the time-dependent structural evolution of single-crystal cubic nanodendrites of Pt_3Cu.[148] Their study provided a detailed, step-by-step description of a four-step growth process, illustrating how intricate control over time and conditions influenced morphology (Fig. 6.2).

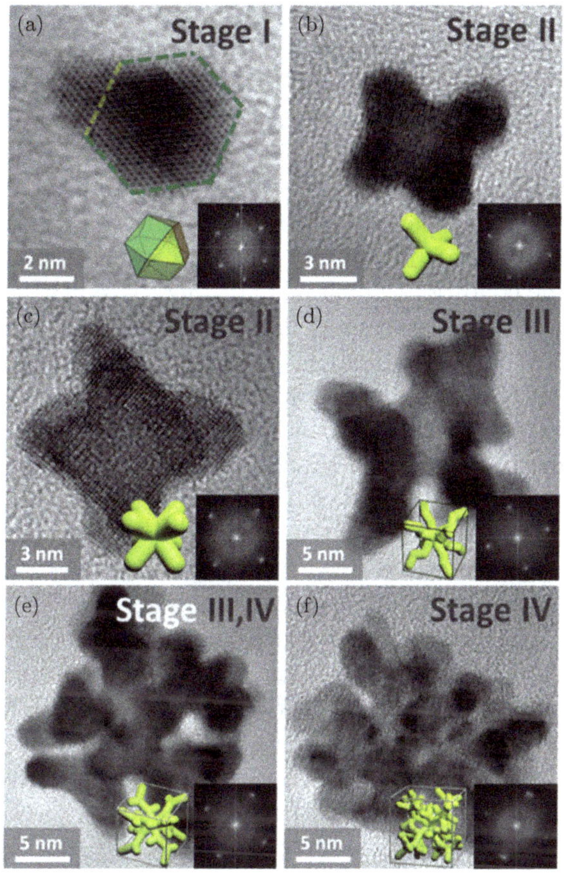

Fig. 6.2. (a)–(f): High-resolution TEM (HRTEM) images of dendritic Pt_3Cu intermediates formed during the synthesis of cubic nanodendrites (e.g., in (f)). Insets show the fast Fourier transform (FFT) patterns of each HRTEM image along with an idealised representation of the imaged particle.
Source: Reproduced from Ref. 148 with permission from the Royal Society of Chemistry.

PVP again featured as a capping agent, with KI introduced to control both reduction rates and facet-selective growth. The formation of the resulting dendritic structures was then summarised as follows. First, initial nucleation of primary cuboctahedral seeds formed after iodide modulation of reduction rates to ensure essentially simultaneous reduction of Cu(II) and Pt(IV). These seeds typically emerged from a supersaturated solution, where the solute concentration and temperature were controlled to initiate crystallisation. Second, seed growth was due to the iodide-controlled selective deposition of material on (100) facets.[149] This led to selective growth at (111) and the formation of tetrapods and then octapods that expressed (100) facets and mirrored the controlled growth conditions seen in the works of Lv[145] and Nosheen,[147] where the chemical environment directed structural assembly. Third, elongation of octapodal structures occurred, likely influenced by the continuous supply of reactants and the stable conditions maintained during the synthesis. This elongation stage proved crucial for defining the ultimate shape and size of the cubic nanodendrites. Fourth, and concurrently with elongation, new branches began to form, complicating the structure. This coincidence of branching with elongation suggested a dynamic interplay between growth kinetics and the available surface sites on the existing crystal structure. While two possibilities presented themselves to account for branching – oxidative etching[150] facilitated by iodide or oriented attachment[151] – some impressive microscopic studies suggested the former and that galvanic reactions were taking place to account for branching. This detailed analysis reinforced the themes of earlier studies,[145,147] emphasising the empirical and experimental nature of nanoparticle synthesis. However, the need to meticulously control facet overgrowth conditions and time, essential for achieving the desired structures, highlighted the challenges of replicating and scaling such precise nanostructures. Each stage of the growth process depended heavily on maintaining specific conditions, underscoring the experimentally iterative approach needed for synthetic control.

Further insights into the formation of Pt–group 11 alloyed nanodendrites were provided in 2017 by Weng *et al.*[152] This work focused on the formation of PtAg nanocrystals, employing L-hydroxyproline (L-Hyp) and, similar to previous studies, PVP in the synthesis. In an initial step, the reduction of silver ions was achieved using ascorbic acid (AA), a common reducing agent in nanoparticle synthesis. This process preceded the addition of hydrazine for the subsequent rapid production of Pt atoms. This latter step, performed in the presence of L-Hyp and PVP, led to the formation of PtAg nuclei. Reference experiments established the need for specific amounts of AA and L-Hyp, with timed studies suggesting nanodendrite formation by oriented growth. The authors proposed a crucial role for L-Hyp, suggesting it adsorbed at the surfaces of nascent nuclei established as hydrazine generated an abundance of Pt atoms that interacted with preformed Ag. This was proposed to reduce the total surface free energy of the seeds, representing a key factor in promoting spontaneous nucleation and epitaxial growth.[153] L-Hyp and PVP were then argued to promote subsequent nanodendrite formation by encouraging growth along the (111) direction. Though the specifics were not explained, the overall process would be well aligned with the nucleation events described in the nanodendritic growth mechanisms proposed by Kuang *et al.*[148]

While the work by Weng *et al.*[152] focused on morphological variations, iterative methodologies in nanoparticle synthesis were further advanced in subsequent studies by Wu *et al.*,[154,155] focusing on the precise control over both morphology and composition. In 2017, this team developed a technique for synthesising core@shell Ag@Pt nanoarchitectures, effectively controlling the overgrowth of a Pt shell on Ag cores (Fig. 6.3).[154] Adjustment of the amount of Pt precursor represented a move aimed at influencing the thickness and uniformity of the shell growing on a preformed group 11 core. Control over the Pt precursor volume not only allowed for the customisation of the nanoarchitecture's final physical properties but also demonstrated the versatility of the method in tailoring nanoparticle architectures. Using 125 μL of H_2PtCl_6, Pt loaded on the surfaces of the Ag nanoparticles roughened the seed surfaces. However, upon adding 250 μL Pt precursor, nanodendrites with irregular morphologies began to emerge, and a further increase in the amount of Pt precursor led to well-developed Ag@Pt nanodendrites. This ability to precisely manage the deposition of Pt on Ag cores echoed the

Fig. 6.3. Core@shell Ag@Pt (Ag = yellow and Pt = blue) made by treating Ag seeds with increasing volumes of 40 mM H_2PtCl_6. From left to right: 125, 250, 375, and 500 μL.
Source: Modified from Ref. 154.

controlled growth processes discussed by Kuang *et al.*,[148] where the overgrowth of predominantly (111) facets was seen.

In 2020, Han *et al.*, interested in the scope of Rh as a catalyst, described the synthesis of PtRh nanodendrites for the HER over a broad pH range.[156,157] They used a process that involved the dispersion of Pt and Rh precursors in oleylamine (OAm), a long-chain amine capable of acting as both a capping and reducing agent.[158] Metal ions were then complexed by cetyltrimethylammonium ions (CTA^+) derived from CTAC and/or by glycine, with the amino- and carboxyl-groups of the latter being closely correlated to the reduction kinetics of metal salts to which it can chelate,[159] as shown in Fig. 6.4. Following OAm reduction to give metallic Pt(0) and Rh(0), these atoms rapidly aggregated to form PtRh nuclei. The presence of CTA^+ and specific levels of glycine proved crucial at this stage; both agents were adsorbed at the surfaces of the forming *fcc* PtRh nuclei, influencing the total surface free energy and inducing anisotropy by promoting overgrowth at the (111) surface facets. Lastly, although not looked at in detail, the involvement of Ostwald ripening, a coarsening effect in which smaller particles dissolve and redeposit onto larger structures,[160] was also argued to contribute to the formation

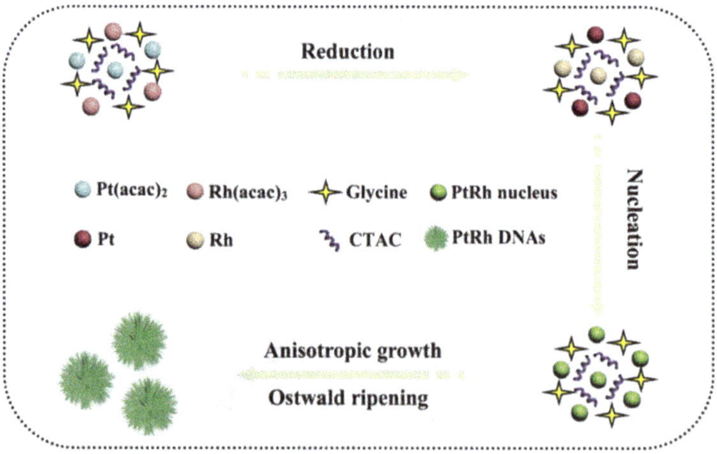

Fig. 6.4. Method of preparing PtRh nanodendrites using OAm coreduction of metal salts, giving rapidly aggregated bimetallic nuclei. CTA^+ (from CTAC) and glycine adsorbed at the surfaces of these nuclei, promoting overgrowth at the (111) facets of *fcc* PtRh.
Source: Reproduced from Ref. 158 with the permission of Elsevier Ltd.

of distinctive nanodendrites. While some details remained obscure, the overall process aligned with the themes observed in the work of Wu *et al.*,[154] where precise control over chemical environments and the use of specific directing agents in exact quantities played pivotal roles in shaping observed nanostructures. Similar to Wu's technique of controlling Ag@Pt architectures through precursor management, Han *et al.*'s method demonstrated how tailored chemical dynamics and the selection of specific molecular interactions could drive the formation of dendritic nanostructures.

Collectively, the studies outlined in this section have illustrated various approaches to NP synthesis, each providing insights into the mechanisms driving the growth of nanodendrites. The techniques highlight how complex chemical dynamics and multiple synthesis steps can be tailored to specific outcomes. However, it is crucial to recognise that the mechanisms described are often specific to the experimental conditions under which they were developed and that even minor deviations can quickly redirect reactions, particularly away from predictable nanodendrite formation. Moreover, conditions seen to promote that formation were generally identified empirically. The resulting specificity indicates that while the methodologies are profoundly insightful, their applicability might be limited to certain environments or material systems, and their generalisation is likely to be extremely complex. The need to better understand these growth mechanisms and the roles of each chemical component in a system requires developments aimed at complete elucidation of the growth processes, verification of their applicability across broader contexts, and refinement of the techniques to enhance their reproducibility in different, ideally straightforward, synthesis environments.

6.3.3 *A systematic protocol for nanodendrite synthesis*

Section 6.3.2 highlights case studies where at least a limited understanding of how to construct PtM nanodendrites has been achieved. In particular, it is noteworthy that the observation of a combination of galvanic and attachment processes demonstrated that nanopods underwent elaboration to form highly branched structures. This suggests that the starting point for systematising nanodendrite formation has the following two basic requirements: a better understanding of the formation and properties of nanopods, and the need to develop a chemical system capable of predictably yielding both nanopods and nanodendrites. Recent work in this direction has utilised the bimetallic combination of Pt with Fe and harnessed a simple dual-capping agent approach. For example, Pazos-Pérez *et al.* reported PtFe nanopods when combining metal salts in a 1:1 mixture of oleic acid (OA) and OAm.[161] This combination offers an enormous range of possible controls in bimetallic nanoparticle formation and will assume some significance in systematising the synthesis of nanodendrites, in particular.[162] The dual use of OA and OAm has been noted[163] to present a mixture of carboxyl anions, protonated oleylamine, and acid–base complexes that are generated depending on the conditions. In particular, electron donation by the carboxyl anions can lend particular stabilisation to evolving nanoparticles.[164] Ligand exchange and *in situ* generation of metal oleates from OA provide further subtle control of reducing conditions.[165] At the end of the day, however, OA is purely a capping agent. It is, moreover, facet-indiscriminate, which promotes the growth of symmetric cuboctahedra. In contrast, the relative growth between the (111) and (100) facets of PtM can be modulated by the presence of OAm,[166] since this capping agent preferentially binds to the (100) facet (cube face). This promotes growth at the (111) facet (cube corners) of cuboctahedra and the formation of anisotropic particles. Such an action has previously been reported in PtFe nanocube formation. Finally, for yet further fine-tuning of particle growth, OAm has already been noted to function as both a capping *and* reducing agent.[103] Despite the practical potential of OA–OAm, compositional control of the product proved difficult in initial PtFe nanopod work;[161] a 2:1 Fe:Pt molar ratio of reagents yielded highly Pt-rich nanopods. This, however, led to an exploration of the effects caused by varying the OA:OAm ratio, resulting in the formation of dumbbells that incorporated an equimolar PtFe nanopod combined with an Fe-based nanocube when employing a 1:3 ratio of OA:OAm.[167] Mechanistic investigation showed that this stoichiometry of pod was strongly favoured, with a deficiency of Fe reagent giving not a depleted yield of dumbbells but isolated 50:50 PtFe nanopods. More interestingly in the current context, using lower temperatures gave dumbbells incorporating not a nanopd but a nanodendrite (Fig. 6.5). These data suggested an intimate relationship

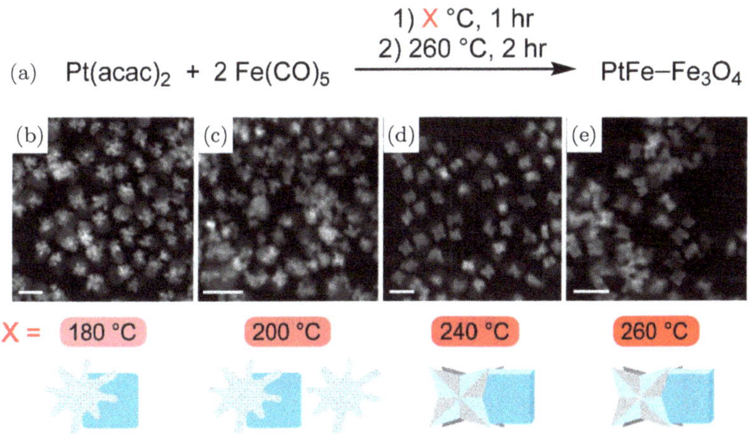

Fig. 6.5. (a) Reaction scheme illustrating the formation of PtFe–Fe_3O_4 nanodumbbells using 1:3 OA:OAm and in which morphology transitions from nanodendritic to nanopodal upon varying the initial reaction temperature. (b)–(e) STEM HAADF images of products obtained with initial temperatures of (b) 180°C, (c) 200°C, (d) 240°C, and (e) 260°C. Scale bars 25 nm.
Source: Modified from Ref. 167.

between the two chemistries and that excessive heating might favour nanopods.

Building on the empirical insights reviewed in Section 6.3.2 and the interplay between nanodendrite and nanopod chemistries outlined above, recent attempts to systematise nanodendrite formation have focused on the idea that reaching a specific nanopod concentration is critical for initiating nanodendrite formation. As suggested in Fig. 6.6, this optimal concentration allows for diffusion-controlled aggregation, with collisions between nanopods expected to lead to coalescence events and nanopod arms beginning to branch, thus facilitating the emergence of nanodendrites.[58]

Before exploring this potentially non-empirical protocol for nanodendrite synthesis, it is essential to understand the relationship between kinetics and thermodynamics, as the principles underpinning this fundamentally influence the formation and growth of nanoparticles, including nanodendrites.[168] Kinetics deals with the rates at which chemical processes occur; in nanoparticle synthesis, kinetic factors are crucial for controlling the formation and characteristics of nanoparticles. One of these factors is the rate of nucleation, which refers to the initial formation of stable clusters from

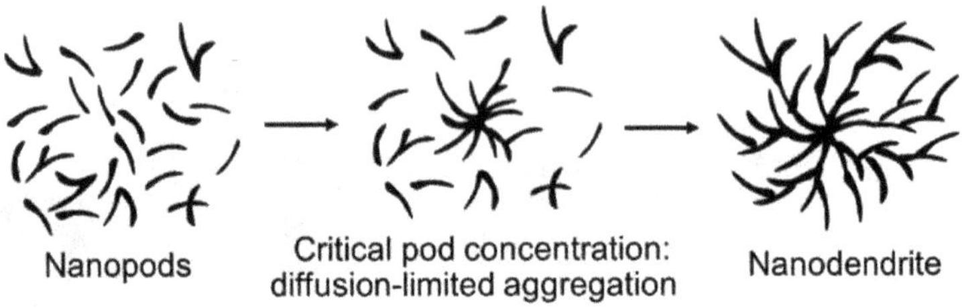

Fig. 6.6. Proposed mechanism for forming polycrystalline nanodendrites. Initially, nanopods form (left). As collisions occur, some nanopod arms begin to branch, leading to the development of nanodendrites (middle). Collisions between further nanopods and the arms of existing nanodendrites then deplete the nanopod supply (right).
Source: Modified from Ref. 168.

the reactants. Nucleation sets the stage for all subsequent growth of nanoparticles.[169] It can dramatically influence both the number of nanoparticles formed and their size distribution depending on how quickly the nuclei are established and stabilised. Once nucleation occurs, growth takes over, where atoms or molecules from the reactants add to the already-formed nuclei. This growth phase can lead to the formation of either single-crystal or polycrystalline nanoparticles, the latter resulting from multiple nucleation events that coalesce. The process of oriented attachment, in which nanoparticles align and join based on specific crystallographic orientations, may also occur during growth, influencing the final morphology and crystallinity of the particles.[169] Subsequently, the dynamics of particle aggregation come into play, affecting how these particles interact and agglomerate, which can further influence the physical properties of the nanoparticle ensemble.[170] These elements – nucleation, growth, and aggregation – collectively determine the rate of particle formation, the ultimate size and morphology of the particles, and their interactions after formation. For example, rapid nucleation combined with slower growth rates might lead to the formation of many small nanoparticles, whereas slower nucleation with faster growth can yield fewer, larger particles. In contrast, thermodynamics concerns the stability and phase behaviour of materials under different conditions. It dictates the energetically favourable states of a system, irrespective of the time taken to reach these states. In nanoparticle synthesis, thermodynamic factors include the solubility of metals in various solvents, the reduction potentials of metal ions, and the surface free energy of particles. These aspects influence the equilibrium size, shape, and composition of nanoparticles, determining the most stable configurations that particles can adopt. In the context of nanodendrite synthesis, the interplay between kinetics and thermodynamics is particularly crucial. Kinetic processes govern the early stages of formation, such as the nucleation and growth of nanopods and their aggregation into more complex structures. Thermodynamic factors then play a role in stabilising these structures, ensuring that they maintain certain sizes and shapes at equilibrium and preventing them from dissolving or reshaping unless the environmental conditions are altered.

Drawing on the empirical insights discussed earlier and recent work on PtFe systems, it has been proposed (Fig. 6.6) that at a certain concentration, nanopods undergo modification resulting in nanodendrite formation. Specifically, this concentration of nanopods allows for diffusion-controlled aggregation, where the movement and collision of nanopods are likely to result in coalescence events. During these interactions, some nanopod arms may start to branch, promoting the development of nanodendrites. This view suggests a critical balance between controlled kinetic conditions and thermodynamic stability that can be harnessed to systematically produce nanodendrites.

This potential approach to induced nanodendrite formation has recently been tested by preparing nanopods at different controlled concentrations to investigate the influence on both the emergence and complexity of nanodendrites.[60] The protocol used included the insertion of an initial reduced-temperature step (200°C, Step 1, Scheme 6.1) into a hitherto established method for generating PtM nanodendrites. This utilised the simple one-pot heating (250°C for 30 min) of metal salts in OA–OAm and treatment with a strong reducing agent, 1,2-hexadencanediol (1,2-HDD), to ensure smooth reduction of all metal ions present. In the modified protocol, Step 1 aimed at moderating the generation of isotropic seed nanoparticles. Subsequently, elevating the temperature (Step 2) allowed for the reduction of feedstock remaining after (temperature-limited) seed formation. The duration of Step 1 significantly impacted outcomes according to three scenarios:

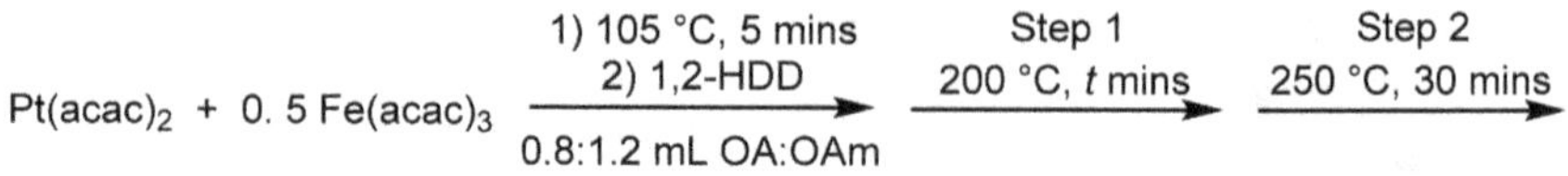

Scheme 6.1. Synthesis process for PtFe alloy nanoparticles incorporating a two-step process, where Step 1 controls the extent of nucleation and targets a nanopod concentration that promotes subsequent nanodendrite formation in Step 2.[58]

(1) Short duration: If Step 1 was brief, few seeds formed, minimally affecting monomer consumption in Step 2. This condition retained extensive amounts of feedstock for subsequent growth of a low concentration of well-formed nanopods in Step 2.
(2) Intermediate duration: Extending Step 1 led to more substantial seed formation, which depleted the feedstock more significantly, but not substantially, and therefore restricted nanopod formation in Step 2, leading to an intermediate concentration of nanopods and potentially triggering nanodendrite formation.
(3) Extended duration: Prolonged nucleation in Step 1 consumed much of the feedstock, reducing nucleation potential in Step 2. However, the already high concentration of nanopods surpassed the threshold at which nanodendrites should collisionally form.

The thesis presented above was tested using simple, commercially available $Pt(acac)_2$ and $Fe(acac)_3$ with a 2:1 Pt:Fe ratio and a 0.8:1.2 OA:OAm volume ratio (for the iteration of these conditions, see Ref. 58). Microscopic analysis of products from experiments that omitted Step 1 ($t = 0$ min) was dominated by nanopod formation (Fig. 6.7(a)). In line with expectations, extending Step 1 to 10 min facilitated a transition to nanodendrite formation (Fig. 6.7(b)), while a further extension to 20 minutes resulted in a return to nanopod formation (Fig. 6.7(c)). These data underscored not only the critical balance between seed concentration and growth conditions necessary for controlling morphology but also the idea that nanodendrite formation could be induced rationally and predictably. A question, however, remained: could these synthetic and resulting morphology changes be made without influencing the *composition* of individual nanoparticles. Within error, the answer appeared to be 'yes' (Fig. 6.7(i)), though the need for clarification was obvious given the apparent trend in Pt content reported by EDX analysis. To

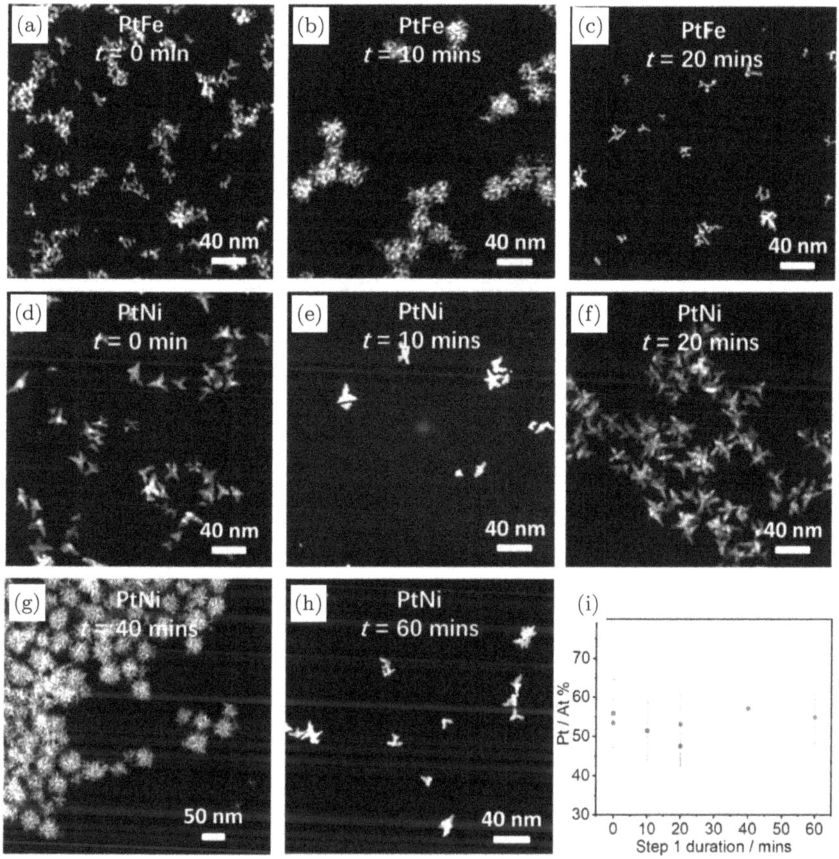

Fig. 6.7. (a)–(c) Representative HAADF images of PtFe nanoparticles synthesised with a 2:1 Pt:Fe reagent ratio and 0.8:1.2 mL OA:OAm volume ratio at 250°C for 30 min, following a mixing period at 200°C ranging from 0 to 20 min (see Scheme 6.1). (d)–(h) Representative HAADF images of corresponding PtNi nanoparticles (1:1 Pt:Ni, 0.4:1.2 mL OA:OAm, 250°C for 30 min after 200°C for 0–60 min). (i) Pt atomic percentage as a function of the 200°C mixing time for the products in (a)–(h); red data PtFe, green data PtNi.
Source: Modified from Ref. 58.

more reliably assess compositional consistency, parallel studies were conducted with equimolar $Pt(acac)_2$ and $Ni(acac)_2$, using a 0.4:1.2 OA:OAm ratio.[58] Adjusting the duration of the 200°C mixing step (from 0 to 60 min) before proceeding to Step 2 (250°C for 30 min) now showed two things clearly: first, increasing the mixing time resulted in the same pattern of morphological change (Fig. 6.7(d)–(h)); second, this was done without significantly altering the elemental composition.

The approach outlined above offers, for the first time, a robust and controlled method for selectively forming nanodendrites across more than one bimetallic system. Importantly, dendrite formation was not only predictable, but its composition was also unaffected. Data revealed the crucial role of seed concentration, and thereby reaction conditions, in influencing the growth dynamics of nanodendrites. Building on this, it is feasible to manipulate both the composition and morphology in Pt-based nanodendrites based on two primary considerations:

(1) Directing reduction: The chemical reduction of metal ions must be steered towards the thermodynamic regime. This provides plentiful atoms for seed formation and circumvents the influence of more positive reduction potential on product composition. In particular, this will enable access to noble metal-poor compositions (Fig. 6.8(a)).
(2) Modulating growth: Adjusting the type and relative concentration of reagents, along with the reaction temperature, directly impacts nanoparticle shape by influencing nucleation and growth rates. This results in different morphologies depending on whether growth is predominantly governed by either thermodynamic or kinetic factors (Fig. 6.8(b)).

While manipulating reduction to ensure that the thermodynamic reaction is straightforward, iteration of growth conditions introduces greater complexity but offers considerable potential. Hence, a low flux of reduced metal monomers typically results in near-pseudospherical thermodynamic growth. In contrast, enhancing the reducing conditions enforces kinetic control and the development of anisotropic nanoparticles due to the rapid evolution of high-energy facets or through oriented attachment when nanocrystallites collide. A further increase in the reduction rate within the kinetic regime promotes more nucleation events, depletes the available feedstock for growth, and leads to the formation of smaller anisotropic particles

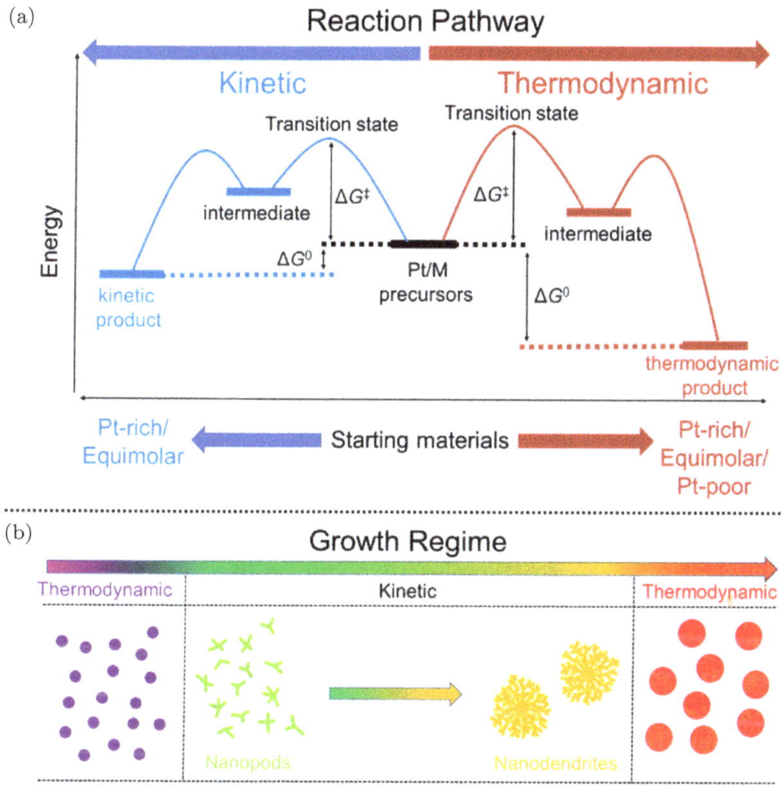

Fig. 6.8. (a) Impact of kinetic and thermodynamic reduction conditions on nanoparticle composition. (b) From left to right, the effects of increasingly chemically-reducing conditions on nanoparticle growth, demonstrating how the supply of feedstock influences morphology and size.

with truncated arms. Lastly, extremely strong reducing conditions revert particles to thermodynamic near-pseudospheres as growth accelerates on all facets.

The combined OA–OAm system facilitates the fine-tuning of the reducing conditions to target a seed concentration that achieves kinetic nanoparticle growth and yields anisotropic products. At its simplest (Scheme 6.1, excluding Step 1), this enables a straightforward, one-pot process at moderate temperatures to quickly render all metal ions zero-valent. Dual control over composition and morphology is then achieved by:

(1) adjusting the reduction rate to initiate kinetic growth, starting with modifications to the OA:OAm ratio while keeping the total reaction volume constant;
(2) expanding the kinetic growth regime by varying the absolute level of OAm, thereby altering the concentration of metal salts and the overall reducing nature of the system;
(3) fine-tuning the reducing conditions using temperature adjustments to increase or decrease the reduction rate;
(4) establishing a new kinetic growth regime by adjusting the concentration of a particular metal salt, thus selectively varying the contribution of that metal in the product.

So far as we are aware, this methodical manipulation of variables provides a first set of principles that can be used for systematically creating bimetallic, anisotropic nanoparticles in which both composition and morphology are predictably varied. Results summarised in Fig. 6.9 extend the ideas presented for PtFe to other bimetallic combinations, demonstrating the reliability of this approach. In this work, identifying the kinetic growth regime initially in PtNi synthesis achieved the formation of nanopods in 1:1 OA:OAm, and then, under more strongly reducing conditions of 0.5:1.5 OA:OAm, naodendrites were obtained. Consistent with the thesis that a concentration window existed at which nanopods generate nanodendrites, even more reducing conditions resulted in a departure from nanodendrite formation. Raising the temperature from 250°C to 275°C (point 3 above) not only enabled straightforward nanodendrite formation but also (at 300°C) proved capable of moving beyond it (including the generation of monometallic nanoparticles). More powerfully (point 4, above), while evidence emerged that nanodendrites formed here were all uniformly intermixed, they could still be made with specific compositions, varying from Pt-rich to Pt-poor. In contrast, Pt-poor nanopods were only achievable by lowering the level of the Pt reagent in the reaction (0.2:1 Pt:Ni). The generality of this approach was finally shown in relation to PtCo systems, where data (illustrated representatively in Fig. 6.9) essentially replicated those for PtFe and PtNi. A cost-effective strategy therefore emerges here for varying Pt

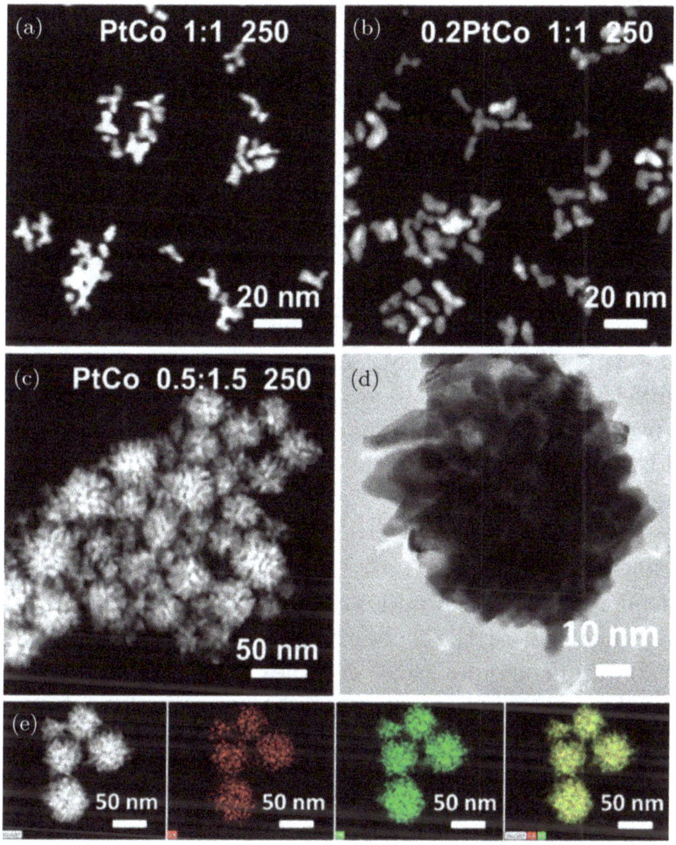

Fig. 6.9. HAADF imaging overview of PtCo nanopods (a)–(b) and nanodendrites (c) synthesised as outlined in Scheme 6.1 ($t = 0$). Each image details the OA:OAm ratio and synthesis temperature (°C), and the reactions used a 1:1 ratio of metal precursors except in (b), where a 0.2:1 ratio was used. (d) Representative HRTEM imaging of a PtCo nanodendrite prepared using 1:1 Pt:Co, 0.5:2 OA:OAm, and 275°C. (e) Representative EDX mapping of the sample imaged in (d).[60]

composition from Pt-rich to Pt-poor within anisotropic nanoparticles in general and nanodendrites in particular. Lastly, EDX spectroscopy (e.g., Fig. 6.9(e)) verified the efficient metal intermixing that is normally targeted for good catalytic properties by virtue of the uniform distribution of surface-active sites. The compositional trends suggested by EDX analysis were also reinforced by atomic emission spectroscopy.

6.4 Computational Insights into Nanodendrite Structure, Composition, and Behaviour

6.4.1 *Introduction*

Although Section 6.3.3 introduced rules for the controllable growth of nanodendrites in Pt-based bimetallic systems under the influence of a specific, yet versatile, set of conditions, transitioning this knowledge more broadly to multimetallic systems will introduce new complications, particularly in achieving the necessary concentration of nanopods.[58] The complexity increases as each metal potentially reacts differently under the same synthesis conditions, complicating control over the reduction rate and the concentration of the resulting nanopods. To address these challenges effectively, a deeper understanding of the reduction rates and synthesis dynamics specific to multimetallic systems is essential.[171] Moreover, understanding the interactions between nanodendrites and their environment will help ensure that morphological and compositional control are maintained during applications or that any evolution is understood. By enhancing our comprehension of these complex dynamics and refining syntheses (including catalyst pretreatment) accordingly, researchers can potentially overcome the barriers currently hindering the generalised production of high-performance, multimetallic dendritic nanocatalysts.

One aid in the development of multimetallic, anisotropic nanoparticles that is suggested by a systematic approach to bimetallic nanodendrites is the use of modelling to understand formation, composition, and reactivity. Certainly, the advancing sophistication of computer programs, propelled by improved computing power and the emergence of more inventive material simulation methodologies in recent decades, have now furnished researchers with a novel toolkit that enables modelling of the properties and behaviours of materials on the nanoscale with increasing confidence.[172] Consequently, computational prediction and rationalisation of nanoparticle growth and the formation (including stability and evolution), structure, and properties of complex nanomaterials have gained traction. Commonly, these have been married to application development and models, either standalone or more often in conjunction with experiments, have provided tools that help to explain the action of new materials. Simulation methods, including those based on molecular dynamics (MD), DFT, and kinetic Monte Carlo (MC), have all been used to gain insight into the behaviour of nanoparticulate materials, and recent advances in the field of PtM nanodendrites are discussed in the following section.

Modelling of the formation of dendrites generally takes one of three primary forms. Continual methods can be applied to homogeneous systems using diffusion and heat equations.[173] However, to probe the internal structure of a dendrite, diffusion-limited aggregation modelling is required.[131] This was developed in the seminal work by Witten and Sander, who used a simple square-lattice model to show how randomly diffusing particles sequentially add to a growing cluster, giving a dendritic aggregate that demonstrates random branching.[174] This approach was subsequently refined to increase its scope, e.g., better handling elevated temperatures.[175] In all cases, a screening effect dictated that the most peripheral tips of the evolving structure dominate the capture of incoming diffusing atoms, with adatom deposition onto a surface facet being prominent in this way, followed by random diffusion before finally attaching to the surface. Meanwhile, structure and growth kinetics can be obtained by kinetic MC modelling,[176–179] and this has allowed structural and compositional effects in near or far proximity to step edges to be evaluated and compared with experiments.

In relation to the theme, as noted above, of modelling the properties of nanomaterials, a recurring aspect in this section will be the use of calculations to study how alloying two metals can tune electronic structure and manipulate (electro)catalytic performance. In the case of Pt, this tends to be with inexpensive 3d transition metals and targets applications in processes such as ORR, HER and methanol oxidation reaction (MOR). This principle was developed in the seminal work of Nørskov published in 2006, which probed the electrocatalytic activity of metal surfaces in the context of the strength of oxygen–metal interactions and their relationship to the position of the metal d-states relative to the Fermi level. Several studies covered will invoke the weaker binding of oxygen at PtM relative to that at Pt or, for nanomaterials with Pt skins, downshifting of the Pt d-states relative to the Fermi level.[180] However, while many studies have appeared that cover a broad range of Pt and Pt-based bimetallic systems on the nanoscale, the modelling of monometallics[181–185] and a range of isotropic bimetallics,[186] such as with Group 11,[187] Pd,[187,188] Ni,[189] Fe,[190] Co,[191] and Bi,[192] will not be covered in detail here. Instead, the focus will be on attempts to explain composition and surface structure in PtM (where M = another metal) in anisotropic, generally nanodendritic contexts.

6.4.2 *Fundamental studies*

In research that sought to study the fundamental aspects of nanodendrite formation using PtCu as a case study, the self-organisation of bimetallic islands on a Cu(111) surface was simulated using different deposition fluxes, relative amounts of Pt, and temperatures by applying a one-layer self-learning kinetic MC model.[179] Nanodendrites consisting of 1,000 and 4,000 atoms were compared, with dendritic branching quantified in terms of fractal dimension[131] (though errors were anticipated since nanodendrites are not strictly fractals) and the ratio of edge atoms to total atoms. Despite the limitations in calculating the fractal dimension, it proved possible to clearly distinguish nanodendrites from non-compact clusters. The main observations were twofold. First, the fractal dimension and the ratio of edge atoms to total atoms decreased and increased, respectively, logarithmically as the deposition flux was raised. Second, the fractal dimension and the ratio of edge atoms to total atoms increased and decreased, respectively, monotonically as temperature was elevated. This approach proved capable of simulating a nanodendrite at 300 K and with a relative number (n) of Pt atoms (n_{Pt}/n_{Cu}) at 0.8, quite similar to that seen experimentally.[178] This done, kinetic MC was applied predictively, simulating (experimentally unknown) dendrites with $n_{Pt}/n_{Cu} \approx 0.5$ at 200 K using the simultaneous deposition of Pt and Cu atoms.

Oriented attachment has recently been reviewed as a mechanism for the colloidal synthesis and growth of anisotropic nanostructures.[193] A variety of experimentally obtained morphologies were summarised, and computational approaches to rationalising oriented attachment as a growth mechanism were discussed. One example featured Pt and bimetallic PtAg nanowires grown through this mechanism, and their formation was modelled computationally. Experimental observation of the composition-dependent formation of PtAg alloy nanowires was rationalised by MD and DFT simulations.[194] Based on the experiments, worm-like nanowires could only be produced at a composition of around $Pt_{50}Ag_{50}$, while other ratios produced nearly spherical nanocrystals. During simulation, the potential energy (E_p), as a function of separation between two out of $Pt_{50}Ag_{50}$, Pt, or Ag nanocrystallites, was found to be dependent on the metal composition. A minimum E_p was achieved when two nanocrystallites were in close contact, and this was the main driving force for attachment, with $Pt_{50}Ag_{50}$ showing more E_p gain than Ag but less than Pt. This suggested that the attachment of Ag nanocrystallites was thermodynamically unfavourable, while a substantial reduction in energy could drive both $Pt_{50}Ag_{50}$ and Pt nanocrystallites to attach. Additionally, however, modelling revealed that difficulties in Pt atom diffusion and reconstruction at the attachment interface inhibited the coalescence of Pt nanocrystallites. As such, only $Pt_{50}Ag_{50}$ nanocrystallites were both thermodynamically and kinetically favoured for generating nanowires via oriented attachment.

In an example of d-/p-block metal alloying – the remarkable rearrangement of branched to hollow nanostructures – $Pt(acac)_2$ was used in conjunction with a strong reducing agent, 1,2-HDD, to generate nanodendritic Pt of approximately 75 nm in size.[195] These were then modified to hollow nanoparticles, defined by single voids (~60 nm wide), encased by polycrystalline *hcp* PtSn shells, by adding $Sn(OAc)_2$ at 270°C. DFT calculations were employed to try and understand this dramatic reconfiguration, which enabled the evaluation of the relative substitution energy (ΔE_{sub}) as a corollary of the thermodynamic driving force for atomic relocation. Results suggested that Pt to Sn substitution occurred in the first layer of the surface, with a slight preference for doing so at Pt(100) and Pt(211) surfaces instead of at a Pt(111) surface (Fig. 6.10(a)). Hence, rather than occupying close-packed sites, Sn was more stable at low-coordinated sites. Moving to the second and third layers in each model, the order of ΔE_{sub} inverted, suggesting that Pt(111) modification was now more favourable (Fig. 6.10(b)). It should be noted, though, that the ratios of calculated differences between ΔE_{sub} values were relatively small given the reaction temperatures, suggesting that all surface facets could likely modify. The overriding preference of Sn adatoms for intermixing at the first layer of the Pt nanodendrite surface was

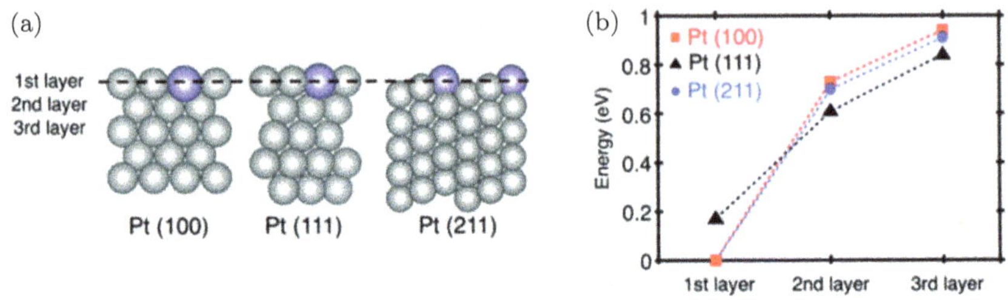

Fig. 6.10. (a) Pt(100), (111), and (211) models showing Sn atoms in the exposed facet (Pt = array of grey spheres, Sn = slightly larger purple spheres interspersed in the first atomic layer) and highlighting how Sn is the low coordinate in the Pt(211) system. (b) Computed relative substitution energies of Sn in the first, second, and third layesr of each model system.
Source: Modified from Ref. 195.

explained by a combination of the elevated temperature at which the reaction was conducted and Sn having a lower surface energy than Pt.[196]

6.4.3 *ORR catalysis*

Pt has been used extensively in oxygen reduction chemistry, and as such, its bimetallic formulations with a range of 3d metals have been investigated for ORR applications. The PtMn alloy nanodendrite catalyst has been prepared using glucose to reduce $Pt(acac)_2$ and $Mn(acac)_3$ in OAm and shown to exhibit ORR durability.[79] Moderate strain was argued to be induced in PtMn by virtue of low-electronegativity Mn contraction and to be responsible for an activity of 0.53 A/mg at 0.9 V versus the RHE. Compared to the well-known high-electronegativity Ni-based Pt alloy counterpart, the PtMn nanodendrite catalyst experienced less transition metal dissolution in acidic solution, underpinning 96% mass activity retention after 10,000 cycles. Fabricated as a membrane electrode assembly, PtMn nanodendrites helped deliver a peak power density of 1.36 W/cm^2 at 0.6 V in a H_2–O_2 hydrogen fuel cell. DFT calculations rationalised the relative thermodynamic stability of the PtMn system in terms of formation enthalpy (ΔH) and cohesive energy (E_{coh}). Following the demonstration that both PtMn and PtNi showed *fcc* structures, a $3 \times 3 \times 3$ supercell of the *fcc* lattice with 108 atoms was used to simulate each system. Based on experimentally derived compositions of $Pt_{60}Mn_{48}$ and $Pt_{38}Ni_{70}$, calculations established that both ΔH and E_{coh} values indicated $Pt_{56}Mn_{44}$ to be easier to generate and more stable than $Pt_{38}Ni_{70}$.

Moving on from Mn to Fe, Wang and co-workers used DFT to investigate the coordination numbers of surface Pt atoms and the energies of different surface facets expressed by pseudospherical, cubic, and dendritic morphologies ({111}, {200}, and {311} facets, respectively) in the context of PtFe systems.[197] Pt atoms on the nanodendrite {311} facet had lower coordination numbers than those on the pseudosphere {111} or cube {200} facets, with the surface energy of {311} being correspondingly the highest. Calculations therefore established how PtFe nanocubes and nanospheres expressed surface facets that were too highly coordinated to allow for facile oxygen adsorption at high operating voltages. This was derived from the recognised correlation between oxygen adsorption/desorption and surface coordination numbers. Notably, highly coordinated Pt atoms at the alloy surface tended to invade the surface and form Pt–O bonds, with a concomitant rearrangement of the interaction of Pt with the surface.[198] Subsequent desorption of oxygen atoms at lower potential then left an active site irreversibly altered. The result of this in the studied system was a gradual disordering of the surface structure and drop in activity. Conversely, high-surface-energy Pt (i.e., that enclosed by high-index facets) tended to adsorb oxygen atoms while preserving surface order. These data were in line with prior first-principles DFT reports, which also showed that the chemisorption energy for O* and OH* (where * denotes adspecies) on Pt nanocrystals increased with a decrease in coordination number, with oxygen adsorption preferentially occurring at edge and vertex atoms.[199,200]

Fe represents one of a triad of metals (Fe, Co, and Ni) that have commonly been tested in experimental ORR on account of their ease of handling and low cost.[72] As such, the (100) surface configuration in Pt_3Co nanodendrites with very low arm densities has been proposed recently to facilitate stable ORR catalysis.[201] Experimental characterisation showed that the Co atoms in empirically Pt_3Co nanodendrites achieved by co-reducing Pt(II) and Co(II) in an OA–OAm mixture were segregated and oxidised at surface-expressed (100) facets. *In situ* X-ray absorption spectroscopy (XAS) verified the prevention of oxygen chemisorption and oxide formation at Pt, and this was used to explain 98% current retention after cycling 5,000 times in acidic media. DFT validation of these data was sought by examining the energetic differences between the surface and subsurface Co (measured as segregation energy, E_{seg}). Subsequent modelling focused on O* to probe the preference for forming surface O* (measured as $E_{seg}(O^*)$) or surface oxides. Results clearly showed the benefits of generating (100) surface facets, with a smaller E_{seg} indicating a preference for Co to be at the surface. Meanwhile, a smaller $E_{seg}(O^*)$ on the (100) surface implied the formation of more surface oxides on Pt_3Co nanodendrites (Fig. 6.11(a)). A corresponding examination of adsorption energies then revealed through $E_{ads}(O^*)$ that O* preferentially bonded to bridge and *fcc* Co sites for the (100) and (111) surfaces, respectively. Adsorption at $Pt_3Co(100)$ was found to be preferred over that at $Pt_3Co(111)$ (Fig. 6.11(b)), with higher reactivity and stronger adsorption at Co being attributed to their elevated DOS and d-band centres relative to those of Pt. A final insight into the system's ORR efficiency came from evaluating the rate-determining ORR step, $O^* + H^* \rightarrow OH^*$.[202] Calculations pointed to compelling drops in reaction enthalpy and activation barrier, which was explained by surface oxides reducing catalyst oxophilicity and interacting with OH* on the bridge sites of the (100) surface, thus promoting the rate-determining step in the four-electron ORR.

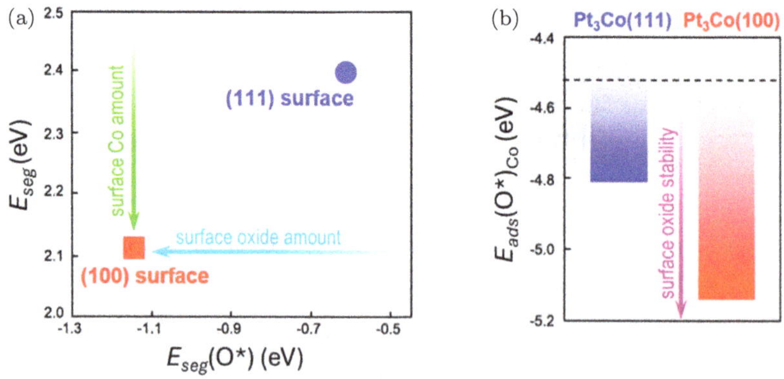

Fig. 6.11. (a) E_{seg} versus $E_{seg}(O^*)$ on (100) and (111) Pt_3Co surfaces, suggesting that the (100) surface encourages more oxide formation. (b) The comparison of $E_{ads}(O^*)$ on the (100) and (111) Pt_3Co surfaces (black dashed line represents adsorption at Pt(111)) points to more stable oxide formation at $Pt_3Co(100)$. *Source*: Modified from Ref. 201.

Moving on to Pt alloying with Cu, Liao *et al.* recently reported high index surface (HIS)-exposed and composition-graded $PtCu_3$@Pt_3Cu@Pt nanodendrites for ORR.[125] Synthetic work focused on the preparation by etching-modulated seeding and growing of $PtCu_3$ nanodendrites that expressed $n(111) \times (110)$ HISs. This was done by oxidatively excavating the faces of nanocubic PtCu seeds, followed by dendritic overgrowth under inert atmosphere conditions. Electrochemical dealloying then achieved $PtCu_3$ nanodendrites coated by Pt_3Cu shells with Pt-rich surfaces based on the proclivity of Pt in bimetallics for surface residence. The observation of excellent mass- and area-specific activities in ORR in perchloric acid led to an interest in using DFT calculations to probe the contribution of alloying to these data. Periodically repeated (111) and (332) slabs of five layers were used to model the catalyst surfaces, and a 20 Å vacuum layer was added in each simulation. The five-layer $PtCu_3$@Pt_3Cu@Pt slab comprised three bottom layers of $PtCu_3$, a subsurface layer of Pt_3Cu, and a top surface layer of Pt. Meanwhile, a $PtCu_3$@Pt_3Cu@Pt(332) stepped facet was used to simulate a HIS that expressed three main types of surface sites: the (111) terrace remote from the step, the (111) terrace adjacent to the step, and the site beneath the step. For comparison, five-layer slabs of Pt(111) and $PtCu_3$@Pt_3Cu@Pt(111) were also modelled. For each, the adsorption energy of atomic oxygen (AE_O) was derived as a corollary of activity.[70,180] $PtCu_3$@Pt_3Cu@Pt(111) revealed an AE_O 0.35 eV more positive than that of Pt(111). Meanwhile, for stepped $PtCu_3$@Pt_3Cu@Pt(332), sites on the (111) terrace remote from the step proved energetically similar to those on the $PtCu_3$@Pt_3Cu@Pt(111) surface, whereas sites on the (111) terrace adjacent to the step were approximately 0.2 eV more negative than those on the $PtCu_3$@Pt_3Cu@Pt(111) surface and roughly 0.15 eV more positive than those for Pt(111). Sites beneath the steps adsorbed O only weakly. Overall, near-step sites were calculated to perform similarly to the best ORR electrocatalysts, while strain induced by alloying with Cu likely caused a compression of the Pt−Pt distances in the catalyst core, which decreased the affinity for oxygen at the $PtCu_3$@Pt_3Cu@Pt(111) surface. Steps were argued to counteract this effect.

Shifting from the alloying of Pt with 3d to 4d metals, a durable PtRh nanodendrite electrocatalyst that evolved a Pt-rich surface

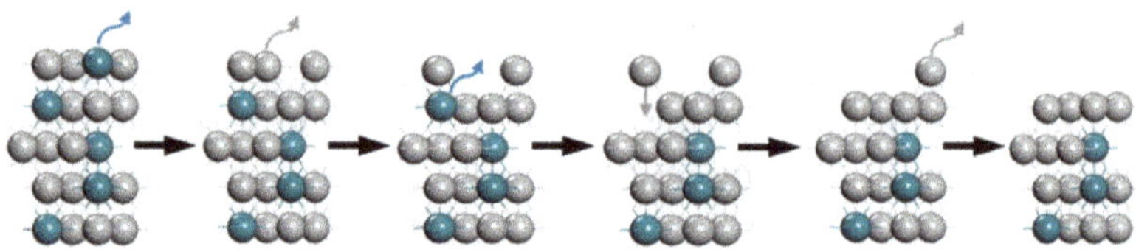

Fig. 6.12. DFT modelling of the possible evolution of PtRh nanodendrites in five time steps by the selective dissolution of either atom type from the initial surface during cycling tests (Pt = rows of light grey spheres, direction of migration of Pt = light grey arrows, Rh = teal spheres interspersed throughout Pt rows, and direction of migration of Rh = teal arrows).
Source: Modified from Ref. 203.

during initial cycling and then retained almost 90% of its mass activity towards ORR after 90,000 cycles (modifying from approximately $Pt_{2.3}Rh$ to approximately Pt_3Rh@Pt-skin according to elemental characterisation) was prepared by reducing $Pt(acac)_2$ and $Rh(acac)_3$ with AA in OAm.[203] Experimental results demonstrated a significant structural evolution to achieve a Pt-rich dendrite surface that strongly resisted dissolution while raising mass activity from 0.026 to 0.271 A/mg_{Pt}. DFT calculations pointed to surface evolution being attributable to the different dissolution energies of Pt and Rh. This conclusion was reached by modelling the dissolution potential of the different metal atoms on a slab. The preferred mode of surface evolution (Fig. 6.12) was found to involve a Rh atom dissolving before the resulting high-energy adjacent Pt atom. Subsequently, the Rh atom in the first sublayer dissolves and is replaced by a Pt atom collapsing down from the original surface. The authors argued that this 'self-healing' process underpinned the generation of a new Pt-rich surface as the final remaining high-energy Pt atom finally dissolves from the original surface. The nature of the modified surface achieved in Pt_3Rh@Pt-skin nanodendrites after durability testing was probed by XPS, with data indicating the transfer of electrons from Pt to Rh after structural evolution. Calculation of the segregation energy of Pt_3Rh under ORR conditions (i.e., with O* on a PtRh(111) slab) pointed to the preservation of the evolved nanodendrite core structure under these conditions. Further, the dissolution potential of surface Pt atoms is another critical parameter, and the Pt skin showed a dissolution potential that suggested good stability. Lastly, the calculation of relatively weak adsorption energies of O* and OH* species at

Pt_3Rh@Pt-skin(111) explained the improved ORR activity relative to the original catalyst.

6.4.4 *HER catalysis*

Many studies have sought to experimentally harness the computational work of Nørskov,[180] who established how both the binding energy of oxygen and the d-band centre of Pt could be modified to promote ORR. In a similar vein, the scope for HER promotion was probed by Guo and colleagues. In contrast to most studies covered here, this did not involve alloying. Rather, they reported the electrocatalytic behaviour of ultrathin Pt nanodendrites post-synthetically treated with NaH_2PO_2 to introduce different levels (0.58–6.5 at%) of electrophilic phosphorus.[204] Experimental work established that the best-performing catalyst contained 0.84 at% of P, achieving an η_{10} of 13.3 mV, and was augmented by studying two models of P inclusion using first-principles DFT calculations. One model located P-doped sites on the surface, and the other did so on the subsurface; the proximity of the Gibbs free energy for hydrogen adsorption (ΔG_{H*}) to zero was then used to evaluate HER performance. Results indicated that atomic-level P doping promoted the desorption of H_2 from the catalyst surface and that the bridge sites were most favourable for HER.

Moving on to bimetallic intermixing, which represents the main thrust of this narrative, facile air oxidative dealloying recently achieved PtCu nanodendrites.[205] A systematic investigation of morphology, composition, and structure revealed three-dimensional open porous nanostructures, and these were subsequently shown to exhibit composition-dependent effectiveness in HER. The best-performing system, $Pt_1Cu_{1.03}$ with a mean nanodendrite size of approximately 40 nm, showed good durability and operated at an overpotential of only 20 mV. This performance was attributed to a combination of the porous dendritic structure and synergistic interactions between Pt and Cu, as suggested by DFT calculations that employed a generalised gradient approximation (GGA)[206] with the Perdew–Burke–Ernzerhof (PBE) function[207,208] used to describe exchange correlation interactions.

The operation of $Pt_1Cu_{1.03}$ in HER was probed to test Pt−Cu interactions, with work functions of Pt(111) and Cu(111) computed

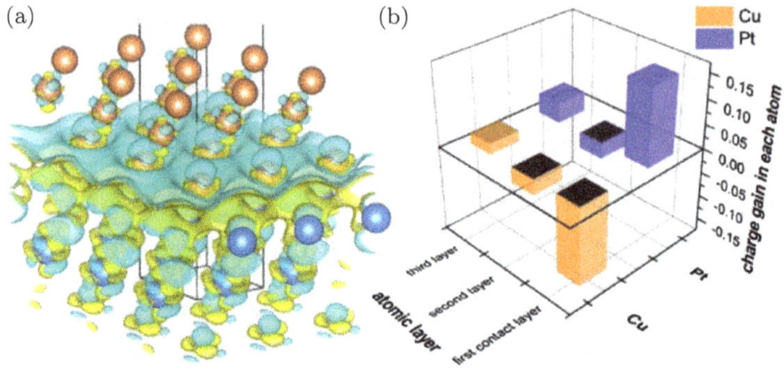

Fig. 6.13. (a) Plot of differential charge densities pointing to electron transfer at the Pt–Cu(111) interface (Pt = blue spheres, Cu = orange spheres, light-blue curved surfaces = areas in which charge density decreases, and light-yellow curved surfaces = areas in which charge density increases). (b) Bader charge analysis quantifying charge transfer and highlighting its ostensible restriction to the Pt–Cu interface (Cu = orange blocks and Pt = dark blue blocks).
Source: Reproduced from Ref. 205 with the permission of Elsevier Ltd.

to be 5.78 and 4.75 eV, respectively, and pointing to electron transfer from Cu to Pt. This drop in charge density at Cu and concomitant growth at adjacent Pt – which would be expected to benefit HER performance[209] – is qualitatively reflected in Fig. 6.13(a). Quantitatively, the interrogation of transferred charge came from a Bader analysis. This showed how the transfer was mainly limited to the interface directly connecting the two different atom types (Fig. 6.13(b)).

DFT calculations have been performed to gain a fundamental understanding of strain effects on the HER.[210] As already noted, ΔG_{H*} can be used to probe HER catalytic activity.[211] In one example of this, curved ultrathin PtPd nanodendrites (CNDs, Fig. 6.14(a)) were prepared along with their compositionally equivalent flat analogues (NDs). This was done by co-reducing Pt(IV) and Pd(II) using AA in the presence of differing levels of $[C_{22}H_{45}N(Me)_2CH_2COOH]^+[Br]^-$ ($C_{22}N$–COOH(Br)) surfactant,[126] resulting in products that differed only in their curvature, which allowed for an exploration of strain effects on electrocatalysis. Experimental HER suggested that introducing compressive strain in the PtPd CNDs underpinned enhanced electrocatalytic performance in that sample, and DFT methods were used to better understand this. To generate ion–electron interactions, $3 \times 4 \times 1$ supercells of

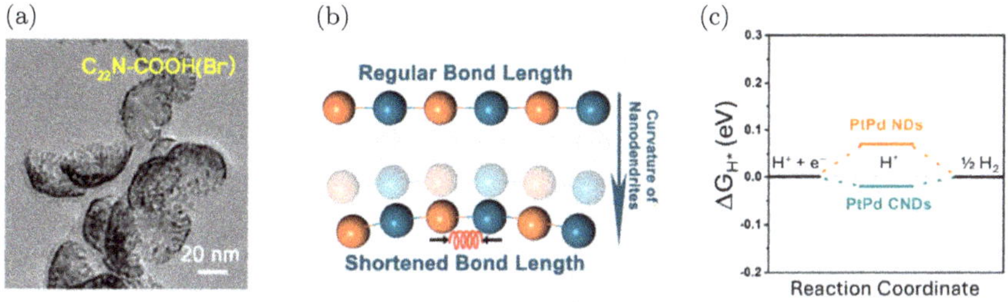

Fig. 6.14. (a) TEM imaging of PtPd CNDs prepared using surfactant C_{22}N-COOH(Br). (b) Side view of ND and CND models illustrating the generation of intrinsic strain (red coiled spring being compressed by small black force arrows) through curvature and bond compression (Pt = orange spheres and Pd = teal spheres). (c) Calculated Gibbs free energy for hydrogen adsorption (ΔG_{H*}), highlighting the enhanced performance of PtPd CNDs (teal energy level) over PtPd NDs (orange energy level) in HER.
Source: Modified from Ref. 210.

PtPd(110) monolayer were modelled using the projector-augmented wave method. Exchange-correlation functionals were described by the GGA method with the PBE functional. The only difference between the CNDs and NDs models was the inclusion of compressive strain in the former, yielding a curvature visible in side view (Fig. 6.14(b)), that induced a lowering in Pt–Pd bond length. The resulting introduction of approximately 2% compressive strain was argued to reduce hydrogen adsorption energy. This achieved a calculated ΔG_{H^*} closer to zero, pointing to improved H atom surface reduction to H* and efficient combination to generate H_2 for desorption (Fig. 6.14(c)).

6.4.5 *Oxidation catalysis*

Though the nanodendritic component is not itself bimetallic, cage-like, quasi-hexagonal Pt nanodendrites have been used to further decorate CoS_2-decorated N-doped graphene oxide (Pt@CoS_2-NrGO).[212] The resulting electrocatalyst was tested for alcohol oxidation in a direct liquid alcohol fuel cell, exhibiting superior performance with higher oxidation peak current densities for (methanol) C1, (ethylene glycol) C2, and (glycerol) C3 electrolytes in acid media. Durability was excellent towards various alcohols, and the surface and adsorption energies of Pt nanodendrites and CoS_2 were calculated using DFT methods. Based on HRTEM analysis, these calculations used the relaxed bulk structures to obtain surface slabs of Pt(111) (9 atomic layers thick) and CoS_2(200) (8 atomic layers thick). The surface energy calculated for Pt(111) was 1.45 J/m^2, which agreed well with previous work.[213] The surface energy calculated for CoS_2(200) was much lower at 0.47 J/m^2, suggesting its greater availability for methanol adsorption. Given that reaction proceeded at the Pt(111) surface, adsorption energies of methanol at this surface and the CoS_2(200) surface were calculated. Results were −0.49 eV and −0.14 eV, respectively, making adsorption feasible on both surfaces, but with the desorption of methanol being slightly easier from CoS_2(200). A better understanding of the adsorption/desorption of methanol at the CoS_2(200) surface came from projected crystal orbital Hamilton population analysis.[214] The calculation of strong Co–O and S–O bonding interactions but significant S–O antibonding interactions meant that the O atoms of methanol were considered to preferentially bind to Co, with the antibonding interactions imparting structural instability and indicating the feasibility of Co–O bond cleavage. Overall, three main theoretical insights emerged. First, CoS_2(200) in Pt@CoS_2-NrGO is available for methanol adsorption. Second, adsorption was more feasible on Pt(111), but a relatively small negative adsorption energy at CoS_2(200) facilitates methanol desorption for further oxidation. Third, antibonding Co–O states contribute to structural instability and promote reaction.

Remaining with alcohol oxidation catalysis, specifically MOR, a bottom-up method has been presented for the stereo-assembly of three-dimensional, porous PtPd nanodendrites on ultrathin $Ti_3C_2T_x$ MXene nanosheets (PtPd/$Ti_3C_2T_x$) for use as anode catalysts in direct methanol fuel cells (DMFCs).[215] The complex composite was prepared by first generating the nanosheets from bulk Ti_3AlC_2 through etching with LiF/HCl. Thereafter, nanosheet functionalisation was necessary (with PVP and hydrazine), before Pd seeds were introduced at the plentiful surface oxygenated functional groups using Pd(II) and formic acid. Finally, Pt(II) and AA were used to grow what turned out to be individually mixed PtPd nanodendrites. A series of spin-polarised DFT calculations were performed to study their electrocatalytic activity and CO adsorption behaviours. These used a simplified array of three Pt atoms and one Pd atom mounted on either carbon or $Ti_3C_2T_x$. The per-atom adsorption energies for

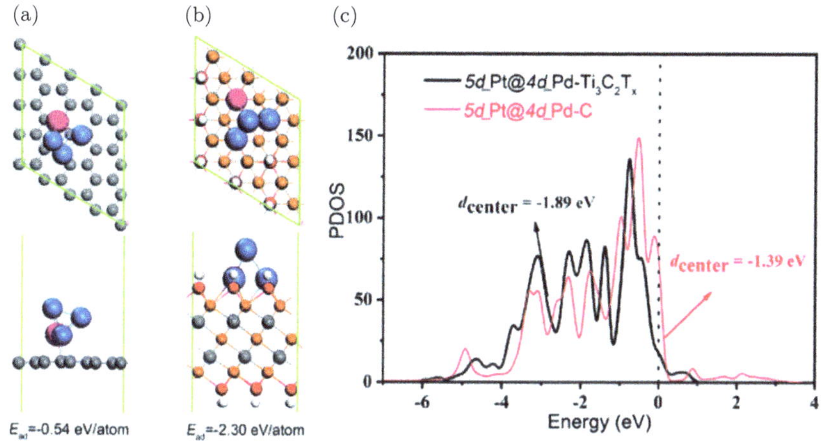

Fig. 6.15. Relaxed structures of PtPd clusters on (a) carbon and (b) $Ti_3C_2T_x$ (Pt = cluster of three large blue spheres, Pd = single large pink spheres, C = array of small grey spheres, Ti = array of small orange spheres, O = array of small red spheres, H = array of small white spheres, limits of model slice = thin green lines, top images = plan views, and bottom images = side-on views). (c) d-PDOS modelling of the downshift in d-band centre for PtPd/$Ti_3C_2T_x$ (pink curve and text) relative to PtPd/C (black curve and text).
Source: Reproduced from Ref. 215 with permission from the Royal Society of Chemistry.

these four-atom PtPd clusters were calculated to be −0.54 eV and up to −2.30 eV (Figs. 6.15(a) and (b)), respectively, implying a significantly strong interaction between nanodendrites and MXene.

The adsorption of CO in supported nanodendrites was modelled next.[216–218] In such scenarios, an upshift of the d-band centre to the Fermi level generally enhances metal–CO binding. In the current case, a map of the partial density of states (PDOS) for PtPd bimetallic clusters on carbon and $Ti_3C_2T_x$ suggested a 0.5 eV downshift in the d-band centre for PtPd/$Ti_3C_2T_x$ and weaker CO adsorption (Fig. 6.15(c)). A DFT estimation of CO adsorption energy corroborated this: −2.30 eV on $Ti_3C_2T_x$ versus −2.71 eV on carbon. Notably, the magnitude of adsorption energy difference correlated well with the downshift in d-band centre. Finally, these determinations were rationalised by locating Mulliken charge distributions near adsorbed CO. With the adsorbate losing 0.188e and 0.141e on carbon and $Ti_3C_2T_x$, respectively, these data confirmed why the MXene-based system was a promising DMFC catalyst; it reduced CO poisoning.

Moving away from alcohol oxidation to that of 3,3′,5,5′-tetramethylbenzidine, Song and co-workers recently reported enhanced peroxidase-like and antibacterial activity in (Ir-decorated) PtPd nanodendrites,[219] using the term 'nanozyme' coined from a landmark paper[220] in 2007 to describe their biomimetic nanomaterials.[221] Both photoluminescence and electron paramagnetic resonance spectra demonstrated their nanozymes' ability to produce relatively high amounts of $^{\bullet}OH$ radical from H_2O_2. The nanodendrite was prepared through simple co-reduction of Pd(II) and Pt(IV) by AA to yield bimetallic nanozyme (D-PP), with Ir decoration carried out post-synthetically by treating D-PP with Na_3IrCl_6, again in the presence of AA, to give DA-PPI. DFT calculations were then employed to analyse the adsorption affinity of both D-PP and DA-PPI towards H_2O_2 and $^{\bullet}OH$. On a representative Pt(111) surface, these species revealed absorption energies of −0.64 and −2.35 eV, respectively. Contrastingly, on a representative Ir(111) surface, the H_2O_2 decomposed to OH^-, the adsorption energy of which was found to be −2.69 eV. This coincided with existing data[222] and indicated that it was a modification of the adsorption properties that induced relatively high peroxidase-like activity in DA-PPI nanozyme but not its bimetallic precursor.

6.5 Summary and Outlook

Achieving optimal catalytic performance by PtM nanodendrites relies heavily on precisely manipulating the composition of multimetallic systems to tailor their electronic properties and surface reactivity. The ability to do this effectively has previously been restricted by the interconnectedness of morphology and composition control; differing kinetic behaviours during metal deposition have typically restricted compositional freedom, while efforts to overcome this have compromised shape and structure. Only recently has a strategy emerged that offers to systematise our approach to this problem, with the control of chemical reduction kinetics in bimetallic systems having been established in the presence of specific surfactants, and a mechanistic relationship shown between nanopods formed by facet-specific overgrowth and nanodendrites formed collisionally. The consequence of this has been an ability to design both types of anisotropic nanomaterials incorporating planned levels of Pt inclusion.[58,168]

The recent advances in understanding the interplay between the factors controlling structure and composition in bimetallic nanodendrites have significant implications. By balancing the reaction rates of precursors with particle growth rates, control over the formation of nanopods has been extended to that of nanodendrites. It has also been established how Pt-poor products can be achieved, maintaining morphological control, while more efficiently consuming Pt reagents, pointing to cost benefits by allowing the scaling back of Pt content in synthetic protocols.[60] Meanwhile, extension of the same strategy to more complex systems, such as co-reducing three metal salts, can be envisaged to yield trimetallic systems that can offer enhanced activity and selectivity,[223] with similar levels of control, seeking to extend recent reports of trimetallic Pt nanodendrites.[224,225] Thereafter, the possibility of improving the accessibility of recently developed high-entropy alloys (HEAs) emerges. These diverge from conventional binary or ternary alloys by incorporating five or more metals in precise proportions in a unified phase. The result is lattice distortions[226] that can instill the alloy with enhanced durability.[227] While these alloys promise significant advances in catalysis, a predictable route to HEA nanostructures is needed. Until recently, methods involved vacuum melting or milling,[228] which were unsuited to nanoscale synthesis. And while particles of approximately 100 nm in size have now been reported, these achieved uniform phase formation only at extremely high temperatures and were uniquely isotropic.[229] As such more complex systems render themselves amenable to experimental study, so the need for more detailed modelling of both nanodendrite synthesis and reactivity will develop, seeking to understand the increasingly multielement surfaces on display. Both approaches will be necessary to develop the more intricate multimetallic systems that will underpin the next generation of Pt-based catalysts targeting net-zero emissions.

References

1. Fankhauser S, Smith SM, Allen M, Axelsson K, Hale T, Hepburn C, *et al.* The meaning of net zero and how to get it right. Nat Clim Chang. 2022;12(1):15–21.

2. Allen MR, Friedlingstein P, Girardin CAJ, Jenkins S, Malhi Y, Mitchell-Larson E, *et al.* Net zero: science, origins, and implications. Ann Rev Environ Res. 2022;47(1):849–887.
3. Darby SJ, Hammond GP, Wu J. Briefing: Stocktaking global warming: the outcomes of the 2023 Dubai Climate Summit (COP28). Proc Inst Civil Eng Energy. 2024;3:4–6.
4. Togo H, Hirata A. Novel health risk alert system for occupational safety in hot environments. IEEE Pulse. 2021;12(4):24–27.
5. Valavanidis A. Extreme Weather Events Exacerbated by the Global Impact of Climate Change. Available online: Chem-tox-ecotox.org/ScientificReviews (accessed on 1 July 2024). 2022;1–40.
6. Rounce DR, Hock R, Maussion F, Hugonnet R, Kochtitzky W, Huss M, *et al.* Global glacier change in the 21st century: Every increase in temperature matters. Science. 2023;379(6627):78–83.
7. Deutch J. Is net zero carbon 2050 possible? Joule. 2020;4(11):2237–2240.
8. Wise M, Dooley J, Luckow P, Calvin K, Kyle P. Agriculture, Land use, energy and carbon emission impacts of global biofuel mandates to mid-century. Appl Energy. 2014;114:763–773.
9. Anika OC, Nnabuife SG, Bello A, Okoroafor ER, Kuang B, Villa R. Prospects of low and zero-carbon renewable fuels in 1.5-degree net zero emission actualisation by 2050: A critical review. Carbon Capture Sci Technol. 2022;5:100072.
10. Pretis F, Schwarz M, Tang K, Haustein K, Allen MR. Uncertain impacts on economic growth when stabilizing global temperatures at 1.5°C or 2°C warming. Phil Trans Roy Soc A. 2018;376(2119):20160460.
11. Nnabuife SG, Oko E, Kuang B, Bello A, Onwualu AP, Oyagha S, *et al.* The prospects of hydrogen in achieving net zero emissions by 2050: A critical review. Sust Chem Clim Act. 2023;2:100024.
12. Osman AI, Mehta N, Elgarahy AM, Hefny M, Al-Hinai A, Al-Muhtaseb AH, *et al.* Hydrogen production, storage, utilisation and environmental impacts: A review. Environ Chem Lett. 2022;20(1):153–188.
13. Shi J, Zhu Y, Feng Y, Yang J, Xia C. A prompt decarbonization pathway for shipping: Green hydrogen, ammonia, and methanol production and utilization in marine engines. Atmosphere. 2023;14(3):584.
14. Midilli A, Kucuk H, Topal ME, Akbulut U, Dincer I. A comprehensive review on hydrogen production from coal gasification: Challenges and opportunities. Int J Hydrogen Energy. 2021;46(50):25385–25412.

15. Boretti A, Banik BK. Advances in hydrogen production from natural gas reforming. Adv Energy Sust Res. 2021;2(11):2100097.
16. Rambhujun N, Salman MS, Wang T, Pratthana C, Sapkota P, Costalin M, *et al.* Renewable hydrogen for the chemical industry. MRS Energy Sust. 2020;7(1):33.
17. Shiva Kumar S, Lim H. An overview of water electrolysis technologies for green hydrogen production. Energy Rep. 2022;8:13793–13813.
18. Panigrahy B, Narayan K, Ramachandra Rao B. Green hydrogen production by water electrolysis: A renewable energy perspective. Mater Today Proc. 2022;67(8):1210–1214.
19. Eicke L, De Blasio N. Green hydrogen value chains in the industrial sector—geopolitical and market implications. Energy Res Soc Sci. 2022;93:102847.
20. Dickson R, Akhtar MS, Abbas A, Park ED, Liu J. Global transportation of green hydrogen via liquid carriers: Economic and environmental sustainability analysis, policy implications, and future directions. Green Chem. 2022;24(21):8484–8493.
21. Pathak PK, Yadav AK, Padmanaban S. Transition toward emission-free energy systems by 2050: Potential role of hydrogen. Int J Hydrogen Energy. 2023;48(26):9921–9927.
22. Izatt SR, McKenzie JS, Bruening RL, Izatt RM, Izatt NE, Krakowiak KE. Selective recovery of platinum group metals and rare earth metals from complex matrices using a green chemistry/molecular recognition technology approach. In: Izatt RM, editor. Metal Sustainability: Global Challenges, Consequences, and Prospects. Hoboken: Wiley; 2016.
23. de Almeida LD, Wang H, Junge K, Cui X, Beller M. Recent advances in catalytic hydrosilylations: Developments beyond traditional platinum catalysts. Angew Chem. 2021;133(2):558–573.
24. Huang L, Zaman S, Tian X, Wang Z, Fang W, Xia BY. Advanced platinum-based oxygen reduction electrocatalysts for fuel cells. Acc Chem Res. 2021;54(2):311–322.
25. Zhu F, Luo L, Wu A, Wang C, Cheng X, Shen S, *et al.* Improving the high-current-density performance of PEMFC through much enhanced utilization of platinum electrocatalysts on carbon. ACS Appl Mater Interfaces. 2020;12(23):26076–2683.
26. Kumar SS, Lim H. Recent advances in hydrogen production through proton exchange membrane water electrolysis – a review. Sust Energy Fuels. 2023;7(15):3560–3583.
27. Liu L. Platinum group metal free nano-catalysts for proton exchange membrane water electrolysis. Curr Opin Chem Eng. 2021;34:100743.

28. Xing W, Yin G, Zhang J, editors. Rotating Electrode Methods and Oxygen Reduction Electrocatalysts. Amsterdam: Elsevier; 2014.
29. Holton DOT, Stevenson JW. The role of platinum in proton exchange membrane fuel cells. Platinum Met Rev. 2013;57(4):259–271.
30. Faraz A, Ambikapathy A, Thangavel S, Logavani K, Arun Prasad G. Battery electric vehicles (BEVs). In: Patel N, Bhoi AK, Padmanaban S, Holm-Nielsen JB, editors. Electric Vehicles: Modern Technologies and Trends. Cham: Springer; 2021.
31. Hosseini SE, Butler B. An overview of development and challenges in hydrogen powered vehicles. Int J Green Energy. 2020;17(1):13–37.
32. Wang Z, Li M, Zhao F, Ji Y, Han F. Status and prospects in technical standards of hydrogen-powered ships for advancing maritime zero-carbon transformation. Int J Hydrogen Energy. 2024;62:925–946.
33. Adler EJ, Martins JRRA. Hydrogen-powered aircraft: Fundamental concepts, key technologies, and environmental impacts. Prog Aerosp Sci. 2023;141:100922.
34. Massaro MC, Biga R, Kolisnichenko A, Marocco P, Monteverde AHA, Santarelli M. Potential and technical challenges of on-board hydrogen storage technologies coupled with fuel cell systems for aircraft electrification. J Power Sources. 2023;555:232397.
35. Gray N, McDonagh S, O'Shea R, Smyth B, Murphy JD. Decarbonising ships, planes and trucks: An analysis of suitable low-carbon fuels for the maritime, aviation and haulage sectors. Adv Appl Energy. 2021;1:100008.
36. Dey S, Mehta NS. Automobile pollution control using catalysis. Res Environ Sust. 2020;2:10006.
37. Izatt RM, Izatt SR, Izatt NE, Krakowiak KE, Bruening RL, Navarro L. Industrial applications of molecular recognition technology to separations of platinum group metals and selective removal of metal impurities from process streams. Green Chem. 2015;17(4):2236–2245.
38. Gastelu G, Saha P, Dyson PJ, Hulla M, Uranga JG. Autocatalysis and CO_2: Carbon capture and utilization (CCU) meets natural processes. ChemCatChem. 2023;15(22):e202300905.
39. Bowker M. Methanol synthesis from CO_2 hydrogenation. ChemCatChem. 2019;11(17):4238–4246.
40. Cieluch M, Podleschny PY, Kazamer N, Wirkert FJ, Rost UW, Brodmann M. Development of a bifunctional Ti-based gas diffusion electrode for ORR and OER by one- and two-step Pt-Ir electrodeposition. Nanomaterials. 2022;12(7):1233.
41. Wang C, Spendelow JS. Recent developments in Pt–Co catalysts for proton-exchange membrane fuel cells. Curr Opin Electrochem. 2021;28:100715.

42. David E, Niculescu VC. Volatile organic compounds (VOCs) as environmental pollutants: Occurrence and mitigation using nanomaterials. Int J Environ Res Pub Health. 2021;18(24):13147.
43. Heck KN, Garcia-Segura S, Westerhoff P, Wong MS. Catalytic converters for water treatment. Acc Chem Res. 2019;52(4):906–915.
44. Pearce AK, Wilks TR, Arno MC, O'Reilly RK. Synthesis and applications of anisotropic nanoparticles with precisely defined dimensions. Nat Rev Chem. 2021;5(1):21–45.
45. Chaudhari NK, Joo J, Kwon H, Kim B, Kim HY, Joo SH, *et al.* Nanodendrites of platinum-group metals for electrocatalytic applications. Nano Res. 2018;11(12):6116–6140.
46. Guo K, Xu D, Xu L, Li Y, Tang Y. Noble metal nanodendrites: Growth mechanisms, synthesis strategies and applications. Mater Horiz. 2023;10(4):1234–1263.
47. Ye E, Regulacio MD, Zhang SY, Loh XJ, Han MY. Anisotropically branched metal nanostructres. Chem Soc Rev. 2015;44(17):6001–6017.
48. Jeyaraj M, Gurunathan S, Qasim M, Kang MH, Kim JH. A comprehensive review on the synthesis, characterization, and biomedical application of platinum nanoparticles. Nanomaterials. 2019;9(12): 1719.
49. Khan MAR, Al Mamun MS, Ara MH. Review on platinum nanoparticles: Synthesis, characterization, and applications. Microchem J. 2021;171:106840.
50. Ren X, Lv Q, Liu L, Liu B, Wang Y, Liu A, *et al.* Current progress of Pt and Pt-based electrocatalysts used for fuel cells. Sust Energy Fuels. 2020;4(1):15–30.
51. Zhang J, Zhang L, Cai Z. Strategies to enhance the electrochemical performances of Pt-based intermetallic catalysts. Chem Commun. 2021;57(1):11–26.
52. Zhang BW, Yang HL, Wang YX, Dou SX, Liu HK. A comprehensive review on controlling surface composition of Pt-based bimetallic electrocatalysts. Adv Energy Mater. 2018;8(20):1703597.
53. Mahata A, Nair AS, Pathak B. Recent advancements in Pt-nanostructure-based electrocatalysts for the oxygen reduction reaction. Catal Sci Technol. 2019;9(18):4835–4863.
54. Zhang J, Yuan Y, Gao L, Zeng G, Li M, Huang H. Stabilizing Pt-based electrocatalysts for oxygen reduction reaction: Fundamental understanding and design strategies. Adv Mater. 2021; 33(20):2006494.
55. Smith BJ, Graziano DJ, Riddle ME, Liu DJ, Sun P, Iloeje C, *et al.* Platinum group metal catalysts – supply chain deep dive assessment.

Available online: https://www.osti.gov/servlets/purl/1871583 (accessed on 15 July 2024). 2022;1–65.
56. Xia J, Ghahreman A. Platinum group metals recycling from spent automotive catalysts: Metallurgical extraction and recovery technologies. Sep Pur Technol. 2023;311:123357.
57. Padamata SK, Yasinskiy AS, Polyakov PV, Pavlov EA, Varyukhin DY. Recovery of noble metals from spent catalysts: A review. Metall Mater Trans B. 2020;51(5):2413–2435.
58. Ming S, Wheatley AEH. Manipulating morphology and composition in colloidal heterometallic nanopods and nanodendrites. Nanoscale. 2023;15(19):8814–8824.
59. She ZW, Kibsgaard J, Dickens CF, Chorkendorff I, Nørskov JK, Jaramillo TF. Combining theory and experiment in electrocatalysis: Insights into materials design. Science. 2017;355(6321):eaad4998.
60. Ming S, Cobb SJ, Rahaman M, Sammy N, Reisner E, Wheatley AEH. Anisotropic heterobimetallic nanomaterials with controlled composition for efficient oxygen reduction at ultralow loading. Adv Funct Mater. 2024;34:2411006.
61. Ding H, Wang P, Su C, Liu H, Tai X, Zhang N, *et al.* Epitaxial growth of ultrathin highly crystalline Pt–Ni nanostructure on a metal carbide template for efficient oxygen reduction reaction. Adv Mater. 2022;34(12):2109188.
62. Jin H, Xu Z, Hu ZY, Yin Z, Wang Z, Deng Z, *et al.* Mesoporous Pt@Pt-skin Pt_3Ni core-shell framework nanowire electrocatalyst for efficient oxygen reduction. Nat Commun. 2023;14(1):1518.
63. Cheng K, Smulders LCJ, van der Wal LI, Oenema J, Meeldijk JD, Visser NL, *et al.* Maximizing noble metal utilization in solid catalysts by control of nanoparticle location. Science. 2022;377(6602):204–208.
64. Strmcnik D, Escudero-Escribano M, Kodama K, Stamenkovic VR, Cuesta A, Marković NM. Enhanced electrocatalysis of the oxygen reduction reaction based on patterning of platinum surfaces with cyanide. Nat Chem. 2010;2(10):880–885.
65. Cao S, Tao FF, Tang Y, Li Y, Yu J. Size- and shape-dependent catalytic performances of oxidation and reduction reactions on nanocatalysts. Chem Soc Rev. 2016;45(17):4747–4765.
66. Stamenkovic VR, Fowler B, Mun BS, Wang G, Ross PN, Lucas CA, *et al.* Improved oxygen reduction activity on $Pt_3Ni(111)$ via increased surface site availability. Science. 2007;315(5811):493–497.
67. Wang YJ, Long W, Wang L, Yuan R, Ignaszak A, Fang B, *et al.* Unlocking the door to highly active ORR catalysts for PEMFC applications: Polyhedron-engineered Pt-based nanocrystals. Energy Environ Sci. 2018;11(2):258–275.

68. Ma Z, Cano ZP, Yu A, Chen Z, Jiang G, Fu X, *et al.* Enhancing oxygen reduction activity of Pt-based electrocatalysts: From theoretical mechanisms to practical methods. Angew Chem Int Ed. 2020;59(42):18334–18348.
69. Kong F, Shi W, Song Y, Ren L, Sun X, Wang J. Surface/near-surface structure of highly active and durable Pt-based catalysts for oxygen reduction reaction: A review. Adv Energy Sust Res. 2021;2(7):2100025.
70. Greeley J, Stephens IEL, Bondarenko AS, Johansson TP, Hansen HA, Jaramillo TF, *et al.* Alloys of platinum and early transition metals as oxygen reduction electrocatalysts. Nat Chem. 2009;1(7): 552–556.
71. Gao G, Zhu G, Chen X, Sun Z, Cabot A. Optimizing Pt-based alloy electrocatalysts for improved hydrogen evolution performance in alkaline electrolytes: A comprehensive review. ACS Nano. 2023;17(21):20804–20824.
72. Lei W, Li M, He L, Meng X, Mu Z, Yu Y, *et al.* A general strategy for bimetallic Pt-based nano-branched structures as highly active and stable oxygen reduction and methanol oxidation bifunctional catalysts. Nano Res. 2020;13(3):638–645.
73. Shifa TA, Wang F, Liu Y, He J. Heterostructures based on 2D materials: A versatile platform for efficient catalysis. Adv Mater. 2019;31(45):1804828.
74. Huang R, Wen YH, Shao GF, Sun SG. Atomic structure and thermal stability of Pt–Fe bimetallic nanoparticles: From alloy to core/shell architectures. Phys Chem Chem Phys. 2016;18(25):17010–17017.
75. Fan F, Chen DH, Yang L, Qi J, Fan Y, Wang Y, *et al.* PtCuFe alloy nanochains: Synthesis and composition-performance relationship in methanol oxidation and hydrogen evolution reactions. J Colloid Interface Sci. 2022;628(B):153–161.
76. Zhao Z, Liu Z, Zhang A, Yan X, Xue W, Peng B, *et al.* Graphene-nanopocket-encaged PtCo nanocatalysts for highly durable fuel cell operation under demanding ultralow-Pt-loading conditions. Nat Nanotechnol. 2022;17(9):968–975.
77. Zhang J, Zhang L, Cui Z. Strategies to enhance the electrochemical performances of Pt-based intermetallic catalysts. Chem Commun. 2021;57(1):11–26.
78. Ye X, Shao RY, Yin P, Liang HW, Chen YX. Ordered intermetallic PtCu catalysts made from Pt@Cu core/shell structures for oxygen reduction reaction. Inorg Chem. 2022;61(38):15239–15246.
79. Nie Y, Sun Y, Song B, Meyer Q, Liu S, Guo H, *et al.* Low-electronegativity Mn-contraction of PtMn nanodendrites boosts

oxygen reduction durability. Angew Chem Int Ed. 2024;63(7): e202317987.

80. Ying J. Atomic-scale design of high-performance Pt-based electrocatalysts for oxygen reduction reaction. Front Chem. 2021;9:753604.
81. Xie C, Niu Z, Kim D, Li M, Yang P. Surface and interface control in nanoparticle catalysis. Chem Rev. 2020;120(2):1184–1249.
82. Pontikakis G. Modeling, reaction schemes and kinetic parameter estimation in automotive catalytic converters and diesel particulate filters [dissertation]. Thessaly: University of Thessaly; 2003.
83. Benck JD, Hellstern TR, Kibsgaard J, Chakthranont P, Jaramillo TF. Catalyzing the hydrogen evolution reaction (HER) with molybdenum sulfide nanomaterials. ACS Catal. 2014;4(11):3957–3971.
84. Liu HL, Nosheen F, Wang X. Noble metal alloy complex nanostructures: Controllable synthesis and their electrochemical property. Chem Soc Rev. 2015;44(10):3056–3078.
85. Gómez-Recio I, Thomas C, Méthivier C, Ruiz-González ML, González-Calbet JM, Portehault D. Metal-support interactions in Pt–WO_3 heterostructures: Role of WO_3 polymorphism. Chem Mater. 2023;35(19):7931–7942.
86. Zhao X, Sasaki K. Advanced Pt-based core-shell electrocatalysts for fuel cell cathodes. Acc Chem Res. 2022;55(9):1226–1236.
87. Zhang XF, Meng HB, Chen HY, Feng JJ, Fang KM, Wang AJ. Bimetallic PtCo alloyed nanodendritic assemblies as an advanced efficient and robust electrocatalyst for highly efficient hydrogen evolution and oxygen reduction. J Alloys Comp. 2019;786:232–239.
88. Vieira LH, Rasteiro LF, Santana CS, Catuzo GL, da Silva AHM, Assaf JM, *et al.* Noble metals in recent developments of heterogeneous catalysts for CO_2 conversion processes. ChemCatChem. 2023;15(14):e202300493.
89. Zhang B, Sun G, Ding S, Asakura H, Zhang J, Sautet P, *et al.* Atomically dispersed Pt_1-polyoxometalate catalysts: How does metal-support interaction affect stability and hydrogenation activity? J Am Chem Soc. 2020;141(20):8185–8197.
90. Bhuvanendran N, Ravichandran S, Lee S, Sanij FD, Kandasamy S, Pandey P, *et al.* Recent progress in Pt-based electrocatalysts: A comprehensive review of supported and support-free systems for oxygen reduction. Coord Chem Rev. 2024;521:216191.
91. Zhou WP, Yang X, Vukmirovic MB, Koel BE, Jiao J, Peng G, *et al.* Improving electrocatalysts for O_2 reduction by fine-tuning the Pt-support interaction: Pt monolayer on the surfaces of a $Pd_3Fe(111)$ single-crystal alloy. J Am Chem Soc. 2009;131(35):12755–12762.

92. Zhong H, Luo X, Chen H, Huang S, Chen Y, Li D. Pt_3Ni alloy catalyst coupled with three-dimensional nitrogen-doped graphene for enhancing the alkaline hydrogen evolution. Electrochim Acta. 2022;429:141030.
93. Lewera A, Timperman L, Roguska A, Alonso-Vante N. Metal-support interactions between nanosized Pt and metal oxides (WO_3 and TiO_2) studied using X-ray photoelectron spectroscopy. J Phys Chem C. 2011;115(41):20153–20159.
94. Cheng Y, Zhao X, Yu Y, Chen L, Cheng T, Huang J, *et al.* Indium oxide supported Pt–In alloy nanocluster catalysts with enhanced catalytic performance toward oxygen reduction reaction. J Power Sources. 2020;446:227332.
95. Akalework NG, Pan CJ, Su WN, Rick J, Tsai MC, Lee JF, *et al.* Ultrathin TiO_2-coated MWCNTs with excellent conductivity and SMSI nature as Pt catalyst support for oxygen reduction reaction in PEMFCs. J Mater Chem. 2012;22(39):20977–20985.
96. Ho VTT, Pan CJ, Rick J, Su WN, Hwang BJ. Nanostructured $Ti_{0.7}Mo_{0.3}O_2$ support enhances electron transfer to Pt: High-performance catalyst for oxygen reduction reaction. J Am Chem Soc. 2011;133(30):11716–11724.
97. Liu J, Jiao M, Mei B, Tong Y, Li Y, Ruan M, *et al.* Carbon-supported divacancy-anchored platinum single-atom electrocatalysts with superhigh Pt utilization for the oxygen reduction reaction. Angew Chem Int Ed. 2019;58(4):1163–1167.
98. Ham K, Shin D, Lee J. The role of lone-pair electrons in Pt–N interactions for the oxygen reduction reaction in polymer exchange membrane fuel cells. ChemSusChem. 2020;13(7):1751–1758.
99. Perini L, Durante C, Favaro M, Perazzolo V, Agnoli S, Schneider O, *et al.* Metal-support interaction in platinum and palladium nanoparticles loaded on nitrogen-doped mesoporous carbon for oxygen reduction reaction. ACS Appl Mater Interfaces. 2015;7(2): 1170–1179.
100. Hammer B, Nielsen OH, Nørskov JK. Structure sensitivity in adsorption: CO interaction with stepped and reconstructed Pt surfaces. Catal Lett. 1997;46(1–2):31–35.
101. Marković NM, Ross PN Jr. Surface science studies of model fuel cell electrocatalysts. Surf Sci Rep. 2002;45(4–6):117–229.
102. Nesselberger M, Ashton S, Meier JC, Katsounaros I, Mayrhofer KJJ, Arenz M. The particle size effect on the oxygen reduction reaction activity of Pt catalysts: influence of electrolyte and relation to single crystal models. J Am Chem Soc. 2011;133(43):17428–17433.

103. Briega-Martos V, Herrero E, Feliu JM. Effect of pH and water structure on the oxygen reduction reaction on platinum electrodes. Electrochim Acta. 2017;241:497–509.
104. Mao X, Wang L, Li Y. Understanding pH-dependent oxygen reduction reaction on metal alloy catalysts. ACS Catal. 2024;14(7): 5429–5435.
105. Calle-Vallejo F, Pohl MD, Reinisch D, Loffreda D, Sautet P, Bandarenka AS. Why conclusions from platinum model surfaces do not necessarily lead to enhanced nanoparticle catalysts for the oxygen reduction reaction. Chem Sci. 2017;8(3):2283–2289.
106. Calle-Vallejo F, Tymoczko J, Colic V, Vu QH, Pohl MD, Morgenstern K, *et al.* Finding optimal surface sites on heterogeneous catalysts by counting nearest neighbors. Science. 2015;350(6257): 185–189.
107. Lim B, Jiang M, Camargo PHC, Cho EC, Tao J, Lu X, *et al.* PdPt bimetallic nanodendrites with high activity for oxygen reduction. Science. 2009;324(5932):1302–1305.
108. Yu T, Kim DY, Zhang H, Xia Y. Platinum concave nanocubes with high-index facets and their enhanced activity for oxygen reduction reaction. Angew Chem Int Ed. 2011;50(12):2773–2777.
109. Xie Y, Cai J, Wu Y, Zang Y, Zheng X, Ye J, *et al.* Boosting water dissociation kinetics on Pt–Ni nanowires by N-induced orbital tuning. Adv Mater. 2019;31(16):1807780.
110. Wu X, Chen XK, Li X, Yan Y, Huang J, Li J, *et al.* A unique ligand effect in Pt-based core–shell nanocubes to boost oxygen reduction electrocatalysis. J Mater Chem A. 2021;9(39):22653–22659.
111. Pang B, Jia C, Wang S, Liu T, Ding T, Liu X, *et al.* Self-optimized ligand effect of single-atom modifier in ternary Pt-based alloy for efficient hydrogen oxidation. Nano Lett. 2023;23(9):3826–3834.
112. Ren J, Tilley RD. Shape-controlled growth of platinum nanoparticles. Small. 2007;3(9):1508–1512.
113. Jenkinson KJ. Manipulation and characterization of morphology and composition in mono- and heterometallic nanoparticles [dissertation]. Cambridge: University of Cambridge; 2019.
114. Khudhayer WJ, Shaikh AU, Karabacak T. Periodic Pt nanorod arrays with controlled porosity for oxygen reduction reaction. Nanosci Nanotechnol Lett. 2012;4(10):999–1007.
115. Takimoto D, Toma S, Suda Y, Shirokura T, Tokura Y, Fukuda K, *et al.* Platinum nanosheets synthesized via topotactic reduction of single-layer platinum oxide nanosheets for electrocatalysis. Nat Commun. 2023;14(1):1–9.

116. Xiao F, Wang Q, Xu GL, Qin X, Hwang I, Sun CJ, *et al.* Atomically dispersed Pt and Fe sites and Pt–Fe nanoparticles for durable proton exchange membrane fuel cells. Nat Catal. 2022;5(6):503–512.
117. He DS, He D, Wang J, Lin Y, Yin P, Hong X, *et al.* Ultrathin icosahedral Pt-enriched nanocage with excellent oxygen reduction reaction activity. J Am Chem Soc. 2016;138(5):1494–1497.
118. Tao L, Huang B, Jin F, Yang Y, Luo M, Sun M, *et al.* Atomic PdAu interlayer sandwiched into Pd/Pt core/shell nanowires achieves superstable oxygen reduction catalysis. ACS Nano. 2020;14(9): 11570–11578.
119. Cao S, Sun T, Li JR, Li QZ, Hou CC, Sun Q. The cathode catalysts of hydrogen fuel cell: From laboratory toward practical application. Nano Res. 2023;16(4):4365–4380.
120. Xia Y, Xiong Y, Lim B, Skrabalak SE. Shape-controlled synthesis of metal nanocrystals: simple chemistry meets complex physics? Angew Chem Int Ed. 2009;48(1):60–103.
121. Lim B, Xia Y. Metal nanocrystals with highly branched morphologies. Angew Chem Int Ed. 2011;50(1):76–85.
122. Cao Y, Yang Y, Shan Y, Huang Z. One-pot and facile fabrication of hierarchical branched Pt–Cu nanoparticles as excellent electrocatalysts for direct methanol fuel cells. ACS Appl Mater Interfaces. 2016;8(9):5998–6003.
123. Gong MX, Jiang X, Xue TY, Shen TY, Xu L, Sun DM, *et al.* PtCu nanodendrite-assisted synthesis of PtPdCu concave nanooctahedra for efficient electrocatalytic methanol oxidation. Catal Sci Technol. 2015;5(12):5105–5109.
124. Huan TN, Ganesh T, Kim KS, Kim S, Han SH, Chung H. A three-dimensional gold nanodendrite network porous structure and its application for an electrochemical sensing. Biosens Bioelectron. 2011;27(1):183–186.
125. Liao Y, Li J, Zhang S, Chen S. High index surface-exposed and composition-graded $PtCu_3$@Pt_3Cu@Pt nanodendrites for high-performance oxygen reduction. Chin J Catal. 2021;42(7):1108–1116.
126. Xu D, Lv H, Jin H, Liu Y, Ma Y, Han M, *et al.* Crystalline facet-directed generation engineering of ultrathin platinum nanodendrites. J Phys Chem Lett. 2019;10(3):663–671.
127. Cao D, Xu H, Cheng D. Branch-leaf-shaped CuNi@NiFeCu nanodendrites as highly efficient electrocatalysts for overall water splitting. Appl Catal B. 2021;298:120600.
128. Chen S, Su H, Wang Y, Wu W, Zeng J. Size-controlled synthesis of platinum-copper hierarchical trigonal bipyramid nanoframes. Angew Chem Int Ed. 2015;54(1):108–113.

129. Lou W, Ali A, Shen PK. Recent development of Au arched Pt nanomaterials as promising electrocatalysts for methanol oxidation reaction. Nano Res. 2022;15(1):18–37.
130. Dong Y, Zhang D, Li D, Jia H, Qin W. Control of Ostwald ripening. Sci China Mater. 2023;66(3):1249–1255.
131. Vicsek T. Fractal Growth Phenomena. Singapore: World Scientific; 1992.
132. Liu L, Nakouzi E, Sushko ML, Schenter GK, Mundy CJ, Chun J, *et al.* Connecting energetics to dynamics in particle growth by oriented attachment using real-time observations. Nat Commun. 2020;11(1):1045.
133. Ahmadi TS, Wang ZL, Green TC, Henglein A, El-Sayed MA. Shape-controlled synthesis of colloidal platinum nanoparticles. Science. 1996;272(5270):1924–1925.
134. Witten TA, Sander LM. Diffusion-limited aggregation, a kinetic critical phenomenon. Phys Rev Lett. 1981;47(19):1400.
135. Siebeneicher S, Reichenberger S, Hengst C, Dornhaus F, Wittek B, Barcikowski S. Activity and durability patterns of 45 binary noble metal alloy nanoparticle variants for commercial diesel exhaust aftertreatment. ChemCatChem. 2023;15(19):e202300563.
136. Sui S, Wang X, Zhou X, Su Y, Riffat S, Liu C. A comprehensive review of Pt electrocatalysts for the oxygen reduction reaction: Nanostructure, activity, mechanism and carbon support in PEM fuel cells. J Mater Chem A. 2017;5(5):1808–1825.
137. Taylor MJ, Beaumont SK, Islam MJ, Tsatsos S, Parlett CAM, Issacs MA, *et al.* Atom efficient PtCu bimetallic catalysts and ultra dilute alloys for the selective hydrogenation of furfural. Appl Catal B. 2021;284:119737.
138. Lopes WA, Jaeger HM. Hierarchical self-assembly of metal nanostructures on diblock copolymer scaffolds. Nature. 2001;414(6865):735–738.
139. Ershov BG. Estimation of the reduction potential of transition metal atoms in aqueous solutions. J Mol Liq. 2023;390(B):123–129.
140. Sriphathoorat R, Wang K, Luo S, Tang M, Du H, Du X, *et al.* Well-defined PtNiCo core–shell nanodendrites with enhanced catalytic performance for methanol oxidation. J Mater Chem A. 2016;4(46):18015–18021.
141. Zhang LY, Zeng T, Zheng L, Wang Y, Yuan W, Niu M, *et al.* Epitaxial growth of Pt–Pd bimetallic heterostructures for the oxygen reduction reaction. Adv Powder Mater. 2023;2(4):100131.
142. Uzio D, Berhault G. Factors governing the catalytic reactivity of metallic nanoparticles. Catal Rev Sci Eng. 2010;52(1):106–131.

143. Zaera F. Shape-controlled nanostructures in heterogeneous catalysis. ChemSusChem. 2013;6(10):1797–1820.
144. Haynes WM. CRC Handbook of Chemistry and Physics. 95th ed. Boca Raton: CRC Press; 2014.
145. Lv JJ, Zheng JN, Li SS, Chen LL, Wang AJ, Feng JJ. Facile synthesis of Pt–Pd nanodendrites and their superior electrocatalytic activity. J Mater Chem A. 2014;2(12):4384–4390.
146. Olenin AY. Mechanisms of metal nanoparticle formation. Nanotechnol Russ. 2012;7(5–6):238–242.
147. Nosheen F, Zhang Z, Xiang G, Xu B, Yang Y, Saleem F, *et al.* Three-dimensional hierarchical PtCu superstructures. Nano Res. 2015;8(3):832–838.
148. Kuang Y, Zhang Y, Cai Z, Feng G, Jiang Y, Jin C, *et al.* Single-crystalline dendritic bimetallic and multimetallic nanocubes. Chem Sci. 2015;6(12):7122–7129.
149. Ma L, Wang C, Gong M, Liao L, Long R, Wang J, *et al.* Control over the branched structures of platinum nanocrystals for electrocatalytic applications. ACS Nano. 2012;6(11):9797–9806.
150. Long R, Zhou S, Wiley BJ, Xiong Y. Oxidative etching for controlled synthesis of metal nanocrystals: atomic addition and subtraction. Chem Soc Rev. 2014;43(17):6288–6310.
151. Taylor E, Chen S, Tao J, Wu L, Zhu Y, Chen J. Synthesis of PtCu nanodendrites through controlled reduction kinetics for enhanced methanol electro-oxidation. ChemSusChem. 2013;6(10): 1863–1867.
152. Weng X, Liu Q, Feng JJ, Yuan J, Wang AJ. Dendrite-like PtAg alloyed nanocrystals: highly active and durable advanced electrocatalysts for oxygen reduction and ethylene glycol oxidation reactions. J Colloid Interface Sci. 2017;504:680–687.
153. Shao FQ, Lin XX, Feng JJ, Yuan J, Chen JR, Wang AJ. Simple fabrication of core-shell AuPt@Pt nanocrystals supported on reduced graphene oxide for ethylene glycol oxidation and hydrogen evolution reactions. Electrochim Acta. 2016;219:321–329.
154. Wu H, Qi W, Peng H, He J. Facile synthesis of Ag@Pt core-shell nanoparticles with different dendrites Pt shells. ChemistrySelect. 2017;2(29):9344–9348.
155. Zhang Q, Li XL, Tao BX, Wang XH, Deng YH, Gu XY, *et al.* CoNi based alloy/oxides@N-doped carbon core-shell dendrites as complementary water splitting electrocatalysts with significantly enhanced catalytic efficiency. Appl Catal B. 2019;254:634–646.
156. Bai J, Huang H, Li FM, Zhao Y, Chen P, Jin PJ, *et al.* Glycerol oxidation assisted electrocatalytic nitrogen reduction: Ammonia

and glyceraldehyde co-production on bimetallic RhCu ultrathin nanoflake nanoaggregates. J Mater Chem A. 2019;7(37):21149–21156.

157. Zhao Y, Bai J, Wu XR, Chen P, Jin PJ, Yao HC, *et al.* Atomically ultrathin RhCo alloy nanosheet aggregates for efficient water electrolysis in broad pH range. J Mater Chem A. 2019;7(27): 16437–16446.

158. Han Z, Zhang RL, Duan JJ, Wang AJ, Zhang QL, Huang H, *et al.* Platinum-rhodium alloyed dendritic nanoassemblies: An all-pH efficient and stable electrocatalyst for hydrogen evolution reaction. Int J Hydrogen Energy. 2020;45(11):6110–6119.

159. Han Z, Wang AJ, Zhang L, Wang ZG, Fang KM, Yin ZZ, *et al.* 3D highly branched PtCoRh nanoassemblies: Glycine-assisted solvothermal synthesis and superior catalytic activity for alcohol oxidation. J Colloid Interface Sci. 2019;554:512–519.

160. Wang L, Yamauchi Y. Facile synthesis of three-dimensional dendritic platinum nanoelectrocatalyst. Chem Mater. 2009;21(15): 3562–3569.

161. Pazos-Pérez N, Rodríguez-González B, Hilgendorff M, Giersig M, Liz-Marzán LM. Gold encapsulation of star-shaped FePt nanoparticles. J Mater Chem. 2010;20(1):61–64.

162. Chen M, Kim J, Liu JP, Fan H, Sun S. Synthesis of FePt nanocubes and their oriented self-assembly. J Am Chem Soc. 2006;128(22): 7132–7133.

163. Mourdikoudis S, Menelaou M, Fiuza-Maneiro N, Zheng G, Wei S, Pérez-Juste J, *et al.* Oleic acid/oleylamine ligand pair: A versatile combination in the synthesis of colloidal nanoparticles. Nanoscale Horiz. 2022;7(9):941–1015.

164. Patsula V, Petrovský E, Kovářová J, Konefal R, Horák D. Monodisperse superparamagnetic nanoparticles by thermolysis of Fe(III) oleate and mandelate complexes. Colloid Polym Sci. 2014;292(9):2097–2110.

165. Guo X, Wang W, Yang Y, Tian Q. Designing a large scale synthesis strategy for high quality magnetite nanocrystals on the basis of a solution behavior regulated formation mechanism. CrystEngComm. 2016;18(47):9033–9041.

166. Chou SW, Zhu CL, Neeleshwar S, Chen GL, Chen YY, Chen CC. Controlled growth and magnetic property of FePt nanostructure: Cuboctahedron, octapod, truncated cube, and cube. Chem Mater. 2009;21(20):4955–4961.

167. Jenkinson KJ, Wagner A, Kornienko N, Reisner E, Wheatley AEH. A one-pot route to faceted FePt-Fe_3O_4 dumbbells: Probing

morphology–catalytic activity effects in O_2 reduction catalysis. Adv Funct Mater. 2020;30(25):2002633.
168. Solov'yov A, Verkhovtsev A, Mason N, Amos R, Bald I, Baldacchino G, *et al.* Condensed matter systems exposed to radiation: Multiscale theory, simulations, and experiment. Chem Rev. 2024;124(13): 8014–8029.
169. Thanh NTK, Maclean N, Mahiddine S. Mechanisms of nucleation and growth of nanoparticles in solution. Chem Rev. 2014;114(15):7610–7630.
170. Zhang W. Nanoparticle aggregation: Principles and modeling. In: Capco DG, Chen Y, editors. Nanomaterial: Impacts on Cell Biology and Medicine. Dordrecht: Springer; 2014.
171. Wu ZP, Shan S, Zang SQ, Zhong CJ. Dynamic core-shell and alloy structures of multimetallic nanomaterials and their catalytic synergies. Acc Chem Res. 2020;53(12):2913–24.
172. Bora T, Dousse A, Sharma K, Sarma K, Baev A, Hornyak GL, *et al.* Modeling nanomaterial physical properties: Theory and simulation. Int J Smart Nano Mater. 2019;10(2):116–143.
173. Liu MX, Wang K, Xia D, Jiang T. Phase field simulation of Al–Si binary dendritic growth and micro-segregation patterns under convection. J Alloys Comp. 2014;589:431–435.
174. Witten TA, Sander LM. Diffusion-limited aggregation. Phys Rev B. 1983;27(9):5686–5697.
175. Ogura S, Fukutani K, Matsumoto M, Okano T, Okada M, Kawamura T. Dendritic to non-dendritic transitions in Au islands investigated by scanning tunneling microscopy and Monte Carlo simulations. Phys Rev B. 2006;73(12):125442.
176. Krumwiede TR, Schulze TP. Dendritic growth shapes in kinetic Monte Carlo models. Model Simul Mat Sci Eng. 2017;25(2):025009.
177. Schulze TP. Simulation of dendritic growth into an undercooled melt using kinetic Monte Carlo techniques. Phys Rev E. 2008;78(2):020601.
178. Soy E, Liang Z, Trenary M. Formation of Pt and Rh nanoclusters on a graphene Moiré pattern on Cu(111). J Phys Chem C. 2015;119(44):24796–24803.
179. Dokukin SA, Kolesnikov SV, Saletsky AM, Klavsyuk AL. Growth of the Pt/Cu(111) surface alloy: Self-learning kinetic Monte Carlo simulations. J Alloys Comp. 2018;763:719–727.
180. Stamenkovic V, Mun BS, Mayrhofer KJJ, Ross PN, Markovic NM, Rossmeisl J, *et al.* Changing the activity of electrocatalysts for oxygen reduction by tuning the surface electronic structure. Angew Chem Int Ed. 2006;45(18):2897–2901.

181. Mei D, Du J, Neurock M. First-principles-based kinetic Monte Carlo simulation of nitric oxide reduction over platinum nanoparticles under lean-burn conditions. Ind Eng Chem Res. 2010;49(21): 10364–10373.
182. Zhdanov VP, Schweinberger FF, Heiz U, Langhammer C. Ostwald ripening of supported Pt nanoclusters with initial size-selected distributions. Chem Phys Lett. 2015;631-632:21–25.
183. Li L, Plessow PN, Rieger M, Sauer S, Sánchez-Carrera RS, Schaefer A, *et al.* Modeling the migration of platinum nanoparticles on surfaces using a kinetic Monte Carlo approach. J Phys Chem C. 2017;121(8):4261–4269.
184. Jørgensen M, Grönbeck H. Scaling relations and kinetic Monte Carlo simulations to bridge the materials gap in heterogeneous catalysis. ACS Catal. 2017;7(8):5054–5061.
185. Bai J, Jia N, Jin P, Chen P, Jiang JX, Zeng JH, *et al.* Metal-organic interface engineering for boosting the electroactivity of Pt nanodendrites for hydrogen production. J Energy Chem. 2020;51:105–112.
186. Angelescu DG, Magno LM, Stubenrauch C. Monte Carlo simulation of the size and composition of bimetallic nanoparticles synthesized in water in oil microemulsions. J Phys Chem C. 2010;114(50): 22069–22078.
187. Yun K, Cho YH, Cha PR, Lee J, Nam HS, Oh JS, *et al.* Monte Carlo simulations of the structure of Pt-based bimetallic nanoparticles. Acta Mater. 2012;60(12):4908–4916.
188. Duan Z, Wang G. Monte Carlo simulation of surface segregation phenomena in extended and nanoparticle surfaces of PtPd alloys. J Phys Cond Matter. 2011;23(47):475301.
189. Lin ZZ, Chen X, Yin C, Tang H, Hu YC, Ning XJ. Theoretical prediction of the growth and surface structure of Pt and Ni nanoparticles. Europhys Lett. 2011;96(6):66005.
190. Müller M, Albe K. Lattice Monte Carlo simulations of FePt nanoparticles: Influence of size, composition, and surface segregation on order-disorder phenomena. Phys Rev B. 2005;72(9):094203.
191. Qin L, Zhang Y, Huang S, Tian H, Wang P. Atomic-scale structure of Co–Pt bimetallic nanoparticles: Monte Carlo simulations. Phys Rev B. 2010;82(7):075413.
192. Perales-Rondón JV, Ferre-Vilaplana A, Feliu JM, Herrero E. Oxidation mechanism of formic acid on the bismuth adatom-modified Pt(111) surface. J Am Chem Soc. 2014;136(38):13110–13113.
193. Chen R, Nguyen QN, Xia Y. Oriented attachment: A unique mechanism for the colloidal synthesis of metal nanostructures. ChemNanoMat. 2022;8(3):e202100474.

194. Peng Z, You H, Yang H. Composition-dependent formation of platinum silver nanowires. ACS Nano. 2010;4(3):1501–1510.
195. Kwon T, Park J, Baik H, Back S, Błasiak B, Cho M, *et al.* Unexpected solution phase formation of hollow PtSn alloy nanoparticles from Sn deposition on Pt dendritic structures. CrystEngComm. 2016;18(32):6019–6023.
196. Li Y, Koel BE. Structural studies of SnPt(100) surfaces: Conditions for alloy formation. Surf Sci. 1995;330(2):193–206.
197. Wang DY, Chou HL, Cheng CC, Wu YH, Tsai CM, Lin HY, *et al.* FePt nanodendrites with high-index facets as active electrocatalysts for oxygen reduction reaction. Nano Energy. 2015;11:631–639.
198. Furuya N, Shibata M. Structural changes at various Pt single crystal surfaces with potential cycles in acidic and alkaline solutions. J Electroanal Chem. 1999;467(1-2):85–91.
199. Han BC, Miranda CR, Ceder G. Effect of particle size and surface structure on adsorption of O and OH on platinum nanoparticles: A first-principles study. Phys Rev B. 2008;77(7):075410.
200. Lim B, Jiang M, Yu T, Camargo PHC, Xia Y. Nucleation and growth mechanisms for Pd-Pt bimetallic nanodendrites and their electrocatalytic properties. Nano Res. 2010;3(1):69–80.
201. Huang TH, Jiang Y, Peng YH, Tseng YT, Yan C, Chien PC, *et al.* Unique (100) surface configuration enables promising oxygen reduction performance for Pt_3Co nanodendrite catalysts. ACS Appl Mater Interfaces. 2023;15(14):18217–18228.
202. Yan SY, Liu CW, Huang TH, Guo YZ, Lee SW, Wang JH, *et al.* Composition effect of oxygen reduction reaction on PtSn nanorods: An experimental and computational study. Int J Hydrogen Energy. 2018;43(31):14427–14438.
203. An Z, Li H, Zhang X, Xu X, Xia Z, Yu S, *et al.* Structural Evolution of a PtRh nanodendrite electrocatalyst and its ultrahigh durability toward oxygen reduction reaction. ACS Catal. 2022;12(6):3302–3308.
204. Guo K, Fan D, Bao J, Li Y, Xu D. Atomic-level phosphorus-doped ultrathin Pt nanodendrites as efficient electrocatalysts. Adv Funct Mater. 2022;32(47):2208057.
205. Du Y, Ni K, Zhai Q, Yun Y, Xu Y, Sheng H, *et al.* Facile air oxidative induced dealloying of hierarchical branched PtCu nanodendrites with enhanced activity for hydrogen evolution. Appl Catal A. 2018;557:72–78.
206. Perdew JP, Chevary JA, Vosko SH, Jackson KA, Pederson MR, Singh DJ, *et al.* Atoms, molecules, solids, and surfaces: Applications of the generalized gradient approximation for exchange and correlation. Phys Rev B. 1992;46(11):6671–6687.

207. Perdew JP, Burke K, Ernzerhof M. Generalized Gradient Approximation Made Simple. Phys Rev Lett. 1996;77(18):3865–3868.
208. Perdew JP, Burke K, Ernzerhof M. Generalized Gradient Approximation Made Simple – erratum. Phys Rev Lett. 1997;78(7):1396.
209. Fan J, Qi K, Zhang L, Zhang H, Yu S, Cui X. Engineering Pt/Pd interfacial electronic structures for highly efficient hydrogen evolution and alcohol oxidation. ACS Appl Mater Interfaces. 2017;9(21):18008–18014.
210. Guo K, Han X, Wei S, Bao J, Lin Y, Li Y, *et al.* Functional surfactant-induced long-range compressive strain in curved ultrathin nanodendrites boosts electrocatalysis. Nano Lett. 2023;23(3): 1085–1092.
211. Zheng Y, Jiao Y, Zhu Y, Li LH, Han Y, Chen Y, *et al.* Hydrogen evolution by a metal-free electrocatalyst. Nat Commun. 2014;5(1):3783.
212. Logeshwaran N, Panneerselvam IR, Ramakrishnan S, Kumar RS, Kim AR, Wang Y, *et al.* Quasihexagonal platinum nanodendrites decorated over CoS_2-N-doped reduced graphene oxide for electro-oxidation of $C1^-$, $C2^-$, and C3-type alcohols. Adv Sci. 2022;9(8):2105344.
213. Singh-Miller NE, Marzari N. Surface energies, work functions, and surface relaxations of low-index metallic surfaces from first principles. Phys Rev B. 2009;80(23):235407.
214. Nelson R, Ertural C, George J, Deringer VL, Hautier G, Dronskowski R. LOBSTER: Local orbital projections, atomic charges, and chemical-bonding analysis from projector-augmented-wave-based density-functional theory. J Comput Chem. 2020;41(21):1931–1940.
215. Yang C, Jiang Q, Liu H, Yang L, He H, Huang H, *et al.* Pt-on-Pd bimetallic nanodendrites stereoassembled on MXene nanosheets for use as high-efficiency electrocatalysts toward the methanol oxidation reaction. J Mater Chem A. 2021;9(27):15432–15440.
216. Xiao BB, Yang L, Liu HY, Jiang XB, Aleksandr B, Song EH, *et al.* Designing fluorographene with FeN_4 and CoN_4 moieties for oxygen electrode reaction: A density functional theory study. Appl Surf Sci. 2021;537:147846.
217. Xiao BB, Jiang XB, Jiang Q. Density functional theory study of oxygen reduction reaction on $Pt/Pd_3Al(111)$ alloy electrocatalyst. Phys Chem Chem Phys. 2016;18:14234–14243.
218. Xiao BB, Jiang XB, Yang XL, Jiang Q, Zheng F. The segregation resistance of the $Pt_2ML/Os/Pd_3Al$ sandwich catalyst for oxygen reduction reaction: A density functional theory study. Phys Chem Chem Phys. 2016;18(43):30174–30182.

219. Song J, He J, Yang L, Wang W, Bai Q, Feng W, *et al.* Enhanced peroxidase-like and antibacterial activity of Ir-coated Pd–Pt nanodendrites as nanozyme. Bioinorg Chem Appl. 2023;2023(1): 1689455.
220. Gao L, Zhuang J, Nie L, Zhang J, Zhang Y, Gu N, *et al.* Intrinsic peroxidase-like activity of ferromagnetic nanoparticles. Nat Nanotechnol. 2007;2(9):577–583.
221. Yang W, Yang X, Zhu L, Chu H, Li X, Xu W. Nanozymes: Activity origin, catalytic mechanism, and biological application. Coord Chem Rev. 2021;448:214170.
222. Shen X, Wang Z, Gao X, Zhao Y. Density functional theory-based method to predict the activities of nanomaterials as peroxidase mimics. ACS Catal. 2020;10(21):12657–12665.
223. Crawley JWM, Gow IE, Lawes N, Kowalec I, Kabalan L, Catlow CRA, *et al.* Heterogeneous trimetallic nanoparticles as catalysts. Chem Rev. 2022;122(6):6795–6849.
224. Eid K, Malgras V, He P, Wang K, Aldalbahi A, Alshehri SM, *et al.* One-step synthesis of trimetallic PtPdRu nanodendrites as highly active electrocatalysts. RSC Adv. 2015;5(39):31147–1152.
225. Huang L, Han Y, Dong S. Highly-branched mesoporous Au–Pd–Pt trimetallic nanoflowers blooming on reduced graphene oxide as an oxygen reduction electrocatalyst. Chem Commun. 2016;52(56): 8659–8662.
226. Tsai MH, Yeh JW. High-entropy alloys: A critical review. Mater Res Lett. 2014;2(3):107–123.
227. Shi Y, Yang B, Liaw PK. Corrosion-resistant high-entropy alloys: A review. Metals. 2017;7(2):43.
228. Alshataif YA, Sivasankaran S, Al-Mufadi FA, Alaboodi AS, Ammar HR. Manufacturing methods, microstructural and mechanical properties evolutions of high-entropy alloys: A review. Metals Mater Int. 2020;26(8):1099–133.
229. Cao G, Liang J, Guo Z, Yang K, Wang G, Wang H, *et al.* Liquid metal for high-entropy alloy nanoparticles synthesis. Nature. 2023;619(7968):73–77.

https://doi.org/10.1142/9781800618183_0007

Chapter 7

Reactive and Irradiation-Driven Molecular Dynamics with MBN Explorer

Alexey V. Verkhovtsev* **and Andrey V. Solov'yov**†

MBN Research Center,
Altenhöferallee 3, 60438 Frankfurt am Main, Germany
**verkhovtsev@mbnresearch.com*
†solovyov@mbnresearch.com

This chapter provides an overview of reactive molecular dynamics (RMD) and irradiation-driven molecular dynamics (IDMD) – the advanced computational methods implemented in the MBN Explorer software package – which allow atomistic simulations of chemical and irradiation-driven transformations in a wide variety of MesoBioNano (MBN) systems. Within the RMD and IDMD frameworks, quantum processes occurring in a system (e.g., bond cleavage and formation) are treated as random, fast, and local transformations that are stochastically incorporated into the classical MD framework, with probabilities elaborated on the basis of quantum mechanics. Major transformations of molecular systems (such as changes in molecular topology, redistribution of atomic partial charges, or changes in interatomic interactions) are simulated using the reactive CHARMM (rCHARMM) force field implemented in MBN Explorer. This chapter presents several exemplary case studies illustrating the use of RMD and IDMD and their validation by comparison with experiments. Particular examples include simulations of the irradiation-induced fragmentation of organometallic and biomolecular systems and the formation of metal-containing nanostructures during focused electron beam induced deposition (FEBID).

7.1 Introduction

Classical molecular dynamics (MD) is a powerful computational method that can provide insights into nanoscale structural features of various molecular, nanoscopic, and condensed matter systems and their thermal, mechanical, and other properties using advanced computer simulations.[1,2] The concept of classical MD is based on the Born–Oppenheimer approximation,[3] which justifies the separation of slow ionic and fast electronic degrees of freedom within a molecular system. Within the classical MD framework, the time evolution of a many-atom system is described by the integration of the classical coupled equations of motion, where interaction potentials or force fields acting between atoms are defined. This simplification of the description of the motion of a quantum system provides significant advantages for computer simulations.

However, the classical MD approach does not describe the electron dynamics and, therefore, most of the quantum transformations that can occur during the system dynamics. These transformations are often induced in the system by external perturbations such as external fields or irradiation by charged particles (electrons, protons, ions, etc.) or photons. The resulting effects may have a global character (electric current, spin ordering, polarisation, magnetisation, etc.) or be local (atomic or molecular excitation, ionisation, dissociation, charge transfer, etc.).

Irradiation-induced local quantum perturbations of a molecular system typically occur on the sub-femtosecond timescale and involve only those atoms that are directly affected by the irradiation. This results in the creation of secondary electrons, ions, reactive species (radicals), and excited molecules that can further interact with the molecular system and cause further chemical transformations. This complex local dynamics typically involves the nearest environment of the targeted molecular site, being a small part of the entire molecular system, and is typically completed within femtoseconds. During this time, some of the initial perturbations of the system, such as quasi-free electrons, electron holes, ionic charges, and interatomic interactions, attain their new characteristics, which practically do not change during the subsequent evolution of the system. The femtosecond timescale is still significantly shorter than the characteristic timescales of the motion of the entire molecular system because in classical molecular dynamics simulations, a typical integration time step is 1–2 fs.

The notable outcome of the process described above is the emergence of bond breaks in the system. These events are most significant as they affect the dynamical behaviour and chemical transformations in the molecular system on the larger timescales, up to nanoseconds and beyond, which is the typical time frame for the classical MD. The bond breaks occur in those parts of the molecular system that are targeted by the irradiation or an external perturbation. They occur randomly with a probability depending on the intensity and the modality of the irradiation. The probabilities of these events are related to the cross-sections of the involved irradiation-induced processes (elastic and inelastic scattering, electronic and vibrational excitation, dissociative electron attachment, collision dissociation, etc.) occurring in the system on the femtosecond timescale and can be elaborated from the collision theory or be obtained from experiment.

The irradiation conditions for a molecular system can vary considerably, depending on the radiation modality, the duration of the system exposure to irradiation, and the system's geometry. Irradiation can be a swift single event, such as a single ion track crossing the molecular system, or it can last over a period of time up to some nanoseconds and even longer. In the latter case, the irradiation-induced bond breaking and charge redistribution in the system occurs throughout the entire irradiation period. Irradiation can be homogeneous within a certain volume or strongly inhomogeneous. The choice of irradiation conditions corresponds to each particular case study.

The scenario described above defines the *irradiation-driven molecular dynamics* (IDMD);[4] see Section 7.3. The IDMD can be introduced as classical MD with the superimposed random process of molecular bond breakage related to the irradiation conditions. Bond breakage is defined as the local alteration of the system force fields, which involves (i) the creation of reactive atomic species (radicals) with dangling bonds, (ii) the possibility of dangling bonds' closure and the creation of new molecular bonds or molecules, (iii) accounting for molecular topology changes (in the cases when it is defined, e.g., molecular mechanics force fields). The characterisation of these modifications of the classical molecular dynamics force fields can be elaborated on the basis of quantum-chemical methods.

The IDMD methodology[4] implemented in the MBN Explorer software package[5] aims to account for the major dissociative transformations of the molecular system induced by its irradiation and possible paths of further reactive transformations. The latter are sensitive to statistical mechanical factors, such as the concentration of the reactive species, their mobility (diffusion), the temperature of the medium, etc.

The concept of IDMD is general and applicable to any kind of molecular system treated with any type of classical force field implemented in MBN Explorer.[6,7] In its current realisation, IDMD can describe systems modelled through pairwise potentials, many-body force fields, molecular mechanics force fields (including the rCHARMM force field;[8] see Section 7.4) and their combinations. The limited number of parameters that determine molecular force fields and their irradiation-driven perturbations results in a countable number of modifications that could occur in a molecular system upon irradiation and makes the method efficient and accurate. This implementation opens up a broad range of possibilities for modelling of irradiation-driven modifications and chemistry of complex molecular systems.

The methodology of reactive force field utilised in IDMD can also be applied to simulate the transformations of MBN systems caused by chemical reactions. Chemical reactions are essentially quantum processes that occur on the femtosecond timescale. They can be treated similarly to the aforementioned irradiation quantum processes in the case of IDMD. The molecular dynamics approach accounting for the possible chemical transformations in the system through the reactive force fields is called *reactive molecular dynamics* (RMD); see Section 7.2.

7.2 Reactive Molecular Dynamics

The reactive molecular dynamics (RMD) approach[8] implemented in MBN Explorer[5] enables classical simulations of chemical reactions and chemistry-based nanoscale phenomena on temporal and spatial scales inaccessible by pure *ab initio* methods. Quantum-mechanical methods such as time-dependent density functional theory (TDDFT) or many-body theory are computationally expensive and limited to systems of a few nanometres in size. On the other hand, conventional (non-reactive) classical MD provides a powerful tool for simulating larger systems on long timescales, but this approach either cannot simulate chemical reactions or can do so only for selected systems using specific interatomic potentials, e.g., reactive empirical bond-order (REBO) potentials for carbon and hydrocarbon systems.[9–13] To circumvent this problem, RMD methods have been developed, which use interatomic force fields capable of locally mimicking the quantum effects due to chemical reactions. One of such force fields is *the reactive CHARMM (rCHARMM) force field*,[8] described in detail in Section 7.4.

The RMD approach[8] implemented in MBN Explorer accounts for fast and local chemical transformations during the MD of molecular or condensed matter systems. Such transformations have a quantum nature and occur probabilistically. In the RMD methodology realised in MBN Explorer, chemical transformations are incorporated into MD on the basis of the Monte Carlo (MC) approach by coupling the chemical transformations having the quantum nature with the classical dynamics of a molecular medium. Such chemical transformations may involve:

- the breaking and formation of covalent bonds in a system,
- change in atomic types and valences,
- change in bond multiplicities,
- redistribution of atomic partial charges,
- change in interatomic interaction potentials,
- formation of specific chemical products associated with specific fragmentation channels,
- changes in the molecular topology of the system, etc.

An extended description of the applications of the RMD methodology can be found in recent reviews[2,14] and the roadmap paper,[15] as well as in this chapter.

The rCHARMM force field enables the description of bond rupture events and the formation of new covalent bonds by chemically active atoms in the system, monitoring its chemical composition and changes in the system's topology that occur during its transformations. The chemically active atoms carry information about their charges and valences, interactions with other atoms in the system, and multiplicities of the bonds that can be formed with other reactive atoms in the system.

As an extension of the 'standard' (non-reactive) CHARMM force field,[16,17] rCHARMM is directly applicable to organic and biomolecular systems.[18,19] Its combination with many other pairwise and many-body force fields enables simulations of thermally driven and post-irradiation chemical transformations in various molecular and condensed matter systems while monitoring their molecular composition and topology changes.[1,2] Due to its versatility and universality, the rCHARMM force field has been applied to study a broad range of processes and phenomena, including collision-induced structural transformations and fragmentation (in the gas phase or after the collision with surfaces);[20] collision-induced fragmentation of molecular and cluster systems in the gas phase and placed in molecular environments;[18,19,21,22] thermally driven and collision-induced chemistry of condensed systems, particularly water;[8,23] and surface chemistry processes lying in the core of modern nanofabrication techniques.[4,24–27] An overview of some of these case studies is given further in this chapter. The input parameters for RMD can be elaborated on the basis of quantum calculations, e.g., DFT (see Section 7.4.1), or they can be obtained from experiments.

7.3 Irradiation-Driven Molecular Dynamics

Atomistic simulations of irradiation-driven transformations in complex molecular and condensed matter systems can be performed using the *Irradiation-Driven Molecular Dynamics (IDMD)* methodology[4] implemented in MBN Explorer.[5] It takes into account the main dissociative transformations of irradiated molecular systems (topological changes, redistribution of atomic partial charges, atomic valences, and bond multiplicities) and possible paths of their further reactive transformations.[4] Such transformations are simulated using the reactive CHARMM (rCHARMM) force field[8] described in Section 7.4. The possibility of combining rCHARMM with many other pairwise and many-body interatomic potentials makes the IDMD approach universal and applicable to many different molecular and condensed matter systems exposed to radiation.

Within the IDMD framework, various quantum collision processes (e.g., ionisation, electronic excitation, bond dissociation via electron attachment, or charge transfer) are treated as random, fast, and local transformations that are incorporated into the classical MD framework in a stochastic manner, with probabilities elaborated based on quantum mechanics.[4] This is possible because the quantum processes take place on sub-femto- to femtosecond timescales (i.e., on timescales comparable to or smaller than a typical time step of MD simulations) and typically involve a relatively small number of atoms.

The probability of each quantum process is equal to the product of the process cross-section and the flux density of the incident particles.[28] Both characteristics should be worked out for each particular case study. Collision process cross-sections can be obtained from any of the following: (i) *ab initio* calculations performed by various dedicated codes, (ii) analytical estimates and models, (iii) experiments, or (iv) atomic and molecular databases. The flux densities of the incident particles are usually specific to the problem and system under consideration. The properties of atoms or molecules (energy, momentum, charge, valence, interaction potentials with other atoms in the system, etc.) involved in such quantum transformations are changed according to their final quantum states in the corresponding quantum processes.

IDMD enables the analysis of rapid energy transfer events into fragmenting covalent bonds caused by quantum processes and the analysis of post-irradiation energy relaxation processes, occurring typically on a picosecond timescale and leading to chemical transformations. The energy transferred to the system through irradiation is absorbed by the involved electronic and ionic degrees of freedom, and chemically reactive sites (atoms, molecules, and molecular sites) in the irradiated system are created.[4] These events lead to changes in the system's molecular topology, the number and type of atomic and molecular species present in the system, and other characteristics that may affect the dynamic behaviour and chemical transformations of the molecular system on longer timescales. The subsequent dynamics of the reactive sites is determined by the classical MD and the thermodynamic state of the system until the system undergoes further irradiation-driven quantum transformations. The chemically reactive sites may also be involved in the chemical reactions leading to changes in their molecular and reactive properties and, ultimately, the formation of chemically stable atomic and molecular species.[2,4,14,24–26]

In IDMD, special attention is paid to the dynamics of secondary electrons and the mechanisms of energy and momentum transfer from the excited electronic subsystem to the vibrational degrees of freedom of the system, i.e., to its heating. For small molecular systems in the gas phase, the ejected electrons can often be decoupled from the system and excluded from the analysis of the post-irradiation dynamics of the system. For extended molecular and condensed phase systems, the interaction of the secondary electrons with the system can be treated within various electron transport theories, such as the diffusion-equation-based approach[29,30] or the MC approach,[31] and considered as an additional irradiation field imposed on the molecular system.[24] Such an analysis provides the spatial distribution of the energy transferred to the medium by the irradiation. Finally, immobilised electrons and electronic excitations transfer the deposited energy to the system's heat via electron–phonon coupling, typically on the picosecond timescale.[32] The IDMD approach accounts for the key results of this relaxation process, determining its duration as well as the temporal and spatial dependence of the amount of energy transferred to the system's heat.

In the absence of irradiation, only reactive transformations become possible at larger timescales, which can be simulated using RMD (see Section 7.2). As such, IDMD, together with RMD, allows a computational analysis of physicochemical processes occurring in the system coupled to radiation and post-irradiation system's dynamics on time and spatial scales far beyond the limits of quantum-mechanics-based computational schemes, such as TDDFT or *ab initio* MD.

IDMD relies on several input parameters, such as bond dissociation energies, molecular fragmentation cross-sections, the amount of energy transferred to the system upon irradiation, the energy relaxation rate, and the spatial region in which the energy is relaxed. These characteristics can be obtained through accurate quantum-mechanical calculations using various computational schemes based on many-body theory, DFT, TDDFT, and collision theory.[15] Alternatively, the required parameters can be obtained from experimental data or analytical models/methods.

The countable number of physically meaningful parameters of IDMD and the reactive molecular force fields, together with a much larger number of different output characteristics accessible for simulations and analysis, make the IDMD method efficient and accurate. As such, IDMD offers unique possibilities for modelling irradiation-driven modifications of complex molecular and condensed matter systems beyond the capabilities of pure quantum or pure classical MD.

The IDMD approach allows the outputs of numerous MC codes simulating radiation and particle transport in different media (e.g., Geant4,[33,34] SEED,[35] and others) to be coupled with the inputs of IDMD, thus achieving the multiscale description of irradiation-driven molecular dynamics, chemistry, and structure formation in many different MBN systems. These important capabilities of MBN Explorer have been demonstrated in a recent work describing the case study of focused electron beam induced deposition.[24] A similar methodology can be used to simulate numerous molecular systems placed in radiation fields of different modalities, geometries, and temporal profiles.

Due to its universality, IDMD can be utilised to study irradiation-induced processes in various molecular systems (in the gas phase or embedded into molecular environments) and condensed matter systems, as discussed in Sections 7.5 and 7.6. Particular areas for the application of this methodology include the analysis of nanoscopic mechanisms of radiation-induced damage of biomolecular systems[29,36] and the nanoscopic mechanisms of the nanostructure formation and growth due to the irradiation with focused beams of charged particles.[4,24–26] An extended description of the applications of this methodology can be found in recent reviews[2,14] and the roadmap paper.[15]

7.4 Generalisation of Standard Molecular Force Fields for RMD and IDMD

MBN Explorer allows the simulation of dissociation and formation of covalent bonds through the rCHARMM force field,[8] which is an extension of the standard CHARMM force field.[16,17] This extension requires the specification of two additional parameters, which define the dissociation energy of a covalent bond and the cutoff radius for bond breakage or formation.

By specifying the additional parameters for the bonded interactions, MBN Explorer considers all molecular mechanics interactions, i.e., bonded, angular, and dihedral, using an alternative parameterisation. If the distance between a given pair of atoms becomes greater than the specified cutoff radius, this particular bonded interaction is removed from the system's topology and is not considered in future calculations.

The standard CHARMM force field[16,17] employs a harmonic approximation to describe the interatomic interactions, thereby limiting its applicability to small deformations of the molecular system. For larger perturbations, the potential should go to zero as the valence bonds break. To allow for the breaking of covalent bonds in the molecular mechanics force field, MBN Explorer uses a modified interaction potential that describes the interaction of atoms connected by chemical bonds. The standard CHARMM force field describing covalent bonds is defined as

$$U^{\mathrm{b}}(r_{ij}) = k^{\mathrm{b}}_{ij}(r_{ij} - r_0)^2 \, , \tag{7.1}$$

where k^{b}_{ij} is the bond stretching force constant, r_{ij} is the distance between atoms i and j, and the parameter r_0 is the covalent bond length. This parameterisation describes well the bond stretching regime for small deviations from r_0 but gives an incorrect result for larger distortions.

For a satisfactory description of the covalent bond breakage, it is reasonable to use the Morse potential. This potential requires one additional parameter compared to the harmonic potential (7.1); this parameter accounts for the energy of the bond dissociation. For a pair of atoms, the Morse potential is given by

$$U^{\mathrm{b}}(r_{ij}) = D_{ij}[e^{-2\beta_{ij}(r_{ij}-r_0)} - 2e^{-\beta_{ij}(r_{ij}-r_0)}], \tag{7.2}$$

where D_{ij} is the dissociation energy of the covalent bond and $\beta_{ij} = (k^{\mathrm{b}}_{ij}/D_{ij})^{1/2}$ determines the steepness of the potential. Figure 7.1 shows the Morse potential for the CN7–CN8B bond, which is one of the covalent bonds in the DNA backbone. For small deviations from r_0, the Morse potential and the harmonic approximation are close to each other.

The breaking of covalent bonds during the simulation automatically employs an improved potential for the valence angles. In the

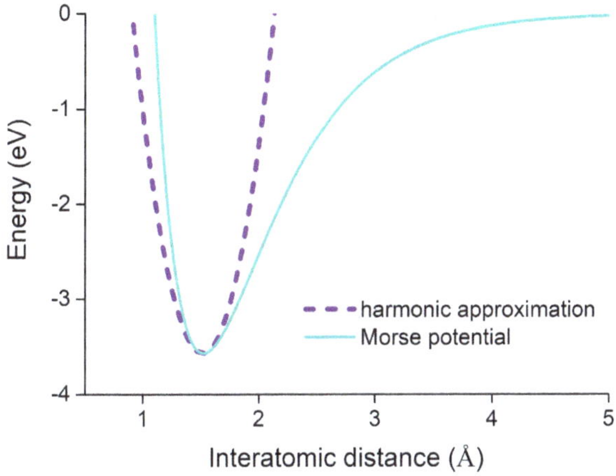

Fig. 7.1. The pairwise carbon–carbon (type CN7–CN8B) interaction potential in harmonic approximation, Eq. (7.1), and modelled with the Morse potential, Eq. (7.2).
Source: Reproduced from Ref. 14 published under an open access Creative Commons CC BY license.

CHARMM force field, the potential associated with the change of a valence angle between bonds with indices ij and jk reads as

$$U^{\mathrm{a}}(\theta_{ijk}) = k^{\theta}_{ijk}(\theta_{ijk} - \theta_0)^2 \ , \tag{7.3}$$

where k^{θ}_{ijk} and θ_0 are parameters of the potential and θ_{ijk} is the actual value of the angle formed by the three atoms. This potential grows rapidly as the angle increases and can lead to nonphysical results when modelling covalent bond breakage. To avoid such cases, the harmonic potential, Eq. (7.3), is replaced in the modified force field with an alternative parameterisation, which reads as

$$U^{\mathrm{a}}(\theta_{ijk}) = 2k^{\theta}_{ijk}\left[1 - \cos(\theta_{ijk} - \theta_0)\right] \ . \tag{7.4}$$

For small variations of the valence angle, this parameterisation is identical to the harmonic approximation (7.3) used in the standard CHARMM force field. For larger values of the angle, the new parameterisation (7.4) defines an energy threshold, which becomes important for an accurate modelling of bond breakage.

The breaking of a covalent bond is accompanied by the breaking of the angular interactions associated with that bond. The effect of bond breakage on the angular potential can be described by a smoothed step function $\sigma(r_{ij})$, defined as

$$\sigma(r_{ij}) = \frac{1}{2}\left[1 - \tanh(\beta_{ij}(r_{ij} - r_{ij}^{*}))\right] , \quad (7.5)$$

where $r_{ij}^{*} = (R_{ij}^{\mathrm{vdW}} + r_0)/2$. This function introduces a correction to the angular interaction potential by assuming that the distance between two atoms involved in an angular interaction increases from the equilibrium value r_0 up to the van der Waals contact value R^{vdW}. Since an angular interaction depends on two bonds connecting the atoms with indices ij and jk, the potential energy describing the valence angular interaction that is subject to rupture is parameterised as

$$\tilde{U}^{\mathrm{a}}(\theta_{ijk}) = \sigma(r_{ij})\,\sigma(r_{jk})\,U^{\mathrm{a}}(\theta_{ijk}) \, . \quad (7.6)$$

As can be seen from Eq. (7.6), the angular potential decreases with increasing bond length between either pair of atoms ij or jk. As an illustration, Fig. 7.2(a) shows the CN8B–ON2–P angular potential which arises, for example, when modelling DNA nucleotides. The shown angular potential is calculated using Eq. (7.6), assuming that the bond between the oxygen and phosphorous atoms is broken. For the sake of illustration, the CN8B–ON2 bond length in this case is taken to be equal to its equilibrium value r_0.

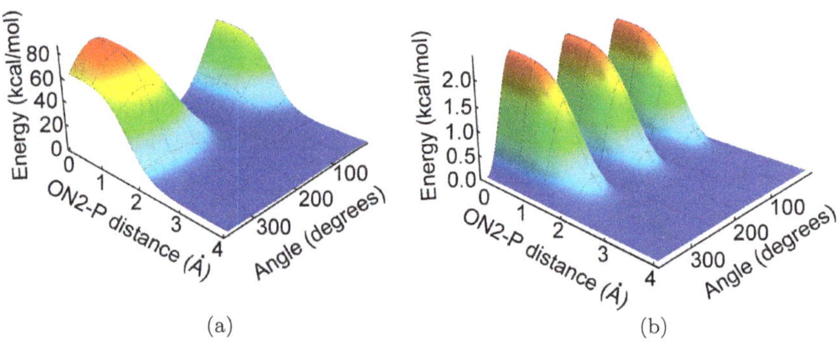

Fig. 7.2. Panel (a): The CN8B–ON2–P angular potential calculated using Eq. (7.6) accounting for the ON2–P bond rupture. Panel (b): The CN4–P–ON2–CN7 dihedral potential calculated using Eq. (7.8) accounting for the ON2–P bond rupture.
Source: Reproduced from Ref. 14 published under an open access Creative Commons CC BY license.

Dihedral interactions arise in the conventional molecular mechanics potential due to the change in the dihedral angles between each of the four topologically defined atoms. Let us consider a quadruple of atoms with indices i, j, k, and l bound by an interaction governed by a change in the dihedral angle. In this case, the dihedral angle is the angle between the plane formed by the atoms i, j, and k and the plane formed by the atoms j, k, and l. In the harmonic approximation, the dihedral energy contribution reads as

$$U^{\mathrm{d}}_{ijkl} = k^{\mathrm{d}}_{ijkl}\left[1 + \cos\left(n_{ijkl}\,\chi_{ijkl} - \delta_{ijkl}\right)\right] , \tag{7.7}$$

where k^{d}_{ijkl}, n_{ijkl}, and δ_{ijkl} are parameters of the potential and χ_{ijkl} is the angle between the planes formed by the atoms i, j, k and j, k, l.

The dihedral interactions are also perturbed by the rupture of a covalent bond; therefore, Eq. (7.7) should be modified to take this effect into account. The rupture of a dihedral interaction between a quadruple of atoms i, j, k, and l should take into account three bonds that contribute to this interaction. Thus, the potential energy describing the dihedral interaction with account for the bond rupture reads as

$$\tilde{U}^{\mathrm{d}}_{ijkl} = \sigma(r_{ij})\,\sigma(r_{jk})\,\sigma(r_{kl})\,U^{\mathrm{d}}_{ijkl} , \tag{7.8}$$

where U^{d}_{ijkl} is the potential in Eq. (7.7) describing the dihedral interaction within the framework of the standard CHARMM force field. The functions $\sigma(r_{ij})$, $\sigma(r_{jk})$, and $\sigma(r_{kl})$ are defined by Eq. (7.5); they are used to limit the dihedral interaction upon increasing the corresponding bond length. Figure 7.2(b) shows a profile of a CN4–P–ON2–CN7 dihedral interaction potential accounting for the rupture of the central ON2–P bond. This dihedral interaction is also important for modelling bond breaks in the DNA molecule.

7.4.1 *Establishing rCHARMM parameters through quantum mechanical calculations*

The rCHARMM parameters (particularly the dissociation energy and the cutoff distance for bond breaking and formation) are commonly determined from density functional theory (DFT) calculations. In Ref. 21, rCHARMM was used to model the irradiation-driven fragmentation of a $W(CO)_6$ molecule (see Fig. 7.3(b)), one

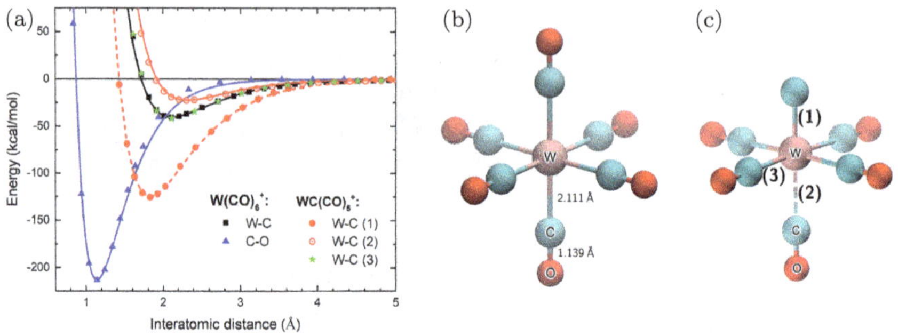

Fig. 7.3. Panel (a): Potential energy curves for W–C and C–O bonds in the $W(CO)_6^+$ and $WC(CO)_5^+$ molecules.[21] Symbols show the results of DFT calculations performed using the B3LYP functional and a LanL2DZ/6-31+G(d,p) basis set. Lines show the fit to these data with the Morse potential, Eq. (7.2). Panels (b) and (c) show, respectively, the optimised structures of $W(CO)_6^+$ and $WC(CO)_5^+$. In the latter case, three different types of W–C bonds, labelled as (1), (2), and (3), appear after one oxygen atom has been removed.
Source: Reproduced from Ref. 21 with permission from Springer Nature.

of the widely used precursors for focused electron beam induced deposition (FEBID).[37–39] DFT calculations were performed to determine the rCHARMM parameters for neutral and singly charged $W(CO)_6$ molecules and their fragments; the results obtained were compared with experimental data.[40] The DFT calculations were performed using the B3LYP exchange-correlation functional and a mixed LanL2DZ/6-31+G(d,p) basis set, the former describing the W atom and the latter applied to C and O atoms. As the IDMD simulations performed in Ref. 21 were aimed at reproducing the appearance energies from electron-impact ionisation experiments,[40] singly charged parent molecule and molecular fragments were considered. The geometry of each molecule was first optimised, and then a potential energy surface scan was performed for different W–C and C–O bonds to calculate equilibrium bond lengths, dissociation energies, and force constants. Atomic partial charges were obtained using the natural bond orbital analysis.[41]

As a benchmark for the method, the equilibrium bond lengths in a neutral $W(CO)_6$ molecule, for which there is a wealth of reference data, were first analysed. The calculated values, $r_0^{\mathrm{W-C}} = 2.07$ Å and $r_0^{\mathrm{C-O}} = 1.15$ Å, are in good agreement with experimental data[42] and with the results of previous DFT calculations.[43] The chosen exchange-correlation functional and the basis set were then used for DFT calculations of a singly charged molecule. The calculated dissociation energies of the W–C bond in the parent cation $W(CO)_6^+$ and in different fragments such as $W(CO)_n^+$ ($n = 1 - 5$) and $WC(CO)_n^+$ ($n = 0 - 5$) are in good agreement with the electron-beam mass spectrometry results in Ref. 40 with a relative discrepancy of a few kilocalories per mole.

Figure 7.3(a) shows potential energy curves for W–C and C–O bonds in the parent $W(CO)_6^+$ molecule as well as in $WC(CO)_5^+$ formed upon removal of an oxygen atom from one of the ligands.[21] Optimised structures of these molecules are shown in Figs. 7.3(b) and 7.3(c), respectively. The symbols illustrate the results of the DFT calculations, while the lines show a fit to these data with the Morse potential, Eq. (7.2). The W–C bond in the parent molecule is significantly weaker (D = 40.7 kcal/mol) than the C–O bond (D = 212.8 kcal/mol), which is a common feature of metal carbonyls.[44] In the case of $WC(CO)_5^+$, three different types of W–C bonds can be distinguished, labelled (1), (2), and (3) in Fig. 7.3(c). When an oxygen atom is removed, the remaining carbon atom of the ligand becomes more strongly bound to the tungsten atom [bond (1)]; the dissociation energy of this W–C bond increases and varies from 119.9 to 143.3 kcal/mol depending on the $WC(CO)_n^+$ fragment considered. The opposite CO group becomes weakly bound to the metal atom [W–C bond (2)], and the equilibrium distance between W and C increases from 2.11 Å to 2.31 Å. Four W–C bonds in the perpendicular plane [bond (3)] remain unaffected and their dissociation energy does not change with respect to that in the parent molecule.

Table 7.1 summarises the rCHARMM parameters used in the calculations.[21] Since the dissociation energy of the W–C bond in

Table 7.1. Parameters of the rCHARMM force field used in the simulations of $W(CO)_6^+$ fragmentation.[21] In the case of $WC(CO)_n^+$, W–C bonds labelled as (1) and (2) refer to the notations in Fig. 7.3(c).

	$W(CO)_n^+$		$WC(CO)_n^+$		
Bond type	W–C	C–O	W–C (1)	W–C (2)	C–O
r_0 (Å)	2.11	1.14	1.83	2.31	1.14
D (kcal/mol)	40.7	212.8	143.3	22.6	212.8
k^b(kcal/mol Å^{-2})	119.9	1493.9	369.4	67.7	1493.9

different $W(CO)_n^+$ ($n = 1-5$) fragments does not vary significantly,[40] the value for the formation of $W(CO)_5^+$, $D = 40.7$ kcal/mol, was used in the simulations. Several DFT potential energy scans were also performed to evaluate the equilibrium angles. It was found that the equilibrium values for the C–W–C and W–C–O angles are 90 and 180 degrees, respectively; see Fig. 7.3(b).

In another illustrative example, rCHARMM was used to model the thermomechanical damage to a DNA molecule induced by a passing ion.[18] It is widely recognised that one of the key events of radiation-induced DNA damage is the formation of single- and double-strand breaks (SSBs and DSBs), as well as more complex damages.[29,36,45] Therefore, the rCHARMM force field was used to describe covalent interactions in the DNA backbone, while covalent interactions in other parts of the DNA molecule were modelled using the standard CHARMM force field.

The dissociation energy and cutoff distance parameters of rCHARMM for the specific bonds were determined from DFT calculations of a guanosine monophosphate (GMP, $C_{10}H_{14}N_5O_8P$) molecule, whose structure is shown in Fig. 7.4. The calculations were performed using the Gaussian 09 software[41] with the B3LYP exchange-correlation functional and the 6-31G(d,p) basis set for wave

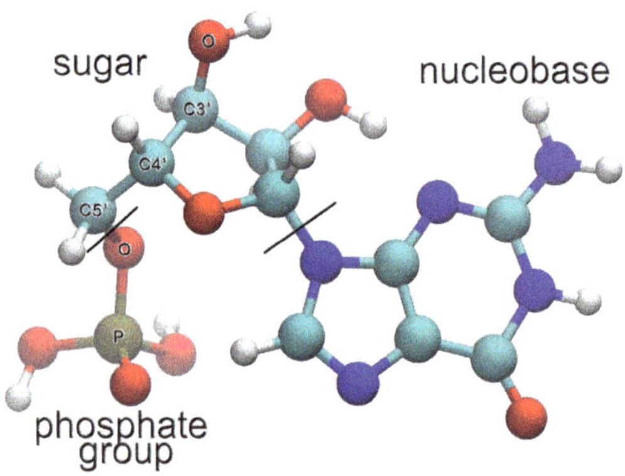

Fig. 7.4. Optimised geometry of guanosine monophosphate (GMP) molecule, which is a structural unit of the DNA.[18] Bonds between the labelled atoms were considered as reactive in the simulations conducted in Ref. 18. The black dashes split the molecule into three parts: the phosphate group, the sugar part, and the nucleobase.
Source: Reproduced from Ref. 18 with permission from Wiley.

Table 7.2. Dissociation energy D, equilibrium distance r_0, and the cutoff distance r_{cutoff} for breakage/formation of covalent bonds in the sugar-phosphate backbone of GMP, calculated at the B3LYP/6-31G(d,p) level of theory.[18] Cutoff distance is defined for each bond as the distances at which the bond energy is equal to $0.1D$.

Bond type	D (kcal/mol)	r_0 (Å)	r_{cutoff} (Å)
C3′–O	160.22	1.415	3.60
C5′–O	160.22	1.415	3.60
C4′–C5′	146.71	1.518	3.90
P–O	146.07	1.609	3.75

function expansion. It was found that the covalent bonds of the DNA backbone, together with the glycosidic bond between the sugar ring and the guanine nucleobase, have the lowest dissociation energies and are therefore most likely to break. The results for the C3′–O, C5′–O, C4′–C5′m and P–O bonds of the DNA backbone (see the labelled atoms in Fig. 7.4) are given in Table 7.2. A rupture of the C4′–C3′ bond does not cause a break in the DNA strand, as the bond is a part of a sugar ring, so it has not been parameterised by the rCHARMM force field in this case. The spring constants for the considered bonds have been taken from the standard CHARMM force field for nucleic acids.[46]

7.5 Radiation-Induced Chemical Transformations of MBN Systems

7.5.1 *Fragmentation of organometallic FEBID precursors*

Molecular fragmentation as a result of energy deposition involves several stages that take place on different timescales. First, the incident radiation (e.g., a photon, electron, or ion) interacts with the molecule and transfers energy to it by means of different mechanisms, e.g., electronic excitation, ionisation, or (in the case of electron irradiation) dissociative electron attachment (DEA). These are fast processes that happen on the sub-femtosecond scale and leave the molecule in an excited electronic state. In the case of ionisation, some fraction of deposited energy is spent in overcoming the ionisation threshold, another fraction is carried away by the ejected electron, while the remaining part is stored in the target in the form of electronic excitations. The latter can involve different molecular orbitals, being of either bonding or antibonding nature.

An excitation involving an antibonding molecular orbital evolves quickly through the cleavage of a particular bond on the femtosecond timescale. The fragmented parent molecule may still retain some amount of the deposited energy, which can lead to the sequential fragmentation of other bonds on the picosecond or even longer timescales.

The excited electronic state may also involve a bonding molecular orbital or may not be localised on a particular bond. In this case, the excess energy can be redistributed over a larger part or even the entire volume of the system and be transferred later into its vibrational degrees of freedom. Relaxation of the deposited energy due to the electron–phonon coupling mechanism[32] leads to an increase in the amplitude of thermal vibrations, which, in turn, leads to evaporation of loosely bound ligands. As was shown in the case of small metal clusters,[32] electron–phonon coupling is a slow process (as compared to a characteristic time of electron–molecule interaction) that happens on a picosecond timescale. The subsequent evaporation process may last up to microseconds.

In Ref. 21, radiation induced fragmentation of a $W(CO)_6$ molecule was studied by means of IDMD simulations with the rCHARMM force field, employing MBN Explorer. Focus was on the events happening from the cleavage of a particular bond to the redistribution of excess energy over the internal degrees of freedom of the molecule, i.e., on the timescales spanning from pico- to microseconds. Although quantum-mechanical calculations can simulate the process of energy deposition, the excitation of specific electronic orbitals, and the initial stage of the fragmentation process, the span of the entire fragmentation process extends far beyond the limits of quantum MD. Thus, classical MD remains the only computational technique that allows us to explore the process within the required time frame at the atomistic level of detail.

The following approach within the framework of classical MD was proposed[21] to describe the aforementioned fragmentation stages, i.e., fast cleavage of an individual bond (referred hereafter as stage I) and

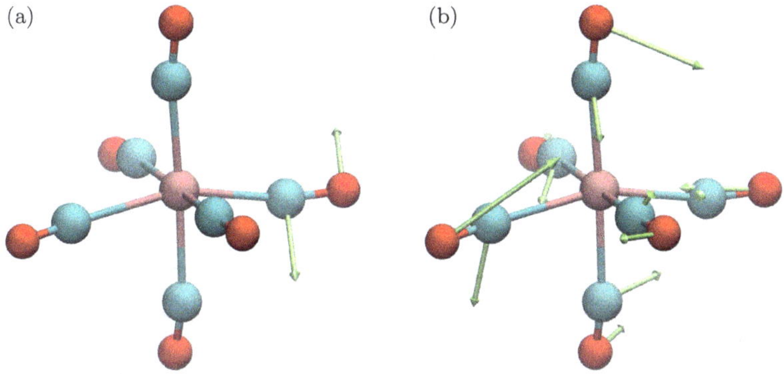

Fig. 7.5. Schematic representation of the two stages of fragmentation of $W(CO)_6$. Panel (a) describes stage I, when the deposited energy is converted into kinetic energy of atoms in a particular bond. The resulting velocities of the atoms are indicated by arrows. Panel (b) describes stage II, when the deposited energy is redistributed over the entire molecule and shared between the kinetic energies of all the atoms.
Source: Reproduced from Ref. 21 with permission from Springer Nature.

slow energy redistribution over all the molecular degrees of freedom (stage II from now on). Both processes result in an increase in the internal energy of the molecule after the energy deposition, which is treated as an initial increase in the kinetic energy of atoms.

In the approach used to model stage I, the energy was locally deposited in a specific covalent bond of the target and converted into the kinetic energy of the two atoms forming the bond; see Fig. 7.5(a). The velocities of these atoms have been defined to obey the total energy and momentum conservation laws:

$$\vec{v}_1 = \frac{\sqrt{2\mu E_{\rm dep}}}{m_1}\vec{u}, \quad \vec{v}_2 = -\frac{\sqrt{2\mu E_{\rm dep}}}{m_2}\vec{u}\,. \tag{7.9}$$

Here, $E_{\rm dep}$ is the amount of deposited energy remaining in the system after ionisation (i.e., excess energy over the ionisation threshold); m_1, m_2, and $\mu = m_1 m_2/(m_1 + m_2)$ are, respectively, the masses and the reduced mass of the atoms forming the bond; and $\vec{u}$ is a unit vector defining the direction of the relative velocity of these atoms upon bond cleavage.

Stage II is governed by the thermal mechanism of fragmentation, where the energy is distributed to all degrees of freedom of the target. In this case, the equilibrium velocities of the atoms corresponding to a given temperature, v_i^{eq}, are scaled by a factor of α depending on the amount of energy deposited; see Fig. 7.5(b). The kinetic energy of N atoms is then given by

$$\sum_i^N \frac{1}{2} m_i (\alpha\, v_i^{\text{eq}})^2 = \frac{3Nk_{\text{B}}T}{2} + E_{\text{dep}} \,. \tag{7.10}$$

The first term on the right-hand side of Eq. (7.10) corresponds to the initial kinetic energy of the atoms at equilibrium (e.g., $T = 300$ K in the simulations), with k_{B} being the Boltzmann constant. The second term on the right-hand side is the excess energy deposited in the molecule during the collision.

Each of the mechanisms described above leads to the formation of a particular group of experimentally observed fragments, while the overall experimental picture is well reproduced when both stages of the fragmentation process are considered. The notations 'stage I' and 'stage II' reflect the different timescales for the processes that precede the bond cleavage, i.e., bond cleavage on the femtosecond timescale after the electronic excitation to an antibonding molecular orbital and fragmentation induced by the relaxation of the deposited energy due to electron–phonon coupling, which occurs on the picosecond timescale. These stages can occur sequentially in the same excited molecule, so that the initial cleavage of a particular bond (stage I) is followed by the redistribution of the remaining energy throughout the molecular fragment (stage II). However, the energy initially deposited in the molecule can be directly redistributed to all degrees of freedom on the picosecond timescale, leading to the fragmentation pattern described by stage II.

The MD simulations of $W(CO)_6^+$ fragmentation were performed using MBN Explorer.[5] First, the structure of the molecule was optimised using the parameters obtained from the DFT calculations; see Section 7.4.1. The molecule was then equilibrated at $T = 300$ K for 100 ns. The equilibration simulation was performed using the Langevin thermostat with a damping time of 2 ps. Atomic coordinates and velocities were recorded every 100 ps. The equilibrated trajectory was sampled to generate random initial geometries and velocity distributions for the simulation of fragmentation. About 1,000 constant-energy simulations of 1 μs duration each were performed for different values of E_{dep} ranging from 0 to 475 kcal/mol.

The upper limit is several times larger than the energy required to break a W–C bond (see Fig. 7.3(a)), allowing the simulation of evaporation of multiple CO ligands. Fragments produced after 1 μs of simulation were analysed, and the corresponding appearance energies were evaluated based on this analysis and compared with experimental data.

In the following, we discuss the fragmentation pathways of a $W(CO)_6$ molecule upon electron impact ionisation using the energy deposition model described above. Simulated appearance energies were compared with the most recent set of experimental data[40] obtained through mass spectrometry for positively charged fragments produced by electron beams of energy $\leq$140 eV, as well as with previously reported experimental values.[47–51]

Figure 7.6 shows the number of CO fragments, n, produced in the dissociation of the parent $W(CO)_6^+$ molecule for a given amount

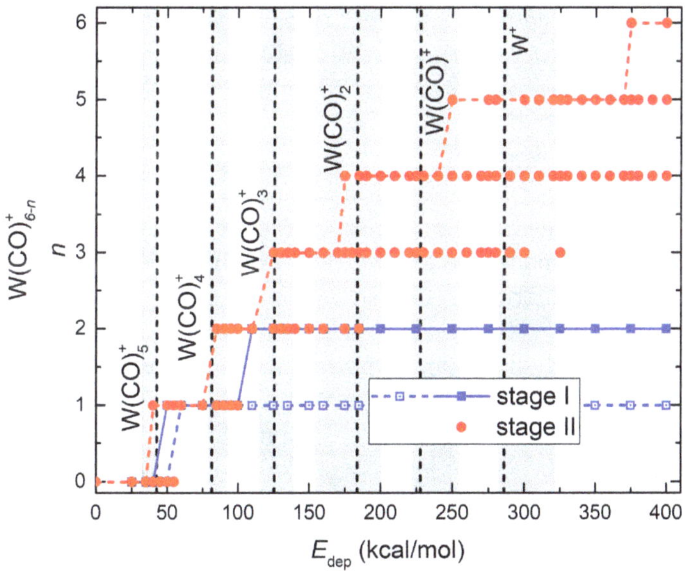

Fig. 7.6. Appearance energies for $W(CO)_{6-n}^+$ molecular fragments produced upon the release of n CO ligands from the parent $W(CO)_6^+$ molecule. The results of the IDMD simulations[21] are shown by symbols. Experimental appearance energies from Ref. 40 are shown by black dashed lines and those from previous studies[47–51] are shown by shaded areas. The dashed red line shows the largest number of CO released at each $E_{\rm dep}$.

Source: Reproduced from Ref. 21 with permission from Springer Nature.

of excess energy E_{dep}. Open blue squares show the results of the simulations of stage I (see Fig. 7.5(a)). In this case, a given amount of energy was deposited in a W–C bond, resulting in a prompt release of one CO group. However, no further fragmentation was observed even at high values of E_{dep}. Due to the large difference in mass between a carbon and a tungsten atom, more than 90% of the deposited energy is transferred to the kinetic energy of the C atom and carried away by the released CO group. Therefore, regardless of the value of E_{dep} given to the molecule by the cleavage of a W–C bond, only a small amount of energy is transferred to the remaining $W(CO)_5^+$ fragment, which is insufficient to observe further fragmentation events.

A single atom or a small fragment produced after the cleavage of a particular bond may hit the remaining large fragment as its escapes, redepositing some amount of energy into the large fragment and thus triggering further fragmentation at stage II (see Fig. 7.5(b)). The filled blue squares describe the situation when the CO group released by the cleavage of a W–C bond collided with the remaining $W(CO)_5^+$ molecule, resulting in the loss of another CO group. In this case, however, no more than two CO ligands escaped from the parent molecule.

The results of simulations describing stage II are shown in Fig. 7.6 by filled red circles. The distribution of the deposited energy over all degrees of freedom of the target leads to the evaporation of several CO fragments. These results are in good agreement with the appearance energies reported in a recent experimental study by Wnorowski *et al.*[40] (black dashed lines) as well as with appearance energies obtained in earlier experiments[47–51] (shaded areas). Note that the first ionisation potential was subtracted from the experimental appearance energies to convert them into the excess deposited energy. Due to the statistical nature of the fragmentation process, the emission of a given number of CO groups was observed at different values of E_{dep}. The lowest E_{dep} values, at which five or fewer CO molecules were recorded after 1 μs, agree well with the experimental results. However, complete fragmentation (i.e, loss of six CO molecules) has been observed only in a few trajectories at $E_{dep} = 375$ and 400 kcal/mol, which are significantly higher than the experimental appearance energy for W^+. This indicates that complete fragmentation takes place on a larger timescale and that longer simulations are needed to observe the complete fragmentation at lower energies.

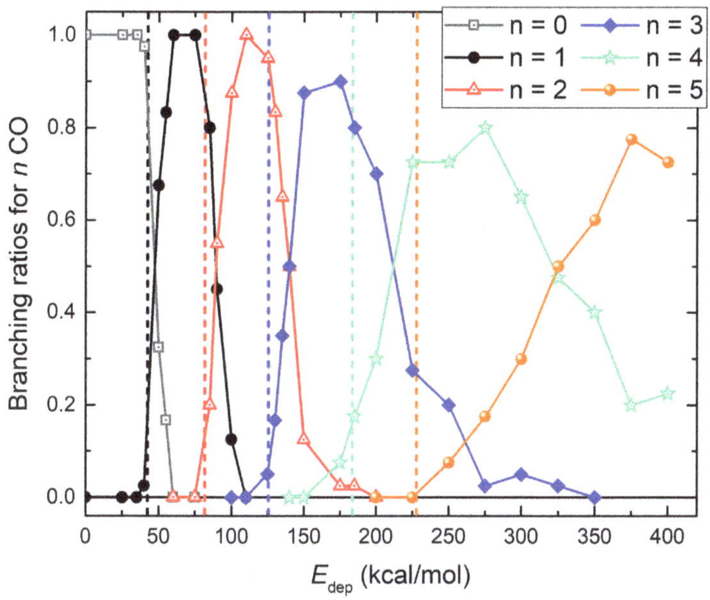

Fig. 7.7. Branching ratios for the formation of n CO ligands upon fragmentation of a $W(CO)_6^+$ molecule during stage II (see Fig. 7.5(b)) as a function of the excess energy E_{dep}. At a given E_{dep}, the sum of branching ratios for all the fragments is equal to 1. Results of the IDMD simulations[21] are shown by symbols. Experimental appearance energies from Ref. 40 are shown by dashed lines. *Source*: Reproduced from Ref. 21 with permission from Springer Nature.

The results of the simulations describing stage II have been used to evaluate the branching ratios for the production of different fragments for a given amount of excess energy E_{dep}; see Fig. 7.7. This analysis provides information on how many carbonyl groups are most likely to be released at a given E_{dep} after 1 μs. Figure 7.7 shows that the emission of one, two, and three CO groups ($n = 1, 2, 3$) takes place in rather well separated energy 'windows'. Fragments corresponding to $n = 1$ and 2 were recorded in these energy ranges, with the maximum probability corresponding to a branching ratio of 1. For the larger numbers of emitted CO, the maximum branching ratios drop down to 0.8, indicating an increased probability of observing different fragments. Note also that the characteristic energy ranges for the emission of n CO groups increase with n. This analysis allows an evaluation of a typical amount of energy that should be deposited in the $W(CO)_6^+$ molecule to observe a particular fragment.

In a more recent study,[22] IDMD simulations were performed to analyse the role of the molecular environment on the irradiation-induced fragmentation of molecular systems. The DI of iron pentacarbonyl, $Fe(CO)_5$, a widely used precursor molecule for FEBID, was considered as a case study. In connection to the earlier experiments,[52] the irradiation-induced fragmentation dynamics of an isolated $Fe(CO)_5^+$ molecule was studied and compared with that of $Fe(CO)_5^+$ embedded in an argon cluster. The metal–ligand separation and the CO ligand dissociation processes were simulated using the rCHARMM force field[8] implemented in MBN Explorer[5] and quantified by analysing the appearance energies for different molecular fragments.

The main outcome of the performed simulations are the fragmentation patterns, i.e., the abundance of different $Fe(CO)_{5-n}^+$ $(n = 0\text{–}5)$ ionic products as a function of the energy deposited to the parent $Fe(CO)_5^+$ ion. Figure 7.8 shows the abundances of the fragment ions for the cases when the excess energy was locally deposited in one of the Fe–C bonds or one of the C–O bonds (Figs. 7.8(a) and 7.8(b), respectively) and when the energy has been redistributed over all the degrees of freedom of the cation (Fig. 7.8(c)). The figure shows that each fragment ion has a specific appearance energy; at larger values of the excess energy E, the abundance distribution reaches a maximum and then decreases as E is further increased. The decrease in the abundance distribution is caused by the fact that if the ion is too 'hot', a larger number of CO ligands are evaporated. The vertical colour bars in Fig. 7.8(c) illustrate the appearance energies of $Fe(CO)_{5-n}^+$ $(n = 1 - 5)$ fragments determined in mass-spectroscopic experiments,[53] which have been converted into the excess energy deposited in $Fe(CO)_5^+$. Figure 7.8(c) shows a good agreement between the fragment appearance energies evaluated from the present simulations and the corresponding experimental values[53] for the gas-phase DI of $Fe(CO)_5$.

In the case of iron pentacarbonyl embedded in an argon cluster (see Fig. 7.9), the RMD simulations[22] reproduced the experimentally observed suppression of $Fe(CO)_5^+$ fragmentation[52] and provided an atomistic understanding of this effect. It was shown[22] that the excess energy deposited to the $Fe(CO)_5^+$ cation as a result of the electron

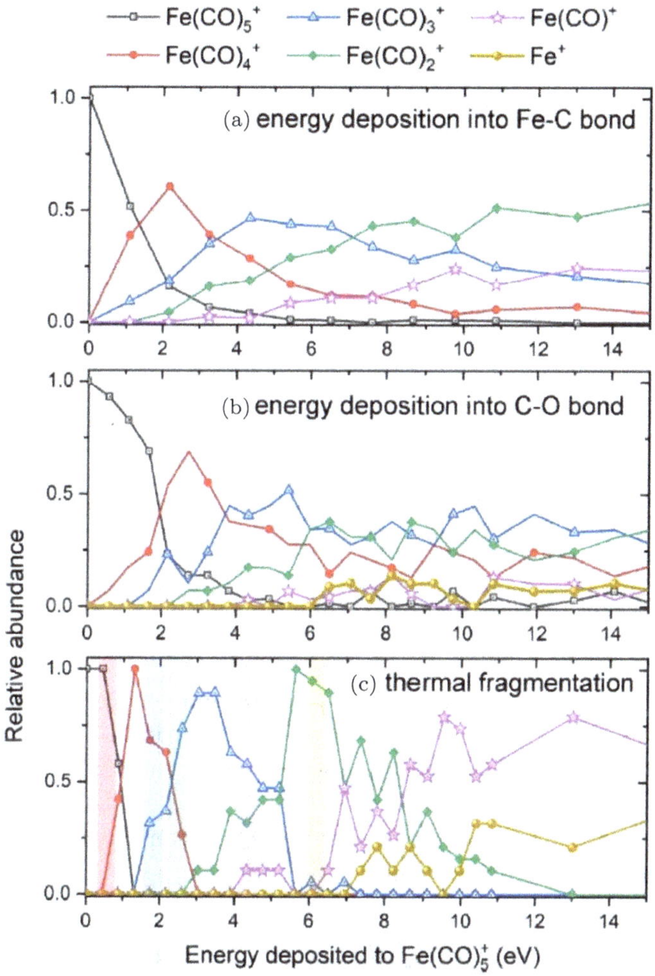

Fig. 7.8. Relative abundance of $Fe(CO)_{5-n}^{+}$ $(n = 0-5)$ fragments produced after the electron-impact-induced fragmentation of isolated $Fe(CO)_5^+$ as a function of energy deposited to the cation.[22] Panels (a) and (b) describe the case of the localised energy deposition into a Fe-C bond and a C-O bond, respectively. Panel (c) describes the thermal mechanism of fragmentation where the energy is distributed over all degrees of freedom of the cation. Vertical colour bars in panel (c) indicate the experimental appearance energies of $Fe(CO)_{5-n}^{+}$ $(n = 1-5)$ from Ref. 53; the width of these bars corresponds to the experimental uncertainty. *Source*: Reproduced from Ref. 22 published under an open access Creative Commons CC BY license.

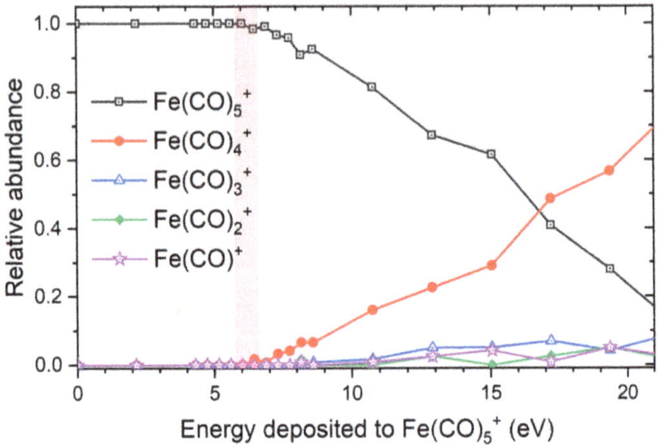

Fig. 7.9. Relative abundance of $Fe(CO)^{+}_{5-n}$ $(n = 0 - 4)$ ionic species produced due to the electron-impact-induced fragmentation of $Fe(CO)^{+}_{5}$ embedded within an argon cluster. Vertical colour bar indicates the experimental appearance energy of $Fe(CO)^{+}_{4}$.[52]
Source: Reproduced from Ref. 22 published under an open access Creative Commons CC BY license.

collision is efficiently quenched by the argon environment, leading to the heating of the cluster and the evaporation of weakly bound argon atoms. A similar process of quenching can take place during the FEBID process (see Section 7.6), when the vibrational energy of precursor molecules physisorbed on a substrate can be efficiently quenched by the environment.

7.5.2 *Fragmentation of the DNA molecule by the ion-induced nanoscale shock wave*

The RMD and IDMD methods can be used to evaluate radiobiological damage caused by heavy ions propagating in various media, including biological ones. An energetic charged particle propagating through the medium loses its energy in inelastic collisions with the medium constituents. The transport of the produced secondary particles and the radiation damage induced by them are the subject of experimental, numerical, and computational studies.[36,54] In particular, the physics and chemistry of radiation damage caused by irradiation with protons and heavier ions has recently become a subject of intense interest due to the use of ion beams in cancer radiotherapy.[29,36,55,56]

Several review papers[29,57] and a book[36] have provided an overview of the main ideas of the multiscale approach (MSA) to the physics of radiation damage with ions. This approach has enabled us to develop knowledge of biodamage at the nanoscale and molecular level and to establish the relationship between the characteristics of the incident particles and the resulting biological damage.

Radiation damage due to ionising radiation is initiated by ions incident on tissue. The initial kinetic energy of the ions ranges from a few to hundreds of MeV per nucleon. As the ions propagate through the tissue, they lose energy due to ionisation, excitation, nuclear fragmentation, etc. Most of the energy loss of the ion is transferred to tissue. For an energetic ion propagating in a medium, the dependence of the energy deposited into the medium on the penetration distance is characterised by a sharp maximum, the Bragg peak, in the region close to the end of ion's trajectory. It is commonly understood that the secondary electrons and free radicals produced in the processes of ionisation and excitation of the medium with ions are largely responsible for the vast portion of biodamage. In the Bragg peak region, the secondary electrons lose most of their energy within 1–2 nm of the ion's path.[58] After that, the electrons continue propagating, elastically scattering with the molecules of the medium until they become bound or solvated electrons are formed.[29,36] Such low-energy electrons remain important agents for biodamage since they can attach to biomolecules such as DNA causing dissociation.[59]

The energy lost by secondary electrons in the processes of ionisation and excitation of the medium is transferred to its heating (i.e., vibrational excitation of molecules) due to the electron–phonon interaction. As a result, the medium within a 1–2 nm region (for ions not heavier than iron) surrounding the ion's path is heated up rapidly.[58,60] The pressure inside this narrow region increases by several orders of magnitude (e.g., by a factor of 10^3 for a carbon ion at the Bragg peak[60]) compared to the pressure in the medium outside that region. This pressure builds up by about 10^{-14}–10^{-13} s, and it is a source of a cylindrical shock wave,[61] which propagates through the medium for about 10^{-13}–10^{-11}) s. Its relevance to biodamage is as follows. If the shock wave is strong enough,[a] it may inflict damage directly by breaking covalent bonds in a DNA molecule.[18,19,58,62–65] Besides, the radial collective motion of the medium induced by the shock wave is instrumental in propagating the highly reactive molecular species, such as hydroxyl radicals and solvated electrons, to large radial distances (up to tens of nanometres), thus increasing the area of an ion's impact.[23,30]

The direct thermomechanical damage of the DNA molecule as a result of interaction with the ion-induced shock wave has been explored using MBN Explorer.[18,19,58,62,63] In earlier investigations,[58,62,63] the DNA damage by ion-induced shock waves was studied by means of classical MD simulations using the standard (non-reactive) CHARMM force field. In those simulations, the potential energy stored in a particular DNA bond was monitored in time as the bond length varied around its equilibrium distance.[58,63] When the potential energy of the bond exceeded a given threshold value, the bond was considered broken.

A more quantitative description of the ion-induced shock wave phenomenon has been made possible[19] by RMD simulations, which allow explicit simulation of covalent bond rupture and formation.[8] A detailed computational protocol for modelling shock-wave-induced DNA damage using the rCHARMM force field[8] was presented in Ref. 18. The target DNA molecule studied in Refs. 18,19 contained 30 complementary DNA base pairs. The molecule was placed in a water box extending 17 nm from the DNA in the x- and y-directions and 8 nm in the z-direction. The total system size was 1,010,994 atoms.

It is widely established that one of the key events of radiation-induced DNA damage concerns the formation of single- and double-strand breaks (SSBs and DSBs) of the sugar-phosphate backbone. Therefore, the rCHARMM force field was used to describe interatomic interactions in the C'_3–O, C'_4–C'_5, C'_5–O, and P–O bonds in the DNA backbone, which connect the sugar ring of one nucleotide and the phosphate group of an adjacent nucleotide; see Fig. 7.4 and

[a]The strength of the shock wave depends on the distance from the ion's path and the value of linear energy transfer (LET), i.e., the amount of energy that a projectile ion transfers to the medium traversed per unit distance.

Table 7.2. Bond dissociation energies and cutoff distances for bond breakage/formation were determined from density functional theory (DFT) calculations.[18] Covalent interactions in other parts of the DNA molecule were modelled using the standard CHARMM force field.

In the MD simulations, the energy lost by the propagating ion is deposited into the kinetic energy of water molecules located inside a 'hot' cylinder of 1 nm radius around the ion's path. The equilibrium velocities of all atoms inside this 'hot' cylinder are increased by a factor α such that the kinetic energy of these atoms reads as[58,62,66]

$$\sum_{i}^{N} \frac{1}{2} m_i \left(\alpha v_i\right)^2 = \frac{3 N k_B T}{2} + S_e\, l. \tag{7.11}$$

Here, S_e is the LET of the projectile ion, l is the length of the simulation box in the z-direction (parallel to the ion's path), and N is the total number of atoms within the 'hot' cylinder. The first term on the right-hand side of Eq. (7.11) is the kinetic energy of the 1 nm radius cylinder at the equilibrium temperature, $T = 300$ K, whereas the second term describes the energy loss by the ion as it propagates through the medium.

The shock wave propagates in the molecular system radially away from the ion track; see Fig. 7.10(a). The range of the shock wave propagation in the aqueous environment can be determined by monitoring the radial density of the water molecules in time. The ion track was, in this case, placed directly through the geometrical centre of the DNA strand. The results of this analysis are shown in Fig. 7.10(b). The wave front moves towards the edge of the simulation box as time passes, and the wave profile becomes lower and broader, showing that the shock wave relaxes as time passes. This indicates that the impact of the shock wave weakens over time. Figure 7.10(b) shows that the maximal density of the wave does not change significantly in the range of 2–6 nm from the ion track; therefore, a DNA strand placed in this range is expected to experience the strongest impact from the shock wave.

The impact of a shock wave on DNA can be characterised through the probability of strand break formation, which can be used to quantify the amount of biodamage induced by the shock wave mechanism.[29] Figure 7.11 shows the number of strand breaks in the DNA

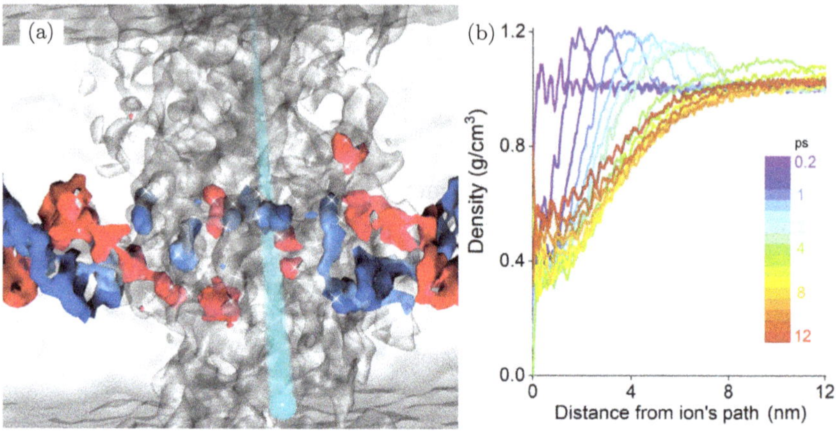

Fig. 7.10. Panel (a): illustration of the propagation of the ion-induced shock wave in the vicinity of the DNA segment containing 30 base pairs.[18] Water molecules inside the 'hot' cylinder surrounding the ion track are highlighted. Panel (b): the density of water in the radial direction from the ion's path is shown at different times from 0.2 ps (the purple curve) to 12 ps (the red curve) after irradiation. The wave front moves towards the edge of the simulation box over time, and the wave profile becomes lower and broader, showing that the shock wave relaxes over time.
Source: Reproduced from Ref. 67 published under an open access Creative Commons CC BY license.

segment caused by a shock wave induced by an argon ion with an LET of 2890 eV/nm as a function of simulation time. The number of breaks rises quickly within the first 7 ps of the simulation, where the shock wave front hits both DNA strands of the target molecule. After this moment, the DNA damage rate becomes lower, lasting until approximately 15 ps; thereafter, the number of breaks remains steady. Even if the number of breaks generally rises with time, some fluctuations in this quantity can be seen locally. This effect might be attributed to the broken bonds, which can be rejoined if the atoms involved get close to each other after the initial bond breakage. This analysis shows that multiple or even complex strand breaks might be induced by the generated shock waves.

The methodology reviewed here can be applied to further computational studies considering irradiation of DNA with different ions and different orientations between the ion's path and the DNA molecule. Such analyses are important for understanding the

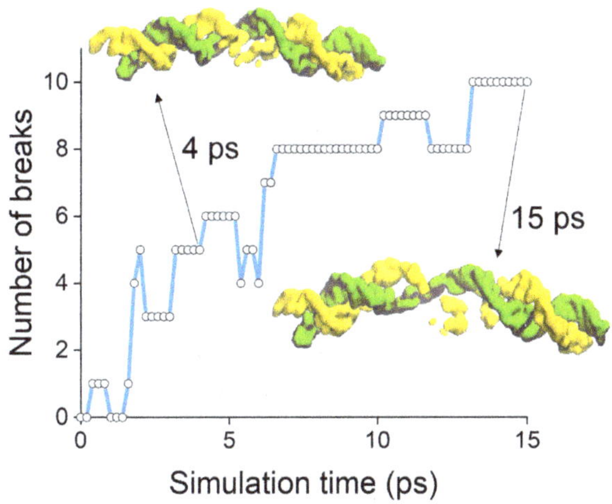

Fig. 7.11. The number of strand breaks in the DNA segment with 30 base pairs induced by the thermomechanical stress due to the argon ion with an LET of 2890 eV/nm as a function of simulation time.[18] Two insets at the top show the breakage of the DNA strands at simulation time instances of 4 and 15 ps. *Source*: Reproduced from Ref. 67 published under an open access Creative Commons CC BY license.

radiation damage with ions on a quantitative level, focusing on particular physical, chemical, and biological effects that bring about lethal damage to cells exposed to ion beams.[29,68,69]

7.6 Atomistic Simulation of the FEBID Process

The controllable fabrication of nanostructures with nanoscale resolution remains a considerable scientific and technological challenge.[70] To address such a challenge, novel techniques have been developed[71] which exploit irradiation of nanosystems with collimated electron and ion beams. One of such techniques, focused electron beam induced deposition (FEBID),[38,71,72] is based on the irradiation of precursor molecules[39] by high-energy electrons while they are being deposited on a substrate. Electron-induced decomposition of the molecules releases its metallic component, which forms a deposit on the surface with size similar to that of the incident electron beam (typically, a few nanometers).[37] As a result, FEBID enables reliable direct-write fabrication of complex, free-standing 3D structures.[37,73]

To date, FEBID has mainly relied on precursor molecules developed for chemical vapor deposition (CVD) – a process mainly governed by thermal decomposition, while dissociation mechanisms in FEBID are predominantly electron-induced reactions. While primary electron (PE) energies during FEBID are typically between 1 and 30 keV, chemical dissociation is most efficient for low-energy (up to several tens of eV) secondary electrons (SEs) created in large numbers when a high-energy PE beam impinges on a substrate. SEs are emitted from the substrate and the deposit, making the electron-induced chemistry that governs FEBID substantially complicated. As a result, as the intended nanostructure resolution falls below 10 nm, FEBID struggles to fabricate structures with the desired size, shape, and chemical composition,[37] which mainly originates from the lack of molecular-level understanding of irradiation-driven chemistry (IDC) underlying nanostructure formation and growth.

FEBID operates through successive cycles of precursor molecules replenishment on a substrate and irradiation by a tightly focused electron beam, which induces the release of metal-free ligands and the growth of metal-enriched nanodeposits. This process involves a complex interplay of phenomena, taking place on different temporal and spatial scales: (i) deposition, diffusion, and desorption of precursor molecules on the substrate; (ii) transport of the primary, secondary, and backscattered electrons; (iii) electron-induced dissociation of the deposited molecules; (iv) follow-up chemistry; (v) relaxation of energy deposited into electronic and vibrational degrees of freedom, and resulting thermomechanical effects. Each of these phenomena requires dedicated computational and theoretical approaches.

Further advances in FEBID-based nanofabrication require a deeper understanding of the relationship between deposition parameters and physical characteristics of fabricated nanostructures (size, shape, purity, crystallinity, etc.). Advanced experiments combined with molecular-level computational modelling can provide the required insights into the fundamental mechanisms of electron-induced precursor fragmentation and the corresponding mechanism of nanostructure formation and growth using FEBID.

Until recently, most computer simulations of FEBID and the nanostructure growth have been performed using an MC approach and diffusion-reaction theory,[71,74,75] which allow for simulations of the average characteristics of the process concerning local growth rates and the nanostructure composition. However, these approaches do not provide any molecular-level details regarding structure (crystalline, amorphous, and mixed) and the IDC involved. At the atomic level, quantum-chemical methods have been used to analyse the adsorption energies and optimised structures of different precursor molecules deposited on surfaces.[76,77] Nevertheless, *ab initio* methods are applicable to relatively small molecular systems with a typical size of up to a few hundred atoms. This makes *ab initio* approaches of limited use to describe the irradiation-induced chemical transformations occurring during the FEBID process.

A breakthrough into the atomistic description of FEBID was achieved recently by means of the IDMD approach.[4] It overcomes the limitations of previously used computational methods and describes FEBID-based nanostructures at the atomistic level by accounting for quantum and chemical transformations of surface-adsorbed molecular systems under focused electron beam irradiation.[4,24–27] Major transformations of irradiated molecular systems are simulated by means of MD with the rCHARMM force field.[8] described in Section 7.4.

In a pioneering study,[4] IDMD was successfully applied for the simulation of FEBID of $W(CO)_6$ precursors on a SiO_2 surface and enabled the prediction of the morphology, molecular composition, and growth rate of tungsten-based nanostructures emerging on the surface during the FEBID process. A snapshot of the MD simulation of the first irradiation phase is shown in Fig. 7.12. The irradiation by an electron beam of the cylindrical shape has been considered. Only those precursors molecules that move inside the cylinder are exposed to radiation that may induce their dissociation. The dissociation rate of the precursor molecules was evaluated from the experimental data.[74]

A follow-up study[24] introduced a novel multiscale computational methodology that couples MC simulations for radiation transport with IDMD for simulating the IDC processes with atomistic resolution. The developed multiscale modelling approach has enabled the simulation of radiation transport and effects in complex systems where all the FEBID-related processes (deposition, irradiation, and replenishment) are accounted for. The spatial and energy distributions of SE and backscattered electrons (BSE) emitted from a SiO_2 substrate were used to simulate electron-induced formation and

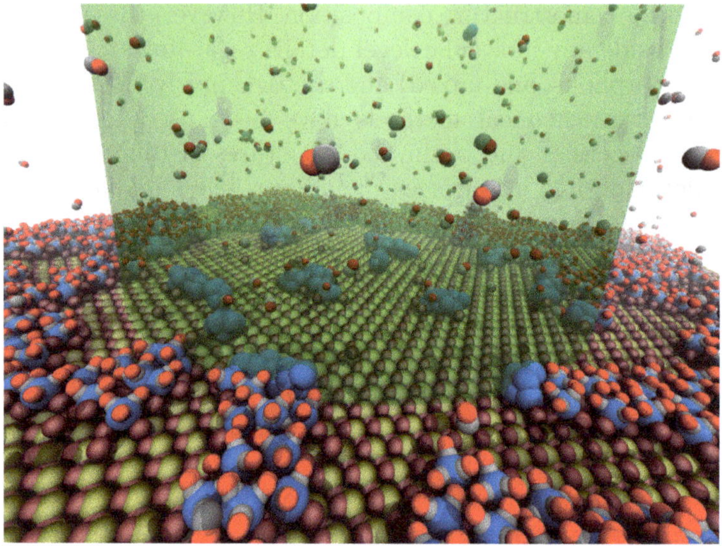

Fig. 7.12. Snapshot of the MD simulation[4] of adsorption of $W(CO)_6$ precursor molecules atop the SiO_2 surface at the early stage of irradiation by an electron beam (a transparent green cylinder). The interaction of deposited precursor molecules with the beam leads to the fragmentation of precursors and to the formation of tungsten clusters, shown in blue.
Source: Reproduced from Ref. 14 published under an open access Creative Commons CC BY license.

growth of metal nanostructures obtained after deposition of $W(CO)_6$ precursors on SiO_2.

Within the IDMD framework, the space-dependent rate for bond cleavage in molecules on the substrate surface is given by

$$P(x, y) = \sigma_{\text{frag}}(E_0)\, J_{\text{PE}}(x, y, E_0) + \sum_i \sigma_{\text{frag}}(E_i)\, J_{\text{SE/BSE}}(x, y, E_i), \tag{7.12}$$

where E_0 is the initial energy of the electron beam; $E_i < E_0$ denote a discrete set of values for the electron energies lower than E_0; $J_{\text{PE/SE/BSE}}(x, y, E_i)$ are space- and energy-dependent fluxes of PE, SE, and BSE per unit area and unit time; and $\sigma_{\text{frag}}(E_i)$ is the energy-dependent molecular fragmentation cross-section. The PE beam flux at the irradiated circular spot of radius R is

$$J_0 = \frac{I_0}{e\, S_0}\,, \tag{7.13}$$

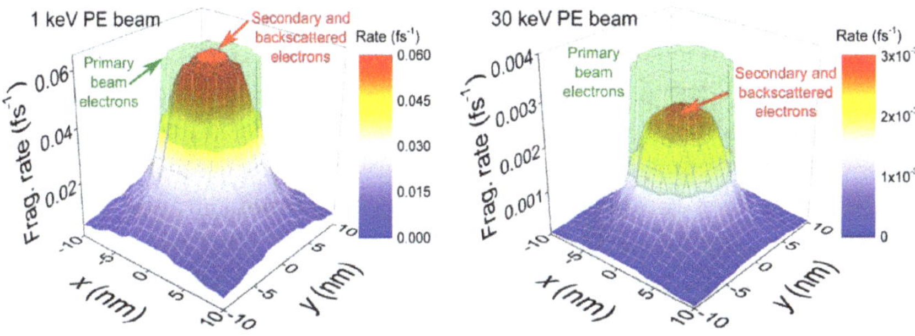

(a) (b)

Fig. 7.13. Electron-induced fragmentation rates for $W(CO)_6$ precursor molecules irradiated with PE beams of $E_0 = 1$ keV (panel (a)) and $E_0 = 30$ keV (panel (b)). The green transparent surface depicts the PE beam area.
Source: Reproduced from Ref. 24 published under an open access Creative Commons CC BY license.

where I_0 corresponds to the PE beam current, $S_0 = \pi R^2$ denotes its area, and e is the elementary charge. The electron distributions were simulated using the MC radiation transport code SEED.[31,35] Molecular fragmentation and further chemical reactions were simulated using MBN Explorer,[5] while MBN Studio[78] was employed to construct the molecular system, perform the precursor molecule replenishment phases, and analyse the IDMD simulation results.

Figure 7.13 illustrates the space-dependent fragmentation rates induced by uniform 1 keV (panel (a)) and 30 keV (panel (b)) beams of unit PE flux $J_0 = 1\ \text{nm}^{-2}\text{fs}^{-1}$ within a circular area of radius $R = 5$ nm. Although the number of BSE/SE electrons for 30 keV is small, their large cross-section (in relation to PE) produces a significant fragmentation probability but less than that due to PE at the beam area. However, for 1 keV, the fragmentation probability due to BSE/SE (∼80 – 90% exclusively due to SE) is very large and significantly extends beyond the PE beam area. These results clearly demonstrate the very different scenarios are to be expected for beams of different energies, which will importantly influence the deposit properties, as well as the prominent role of low-energy SE on molecular fragmentation.

Each irradiation phase lasts for a time known as dwell time τ_d, whose typical duration in experiment ($\geq \mu$s) is still computationally demanding for MD. To address this challenge, the irradiation phase was simulated in Ref. 4 for 10 ns, and simulated PE fluxes J_0 (and hence PE beam currents I_0) were then scaled to match the same number of PE per unit area and per dwell time as in experiments. As for replenishment, its characteristic times are also typically very long (~ms). In simulations, the CO molecules desorbed to the gas phase are simply removed during the replenishment stages, and new $W(CO)_6$ molecules are deposited.

As described above, the tabulated space-resolved fragmentation probability per primary electron is used in the IDMD simulation of the irradiation phase to link the bond dissociation rate to the electron flux. As the realistic experimental timescale for τ_d is challenging for MD, the simulated PE fluxes J_0 (and hence PE beam currents I_0) are rescaled to match the same number of PE per unit area and per dwell time as in experiments. The correspondence of simulated results to experimental ones is established through the correspondence of the electron fluence per dwell time per unit area in simulations and experiments.[4] Such an approach is valid in the case when different fragmentation events occur independently and do not induce a collective effect within the system. In this case, the irradiation conditions for the deposited precursor molecules are the same in simulations and in experiments. This correspondence condition gives

$$I_{\rm exp} = I_{\rm sim} \frac{S_{\rm exp}}{\lambda S_{\rm sim}} = I_{\rm sim} \frac{R_{\rm exp}^2}{\lambda R_{\rm sim}^2}, \tag{7.14}$$

$$\lambda = \frac{\tau_d^{\rm exp}}{\tau_d^{\rm sim}}, \tag{7.15}$$

where $S_{\rm exp}$ and $S_{\rm sim}$ are the electron beam cross-sections used in experiments and simulations, respectively, and $R_{\rm exp}$ and $R_{\rm sim}$ are the corresponding beam spot radii.

The rCHARMM force field used to model the structure and dynamics of irradiated $W(CO)_6$ molecules atop the hydroxylated SiO_2 surface requires the specification of several parameters (the equilibrium bond lengths, bond stiffness values, and dissociation energies), as described in Section 7.4.1. Additionally, we need to define the dissociative chemistry of precursors, including the definition of the molecular fragments and atomic valences. In the model considered in Ref. 4, only the dissociation and formation of the W–C and W–W bonds was permitted, while the C–O bonds were

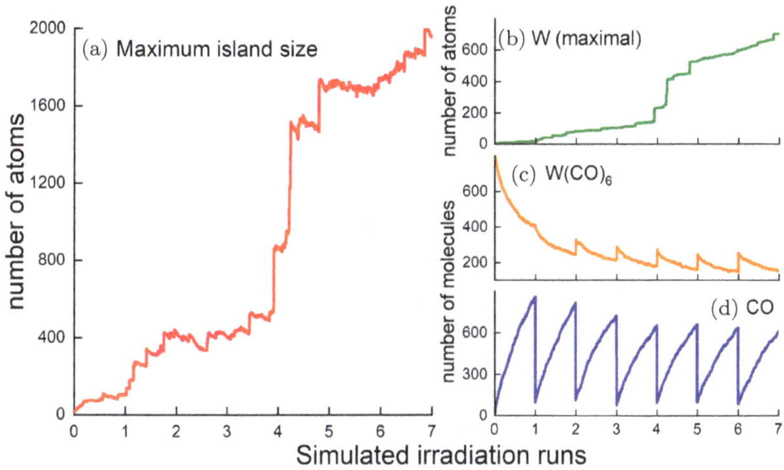

Fig. 7.14. Time evolution of the size of the largest W-enriched island (a), the number of W atoms (b), and the number of $W(CO)_6$ (c) and CO (d) molecules in the system during the irradiation periods.[4] The irradiation periods are marked by the successive numbers. The duration of each period is 10 ns. In simulations, the electron beam radius R is equal to 5 nm, and the beam current $I_0 = 4$ μA. *Source*: Reproduced from Ref. 14 published under an open access Creative Commons CC BY license.

treated within the harmonic approximation, Eq. (7.1), preventing those bonds from breakage.

The FEBID process was modelled[4] with the rescaled computationally accessible parameters (the irradiation time and the beam current). These parameters may differ from experimental values, but they have to be chosen so that they correspond to a given (as in the experiment) number of electrons N_e targeting the system (the electron fluence) and thus producing the irradiation induced effects on the same scale as in the experiment. Following this idea, the irradiation time in IDMD simulations is typically decreased as compared to the corresponding experimental values.

Figure 7.14(a) shows the time dependence of the size of the largest W-enriched island emerged in the simulations. Panels (b), (c), and (d) show this evolution for the numbers of W atoms, $W(CO)_6$, and CO molecules in the system during the irradiation periods, respectively. The irradiation periods are marked in the plots by successive numbers. The duration of each period is 10 ns. The replenishment periods are excluded from the plots. However, the drops in the number of CO molecules and the number of $W(CO)_6$, corresponding to changes in the system that occur during the replenishment periods, are clearly seen. Both the size and number of W atoms in the islands grow due to the attachment of new atoms to the islands in the course of the FEBID process. Figures 7.14(a) and 7.14(b) indicate the coalescence of smaller islands into a larger single nanostructure (ripening process) during the initial stage of the FEBID process. The irregular spikes on the curves arise at the instants when separate islands merge together.

The W–C bonds in precursors dissociate during the irradiation periods, leading to the appearance of the CO molecules. Most of these are created in the vicinity of the surface and are later evaporated into the vacuum chamber. The evaporation process continues during the replenishment periods. Figure 7.14(d) shows that, during each irradiation period, the number of CO molecules grows nonlinearly. To account for the evaporation process of CO during the replenishment periods, the CO molecules have been removed from the simulation box after each irradiation phase. After that, the new $W(CO)_6$ molecules are deposited on the surface according to the chosen deposition rate and the duration of the deposition process. This results in an abrupt decrease in the CO molecules and an increase in $W(CO)_6$ numbers before the start of each new irradiation cycle.

Figure 7.15 shows the results of the IDMD simulations conducted in Ref. 24. The number of atoms (one W, C, or O) in the largest island is shown for three simulation conditions close to the experimentally reported conditions:[79] 30 keV at $I_0 = 0.3$ nA, 10 keV at $I_0 = 2.3$ nA, and 1 keV at $I_0 = 3.7$ nA. Smaller clusters tend to merge with time, giving rise to larger structures and, eventually, to the largest island displayed in the figure. The jumps in the island size observed with some frequency are due to the merging of independent clusters that grow on the substrate.

Experimental measurements performed to date have been limited to particular values of energy and current due to the characteristics of the electron source.[79] In contrast, the IDMD simulation method permits the exploration of a much broader range of electron beam parameters. Full symbols in Fig. 7.16(a) depict the simulated metal contents of the deposits as a function of the experimentally equivalent current I_{exp}. Error bars show the standard deviations obtained from

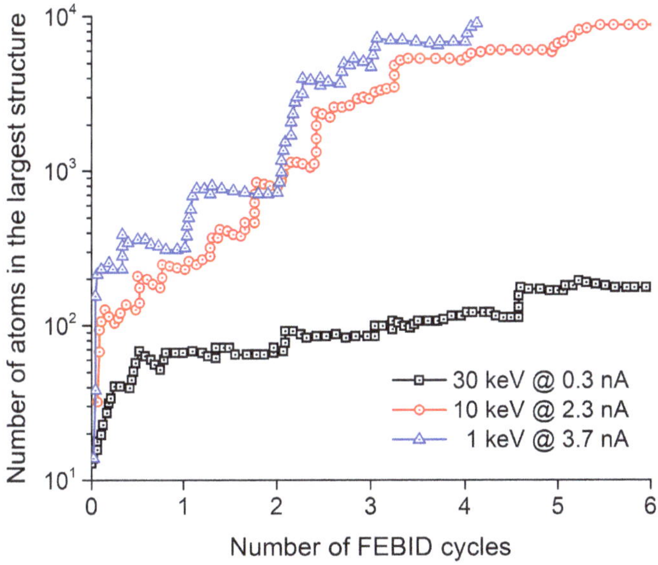

Fig. 7.15. Evolution of the number of atoms in the largest simulated islands for PE beams of energies 1, 10, and 30 keV for different currents, as indicated.[24] *Source*: Reproduced from Ref. 14 published under an open access Creative Commons CC BY license.

three independent simulations for each case. Experimental results[79] are shown by open symbols. Numbers next to symbols represent the primary beam energies in keV. It is clearly seen that the results from simulations are within the range of experimental uncertainties, which indicates the predictive capabilities of the simulations.

This analysis provides a detailed 'map' of the attainable metal content in the deposits as a function of the beam parameters, which is a valuable outcome for the optimisation of FEBID with $W(CO)_6$ on SiO_2. Dashed lines in Fig. 7.16(a) correspond to the limiting values of the PE beam energy and current studied in Ref. 24. These results clearly show that, within the analysed energy domain, a decrease in the beam energy and an increase in the current promote a faster growth of the deposit and also augments its metal content. Simulation results provide the grounds for clearly understanding such trends: an increment in the current means a larger number of PE per unit time, while a reduction in the energy produces an increase in the SE yield. These lead to both the greater size of the deposit and

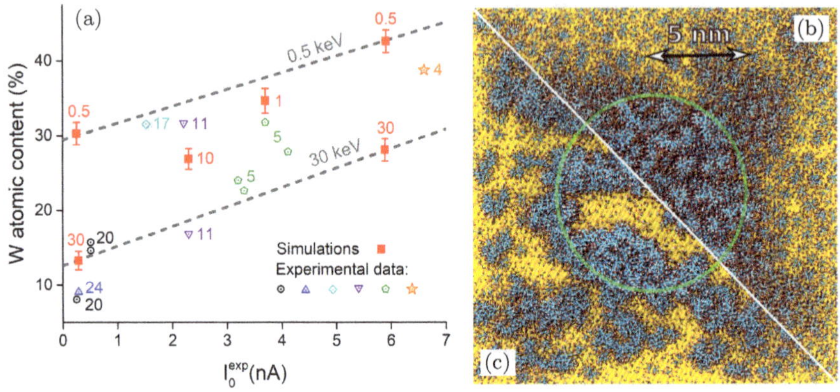

Fig. 7.16. Compositions and morphologies of the deposits created by FEBID.[24] Panel (a): dependence of the deposited metal content on the beam energy E_0 and current I_{exp} from experiments (open symbols of different colours)[79] and simulations (full red symbols).[24] Numbers next to symbols represent the beam energy in keV for each case. Panels (b) and (c) show the top views of the deposits produced by 10 keV@2.3nA and 1 keV@3.7nA beams, respectively. The green circular area in the central part of the figure marks the PE beam spot. Light-blue spheres represent W atoms, while darker white and red spheres represent C and O atoms, respectively; the SiO_2 substrate is represented by a yellow surface beneath the deposit.

Source: Reproduced from Ref. 24 published under an open access Creative Commons CC BY license.

its larger metal content due to the increased probability for bond cleavage (see Fig. 7.13).

Figures 7.16(b) and 7.16(c) show top views of the simulated deposits for 1 keV at 3.7 nA and 10 keV at 2.3 nA, after five and seven irradiation cycles, respectively (the number of atoms in the largest island is similar in these cases, ~12000). The green circular line marks the area covered by the PE beam (having a radius of 5 nm). These figures show that different energy–current regimes lead to distinct deposit microstructures and edge broadenings. While the more energetic 10 keV energy beam produces a deposit almost entirely localised within the PE beam area, the 1 keV beam produces a more sparse deposit (at least during the early stage of the FEBID process) that significantly extends beyond the PE beam area, producing an undesired edge broadening of the structure.

As described above, investigating the physicochemical phenomena that govern the formation and growth of nanostructures coupled to radiation is a complex multi-parameter problem. Indeed, a vast number of parameters (e.g., different precursor molecules, substrate types, irradiation and replenishment regimes, and additional molecular species that may facilitate precursor decomposition) can be varied with the aim of improving the purity of the grown deposits and increasing their deposition rate. It is therefore essential to develop a comprehensive computational protocol for atomistic simulations of the FEBID process.

In Ref. 25, a detailed computational methodology for modelling the formation and growth of metal-containing nanostructures during FEBID by means of IDMD has been formulated. Different computational aspects of the methodology, as well as the key input parameters describing the precursor molecules, the substrate, and the irradiation conditions have been systematically described. A step-by-step simulation and parameter determination workflow represented a comprehensive computational protocol for simulating and characterising a broad range of nanostructures created by means of FEBID.

The formulated computation protocol was applied to simulate the FEBID of $Pt(PF_3)_4$ – a widely studied precursor molecule[80–84] – on a SiO_2 surface. As such, the study[25] extended the above-described IDMD-based studies[4,24] of the FEBID of $W(CO)_6$ towards another precursor molecule which has been commonly used to fabricate platinum-containing nanostructures. In contrast to the earlier studies,[4,24] the study performed in Ref. 25 has focused on the case of low precursor surface coverage (below one monolayer), which is of interest for the FEBID-based fabrication of free-standing 3D nanostructures.[85,86] Particular focus has been placed on the atomistic characterisation of the initial stage of the FEBID process, including the nucleation of Pt atoms, the formation of small metal clusters on the surface followed by their aggregation, and, eventually, the formation of dendritic platinum nanostructures.

IDMD simulations have been performed for 23 cycles of the FEBID process with a total simulation time of 230 ns. The accumulated fluence of PE is $\sim 7.5 \times 10^{18}$ cm^{-2}. This corresponds to the equivalent total experimental irradiation time of 20 ms with the following beam parameters taken from Ref. 80: electron current $I_{\text{exp}} =$ 2.8 nA and the estimated beam spot radius $R_{\text{exp}} = 40$ nm. Although the experimental irradiation was performed with a continuous beam, the dwell time of a single irradiation cycle was set to $\tau_d^{\text{exp}} = 1$ ms.

This corresponds to a PE flux of $J_0^{\rm exp} \approx 3480$ nm^{-2}ms^{-1}. Following the earlier studies of FEBID using IDMD,[4,24] a dwell time value of $\tau_d = 10$ ns is used in the simulations. The rescaled electron current in the simulations is $I_{\rm sim} \approx 4.4$ μA.

A snapshot of the system at the end of the 20th FEBID cycle is presented in Fig. 7.17. Panels (a)–(d) correspond to different amounts of energy transferred to the medium via the Pt–P bonds fragmentation. The energy parameter $E_{\rm dep}$ governing the bond fragmentation is equal to 300 kcal/mol and 205 kcal/mol for the cases presented in panels (a) & (c) and panels (b) & (d), respectively. As shown in Figs. 7.17(a) and 7.17(b), three spatial regions can be distinguished where different structures are formed on the surface depending on the spatial distributions of the fragmentation probability and deposited precursor molecules. Inside the beam spot area with the diameter of 10 nm (indicated by the blue circle in Figs. 7.17(a) and 7.17(b)), a high probability of Pt–P bond fragmentation leads to the dissociation of $Pt(PF_3)_4$ molecules and the formation of metal clusters. The clusters grow, merge, and interconnect during the irradiation process, forming a network of thread-like metallic nanostructures. The transition region of 1 nm radius outside the beam spot area contains smaller metal clusters with a larger number of PF_3 ligands attached due to a lower fragmentation probability in this spatial region. The spatial region near the simulation box boundaries contains mostly intact or less fragmented precursor molecules.

The largest cluster formed at the end of the simulation was selected to study its evolution during the FEBID process. The cluster's evolution is tracked back to the initial nucleation stage using the coordinates of the centre of mass of the metal core. The cluster structure and the number of Pt atoms in the cluster at the end of each FEBID cycle are shown in Fig. 7.18. The platinum-containing nanostructure grows via coalescence of the neighboring metal clusters of different sizes. The coalescence takes place via an interplay of the following mechanisms: the addition of a fragmented precursor molecule with a single Pt atom (see the evolution of the cluster structure at FEBID cycles 3–4 and 8–9 in Fig. 7.18), the merging of two clusters of comparable size (see the cluster structure at cycles 4–5, 9–10, and 10–11), and a combination of both. Metal clusters containing up to about 30 Pt atoms preserve a spherical shape and tend to rearrange after an elongation caused by the merging of clusters of

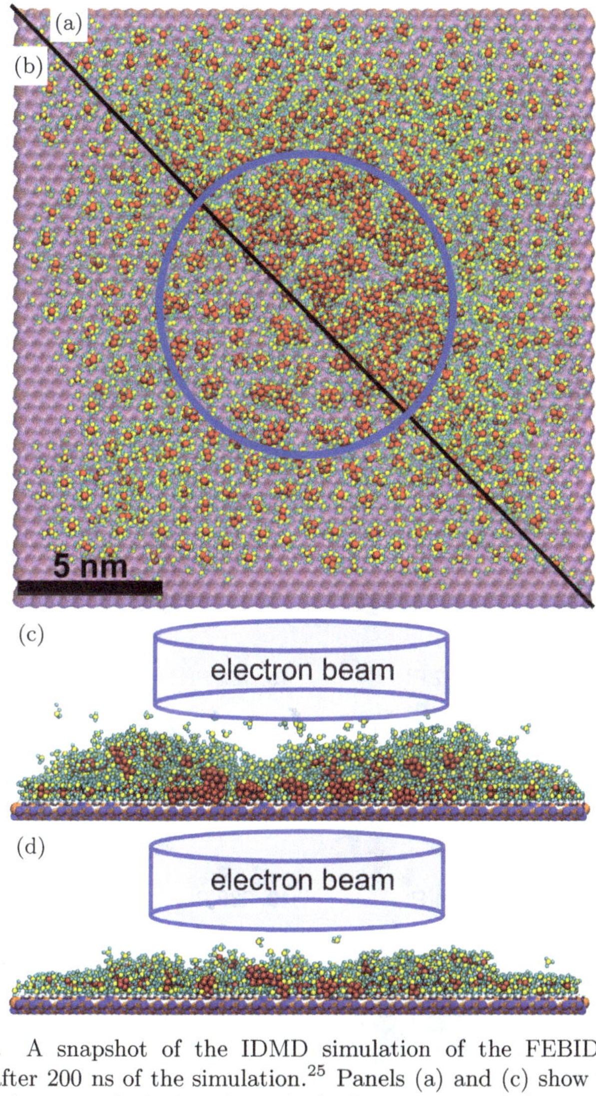

Fig. 7.17. A snapshot of the IDMD simulation of the FEBID process of $Pt(PF_3)_4$ after 200 ns of the simulation.[25] Panels (a) and (c) show the top and side views of a 6 nm thick slice through the beam centre of the grown structure for the energy $E_{\mathrm{dep}} = 300$ kcal/mol transferred to the medium during fragmentation. Panels (b) and (d) show the corresponding views for the deposition energy of $E_{\mathrm{dep}} = 205$ kcal/mol. The scale bar is the same for all figures. The blue cylinder indicates the PE beam spot.
Source: Reproduced from Ref. 25 published under an open access Creative Commons CC BY license.

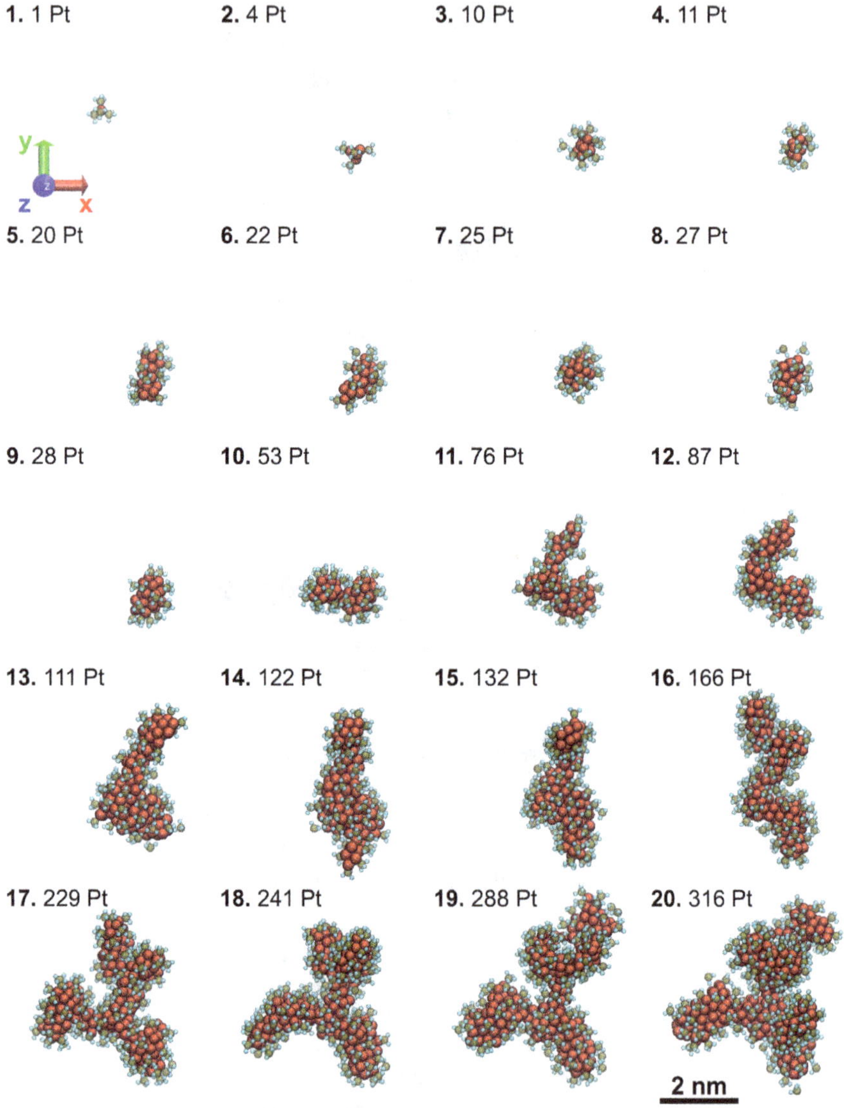

Fig. 7.18. Evolution of the largest cluster in the course of simulation at the end of each FEBID simulation cycle (shown in bold). The number of Pt atoms is also indicated for each structure.

Source: Reproduced from Ref. 25 published under an open access Creative Commons CC BY license.

comparable size (see cycles 1–9). The sequential coalescence of larger clusters containing several tens of Pt atoms results in the formation of randomly oriented branched structures (see cycles 10–20). The coalescence of all the isolated Pt-containing clusters into a bigger structure continues after the 20 simulated cycles. The continuation of the irradiation and replenishment processes can lead to further growth and interconnection of the branched clusters into a single metal network.

The IDMD approach enables us to characterise on the atomistic level the morphology of FEBID nanostructures. The morphology of deposits governs many physical properties, including electrical and thermal conductivity and magnetic properties.[87,88] In Ref. 25, the morphology, size, and metal content of a Pt-containing deposit were analysed as functions of electron fluence and adsorbate concentration. Particular focus was placed on analysing the morphological transition from isolated metal-enriched islands formed on the substrate into a single metal nanostructure. A simple estimate was made to evaluate the threshold island size at which the morphological transition will be completed at given conditions and the number of adsorbed precursor molecules enabling such a transition. In general, these parameters depend on the irradiation and replenishment conditions.

The characteristics of the FEBID nanostructures (such as lateral size, height, metal concentration, and growth rate) are determined by the precursor fragmentation rate (which depends on the energy and flux of PEs, BSEs, and SEs) and the surface density of adsorbed precursors. Two different FEBID regimes can be distinguished depending on the number of precursor molecules adsorbed on the surface and the number of fragmented molecules during the irradiation period at a given electron current density – the electron-limited regime (ELR) and the precursor-limited regime (PLR).[38,71] In the ELR, only a fraction of all the available precursor molecules on the surface exposed to irradiation by the PEs, BSEs, and SEs become fragmented during the irradiation phase. In the PLR, all the precursor molecules available on the surface become fragmented during the specific irradiation period. In this case, the high degree of molecular dissociation and subsequent irradiation-driven chemical reactions determine the properties of the growing metal-containing nanostructure.

In Ref. 26, the IDMD method was used to study the growth of metal nanostructures under different irradiation and replenishment conditions corresponding to the ELR and PLR regimes of the FEBID process. $Fe(CO)_5$, one of the most common FEBID precursors used to fabricate magnetic nanostructures,[89–95] was chosen as a case study. The dependence of the nanostructures' morphology and metal content on the number of adsorbed molecules and electron flux was investigated on the atomistic level. It was found that varying the electron current during the FEBID process significantly changes the deposit's morphology (Fig. 7.19(a)) and elemental composition (Fig. 7.19(b)). At a low beam current, corresponding to a low degree of precursor fragmentation, the deposits consist of isolated small-size iron clusters surrounded by organic ligands (see the top panel in Fig. 7.19(a)). In this irradiation regime, metal clusters formed as a result of electron-irradiation-induced fragmentation of precursor molecules do not agglomerate with increasing electron fluence, resulting in a relatively low metal content in the deposit. Higher beam current facilitates the precursor fragmentation, and the metal clusters coalesce into dendrite-like structures with the size corresponding to the PE beam (see the bottom panel of Fig. 7.19(a)). In this regime, the deposit's metal content increases twofold compared to the case of low current, as shown in Fig. 7.19(b).

In Ref. 27, the role of thermal effects in the FEBID of $Me_2Au(tfac)$ precursor molecule was studied by means of IDMD simulations. $Me_2Au(tfac)$ is a commonly used precursor molecule for the fabrication of gold-containing nanostructures. The FEBID process with this precursor molecule was simulated at different temperatures in the 300–450 K range. The deposit's structure, morphology, growth rate, and elemental composition at different temperatures were analysed. The simulations[27] confirmed the experimental observation that the deposits consist of small gold clusters embedded into a carbon-rich organic matrix; see Fig. 7.20.

The deposit's growth rate, defined as the average deposit height per accumulated electron fluence, was analysed at different temperatures, ranging from 300 to 450 K. The results of this analysis are presented in Fig. 7.21. The decrease in the growth rate with temperature was explained by (i) the increased mobility of atoms and

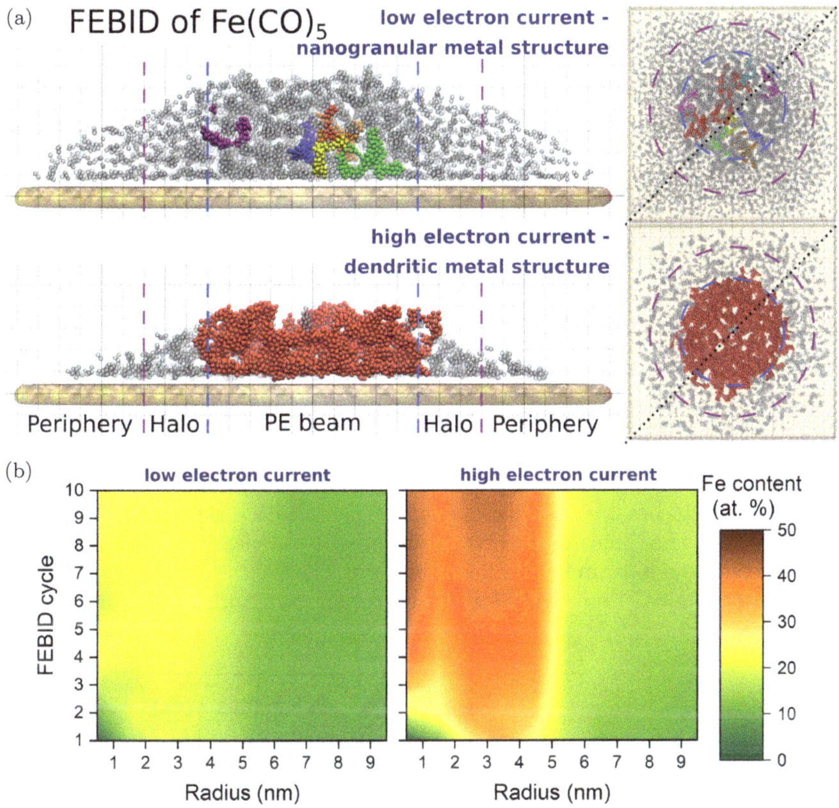

Fig. 7.19. Results of the IDMD simulations[26] of the FEBID process for $Fe(CO)_5$. (a) Snapshots of the simulated iron-containing nanostructures: side view of diagonal cross-sections indicated by dotted black lines (left) and their top view (right). The top and bottom snapshots in panel (a) correspond to electron currents of 1 and 4 nA, respectively. Only iron atoms are shown for clarity. Topologically disconnected metal clusters containing more than 100 iron atoms are shown in different colours (red, blue, green, yellow, orange, and purple) in the central part of the nanostructures. Smaller clusters containing less than 100 iron atoms are shown as pale grey spheres. Boundaries of the primary electron beam, halo, and peripheral regions are indicated by dashed blue and purple lines in the left column and by circles in the right column. Grid line spacing is equal to 1 nm in all dimensions. (b) Atomic Fe content of the grown iron-containing structures as a function of the number of simulated irradiation–replenishment FEBID cycles for electron currents of 1 (left) and 4 nA (right).

Source: Adapted from Ref. 26 with permission from the Royal Society of Chemistry.

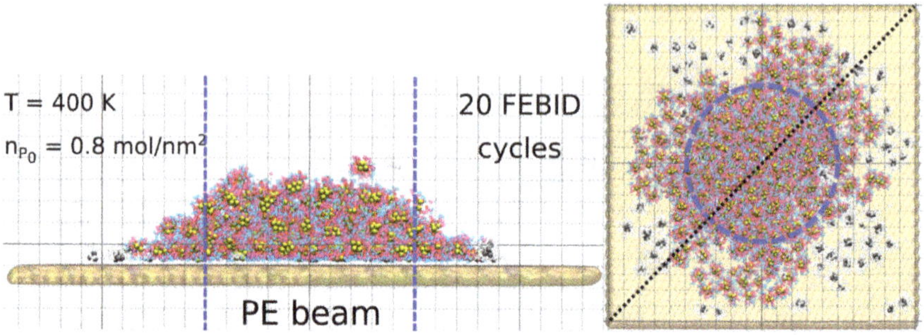

Fig. 7.20. Snapshots of the IDMD simulations[27] of the FEBID process for $Me_2Au(tfac)$ with electron current of 4 nA at temperature $T = 400$ K and the corresponding steady-state concentration of adsorbed precursors n_{p_0}. Left panel shows the system's side view of diagonal cross-sections indicated by dotted lines; the right panel shows the system's top view. The largest topologically connected gold-containing compound is visualised in colour. Isolated precursor molecules and small fragments are shown in greyscale. The primary electron beam spot is depicted by dashed lines in the left column and by circles in the right column. Grid line spacing is 1 nm in all dimensions.
Source: Reproduced from Ref. 27 with permission from Springer Nature.

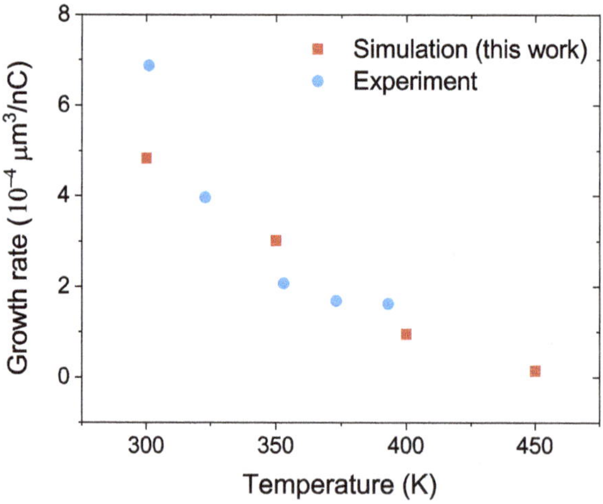

Fig. 7.21. Growth rate of the gold-containing deposit (red squares) defined as its height in the beam spot area per accumulated electron fluence at different temperatures in the range of 300–450 K, simulated using IDMD.[27] Blue circles depict the experimental data[96] on the deposit's growth rate during the FEBID of $Me_2Au(acac)$.
Source: Reproduced from Ref. 27 with permission from Springer Nature.

molecules at elevated temperatures, causing a faster formation of larger clusters, and (ii) a limitation of the volumetric growth of the deposit by a smaller number of precursor molecules adsorbed on a surface at elevated temperatures. The evaluated dependence of the deposit's growth rate on temperature agreed with the experimentally measured dependence for a structurally similar $Me_2Au(acac)$ molecule.[96]

7.7 Conclusion

This chapter has provided an overview of the RMD[8] and IDMD[4] methodologies, which enable atomistic simulations of the chemical and irradiation-driven transformations of complex MBN systems exposed to various conditions and radiation modalities. RMD and IDMD rely on several input parameters with a clear physical meaning, such as the bond dissociation energies, molecular fragmentation cross sections, amount of energy transferred to the system upon irradiation, energy relaxation rate, and spatial region wherein the energy is relaxed. Major transformations of irradiated molecular systems (such as topological changes, redistribution of atomic partial charges, and alteration of interatomic interactions) are simulated with the reactive rCHARMM force field[8] implemented in the MBN Explorer[5] software package.

The utilisation of RMD and IDMD methods has been illustrated through several exemplary case studies related to the irradiation-induced fragmentation of molecular and biomolecular systems and to the controlled fabrication of nanostructures using FEBID. The results presented in this chapter demonstrate that RMD and IDMD provide powerful computational tools to model chemical and irradiation-induced transformations of complex MBN systems at the atomistic level of detail. The IDMD simulations of the FEBID process demonstrate a great predictive power, yielding the morphology of the simulated metal nanostructures, their composition, and growth characteristics in good agreement with available experimental data.

The outcomes of atomistic simulations of the FEBID process using IDMD can be used to construct stochastic models of FEBID using the SD methodology[97] implemented in MBN Explorer. An illustration of the application of SD for modelling the FEBID process was presented recently.[98] It was demonstrated that the FEBID process can be described through step-by-step transformations occurring to particles of different types, representing intact and fragmented precursor molecules, ligands, isolated metal atoms, and the substrate. The probabilities of the underlying processes occurring in the system can be determined through atomistic MD simulations and track-structure MC calculations.

The RMD and IDMD methods presented in this chapter open up a broad range of possibilities for modelling of irradiation-driven modifications and chemistry of complex molecular systems. This methodology provides a wide range of possibilities for atomistic-level study of FEBID and many other processes in which the irradiation of molecular systems and irradiation-driven chemistry play the key role.

References

1. Solov'yov IA, Korol AV, Solov'yov AV. Multiscale modeling of complex molecular structure and dynamics with MBN Explorer. Cham: Springer International Publishing. 2017.
2. Solov'yov IA, *et al.*, editors. Dynamics of systems on the nanoscale. Cham: Springer International Publishing. 2022.
3. Born M, Oppenheimer JR. Zur Quantentheorie der Molekeln. Ann Phys. 1927;389:457–484.
4. Sushko GB, Solov'yov IA, Solov'yov AV. Molecular dynamics for irradiation driven chemistry: Application to the FEBID process. Eur Phys J D. 2016;70:217.
5. Solov'yov IA, *et al.* MesoBioNano Explorer – A universal program for multiscale computer simulations of complex molecular structure and dynamics. J Comput Chem. 2012;33:2412.
6. Solov'yov IA, Sushko G, Solov'yov AV. MBN Explorer User's Guide (Version 3.0). Frankfurt am Main: MesoBioNano Science Publishing. 2017.
7. Verkhovtsev A, *et al.* MBN Explorer and MBN Studio Tutorials: Version 5.0. Frankfurt am Main: MesoBioNano Science Publishing. 2024.
8. Sushko GB, *et al.* Studying chemical reactions in biological systems with MBN Explorer: Implementation of molecular mechanics with dynamical topology. Eur Phys J D. 2016;70:12.
9. Tersoff J. Empirical interatomic potential for carbon, with applications to amorphous carbon. Phys Rev Lett. 1988;61:2879.

10. Tersoff J. New empirical approach for the structure and energy of covalent systems. Phys Rev B. 1988;37:6991.
11. Brenner DW. Empirical potential for hydrocarbons for use in simulating the chemical vapor deposition of diamond films. Phys Rev B. 1990;42:9458.
12. Stuart SJ, Tutein AB, Harrison JA. A reactive potential for hydrocarbons with intermolecular interactions. J Chem Phys. 2000;112:6472.
13. Brenner DW, *et al.* A second-generation reactive empirical bond order (REBO) potential energy expression for hydrocarbons. J Phys Condens Matter. 2002;14:783.
14. Verkhovtsev AV, Solov'yov IA, Solov'yov AV. Irradiation-driven molecular dynamics: A review. Eur Phys J D. 2021;75:213.
15. Solov'yov AV, *et al.* Condensed matter systems exposed to radiation: Multiscale theory, simulations, and experiment. Chem Rev. 2024;124:8014–8129.
16. MacKerell AD Jr, *et al.* All-atom empirical potential for molecular modeling and dynamics studies of proteins. J Phys Chem B. 1998;102:3586.
17. MacKerell AD Jr, Feig M, Brooks CL III. Extending the treatment of backbone energetics in protein force fields: Limitations of gas-phase quantum mechanics in reproducing protein conformational distributions in molecular dynamics simulations. J Comput Chem. 2004;25:1400.
18. Friis I, *et al.* Modeling the effect of ion-induced shock waves and DNA breakage with the reactive CHARMM force field. J Comput Chem. 2020;41:2429.
19. Friis I, *et al.* Lethal DNA damage caused by ion-induced shock waves in cells. Phys Rev E. 2021;104:054408.
20. Verkhovtsev AV, Korol AV, Solov'yov AV. Classical molecular dynamics simulations of fusion and fragmentation in fullerene–fullerene collisions. Eur Phys J D. 2017;71:212.
21. de Vera P, *et al.* Reactive molecular dynamics simulations of organometallic compound $W(CO)_6$ fragmentation. Eur Phys J D. 2019;73:215.
22. Andreides B, *et al.* Role of the molecular environment in quenching the irradiation-driven fragmentation of $Fe(CO)_5$: A reactive molecular dynamics study. J Phys Chem A. 2023;127:3757.
23. de Vera P, *et al.* Simulation of the ion-induced shock waves effects on the transport of chemically reactive species in ion tracks. Eur Phys J D. 2018;72:147.
24. de Vera P, *et al.* Multiscale simulation of the focused electron beam induced deposition process. Sci Rep. 2020;10:20827.

25. Prosvetov A, *et al.* Irradiation-driven molecular dynamics simulation of the FEBID process for $Pt(PF_3)_4$. Beilstein J Nanotechnol. 2021;12:1151.
26. Prosvetov A, *et al.* Atomistic simulation of the FEBID-driven growth of iron-based nanostructures. Phys Chem Chem Phys. 2022;24:10807.
27. Prosvetov A, *et al.* Atomistic modeling of thermal effects in focused electron beam-induced deposition of Me_2Au(tfac). Eur Phys J D. 2023;77:15.
28. Landau LD, Lifshitz EM. Quantum mechanics: Non-relativistic theory. 3rd ed. Oxford: Butterworth-Heinemann. 1981.
29. Surdutovich E, Solov'yov AV. Multiscale approach to the physics of radiation damage with ions. Eur Phys J D. 2014;68:353.
30. Surdutovich E, Solov'yov AV. Transport of secondary electrons and reactive species in ion tracks. Eur Phys J D. 2015;69:193.
31. Dapor M. Transport of energetic electrons in solids. 3rd ed. Cham: Springer International Publishing. 2020.
32. Gerchikov LG, *et al.* Non-adiabatic electron-ion coupling in dynamical jellium model for metal clusters. J Phys B At Mol Opt Phys. 2000;33:4905.
33. Agostinelli S, *et al.* Geant4 – a simulation toolkit. Nucl Instrum Meth A. 2003;506:250–303.
34. Allison J, *et al.* Recent developments in Geant4. Nucl Instrum Meth A. 2016;835:185–225.
35. Azzolini M, *et al.* Secondary electron emission and yield spectra of metals from Monte Carlo simulations and experiments. J Phys Condens Matter. 2019;31:055901.
36. Solov'yov AV, editor. Nanoscale insights into ion-beam cancer therapy. Cham: Springer International Publishing. 2017.
37. Utke I, Hoffmann P, Melngailis J. Gas-assisted focused electron beam and ion beam processing and fabrication. J Vac Sci Technol B. 2008;26:1197–1276.
38. Huth M, *et al.* Focused electron beam induced deposition: A perspective. Beilstein J Nanotechnol. 2012;3:597–619.
39. Barth S, Huth M, Jungwirth F. Precursors for direct-write nanofabrication with electrons. J Mater Chem C. 2020;8:15884–15919.
40. Wnorowski K, *et al.* Electron ionization of $W(CO)_6$: Appearance energies. Int J Mass Spectrom. 2012;314:42.
41. Frisch MJ, *et al.* Gaussian 09, Revision E.01. Wallingford: Gaussian Inc. 2016.
42. Arnesen SV, Seip HM. Studies on the failure of the first Born approximation in electron diffraction. V. Molybdenum- and tungsten hexacarbonyl. Acta Chem Scand. 1966;20:2711.

43. Szilagyi RK, Frenking G. Structure and bonding of the isoelectronic hexacarbonyls $[Hf(CO)_6]^{2-}$, $[Ta(CO)_6]^{-}$, $W(CO)_6$, $[Re(CO)_6]^{+}$, $[Os(CO)_6]^{2+}$, and $[Ir(CO)_6]^{3+}$: A theoretical study. Organometallics. 1997;16:4807.
44. Diefenbach A, Bickelhaupt FM, Frenking G. The nature of the transition metal-carbonyl bond and the question about the valence orbitals of transition metals. A bond-energy decomposition analysis of $TM(CO)_6^{\varphi}$ ($TM^{\varphi} = Hf^{2-}$, Ta^{-}, W, Re^{+}, Os^{2+}, Ir^{3+}). J Am Chem Soc. 2000;122:6449.
45. Ward JF. DNA damage produced by ionizing radiation in mammalian cells: Identities, mechanisms of formation, and reparability. Prog Nucleic Acid Res Mol Biol. 1988;35:95–125.
46. MacKerell AD Jr, Banavali N, Foloppe N. Development and current status of the CHARMM force field for nucleic acids. Biopolymers. 2001;56:257.
47. Michels GD, Flesch GD, Svec HJ. Comparative mass spectrometry of the Group 6B hexacarbonyls and pentacarbonyl thiocarbonyls. Inorg Chem. 1980;19:479.
48. Winters RE, Kiser RW. Mass spectrometric studies of chromium, molybdenum, and tungsten hexacarbonyls. Inorg Chem. 1965;4:157.
49. Bidinosti DR, McIntyre NS. Electron-impact study of some binary metal carbonyls. Can J Chem. 1967;45:641.
50. Foffani A, *et al.* Mass spectra of metal hexacarbonyls. Z Phys Chem. 1965;45:79.
51. Qi F, *et al.* Vacuum ultraviolet photoionization and dissociative photoionization of $W(CO)_6$. J Chem Phys. 1997;107:10391.
52. Lengyel J, Fedor J, Fárník M. Ligand stabilization and charge transfer in dissociative ionization of $Fe(CO)_5$ aggregates. J Phys Chem C. 2016;120:17810.
53. Lacko M, *et al.* Electron-induced ionization and dissociative ionization of iron pentacarbonyl molecules. Eur Phys J D. 2015;69:84.
54. García Gómez-Tejedor G, Fuss MC, editors. Radiation damage in biomolecular systems. Dordrecht: Springer Science + Business Media B.V. 2012.
55. Schardt D, Elsässer T, Schulz-Ertner D. Heavy-ion tumor therapy: Physical and radiobiological benefits. Rev Mod Phys. 2010;82: 383–425.
56. Linz U, editor. Ion beam therapy: Fundamentals, technology, clinical applications. Berlin Heidelberg: Springer-Verlag. 2012.
57. Surdutovich E, Solov'yov AV. Multiscale modeling for cancer radiotherapies. Cancer Nanotechnol. 2019;10:6.

58. Surdutovich E, Yakubovich AV, Solov'yov AV. Biodamage via shock waves initiated by irradiation with ions. Sci Rep. 2013;3:1289.
59. Park Y, *et al.* DNA damage induced by low-energy electrons: Conversion of thymine to 5,6-dihydrothymine in the oligonucleotide trimer TpTpT. Radiat Res. 2011;175:240.
60. Toulemonde M, Surdutovich E, Solov'yov AV. Temperature and pressure spikes in ion-beam cancer therapy. Phys Rev E. 2009;80:031913.
61. Surdutovich E, Solov'yov AV. Shock wave initiated by an ion passing through liquid water. Phys Rev E. 2010;82:051915.
62. Yakubovich AV, Surdutovich E, Solov'yov AV. Atomic and molecular data needs for radiation damage modeling: Multiscale approach. AIP Conf Proc. 2011;1344:230.
63. de Vera P, *et al.* Molecular dynamics study of accelerated ion-induced shock waves in biological media. Eur Phys J D. 2016;70:183.
64. Fraile A, *et al.* First principles simulation of damage to solvated nucleotides due to shock waves. J Chem Phys. 2019;150:015101.
65. Bottländer D, Mücksch C, Urbassek HM. Effect of swift-ion irradiation on DNA molecules: A molecular dynamics study using the REAX force field. Nucl Instrum Meth B. 2015;365:622.
66. Yakubovich AV, Surdutovich E, Solov'yov AV. Thermomechanical damage of nucleosome by the shock wave initiated by ion passing through liquid water. Nucl Instrum Meth B. 2012;279:135.
67. Verkhovtsev AV, Solov'yov IA, Solov'yov AV. Advances in multiscale modeling for novel and emerging technologies. Eur Phys J D. 2021;75:207.
68. Verkhovtsev A, Surdutovich E, Solov'yov AV. Multiscale approach predictions for biological outcomes in ion-beam cancer therapy. Sci Rep. 2016;6:27654.
69. Verkhovtsev A, Surdutovich E, Solov'yov AV. Phenomenon-based evaluation of relative biological effectiveness of ion beams by means of the multiscale approach. Cancer Nanotechnol. 2019;10:4.
70. Cui Z. Nanofabrication: Principles, capabilities and limits. Cham: Springer International Publishing. 2017.
71. Utke I, Moshkalev S, Russell P, editors. Nanofabrication using focused ion and electron beams. Oxford: Oxford University Press. 2012.
72. van Dorp WF, Hagen CW. A critical literature review of focused electron beam induced deposition. J Appl Phys. 2008;104:081301.
73. Huth M, Porrati F, Dobrovolskiy OV. Focused electron beam induced deposition meets materials science. Microelectron Eng. 2018;185–186:9–28.

74. Fowlkes JD, Rack PD. Fundamental electron-precursor-solid interactions derived from time-dependent electron-beam-induced deposition simulations and experiments. ACS Nano. 2010;4:1619.
75. Sanz-Hernández D, Fernández-Pacheco A. Modelling focused electron beam induced deposition beyond Langmuir adsorption. Beilstein J Nanotechnol. 2017;8:2151.
76. Muthukumar K, *et al.* Spontaneous dissociation of $Co_2(CO)_8$ and autocatalytic growth of Co on SiO_2: A combined experimental and theoretical investigation. Beilstein J Nanotechnol. 2012;3:546.
77. Muthukumar K, Jeschke HO, Valentí R. Dynamics and fragmentation mechanism of $(C_5H_4CH_3)Pt(CH_3)_3$ on SiO_2 surfaces. Beilstein J Nanotechnol. 2018;9:711.
78. Sushko GB, Solov'yov IA, Solov'yov AV. Modeling MesoBioNano systems with MBN Studio made easy. J Mol Graph Model. 2019;88:247.
79. Porrati F, Sachser R, Huth M. The transient electrical conductivity of W-based electron-beam-induced deposits during growth, irradiation and exposure to air. Nanotechnology. 2009;20:195301.
80. Barry JD, *et al.* Electron beam induced deposition of low resistivity platinum from $Pt(PF_3)_4$. J Vac Sci Technol B. 2006;24:3165.
81. Botman A, *et al.* Electron postgrowth irradiation of platinum-containing nanostructures grown by electron-beam-induced deposition from $Pt(PF_3)_4$. J Vac Sci Technol B. 2009;27:2759.
82. Landheer K, *et al.* Low-energy electron-induced decomposition and reactions of adsorbed tetrakis(trifluorophosphine)platinum [$Pt(PF_3)_4$]. J Phys Chem C. 2011;115:17452.
83. May O, Kubala D, Allan M. Dissociative electron attachment to $Pt(PF_3)_4$– a precursor for Focused Electron Beam Induced Processing (FEBIP). Phys Chem Chem Phys. 2012;14:2979.
84. Zlatar M, Allan M, Fedor J. Excited states of $Pt(PF_3)_4$ and their role in focused electron beam nanofabrication. J Phys Chem C. 2016;120:10667.
85. Fowlkes JD, *et al.* Simulation informed CAD for 3D nanoprinting. Micromachines. 2020;11:8.
86. Fowlkes JD, *et al.* High-fidelity 3D-nanoprinting via focused electron beams: Computer-aided design (3BID). ACS Appl Nano Mater. 2018;1:1028.
87. Huth M, *et al.* Conductance regimes of W-based granular metals prepared by electron beam induced deposition. New J Phys. 2009;11:033032.

88. Huth M, *et al.* Temperature-dependent growth characteristics of Nb- and CoFe-based nanostructures by direct-write using focused electron beam-induced deposition. Micromachines. 2020;11:28.
89. Takeguchi M, Shimojo M, Furuya K. Fabrication of magnetic nanostructures using electron beam induced chemical vapour deposition. Nanotechnology. 2005;16:1321.
90. Gavagnin M, *et al.* Magnetic force microscopy study of shape engineered FEBID iron nanostructures. Phys Stat Solidi A. 2014;211:368.
91. Serrano-Ramon L, *et al.* Ultrasmall functional ferromagnetic nanostructures grown by focused electron-beam-induced deposition. ACS Nano. 2011;5:7781.
92. Lukasczyk T, *et al.* Electron-beam-induced deposition in ultrahigh vacuum: Lithographic fabrication of clean iron nanostructures. Small. 2008;4:841.
93. Gavagnin M, *et al.* Synthesis of individually tuned nanomagnets for nanomagnet logic by direct write focused electron beam induced deposition. ACS Nano. 2013;7:777.
94. Porrati F, *et al.* Magnetotransport properties of iron microwires fabricated by focused electron beam induced autocatalytic growth. J Phys D Appl Phys. 2011;44:425001.
95. Lavrijsen R, *et al.* Fe:O:C grown by focused-electron-beam-induced deposition: Magnetic and electric properties. Nanotechnology. 2011;22:025302.
96. Mulders JJ, Belova LM, Riazanova A. Electron beam induced deposition at elevated temperatures: Compositional changes and purity improvement. Nanotechnology. 2011;22:055302.
97. Solov'yov IA, *et al.* Multiscale modeling of stochastic dynamics processes with MBN Explorer. J Comput Chem. 2022;43:1442.
98. Solov'yov IA, *et al.* Stochastic dynamics simulation of the focused electron beam induced deposition process. 2025;arXiv:2506.18163 [physics.comp-ph].

https://doi.org/10.1142/9781800618183_0008

Chapter 8

Molecular Mechanisms of Radiation Action on Peptides, Proteins, DNA Strands and Antibiotics Studied in the Gas Phase

Thomas Schlathölter[*,†,§] **and Jean-Christophe Poully**[‡,¶]

[*] *Zernike Institute for Advanced Materials, University of Groningen, Nijenborgh 4, 9747 AG Groningen, The Netherlands*

[†] *University College Groningen University of Groningen, Hoendiepskade 23/24, 9718 BG Groningen, The Netherlands*

[‡] *CIMAP UMR 6252, CEA, CNRS, ENSICAEN, Université de Caen Normandie, Bd Becquerel, 14070 Caen, France*

[§] *t.a.schlatholter@rug.nl*

[¶] *poully@ganil.fr*

Gas-phase studies are an ideal tool for the investigation of the molecular mechanisms underlying ionising radiation action upon biological systems. The field has strongly benefitted from the advent of electrospray ionisation as a tool for gentle transfer of complex biomolecular ions from solution into the gas phase. Radiofrequency techniques are commonly used for low-energy ion transport, manipulation, and trapping. The trapped molecular ions are then exposed to ionising radiation, e.g., from synchrotrons or heavy ion beamlines. The biomolecular interaction products are analysed by means of mass spectrometry. In this chapter, we provide a review of experimental investigations that focus on three different classes of biomolecular systems: peptides and proteins, DNA strands, and antibiotics. Molecular processes such as ionisation, fragmentation, and H transfer are discussed. The chapter closes with a brief description of current experimental developments towards conformational control through ion mobility spectrometry, electron spectroscopy in crossed-beam experiments, and time-resolved measurements, e.g., using ultrashort pulses from free electron lasers.

8.1 Introduction

When biological matter is irradiated by ionising particles, such as energetic photons, ions, and electrons, the first processes to occur are purely physical and often involve the interaction with electrons from water molecules and from biomolecules, including proteins, DNA, carbohydrates, and lipids. These initial molecular electronic excitation and ionisation processes take place at the attosecond time scale. The resulting altered molecular electronic structure then induces dynamics of charge, energy, and atoms, leading to noncovalent or/and covalent bond cleavage, as well as cross-linking between biomolecules. All these so-called *direct effects* can thus create new chemical species. Among the latter, free radicals are particularly important due to their high chemical reactivity; for instance, hydroxyl from water radiolysis induces secondary processes in proteins, such as backbone cleavage and in some cases aggregation of the fragments formed,[1] cross-linking between proteins to form larger systems, or even nanoparticles with diameter in the 10 nm range[2] and generally quenching but also enhancement of biological activity.[3]

These *indirect effects* generally require the diffusion of free radicals at rates that decrease by several orders of magnitude when cooling samples from room temperature to cryogenic temperatures.[4] Therefore, irradiation of frozen samples allows for enhancing the relative contribution of direct effects, although indirect effects from molecules in the close vicinity of a given molecular system are still present. In many situations, direct effects represent a non-negligible fraction of the total effects of the ionising radiation, for instance, at around 50% regarding DNA in the cell nucleus, where DNA is tightly packed.[5] The same applies to connective tissues, such as skin, cartilage, and tendons, which are primarily made up of a dense extracellular matrix containing structural macromolecules, such as proteins (collagen, elastin, etc.) and glycosaminoglycans. Early studies aiming at understanding these effects have been reviewed more than 30 years ago.[4,6]

More recently, we have surveyed the efforts aimed at a description of the direct effects of radiation on the structure and stability of collagen and other proteins, with emphasis on the irradiation of molecular systems in the gas phase.[7] The gas phase has several advantages for studying the radiation-induced processes in biomolecular systems. First of all, the influence of a molecular environment is absent in the gas phase, leaving the intrinsic properties of the systems for direct experimental investigation. The absence of a chemical environment is also ideal for comparing experimental data with theoretical calculations, most often obtained by using quantum-chemical methods. Furthermore, in particular, gas-phase molecular ions allow for precise control of molecular properties: mass spectrometric techniques allow for defining the initial state of a molecular system in terms of size, stoichiometry, temperature, charge, and tautomeric state. Techniques such as ion mobility spectrometry or infrared (IR)/ultraviolet (UV) hole burning spectroscopy are sensitive to the geometrical structure of molecular systems, and thus allow for selecting a given conformer (e.g., before interaction with a laser[8,9]) or tracing specific ionic products to one conformer.[10] In addition, large biological systems such as intact viruses can be brought to the gas phase thanks to techniques such as electrospray (or nanospray) ionisation (ESI) or matrix-assisted laser desorption ionisation (MALDI).[11] Last but not least, mass spectrometry is also an important tool to probe the mass-to-charge (m/z) ratio of ionic products formed following the interaction between the molecular system and the ionising radiation.

Recently, several roadmaps and reviews have surveyed important technological as well as scientific advances in the field of biomolecular radiation damage.[12–17] In the current chapter, we focus on the direct physical and chemical effects of the radiation-induced removal of one or several electrons from biomolecular ions produced by ESI or MALDI. We describe the existing knowledge, including our own contributions to the field, regarding peptides, proteins, and DNA strands, as well as antibiotics and their noncovalent complexes with receptors, with the latter being relevant to specific molecular recognition. The conclusion includes perspectives on experimental efforts towards describing the radiation-induced processes in time (on the femtosecond to picosecond time scale), not only measuring the energy of electrons ejected from biomolecular ions (especially deprotonated DNA strands) but also understanding the influence of geometrical structure on these processes. The latter is highly relevant for biological molecules due to the structure–function relationship.

8.2 Experimental Techniques

The first generation of gas-phase experiments for investigation of radiation action on biomolecular systems have relied on the use of effusive beams, commonly from heated solid samples. Most complex biomolecular systems such as DNA and peptides are subject to significant structural and chemical changes such as thermal decomposition at temperatures that would be required to yield sufficient vapor pressures. As a consequence, for a considerable time, experiments have been constrained to relatively basic building blocks of DNA and proteins, such as nucleobases and amino acids.[18–20] The most common methods for bringing larger molecules into the gas phase are matrix-assisted laser desorption ionisation (MALDI) and electrospray ionisation (ESI). Both techniques have been developed in the mass spectrometry community.

In the ESI process, a liquid containing the analyte is dispersed into an aerosol by an electric field (1 kV/mm). The charged droplets shrink as the solvent evaporates, undergoing fission until 'naked' analyte ions form. Depending on solution properties (e.g., pH), multiple protonated or deprotonated ions are produced at various charge states. The Zubarev group, for instance, interfaced an ESI Fourier-transfrom ion cyclotron resonance mass spectrometer with a low-energy electron beam to study electron impact ionisation of protonated peptides.[21] Nielsen *et al.* pioneered using ESI for high-energy collision experiments with complex biomolecules such as adenosine monophosphate.[22] Around 2010, the Schlathölter group[23,24] and the team led byGiuliani[25] conducted the first experiments where electrosprayed biomolecules were exposed to ionising radiation. Since then, the number of groups utilising ESI for studies of complex gas-phase molecular ions has strongly increased. Early work was hindered by

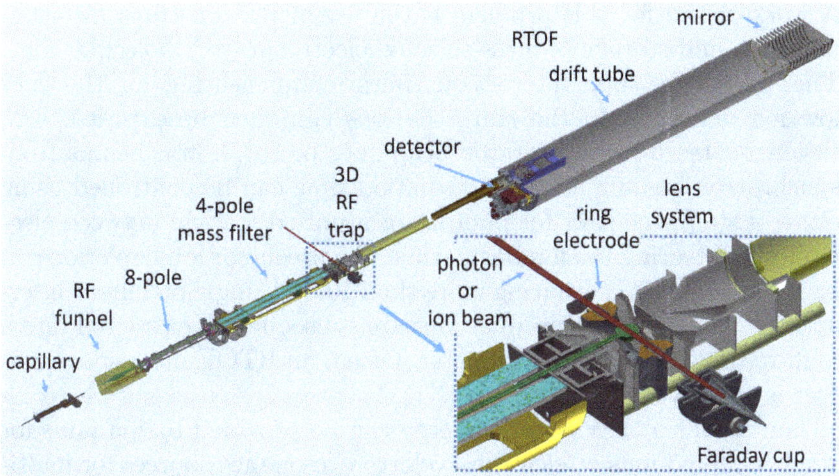

Fig. 8.1. An implementation of an ESI-based ion trap mass spectrometer developed by the Schlathölter group. The electrosprayed molecular ions undergo phase-space compression in an ion funnel, transported by an RF eight-pole and mass filtered by an RF four-pole. In the 3D RF ion trap, the ions are exposed to ionising radiation, mostly MeV ions or photons. The interaction products are mass analysed using a reflectron time-of-flight (RTOF) mass spectromter. A Faraday cup is used to measure MeV ion currents. For photon experiments, it is replaced with a photodiode.

low ion beam currents, leading many researchers to adopt radiofrequency (RF) ion funnels for improved ion collection. Currently, most experimental approaches (see Fig. 8.1 for a schematic drawing of a typical apparatus, developed by the Schlathölter group) are based on systems that feature RF ion funnels as phase-space compressors, often in the form of printed-circuit-based devices based on the concept introduced by the Smith group.[26] Funnels can either be used as standalone devices or in the form of two subsequent funnels.[27] In the latter case, biomolecular ion beam currents of hundreds of pA can be reached.

As the electrosprayed molecules are produced in ionic form, they can much more efficiently be further transported, manipulated, and stored using RF techniques. RF quadrupoles can be used as mass filters to isolate the desired mass-to-charge ratio. However, ESI fluxes are relatively low, with currents often in the few picoampere range, which is a limitation for most straightforward collision experiments.

A workaround for this problem is the use of RF ion traps for accumulation and storage of mass-selected electrosprayed molecular ions. This allows for long interaction times, compensating for the very low ion densities. In the trap, the ions can then be exposed, e.g., to ion beams from accelerator beamlines or to photon beams from synchrotron beamlines. The irradiation time can be controlled using either a simple shutter for photons or a pulsed voltage between electrodes for steering the ion beam; thus, the precursor ion depletion can be kept below 10%, ensuring more than 10% of single precursor interactions. The resulting product ions are subsequently extracted into a high-resolution mass spectrometer. Often, an RTOF mass spectrometer is employed.

Ion beams at keV kinetic energy can be provided by ion guns for singly charged ions or electron-cyclotron resonance sources for multiply charged ions, both using high-voltage electric fields to accelerate the ions. Reaching the MeV range requires large-scale accelerators, such as cyclotrons. This kinetic energy is more relevant to the conditions encountered by biomolecules in tumour cells during a hadrontherapy, a promising cancer treatment technique using heavy-ion beams instead of hard X-rays. Ionising photons in the vacuum ultraviolet (VUV) or soft X-ray ranges with sufficient flux for gas-phase experiments, high resolution, and broad tunability are provided by many synchrotron radiation facilities around the world. Furthermore, most of the times, they include one or several beamlines dedicated to atomic and molecular physics, often with experimental stations available for users. Technical developments are also performed, allowing significant advances in the field, for instance, regarding filters that remove higher-order harmonics from VUV beams to determine precise ionisation energies of biomolecules in the gas phase.

In 2010, the Groningen group was the first to study collisions of keV ions with protonated peptides held in an RF ion trap[23,24] (see Fig. 8.1).

The first VUV photoionisation data obtained by exposing RF trapped protonated proteins to synchrotron radiation was published by the SOLEIL group.[25] Independently and in parallel, the Groningen team published the first study on the protonated peptide leucine enkephalin, with data recorded at the BESSY II synchrotron in Berlin.[24]

Since then, the combination of ESI and RF ion trapping has been employed by many groups – often in collaboration – to experimentally study the interaction of energetic ions and photons with biomolecular systems.

8.3 Peptides and Proteins

8.3.1 *General considerations*

Studying the consequences of ionising radiation on proteins is highly important because these biomolecules are essential to life: they are not only involved in a multitude of cellular functions but they are also responsible for the crucial mechanical properties of connective tissues such as skin, tendons, or cartilage. These tissues are mainly composed of a dense extracellular matrix where proteins such as collagen, elastin, and fibronectin dominate. In the extracellular matrix, collagen provides most of its frame and stiffness via collagen supramolecular structures made of assemblies of fibres. Each fibre is made of many parallel triple helices bound by noncovalent interactions, and three collagen proteins are wound around each other in a triple helix. We illustrate these aspects in Fig. 8.2 for the example of cartilage, a tissue located between the synovial fluid and bone. Rare chondrocyte cells are embedded in a dense extracellular matrix mainly made of collagen fibers but also featuring aggrecans (assemblies of glycosylated proteins).

The triple-helix secondary structure is typical for collagen, and due to its peculiar primary structure, every third amino acids is glycine, and the proline content is very high (around 20%). The collagen triple helix is also intrinsically highly stable, thanks to strong intermolecular H bonds linking the three protein backbones. This stability is further enhanced by an unusually abundant post-translational modification, i.e., proline hydroxylation. Being the most abundant protein in mammals, collagen is extremely relevant for studies aiming at describing the consequences of radiation action on biological tissues, especially in the context of cancer radiotherapy, where tissues are irradiated by X-rays, electrons, or ion beams in order to deposit energy and kill malignant cells. For instance, in the case of chondrosarcoma, a cancer of cartilage, ionising particles

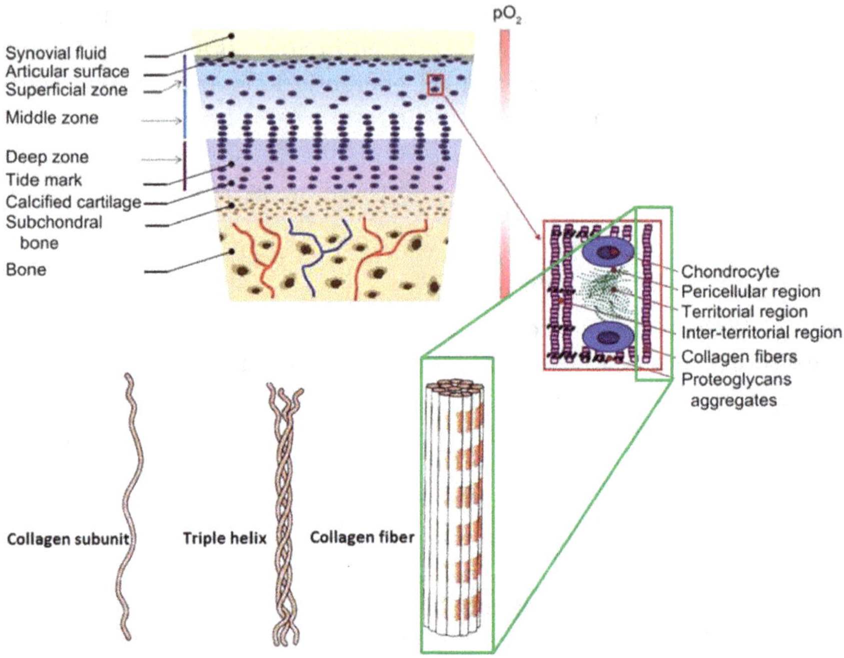

Fig. 8.2. Top: scheme of cartilage with the partial oxygen pressure indicated as a shaded red bar on the right-hand side; bottom: drawing of a collagen fibre made of many triple helices.
Source: Adapted from Saintigny *et al.*[28]

mainly go through the extracellular matrix and directly impact collagen (see Fig. 8.2), with indirect effects due to water radiolysis being of minor importance. A particularly interesting and important question is whether collagen irradiation is able to induce fragmentation of the protein backbone and create small peptides that might act as messengers for cells (for cartilage, it is clear from Fig. 8.2 that chondrocytes are directly surrounded by collagen). These messengers could then induce cell stress or even senescence and, in the case of cartilage, ultimately contribute to arthrosis, normally a degenerative process and part of the ageing process. We will return to this point later in this chapter. Another highly relevant context is radiation protection of astronauts, who are exposed to ionising radiation originating from galactic cosmic rays and solar events, both composed of high-energy protons.

DNA mutations can change the primary structure of proteins and trigger diseases. A good example is *osteogenesis imperfecta*, which is caused by the substitution of glycine in collagen.[29] In order to systematically investigate the role of mutations on the higher-order structure and stability of collagen, straightforward techniques such as X-ray diffraction and thermal melting require crystallisation, which is extremely difficult for collagen. For this reason, much shorter collagen mimetic peptides (CMPs) have been widely used, with the advantage of allowing on-demand synthesis of controlled sequences. Popular CMP are $(PPG)_{10}$ and $(POG)_{10}$ but also $(GPO)_9$ (P, O, and G being proline, hydroxyproline, and glycine, respectively), which form stable triple helices in crystals (see Fig. 8.3) and in aqueous solutions. Hydroxylation increasing both the melting temperature and the enthalpy of denaturation of the triple helix.[30,31] As stated by Kotch *et al.*,[32] 'the means by which hydroxyproline stabilizes collagen has engendered dispute.' Thanks to CMP, investigations in the condensed phase brought arguments supporting intrinsic stereoelectronic against water-mediated effects.[33,34] Stereoelectronic effects arise from the spatial interaction between molecular orbitals, usually in a donor/acceptor fashion, involving the hydroxyl group of hydroxyproline in the case of CMP. However, only gas-phase studies allow for completely ruling out the influence of the molecular environment: such experiments have been performed with protonated $(PPG)_{10}$ and $(POG)_{10}$ by means of ion mobility spectrometry (IMS) coupled to mass spectrometry.[35] IMS separates molecular ions of different geometrical structures thanks to soft collisions with an inert gas (often helium). In these gas-phase experiments, in the absence of water, IMS measurements demonstrate a conservation of the triple helix for trimers of $(PPG)_{10}$ and $(POG)_{10}$ with seven protons, as well as a conservation of two parallel triple helices (see Fig. 8.3) for the hexamers of $(PPG)_{10}$ with nine protons. When the number of protons is low, the condensed-phase structure is lost since the molecular system folds, which can be explained by the decreased Coulomb repulsion between protons. Furthermore, the dissociation of the triple helix upon gentle heating resulting from soft collisions with helium requires a higher collision energy for the $(POG)_{10}$ trimer than for the $(PPG)_{10}$ trimer, showing that the stabilisation of the triple helix by hydroxyprolines does not require water molecules.

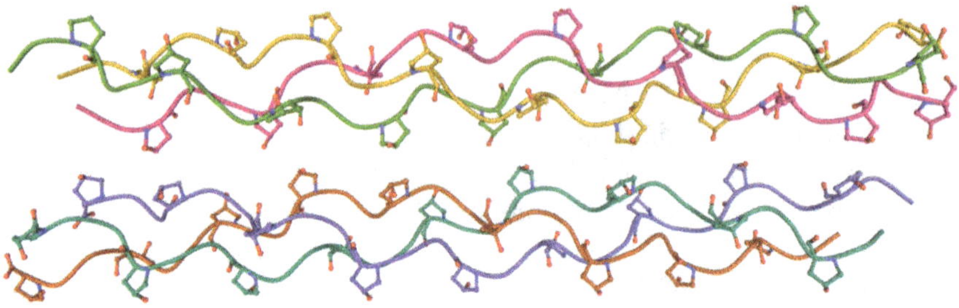

Fig. 8.3. Crystal structure of the (GPO)$_9$ CMP hexamer (PDB 3B0S). Peptide backbones are depicted as ribbons, and only carbon, oxygen (red dots), and nitrogen atoms (blue dots) of hydroxyproline side-chains are shown. Water molecules are not displayed.

In the following, the direct effects of ionising radiation on peptides and proteins are described at the molecular level based on experiments where molecular systems are brought to the gas phase. Emphasis will be placed on collagen, following a bottom-up approach from isolated peptides to assemblies of triple helices, the latter serving as model systems of collagen fibrils.

8.3.2 *Isolated peptides and proteins*

8.3.2.1 *Neutral molecular systems*

Much work has been dedicated to the experimental but also theoretical investigation of the effects of both ionising particles, such as electrons and photons, and atomic ions interacting with isolated peptides or proteins. These thermally fragile and reactive molecules can be brought into the gas phase by several techniques, with electrospray ionisation (ESI) and matrix-assisted laser desorption ionisation (MALDI) being the most common. Both produce protonated or deprotonated species, but MALDI can also form isolated neutral peptides and even proteins, e.g., when coupled to supersonic jet cooling. After interaction with ionising particles, analysis of the radiation-induced molecular ionic products is often performed using mass spectrometry, for instance, by means of time-of-flight spectrometry, Fourier-transform ion cyclotron resonance (FT-ICR) cells, or RF multipoles. With ionising particles, one expects ionisation of the target molecular system. The most straightforward way of experimentally demonstrating that this process occurs in the gas phase is to irradiate neutral molecules and to detect positive ions resulting from the removal of one or several electrons. The Schlag group applied this method to several peptides and proteins.[36–38] For instance, the group demonstrated that angiotensin-I could be brought intact to the gas phase by IR laser-induced desorption coupled to supersonic jet cooling and singly ionised by UV laser multiphoton absorption.[39] At low laser fluence, only the intact molecular ion is observed, whereas fragments coming from cleavage of the peptide backbone covalent bonds appear when laser fluence increases. The latter phenomenon is explained by a higher number of photons absorbed by angiotensin-I, thus inducing not only the removal of one electron but also electronic excitation: in most cases, this excitation energy is then converted into vibrational excitation energy after internal conversion to the electronic ground state, with molecular fragmentation as a final result (as detailed in the following).

Technically, a simple way of producing ionising particles in the gas phase is by heating a filament to thermally emit electrons with near-zero kinetic energy and accelerating these electrons to the required energy by means of electric fields. As the ionisation energy (IE) of most molecular systems under study is around 10 eV, a matching electron kinetic energy is then sufficient for ionisation. However, as ionisation cross-sections are usually low close to the ionisation threshold, often a significantly higher electron kinetic energy of 70 eV is chosen, which leads to efficient ionisation and delivers mass spectra where a high abundance of fragments is observed. This is because stronger bound electrons (up to inner valence electrons) can be ejected, and the resulting molecular cation can reach an excited state.[40,41]

For neutral molecules containing UV chromophores, ionisation is possible via laser-induced UV multiphoton absorption and ionisation. Regarding peptides, depending on the sequence and the laser fluence (and thus the number of absorbed photons), the probability of intramolecular hole transfer between amino acids leading to cleavage of the peptide backbone can be controlled.[37,38,42,43] Useful information about the primary sequence of the peptide can even be obtained, as illustrated by fragmentation of angiotensin-I[39] Using femtosecond instead of nanosecond UV laser pulses, much more efficient ionisation of peptides can be achieved, and more localised fragmentation close to the chromophore is observed because multiphoton absorption occurs before internal conversion and before the excitation energy dissipates over the system.[44]

When photon energy rises above the ionisation energy of molecules, single-photon absorption can lead to ejection of one electron; thus, neutral peptides without UV chromophore can still be ionised with a laser beam, for instance, following the irradiation by VUV photons of 10.5 eV energy (these photons can be produced by frequency tripling of a 355 nm laser beam focused in a cell filled with rare gases).[45] To obtain more information, one would like to scan the photon energy continuously, for instance, using VUV synchrotron light, but the photon flux is several orders of magnitude lower than that of a laser. To circumvent this, the molecular density can be increased by using a thermally heated oven. However, this is not a viable option for peptides or proteins, as they tend to thermally decompose or/and react in the oven. Another solution is to irradiate the molecules for a much longer time; this is possible by using RF traps such as 3D Paul traps or linear multipoles where molecular ions can be stored for up to several seconds. Much progress has been achieved by following this strategy.

8.3.2.2 *Ionic molecular systems: Ionisation processes*

In biological tissues, peptides and proteins are most of the time charged, as they very often contain residues with acidic or basic sidechains. It is thus relevant to study protonated or deprotonated peptides and proteins. Furthermore, the additional charge allows for the transportation, manipulation, and selection of mass and geometrical structure by means of electric and magnetic fields. For the particular cases discussed here, it is possible to select the molecule of interest in a single well-defined charge state from a broad distribution of protonation/deprotonation states, delivered by ESI. Subsequently, the ions can be injected into an RF ion trap (see the experimental section) or an FT-ICR cell, where they can be stored sufficiently long to allow exposure to ionising radiation for periods of several milliseconds to few seconds. This method has early been applied to protonated peptides $[P+nH]^{n+}$ (with n being the number of protons) by R. Zubarev's group, who reported mass spectra resulting from electron-impact ionisation and high-resolution analysis in an FT-ICR spectrometer.[46,47]

Peaks at m/z corresponding to $[P+nH]^{(n+1)+}$ were unambiguously assigned to non-dissociative single ionisation of the precursor ion, allowing the determination of a cross-section in the 10^{-15} cm^{-2} range for single ionisation at 20 eV electron kinetic energy. By scanning this kinetic energy, the authors also obtained threshold ionisation energies of around 11 eV for several singly protonated peptides, indicating negligible effects of amino acid sequence on peptide ionisation energies.

An interesting question is then, when considering a particular peptide charge state, whether this ionisation energy obtained by electron impact the lowest energy required for ionising the system. To answer this question, Giuliani and co-workers studied single-photon absorption in the vacuum ultraviolet (VUV) range using a synchrotron beamline coupled to a linear quadrupole mass spectrometer. Protonated peptides were irradiated, and product cations were analysed. The results show that for singly protonated substance P, within the experimental uncertainty, the vertical ionisation energy of 10.5 eV determined for photoabsorption matches the value obtained by electron impact.[48] The Zubarev and Giuliani groups also showed that ionisation energy increases linearly with the protonation state of the peptide (about 1.1 eV/proton) for both electron-impact and single-photon ionisation. The linear increase was attributed to Coulomb attraction between the ejected electron and the resulting $[P+nH]^{(n+1)+}$ ion.[49,50] Interestingly, different results were obtained for larger proteins such as cytochrome C or ubiquitin upon photoionisation. This was explained by the fact that these proteins change their geometrical structure as the protonation state increases, more precisely by unfolding due to Coulomb repulsion between the positive charges.[50] The authors conclude that instead of the net charge of the system, it is the charge density that governs the ionisation energy of peptides and proteins. In addition, since this quantity is governed by Coulomb interaction between the ejected electron and the remaining isolated species, a straightforward direction of research is to study molecular anions as precursors, as they should require much lower electron detachment energies. To detect the anionic products of this process in mass spectrometers, one has to choose a multiply deprotonated precursor since electron detachment will reduce the charge state of the anion, and to be experimentally observable, the product needs to remain negatively charged. Early studies using lasers demonstrated that one electron could be detached from multiply deprotonated peptides by single-photon absorption in the UV range, with photon energies as low as 4 eV, confirming predictions from the Coulomb picture.[51,52] At these photon energies, electron detachment was explained by electronic excitation via strong π–π^* transitions in aromatic amino acid side-chains followed by internal conversion to an autoionising state. The electronic nature of electron detachment was later confirmed by the observation of efficient electron photodetachement from larger multiply deprotonated proteins.[53] In the VUV range, single-photon absorption can induced multiple-electron detachment. In synchrotron studies, a threshold value of 11.4 eV was measured for double-electron detachment from insulin^{5-} anions.[54] Multiple-electron detachment from proteins has also been observed for single-electron impact at kinetic energies above 40 eV[55] and for multiple-photon absorption using femtosecond pulses in the VUV range.[56] Furthermore, the capture of several electrons from multiply protonated or deprotonated proteins has been observed in collisions with keV multiply charged ions.[57,58] Capture of up to six electrons was observed for keV Xe^{25+} ions interacting with multiply deprotonated cytochrome c.[59]

In all the cases discussed above, valence electrons were the subject of ionisation or detachment. It is a logical next step to study the ejection of much stronger-bound core electrons. Depending on photon energy, core electrons can either be excited to an unoccupied molecular orbital or ejected into the continuum. For low atomic number atoms, the resulting core-excited sites decay predominantly non-radiatively by means of Auger decay, in which an electron of high kinetic energy is ejected (see Section 8.5). As a consequence, core-level *excitation* leads to single ionisation and core-level *ionisation* leads to double ionisation. Before experiments on ions were conducted, the group led by K. Prince performed soft X-ray photoelectron spectroscopy close to the atomic K-edges on neutral gas-phase amino acids.[60–62]

Near-edge X-ray absorption spectra of protonated proteins were later performed by monitoring the yield of non-dissociative single ionisation following single-photon absorption using a mass spectrometer coupled to a synchrotron beamline. Spectral features were found to be very similar to those obtained in the condensed phase.[63] Ionisation threshold energies of 1s orbitals were also determined by studying the yield of non-dissociative double ionisation versus photon energy. In addition, the protonation state of multiply protonated melittin was found to influence the photon-energy dependence of non-dissociative single versus double ionisation upon X-ray absorption and electronic excitation. This was interpreted as an indirect proof of electron-impact ionisation of a protein upon impact of an Auger electron emitted from the same protein. The probability of this process strongly depends on the conformation on the protein, which in turn depends on the protonation state: higher protonation states lead to Coulomb-repulsion-induced protein unfolding.[64]

All these results show that when targeting electrons, different types of ionising radiation (photons, electrons, and ions) and energy can lead to single- and also multiple-electron emission from peptides and proteins, akin to the case of oligonucleotides (see Section 8.5 of this chapter). This is highly relevant for radiobiology because these electrons can lead to secondary damage, for instance, when dissociatively attaching to other biomolecules located in the close vicinity of the ionised molecule.[65]

8.3.2.3 *Ionic molecular systems: Fragmentation processes*

We have mentioned above that ionisation of peptides by electron or laser beams can lead to fragmentation due to covalent bond cleavage. In particular, for larger molecular ions, fragmentation requires substantial excitation energy, and experiments are often performed at large-scale facilities, such as accelerators or synchrotrons, using both ion or photon beams. Studies on protonated peptides have been pioneered by the Schlathölter group, who reported the irradiation of protonated leucine-enkephalin by ion beams (proton and helium) with kinetic energy in the keV range.[23,24] They recorded the mass spectrum of product cations as a function of the ion kinetic energy, element and charge state (for helium), and observed extensive fragmentation of the peptide, producing mainly small ions from amino acid side-chains (immonium ions) but also, in lower abundance, larger fragments following backbone peptide bond cleavage. This is very different from the results of collision-induced dissociation (CID) at low kinetic energies (on the order of 10 eV), where sequential soft collisions with an inert gas (often helium or nitrogen) induce vibrational excitation of the molecular ion which leads to gentle heating. Therefore, in CID, only the lowest-energy statistical fragmentation pathways in the electronic ground state are probed. For protonated peptides, these pathways include peptide bond cleavage or loss of neutral molecules such as H_2O.[66]

In the case of keV ion projectile beams, the dominating processes underwent by molecular targets in the gas phase are electron capture (mainly for highly charged ions) and electronic excitation (mainly for low ion charge states). In both cases, the resulting electronic excitation of the peptide ion can be followed by direct fragmentation if the excited state is repulsive. The fingerprint of repulsive states are often very specific fragment ions. Moreover, even when internal conversion brings the system back to the electronic ground state, the amount of vibrational energy in the molecular system can be very high, leading to the fast rupture of multiple bonds. In addition, the ionised and excited biomolecular ion can be subject to charge, atoms, and/or energy transfer, which can strongly influence fragmentation.[67]

In order to get more insight on the fragmentation processes occurring in the same system (protonated leucine-enkephalin) after interaction with ionising radiation, the Schlathölter group studied the influence of the type of radiation and extended their studies to VUV photoionisation in the 8–40 eV energy range.[68] The highly resolved and continuously tunable synchrotron beam allowed to precisely control and scan the photon energy over this range, which encompasses the ionisation energy of the protonated peptide (see Section 8.3.2.2). When recording mass spectra of the single-photon absorption product ions, a progressive change in fragmentation pattern with photon energy was observed and was explained by the transition between photoexcitation and photoionisation. In the 20–40 eV range, well above the ionisation energy, the mass spectra were similar to those obtained by ion impact, suggesting that ionisation dominates in both cases but is followed by extensive fragmentation. Upon exposing the same protonated peptide to UV femtosecond laser pulses, again very similar mass spectra were observed,[69] indicating that for all radiation modes under study, electronic excitation is quickly transformed into vibration excitation via internal conversion. As a consequence, the fragmentation pattern exhibits little memory of the radiation mode.

Increasing photon energy up to the soft X-ray range did not produce fundamentally new fragment ions. However, partial ion yields were found to have different spectroscopic signatures around K-edges. For instance, site-specific photoabsorption in aromatic side-chains caused little fragmentation, whereas excitation along the peptide backbone caused strong fragmentation. This proved a strong influence of the excited electronic states reached after photoabsorption on the fragmentation process.[70]

The interaction of a number of other protonated peptides with various types of ionising radiation has been investigated as well. Qualitatively very similar fragmentation patterns were observed, e.g., for substance P[24,48,71] suggesting little or no influence of the amino acid sequence on fragmentation. Electron detachment and fragmentation induced by KeV ion collision have been further developed by the Jackson group into the novel mass spectrometric technique charge transfer dissociation (CTD).[72–74]

In sharp contrast to protonated peptides, fragmentation is weak or even absent for the case of protonated proteins. There are a number of parameters which may contribute to this difference: due to the larger number of atoms, proteins have more vibrational degrees of freedom and generally a higher density of electronic states. Schlathölter and co-workers therefore performed systematic studies on protonated peptides and proteins as a function of size, i.e., the number of vibrational degrees of freedom, under single-photon absorption conditions both in the VUV and X-ray ranges. In the X-ray range, the yield of non-dissociative ionisation was found to decrease linearly as a function of the amount of internal energy per degree of freedom (DOF) of the protein (see Fig. 8.4), indicating that statistical fragmentation after redistribution of the excitation energy over all vibrational DOF dominates.[75] The relationship between fragmentation and the number of DOF is less obvious in the VUV range,[76] presumably because geometrical structure and, in particular, the location of protons are more relevant. In the low internal energy regime, proton transfer is crucial for fragmentation.[66] For both VUV and soft X-ray photon energies, however, fragmentation yields generally decrease with the system size and vanish at around 5 meV per DOF. The heat capacity of protonated peptides and proteins is therefore a highly important parameter for fragmentation following ionisation.

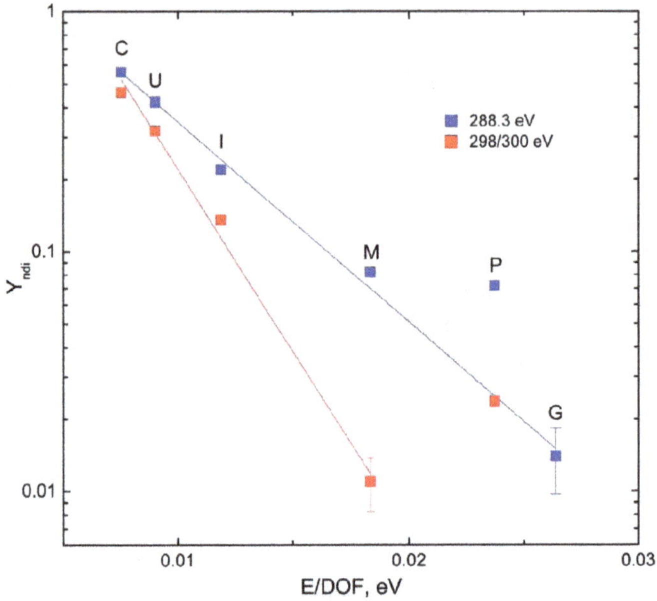

Fig. 8.4. Yield of non-dissociative ionisation (NDI) after absorption of one photon plotted as a function of the deposited energy E per degree of freedom (DOF) for protonated peptides and proteins ranging from 1.88 to 12.3 kDa in mass (gramicidin: G, substance P: P, melittin: M, insulin: I, ubiquitin: U and cytochrome c: C), assuming E = 18.5 eV. The photon energies (288.3 eV and 298/300 eV) are indicated for the two sets of data.
Source: Reproduced from Ref. 75 with permission from the PCCP Owner Societies.

In this picture, non-covalent interactions such as H bonds, present in most peptides in the gas phase, are expected to be more sensitive to an increase in vibrational energy (or temperature) than covalent bonds. However, in the case of gas-phase single molecules, the dissociation of such non-covalent bonds only changes the conformation and not the mass of the molecule, which is why mass spectrometry is insufficient for the characterisation of the process. In the following sections, we see how biologically relevant noncovalently bound assemblies of several peptides respond to ionising radiation in the gas phase.

8.3.2.4 *Triple helices of collagen mimetic peptides*

As discussed in the introduction of this section (see Section 8.3.1) Poully and co-workers have used ESI, mass spectrometry, and IMS to show that gas-phase complexes of CMPs ($(PPG)_{10}$ and $(POG)_{10}$) bound by noncovalent interactions (mainly H bonds between the backbones) adopt a triple-helical geometrical structure as in the condensed phase if the number of peptides is three and the protonation state of the system is seven (for instance, $[((PPG)_{10})_3{+}7H]^{7+}$).[35] This triple helix is a typical feature of collagen in biological tissues, which is why these $(PPG)_{10}$- and $(POG)_{10}$-based complexes are ideal model systems regarding radiation biology.

In VUV and soft X-ray synchrotron radiation studies on gas-phase $[((PPG)_{10})_3{+}7H]^{7+}$, it was shown that single-photon absorption leads to three main processes: (i) non-dissociative ionisation (NDI), (ii) disruption of the helix after loss of noncovalent interactions and formation of three isolated peptides, and (iii) subsequent fragmentation of these isolated peptides via backbone bond cleavage (see Fig. 8.5).[77] In the photon energy range between 14 and 22 eV, NDI (i) relatively increases with photon energy, whereas dissociation

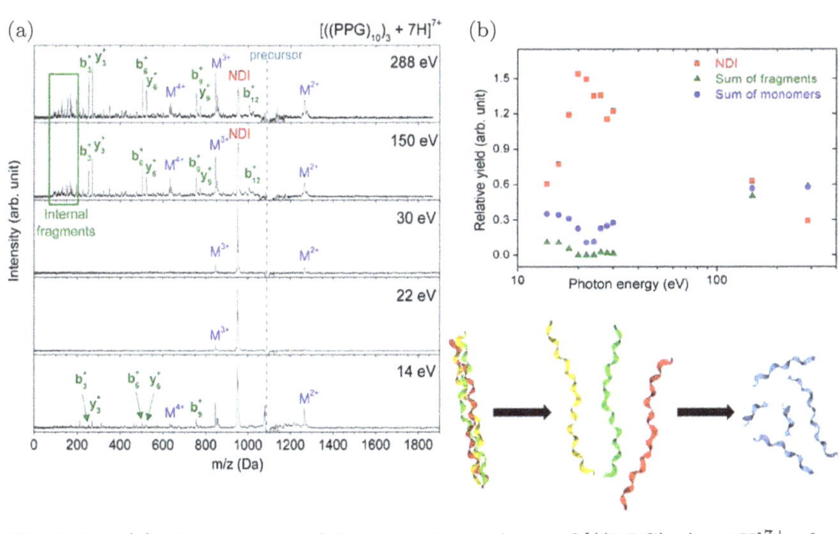

Fig. 8.5. (a) Mass spectra of the cationic products of $[((PPG)_{10})_3{+}7H]^{7+}$ after single-photon absorption. The usual nomenclature is used to name peptide backbone fragment ions (labelled in green), peptide monomers with n positive charges (M^{n+}) are labelled blue, and NDI stands for non-dissociative ionisation of the precursor ion, whose peak has negative intensity. (b) Evolution of the yields of NDI, monomer formation, and peptide backbone fragmentation as a function of photon energy.
Source: Reproduced from Ref. 77 with permission from the PCCP Owner Societies.

of noncovalent bonds (ii) and covalent bonds (iii) surprisingly drops down to almost zero. As the ($[((PPG)_{10})_3{+}7H]^{7+}$) ionisation energy is expected to be around 13 eV,[50] these results reveal a progressive transition from photoexcitation to photoionisation of the triple helix. From 22 to 30 eV, NDI decreases while dissociation of noncovalent interactions reappears, but fragmentation of the peptide backbone is absent. Since the latter process is observed at 150 eV and dominates at 288 eV, these results were explained by the fact that after photoionisation, the system is left with an amount of internal energy, which progressively increases with photon energy. Moreover, the fragment ions formed after peptide backbone bond cleavage are the same as those formed in the case of photoionisation of isolated protonated $(PPG)_{10}$ and $(POG)_{10}$ peptides and were attributed to the cleavage of the same peptide bonds as in CID, where statistical fragmentation dominates. Interestingly, in the case of the hydroxylated triple helix $[((POG)_{10})_3{+}7H]^{7+}$, no backbone bond cleavage is observed in the 14–30 eV range, consistent with the higher binding energy of the triple helix due to stereoelectronic effects (see the introduction of this section).[77]

When protonated $(PPG)_{10}$ and $(POG)_{10}$ CMPs and triple helices are irradiated by carbon ions at the Bragg peak energy, the same processes as those triggered by soft X-ray photoabsorption occur since the obtained mass spectra contain the same peaks.[78] Interestingly, the data suggest that the amount of excitation energy increases with the size of the system, presumably because of the higher number of electrons and thus a higher probability of multiple interactions of one carbon ion with the system.

From these data, it follows that the main mechanism in the relaxation of a single-photon ionised gas-phase protonated CMP triple helix is internal conversion to the hot ground state of the system followed by inter- and intra-molecular vibrational energy redistribution. More generally, in light of previous investigations discussed earlier in this section, this mechanism also plays an important role in protonated peptides and proteins.

8.3.2.5 *Assemblies of triple helices*

As a first step towards studying the response of collagen fibrils upon ionising radiation, hexamers of collagen-mimetic peptides with nine protons ($[((PPG)_{10})_6+9H]^{9+}$ and $[((POG)_{10})_6+9H]^{9+}$) have been irradiated by X-rays in the gas phase.[79] The native structure of these systems, a parallel dimer of two triple helices, is conserved in the gas phase (see the introduction of this section and Fig. 8.3). Absorption of one 100 eV photon by $[((PPG)_{10})_6+9H]^{9+}$ induces NDI and/or separation of the two triple helices with a small probability, with the most abundant process being the dissociation of the helices into isolated peptides. These peptides also undergo minor fragmentation processes with the same pathways as observed in regular triple helices and in isolated peptides of $(PPG)_{10}$ but with a lower relative abundance that, however, increases with photon energy up to 288 eV. In $[((POG)_{10})_6+9H]^{9+}$), this relative abundance is lower at any photon energy under study, which is consistent with the stabilisation of the two triple helices in the hexamer due to stereoelectronic effects (see Section 8.3.2.4). Therefore, these results are in line with the main mechanism proposed to account for the behaviour of protonated peptides and proteins upon ionising radiation (see Section 8.3.2.4).

However, this mechanism cannot explain the notable differences between $[((POG)_{10})_6+9H]^{9+}$) and $[((PPG)_{10})_6+9H]^{9+}$: in the former, NDI is much more abundant and non-dissociative double ionisation is observed. To understand this, it is important to realise that in the studied hexamers, the two triple helices are bound by a dipole–dipole interaction that does not involve hydroxyproline side-chains in the first place. However, in $[((POG)_{10})_6+9H]^{9+}$), the side-chains contain hydroxyl groups pointing out of the helix. From the structure of the hydroxylated hexamer shown in Fig. 8.3, one can see pairs of hydroxyl groups facing each other and being close (oxygen atoms only 5 Å apart). After H loss induced by VUV or soft X-ray, these two hydroxyl groups can turn into radicals. Subsequently, the radical sites may recombine and create a covalent bond. The resulting cross-link would strongly stabilise the hexamer with respect to radiation-induced vibrational excitation. Quantum-chemical calculations on model systems using density-functional methods support this process. The same mechanism is widely accepted to explain cross-linking between tyrosine side-chains in proteins in the condensed phase following UV or gamma photon irradiation,[80] and collagen cross-linking has recently been employed to produce a biomaterial for cell encapsulation.[81]

Here, H loss from each of the two hydroxyl groups has to be triggered by photoionisation and can proceed via different processes. First, direct ionisation of hydroxyl can be followed by proton transfer to another part of the system, a probable process since ionisation-induced proton transfer is a general process in biomolecular systems.[67] Second, ionisation at another site can also occur and lead to energy transfer and O–H bond cleavage. More work is needed to establish the precise mechanism responsible for the cross-linking of the two triple helices in the gas phase, but in any case these results constitute the first evidence of ionising radiation-induced cross-linking between biomolecules in the absence of a molecular environment.

8.3.3 *Other collagen-related peptides*

Processes induced by UV and soft X-ray in two collagen-related peptides have been studied by Poully *et al.*: PGGPPGPKGNSGEPGAPGSKGDTGAK (in short, PK26) is a 26 amino-acid sequence found in the triple-helical region of human collagen and PGGPOGPKGNSGEOGAOGSKGDTGAK (in short, PK26-Hyp) is an identical sequence but with the three prolines in the Yyy position of the XxxYyyGly repeats being substituted by hydroxyprolines. Upon single VUV or X-ray single-photon absorption, it is expected that these peptides undergo the same processes as the CMPs mentioned earlier in this section, namely excitation, ionisation, energy redistribution, and fragmentation. This is confirmed by the observation of peaks corresponding to non-dissociative ionisation and backbone fragmentation in the respective mass spectra in the 14–531 eV range.[82]

Since the fragmentation yield increases with photon energy, the same mechanism as invoked in Section 8.3.2.4 can be assumed. Interestingly, in the VUV range, additional peaks, absent from the mass spectra of $(PPG)_{10}$-containing systems, are observed at m/z slightly below the precursor or the NDI peak. These peaks are due to neutral losses from the precursor or the singly ionised intact peptide, respectively. What makes protonated PK26 and PK26-Hyp peptides different from $(PPG)_{10}$ is the presence of residues other than glycine and proline. The neutral groups were shown to originate from the side-chains of these amino acids. Importantly, loss of H_2O is attributed to the hydroxyproline side-chain, which is confirmed by the observation of this process in $(POG)_{10}$-containing systems. A radical-mediated mechanism explains these neutral losses, which, in addition to the cross-linking mechanism evoked in the previous section, presents further arguments supporting the role of radicals in the reactivity of peptide-containing systems upon ionising radiation.

In the experiments discussed above, backbone fragmentation is not random in the studied peptides: it appears that the Gly–Pro peptide bond is highly prone to cleavage, especially in CMPs. This specific cleavage site has also been found after UV irradiation of the same CMPs in solution.[83] In the gas phase, electronic excitation after UV irradiation of protonated or deprotonated PK26 and PK26-Hyp peptides also leads to preferential Gly–Pro peptide bond cleavage, which indicates that radiation-induced Gly–Pro bond cleavage is an intrinsic property of collagen-related peptides.[84] Since collagen is the main constituent of the extracellular matrix of connective tissues, such cleavage could be abundant in irradiated tissue, and the resulting small peptidic collagen fragments might diffuse to surrounding cells and trigger a biological response. Interestingly, such responses have been observed for chondrocytes (cartilage cells) when exposed to peptides with sequences selected from human collagen, e.g., fragments formed after Gly–Pro peptide bond cleavage.[84]

8.4 Antibiotics and Noncovalent Complexes Involved in Molecular Recognition

8.4.1 *Rifamycin*

Rifamycin is the name of a family of antibiotics that are naturally produced by *Amycolatopsis mediterranei* and have a broad spectrum of activity against Gram-positive and -negative bacteria. Their mechanism of action involves highly selective inhibition of the bacterial DNA-dependent RNA polymerase via noncovalent binding.[85] Rifamycin SV has a cyclic structure, as shown in Fig. 8.6: it contains an aromatic naphthofuran group with a sodium cation bound to an oxygen anion. CID experiments demonstrated that this ionic bond survives when protonated rifamycin SV is brought into the gas phase by ESI, mostly because the naphthofuran fragment contains sodium.[86]

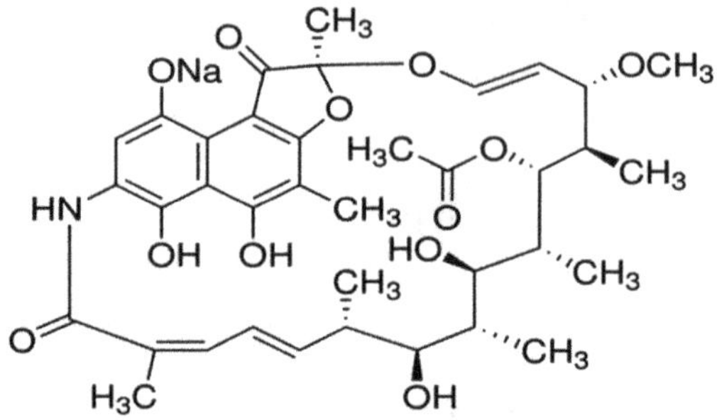

Fig. 8.6. Chemical structure of sodiated rifamycin SV.

This molecular salt has been irradiated by X-rays, and the cationic products were analysed through mass spectrometry. It reveals ionisation and extensive fragmentation after single photon absorption. Moreover, a peak absent from CID spectra has been assigned to the fragment complementary to naphthofuran but containing sodium. This fragment is due to photoabsorption-induced sodium transfer. The same fragment is also observed in the mass spectra of the protonated rifamycin noncovalent dimer, along with another peak attributed to intermolecular proton transfer. We have already seen that the latter process is quite common in photoionisation of gas-phase biomolecular systems, but sodium transfer had been never reported earlier.[86] The proposed mechanism involves the ejection of an electron from the negatively charged oxygen bound to sodium, inducing a decrease in binding energy and thus the release of sodium. Sodium ions are required in animals for several key functions, for which sodium often binds negatively charged glutamic and aspartic acid residues in proteins. Removal of electrons from these residues by means of ionising radiation might thus trigger sodium transfer, leading to perturbations in regular enzyme function.

Another interesting mass spectrometric observation in single-photon X-ray absorption in rifamycin monomers and dimers is the presence of groups of intense peaks starting from $m/z = 273$ and going down below 100 by steps of approximately 14. Rifamycin contains carbon, nitrogen, and oxygen as heavy atoms, meaning that these peaks are likely formed by sequential dissociation of the naphthofuran fragment into singly charged cations. Fragment peaks in this mass range are grouped, and within a given group, peaks are separated by $m/z = 1$, which implies singly charged cations and extensive

H scrambling. The intensity of these peaks rises with photon energy in the 100–300 eV range, indicating that considerable vibrational energy is needed to form them. We discuss in the following section another example of this feature.

8.4.2 *Vancomycin*

Vancomycin is a last-resort glycopeptide antibiotic against infections caused by Gram-positive bacteria, blocking the renewal of their cell wall by acting as a ligand for a precursor of the main cell-wall component. The seven-residue peptidic chain of vancomycin contains five tyrosine side-chains that are heavily modified by post-translational modifications involving the addition of OH groups, cross-linking, the addition of chlorine atoms to two of these side-chains, and the glycosylation of the central side-chain (see Fig. 8.7). The numerous and

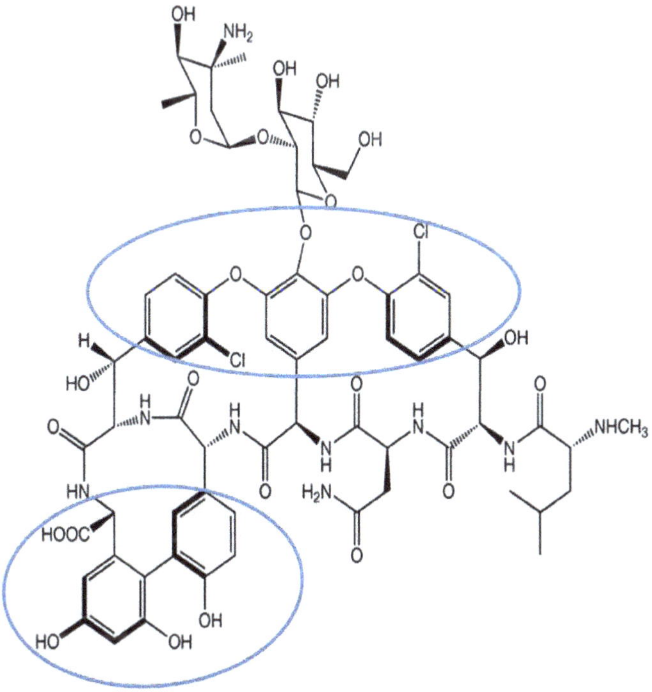

Fig. 8.7. Chemical structure of vancomycin. The groups forming the fragments constituting the signature of post-translational modifications are encircled.

dense cross-linkages in the resultant peptide create a relatively rigid binding pocket, allowing for stereospecific recognition of the receptor via noncovalent binding. The behaviour of doubly protonated vancomycin upon single VUV and X-ray photoabsorption has been probed using mass spectrometry, revealing extensive fragmentation in the whole photon energy range. Moreover, plotting the relative yield of the fragments coming from glycosidic bond cleavage after ionisation led to the conclusion that this process has a very low activation energy.[87,88] Charge state is a crucial parameter regarding the response of vancomycin upon UV or VUV photoabsorption since singly and doubly deprotonated vancomycin mainly undergo CO_2 loss, with or without electron detachment.[89]

In the low-m/z range of the mass spectra containing cationic products of X-ray photoabsorption of doubly protonated vancomycin, several groups of peaks are observed: they are similar to those observed for rifamycin (see the previous section), and their intensity also increases with photon energy. Thus, they most probably originate from sequential fragmentation with extensive H scrambling, and their formation requires a high amount of vibrational energy. Interestingly, all these peaks can be explained by only three pathways involving the modified tyrosine side-chains, and each of these side-chains can be identified individually thanks to the mass differences. Thus, these groups of peaks can be seen as a signature of the complex post-translational modifications of tyrosine residues of vancomycin. Moreover, this signature is robust with respect to photon energy and even molecular environment, which makes it appealing for applications in analytical chemistry.[87]

8.4.3 *Vancomycin–$Ac_2{}^{L}K^{D}A^{D}A$ noncovalent complexes*

Molecular recognition of a receptor by a drug is needed for medical efficiency and specificity, with the latter being crucial for avoiding side-effects. Drug-receptor-specific binding is governed by noncovalent links, such as H bonds or charge–dipole interactions. Despite their relevance, little is known about radiation effects on noncovalent complexes involved in molecular recognition: gamma rays have been shown to damage a subtype of the estrogen receptor protein and destabilise its complex with a DNA sequence,[90] and bond cleavages within a protein–DNA complex have been observed by means of X-ray crystallography.[91] However, such condensed-phase investigations often suffer from a lack of knowledge about the stoichiometry of noncovalent complexes, which is not the case when using gas-phase mass spectrometry techniques, which allow studying a model system for specific molecular recognition, i.e., the vancomycin–$Ac_2{}^LK^DA^DA$ noncovalent complex (denoted 'V+R' in the following). The native molecular interactions in the V+R complex have been unravelled by structural investigation using X-ray diffraction[92] (see Fig. 8.8), with emphasis on the H bonds involving the CO_2^- group of the receptor. IR spectroscopy was employed to show that this biologically active structure is conserved for deprotonated complexes in the gas phase[93] but not for protonated species.[94] Indeed, in the doubly protonated complex, $Ac_2{}^LK^DA^DA$ is neutral, and the absence of negative charge strongly disturbs the intermolecular H-bonding network, causing it to leave the binding pocket of vancomycin and being linked to the disaccharide moiety in a non-specific way.

VUV and soft X-ray single-photon absorption was studied using mass spectrometry for the doubly protonated $[V+R+2H]^{2+}$ complex over a wide photon energy range of 14–32 eV and 100–531.5 eV.[87,88] No intact ionised complex was observed, even at the lowest photon energy, which is consistent with a very low binding energy and/or barrier for dissociation after ionisation. The latter process mainly leads to loss of noncovalent interactions and the formation of the complementary cations of vancomycin and $Ac_2{}^LK^DA^DA$, which also undergo subsequent fragmentation via the same pathways as the isolated protonated molecules (glycosidic bond cleavage for vancomycin and backbone bond cleavage for $Ac_2{}^LK^DA^DA$; see Fig. 8.8), most probably in the hot ground electronic state of the ionised system. Interestingly, the glycosidic bond of vancomycin can be broken without dissociating the noncovalent bond between $Ac_2{}^LK^DA^DA$ and the sugar group, forming the $[B_0+R]^+$ fragment (see Fig. 8.8). The evolution of the yields of $[B_0+R]^+$ and its complementary fragment $[Y_0]^+$ (see Fig. 8.9) shows that their common threshold must be around 12 eV, very close to the probable ionisation energy of the precursor ion, given the values measured for protonated peptides (see Section 8.3.2.2). Thus, it confirms the very low potential energy barrier for the cleavage of the glycosidic bond of vancomycin upon ionisation. Moreover, ionisation also triggers proton transfer from vancomycin

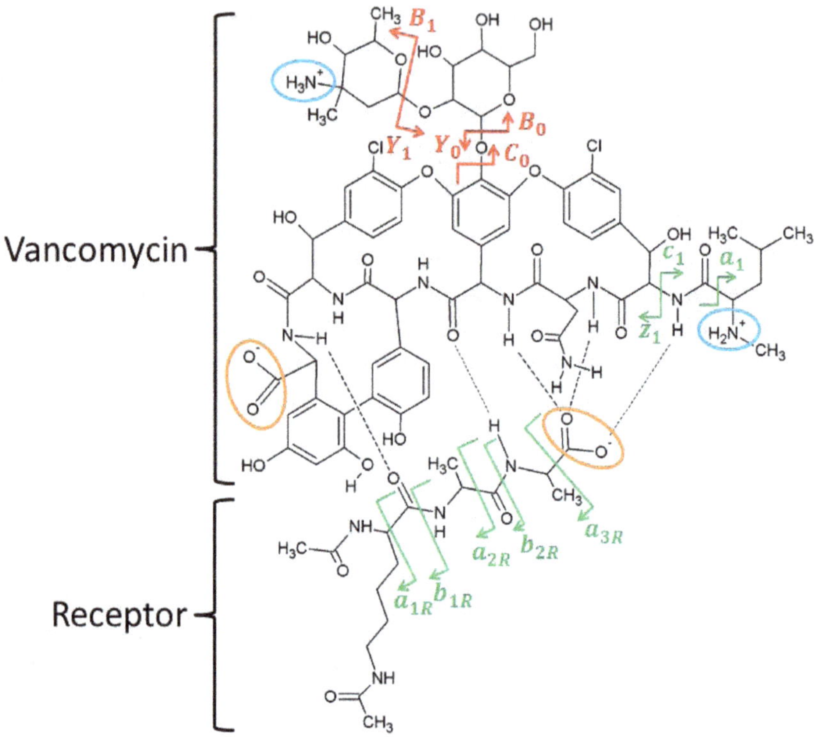

Fig. 8.8. Chemical structure of the V+R complex in the condensed phase, showing native intermolecular H bonds as dashed lines. The most basic and acidic sites are protonated and deprotonated and are encircled with light-blue and orange solid lines, respectively. The usual nomenclature for peptide backbone and oligosaccharide fragmentation is used for the main fragments observed after photoabsorption. Arrows indicate the side of the cleaved bond on which the charge is located.
Source: Reproduced from Ref. 88 with permission from Wiley on behalf of Chemistry Europe.

to $Ac_2{}^{L}K^{D}A^{D}A$ since a high amount of $[R+H]^+$ is detected. In the X-ray range, the latter process becomes less likely as photon energy and thus vibrational energy deposition rise, presumably due to faster dissociation of the complex, which would quench intermolecular proton transfer.

In the negative ESI mode, deprotonated V+R complexes featuring the native binding interactions are produced. Synchrotron radiation has been employed to probe the response of these complexes to

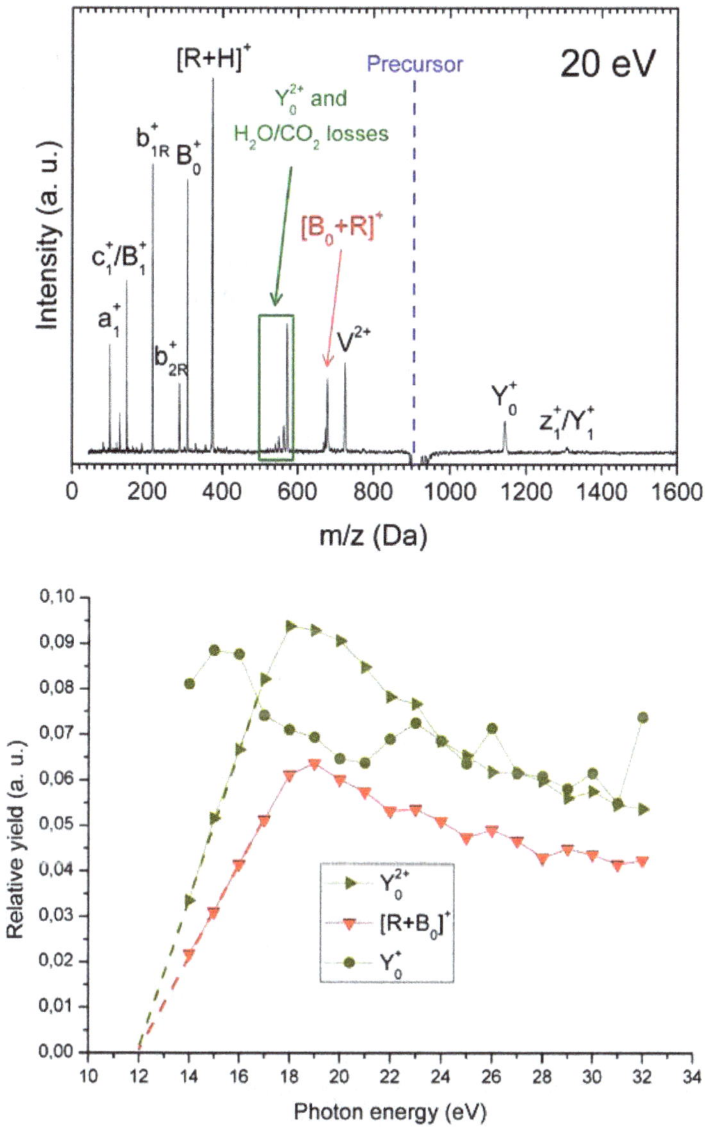

Fig. 8.9. Top: mass spectrum of $[V+R+2H]^{2+}$ after absorption of one 14 eV photon. The same nomenclature as in Fig. 8.8 has been used for the fragments. Bottom: relative yield of the main species formed after single photoabsorption of $[V+R+2H]^{2+}$ as a function of photon energy, normalised by the detector efficiency and the sum of all peak areas. The dashed lines are added to guide the eye. *Source*: Reproduced from Ref. 88 with permission from Wiley on behalf of Chemistry Europe.

single-photon absorption in the 4.5–30 eV photon energy range.[87,89] In sharp contrast with the protonated complex, the deprotonated complex retains noncovalent interactions to a significant extent after photoabsorption and covalent bond cleavage. Electron detachment from $[V+R-2H]^{2-}$ dominates in the VUV range and is often followed by CO_2 loss and/or dissociation of the complex. The precursor is composed of $[V-H]^-$ bound to $[R-H]^-$ (see Fig. 8.8); therefore, the electron can be detached from either molecules. If it is detached from $[R-H]^-$, the carboxylate group becomes neutral, and thus the binding energy of the complex drops, triggering facile dissociation. If it is detached from vancomycin, the deprotonated group of $[R-H]^-$ allows the native interactions to be conserved, and it most probably leads to not only the sole CO_2 loss from vancomycin but also – to a much lesser extent – the dissociation of the complex and the formation of $[R-H]^-$. At 16 eV photon energy, $[R-H]^-$ is released intact after electron detachment from $[V-H]^-$, whereas it mainly loses CO_2 at 24 eV, indicating that vibrational energy deposition increases with photon energy in this range. Another demonstration of the role of the charge state of vancomycin in the processes induced by photoabsorption is given by the study performed on [V+R-H]-, where vancomycin is neutral.[89] In this case, single-photon absorption in the 4.5–10 eV range mainly triggers HCl loss from vancomycin, presumably after a rearrangement. Since the precursor is singly charged, photoexcitation and not electron detachment is responsible for this process because neutrals are lost in the mass spectrometer. It indicates that the photoexcited states involve the modified tyrosine side-chains of vancomycin, to which both Cl atoms are bound and which strongly absorb UV light. An interesting process also occurs in highly photoexcited $[V+R-H]^-$: proton transfer from neutral vancomycin to $[R-H]^-$, followed by dissociation of the complex together with CO_2 loss from vancomycin to release $[V-H-CO_2]^-$. The lowest-energy pathway is probably proton transfer to the carboxylate group of $[R-H]^-$ because it neutralises this group. This demonstrates that proton transfer can be induced from an electronically excited large neutral molecule by VUV photoabsorption in the absence of ionisation.

8.5 DNA

8.5.1 *Ion collisions with gas-phase oligonucleotide ions*

The molecular mechanisms behind hadrontherapy are still not fully understood. Especially in the Bragg peak range, proton and heavy-ion interactions with DNA become particularly complex. For instance, a single 1 MeV/u carbon ion interacting with a short DNA segment can trigger multiple electronic excitations, valence ionisations, and inner-shell ionisations. Following these initial events, the DNA undergoes processes such as Auger decay, charge and energy transfer, and internal conversion of electronic excitation into vibrational energy. These processes ultimately lead to DNA damage, including strand breaks and base losses. To gain deeper insights into these molecular processes, one strategy is the elimination of the DNA chemical environment in gas-phase experiments.

Numerous studies have focused, for instance, on keV ion collisions with neutral DNA building blocks, such as DNA bases,[18,19,95] nucleosides,[96] or deoxyribose,[95,97] revealing key phenomena, such as the structural fragility of deoxyribose compared to nucleobases. However, keV ion collisions with these relatively small systems mostly induce multifragmentation, which is in stark contrast to DNA damage in biological systems, where single- and double-strand breakage (i.e., scission of one or very few bonds in a longer DNA sequence) are considered most relevant. The first experiments on more complex systems focused on nucleobase clusters, in which fragmentation channels opened up but were closed in the isolated molecule.[98]

However, it is only in the past decade that electrospray ionisation and radiofrequency ion trapping have enabled the study of protonated and deprotonated oligonucleotides, i.e., actual strands of DNA. In 2013, the Schlathölter team published results from the first collision study of keV carbon ions with doubly protonated oligonucleotides $[dGCAT+2H]^{2+}$. Figure 8.10 shows a typical mass spectrum of the product ions. The precursor ion is shown as an inset. The spectrum is dominated by peaks in the low mass region, with the

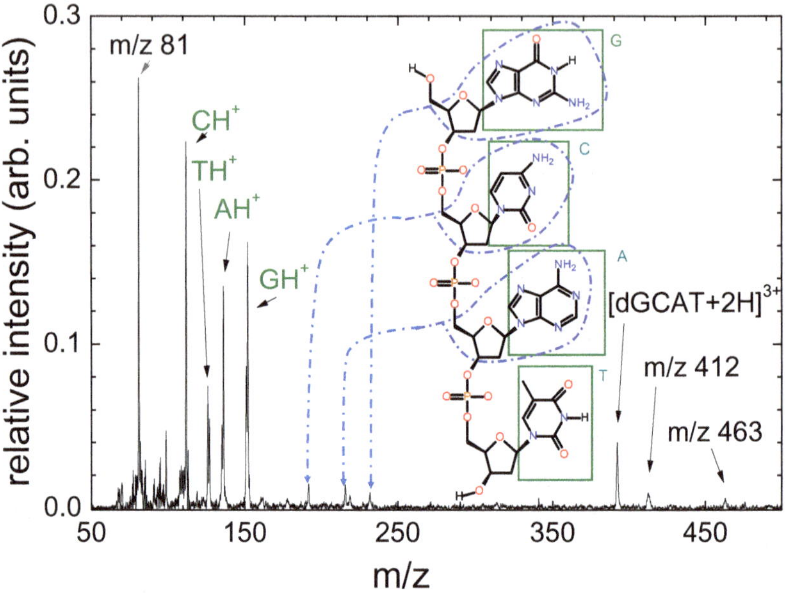

Fig. 8.10. Mass spectrum of interaction products from 40 keV C^{4+} collisions with $[dGCAT + 2H]^{2+}$. The intensity was normalised to the total loss of the parent peak after irradiation.
Source: Figure reprinted with permission from Ref. 70. Copyright (2013) the American Physical Society.

$m/z = 81$ peak being due to deoxyribose fragmentation, whereas the peaks CH^+, TH^+, AH^+, and GH^+ are due to protonated nucleobases, formed by glycosidic bond cleavage, a process also observed in biological systems. A closer inspection of the nucleobase peaks reveals that they are accompanied by peaks of the radical cations C^+, T^+, A^+, and G^+. Furthermore, formation of larger nucleoside-related fragments is observed but at much weaker intensities. NDI into $[dGCAT{+}2H]^{3+}$ is weak but observable.

The initial experiments with keV carbon ion collisions on $[dGCAT{+}2H]^{2+}$ had two key limitations. First, while relevant to medicine (keV energies actually occur at the end of the Bragg peak in heavy ion therapy), higher ion energies in the MeV range would better represent the conditions in the bulk of the Bragg peak, especially near its maximum. Second, in biological cells, DNA is mostly present in deprotonated form, typically with about one negative charge per three nucleotides.

Over the past few years, Schlathölter and Poully have conducted a series of experimental campaigns where the apparatus shown in Fig. 8.1 was interfaced with the IRRSUD beamline at GANIL (Caen, France), which can deliver heavy ion (in particular, carbon) beams at Bragg peak energies. In most of these experiments, the ESI source was operated in negative mode in order to produce deprotonated oligonucleotide anions. A typical mass spectrum of the collision products is shown in Fig. 8.11. The black line is a difference spectrum between trapped anions without ion collisions and those with ion collisions. The negative peak, labelled $[d^{Br}UAAAAG\text{-}2H]^{2-}$ thus corresponds to collision-induced loss of precursor ions from the trap. However, only to a very small extent

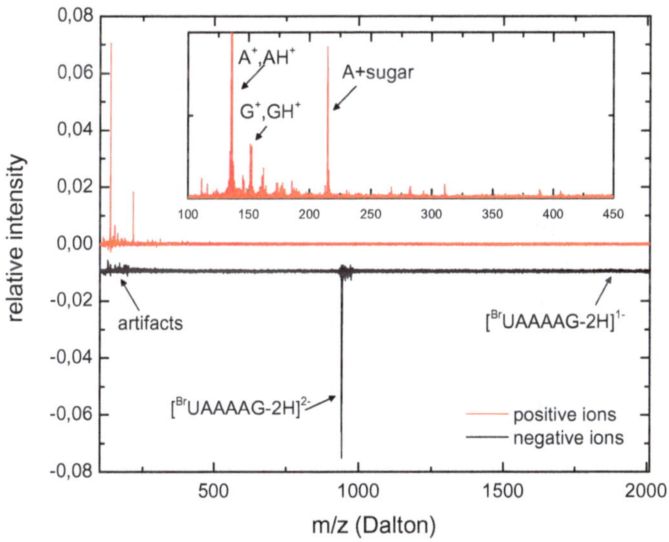

Fig. 8.11. Mass spectrum of interaction products from 10.4 MeV $^{13}C^{3+}$ collisions with $[d^{Br}UAAAAG{+}2H]^{2-}$ (^{Br}U is bromouracil). The intensity was normalised to the total loss of the parent peak after irradiation. The black (lower) spectrum represents negative ions and features a strong negative peak due to precursor ion loss from the trap, as well as positive peaks due to fragment anions. The red (upper) spectrum represents positive ions and only features positive peaks due to fragment cations. The negative ion data were shifted by 0.005 for better visualisation.

do these ions suffer non-dissociative single-electron detachment into $[d^{Br}UAAAAG\text{-}2H]^{1-}$. Negative fragments are not observed either (the weak peaks at smaller m/z are artefacts, probably due to MeV ions hitting diaphragms). It is obvious that the vast majority of the collisions must lead to neutral or positively charged species.

The 3D ion trap used in the experiments consists of two hyperbolic electrodes for ion injection (from the ESI source side) and ion extraction (into the reflectron time of flight mass spectrometer, RTOF) and a hyperbolic ring electrode, carrying the RF (see inset of Fig. 8.1). RF quadrupole traps of this type hold positive and negative ions at the same time. Positively charged collision products can thus simply be studied through a polarity reversal of the voltages of the extraction system and RTOF. The red spectrum in Fig. 8.11 shows the respective data. Mostly small fragments are observed, qualitatively similar to what is observed in collisions of keV carbon ions with protonated oligonucleotides (see Fig. 8.10). The strongest peaks are observed for radical nucleobase ions and for protonated nucleobases (here, mostly A^+, AH^+, G^+, and GH^+) and for the adenine + sugar nucleoside, also observed in Fig. 8.10.

The very efficient production of small, positively charged fragments from a doubly deprotonated precursor implies the removal of at least three and probably more electrons in a single collision process. In the context of biological radiation damage, the transfer of MeV ion kinetic energy to a target system is usually described in terms of electronic stopping, i.e., inelastic collisions of the ion with bound electrons of the target. In the process, electrons can either be excited or ionised into the continuum, and both valence and inner-shell electrons can be involved. The Tribedi group has shown, with the example of the DNA base adenine, that inner-shell ionisation only contributes a few percent to the total ionisation cross-section;[99] however, the resulting inner-shell vacancies are typically filled by the Auger type of decay, leading to the emission of one or more additional electrons. The amount of electronic stopping, and thus the number of ejected electrons as well, naturally depends on the MeV ion trajectory through the target molecule, with only glancing collisions leading to little energy deposition and single ionisation. Even for identical trajectories through the oligonucleotide, the stopping process is of statistical nature. To investigate the mechanism of multiple-electron removal in more detail, it is thus straightforward to employ a more controlled way of creating inner-shell vacancies, which is described in the following section. It is important to mention that multiple-electron detachment and glycosidic bond cleavage leading to the formation of radical nucleobase cations and protonated nucleobase cations are very robust processes, observed for all small deprotonated oligonucleotides that we have investigated so far (see, e.g., Ref. 100 for the example of $[dTGGGT\text{-}2H]^{2-}$).

8.5.2 *VUV and soft X-ray photoabsorption in gas-phase oligonucleotide ions*

In the pioneering paper by Gonzalez-Magaña *et al.*, keV C^{q+} collisions with the doubly-protonated oligonucleotide $[dGCAT{+}2H]^{2+}$ have already been compared with VUV and soft X-ray photoexcitation and photoionisation at selected energies.[70] The results are shown in Fig. 8.12 and should be compared with the mass spectra for keV ion impact in Fig. 8.10.

At all photon energies, the same small-mass singly charged fragment cations dominate, i.e., the radical and protonated nucleobases B^+ and BH^+ (B = C, T, A, G), formed by glycosidic bond cleavage, as well as the deoxyribose fragment at $m/z = 81$, indicating backbone fragmentation. Only for VUV photons at 40 eV and even more at 15 eV, the somewhat larger fragments and $[dGCAT + 2H]^{3+}$ due to NDI are observed. These channels are also open in the case of keV C^{q+} ions (see Fig. 8.10). As discussed in previous sections, NDI is usually only observed for low excitation energy processes, which is why it contributes most for 15 eV photons, only a couple of eV above the ionisation threshold where only outer valence electrons can be ejected. NDI is weaker at 30 eV, where inner valence electrons (e.g., carbon 2s) are also accessible.

One can now once again ask the question:what about deprotonated oligonucleotides? Li *et al.* have investigated VUV photoabsorption in multiply deprotonated oligonucleotides using different systems, all containing the human telomere sequence TTAGGG.[101] Mass spectra obtained for 30 eV photoaborption by $[dTTAGGG\text{-}3H]^{3-}$, $[dTTAGGG(CCG)_2\text{-}5H)^{5-}$, and $[dTTAGGG(CCG)_4\text{-}7H)^{7-}$ are shown in Fig. 8.13. In contrast to what we have seen up to now for keV ion impact and VUV photoionisation of protonated

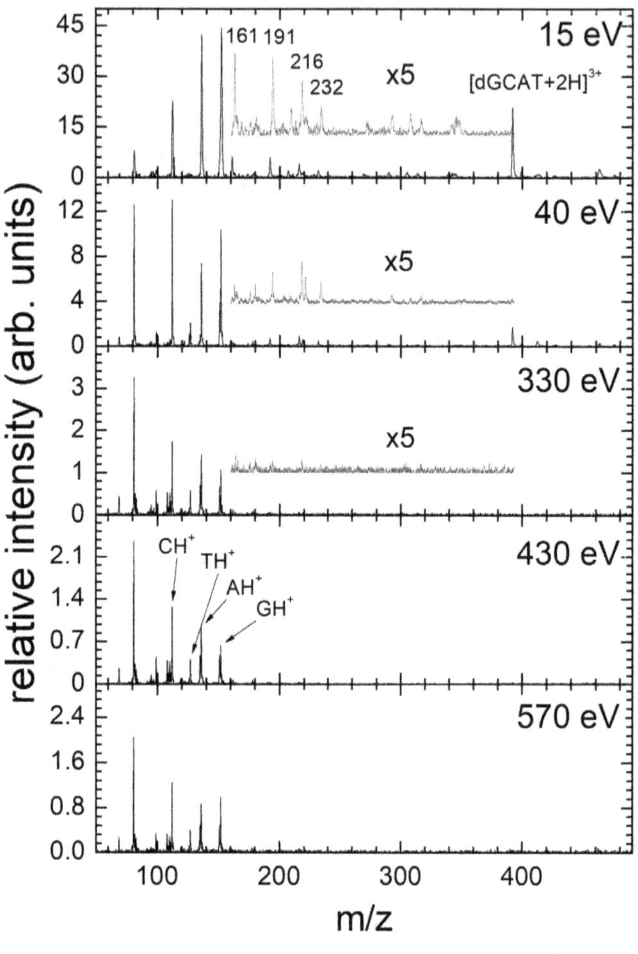

Fig. 8.12. Mass spectra of molecular cations formed upon $[\text{dGCAT} + 2\text{H}]^{2+}$ photofragmentation at 15, 40, 330 (carbon K-edge), 430 (nitrogen K-edge), and 570 eV (oxygen K-edge).
Source: Figure reprinted with permission from Ref. 70. Copyright (2013) the American Physical Society.

oligonucleotides and for MeV ion impact on deprotonated oligonucleotides, neither small fragments nor positive photoabsorption products are observed at all. Instead, the data reveal an interesting trend that had earlier been observed for fragmentation as a function of peptide size (see Section 8.3.2.3 and Ref. 76): for a given photon energy, fragmentation generally decreases when the size of

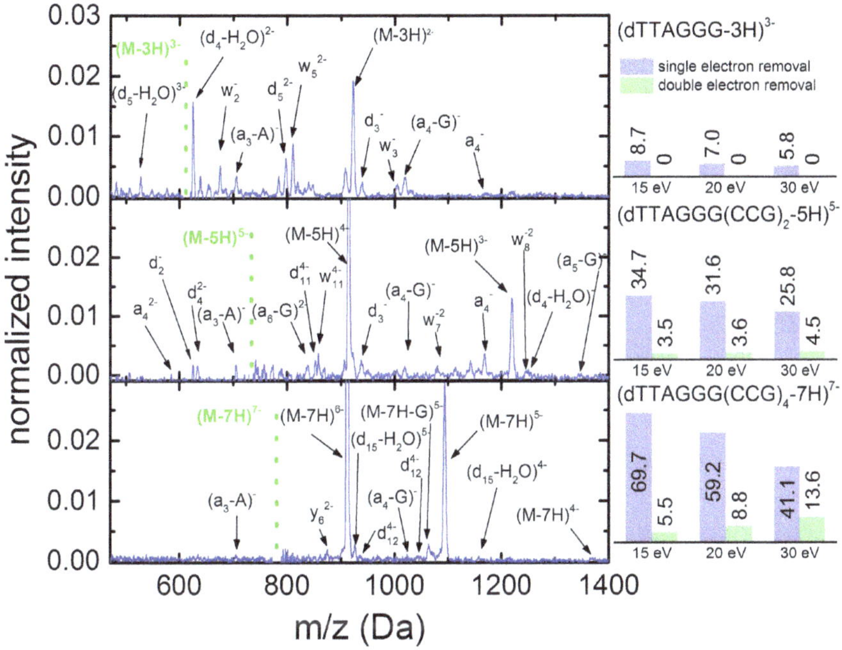

Fig. 8.13. VUV photofragmentation spectra (30 eV) for the deprotonated oligonucleotides [dTTAGGG-3H]$^{3-}$ (top), [dTTAGGG(CCG)$_2$-5H)$^{5-}$ (middle), and [dTTAGGG(CCG)$_4$-7H)$^{7-}$ (bottom). The position of the precursor m/z is always indicated by a dashed green line. All spectra are normalised to the height of the precursor peak. Single- and double-electron detachment yields are shown in the right panel. Single-electron detachment yields (shown in purple) have a higher intensity, while double-electron detachment yields (shown in green) have a lower intensity.
Source: Reproduced from Ref. 101 with permission from Wiley on behalf of Chemistry Europe.

the system increases, i.e., with the internal temperature, if we assume that fragmentation occurs after vibrational energy redistribution. For [dTTAGGG-3H]$^{3-}$, fragmentation into intermediate-size fragments is dominating, and non-dissociative single electron detachment into [dTTAGGG-3H]$^{2-}$ only contributes on the few percent level. For larger systems such as [dTTAGGG(CCG)$_4$-7H)$^{7-}$, non-dissociative detachment into [dTTAGGG(CCG)$_4$-7H)$^{q-}$ prevails, with $q = 6, 5$, and even 4. The experiments also confirmed[102] that – due to their low vertical ionisation energy – repetitive G sequences function as traps for the photoinduced holes, leading to predominant bond scission in the GGG part of the sequence.[101]

For a deeper understanding of the molecular physics underlying VUV photoionisation of deprotonated oligonucleotides (and other biomolecules), it would be instrumental to measure PhotoElecton(s) and PhotoIon(s) in COincidence (PEPICO). Maclot *et al.*, for instance, have applied this technique to VUV (50 eV) photoionisation of the nucleoside thymidine[96] and recorded photofragmentation mass spectra as a function of the initial binding energy of the ejected electron. It was convincingly shown that the ejection of weakly bound electrons predominantly leads to glycosidic bond cleavage, whereas the ejection of deeper valence electrons leads to extensive fragmentation. Unfortunately, the use of RF ion traps rules out such coincidence studies for deprotonated oligonucleotides at this point.

Finally, Li *et al.* have also investigated soft X-ray photoionisation of deprotonated oligonucleotides.[100] In contrast to the case of VUV photoionisation, where solely negatively charged photoions are formed, soft X-ray photoionisation of $[dTGGGGT\text{-}2H]^{2-}$ gives qualitatively similar results to what is observed in MeV ion collisions (see Fig. 8.11). In the negative mode, only non-dissociative single-electron detachment into $[dTGGGGT\text{-}2H]^{1-}$ is observed as a weak channel, whereas fragmentation into small nucleobase fragments T^+/TH^+ and G^+/GH^+ indicates efficient multiple-electron removal and glycosidic bond cleavage (see Fig. 8.14). Different from the MeV ion collisions case, however, this experiment narrows down the initial ionisation/excitation process, as the photon energy was scanned from 280 eV to 310 eV. Figure 8.14(a) shows the results for 292 eV carbon 1s *excitation* into an unoccupied molecular orbital (see Fig. 8.14(e)). This process does not affect the intermediate charge state but creates a carbon 1s vacancy. Figure 8.14(b) shows the results for 300 eV carbon 1s *ionisation* into the continuum (see Fig. 8.14(e)). This process decreases the charge state by one and creates a carbon 1s vacancy. In both cases, the final charge state is reached after Auger decay of the inner-shell vacancy (see Fig. 8.14(f)). In atoms and small molecules, Auger decay predominantly leads to emission of a single Auger electron. With increasing size of the molecule, more valence electrons are available in the vicinity of the core vacancy, and the probability of double Auger decay was found to range from 6.3% in CH_4 to 29% in CCl_4 for a carbon 1s hole.[103] Auger decay that involves detachment

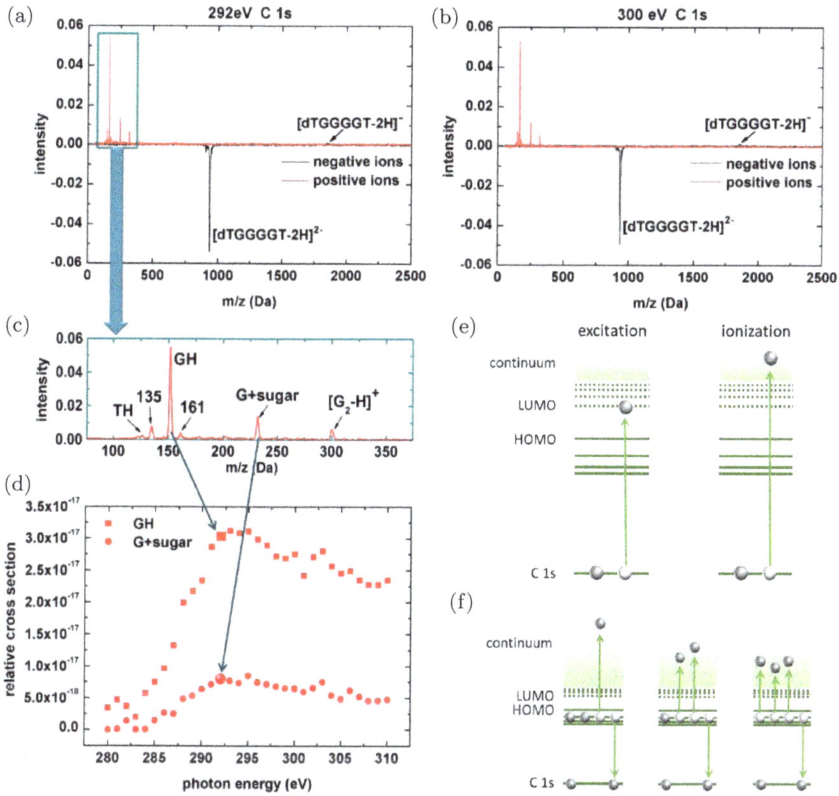

Fig. 8.14. Top panel: X-ray photoabsorption mass spectra of [dTGGGGT–2H]$^{2-}$ anions (m/z= 930.6) for $E_{\text{X-ray}}$ = 292 eV (a) and 300 eV (b) (black: negative ions, red: positive ions). (c) Zoomed-in image of the cationic fragment region 75–375 Da. (d) Relative total photoion yields as a function of photon energy for two fragments. (e) Schematic diagram of C 1s excitation and ionisation. (f) Sketch for single-, double-, and triple-valence electron detachment, associated with Auger decay.

Source: Reproduced from Ref. 100 with permission from the Royal Society of Chemistry.

of several electrons is generally much stronger in negative ions than in neutral or positive systems.[104] It is therefore feasible that multi-electron Auger decay in the very [dTGGGGT-2H]$^{2-}$ anions is pronounced, but it is remarkable that multiple-electron detachment of at least three electrons to reach charge reversal appears to dominate.[100]

Multiple-electron Auger decay is not only interesting from a fundamental point of view but also has implications for understanding and modelling DNA damage. K-shell vacancies are commonly induced by ionising radiation in biological tissue, and it is well established that the action of Auger electrons, emitted in their decay, needs to be considered. Single Auger decay of, e.g., a carbon 1s vacancy leads to the emission of an electron with about 300 eV of kinetic energy, which has about 10 nm range in tissue. In contrast, the emission of (multiple) electrons at energies far below 100 eV implies ranges in the few nanometres regime. The latter is much closer to the characteristic dimension of DNA (the double helix has a diameter of about 2 nm), i.e., in the case of multiple-electron Auger decay the emitted electrons will likely lead to DNA damage in the very close vicinity of the initial carbon 1s vacancy, allowing for the induction of double-strand breaks or even clustered DNA damage.[105]

8.5.3 *Soft X-ray spectroscopy of deprotonated oligonucleotides*

Soft X-ray photoabsorption of biomolecular ions is ideally suited for spectroscopic studies. To this end, photoionisation/photofragmentation mass spectra are recorded as a function of photon energy around the relevant absorption atomic edges. Partial ion yields of characteristic photoions are then plotted as a function of photon energy. An example, recorded with a relatively low energy resolution, is shown in Fig. 8.14(d) for GH^+ from $[dTGGGGT\text{-}2H]^{2-}$ at the carbon K-edge.

The Schlathölter group has performed a series of experiments at the IonTrap endstation[107–109] at the BESSY II synchrotron (Helmholtz Zentrum Berlin, Germany), focusing on the short protonated oligonucleotide $[d(^{F}UAG{+}H]^{+}$.[106] The linear ion trap of this endstation can hold two orders of magnitude more ions than the Groningen setup, which makes high-energy-resolution measurements much easier. At the same time, the mass resolution of the time-of-flight mass spectrometer is markedly lower. $d(^{F}UAG$ is an oligonucleotide that features the nucleobases adenine (A) and guanine (G) and the nucleobase analog 5-fluorouracil (^{F}U). 5-fluorouracil has a high ionisation potential and very low proton affinity. It is thus not expected to be involved in protonation and positive charge transport, but the fluorine K shell can be addressed selectively, allowing for site-selective ionisation/excitation. A and G have very similar proton affinities, and the protonation site therefore needs to be determined experimentally. For both nucleobases, the highest proton affinities are found for protonation of nitrogen atoms with imine character (one double bond, =N-). Protonation of an imine nitrogen turns it into an amine nitrogen (three single bonds, >N-). 1s electrons are weaker bound in imine N than in amine N.[110] In nucleobase nitrogen K-edge X-ray absorption spectra, imine nitrogen transition energies (1s-π_1^*) fall into a lower band than amine transition energies (1s-π_2^*).[110] The protonation of an imine nitrogen in either A or G is therefore expected to influence the ratio of the two bands and possibly their spectral shape.

The obtained mass spectra for nitrogen K-edge X-ray photoionisation of $[d(^FUAG+H]^+$ are qualitatively very similar to what is observed for the oligonucleotide $[dGCAT+2H]^{2+}$ (see Fig. 8.12): mostly, radical and protonated nucleobase cations are formed, and no non-dissociative ionisation is observed. The top panel in Fig. 8.15 shows the partial ion yields as a function of photon energy for the G, A, and ^{F}U fragments (either protonated or as radical cations) as a function of photon energy around the nitrogen K-edge. The bottom panel displays results from time-dependent density functional theory (TD-DFT) calculations using a short-range corrected potential that is particularly suited for describing soft X-ray absorption.[111,112] The TD-DFT data have been calculated for protonation on the A (A-N1 protonated) and G (G-N7 protonated) nucleobases. For comparison, data for a neutral oligonucleotide are added.

A comparison between theory and experiment only shows good agreement for A-N1 protonation, indicating that this is the preferred binding site of the proton in this gas-phase oligonucleotide. This finding proves two things: (i) for large molecular systems, the interpretation of gas-phase soft X-ray absorption spectra requires input from high-level quantum chemical calculations; (ii) the combination of experiment and theory allows for astonishingly deep insights into the geometric and electronic structure of complex gas-phase biomolecules.

In a subsequent study, Wang and coworkers[113] have therefore set out to employ the same combination of experiment and theory to investigate the mechanism of formation of B^+ and BH^+ ions

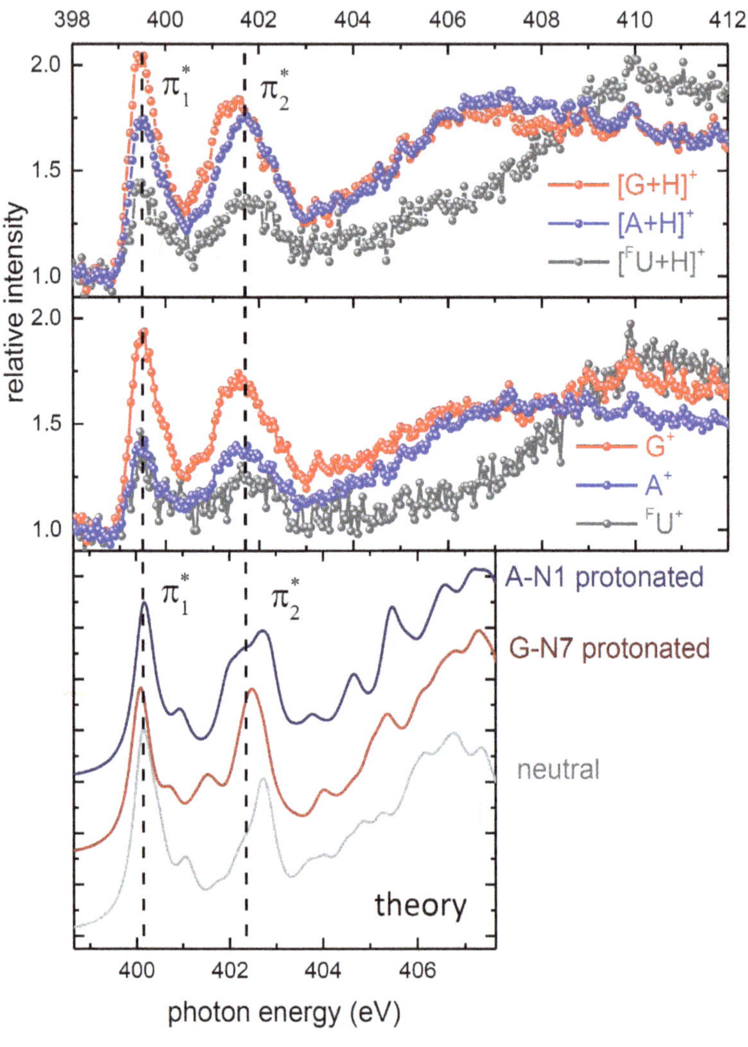

Fig. 8.15. Top panel: experimental N K-edge X-ray absorption spectra for $[\mathrm{d}(^{\mathrm{F}}\mathrm{UAG})+\mathrm{H}]^{+}$ for the fragments $[\mathrm{G+H}]^{+}$ (red), $[\mathrm{A+H}]^{+}$ (blue), and $[^{\mathrm{F}}\mathrm{U+H}]^{+}$ (grey). Middle panel: corresponding data for G^{+} (red), A^{+} ^{F} (blue), and U^{+} (grey). The intensities are relative to the non-resonant intensity just below the N K-edge. Bottom panel: TD-DFT data, convoluted with the experimentally observed resolution (0.5 eV) for the neutral oligonucleotide, as well as for A-N1 and G-N7 protonation. Note that the axis of the theoretical data is shifted with respect to the experimental data by -0.65 eV.
Source: Reproduced from Ref. 106 with permission from the PCCP Owner Societies.

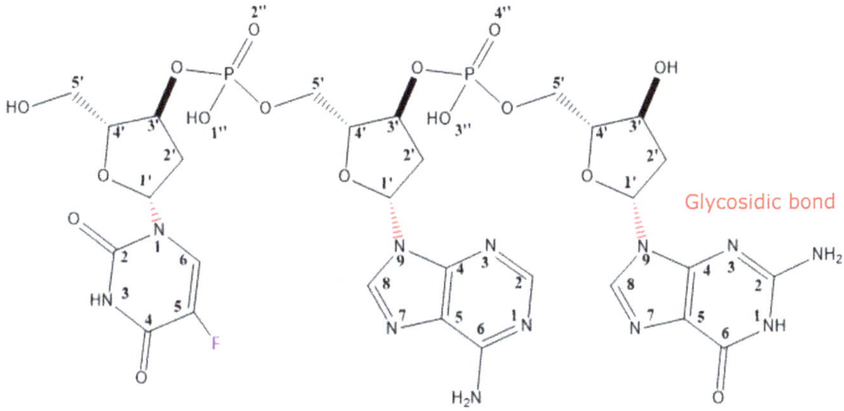

Fig. 8.16. Structure of d(^{F}UAG) and location of the glycosidic bonds between nucleobase and deoxyribose (marked in red).
Source: Reproduced from Ref. 113 with permission from the PCCP Owner Societies.

(B: nucleobase) formed in soft X-ray photoionisation of the same protonated oligonucleotide. As mentioned earlier, these fragments are formed by glycosidic bond cleavage (see Fig. 8.16), and their formation is a dominating feature of most interaction processes of ionising radiation with gas-phase oligonucleotides. It is important to realise that a mere glycosidic bond cleavage does not explain the formation of B^+ and BH^+, as the cleaved-off group has B-H character! Only from a protonated nucleobase is B^+ formation by bond cleavage possible. In all other cases, i.e., B^+ and BH^+ formation from a neutral base or BH^+ formation from a protonated base, hydrogen and/or proton transfer needs to be involved. This transfer most likely needs to occur on very short timescales, as it has to compete with the bond cleavage process.

Figure 8.17 shows experimental and theoretical X-ray absorption data for the O K-edge. The bottom panel shows oscillator strengths determined by TD-DFT calculations. The green sticks denote transitions involving O 1s sites on the sugar phosphate backbone while the orange sticks denote transitions involving the three oxygen atoms in ^{F}U and G. Clearly, In the spectrum obtained by the convolution of the stick spectra with a Gaussian of 0.4 eV fwhm (middle panel), the latter transitions lead to a peak centred at a photon energy of 532 eV. Targeting this peak implies localised 1s excitation on one of these two

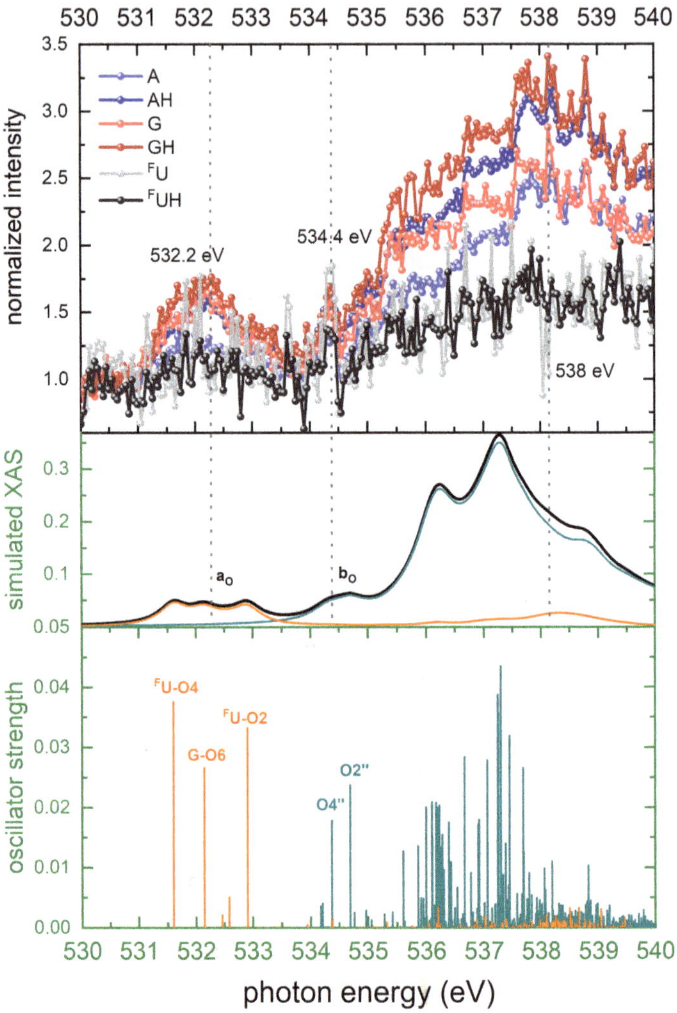

Fig. 8.17. Top panel: normalised ion yields of A^+ (light blue), G^+ (light red), and ${}^{F}U^+$ (grey) and AH^+ (blue), GH^+ (red), and ${}^{F}UH^+$ (black) followed by $[d({}^{F}UAG) + H]^+$ X-ray absorption near the O K-edge. Bottom panel: stick spectrum of TD-DFT oscillator strengths for O 1s transitions. The orange lines represent transitions occurring in the nucleobase, while the dark-green lines represent transition on the sugar-phosphate backbone. Five particular transitions are labelled. Middle panel: convoluted TD-DFT data (Gaussian FWHM = 0.4 eV). The black curve is the sum spectrum of all core transitions based either on the nucleobases (orange) or on the backbone (green).

Source: Reproduced from Ref. 113 with permission from the PCCP Owner Societies.

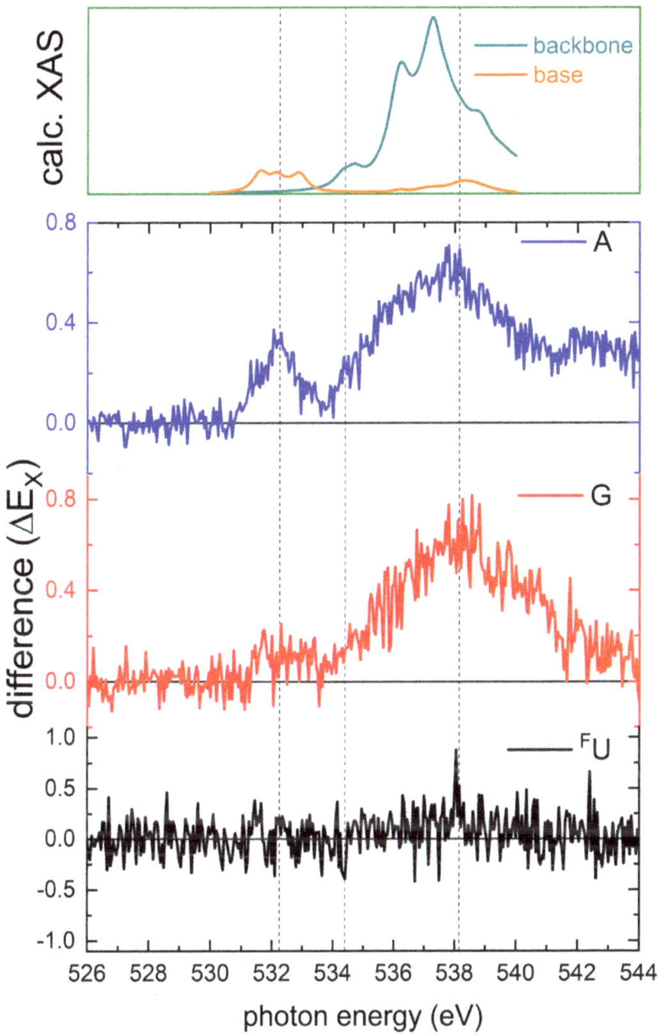

Fig. 8.18. Difference in protonated nucleobase BH^+ and non-protonated nucleobase B^+ ion yields, $\Delta(E_X)$, as a function of photon energy at the O K-edge for all three nucleobases: adenine (A, blue), guanine (G, red), and fluorouracil (^{F}U, black). The vertical dotted grey lines indicate the resonant photon absorption maxima observed in Fig. 8.17. The top panel shows again the theoretical data from the middle panel of Fig. 8.17.
Source: Reproduced from Ref. 113 with permission from the PCCP Owner Societies.

nucleobases. The top panel in Fig. 8.17 contains the experimentally obtained partial ion yields for the six nucleobase fragments, formed after glycosidic bond cleavage. Besides A^+ and G^+, these fragments all require H transfer. It is important to note that H transfer to the A moiety is observed, even though A does not contain an O X-ray chromophore. This entails the question of whether the observed H transfer merely requires excitation energy from the X-ray absorption at any atomic site (O, N or C) or whether it is affected by the excitation site. In order to answer that question, it is helpful to introduce the difference $\Delta(E_X) = Y_{\mathrm{BH}}^{\mathrm{norm}}(E_X) - Y_{\mathrm{B}}^{\mathrm{norm}}(E_X)$ of the normalised BH^+ and B^+ partial ion yields as a function of photon energy E_X and plot this quantity as a function of photon energy. This quantity is a measure for H-transfer to nucleobases when glycosidic bond cleavage occurs, as BH^+ can only be formed by H transfer. The data for the O K-edge are shown in Fig. 8.18, with the theoretical spectrum in the top panel and the experimental data for A, G, and ^{F}U below. The most obvious observation is the absence of signal in the difference spectrum for ^{F}U, implying that X-ray absorption is not substantially influencing hydrogen transfer to this nucleobase. On the other hand, for both A and G, a strong effect is observed, indicating that for these nucleobases, resonant soft X-ray absorption facilitates H transfer. The effect is observed for both: absorption on the backbone (photon energy range above 534 eV) or on the O-containing nucleobases ^{F}U and G (resonance centred around 532 eV). For G, the peak at 532 eV is observed to be quenched. This counterintuitive finding could indicate that O K-edge absorption in G often leads to fragmentation of the G moiety rather than mere glycosidic bond cleavage.

8.6 Conclusions and Outlook

Interaction between ionising particles and trapped biomolecular ions in the gas phase leads to ejection of one or several electrons in the continuum and leaves an electronically excited molecular system that, in most cases, undergoes fast internal conversion to the electronic ground state, with a corresponding amount of vibrational excess energy. The latter triggers a number of processes that can be experimentally studied by analysing the ionic products of the interaction thanks to mass spectrometry. For instance, photoabsorption by (multiply-)protonated peptides and antibiotics, as well as their noncovalent complexes, leads to noncovalent and/or covalent bond cleavage, and the amount of vibrational excitation energy has been found to increase with photon energy from the ionisation threshold (in the 10–15 eV range) to soft X-rays because electrons of deeper orbitals can be ejected, creating highly electronically excited systems. Interestingly, the response of CMP assemblies to ionising radiation is influenced by typical collagen properties such as stabilisation of the triple helix by stereoelectronic effects resulting from proline hydroxylation. In addition to fragmentation, other processes such as proton or sodium transfer and cross-linking are triggered by photoionisation. Since these processes are expected to be strongly dependent on the geometrical structure of the biomolecular ion, it would be highly valuable to irradiate only one structure at a time to be able to compare the results for several structures, for instance, a folded and an extended structure of a protein of a given charge state. Experimental developments are in progress in this direction by using an ion mobility filter prior to the injection of biomolecular ions in the RF trap, following what has already been achieved in the mass spectrometry community, for instance, to measure the unfolding signature of proteins upon CID.[114]

For protonated and deprotonated oligonucleotide ions, we have found that glycosidic bond cleavage accompanied by fast H transfer appears to be a characteristic feature for all interactions that lead to high excitation energies. An improved understanding of X-ray-induced fragmentation in oligonucleotide ions, with a focus on X-ray-induced H-transfer and charge migration, will require time-resolved experiments on femtosecond and picosecond timescales. A straightforward approach would be soft X-ray/soft X-ray or soft X-ray/IR pump-probe experiments. In such experiments, the pump pulse would be a soft X-ray, tuned in energy to photoabsorption predominantly on selected nucleobases, on the deoxyribose moieties, or on the phosphate groups. After the initial photoabsorption process, the molecules will be left to dynamically respond until – after a time-delay Δt – the soft X-ray or IR probe pulse would further ionise and potentially fragment the system. Processes such as H migration could then be studied in a time-resolved manner by investigating the Δt dependence of characteristic fragment ion yields. Such experiments could be implemented by interfacing existing ESI/ion-trap setups with a soft X-ray free-electron-laser beamlines, ideally operating in 'seeded' mode with reproducible pulse shapes. The first experiments in this direction are currently ongoing.

Soft X-ray and MeV ion-induced multiple-electron detachment from deprotonated oligonucleotides is another key experimental finding that remains to be fully understood. Here, electron spectroscopy experiments would be required to characterise the properties of the emitted electrons. Such experiments cannot be performed in the IMS/ion-trap systems currently used in the field. Instead, crossed-beam experiments, where a beam of oligonucleotide ions is interacting with, say, a soft X-ray photon beam would be required. In that geometry, electron spectrometers can be used to study emitted electrons, but very high biomolecular ion beam currents are needed.

References

1. Vuckovic M, Radojcic M, Milosavljevic B. Gamma-radiation induced damage of proteins in the thick fraction of egg white. J Serb Chem Soc. 2005;70:1255–1262.
2. Soto Espinoza SL, *et al.* Radiation synthesis of seroalbumin nanoparticles. Radiat Phys Chem. 2012;81:1417–1421.
3. Lee SS, *et al.* Enhancement of chaperone activity of plant-specific thioredoxin through γ-ray mediated conformational change. Int J Mol Sci. 2015;16:27302–27312.
4. Kempner ES. Damage to proteins due to the direct action of ionizing radiation. Q Rev Biophys. 1993;26:27–48.
5. Krisch RE, Flick MB, Trumbore CN. Radiation chemical mechanisms of single- and double-strand break formation in irradiated SV40 DNA. Radiat Res. 1991;126:251–259.
6. Kempner E. Effects of high-energy electrons and gamma rays directly on protein molecules. J Pharm Sci. 2001;90:1637–1646.
7. Lalande M, *et al.* Direct radiation effects on the structure and stability of collagen and other proteins. ChemBioChem. 2019;20:2972–2980.
8. Bull JN, *et al.* Ion mobility action spectroscopy of flavin dianions reveals deprotomer-dependent photochemistry. Phys Chem Chem Phys. 2018;20:19672–19681.

9. Choi CM, *et al.* Ion mobility resolved photo-fragmentation to discriminate protomers. Rapid Commun Mass Spectrom. 2019; 33:28–34.
10. Bhattacharya A, *et al.* Conformation specific and charge directed reactivity of radical cation intermediates of α-substituted (amino, hydroxy, and keto) bioactive carboxylic acids. Phys Chem Chem Phys. 2010;12:9700–9712.
11. Milewska A, *et al.* Mass spectrometry in virological sciences. Mass Spectrom Rev. 2020;39:499–522.
12. Barreiro-Lage D, *et al.* "Smart decomposition" of cyclic alanine-alanine dipeptide by VUV radiation: A seed for the synthesis of biologically relevant species. J Phys Chem Lett. 2021;12:7379–7386.
13. Zettergren H, *et al.* Roadmap on dynamics of molecules and clusters in the gas phase. Eur Phys J D. 2021;75:152.
14. Ueda K, *et al.* Roadmap on photonic, electronic and atomic collision physics: I. Light-matter interaction. J Phys B At Mol Opt Phys. 2019;52:171001.
15. Aumayr F, *et al.* Roadmap on photonic, electronic and atomic collision physics: III. Heavy particles: With zero to relativistic speeds. J Phys B At Mol Opt Phys. 2019;52:171003.
16. Giuliani A, *et al.* Contribution of synchrotron radiation to photoactivation studies of biomolecular ions in the gas phase. Mass Spectrom Rev. 2014;33:424–441.
17. Solov'yov AV, *et al.* Condensed matter systems exposed to radiation: Multiscale theory, simulations, and experiment. Chem Rev. 2024;124:8014–8129.
18. Coupier B, *et al.* Inelastic interactions of protons and electrons with biologically relevant molecules. Eur Phys J D. 2002;20:459–468.
19. de Vries J, *et al.* Cq+-induced excitation and fragmentation of uracil: Effects of the projectile electronic structure. J Phys B At Mol Opt Phys. 2002;35:4373–4381.
20. Schlathölter T. Ion-induced radiation damage in biomolecular systems. In: García Gómez-Tejedor G, Fuss MC, editors. Radiation damage in biomolecular systems. Dordrecht: Springer Netherlands. 2012. pp. 177–190.
21. Fung YME, Adams CM, Zubarev RA. Electron ionization dissociation of singly and multiply charged peptides. J Am Chem Soc. 2009;131:9977–9985.
22. Liu B, *et al.* Collision-induced dissociation of hydrated adenosine monophosphate nucleotide ions: Protection of the ion in water nanoclusters. Phys Rev Lett. 2006;97:133401.

23. Bari S, Hoekstra R, Schlathölter T. Peptide fragmentation by keV ion-induced dissociation. Phys Chem Chem Phys. 2010;12:3376–3383.
24. Bari S, Hoekstra R, Schlathölter T. Fast side-chain losses in keV ion-induced dissociation of protonated peptides. Int J Mass Spectrom. 2011;299:64–70.
25. Milosavljevic AR, *et al.* Photoionization of a protein isolated in vacuo. Phys Chem Chem Phys. 2011;13:15432–15436.
26. Shaffer SA, *et al.* An ion funnel interface for improved ion focusing and sensitivity using electrospray ionization mass spectrometry. Anal Chem. 1998;70:4111–4119.
27. Ibrahim Y, *et al.* Improving mass spectrometer sensitivity using a high-pressure electrodynamic ion funnel interface. J Am Soc Mass Spectrom. 2006;17:1299–1305.
28. Saintigny Y, *et al.* Impact of therapeutic irradiation on healthy articular cartilage. Radiat Res. 2015;183:135–146.
29. Uitto J, Lichtenstein JR. Defects in the biochemistry of collagen in diseases of connective tissue. J Invest Dermatol. 1976;66:59–79.
30. Berisio R, *et al.* Crystal structure of the collagen triple helix model $[(\text{Pro-Pro-Gly})_{10}]_3$. Protein Sci. 2002;11:262–270.
31. Okuyama K, *et al.* Crystal structure of $(\text{Gly-Pro-Hyp})_9$: Implications for the collagen molecular model. Biopolymers. 2012;97:607–616.
32. Kotch FW, Guzei IA, Raines RT. Stabilization of the collagen triple helix by o-methylation of hydroxyproline residues. J Am Chem Soc. 2008;130:2952–2953.
33. Shoulders MD, Raines RT. Collagen structure and stability. Annu Rev Biochem. 2009;78:929–958.
34. Bella J. Collagen structure: New tricks from a very old dog. Biochem J. 2016;473:1001–1025.
35. Lalande M, *et al.* Isolated collagen mimetic peptide assemblies have stable triple-helix structures. Chem Eur J. 2018;24:13728–13733.
36. Grotemeyer J, Schlag E. Biomolecules in the gas-phase. 4. multiphoton ionization of bovine insulin. Org Mass Spectrom. 1987;22:758–760.
37. Grotemeyer J, Schlag E. Peptides investigated by laser desorption-multiphoton ionization mass-spectrometry. Org Mass Spectrom. 1988;23:388–396.
38. Walter K, *et al.* Biomolecules in the gas phase. v. small peptides investigated by multiphoton mass spectrometry. Chem Phys. 1988;125:155–164.

39. Grotemeyer J, *et al.* Biomolecules in the gas-phase. 2. multiphoton ionization mass-spectrometry of angiotensin-i. Org Mass Spectrom. 1986;21:595–597.
40. Björkman S. Electron impact and methane, isobutane and ammonia chemical ionization mass spectra of peptide amides related to melanostatin (Pro-Leu-Gly-NH_2). Biomed Mass Spectrom. 1982;9:315–322.
41. Svec HJ, Junk GA. The mass spectra of dipeptides. J Am Chem Soc. 1964;86:2278–2282.
42. Weinkauf R, *et al.* Elementary processes in peptides: Electron mobility and dissociation in peptide cations in the gas phase. J Phys Chem. 1995;99:11255–11265.
43. Remacle F, *et al.* Electronic control of site selective reactivity: A model combining charge migration and dissociation. J Phys Chem A. 1999;103:10149–10158.
44. Weinkauf R, *et al.* Femtosecond versus nanosecond multiphoton ionization and dissociation of large molecules. J Phys Chem. 1994;98:8381–8391.
45. Köster C, Grotemeyer J. Single-photon and multi-photon ionization of infrared laser-desorbed biomolecules. Org Mass Spectrom. 1992;27:463–471.
46. Zubarev R, Nielsen M, Budnik B. Tandem ionization mass spectrometry of biomolecules. Eur J Mass Spectrom. 2000;6:235–240.
47. Budnik BA, Zubarev RA. $MH^{2+\cdot}$ ion production from protonated polypeptides by electron impact: Observation and determination of ionization energies and a cross-section. Chem Phys Lett. 2000;316:19–23.
48. Canon F, *et al.* Action spectroscopy of a protonated peptide in the ultraviolet range. Phys Chem Chem Phys. 2015;17:25725–25733.
49. Budnik BA, *et al.* Ionization energies of multiply protonated polypeptides obtained by tandem ionization in fourier transform mass spectrometers. J Mass Spectrom. 2002;37:1141–1144.
50. Giuliani A, *et al.* Structure and charge-state dependence of the gas-phase ionization energy of proteins. Angew Chem Int Ed. 2012;51:9552–9556.
51. Antoine R, *et al.* Photo-induced formation of radical anion peptides. electron photodetachment dissociation experiments. Rapid Commun Mass Spectrom. 2007;21:265–268.
52. Antoine R, *et al.* Electron photodetachment of trapped doubly deprotonated angiotensin peptides. UV spectroscopy and radical recombination. Eur Phys J D. 2009;51:117–124.

53. Antoine R, Dugourd P. Visible and ultraviolet spectroscopy of gas phase protein ions. Phys Chem Chem Phys. 2011;13:16494–16509.
54. Brunet C, *et al.* Valence shell direct double photodetachment in polyanions. New J Phys. 2013;15:063024.
55. Zubarev RA, Yang H. Multiple soft ionization of gas-phase proteins and swift backbone dissociation in collisions with $\leq$99 eV electrons. Angew Chem Int Ed. 2010;49:1439–1441.
56. Schlathölter T, *et al.* Multiple ionization of free ubiquitin molecular ions in extreme ultraviolet free-electron laser pulses. Angew Chem Int Ed. 2016;55:10741–10745.
57. Martin S, *et al.* Electron capture and deprotonation processes observed in collisions between Xe^{8+} and multiply protonated cytochrome-c. Phys Rev A. 2014;89:012707.
58. Li P, Kreft I, Jackson GP. Top-down charge transfer dissociation (CTD) of gas-phase insulin: Evidence of a one-step, two-electron oxidation mechanism. J Am Soc Mass Spectrom. 2018;29:284–296.
59. Milosavljević AR, *et al.* Multiple electron capture from isolated protein poly-anions in collision with slow highly charged ions. Phys Chem Chem Phys. 2017;19:19691–19698.
60. Zhang W, *et al.* Electronic structure of aromatic amino acids studied by soft x-ray spectroscopy. J Chem Phys. 2009;131:035103.
61. Plekan O, *et al.* An x-ray absorption study of glycine, methionine and proline. J Electron Spectros Relat Phenom. 2007;155:47–53.
62. Plekan O, *et al.* Investigation of the amino acids glycine, proline, and methionine by photoemission spectroscopy. J Phys Chem A. 2007;111:10998–11005.
63. Milosavljević AR, *et al.* Gas-phase protein inner-shell spectroscopy by coupling an ion trap with a soft x-ray beamline. J Phys Chem Lett. 2012;3:1191–1196.
64. Bari S, *et al.* Soft x-ray spectroscopy as a probe for gas-phase protein structure: Electron impact ionization from within. Chem Eur J. 2018;24:7631–7636.
65. Pan X, Sanche L. Mechanism and site of attack for direct damage to DNA by low-energy electrons. Phys Rev Lett. 2005;94:198104.
66. Paizs B, Suhai S. Fragmentation pathways of protonated peptides. Mass Spectrom Rev. 2005;24:508–548.
67. Chevalier F, Schlathölter T, Poully J-C. Radiation-induced transfer of charge, atoms, and energy within isolated biomolecular systems. ChemBioChem. 2023;24:e202300543.
68. Bari S, *et al.* Photodissociation of protonated leucine-enkephalin in the VUV range of 8–40 eV. J Chem Phys. 2011;134:024314.

69. Reitsma G, *et al.* Femtosecond laser induced ionization and dissociation of gas-phase protonated leucine enkephalin. Int J Mass Spectrom. 2014;365-366:365–371.
70. González-Magaña O, *et al.* Near-edge x-ray absorption mass spectrometry of a gas-phase peptide. J Phys Chem A. 2012;116:10745–10751.
71. Li P, Jackson GP. Charge transfer dissociation (CTD) mass spectrometry of peptide cations: Study of charge state effects and side-chain losses. J Am Soc Mass Spectrom. 2017;28:1271–1281.
72. Hoffmann WD, Jackson GP. Charge transfer dissociation (CTD) mass spectrometry of peptide cations using kiloelectronvolt helium cations. J Am Soc Mass Spectrom. 2014;25:1939–1943.
73. Edwards HM, *et al.* Differentiation of leucine and isoleucine residues in peptides using charge transfer dissociation mass spectrometry (CTD-MS). Rapid Commun Mass Spectrom. 2022;36:e9246.
74. Sasiene ZJ, Jackson GP. Installation protocol for charge transfer dissociation mass spectrometry on ion trapping mass spectrometers. Rapid Commun Mass Spectrom. 2018;32:964–972.
75. Egorov D, *et al.* Near edge x-ray absorption mass spectrometry of gas phase proteins: The influence of protein size. Phys Chem Chem Phys. 2016;18:26213–26223.
76. Egorov D, Hoekstra R, Schlathölter T. A comparative VUV absorption mass-spectroscopy study on protonated peptides of different size. Phys Chem Chem Phys. 2017;19:20608–20618.
77. Schwob L, *et al.* Single-photon absorption of isolated collagen mimetic peptides and triple-helix models in the VUV-X energy range. Phys Chem Chem Phys. 2017;19:18321–18329.
78. Lalande M, *et al.* Irradiation of isolated collagen mimetic peptides by x-rays and carbon ions at the bragg-peak energy. Phys Rev A. 2018;98:062701.
79. Abdelmouleh M, *et al.* Ionizing radiation induces cross-linking of two noncovalently bound collagen mimetic peptide triple helices in the absence of a molecular environment. Phys Chem Chem Phys. 2023;25:29249–29256.
80. Giulivi C, Traaseth N, Davies K. Tyrosine oxidation products: Analysis and biological relevance. Amino Acids. 2003;25:227–232.
81. Partlow BP, *et al.* Dityrosine cross-linking in designing biomaterials. ACS Biomater Sci Eng. 2016;2:2108–2121.
82. Schwob L, *et al.* Radical-driven processes within a peptidic sequence of type I collagen upon single-photon ionisation in the gas phase. Phys Chem Chem Phys. 2017;19:22895–22904.

83. Jariashvili K, *et al.* UV damage of collagen: Insights from model collagen peptides. Biopolymers. 2012;97:189–198.
84. Montanari J, *et al.* Pilot screening of potential matrikines resulting from collagen breakages through ionizing radiation. Radiat Environ Biophys. 2024;63:337–350.
85. Floss HG, Yu T-W. Rifamycin mode of action, resistance, and biosynthesis. Chem Rev. 2005;105:621–632.
86. Abdelmouleh M, *et al.* X-ray photoabsorption-induced processes within protonated rifamycin sodium salts in the gas phase. Eur Phys J D. 2021;75:74.
87. Abdelmouleh M, *et al.* Mass spectral signatures of complex post-translational modifications in proteins: A proof-of-principle based on x-ray irradiated vancomycin. J Am Soc Mass Spectrom. 2020;31:1738–1743.
88. Abdelmouleh M, *et al.* Photoinduced processes within noncovalent complexes involved in molecular recognition. Chem Eur J. 2020;26:2243–2250.
89. Liu M, *et al.* UV-VUV photofragmentation spectroscopy of isolated neutral fragile macromolecules: A proof-of-principle based on a deprotonated vancomycin-peptide noncovalent complex. J Phys Chem A. 2022;126:9042–9050.
90. Štísová V, *et al.* Radiation damage to DNA-protein specific complexes: Estrogen response element-estrogen receptor complex. Radiat Prot Dosim. 2007;122:106–109.
91. Bury CS, *et al.* Radiation damage within nucleoprotein complexes studied by macromolecular x-ray crystallography. Radiat Phys Chem. 2016;128:118–125.
92. Nitanai Y, *et al.* Crystal structures of the complexes between vancomycin and cell-wall precursor analogs. J Mol Biol. 2009;385:1422–1432.
93. Poully JC, *et al.* Probing the specific interactions and structures of gas-phase vancomycin antibiotics with cell-wall precursor through irmpd spectroscopy. Phys Chem Chem Phys. 2010;12:3606–3615.
94. Poully JC, *et al.* Combining ion mobility mass spectrometry and infrared multiphoton dissociation spectroscopy to probe the structure of gas-phase vancomycin–$Ac_2^{L}K^{D}A^{D}A$ non-covalent complex. Int J Mass Spectrom. 2010;297:28–35.
95. Alvarado F, *et al.* Precise determination of 2-deoxy-d-ribose internal energies after kev proton collisions. Chem Phys Chem. 2008;9:1254–1258.
96. Maclot S, *et al.* Determination of energy-transfer distributions in ionizing ion-molecule collisions. Phys Rev Lett. 2016;117:073201.

97. Alvarado F, *et al.* Quantification of ion-induced molecular fragmentation of isolated 2-deoxy-d-ribose molecules. Phys Chem Chem Phys. 2006;8:1922–1928.
98. Schlathölter T, *et al.* Ion-induced biomolecular radiation damage: From isolated nucleobases to nucleobase clusters. Chem Phys Chem. 2006;7:2339–2345.
99. Bagdia C, *et al.* 1s-1s electron transfer in collisions of fast C and O ions with adenine. Nucl Instrum Meth B. 2020;462:68–74.
100. Li W, *et al.* Multiple valence electron detachment following Auger decay of inner-shell vacancies in gas-phase DNA. Chem Sci. 2021;12:13177–13186.
101. Li W, *et al.* Hole migration in telomere-based oligonucleotide anions and G-quadruplexes. Chem Eur J. 2019;25:16114–16119.
102. Cauët E. Unique hole-trapping property of the human telomere sequence. J Biomol Struct Dyn. 2011;29:557–561.
103. Roos AH, *et al.* Abundance of molecular triple ionization by double Auger decay. Sci Rep. 2018;8:16405.
104. Perry-Sassmannshausen A, *et al.* Multiple photodetachment of carbon anions via single and double core-hole creation. Phys Rev Lett. 2020;124:083203.
105. Surdutovich E, Solov'yov AV. Double strand breaks in DNA resulting from double ionization events. Eur Phys J D. 2012;66:206.
106. Wang X, *et al.* Site-selective soft X-ray absorption as a tool to study protonation and electronic structure of gas-phase DNA. Phys Chem Chem Phys. 2021;23:11900–11906.
107. Hirsch K, *et al.* X-ray spectroscopy on size-selected clusters in an ion trap: From the molecular limit to bulk properties. J Phys B At Mol Opt Phys. 2009;42:154029.
108. Niemeyer M, *et al.* Spin coupling and orbital angular momentum quenching in free iron clusters. Phys Rev Lett. 2012;108:057201.
109. Langenberg A, *et al.* Spin and orbital magnetic moments of size-selected iron, cobalt, and nickel clusters. Phys Rev B. 2014;90:184420.
110. Plekan O, *et al.* A theoretical and experimental study of the near edge x-ray absorption fine structure (NEXAFS) and x-ray photoelectron spectra (XPS) of nucleobases: Thymine and adenine. Chem Phys. 2008;347:360–375.
111. Besley NA, Peach MJ, Tozer DJ. Time-dependent density functional theory calculations of near-edge x-ray absorption fine structure with short-range corrected functionals. Phys Chem Chem Phys. 2009;11:10350–10358.

112. Besley NA, Asmuruf FA. Time-dependent density functional theory calculations of the spectroscopy of core electrons. Phys Chem Chem Phys. 2010;12:12024–12039.
113. Wang X, *et al.* Intramolecular hydrogen transfer in DNA induced by site-selective resonant core excitation. Phys Chem Chem Phys. 2022;24:7815–7825.
114. Dixit SM, Polasky DA, Ruotolo BT. Collision induced unfolding of isolated proteins in the gas phase: Past, present, and future. Omics. 2018;42:93–100.

https://doi.org/10.1142/9781800618183_0009

Chapter 9

Radiation Damage of Biomolecular Systems: A Multiscale Approach

Alexey V. Verkhovtsev* **and Andrey V. Solov'yov**†
MBN Research Center,
Altenhöferallee 3, 60438 Frankfurt am Main, Germany
**verkhovtsev@mbnresearch.com*
†*solovyov@mbnresearch.com*

This chapter provides a comprehensive review of the multiscale approach (MSA) to the physics of radiation damage with ions and its application in ion beam cancer therapy. The assessment of radiation damage is based on a series of effects that occur on different time, space, and energy scales, starting with ion entering the tissue and ending with the analysis of the irreparable DNA damage events that lead to cell inactivation. The MSA allows the prediction of survival probabilities for cells irradiated with ions. These probabilities depend on the cell type, type of ions, the fluence of the beam, the depth in the tissue, the initial energy of ions, the concentration of oxygen in the tissue, and other external conditions. According to the MSA, the scenario of radiation damage with ions includes the ion-induced shock waves. Predictions of the calculated quantities are compared with different sets of experiments, including those involving different cell lines. The chapter discusses applications of the MSA to the analysis of the dependence of cell survival probability on depth in the case of spread-out Bragg peak and the dependence of the relative biological effectiveness on the linear energy transfer, in particular the overkill effect.

9.1 Introduction

The physics and chemistry of radiation damage caused by ion irradiation have been studied intensively in recent decades due to the use of proton and carbon ion beams in cancer therapy.[1–6] Ion-beam cancer therapy (IBCT) was first realised in the 1950s as proton therapy due to the favourable shape of the depth–dose distribution resulting from the fundamental difference in the energy deposition profile between charged projectiles and photons. For ions, this shape is characterised by the Bragg peak, which is a sharp maximum in the dose deposition close to the end of their trajectories, corresponding to the maximum in the linear energy transfer (LET). This key feature enables high doses of radiation to be delivered to tumours during IBCT, maximising the destruction of cancer cells while minimising the radiation damage to surrounding healthy tissue.

Over the past 30 years, technological and clinical advances of IBCT have developed more rapidly than the understanding of radiation damage with ions. Although empirical approaches have so far produced exciting results for several hundred thousand patients,[7] many questions about the mechanisms involved in radiation damage with ions remain open, and the fundamental quantitative scientific knowledge of the involved physical, chemical, and biological effects is, to a significant extent, missing. In fact, a series of studies[8–10] that elucidated the importance of low-energy (below the ionisation threshold) electrons appeared around 2000–2003, while the treatment of patients at GSI (Darmstadt, Germany) started in 1997.[11] The dominant molecular mechanism of a double-strand break (DSB), the most important DNA lesion,[12,13] is still unknown. Even the significance of the relation of DNA damage (including DSBs) compared to the damage of other cellular components to the cell inactivation or sterilisation is not entirely clear. In addition to IBCT, the mechanisms of radiation damage (including biodamage) induced by protons and heavier ions have attracted attention in the context of many important applications,[14] such as radiation protection of humans, electronic devices, and materials, e.g., from galactic cosmic rays during potential long-term human space missions.[15,16]

Over many decades of using radiation with photons, a large body of data relating the radiation damage to deposited dose has been accumulated. These data are currently used to describe the biological damage caused by ions.[2] However, there are significant qualitative and quantitative differences between the effects of ions and photons on tissue. The first difference is in the localisation of the dose distribution for ions, as opposed to the mostly uniform dose distribution for photons. This feature is manifested longitudinally (along the ion's path) as the Bragg peak. Radially (with respect to the ion's path) it appears as a sharply decreasing (within several tens of nanometres) radial dose distribution, whereas the average distance between adjacent ions in clinically used beams is several hundreds nanometres.

The second difference is a consequence of the first. Secondary particles, such as electrons and free radicals produced as a result of the interaction (ionisation and excitation) of ions with the medium, emerge at the location of the Bragg peak in much larger number densities than those produced by photons, and their distribution is also non-uniform. These secondary particles are largely responsible for biological damage. In order to assess the damage, it is therefore important to distinguish the biological effects of the locally deposited dose and the local number density of the secondary particles. For instance, clustered damage, which is considered more lethal than isolated damage, may be caused by several low-energy electrons, which are not associated with a large dose deposition. This changes the effect of radiation qualitatively and quantitatively.[2,17]

There are also differences in the chemical interactions related to a different balance between free electrons, free radicals, and other agents for ion versus photon irradiation. These differences affect, for example, the resistivity of cells to radiation and are therefore important for the assessment of radiation damage. Finally, the Bragg peak leads to thermomechanical effects resulting from large gradients in the radial dose deposition.

One of the most important questions in the foundation of the science devoted to radiation damage with ions is the question of the molecular mechanisms leading to DNA damage or, more generally, biodamage. While 'whether the biodamage leads to cell sterilisation?' is a biological question, that about the mechanisms of biomolecular damage belongs to the realm of physics and chemistry.

The phenomenon-based multiscale approach (MSA) to the physics of radiation damage with ions has been formulated and elaborated over the past 15 years (see Refs. 4,5,18,19 and references therein). This approach considers relevant physical, chemical, and biological

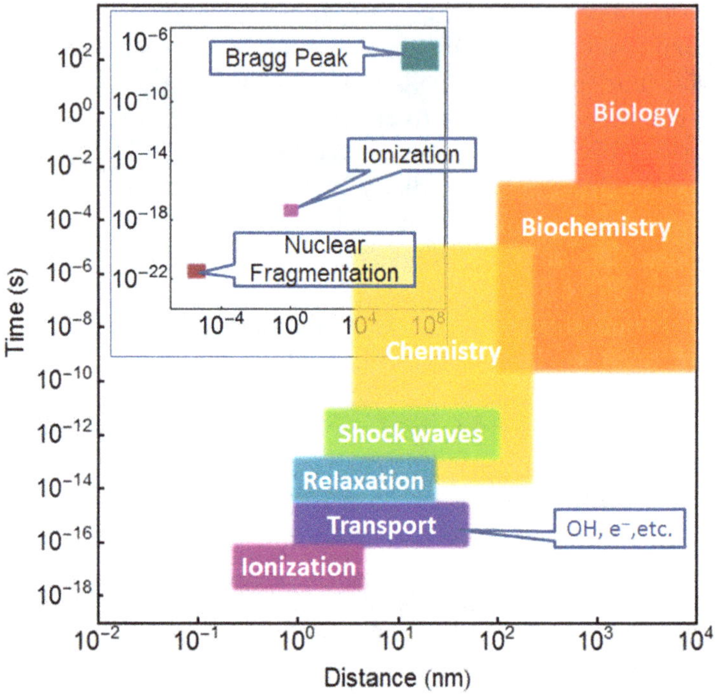

Fig. 9.1. Features, processes, and disciplines, associated with ion beam therapy shown in a space–time diagram, depicting approximate temporal and spatial scales of the phenomena. The history from ionisation/excitation to biological effects on the cellular level is shown in the main figure, and the features of ion propagation are shown in the inset.
Source: Reproduced from Ref. 4 with permission from Springer Nature.

effects occurring on different scales in space, time, and energy, and it explores their manifestation as biological damage. Temporal and spatial scales are shown schematically in Fig. 9.1. The distinguishing feature of the MSA, compared tp many other methods devoted to the assessment of biological damage,[20–25] is its inclusiveness with respect to effects occurring on all scales relevant to radiation damage.

The key phenomena and processes addressed by the MSA are ion stopping in the medium, the production of secondary electrons and free radicals as a result of ionisation and excitation of the medium, the transport of secondary electrons and reactive molecular species, the interaction of secondary particles with biomolecules, radiation chemistry, thermomechanical effects caused by nanoscale shock waves induced by ions, and the analysis of induced biodamage. The important outcome of the MSA is the prediction of cell response to ion irradiation based on the assessment of complex DNA damage produced by a cascade of the aforementioned processes. The MSA also demonstrated great success in predicting cell survival probabilities as a function of the radiation dose in a wide range of system parameters, including different cell types, ions with different values of LET, and oxygenation levels, as well as different cell repair conditions.[4,5,26–28] An overview of some of these main achievements is given in this chapter.

9.2 Multiscale Scenario of Radiation Damage

Biological damage due to ionising radiation is initiated by ions incident on tissue. For ion beams used clinically in IBCT, the initial energy of ions ranges from a few MeV to hundreds of MeV. In the process of propagation through tissue, ions lose their energy in the processes of ionisation, excitation, nuclear fragmentation, etc. Most of the energy lost by an ion is transferred to tissue.[a] Naturally, radiation damage is associated with this transferred energy, and the dose (i.e., deposited energy density) is a common indicator for the assessment of the damage.[2,4,29] The profile of the LET along the ion's path is characterised by a plateau followed by a sharp Bragg peak. The position of this peak depends on the initial energy of the ion and marks the location of the maximum radiation damage. In the process of radiation therapy, a tumour is being 'scanned' with the Bragg peak both laterally and longitudinally in order to deposit a large dose to the target and spare healthy tissues surrounding it. The longitudinal scanning is achieved either by changing the initial energy of projectiles or by using scatterers of variable thickness in front of the target.

However, the deposition of large doses in the vicinity of the Bragg peak does not explain how the radiation damage occurs since projectiles themselves only interact with a few biomolecules along their trajectory, and this direct damage is only a small fraction of the

[a]The only part that is not transferred is emitted as radiation. This part, in the case of ions interacting with tissue, is deemed to be insignificant.

overall damage. It is commonly understood that the secondary electrons and free radicals produced in the processes of ionisation and excitation of the medium with ions are largely responsible for the vast portion of the biodamage.

Secondary electrons are produced during a rather short duration of 10^{-18}–10^{-17} s following the ion's passage. The energy spectrum of these electrons has been extensively discussed in the literature.[30–34] It was found that most secondary electrons have energy below 50 eV (more than 80% for an ion energy of 0.3 MeV/u, corresponding to the kinetic energy of ions near the Bragg peak), and only a few secondary electrons (less than 10% for 0.3 MeV/u ions) have energy higher than 100 eV. This has several important consequences. First, the ranges of propagation of these secondary electrons in tissue are rather small, around 10 nm.[35] Second, the angular distribution of their velocities as they are ejected from their original host and as they scatter further is largely uniform;[23] this allows one to consider their transport using a random walk approach.[17,18,36–38]

The next time scale of 10^{-16}–10^{-15} s corresponds to the propagation of secondary electrons in tissue. These electrons (which start with about 45–50 eV energy) are called ballistic. Such electrons lose most of their energy within 1–1.5 nm of the ion's path.[39] After that, they continue propagating, elastically scattering with the molecules of the medium until they get bound or solvated electrons are formed. It is important to note that these low-energy electrons remain important agents for biodamage since they can attach to biomolecules such as DNA, causing dissociation.[40] The solvated electrons may also play an important role in the scenario of radiation-induced biodamage.[13,41,42]

Additionally, the energy lost by electrons during the previous stage in the processes of ionisation, excitation, and electron–phonon interaction is transferred to the medium. As a result of this relaxation, the medium within about a 1–1.5 nm cylinder (for ions not heavier than iron) around the ion's path becomes very hot.[39,43] This cylinder is referred to as the 'hot cylinder'. The pressure inside this cylinder increases by a factor of about 10^3 compared to the pressure in the medium outside the cylinder. This pressure builds up by about 10^{-14}–10^{-13} s and it is a source of a cylindrical shock wave.[44] This shock wave propagates through the medium for about 10^{-13}–10^{-11} s. Its relevance to biodamage is as follows. If the shock wave is strong enough (where the strength depends on the distance from the ion's path and the LET), it may inflict damage directly by breaking covalent bonds in a DNA molecule.[39] Besides, the radial collective motion that takes place during this time is instrumental in propagating highly reactive species (such as hydroxyl radicals), just

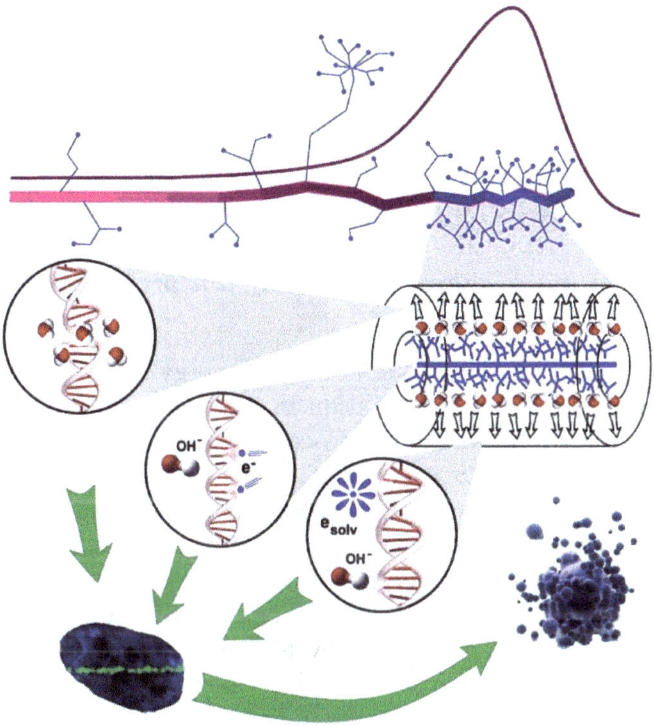

Fig. 9.2. The scenario of biological damage with ions. Ion propagation ends with a Bragg peak, shown in the top-right corner. A segment of the track at the Bragg peak is shown in more detail. Secondary electrons and radicals propagate away from the ion's path damaging biomolecules (central circle). They transfer the energy to the medium within the hot cylinder. This results in a rapid temperature and pressure increase inside this cylinder. The shock wave (shown in the expanding cylinder) due to this pressure increase may damage biomolecules by mechanical stress (left circle), but it also effectively propagates reactive species, such as radicals and solvated electrons to larger distances (right circle). A living cell responds to all the shown DNA damage by creating foci (visible in the stained cells), in which enzymes attempt to repair the induced lesions. If these efforts are unsuccessful, the cell dies; an apoptotic cell is shown in the lower-right corner. *Source*: Reproduced from Ref. 4 with permission from Springer Nature.

formed solvated electrons, etc., to a larger radial distance (up to tens of nanometres), thus increasing the area of an ion's impact.

The assessment of the primary damage to DNA molecules and other parts of cells due to the above effects has been done within the MSA. This damage happens within 10^{-5} s from the ion's passage and consists of various lesions on DNA and other biomolecules. Some of these lesions may be repaired by the living system, but some may not, and the latter may lead to cell sterilisation. The scenario described above is illustrated in Fig. 9.2.

9.3 Propagation of Ions in Tissue and Primary Ionisation of the Medium

9.3.1 *The main characteristics of ion propagation in the medium*

Ions propagating in the tissue lose their energy due to ionisation of the medium, nuclear fragmentation in collisions with nuclei, excitations of the medium, etc. This process is described by the stopping power of the medium, $S = -dE/dx$, where E is the kinetic energy of the ion and x is the longitudinal coordinate. In quantitative terms, this characteristic is often considered synonymous to the LET, which describes the amount of energy absorbed by the medium per unit length of the projectile's trajectory. The LET profile for ions is characterised by a plateau followed by a sharp Bragg peak where the LET reaches its maximum. The tail is caused by the energy loss of the lighter products of nuclear fragmentation, such as protons, neutrons, and α-particles.

The behaviour of the LET is explained by features of inelastic cross-sections (CSs) of the projectile in the medium. The Bragg peak in the stopping power of massive charged particles is described by the Bethe–Bloch formula:[45–47]

$$-\frac{dE}{dx} = \frac{4\pi n_e z^2 e^4}{mV^2}\left[\ln\frac{2mV^2}{\langle I\rangle(1-\beta^2)} - \beta^2\right], \qquad (9.1)$$

where m and e are the mass and charge of electron, V is the velocity of the projectile, $\beta = V/c$ (c is the speed of light in vacuum), $z|e|$ is the charge of the projectile, n_e is the number density of electrons in the target, and $\langle I \rangle$ is the mean excitation energy of its molecules.

Equation (9.1) provides the dependence of the stopping power on the energy of the ion and practically depends on a single parameter, the mean excitation energy $\langle I \rangle$. This parameter for liquid water is usually chosen empirically somewhere between 70 and 80 eV.[48,49] The use of such a parameter is sufficient for the calculations of the position of the Bragg peak and its shape, and Eq. (9.1) has been used in many Monte Carlo (MC) simulations for that purpose.[48] However, this parameter hides all physical processes such as ionisation and excitation of the medium, even though these processes are important for understanding the scenario of radiation damage. In Refs. 18,30,31, a different approach has been used: the singly differentiated (with respect to the secondary electron energy) ionisation CSs of water molecules in the medium have been employed as a physical input. This uncovers the physics integrated in the empirical parameter $\langle I \rangle$ and allows for not only describing the features of the Bragg peak but also obtaining the energy spectrum of secondary electrons, which are very much involved in subsequent radiation damage.

9.3.2 *Singly differentiated cross-sections of ionisation*

The total ionisation CS, σ_t, differentiated with respect to secondary electron kinetic energy, W, i.e., singly differentiated cross-section (SDCS) is the main quantity of interest. Besides the kinetic energy of secondary electrons and the properties of target molecules, the SDCS depends on the velocity V of the projectile and its charge, $z|e|$.

In Refs. 30,31,50, the semi-empirical Rudd's expression[51] for the calculation of SDCS has been used:

$$\frac{d\sigma_t}{dW} = z^2 \sum_i \frac{4\pi a_0^2 N_i}{I_i} \left(\frac{I_0}{I_i}\right)^2 \frac{F_1(v_i) + F_2(v_i)\omega_i}{(1+\omega_i)^3 \left(1 + \exp(\alpha(\omega_i - \omega_i^{\max})/v_i)\right)}. \tag{9.2}$$

This analytic expression, containing a number of fitting parameters, is a combination of the experimental data and calculations within the plane-wave Born approximation and other theoretical models.[51] In Eq. (9.2), the sum is taken over the electron shells of a target molecule (e.g., a water molecule), $a_0 \approx 0.529$ Å is the Bohr radius, $I_0 \approx 13.6$ eV, N_i is the shell occupancy, I_i is the ionisation potential of the shell, $\omega_i = W/I_i$ is the dimensionless normalised kinetic energy of the ejected electron, $v_i = \sqrt{mV^2/(2I_i)}$ is the dimensionless normalised projectile velocity, and $\omega^{\max}$ is the cut-off energy which depends on v_i and I_i.[4,51]

Functions F_1 and F_2 in Eq. (9.2) contain a number of fitting parameters, which depend on the medium. The parameters of Rudd's model have been developed to calculate the SDCS for the ionisation of water vapour,[51] liquid water,[31] and DNA nucleobases.[52] Rudd's model was initially developed for non-relativistic protons[51] and was later modified to include heavier ions at relativistic velocities.[30]

An alternative method has been used in Ref. 53, where the dielectric formalism is based on the experimental measurements of the energy-loss function of the target medium, $\mathrm{Im}\,(-1/\epsilon(\mathcal{E},q))$, where $\epsilon(\mathcal{E},q)$ is the complex dielectric function and $\hbar q$ and $\mathcal{E}$ are the momentum and energy transferred in the electronic excitation, respectively.[54,55] This formalism allows us to obtain the SDCS not only for liquid water but also for a real biological medium containing sugars amino acids, etc.[56,57] If the energy-loss function is experimentally known, many-body interactions and target physical state effects are naturally included in these calculations.

9.3.3 *The position of the Bragg peak*

The stopping CS, defined as

$$\sigma_{\mathrm{st}} = \sum_i \int_0^\infty (W + I_i) \frac{d\sigma_{t,i}}{dW} dW, \tag{9.3}$$

where the sum is taken over all electrons of the target, gives the average energy lost by a projectile in a single collision, which can be further translated into energy loss within an ion's trajectory segment dx:

$$S(E) = -\frac{dE}{dx} = n\sigma_{st}(E), \tag{9.4}$$

where n is the number density of molecules of the medium. This quantity is known as the stopping power.[29,49] As discussed in Section 9.3.1, for ions, this quantity is similar to the LET.

The LET found from Eq. (9.4) is a function of the kinetic energy of the ion rather than the ion's position along the path in the medium. However, the dependence of LET (and other quantities) on this position is more suitable for IBCT applications. Integrating the inverse LET, given by Eq. (9.4), yields

$$x(E) = \int_{E}^{E_0} \frac{dE'}{|dE'/dx|}, \tag{9.5}$$

where E_0 is the initial energy of the projectile. The obtained correspondence between the position of the ion along the path and its energy allows one to obtain all quantities of interest in terms of x rather than E. As an illustration, the depth dependence of the average LET as a function of x is shown in Fig. 9.3.

The calculations of the LET include the effects that were discussed above, such as SDCS calculated using the semi-empirical Rudd's model (9.2), modified for relativistic energies using the effective charge described in Section 9.3.4. The effect of energy straggling

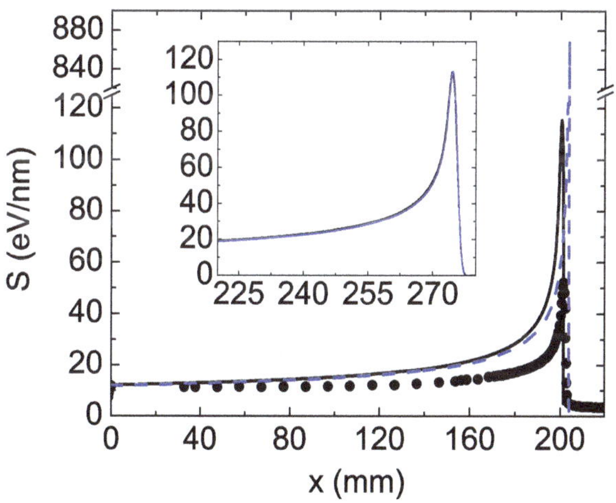

Fig. 9.3. The dependence of the LET on depth for carbon ions in liquid water. The calculations (solid line) are done for ions with an initial energy of 330 MeV/u and using Eqs. (9.3) and (9.4). Experimental results[58] for the same energy are shown with dots. The dashed line depicts the LET dependence without the effect of energy straggling. In the inset, two almost coinciding curves show the agreement between the analytical calculations and MC simulations[48] for 420 MeV/u carbon ion projectiles with straggling being included.
Source: Reproduced from Ref. 4 with permission from Springer Nature.

due to multiple ion scattering, described in Section 9.3.5, is also taken into account. This effect explains why the height of the Bragg peak decreases with the increasing initial energy of ions and thus the increasing depths of the corresponding Bragg peaks.

In Fig. 9.3, the LET calculated using the methodology described above is compared with the experimental results.[58] As can be seen from the figure, the experimental dots at the Bragg peak are systematically lower than the calculated curve, the difference being due to the nuclear fragmentation component, which has not been included in the analytical calculations. It is feasible to include it, as has been done in Ref. 59 for protons, if the appropriate fragmentation CSs are known. As a result of nuclear fragmentation reactions, the beam is attenuated, and new projectiles such as protons, neutrons, and α-particles are formed. Since these particles are lighter than the incident ions, after fragmentation they carry a larger portion of the energy, and their penetration depths are larger than that of the original ions.[58] This results in a tail after the Bragg peak, which can also be seen in Fig. 9.3.

9.3.4 *Charge transfer effect*

The incident ions are usually stripped of all electrons; however, as they slow down, they pick electrons off, and their charge reduces. In order to find the stopping power and estimate the secondary electron spectra (in the first approximation), the charge z of the stripped ion in Eq. (9.2) should be replaced with an effective charge z_{eff} which decreases with decreasing energy, making the ionisation CS effectively smaller. In Refs. 30,31, the following empirical formula for z_{eff}, suggested by Barkas,[60] was used:

$$z_{\text{eff}} = z(1 - \exp(-125\beta z^{-2/3})). \tag{9.6}$$

The effective charge given by this expression slowly changes at high projectile velocities but rapidly decreases in the vicinity of the Bragg peak. As a result, charge transfer significantly affects the height of the Bragg peak and only slightly shifts its position towards the projectile's entrance. This happens due to the stopping CS as a function of velocity has a sharp peak as velocity decreases. At the same time, the stopping CS σ_{st} is proportional to z_{eff}^2. If the latter decreases with decreasing V, the Bragg peak shifts towards the direction of the beam's entrance to the tissue. For instance, accounting for charge transfer, the Bragg peak for carbon ions occurs at $E = 0.3$ MeV/u rather than at $E = 0.1$ MeV/u.

9.3.5 *The effect of ion scattering*

Tracks of ions emerging from clinically used accelerators do not interfere, i.e., the effects of a single ion do not spread far enough to reach the area affected by adjacent ions. Therefore, it is usually sufficient to study a single ion interacting with tissue and then combine these effects to relate the action of the beam to the dose. Although the Bragg peak is a feature of every ion's LET, each peak cannot be observed separately. Since each projectile in the beam experiences its own multiple scattering sequence, peaks for different ions occur at slightly different spatial locations, and only the Bragg peak, averaged over the whole beam, is observed experimentally. Therefore, in order to compare the shape of the Bragg peak with experiments, the whole ion beam should be considered.

In Ref. 30, the Bragg peak for an ion beam was obtained by introducing the energy-loss straggling due to ion scattering. The energy straggling, described using a semi-analytical model,[61] is given by

$$\left\langle \frac{dE}{dx}(x) \right\rangle = \frac{1}{\lambda_{\rm str}\sqrt{2\pi}} \int_0^{x_0} \frac{dE}{dx}(x') \exp\left[-\frac{(x'-x)^2}{2\lambda_{\rm str}^2} \right] dx', \quad (9.7)$$

where x_0 is the maximum penetration depth of the projectile and $\lambda_{\rm str} = 0.8$ mm is the standard deviation of the longitudinal straggling calculated for a carbon ion of this energy range.[62] The Bragg peak shown in Fig. 9.3 was calculated using Eq. (9.7).

9.3.6 *Energy spectra of secondary electrons*

The emission of secondary electrons in collisions of protons and heavier ions with atoms and molecules has been studied theoretically and experimentally for decades.[34,51,63,64] The quantity of interest is the probability of producing N_e secondary electrons with kinetic energy W, in the interval dW, emitted from a segment Δx of the trajectory of a single ion at the depth x, corresponding to the kinetic energy of the ion E. This quantity is proportional to the SDCS discussed in Section 9.3.2:

$$\frac{dN_e(W,E)}{dW} = n\Delta x \frac{d\sigma_t}{dW}, \tag{9.8}$$

where n is the number density of molecules of the medium (for water at standard conditions, $n \approx 3.3 \times 10^{22}$ cm^{-3}). Equation (9.8) relates the energy spectrum of secondary electrons to the SDCS regardless of the method used to obtain it.

One important characteristics that can be obtained from the SDCS is the average energy of the secondary electrons, $\langle W \rangle$, which is given by

$$\langle W \rangle(E) = \frac{1}{\sigma_t} \int_0^{\infty} W \frac{d\sigma_t}{dW} dW. \tag{9.9}$$

The dependence of $\langle W \rangle$ on the ion's energy E indicates that the energy of secondary electrons is somewhere below 50 eV for the whole range of the ion's energy, and it levels out as the energy of the projectiles increases. There are several implications of this fact:

(1) Since the dependence $\langle W \rangle(E)$ over a relevant range of projectile energies ($E = 0.3 - 400$ MeV/u) is weak, the number of produced secondary electrons is largely proportional to the value of the LET, or, more precisely, to the electronic component of the LET, S_e, which excludes nuclear stopping. If the ion is destroyed in a nuclear collision, the ionisation due to its debris should be considered instead; if it survives, then its ionising capabilities do not change too much, unless it slows down considerably; then, its stopping power may change accordingly.
(2) The expression for $\langle W \rangle$ is independent of the charge of the projectile, e.g., the difference between protons and heavier ions lies in their values of S_e, i.e., in the number of secondary electrons, but not in their relative energy spectra. The difference between the effects of these different ions therefore lies in the number of secondary electrons produced by these ions per unit path length.
(3) Most secondary electrons are only capable of ionising one or two water molecules, so there is no significant avalanche ionisation effect.[30] Since the average energy of secondary electrons in the vicinity of the Bragg peak is about 40 eV, the maximum average energy that can be transferred to the next-generation secondary

electron is $(40\ \text{eV} - I_i)/2$, which is about 15 eV for the outermost electrons – a quantity of energy barely sufficient to cause further ionisation.

(4) Finally, what is crucial for considering the next scale of electron propagation is that at energies below 50 eV, the CSs of the electrons are nearly isotropic,[23,65] and it is possible to use the random walk approximation in order to describe their transport.[17,18,37,38]

9.4 Transport of Secondary Electrons in Ion Tracks

The next stage of the multiscale scenario is related to the emission of secondary electrons (most of which have energies below 50 eV) from the molecules of the medium as a result of ionisation. These electrons are called ballistic until their energy becomes sufficiently small and coupling with phonons, recombination, and other quantum processes begin to dominate their transport. While electrons are ballistic, their interactions with molecules can be described as a sequence of elastic and inelastic collisions. This section gives an overview of an analytical approach to analysing the propagation of ballistic electrons and the comparison of the results obtained with those of MC simulations.

The main mechanism of radiation damage by ballistic electrons is inelastic collisions with targets. A target in this discussion is a biomolecule, such as DNA. Therefore, the probability of biodamage is a combination of the number of electrons (or other secondary particles) colliding with a given segment of a biomolecule and the probability of a certain inelastic process on impact. The first part is described by the fluence of electrons or other particles on the target. Fluence is the integral of the flux of particles (the number of particles hitting a part of the target's surface per unit time) over the entire duration after the ion's passage and over the surface of the target. In general, the fluence depends on the distance of the target from the ion's path and its geometrical orientation. This part can be calculated analytically with sufficient accuracy for understanding the scenario of radiation damage. The second part, i.e., the probability of a certain inelastic process on impact, is more difficult to assess mainly because of the diverse variety of possible processes. However, there are plenty of data that allows one to make reasonable quantitative estimates for this probability.

This section starts with the analysis of the transport of secondary electrons, while the production and transport of reactive species, such as solvated electrons and radicals, are also considered. Then, the fluence for relevant configurations is calculated. Important characteristics of the track structure, such as radial dose, can also be calculated as a byproduct of transport analysis. The random walk approach[66,67] used for these problems allows one to make simple analytical calculations. The main requirement for the use of this approach is that the elastic and inelastic scattering of secondary electrons is isotropic. The anisotropy in the angular dependence of the CSs for sub-50-eV electrons appears to be insignificant.[23] As noted above, in the Bragg peak region, more than 80% of secondary electrons satisfy this condition, and only for less than 10% of δ-electrons with energies higher than 100 eV is this condition violated significantly.

This section is largely devoted to the transport of sub-50-eV electrons. Moreover, unless specifically stated to the contrary, these secondary electrons are produced by carbon ions in the vicinity of the Bragg peak in liquid water. At this part of the ion's trajectory, while a 0.3 MeV/u carbon ion passes 1 μm along the path, a typical radius within which the secondary electrons propagate is about 2–3 nm.[66] This allows one to assume that the electron diffusion is cylindrically symmetric with respect to the ion's path. The electronic component of the LET, S_e, remains nearly constant along this 1 μm of ion's path described by the coordinate ζ. Therefore, the number of ejected secondary electrons per unit length $\frac{dN_e}{d\zeta}$ is independent of ζ. A typical elastic mean free path of sub-50-eV electrons ranges between 0.1 and 0.45 nm.[23,65] Since the scale along the Bragg peak is measured in tens of μm, while the radial scale is only in tens of nm, one can assume ζ to be ranging from $-\infty$ to $+\infty$.

As an ion passes through tissue, it ionises molecules, and the ejected secondary electrons (first generation) start with the average energy of $\sim$45 eV, that is, the average energy of secondary electrons formed in the vicinity of the Bragg peak.[50] These electrons lose most of this energy within 1–1.5 nm of the ion's path, ionising more molecules (including biomolecules). The second generation of slower electrons is thus formed. The transport of these two generations can be represented by coupled diffusion equations:[66]

$$\begin{aligned}\frac{\partial n_1(\mathbf{r},t)}{\partial t} &= D_1\nabla^2 n_1(\mathbf{r},t) - \frac{n_1(\mathbf{r},t)}{\tau_1},\\ \frac{\partial n_2(\mathbf{r},t)}{\partial t} &= D_2\nabla^2 n_2(\mathbf{r},t) + 2\frac{n_1(\mathbf{r},t)}{\tau_1} - \frac{n_2(\mathbf{r},t)}{\tau_2}.\end{aligned} \tag{9.10}$$

Here, the index '1' marks the first generation of secondary electrons; D_1 is their diffusion coefficient, τ_1 is their lifetime (before they convert to the second generation); accordingly, $D_1 = vl/6 = 0.265\ \mathrm{nm}^2\mathrm{fs}^{-1}$ and $\tau_1 = l_{\mathrm{ion}}/v = 0.64$ fs, using the mean free path data.[65,68] These electrons can ionise another water molecule. In this process, these electrons lose ionisation energy to the medium and share the rest of their energy with electrons of the second generation. It is assumed that the remaining energy of the first-generation electrons is equal to the energy acquired by the second-generation electrons, so both types of electrons are referred to as second-generation electrons. Thus, the factor of two appears in the second term on the right-hand side (RHS) of the second equation in (9.10). The index '2' in this equation corresponds to the second-generation electrons; their energy is $\sim$15 eV; accordingly, $D_2 = 0.057\ \mathrm{nm}^2\mathrm{fs}^{-1}$ and $\tau_2 = 15.3$ fs. The solutions of the above equations are[66]

$$\begin{aligned}n_1(t,\rho) &= \frac{dN_e}{d\zeta}\frac{1}{4\pi D_1 t}\exp\left(-\frac{\rho^2}{4D_1 t} - \frac{t}{\tau_1}\right),\\ n_2(t,\rho) &= \frac{1}{2\pi\tau_1}\frac{dN_e}{d\zeta}\int_0^t \frac{1}{D_1 t' + D_2(t-t')}\\ &\quad\times \exp\left(-\frac{\rho^2}{4(D_1 t' + D_2(t-t'))} - \frac{t-t'}{\tau_2} - \frac{t'}{\tau_1}\right) dt'.\end{aligned} \tag{9.11}$$

From solutions (9.11), it follows[66] that a few femtoseconds after the ion's passage, all secondary electrons lose energy, and the number density of secondary electrons is mainly given by $n_2(\rho,t)$. The distribution becomes slightly broader with time, but the main effect is the exponential decrease with time. As a result of this decrease, the so-called pre-solvated electrons are formed. A pre-solvated stage of electrons is a transition stage between low-energy ballistic electrons and a relatively stable compound of electrons with water molecules, known as solvated electrons. This transition takes about 1 ps.

It is also possible to assess the energy deposition density (dose), $\varepsilon(\rho, t)$, in the medium by ions and secondary electrons. In order to do this, it is assumed that the average energy, $\bar{w}$, is deposited to the medium with each ionisation. Moreover, since the first-generation electrons start with about 45 eV and $\bar{w} \approx 15$ eV, the second-generation electrons energy is also approximately $(45 - 15)/2 = 15$ eV. Then, the rate of energy deposition is proportional to the rate of inelastic events:

$$\frac{\partial \varepsilon(\rho, t)}{\partial t} = \bar{w} \left(\frac{dN_e}{d\zeta} \delta^{(2)}(\rho)\delta(t) + \frac{n_1(\rho, t)}{\tau_1} + \frac{n_2(\rho, t)}{\tau_2} \right). \quad (9.12)$$

The first energy deposition occurs right at the ion's path where the molecules are ionised by the ion; this corresponds to the first term on the RHS of Eq. (9.12), where δ's are the corresponding δ-functions. The second deposition (the second term on the RHS) is the ionisation by secondary electrons at the end of their ionisation mean free paths. Finally, the third deposition is due to the remaining energy loss from the excitation of molecules by the second-generation electrons. After that, the electrons enter the pre-solvated stage.

Time integration of Eq. (9.12) gives the dependence of the radial dose on time. Figure 9.4 shows the radial dose distributions at 5, 10, 20, and 50 fs instants. At small radii, the distribution is due to primary ionisation, and it does not change with time. At larger radii, the dose slowly increases because of energy loss by the second generation of electrons.

The solid curve in Fig. 9.4 gives the upper estimate of pressure near the ion's path. Indeed, the energy $U(\rho)$ inside a cylinder of radius ρ and length L, coaxial with the ion's path, is given by

$$U(\rho) = \int_0^{\rho} \varepsilon(\rho')2\pi\rho' d\rho' L, \quad (9.13)$$

where $\varepsilon(\rho)$ is the radial dose given by

$$\varepsilon(\rho) = \int_0^{t_r} \frac{\partial \varepsilon(\rho, t)}{\partial t} dt. \quad (9.14)$$

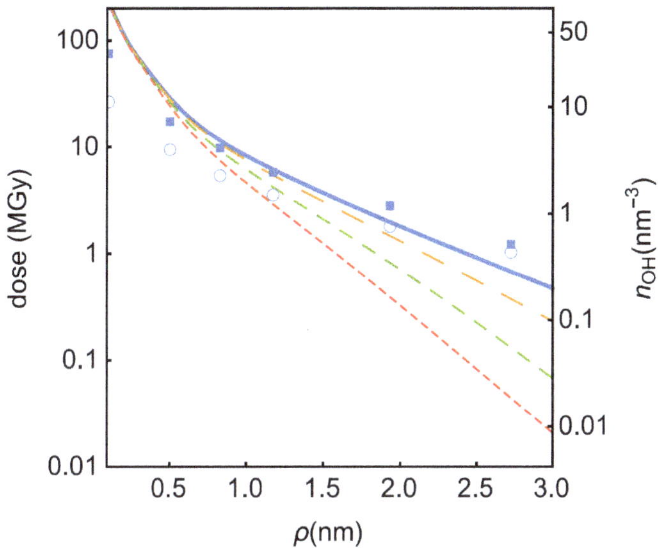

Fig. 9.4. The radial dose distributions as functions of the distance from the ion's path at times 50, 20, 10, and 5 fs are shown with solid and dashed lines with a diminishing dash size, correspondingly. The solid line also represents the initial radial distribution of hydroxyl radicals (the scale is on the right). Solid squares mark the radial dose data[69] for a 2 MeV/u carbon ion, multiplied by a factor of four. Empty circles represent the radial dose by a single generation of secondary electrons, as calculated in Ref. 37.
Source: Reproduced from Ref. 66 with permission from Springer Nature.

Then, the force normal to the surface of this cylinder is given by

$$F(\rho) = -\frac{\partial U(\rho)}{\partial \rho} = \varepsilon(\rho)2\pi\rho L, \tag{9.15}$$

and the pressure $P(\rho)$ is

$$P(\rho) = \frac{F(\rho)}{2\pi\rho L} = \varepsilon(\rho). \tag{9.16}$$

Thus, the radial dose is equal to the pressure in the medium. This notion and the results shown in Fig. 9.4 justify the idea of the ion-induced shock waves discussed in Section 9.5. The time $\leq$50 fs within which the radial dose distribution evolved is sufficient for the temperature and pressure distributions to form around the ion's path, following the process of energy relaxation, in which the energy stored in electronic excitations is transferred to vibrational excitations and then to translational degrees of freedom.

9.5 Thermomechanical Effects

The energy lost by secondary electrons in the processes of ionisation and excitation of the medium is transferred to the vibrational excitation of molecules of the medium due to the electron–phonon interaction. As a result, the medium within the cylinder of the ∼1–2 nm radius surrounding the ion's path is heated up rapidly, and the pressure inside this cylinder increases by several orders of magnitude (e.g., by a factor of 10^3 for a carbon ion at the Bragg peak[43]) compared to the pressure in the medium outside the cylinder. High local temperature and pressure around the ion's path initiate a strong cylindrical explosion of the excited medium, resulting in the formation of an ion-induced shock wave.[44]

Different aspects of this effect have been analysed in the framework of MSA,[28,39,44,70,71] resulting in the theoretical prediction of the ion-induced shock waves. Thermomechanical phenomena induced by ion beams also occur in solid targets, where their effects are easier to detect and which present important technological applications.[72,73] Experimentally, the ion-induced shock waves have not yet been directly observed, but there are several indirect observations, suggesting that their effect is significant for any assessment of ion-induced biodamage.

9.5.1 *Hydrodynamic expansion on the nanometer scale*

The problem of the expansion of the medium driven by the high pressure inside the hot cylinder was analysed in detail in Ref. 44. It was shown that the expansion is cylindrically symmetric. If the ratio of pressures inside and outside the hot cylinder is high enough, as happens for large values of LET, the cylindrical expansion of the medium is described as a cylindrical shock wave, driven by a 'strong explosion'.[74] For an ideal gas, this condition holds until about $t = 1$ ns, but in liquid water, the shock wave relaxes much sooner. Classical molecular dynamics (MD) simulations of liquid water expansion showed[70] that the shock wave weakens by about 0.5 ps after the ion's passage. The solution of the hydrodynamic problem describing the strong explosion regime of the shock wave, as well as its mechanical features and limitations, are described in Refs. 74–76. The solution for the cylindrical case has been reproduced and analysed in Ref. 44 in order to apply it for the nanometre-scale dynamics of the DNA surroundings.

A self-similar flow of water and heat transfer depend on a single variable, ξ. This variable is a dimensionless combination of the radial distance ρ from the ion's path, the time t after the ion's passage, the ion's LET S_e, and the density of undisturbed water, $\varrho = 1$ g/cm^3. This combination is given by

$$\xi = \frac{\rho}{\beta\sqrt{t}}\left(\frac{\varrho}{S_e}\right)^{1/4}, \tag{9.17}$$

where β is a dimensionless parameter equal to 0.86 for $\gamma = C_P/C_V = 1.222$ (where C_P and C_V are molar heat capacity coefficients at constant pressure and volume, respectively).[44] The radius R and the speed u of the wave front are given by

$$R = \rho/\xi = \beta\sqrt{t}\left(\frac{S_e}{\varrho}\right)^{1/4} \tag{9.18}$$

and

$$u = \frac{dR}{dt} = \frac{R}{2t} = \frac{\beta}{2\sqrt{t}}\left(\frac{S_e}{\varrho}\right)^{1/4}, \tag{9.19}$$

respectively. The position of the wave front calculated according to Eq. (9.18) for different projectile ions in the Bragg peak region is illustrated in Fig. 9.5. Pressure P at the wave front can be obtained as[44]

$$P = \frac{2}{\gamma+1}\varrho u^2 = \frac{1}{\gamma+1}\frac{\beta^4}{2}\frac{S_e}{R^2}. \tag{9.20}$$

Then, one can solve the hydrodynamic equations in order to obtain the expressions for speed, pressure, and density in the wake of the shock wave, i.e., behind the wave front.[44]

The forces acting on DNA segments were predicted to be as large as 2 nN, which is more than enough to break a covalent

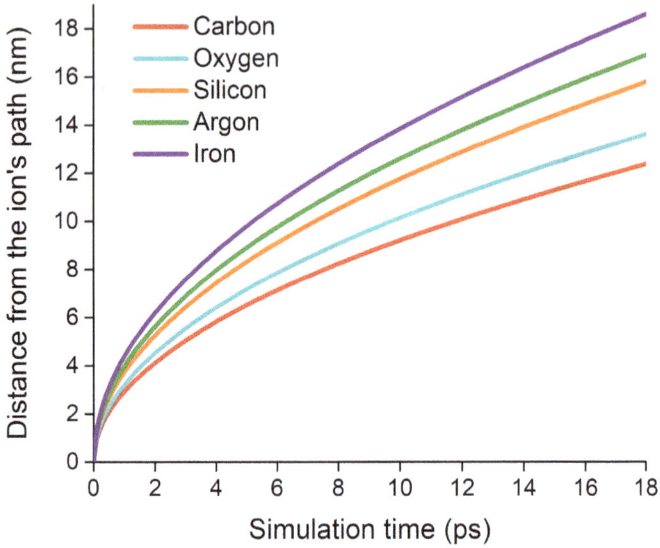

Fig. 9.5. Radial distance travelled by the front of the shock wave induced by the five studied ions in the Bragg peak region. The radial distance increases with the LET of the ion, being smallest for the carbon ion (red curve) and largest for the iron ion (purple curve). The results are obtained using Eq. (9.18).
Source: Reproduced from Ref. 28 with permission from the American Physical Society.

bond; however, these forces are only acting for a short time, and it remained unclear whether this is sufficient to cause severe damage to DNA molecules. A series of papers was devoted to the study of a shock-wave-induced biodamage (e.g., a strand break) by mechanical force.[28,39,71] It was found that the probability of the direct mechanical bond breaking due to shock waves depends on the value of LET. While this effect is negligible for protons and α-particles, it becomes dominant for heavy ions such as iron.[28] The direct DNA damage due to the ion-induced shock waves is discussed in this section.

Another set of questions concerns the transport of reactive species (such as hydroxyl radicals) by the shock wave and its impact on the initial conditions for the chemical phase of radiation damage. These questions were studied in Refs. 28,39,77,78. It turned out that this effect is important even for light ions such as protons. This effect is reviewed in Section 9.5.2.

9.5.1.1 *Quantification of the number of bond breaks in the DNA double twist*

In Ref. 28, reactive MD simulations of the shock-wave-induced damage of a DNA molecule with 30 base pairs have been conducted using the MBN Explorer software package.[79] The simulations employed the reactive CHARMM (rCHARMM) force field,[80] which is also described in detail in Refs. 14,81–83. The simulation geometry is illustrated in Fig. 9.6(a).

The simulations[28] revealed that the projectile ion propagating in close proximity to the geometrical centre of the molecule produces significant damage within the central segment containing 20 DNA base pairs. The DNA damage produced in segments of such size may lead to complex irreparable lesions in a cell.[4,26,84] The number of bond breaks in the DNA double twist was counted after each completed MD simulation and analysed as a function of the distance d_{geo} from the ion's path to the principal axis of inertia of the DNA molecule; see Fig. 9.6(a).

Figure 9.7 shows that the number of bond breaks produced in the DNA double twist by the ion-induced shock wave increases with the LET of a projectile ion. Simulation results obtained for the scaled bond dissociation energies $D_e/6$ (Fig. 9.7(a)) reveal that up to two DNA backbone bonds break due to the shock wave induced by the carbon ion, whereas up to 50 bonds may be broken due to the iron ion impact. For every combination of the bond dissociation energy and ion's LET, the average number of bond breaks, $\mathcal{N}_{\text{SW}}(D_e, S_e)$, fluctuates around certain values within a certain distance range from the ion's path of ~1 nm.[28] As the ion passes at larger distances from the target DNA molecule, the average number of bond breaks within the DNA double twist gradually decreases. Figure 9.7 shows that the shock-wave-induced thermomechanical stress of the DNA mostly occurs at $\lesssim$1 nm from the ion's path for ions lighter than iron.

9.5.1.2 *Force exerted by the shock wave on the DNA*

The characteristic range of the shock-wave-induced thermomechanical damage, R_{SW}, can be evaluated by analysing the pressure on the shock wave front, Eq. (9.20), and the corresponding force exerted by the shock wave on covalent bonds in the DNA backbone.[28] In the

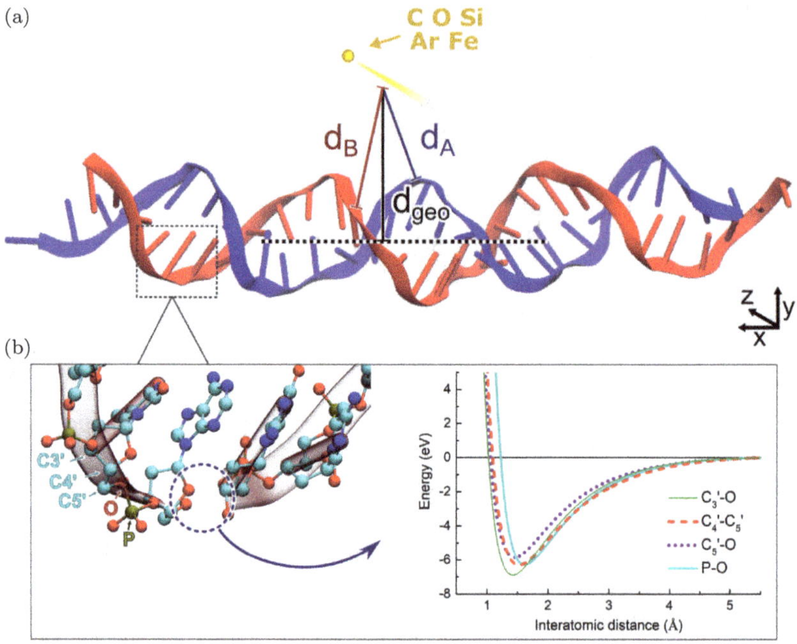

Fig. 9.6. Geometry of the DNA molecule considered in the reactive MD simulations.[28] Panel (a) shows an ion (C, O, Si, Ar, and Fe) propagating in close proximity to the DNA molecule consisting of 30 complementary base pairs. The ion track is oriented along the z-axis; the x-axis is oriented along one of the principal axis of inertia of the chosen DNA molecule, and the y-axis is along the line defining the shortest distance between the ion track and the selected principal axis of inertia. The collision parameter $d_{\rm geo}$ is defined as the displacement of the ion's path along the y-axis with respect to the principal axis of inertia. The specific collision parameters $d_{\rm A}$ and $d_{\rm B}$ are defined as the shortest distances from the ion's path to DNA strand A and strand B, respectively. Panel (b) illustrates C_3'–O, C_4'–C_5', C_5'–O, and P–O bonds in the DNA sugar-phosphate backbone and the corresponding potential energy curves obtained by means of DFT calculations.[85] *Source*: Reproduced from Ref. 28 with permission from the American Physical Society.

cited study, the analysis was performed for the C_3'–O, C_4'–C_5', C_5'–O, and P–O bonds in the DNA backbone, as shown in Fig. 9.6(b). For the sake of simplicity, the model presented in Ref. 28 considered a molecular bond oriented parallel to the direction of the shock-wave propagation. The force stretching the bond is given by

$$F = (\pi a_0^2)\,\frac{\partial P}{\partial r}\,l, \tag{9.21}$$

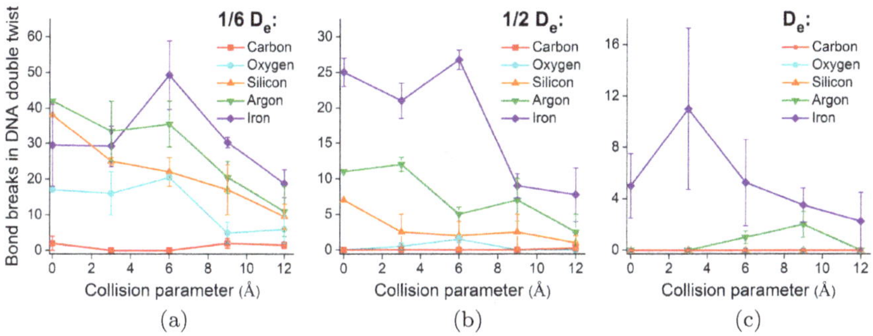

Fig. 9.7. Average number of bond breaks in the DNA double twist calculated as a function of the collision parameter $d_{\rm geo}$ for the five studied ions. $d_{\rm geo}$ is the distance from the ion track to the principal axis of inertia of the DNA segment, as shown in Fig. 9.6(a). Panels (a), (b), and (c) show the results of simulations employing the bond dissociation energies D_e derived from the DFT calculations (see Fig. 9.6(b)) and the values of D_e scaled by the factors of 1/2 and 1/6. Two independent MD simulations were performed for each ion and each collision geometry; error bars indicate the corresponding standard deviation.
Source: Reproduced from Ref. 28 with permission from the American Physical Society.

where πa_0^2 is the transverse area of the molecular bond exposed to the pressure created by the shock wave front, $\frac{\partial P}{\partial r}$ is the pressure gradient, and l is a characteristic interatomic distance in the medium on which the pressure gradient is evaluated. The calculations in Ref. 28 were performed using the value $a_0 = 0.15$ nm corresponding to the van der Waals radius for atoms forming the DNA backbone[86] and $l \approx 0.15$ nm, which is a characteristic length of covalent bonds in the DNA backbone.

It was found that the threshold distance from the ion track, $R_{\rm SW}$, below which the bonds in the DNA backbone can be broken by the shock-wave-induced thermomechanical stress has the following dependence on LET:

$$R_{\rm SW} = b\, S_e^{1/3}, \tag{9.22}$$

where b is a pre-factor depending on the parameters of a specific covalent bond (dissociation energy and stiffness) and on the ion's LET. The complete derivation of this dependence can be found in Ref. 28.

Figure 9.8 shows the dependence of $R_{\rm SW}$ on S_e for the five studied ions in the Bragg peak region.[28] Symbols show the $R_{\rm SW}$ values

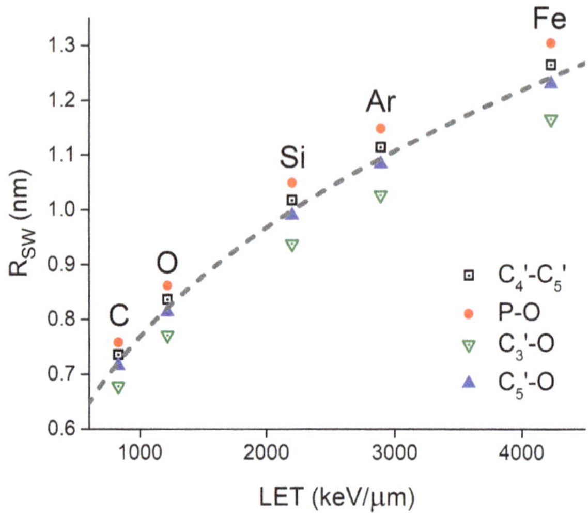

Fig. 9.8. The threshold distance R_{SW} for cleavage of the C_3'–O, C_4'–C_5', C_5'–O, and P–O bonds due to the shock-wave-induced thermomechanical stress by the five studied ions at the Bragg peak region.[28] The R_{SW} values are calculated according to Eq. (9.22) (symbols). The coefficient b in Eq. (9.22) varies in the range (0.072–0.081) $\text{nm}^{4/3}\,\text{eV}^{-1/3}$ for the four bonds considered. Dashed line shows the $R_{\text{SW}} = b\,S_e^{1/3}$ dependence with the average value $b = 0.077\ \text{nm}^{4/3}\,\text{eV}^{-1/3}$.

Source: Reproduced from Ref. 28 with permission from the American Physical Society.

for cleavage of the C_3'–O, C_4'–C_5', C_5'–O, and P–O bonds in the DNA backbone, calculated according to Eq. (9.22). The coefficient b varies in the range (0.072–0.081) $\text{nm}^{4/3}\,\text{eV}^{-1/3}$ for the four bonds considered. The $R_{\text{SW}} = b\,S_e^{1/3}$ dependence with the average value $b = 0.077\ \text{nm}^{4/3}\,\text{eV}^{-1/3}$ is shown in Fig. 9.8 by the dashed line.

The results presented in Fig. 9.8 indicate that the threshold distance from the ion's path for the bond rupture by the pressure gradient on the shock wave front varies from 0.7 nm for a carbon ion at the Bragg peak to 1.3 nm for an iron ion at the Bragg peak. The estimated R_{SW} values are consistent with the results of MD simulations shown in Fig. 9.7. However, no strand breaks have been observed in the simulations for carbon and oxygen projectile ions for the default bond dissociation energies D_e (Fig. 9.7(c)), which can be attributed to a small number of simulated trajectories and a low number of such events. Accounting for different possible orientations of the molecular bonds in the DNA backbone should also increase their average stability. Simulations[28] performed with the scaled bond dissociation energies $D_e/2$ and $D_e/6$ (Fig. 9.7(b) and Fig. 9.7(a), respectively) indicate the formation of bond breaks in the DNA backbone by the shock wave induced by carbon and oxygen ions within the range of distances from the ion track, which are consistent with the $R_{\rm SW}$ values determined by Eq. (9.22).

9.5.2 *Transport of reactive species by the radial collective flow*

The analysis done in Ref. 44 suggests that a considerable collective radial flow emerges from the hot cylinder region of medium. The maximum mass flux density carried by the cylindrical shock wave is given by $\varrho_{\rm f} u$, where $\varrho_{\rm f} = \frac{\gamma+1}{\gamma-1}\varrho$ is the matter density on the wave front. This flux density is inversely proportional to radius ρ and is linear with respect to $\sqrt{S_e}$. It sharply drops to zero in the wake of the wave along with the density. A sharp rarefaction of the volume in the wake of the wave follows from the results in Ref. 44. This is the effect of cavitation on a nanometre scale, and due to this effect, the water molecules of the hot cylinder along with all reactive species formed in this cylinder are pushed out by the radial flow. Such a mechanism of propagation of reactive species, formed within the hot cylinder, is competitive (and becomes dominant at relatively small values of LET) with the diffusion mechanism, studied in track-structure MC simulations.[23]

First estimates of the transport were done in Ref. 39. The time at which the wave front reaches a radius ρ can be derived from Eq. (9.18), as it is equal to $(\rho^2/\beta^2)\sqrt{\varrho/S_e}$. This time has to be compared with diffusion times, which can be estimated for different reactive species as ρ^2/D, where ρ is the distance from the ion's path and D is the corresponding diffusion coefficient. The ratio of these times is equal to $(\sqrt{\varrho/S_e})D/\beta^2$. For all relevant species, the diffusion coefficient is less than 10^{-4} cm^2/s.[87] Therefore, the above ratio is less than $10^{-3}/\sqrt{S_e(\text{keV/nm})}$, which is much less than unity even for protons. For instance, for carbon ion projectiles, the wave front reaches 5 nm from the path in 2.8 ps after the ion's traverse, while hydroxyl radicals reach the same distance via the diffusion mechanism in about 9 ns, a more than 3,000 times longer time. In fact, the lifetime of hydroxyl free radicals is shorter than 5 ns;[13,29,87] therefore, the shock wave transport may be the only means to deliver hydroxyl radicals to distances farther than 3.5 nm of the ion's path.

The next stage of this study was a series of works that were by and large based on MD simulations of shock wave propagation. In Ref. 88, δ-electrons were included in the dynamics following the ion's traverse. These radial doses were obtained at different values of LET for carbon ions, far away and in the vicinity of the Bragg peak. δ-electrons take energy away from the track and, consequently, from the shock waves. The radial dose (or pressure) distributions look similarly to those in Fig. 9.4 when ions are in the Bragg peak region, but they differ significantly if the ion is far from the Bragg peak. These radial dose distributions were used for initial conditions for MD simulations. Depending on these initial conditions, the shock wave was stronger or weaker, sharper or more dispersed. Expectedly, in the vicinity of the Bragg peak (where δ-electrons are suppressed) the results were similar to the step function (a.k.a. 'hot cylinder') initial conditions. All simulations were performed using the MBN Explorer software package.[79]

The following work in this series[77] was devoted to reactive MD simulations performed with MBN Explorer.[79] The rCHARMM force field[80] was used, which enables us to simulate the main chemical reaction involving hydroxyl radicals,

$$OH^{\bullet} + OH^{\bullet} \longrightarrow H_2O_2. \tag{9.23}$$

The transport of radicals in a track was studied at LET away from the Bragg peak and in its region. Figure 9.9 shows the simulation results for 500 keV protons (with low LET), where shock waves were not present/[77] Fig. 9.9(a) shows the mean square displacement of OH radicals, with the transport being purely due to diffusion. A hydroxyl diffusion coefficient $D_{\mathrm{OH}} = 0.22$ Å^2/ps was obtained from these simulations,[77] which agrees with the results of other MC simulations[89] (0.3 Å^2/ps) and with the value of 0.28 Å^2/ps used in the popular MC simulation codes PARTRAC[90] and GEANT4-DNA.[91] The numbers of OH and H_2O_2 molecules during simulations are shown by histograms with error bars in Fig. 9.9(b), representing the average and standard deviations of three independent runs. These are compared with results reported using the GEANT4-DNA code:[92] dashed lines

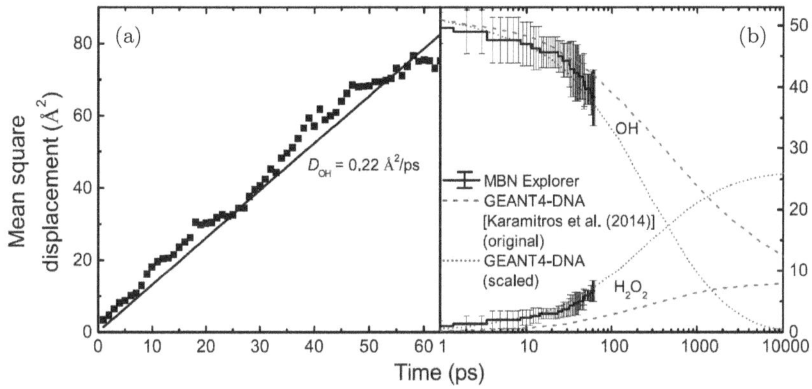

Fig. 9.9. Diffusion and reactivity of the OH radicals produced by 500 keV protons in liquid water, where a shock wave is not expected to occur due to the very low LET. Mean square displacement of OH radicals (panel (a)) and the number of OH and H_2O_2 molecules (panel (b)) as a function of simulation time. The results of reactive MD simulations[77] using MBN Explorer[79] (histograms with error bars) are compared with MC simulations using GEANT4-DNA[92] (dashed and dotted lines).
Source: Reproduced from Ref. 78 published under an open access Creative Commons CC BY license.

depict the original results, in which many reactions among reacting species are possible, while dotted lines are scaled in a way in which only the reaction (9.23) is included (see Ref. 77 for more details). Even though it is computationally expensive to reach long times with MD simulations, it is clear that the results of MD simulations at low LET (and no shock wave) agree with the evolution predicted by MC simulations, where shock waves are not included. These findings presented in Fig. 9.9 validate the MD-based model used in Ref. 77.

Similar quantities were found for the case of a carbon ion in the Bragg peak region, where the shock waves are substantial. The shock waves develop naturally in the MD simulations as a result of the energy deposited along the ion's track. However, computer simulations allow one to artificially 'switch off' this process in order to distinguish its influence in the evolution of the radiation chemistry scenario. Such an artificial switching off was used in Ref. 77 to assess how the presence of shock wave modifies the radiation chemistry as compared to the traditional picture of transport of reactive species where shock waves is not included.

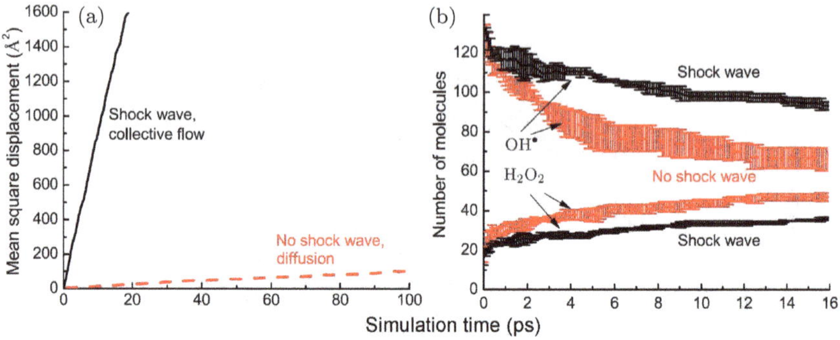

Fig. 9.10. Panel (a): Mean square displacement of the OH radicals produced around a 200 keV/u carbon ion's path. Panel (b): Time evolution of the number of OH radicals and produced H_2O_2 molecules. Results of simulations where the shock wave is allowed to develop (transport by the collective flow, solid lines) and where it is artificially 'switched off' (transport by diffusion, dashed lines) are shown, which demonstrate the shock wave effects on the radiation chemistry. *Source*: Reproduced from Ref. 78 published under an open access Creative Commons CC BY license.

Figure 9.10(a) compares the mean square displacement of radicals in simulations with the wave included (i.e., in presence of a substantial collective flow) versus that switched off (i.e., with the transport being purely due to diffusion). The effective diffusion coefficient for the case of shock-wave-induced collective flow is about 80 times larger than the diffusion coefficient obtained for the simulation with the shock wave switched off.

The reactivity of the OH radicals in the presence of the shock wave is illustrated in Fig. 9.10(b). It depicts the average number of OH and H_2O_2 molecules, together with the standard deviations obtained after three independent runs, in the two cases where the shock wave is artificially switched off and naturally allowed to develop, respectively. As is clearly seen, the evolution of the number of molecules is different for the simulations with the shock waves on and off. The transport of radicals by the collective flow does not only propagate the radicals much faster than diffusion but also prevents their recombination, both by their spreading and by creating harsh conditions for the formation of O–O bonds.[77] These results confirm the evaluations made in Ref. 66 that, in the absence of ion-induced shock waves, the presence of hydroxyl radicals outside of ion tracks would be strongly suppressed. These results also allow one to infer the radial range of spread of reactive species.

A further step has been made in Ref. 28, where the range of shock wave propagation in liquid water and hence the range of shock-wave-driven propagation of reactive species have been evaluated from MD simulations by means of MBN Explorer. A shock wave propagation was simulated in a pure water box with dimensions of 49.5 nm $\times$ 49.5 nm $\times$ 8.0 nm. The evolution of radial density of water around the tracks of the five projectile ions (carbon to iron at the Bragg peak) is shown in Fig. 9.11. Water molecules located in the vicinity of the ion's path are transported away from their initial positions, which results in the formation of a cylindrical cavity around the ion's path. The radius of the cavity grows with time up to the values of about 6 nm for carbon and oxygen ions, while the density of water increases at larger distances from the ion's path. Following the mass conservation law, the mass of water molecules transported from the region in the vicinity of the ion track is equal to the mass of excess water molecules at larger distances from the track.[28]

The MD simulation of the propagation of the shock wave induced by carbon ions revealed[28] that at a certain time instance ($\sim$16.7 ps), the shock wave stopped propagating away from the ion's path and started to move slowly in the inward direction. This happens when the pressure of the shock wave front drops below a certain value determined by the balance of the pressure at the wave front and the water surface tension pressure.[93,94]

According to Eq. (9.18), the front of the carbon-ion-induced shock wave propagates by the time $t = 16.7$ ps to the distance $R = 11.9$ nm from the ion track. This characteristic distance defines the propagation range of free radicals, R_r, which are transported by the shock-wave-driven collective flow (see also Section 9.6.3). To evaluate the range of shock wave propagation for heavier ions, one would need to run longer simulations and consider much larger simulation boxes than the one used in Ref. 28. Alternatively, the range of the shock wave propagation induced by high-LET ions can be estimated from the analysis of the pressure on the shock wave front according to Eq. (9.20). According to the analysis performed in Ref. 28, the pressure induced by the shock wave front is $P = 0.115$ GPa at the distance $R = 11.9$ nm from the ion track, which is the maximum distance of the wave front propagation for a carbon ion.

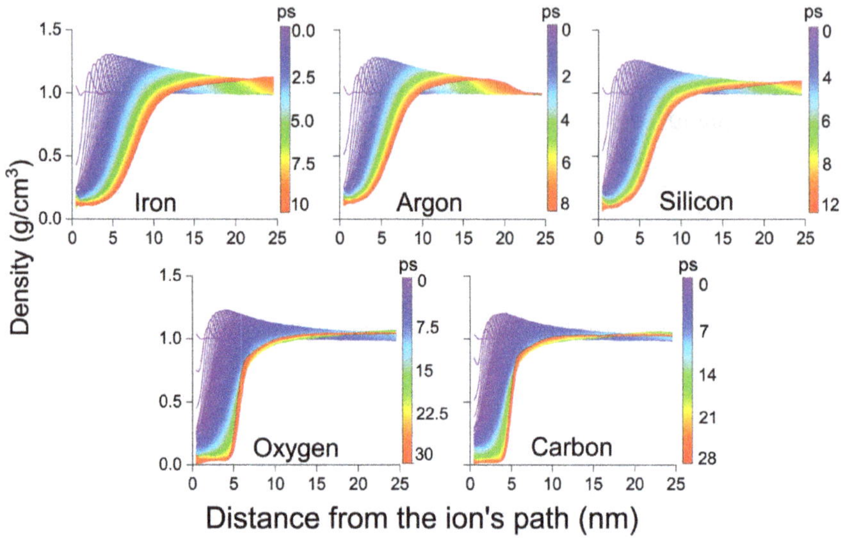

Fig. 9.11. The average density of water as a function of radial distance from the ion track at different time instants. The dynamics of the water medium is caused by propagation of the shock wave induced by iron, argon, silicon, oxygen, and carbon ions inside a 49.5 nm $\times$ 49.5 nm $\times$ 8.0 nm water box. The simulation time (measured in ps) is depicted as a colour scale. Water molecules located in the vicinity of the ion's path are transported away from their initial positions, which results in the formation of a cylindrical cavity around the ion's path (see the purple curves). The radius of the cavity grows with time up to the values of about 6 nm for carbon and oxygen ions, while the density of water increases at larger distances from the ion's path (see the red curves). The shock wave induced by iron, argon, and silicon ions (top row) reached the simulation box boundary much faster than the shock wave induced by oxygen and carbon ions (bottom row). Therefore, the simulation time for iron, argon, and silicon ions is about three times shorter than for the lighter ions.
Source: Reproduced from Ref. 28 with permission from the American Physical Society.

The shock wave propagation in the radial direction away from the ion's path causes cavitation in its wake, leading to the formation of a rarefied cylindrical region.[44,93] This effect was observed in MD simulations described in Fig. 9.11. During the shock wave propagation, the pressure at the shock wave front becomes balanced by the surface tension pressure building up at the border of the wake region. As a result, the growth of the wake region stops, and this region shrinks after the instant when the pressure of the wave front becomes equal to the water surface tension pressure on the surface of the wake region. The latter can be estimated as

$$P = \frac{\xi}{R}, \tag{9.24}$$

where ξ is the coefficient of surface tension and R is the distance from the ion track. Using the aforementioned values $P = 0.115$ GPa and $R = 11.9$ nm for the carbon ion at the Bragg peak, one obtains the surface tension coefficient $\xi = 1.37$ N/m. Note that the medium in the vicinity of the shock wave front is far from equilibrium, and the density of the medium is significantly higher than that of water at ambient conditions. The high water density in the vicinity of the shock wave front and the large amount of energy deposited into the medium explain the large value of the corresponding surface tension coefficient.

Assuming that ξ depends weakly and smoothly on LET (or does not depend at all) at the pressures balance point, the radii $R_{\rm r}$ for different ions can be evaluated. The radii $R_{\rm r}$ define the propagation ranges of free radicals transported by the shock wave induced by different ions. Equating the pressure on the shock wave front, Eq. (9.20), and the surface tension pressure, Eq. (9.24), one obtains a linear dependence of $R_{\rm r}$ on LET:

$$R_{\rm r} = \frac{\beta^4}{2(\gamma + 1)\,\xi} S_e. \tag{9.25}$$

The calculated values of $R_{\rm r}$ for the five studied ions at the Bragg peak region are summarised in Table 9.1. The results indicate that, for an iron ion, the free radicals are transported by the shock wave

Table 9.1. The maximum radii of the shock wave wake region for the five ions from carbon to iron at their Bragg peak energies.[28] These characteristic radii define the propagation range of free radicals $R_{\rm r}$, Eq. (9.25), which are transported by the propagating shock wave.

	Carbon	Oxygen	Silicon	Argon	Iron
$R_{\rm r}$ (nm)	11.9	17.5	31.6	41.5	60.8

to a distance of ∼60 nm. This value exceeds by an order of magnitude the typical distances that radicals can diffuse in the medium at equilibrium during the time corresponding to the duration of the formation of the shock wave wake region with the maximum radius.

9.6 Assessment of Radiation Damage Using MSA

9.6.1 *Calculation of the number of secondary electrons incident on a DNA target*

As shown in Ref. 66, the number densities of the first and second generations of secondary electrons are given by Eqs. (9.11). A target is chosen to be a rectangle of area $\xi\eta$, where $\xi = 6.8$ nm and $\eta = 2.3$ nm are the length of two twists and the diameter of a DNA molecule, respectively. Thus, electrons or radicals hitting such a target would be hitting two rungs of a DNA molecule masked by this target. The plane of the target is chosen to be parallel to the ion's path with dimension ξ along and η perpendicular to the path; see Fig. 9.12. Then, the angle $\phi = 2\arctan\frac{\eta/2}{r}$ inscribes the target in a plane perpendicular to the ion's path, where r is the distance between the target and the path.

The number of first-generation electrons hitting the described target segment of area $r\phi\xi \approx \xi\eta$ parallel to the ion's path per unit time is

$$\Phi_1 = -\phi\xi r D_1 \frac{\partial n_1(r,t)}{\partial r} = \frac{\phi}{2\pi}\frac{dN_1}{dx}\frac{r^2\xi}{4D_1 t^2}\exp\left(-\frac{r^2}{4D_1 t} - \frac{t}{\tau_1}\right). \tag{9.26}$$

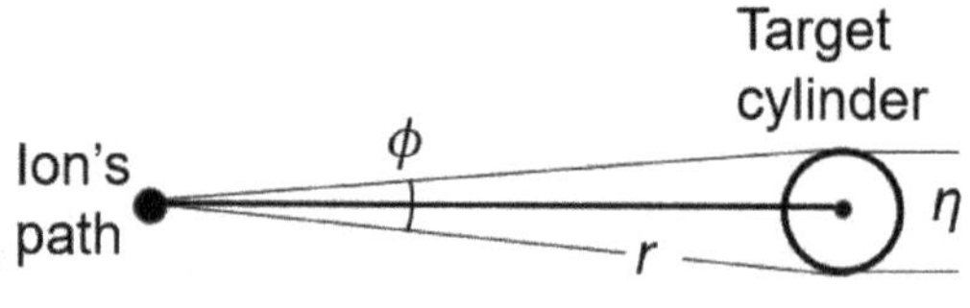

Fig. 9.12. The geometry of the problem in the plane perpendicular to ion's path. The target cylinder that encloses a DNA twist is shown as a circle with a diameter η. The dimension ξ is perpendicular to the plane of the figure. *Source*: Reproduced from Ref. 96 with permission from Springer Nature.

Its integral over time,

$$\int_0^\infty \Phi_1 dt = \frac{\phi}{2\pi}\int_0^\infty \frac{dN_1}{dx}\frac{r^2\xi}{4D_1t^2}\exp\left(-\frac{r^2}{4D_1t}-\frac{t}{\tau_1}\right)dt$$
$$= \frac{\phi}{2\pi}\frac{dN_1}{dx}\frac{r\xi}{\sqrt{D_1\tau_1}}K_1\left(\frac{r}{\sqrt{D_1\tau_1}}\right), \tag{9.27}$$

where K_1 is the Macdonald function (modified Bessel function of the second kind),[95] gives the total number of first-generation secondary electrons that hit this area. The contribution from the second generation of electrons is obtained similarly:

$$\Phi_2(t,r) = -r\phi\xi D_2\frac{\partial n_2(r,t)}{\partial r}$$
$$= \phi\frac{\xi r^2 D_2}{4\pi\tau_1}\frac{dN_1}{dx}\int_0^t \frac{1}{(D_1t' + D_2(t-t'))^2}$$
$$\times \exp\left(-\frac{r^2}{4(D_1t' + D_2(t-t'))}-\frac{t-t'}{\tau_2}-\frac{t'}{\tau_1}\right)dt', \tag{9.28}$$

and then

$$\int_0^\infty \Phi_2 dt = \phi\frac{\xi r^2 D_2}{4\pi\tau_1}\frac{dN_1}{dx}\int_0^\infty\int_0^t \frac{1}{(D_1t' + D_2(t-t'))^2}$$
$$\times \exp\left(-\frac{r^2}{4(D_1t' + D_2(t-t'))}-\frac{t-t'}{\tau_2}-\frac{t'}{\tau_1}\right)dt'dt \tag{9.29}$$

gives the number of second-generation secondary electrons that hit the same area.

The average number of simple lesions due to a single ion, $\mathcal{N}_e(r)$, can be obtained as the sum

$$\mathcal{N}_e(r) = \mathcal{N}_1(r) + \mathcal{N}_2(r) = \Gamma_e(S_e)\int_0^\infty \Phi_1 dt + \Gamma_e(S_e)\int_0^\infty \Phi_2 dt, \tag{9.30}$$

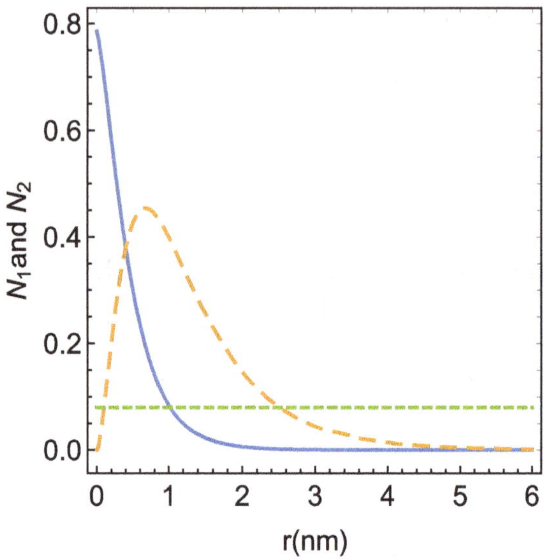

Fig. 9.13. The average numbers of simple lesions due to a single carbon ion with a Bragg-peak energy propagating through a uniform chromatin as functions of radial distance from the ion's path. The lesions are produced by secondary electrons of the first (solid line) and second (dashed line) generations, $\mathcal{N}_1(r)$ and $\mathcal{N}_2(r)$, respectively. These dependencies are calculated using the corresponding number of hits, Eqs. (9.27) and (9.29), multiplied by the probability of the production of a simple lesion per hit, $\Gamma_e = 0.03$ (used in Ref. 4). A straight green line shows the value for reactive species, $\mathcal{N}_r(r)$, calculated using Eq. (9.37) with numbers from Ref. 26.
Source: Reproduced from Ref. 96 with permission from Springer Nature.

where $\mathcal{N}_1(r)$ and $\mathcal{N}_2(r)$ are the average numbers of simple lesions produced by secondary electrons of the first and second generations, respectively, and $\Gamma_e(S_e)$ is the average probability for an electron to induce a simple lesion on a hit. The probability Γ_e depends on the kinetic energy of secondary electrons incident on the DNA double twist and hence on the ion's LET. The functions $\mathcal{N}_1(r)$ and $\mathcal{N}_2(r)$ are plotted in Fig. 9.13.

Equation (9.30) gives the average number of simple DNA lesions due to secondary electrons of the first and second generations as a function of the distance of the target DNA segment from the ion's path. The next step is the addition of the contribution of reactive species, which is a product of the average number of hits on the chosen area by reactive species N_r (whose value is similar to $\int_0^\infty \Phi_1 dt$

and $\int_0^\infty \Phi_2 dt$) and the probability of lesion production per hit, Γ_r. The number N_r depends on the value of the LET since at small values of LET, the transport of radicals is defined by diffusion, while at high values, the collective flow is expected to dominate this process.

9.6.2 *Calculation of the reactive species contribution for small values of LET*

The number of produced reactive species, such as free radicals and solvated electrons, depends on the LET. If the LET is not very high, it is expected that the number of reactive species is proportional to the secondary electron production, dN_1/dx, and, therefore, increases nearly linearly with the LET.[4] At sufficiently high values of the LET, further production of radicals is possible due to water radiolysis at locations adjacent to the ion's path, but the quantification of this effect is a subject of future studies. In Ref. 66, a linear dependence between the number of reactive species and the LET was assumed, and the difference between high and low values of the LET was defined only by the mechanism of transport of the reactive species; at low LET, this transport is defined by diffusion. Moreover, this means that chemical reactions such as $2OH^\bullet \to H_2O_2$ and $e^-_{aq} + OH \to OH^-$ are rare, and their frequency can be neglected compared to the diffusion term.[66] Thus, the transport of reactive species in the low-LET case can be calculated by solving a diffusion equation,

$$\frac{\partial n_r}{\partial t} = D_r \nabla^2 n_r, \tag{9.31}$$

where n_r is the number density and D_r is the diffusion coefficient for reactive species.

The initial conditions for this equation can be taken from Ref. 66:

$$\frac{\partial n_r(r,t)}{\partial t} = \frac{dN_1}{dx}\delta^{(2)}(r)\delta(t) + \frac{n_1(r,t)}{\tau_1} + \frac{n_2(r,t)}{\tau_2}, \tag{9.32}$$

where the first term describes the species formed at sites of original ionisations by the projectile, while other two terms are due to inelastic processes involving secondary electrons of the first and secondary generations, respectively. Ionisations and excitations that lead to the production of reactive species, $n_r(r,t)$, through the mechanism described by Eq. (9.32) take place by about 50 fs.[66] By that time, the forming reactive species are localised within 3 nm of the ion's path. These are the initial conditions for the following propagation of reactive species by the diffusion and/or collective flow, which happen on much larger scales of up to 100 ps in time and 50 nm in distance. Therefore, in Ref. 66, a simplified initial condition was used:

$$\frac{\partial n_r(r,t)}{\partial t} = K\frac{dN_1}{dx}\delta^{(2)}(r)\delta(t), \tag{9.33}$$

where K is the number of reactive species produced due to each secondary electron of the first generation ejected by an ion. The value of $K \approx 6$ can be evaluated as follows. The primary ionisation produces H_2O^+, which is likely to produce a hydroxyl radical.[13] The same thing happens when the secondary electron of the first generation ionises a water molecule (and thus becomes a second-generation electron). Then, two electrons of the second generation (the ionising and ejected ones) can produce about four reactive species: two as a result of further energy loss in inelastic processes and another two if they become solvated electrons. A more accurate number for K can be obtained if the probabilities of the above processes are combined following a comprehensive radiochemical analysis.

The solution to Eq. (9.31) with the initial condition (9.33) is given by

$$n_r(r,t) = K\frac{dN_1}{dx}\frac{1}{4\pi D_r t}\exp\left(-\frac{r^2}{4D_r t}\right). \tag{9.34}$$

The next step is to find the number of reactive species, Φ_r, incident on the target at a distance r from the ion's path per unit time. This can be done similarly to Eqs. (9.26) and (9.27):

$$\Phi_r = -\phi\xi r D_r\frac{\partial n_r(r,t)}{\partial r} = \frac{\phi}{2\pi}K\frac{dN_1}{dx}\frac{r^2\xi}{4D_r t^2}\exp\left(-\frac{r^2}{4D_r t}\right), \tag{9.35}$$

and its integral over time then reads as

$$\int_0^\infty \Phi_r dt = \frac{\phi\xi}{2\pi}K\frac{dN_1}{dx} = K\frac{dN_1}{dx}\frac{\xi}{\pi}\arctan\frac{\eta/2}{r}. \tag{9.36}$$

9.6.3 *Calculation of the reactive species contribution for large values of LET*

If the reactive species are formed in large quantities as a result of a high-LET ion's traverse, the collective flow due to the shock wave is the main instrument for the transport of these species away from the ion's path. The reactive MD simulations[28,77] performed using the MBN Explorer software package[79,80] have shown that, in the case of high LET, the range depends on the value of the LET, but a more extensive investigation is still needed to obtain a more detailed dependence.

In Ref. 26, this transport was described using the following model. The value of the average number of lesions at a distance r from the ion's path, $\mathcal{N}_r = \Gamma_r N_r$, was considered to be a constant within a certain LET-dependent range, $R_r(S_e)$, defined by the radius of shock wave propagation:

$$\mathcal{N}_r(r, S_e) = \mathcal{N}_{r,0}(S_e)\,\theta[R_r(S_e) - r]. \tag{9.37}$$

$\theta(x)$ on the RHS of Eq. (9.37) is the Heaviside function. A linear dependence, $R_r \propto S_e$, was explored in an earlier study,[93] and a conservative estimate of $R_r \approx 10$ nm was derived for carbon ions in the Bragg peak region.[4] As described in Section 9.5.2, the value of R_r for carbon ions was recently evaluated more precisely on the basis of MD simulations,[28] and the R_r values for heavier ions were estimated from the $R_r \propto S_e$ dependence by analysing the pressure at the shock wave front; see Table 9.1.

The value $\mathcal{N}_{r,0}(S_e)$ depends on the number of formed radicals, which is in turn proportional to the number of generated secondary electrons and hence proportional to the LET. $\mathcal{N}_{r,0}(S_e)$ also depends on the level of oxygenation of the medium since the concentration of oxygen dissolved in the medium affects the number of formed radicals and, consequently, the creation of DNA lesions. The value of $\mathcal{N}_{r,0}(S_e) = 0.08$ for the environment with a normal concentration of oxygen was derived earlier[26] by comparing the experimental results[97] for plasmid DNA, dissolved in pure water and in a scavenger-rich solution, and irradiated with carbon ions at Bragg peak energies. A large number of cell survival experiments performed at hypoxic conditions were reproduced with the value of $\mathcal{N}_{\mathrm{r},0} = 0.04$.[5,26] More information about $\mathcal{N}_r$ can be obtained from reactive and irradiation-driven MD simulations[14,80,81,98] using the MBN Explorer package,[79] which is capable of resolving the corresponding temporal and spatial scales.

A comprehensive picture of the transport of reactive species includes diffusion (dominant at low values of the LET), collective flow (dominant at high values of the LET), and chemical reactions. With this understanding, Eq. (9.36) should gradually transform into Eq. (9.37) as the LET increases. In addition to these equations, the effective range of reactive species is limited by the criterion of lethality (see Section 9.6.4), which requires a minimal fluence at each site. Further discussion on this topic was presented in Ref. 27.

9.6.4 *Evaluation of the number of ion-induced DNA lesions and cell survival probability*

This section outlines the procedure for evaluating the number of lesions of the DNA molecule produced upon its irradiation with ions and the corresponding cell survival probabilities. The previously developed formalism described in detail in Refs. 4,5,19 has been extended[28] towards accounting for the DNA lesions produced by the thermomechanical stress imposed on the DNA molecule by the propagating shock wave.

This formalism starts with the calculation of $\mathcal{N}(r)$ — the total average number of simple lesions, i.e., single-strand breaks (SSBs), produced in a DNA double convolution (a DNA double twist) located at distance r from the ion's path. According to the MSA analysis,[4,5] this number is equal to

$$\mathcal{N}(r, S_e) = \mathcal{N}_e(r, S_e) + \mathcal{N}_r(r, S_e) + \mathcal{N}_{\rm SW}(r, S_e), \tag{9.38}$$

where $\mathcal{N}_e(r, S_e)$, Eq. (9.30), is the number of lesions produced by secondary electrons; $\mathcal{N}_r(r, S_e)$, Eq. (9.37), is the number of lesions produced by free radicals; and $\mathcal{N}_{\rm SW}(r, S_e)$ is the number of lesions produced by the thermomechanical stress imposed on DNA by the propagating shock wave.

The creation of DNA lesions by secondary electrons, free radicals, and the shock wave are statistically independent events taking place at different time scales after the ion passage.[4,5] Therefore, the total average number of simple lesions in a DNA double twist, $\mathcal{N}(r, S_e)$, is a cumulative quantity derived by integrating all the events over time. $\mathcal{N}_e(r, S_e)$ and $\mathcal{N}_r(r, S_e)$ were worked out earlier within the MSA,[4,5,19,26,66] whereas $\mathcal{N}_{\rm SW}(r, S_e)$ has been quantified recently by means of MD simulations.[28]

Knowing $\mathcal{N}(r, S_e)$ at a given distance r, one can use the Poisson statistics to calculate probabilities of different independent events. The probability to produce k lesions in a DNA double twist placed at a distance r from the ion track is equal to

$$\mathcal{P}_k(r, S_e) = \frac{\mathcal{N}^k(r, S_e)}{k!}\, e^{-\mathcal{N}(r, S_e)}. \tag{9.39}$$

A lethal DNA lesion is defined within the MSA framework as one double-strand break (DSB) plus at least two additional single lesions (SSBs) occurring within a DNA double twist.[4] This definition relies on earlier findings[84,99] that complex DNA damage is irreparable for a cell if the damage occurs in a localised DNA segment, which typically consists of two helical turns containing 20 base pairs.

Lesions within the DNA double twist may occur on one DNA strand or be present on both strands. Let us define n as the number of vulnerable covalent bonds in one strand in a DNA double twist. In Ref. 28, this number was set equal to 80, as four vulnerable covalent bonds (C_3'–O, C_4'–C_5', C_5'–O, and P–O) in the sugar-phosphate backbone correspond to each nucleotide; see Fig. 9.6(b). Therefore, the total number of such bonds in both strands in the DNA double twist is $2n = 160$.

According to Ref. 28, the probability $\mathcal{P}_\nu^{(1)}$ of creating ν SSBs located on one DNA strand within the double twist is equal to

$$\mathcal{P}_\nu^{(1)} = \begin{cases} 1, & \nu = 0 \\ 2\,\dfrac{n!}{(n-\nu)!}\,\dfrac{(2n-\nu)!}{(2n)!}, & \nu = 1, 2, \ldots, n \\ 0, & \nu = n+1, n+2, \ldots, 2n, \end{cases} \tag{9.40}$$

and the probability $\mathcal{P}_\nu^{(2)}$ that ν lesions result in at least one DSB within the DNA double twist reads as[28]

$$\mathcal{P}_\nu^{(2)} = \begin{cases} 0, & \nu = 0, 1 \\ 1 - \mathcal{P}_\nu^{(1)} \\ \quad \equiv 1 - 2\,\dfrac{n!}{(n-\nu)!}\,\dfrac{(2n-\nu)!}{(2n)!}, & \nu = 2, 3, \ldots, n \\ 1, & \nu = n+1, n+2, \ldots, 2n. \end{cases} \tag{9.41}$$

Following the above-introduced criterion for a lethal DNA lesion, one can derive the probability of such event as follows:

$$\begin{aligned} \mathcal{P}_l(r, S_e) = {} & \lambda \sum_{\nu=3}^{\nu_{\max}} \mathcal{P}_\nu^{(1)} \frac{\mathcal{N}^\nu(r, S_e)}{\nu!} e^{-\mathcal{N}(r,S_e)} \\ & + \lambda \mathcal{P}_3^{(2)} \frac{\mathcal{N}^3(r, S_e)}{3!} e^{-\mathcal{N}(r,S_e)} \\ & + \sum_{\nu=4}^{\nu_{\max}} \mathcal{P}_\nu^{(2)} \frac{\mathcal{N}^\nu(r, S_e)}{\nu!} e^{-\mathcal{N}(r,S_e)}. \end{aligned} \tag{9.42}$$

Here, λ is the probability that an SSB can be converted to a DSB and $\nu_{\max} = 2n$. Accounting for λ relies on the experimental findings[9,10] that the DSBs caused by low-energy electrons with energies higher than ~5 eV happen in one hit. In this case, the subsequent break in the second DNA strand occurs due to the action of debris generated by the first SSB. Following Refs. 9,10, λ is set equal to 0.15 within the MSA framework.[4]

The first term on the RHS of Eq. (9.42) describes the sum of probabilities of having all ν ($\nu = 3, 4, \ldots, 2n$) lesions on one DNA strand with the subsequent conversion of one SSB into a DSB. The second term is the probability of three lesions with at least one DSB among them and the subsequent conversion of one SSB into a DSB, i.e., creating two DSBs. The third term is the sum of probabilities of ν lesions ($\nu = 4, 5, \ldots, 2n$) with creation of at least one DSB.

After simple algebraic transformations Eq. (9.42) can be rewritten in the following form:

$$\mathcal{P}_l(r,S_e) = \lambda \sum_{\nu=3}^{\nu_{\max}} \frac{\mathcal{N}^\nu(r,S_e)}{\nu!}\, e^{-\mathcal{N}(r,S_e)} + (1-\lambda)\sum_{\nu=4}^{\nu_{\max}} \mathcal{P}_\nu^{(2)}\, \frac{\mathcal{N}^\nu(r,S_e)}{\nu!}\, e^{-\mathcal{N}(r,S_e)}, \tag{9.43}$$

with $\mathcal{P}_\nu^{(2)}$ defined above in Eq. (9.41).

At small LET values when the number of lesions in a DNA double twist $\mathcal{N}(r,S_e) \lesssim \lambda \ll 1$, the probability of lethal events $\mathcal{P}_l(r,S_e)$ is simplified to

$$\mathcal{P}_l(r,S_e) \simeq \lambda\, \frac{\mathcal{N}^3(r,S_e)}{3!} + \frac{7}{8}\, \frac{\mathcal{N}^4(r,S_e)}{4!}. \tag{9.44}$$

In the region $1 \ll \mathcal{N}(r,S_e) \ll \nu_{\max}$, the probability of a lethal lesion is given by

$$\mathcal{P}_l(r,S_e) \simeq 1 - 2(1-\lambda)\, e^{-\frac{\mathcal{N}(r,S_e)}{2}} - \frac{3}{24}(1-\lambda)\, e^{-\mathcal{N}(r,S_e)}\, \mathcal{N}^3(r,S_e). \tag{9.45}$$

This means that the probability of lethal lesions $\mathcal{P}_l(r,S_e) \to 1$ within the entire region, where $1 \ll \mathcal{N}(r,S_e) \ll \nu_{\max}$.

Knowing $\mathcal{P}_l(r,S_e)$, one can calculate the number of lethal events in a cell nucleus traversed by a projectile ion. Equation (9.43) represents the probability of creating a lethal lesion in a DNA double twist located at a distance r from the ion track. Integrating $\mathcal{P}_l(r,S_e)$ over the area perpendicular to the ion's trajectory and convoluting the result with the number density of DNA double twists in a cell nucleus, one derives the average number of lethal lesions per unit length of the ion's trajectory:

$$\frac{dN_l}{dx} = n_s \int_0^\infty \mathcal{P}_l(r,S_e)\, 2\pi r\, dr \equiv n_s\, \sigma_l(S_e). \tag{9.46}$$

Here, n_s is the number density of DNA double twists in a cell nucleus, which is equal to the number of DNA base pairs accommodated in a cell nucleus, $N_{\rm bp}$, divided by the number of DNA base pairs in one double twist and by the nuclear volume V_n:[26]

$$n_s = \frac{N_{\rm bp}}{20\,V_n}. \tag{9.47}$$

The function $\sigma_l(S_e)$ is the cross-section of producing lethal DNA damage in a cell nucleus, which depends on LET and the concentration of oxygen in the target. The $\sigma_l(S_e)$ dependence originates from the dependence of $\mathcal{N}(r, S_e)$ on LET; this dependence is discussed later in this section.

The number of lethal events in a cell nucleus at a given dose d produced by $N_{\rm ion}$ ions is equal to[4]

$$Y_l(S_e) = \frac{{\rm d}N_l(S_e)}{{\rm d}x}\,\bar{z}\,N_{\rm ion}(S_e), \tag{9.48}$$

where $\bar{z}$ is the average distance traversed by $N_{\rm ion}$ ions through the cell nucleus. The average number of ions hitting the nucleus, $N_{\rm ion}$, depends on the nucleus area A_n, the dose and LET:

$$N_{\rm ion}(S_e) = A_n \frac{\varrho\, d}{S_e}, \tag{9.49}$$

where ϱ is the mass density of the irradiated medium, taken to be equal to the density of liquid water ($\varrho = 1$ g/cm^3). The probability of cell survival is given by the probability of zero lethal lesions occurrence.[4] According to the Poisson statistics, it is equal to

$$\Pi_{\rm surv} = e^{-Y_l(S_e)}. \tag{9.50}$$

Substituting $N_{\rm ion}$ into Y_l and taking the logarithm of $\Pi_{\rm surv}$, one obtains

$$\ln \Pi_{\rm surv} = -Y_l(S_e) = -\frac{{\rm d}N_l(S_e)}{{\rm d}x}\,\bar{z}\,A_n \frac{\varrho\, d}{S_e}, \tag{9.51}$$

where $\left(\frac{{\rm d}N_l}{{\rm d}x}\,\bar{z}\right)$ is the average number of lethal events created by a single ion in a cell nucleus. A successful comparison of calculated survival curves at different LET values for a number of different cell lines has been demonstrated in Refs. 26,27. A brief overview of this analysis is given in Section 9.7.1.

The dependence of the cross-section of a DNA lethal lesion, σ_l, and the number of lethal lesions in a cell nucleus, Y_l, on LET was analysed in Ref. 28. For the low-LET irradiation, the number of lesions in a DNA double twist is $\mathcal{N}(r, S_e) \ll 1$, and hence $\mathcal{P}_{l,S_e}(r) \sim \mathcal{N}^3(r, S_e)$, according to Eq. (9.44).

At low LET values, e.g., for protons in the Bragg peak region, the number of lesions in a DNA double twist is linearly proportional to LET, $\mathcal{N}(r, S_e) \propto S_e$. This dependence arises because (i) the number of secondary electrons incident on a DNA double twist, and hence $\mathcal{N}_e(r)$, is proportional to LET;[4] and (ii) the number of formed free radicals is proportional to the number of secondary electrons.[66] The shock wave induced by protons at the Bragg peak does not transport free radicals and other reactive species to the distances much larger than the secondary electron propagation range R_e. However, the diffusion of free radicals on the picosecond timescale might be affected by the temperature increase in the vicinity of the ion tracks. Following the analysis described above, the free radicals propagation range R_r is smaller than $R_\mathrm{e} \sim 1 - 2$ nm in the S_e region up to 70–140 keV/μm. Combining Eqs. (9.44) and (9.46) and using the $\mathcal{N}(r, S_e) \propto S_e$ dependence, one obtains the following dependence of σ_l on LET:

$$\sigma_l(S_e) \propto S_e^3. \tag{9.52}$$

The number of lethal lesions in a cell nucleus Y_l, Eq. (9.48), thus increases with LET as

$$Y_l(S_e) \propto \frac{\sigma_l(S_e)}{S_e} \sim S_e^2. \tag{9.53}$$

The quantity $\mathcal{N}_\mathrm{r}(r, S_e)$ might grow with LET due to the increase in the SW radius and, correspondingly, R_r. The growth of R_r results in lowering the density of free radicals and thus their recombination rate constant. The additional growth of $\mathcal{N}_\mathrm{r}(r, S_e)$ with S_e will result in a faster growth of σ_l and Y_l with increasing S_e. Even steeper dependencies of σ_l and Y_l on LET may arise at higher S_e values when the number of lesions in a DNA double twist $\mathcal{N}(r, S_e) \lesssim 1$ due to a steeper dependence of $\mathcal{P}_l(r, S_e)$ on $\mathcal{N}(r, S_e)$.

Finally, let us consider the case of $\mathcal{N}(r, S_e) \gg 1$ when the probability $\mathcal{P}_l(r, S_e) \rightarrow 1$. This case describes iron and heavier ions at the Bragg peak.[28] In this case, multiple lesions are created by the shock-wave-induced thermomechanical stress of the DNA double twist within the distance range of $r < R_{\rm SW}(S_e)$ from the ion track. The number of lesions produced by the shock wave in the region $r < R_{\rm SW}(S_e)$ is much larger than the number of lesions produced by secondary electrons and free radicals, i.e., $\mathcal{N}_{\rm SW}(r, S_e) \gg \mathcal{N}_{\rm e}(r, S_e)$ and $\mathcal{N}_{\rm SW}(r, S_e) \gg \mathcal{N}_{\rm r}(r, S_e)$. As described in detail in Section 9.5.1.1, the number of lesions $\mathcal{N}_{\rm SW}(r, S_e)$ has been evaluated through the MD simulations for the five ions with different LET values at the Bragg peak region.[28] These results suggest the following stepwise dependence of $\mathcal{N}(r, S_e)$ on distance r from the ion track:

$$\mathcal{N}(r, S_e) = \mathcal{N}_{\rm SW}(S_e)\, \theta(R_{\rm SW}(S_e) - r). \tag{9.54}$$

Since at large LET values $\mathcal{N} \approx \mathcal{N}_{\rm SW} \gg 1$ is within the range $r < R_{\rm SW}(S_e)$, the probability $\mathcal{P}_l(r, S_e) \rightarrow 1$ at $r < R_{\rm SW}(S_e)$. Then, for high-LET irradiation, one obtains

$$\sigma_l(S_e) = \int_0^\infty \mathcal{P}_l(r, S_e)\, 2\pi r\, dr = \pi R_{\rm SW}^2(S_e). \tag{9.55}$$

In this case, the number of lethal events in a cell nucleus at a given dose d produced by $N_{\rm ion}$ ions, Eq. (9.48), transforms into

$$Y_l(S_e) = \pi R_{\rm SW}^2(S_e)\, n_s\, \bar{z}\, A_n \frac{\rho\, d}{S_e}, \tag{9.56}$$

with the probability of cell survival being given by Eq. (9.50).

The characteristic range for inducing bond breakage by the shock-wave-induced thermomechanical stress, $R_{\rm SW}$, has been discussed above in Sections 9.5.1.1 and 9.5.1.2. As demonstrated in Ref. 28, $R_{\rm SW}$ depends on LET as $R_{\rm SW} = b\, S_e^{1/3}$, where b is the proportionality factor. In the case of large LET values (where the condition $\mathcal{N} \approx \mathcal{N}_{\rm SW} \gg 1$ is fulfilled), the number of lethal events in a cell nucleus can be written as

$$Y_l(S_e) = \alpha\, S_e^{-1/3}, \tag{9.57}$$

where

$$\alpha = \pi\, b^2\, n_s\, \bar{z}\, A_n \rho\, d. \tag{9.58}$$

This means that the number of lethal events in a cell nucleus at a given dose d decreases slowly with high LET, which corresponds to the experimental observations for iron and heavier ions at the Bragg peak region; see Section 9.7.3.

Now, let us return to the yield of lethal lesions in a cell nucleus Y_l, Eq. (9.48). That expression can be rewritten in several ways:

$$Y_l(S_e) = \frac{dN_l}{dx}\,\bar{z}\,N_{\text{ion}}(d) = \frac{\pi}{16}N_g\frac{\sigma(S_e)}{S_e}d = \frac{\pi}{16}N_g\sigma(S_e)F_{\text{ion}}, \tag{9.59}$$

where F_{ion} is the ion fluence. Let us comment on the universality and versatility of this expression. Its first representation, $\frac{dN_l}{dx}\,\bar{z}\,N_{\text{ion}}(d)$, indicates that the yield of lethal lesions is a product of only two quantities, the number of lethal lesions per unit length of ion's trajectory and the total length of ion's path through the cell nucleus, which can be broken into the average length of a traverse by the number of ions passing through the nucleus. This number depends on the dose d. However, in the case of ions, the dose is not an independent parameter but is regulated by the ion fluence:

$$d = \frac{S_e\bar{z}\,N_{\text{ion}}}{\varrho V} = \frac{S_e\,F_{\text{ion}}}{\varrho}, \tag{9.60}$$

where ϱ is the mass density of the nucleus.

Let us discuss a number of effects and limits.

First, it is interesting to analyse the limits of N_{ion}; its minimum (non-zero) value is one. Then, both the dose and the yield are defined by LET, with the former being linear with respect to it and the latter being linear when the LET is small, but may be quadratic when the LET is larger. This enhancement is expected as a result of the transport of reactive species to larger distances by ion-induced shock waves. If the LET is too large, the lethal damage may occur already on a fraction of $\bar{z}$. This means that the 'rest' of the dose is wasted, the relative biological effectiveness is reduced, and the so-called 'overkill' effect is observed. On the other hand, when N_{ion} is very large, the ion tracks are likely to overlap. This corresponds to the case of large ion fluences, which was discussed in Ref. 96. This limit may be important for applications of laser-driven proton beams.[100]

Second, Eq. (9.60) is only valid when the LET is the same for all ions; when it is not, e.g., in the case of a spread-out Bragg Peak (SOBP), then

$$d = \sum_j \frac{S_{ej} \bar{z} F_j}{\rho}, \tag{9.61}$$

where a subscript j indicates a corresponding component of the ion beam. This dependence was exploited in Ref. 101, and it is discussed in Section 9.7.2 because the SOBP is used clinically and in many experiments.

Third, more intriguing effects are seen in the second representation, $\frac{\pi}{16} N_g \frac{\sigma(S_e)}{S_e} d$. As mentioned earlier, at relatively small values of LET, $\frac{dN_l}{dx}$ is linear with LET. Then, the yield is linear with the dose. However, when LET increases, $\sigma(S_e) = \kappa S_e^{\delta}$, where κ is a coefficient and δ is some positive power that depends on the range of LET, as explained above. As a result, the yield is proportional to $S_e^{\delta-1} d$. This effect is purely due to the ion-induced shock wave effect due to reactive species propagation as well as due to direct bond breaking effect at large LET values.

Fourth, N_g in this representation is the number of base pairs in the whole cell nucleus, which is derived from the expectation that the cell is in interphase and that chromatin is uniformly distributed throughout the nucleus. In particular, this means that the yield would be the same for all human cells. While this may be true for healthy cells of normal tissue, it may not be true for cancer cells. More research is needed to clarify this.

Fifth, the oxygen concentration dependence is 'hidden' in the value of $\frac{dN_l}{dx}$. It affects the reactive species effect through the value of $\mathcal{N}_r$ that enters Eq. (9.37). In the case of hypoxic conditions, the damage induced by secondary species may be repaired chemically through reduction of DNA radicals by endogenous thiols such as glutathione or other sulphur-containing cellular constituents,[102] thus decreasing the number of individual and clustered DNA lesions processed by enzymatic repair mechanisms. The map of oxygen concentration automatically produces the map of the oxygen enhancement ratio (OER), which is the ratio of doses required to achieve the same biological effect with a given oxygen concentration to that at normal aerobic conditions. The analysis of OER performed within the MSA framework was presented in Ref. 26.

Sixth, if the LET is fixed, Eq. (9.59) suggests that the yield and therefore the logarithm of survival probability is linear with dose, so that survival curves are straight lines in their traditional coordinates. A comparison of the number of survival curves across a range of LET values supports this observation, but there are experiments where the so-called shouldered survival curves have been observed (see Section 9.7.1). At this point, it is worth mentioning that a large body of research on X-rays survival curves[29] shows that straight survival curves indicate a single-hit scenario of radiation damage. This means that a single hit of a target (in our case, with an ion) leads to cell inactivation with a certain probability. This probability includes the probability of DNA damage repair. In the framework of molecular theories developed from the 1950s to the 1990s,[29] including the microdosimetric kinetic model (MKM),[103,104] shouldered survival curves are the result of either nonlinear damage or repair. Some studies in this direction within the MSA are briefly described in Section 9.7.1.

9.7 Examples of Application of MSA

9.7.1 *Survival probability for different cell lines irradiated with ions*

The above-presented formalism has been utilised in Refs. 26,27 to evaluate survival probability for various cell lines irradiated with ions. Figures 9.14(a) and 9.14(b) show the survival curves for human adenocarcinomic A549 cells and normal fibroblasts AG1522, irradiated with protons and alpha-particles at different values of LET. The calculated curves (lines) are compared with the experimental data (symbols) in terms of the survival of the same cells in the same conditions. Different cell lines have different cross-sectional areas of their nuclei and, thus, the average distance $\bar{z}$ of the ion's traverse through the nucleus.[26] This results in different slopes of the survival curves calculated for A549 and AG1522 cell lines at comparable values of LET. A comparison of calculated survival curves for other human cell lines, such as glioblastoma A172 cells and normal skin

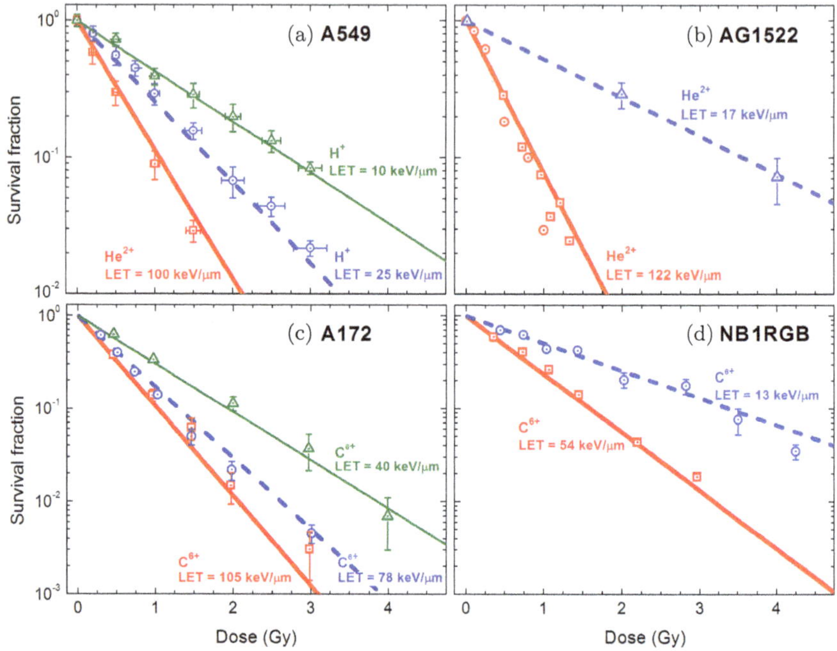

Fig. 9.14. Survival curves for different human cell lines: (a) adenocarcinomic A549 cells, (b) normal fibroblasts AG1522, (c) glioblastoma A172 cells, and (d) normal skin fibroblasts NB1RGB. The calculated survival probabilities are shown by lines and experimental data from Ref. 105,106 (A549), Ref. 107–109 (AG1522), Ref. 110,111 (A172), and Ref. 110,112 (NB1RGB) are shown by symbols.
Source: Reproduced from Ref. 26 published under an open access Creative Commons CC BY license.

fibroblasts NB1RGB, with experiments is presented in Figs. 9.14(c) and 9.14(d).

The criterion of lethality and Eq. (9.59) produce linear (on a semi-log plot) survival curves for cells irradiated with ions, as shown in Fig. 9.14. This model includes the probability of enzymatic repair embedded into the criterion. The criterion itself can be different for different cell lines, but it will lead to straight lines nevertheless. The 'shoulderness through damage' translates into MSA language as tracks overlap.[96] In this case, $\frac{dN_l}{dx}$ depends on fluence and therefore dose, and Eq. (9.59) becomes nonlinear with respect to dose, resulting in shouldered survival curve.[96] However, this happens at very large values of fluence and dose, far larger than those used clinically.

Therefore, it is more likely that a shouldered curve in ion therapy may result due to the repair process. The solution to this problem was suggested in Ref. 26.

This solution does not change the expression for the yield Y_l, Eq. (9.59), except for a constant coefficient. What is changed is the logarithm of survival probability (9.51); instead of being linear with respect to the yield, it becomes a quadratic function,

$$-\ln \Pi_{\text{surv}} = Y_l - (\chi_0 - \chi_1 Y_l)Y_l = (1 - \chi_0)Y_l + \chi_1 Y_l^2, \qquad (9.62)$$

where χ_0 and χ_1 are positive constants. The first representation can be phenomenologically interpreted in such a way that the cell lines for which the survival curves are shouldered are more resistive than those for which the survival is linear. At small values of yield, the RHS of Eq. (9.62) is linear with respect to Y_l with a coefficient $1 - \chi_0 < 1$. However, as the yield increases, the resistivity decreases linearly, and when $(\chi_0 - \chi_1 Y_l)$ turns to zero, the survival becomes 'normal'. This is formalised as

$$-\ln \Pi_{\text{surv}} = (1 - \chi)Y_l = Y_l - \Theta(\chi_0 - \chi_1 Y_l)\,(\chi_0 - \chi_1 Y_l)\,Y_l. \quad (9.63)$$

The coefficient

$$\chi = (\chi_0 - \chi_1\, Y_l)\ \Theta(\chi_0 - \chi_1\, Y_l) \qquad (9.64)$$

gradually approaches zero with increasing number of lesions until it becomes equal to zero at a critical value, $\tilde{Y}_l = \chi_0/\chi_1$, which depends, in particular, on dose and LET. Above this critical value, Eq. (9.51) remains valid. Thus, the critical yield $\tilde{Y}_l$ is the transition point in the survival curve from the linear-quadratic to the linear regime. Examples of application of this model are shown in Fig. 9.15.

For $Y_l < \chi_0/\chi_1$, the survival probability given by Eq. (9.63) can be rewritten as[26]

$$-\ln \Pi_{\text{surv}} = (1 - \chi_0)\,\frac{\pi}{16}\sigma\, N_{\text{g}}\,\frac{d}{S_e} + \chi_1\left(\frac{\pi}{16}\sigma\, N_{\text{g}}\right)^2 \frac{d^2}{S_e^2}. \qquad (9.65)$$

At this point, the famous empirical parameters α and β of the linear-quadratic model[29] given by

$$-\ln \Pi_{\text{surv}} = \alpha d + \beta d^2 \qquad (9.66)$$

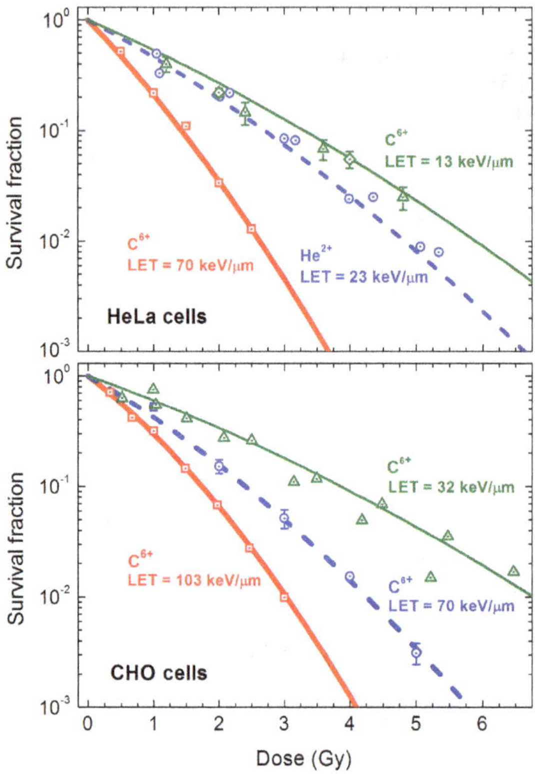

Fig. 9.15. Survival curves for repair-efficient HeLa and CHO cell lines. The calculated survival probabilities[26] are shown by lines and experimental data from Refs. 113,114 are shown by symbols. The survival curves are calculated using Eq. (9.63), with $\chi_0 = 0.08$ and $\chi_1 = 0.07$ for HeLa cells and $\chi_0 = 0.4$ and $\chi_1 = 0.045$ for CHO cells.
Source: Reproduced from Ref. 5 with permission from Springer International Publishing.

can be introduced. Equation (9.65) provides the molecular-level expressions for these parameters at doses $d \leq \frac{16}{\pi}\frac{S_e}{\sigma N_g}\frac{\chi_0}{\chi_1}$:

$$\alpha = (1 - \chi_0)\,\frac{\pi}{16}\sigma\,N_{\rm g}\,\frac{1}{S_e}, \quad \beta = \chi_1\left(\frac{\pi}{16}\sigma\,N_{\rm g}\right)^2\frac{1}{S_e^2}. \tag{9.67}$$

At $Y_l > \chi_0/\chi_1$, i.e., for $d > \frac{16}{\pi}\frac{S_e}{\sigma N_g}\frac{\chi_0}{\chi_1}$, the survival curves are linear, and the parameter α is given by

$$\alpha = \frac{\pi}{16}\frac{\sigma\,N_{\rm g}}{S_e}. \tag{9.68}$$

9.7.2 Survival curves along a spread-out Bragg peak

In Ref. 101, an algorithm was suggested for choosing the energy distribution of ion fluence at the entrance in order to achieve the uniform cell survival distribution throughout the SOBP. In the beginning, it was shown that the uniform dose distribution leads to an increase in cell inactivation along the SOBP towards a sharp maximum at its distal end. This section shows the application of this algorithm to achieve the uniform cell survival at a constant oxygen concentration along the SOBP.

Let the maximum initial energy at the entrance be E_0, and let it change by steps of ΔE to construct the SOBP; the depth of each pristine Bragg peak can be denoted by x_j, where $j = 0, 1, 2, \ldots, J$. According to Eqs. (9.59) and (9.61), at a given depth x, the yield is

$$Y_l = \frac{\pi}{16} N_g \sum_j \sigma(S_j(x)) F_j = Y_0, \tag{9.69}$$

where Y_0 is the target yield throughout the SOBP. The goal is to obtain the distribution of F_j. Clearly,

$$F_0 = Y_0 \frac{16}{\pi N_g \sigma(S_0(x_0))}, \tag{9.70}$$

the fluence at maximum energy corresponds to the desired yield at the distal end of the Bragg peak. Then,

$$\frac{\pi}{16} N_g (\sigma(S_1(x_1)) F_1 + \sigma(S_0(x_1)) F_0) = \frac{\pi}{16} N_g \sigma(S_0(x_0)) F_0, \tag{9.71}$$

which gives

$$F_1 = \frac{\sigma(S_0(x_0)) - \sigma(S_0(x_1))}{\sigma(S_1(x_1))} F_0. \tag{9.72}$$

In the next step, we find F_2 from

$$\begin{aligned} &\frac{\pi}{16} N_g (\sigma(S_2(x_2)) F_2 + \sigma(S_1(x_2)) F_1 + \sigma(S_0(x_2)) F_0) \\ &\quad = \frac{\pi}{16} N_g \sigma(S_0(x_0)) F_0, \end{aligned} \tag{9.73}$$

and so on. If the oxygen concentration depends on x, this affects all $S_j(x)$ and can be easily included in the algorithm. Figure 9.16 shows the application of the algorithm for a proton SOBP example.

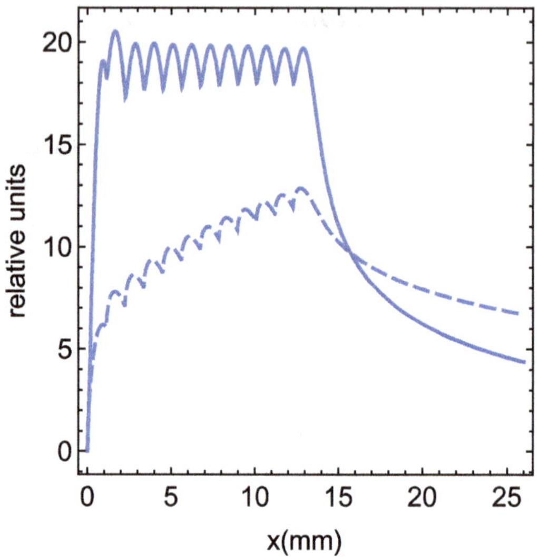

Fig. 9.16. The solid line shows the profile of the yield dependence of lethal lesions in cells along the SOBP as a function of distance on the distal end of the SOBP. The dashed line shows the profile of the depth-dose curve that produced the above result.
Source: Reproduced from Ref. 101 with permission from Springer Nature.

This example demonstrates the potential of using the MSA for treatment planning and optimisation purposes. It also illustrates the importance of the ion-induced shock wave concept. In a uniform beam, the dose is linear with respect to the LET (and the production of reactive species). In order for the *observed* effect of the increase in cell inactivation probability with the depth along the SOBP to take place in the case of the uniform dose throughout the SOBP, there should be some dependence of the cell survival probability on LET beyond the linear dependence. The only physical reason for this dependence can be the dependence of *radial range* of reactive species propagation on LET, $R_r(S_e)$, and only shock waves can make such a dependence possible.

9.7.3 *Shock-wave-induced DNA lethal damage of cells irradiated with high-LET ions*

This section describes the results of the evaluation of survival probabilities of cells irradiated with high-LET ions within the MSA

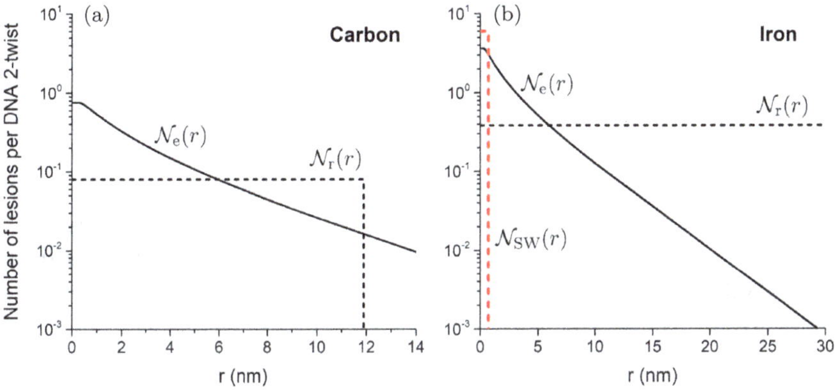

Fig. 9.17. Average number of simple lesions per DNA double twist due to a single carbon ion (a) and iron ion (b) at their Bragg peak energies, as a function of radial distance from the ion's path. $\mathcal{N}_e(r)$ and $\mathcal{N}_r(r)$ are the numbers of simple lesions produced by secondary electrons and free radicals, respectively. $\mathcal{N}_{SW}(r)$ is the average number of lesions produced due to direct thermomechanical damage by the ion-induced shock wave. The value $\mathcal{N}_{SW} = 5.4$ at $r \leq R_{SW} = 0.4$ nm was obtained from MD simulations.[28]

Source: Reproduced from Ref. 28 with permission from the American Physical Society.

formalism.[28] This analysis reveals the significant role of the shock-wave-induced thermomechanical mechanism of DNA damage in the cell inactivation.

Figure 9.17 shows the average number of simple lesions per a DNA double twist as a function of radial distance from the ion's path for irradiation with a carbon ion (Fig. 9.17(a)) and with an iron ion (Fig. 9.17(b)) in the vicinity of the corresponding Bragg peaks. The average number of simple lesions created by secondary electrons and free radicals ($\mathcal{N}_e(r, S_e)$ and $\mathcal{N}_r(r, S_e)$) is calculated according to Eqs. (9.30) and (9.37), respectively. The average number of lesions created by the shock-wave-induced thermomechanical stress of the DNA, $\mathcal{N}_{SW}$, is taken from the MD simulations described above in Section 9.5.1.1. The number of breaks corresponds to the bond dissociation energies D_e obtained from the DFT calculations.[85]

Following the MD simulations (see Fig. 9.7), the thermomechanical stress by the carbon-ion-induced shock wave does not produce any lesions within the DNA double twist for the bond dissociation energies D_e. In the case of irradiation with a carbon ion, the lesions are created by secondary electrons, free radicals, and other reactive species, which are spread over the large distance range by the shock wave; see Fig. 9.17(a). This is in agreement with the results of earlier studies[4,39] which demonstrated that at the values of LET typical for a single carbon ion at the Bragg peak (S_e = 830 keV/μm), most of the ion-induced DNA damage occurs via chemical effects involving interactions of DNA molecules with secondary electrons, free radicals, solvated electrons, etc. In contrast, the number of lesions produced by the thermomechanical stress caused by the iron-ion-induced shock wave outweighs the number of lesions produced by the chemical effects at distances $r \leq R_{\rm SW}$ from the ion's path, as shown in Fig. 9.17(b). The analysis described in the following has been performed using the effective radius $R_{\rm SW}$ = 0.4 nm for the iron ion at the Bragg peak, which is lower than the values reported in Section 9.5.1.2. One should stress that the model presented in Section 9.5.1.2 gives the maximal values of $R_{\rm SW}$ for the five studied ions at the Bragg peak, corresponding to the ideal orientation of the molecular bond parallel to the direction of the shock wave propagation. Accounting for different orientations of the bonds in the DNA backbone with respect to the direction of a shock wave propagation should lead reduced $R_{\rm SW}$ values. The value $R_{\rm SW}$ = 0.4 nm has been obtained by averaging the number of lesions produced due to direct thermomechanical damage by the iron-ion-induced shock wave, $\mathcal{N}_{\rm SW}$, over the range of distances from the ion track to the principal axis of the DNA molecule; see Fig. 9.7(c).

On the basis of the non-reactive MD simulations and subsequent estimates for the energy deposited into the DNA backbone bonds, it was concluded earlier[39] that the bond breaking due to the shock-wave-induced thermomechanical stress becomes dominant for ions heavier than argon, propagating in liquid water. This result has been confirmed in Ref. 28 by means of reactive MD simulations.

The calculated probabilities $\mathcal{P}_l(r, S_e)$ of producing the lethal DNA damage in a DNA double twist located at a distance r from the ion's path, Eq. (9.43), are shown in Fig. 9.18 for carbon and iron ions. In the case of irradiation with iron ions (see Fig. 9.18(b)), accounting for the shock-wave-induced thermomechanical stress results in a significant increase of the probability of lethal DNA damage within the characteristic distance $r \leq R_{\rm SW}$ from the ion track. A conservative estimate for the number of bond breaks produced by the

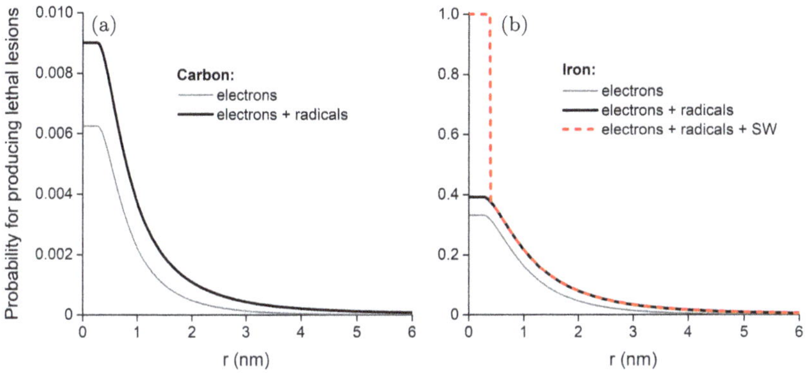

Fig. 9.18. Probability for producing lethal lesions in a DNA double twist as a function of radial distance from the ion's path for irradiation with a carbon ion (panel (a)) and with an iron ion (panel (b)) in the vicinity of the corresponding Bragg peaks. Solid grey, solid black and dashed red curves show, respectively, the contribution of only secondary electrons, secondary electrons, and free radicals, as well as these agents together with the thermomechanical stress of the DNA induced by the shock wave (SW).
Source: Reproduced from Ref. 28 with permission from the American Physical Society.

iron-ion-induced shock wave thermomechanical stress, corresponding to the largest bond dissociation energy D_e (see Fig. 9.7), reveals that five or more bond breaks within the DNA double twist are created when the iron ion propagates at distances smaller than $R_{\rm SW} = 0.4$ nm from the principal axis of inertia of the DNA molecule. The indicated number of breaks exceeds the minimal number of lesions needed to produce lethal DNA damage, and hence the probability $\mathcal{P}_l(r, S_e) = 1$ at $r \leq 0.4$ nm from the iron ion track. This means that even a single hit of a cell nucleus by a high-LET ion will be sufficient to inactivate the cell.

The MSA-based calculations of the survival probabilities for human fibroblasts irradiated with carbon ions at high values of LET (see Ref. 28 and Fig. 9.14) indicated that the shock wave mechanism enhances transport of radicals and thus reduces their fast recombination, thereby increasing the damaging effect of projectile ions. The cell survival probabilities obtained with accounting for the DNA damage produced by the secondary electrons, free radicals, and the shock wave mechanism were found to be in good agreement with the calculated survival probabilities using experimental data.

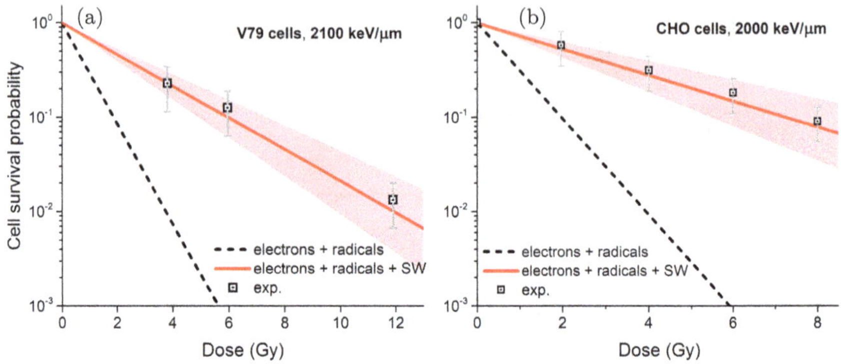

Fig. 9.19. Survival probability as a function of deposited dose for normal rodent cells, (a) V79 and (b) CHO, irradiated with iron ions at the indicated values of LET in the vicinity of the Bragg peak. Solid red lines show the probabilities calculated within the MSA framework accounting for the shock-wave-induced thermomechanical damage. Shaded areas illustrate variation of cell survival probabilities due to variation in the cell nucleus area (see text for details). Dashed lines show the cell survival probabilities calculated while accounting for the DNA damage produced only by secondary electrons and free radicals. Symbols denote experimental data for irradiation of the V79[115] and CHO[116] cells.
Source: Reproduced from Ref. 28 with permission from the American Physical Society.

The shock wave mechanism plays an even bigger role in producing lethal damage to cells by high-LET ions, as demonstrated in Fig. 9.18(b). Figure 9.19 shows survival probabilities for two normal rodent cells, V79 and CHO, irradiated with high-LET iron ions in the vicinity of the Bragg peak. Solid red lines show the probabilities calculated while accounting for the shock-wave-induced thermomechanical damage. These probabilities were calculated within the MSA using the number of lethal lesions Y_l, defined by Eq. (9.56), and $R_{\rm SW} = 0.4$ nm, as discussed above. Shaded areas demonstrate the variation of the calculated cell survival probabilities related to the variation of the nucleus size.[28] Dashed black lines show the probabilities calculated while accounting for DNA damage produced only by secondary electrons and free radicals.

It is apparent that for the irradiation with high-LET ions, the shock-wave-induced thermomechanical stress of the DNA has a significant impact on the cell survival probabilities. If this mechanism is not taken into consideration, the calculated survival probabilities deviate by orders of magnitude from the experimental values.[115,116] Indeed, according to Eqs. (9.46)–(9.51), the number of lethal lesions Y_l produced by secondary electrons and free radicals in a cell nucleus grows with an increase in LET. As a consequence, the slope of cell survival curves would monotonically increase with an increase in LET. This behaviour contradicts with the experimentally observed phenomenon known as the 'overkill' effect, which manifests itself when cells are irradiated with high-LET ions. At higher LET, a given dose can be delivered with a smaller number of ions. This increases the chances of some cells remaining non-targeted, i.e., the cell survival probability should increase. This leads to a less steep dependence of cell survival probability on the deposited dose.

Different approaches have been adopted in existing radiobiological models to account for the 'overkill' effect. For instance, empirical saturation corrections due to non-Poisson distribution of lethal lesions in the cell nucleus were introduced in the commonly used LEM and MKM models to describe the radiobiological response to high-LET irradiation.[117,118] In contrast to other models, the MSA describes quantitatively the 'overkill' effect by accounting for the shock-wave-induced thermomechanical stress of the DNA. Within the experimental uncertainties, the calculated survival probabilities for cells irradiated with iron ions are in a good agreement with the experimental results.[115,116] This agreement provides strong experimental evidence for the biodamage effects caused by ion-induced shock waves upon irradiation of biological targets with high-LET ions.

9.8 Conclusions and Outlook

This chapter has provided an overview of the major methodological concepts of the MSA to the physics of ion beam therapy and has demonstrated that the whole approach converges to a single formula that calculates the yield of lethal lesions in a cell irradiated with ions. This yield, equal to the logarithm of the inverse probability of cell survival, depends on the depth, the composition of the tissue in front of the cell, oxygen concentration, and the type of the cell. It has been shown that the MSA can be used to calculate the probability of cell survival in a variety of conditions, such as high and low LET values, high and low ion fluences, and aerobic and hypoxic environments. The MSA generally predicts linear survival curves, but it can also explain shouldered curves in special cases. This makes it a truly universal and robust method for assessing radiation biodamage with ions. In addition to its effectiveness, the method answers many questions about the nature of effects, which occur on a variety of time, space, and energy scales.

The MSA has been designed as an inclusive scientific approach. Its additional strength lies in its ability for adjustment to changing external conditions, e.g., the presence of radiosensitising nanoparticles of given composition, size, and density.[119–121] In such cases, additional effects can be included in the scenario. The ion-induced shock wave phenomenon may change the initial conditions for the chemical phase of radiation damage. This prediction could be compared with the track structure simulations if the shock waves were effectively included in their scenario (e.g., by increasing the diffusion coefficients for reactive species depending on their positions in the track for some time on the picosecond scale).

To summarise, the MSA methodology provides a comprehensive picture of radiation biodamage with ions by integrating all up-to-date known effects and phenomena. As such, the MSA enables us to describe different irradiation scenarios, including irradiation with low-, medium-, and high-LET ions, spread-out Bragg peaks, the LET dependence of the oxygen enhancement ratio, and the overkill effect. The future directions for the MSA development may concern the following:

(1) The elaboration of different elements of the approach, such as quantum processes induced by low-energy secondary electrons and chemically reactive species, the formation of strand breaks, and repair mechanisms.
(2) The experimental observation of ion-induced shock waves, predicted and already included in the scenario of radiation damage would be the most significant step towards the recognition of the MSA. Some steps were made towards designing experiments in which the ion-induced shock waves could be observed directly.[93,94]
(3) A more elaborated scenario of the transport of reactive species, including the collective flow due to the shock waves as a function of LET and its comparison with MC simulations, will also be an important development.

(4) A comprehensive study of survival curves for a large variety of cell lines and conditions is definitely desired.
(5) Experiments with high fluences and disabled DNA repair function could explore the effects of tracks overlap and measure the effective radii of ion tracks, which can help in better understanding the transport of reactive species.
(6) A better understanding of SOBP features will certainly improve the optimisation of therapy planning and bring it to a more scientific level.
(7) The radiosensitising effect of nanoparticles should be further explored in collaboration with experimentalists.
(8) Finally, the MSA should be applied on the next, larger, scale to optimise the achievement of tumour control as a function of relevant external and internal conditions.

References

1. Amaldi U, Kraft G. European developments in radiotherapy with beams of large radiobiological effectiveness. J Radiat Res. 2007;48:A27–A41.
2. Schardt D, Elsässer T, Schulz-Ertner D. Heavy-ion tumor therapy: Physical and radiobiological benefits. Rev Mod Phys. 2010;82: 383–425.
3. Durante M, Loeffler J. Charged particles in radiation oncology. Nat Rev Clin Oncol. 2010;7:37.
4. Surdutovich E, Solov'yov AV. Multiscale approach to the physics of radiation damage with ions. Eur Phys J D. 2014;68:353.
5. Solov'yov AV, editor. Nanoscale insights into ion-beam cancer therapy. Cham: Springer International Publishing. 2017.
6. García Gómez-Tejedor G, Fuss MC, editors. Radiation damage in biomolecular systems. Dordrecht: Springer Science + Business Media B.V. 2012.
7. Particle Therapy Co-Operative Group. 2024. Available from: https://www.ptcog.site/.
8. Boudaïffa B, *et al.* Resonant formation of DNA strand breaks by low-energy (3 to 20 eV) electrons. Science. 2000;287:1658.
9. Huels MA, *et al.* Single, double, and multiple double strand breaks induced in DNA by 3–100 eV electrons. J Am Chem Soc. 2003;125:4467.

10. Sanche L. Low energy electron-driven damage in biomolecules. Eur Phys J D. 2005;35:367–390.
11. Jäkel O, Kraft G, Karger CP. The history of ion beam therapy in Germany. Z Med Phys. 2022;32:6–22.
12. Chatterjee A, Holley WR. Computer simulation of initial events in the biochemical mechanisms of DNA damage. Adv Radiat Biol. 1993;17:181–226.
13. von Sonntag C. The chemical basis of radiation biology. London: Taylor & Francis. 1987.
14. Solov'yov AV, *et al.* Condensed matter systems exposed to radiation: Multiscale theory, simulations, and experiment. Chem Rev. 2024;124:8014–8129.
15. Durante M, Cucinotta FA. Physical basis of radiation protection in space travel. Rev Mod Phys. 2011;83:1245–1281.
16. Kronenberg A, Cucinotta FA. Space radiation protection issues. Health Phys. 2012;103:556–567.
17. Surdutovich E, Gallagher DC, Solov'yov AV. Calculation of complex DNA damage induced by ions. Phys Rev E. 2011;84:051918.
18. Solov'yov AV, *et al.* Physics of ion beam cancer therapy: A multiscale approach. Phys Rev E. 2009;79:011909.
19. Surdutovich E, Solov'yov AV. Multiscale modeling for cancer radiotherapies. Cancer Nanotechnol. 2019;10:6.
20. Friedland W, *et al.* Comprehensive track-structure based evaluation of DNA damage by light ions from radiotherapy-relevant energies down to stopping. Sci Rep. 2017;7:45161.
21. Frese MC, *et al.* A mechanism-based approach to predict the relative biological effectiveness of protons and carbon ions in radiation therapy. Int J Radiat Oncol. 2012;83:442.
22. Semenenko VA, Stewart RD. Fast Monte Carlo simulation of DNA damage formed by electrons and light ions. Phys Med Biol. 2006;51:1693.
23. Nikjoo H, *et al.* Track-structure codes in radiation research. Radiat Meas. 2006;41:1052–1074.
24. Nikjoo H, *et al.* Modelling of DNA damage induced by energetic electrons (100 eV to 100 keV). Radiat Prot Dosim. 2002;99:77.
25. Plante I, Cucinotta FA. Energy deposition and relative frequency of hits of cylindrical nanovolume in medium irradiated by ions: Monte Carlo simulation of tracks structure. Radiat Environ Biophys. 2010;49:5.
26. Verkhovtsev A, Surdutovich E, Solov'yov AV. Multiscale approach predictions for biological outcomes in ion-beam cancer therapy. Sci Rep. 2016;6:27654.

27. Verkhovtsev A, Surdutovich E, Solov'yov AV. Phenomenon-based evaluation of relative biological effectiveness of ion beams by means of the multiscale approach. Cancer Nanotechnol. 2019;10:4.
28. Friis I, *et al.* Lethal DNA damage caused by ion-induced shock waves in cells. Phys Rev E. 2021;104:054408.
29. Alpen EL. Radiation biophysics. San Diego: Academic Press. 1998.
30. Surdutovich E, *et al.* Ion-induced electron production in tissue-like media and DNA damage mechanisms. Eur Phys J D. 2009;51:63.
31. Scifoni E, Surdutovich E, Solov'yov AV. Spectra of secondary electrons generated in water by energetic ions. Phys Rev E. 2010;81: 021903.
32. Pimblott SM, Siebbeles LDA. Energy loss by non-relativistic electrons and positrons in liquid water. Nucl Inst Meth B. 2002; 194:237.
33. Pimblott SM, LaVerne JA, Mozumder A. Monte Carlo simulation of range and energy deposition by electrons in gaseous and liquid water. J Phys Chem. 1996;100:8595.
34. Pimblott SM, LaVerne JA. Production of low-energy electrons by ionizing radiation. Rad Phys Chem. 2007;76:1244.
35. Meesungnoen J, *et al.* Low-energy electron penetration range in liquid water. Radiat Res. 2002;158:657.
36. Surdutovich E, Solov'yov AV. Double strand breaks in DNA resulting from double ionization events. Eur Phys J D. 2012;66:206.
37. Surdutovich E, Solov'yov AV. Random walk approximation for the radial dose dependence. Eur Phys J D. 2012;66:245.
38. Bug M, *et al.* Nanoscale characterization of ion tracks: MC simulations versus analytical approach. Eur Phys J D. 2012;66:291.
39. Surdutovich E, Yakubovich AV, Solov'yov AV. Biodamage via shock waves initiated by irradiation with ions. Sci Rep. 2013;3:1289.
40. Park Y, *et al.* DNA damage induced by low-energy electrons: Conversion of thymine to 5,6-dihydrothymine in the oligonucleotide trimer TpTpT. Radiat Res. 2011;175:240.
41. Smyth M, Kohanoff J. Excess electron interactions with solvated DNA nucleotides: Strand breaks possible at room temperature. J Am Chem Soc. 2012;134:9122.
42. Becker D, Adhikary A, Sevilla MD. Physicochemical mechanisms of radiation-induced DNA damage. In: Hatano Y, Katsumura Y, Mozumder A, editors. Charged particle and photon interactions with matter recent advances, applications, and interfaces. Boca Raton: CRC Press. 2010. pp. 503–542.
43. Toulemonde M, Surdutovich E, Solov'yov AV. Temperature and pressure spikes in ion-beam cancer therapy. Phys Rev E. 2009;80:031913.

44. Surdutovich E, Solov'yov AV. Shock wave initiated by ion passing through liquid water. Phys Rev E. 2010;82:051915.
45. Bethe H. Zur Theorie des Durchgangs schneller Korpuskularstrahlen durch Materie. Ann Phys. 1930;397:325.
46. Bloch F. Bremsvermögen von Atomen mit mehreren Elektronen. Z Physik. 1933;81:363.
47. Bloch F. Zur Bremsung rasch bewegter Teilchen beim Durchgang durch Materie. Ann Phys. 1933;408:285.
48. Pshenichnov I, Mishustin I, Greiner W. Comparative study of depth-dose distributions for beams of light and heavy nuclei in tissue-like media. Nucl Inst Meth B. 2008;266:1094.
49. Abril I, *et al.* Energy loss of hydrogen- and helium-ion beams in DNA: Calculations based on a realistic energy-loss function of the target. Radiat Res. 2011;175:247.
50. Obolensky O, *et al.* Ion beam cancer therapy: Fundamental aspects of the problem. Nucl Inst Meth B. 2008;266:1623.
51. Rudd ME, *et al.* Electron production in proton collisions with atoms and molecules: Energy distributions. Rev Mod Phys. 1992; 64:441.
52. Francis Z, *et al.* Calculation of lineal energies for water and DNA bases using the Rudd model cross sections integrated within the Geant4-DNA processes. J Appl Phys. 2017;122:014701.
53. de Vera P, *et al.* Semiempirical model for the ion impact ionization of complex biological media. Phys Rev Lett. 2013;110:148104.
54. Landau LD, Lifshitz EM, Pitaevskii LP. Electrodynamics of continuous media. 2nd ed. Oxford: Elsevier Butterworth-Heinemann. 1984.
55. Lindhard J. On the properties of a gas of charged particles. Dan Mat Fys Medd. 1954;28:1–57.
56. de Vera P, *et al.* Ionization of biomolecular targets by ion impact: Input data for radiobiological applications. J Phys Conf Ser. 2013;438:012015.
57. de Vera P, *et al.* Analytical model of ionization and energy deposition by proton beams in subcellular compartments. Eur Phys J D. 2014;68:96.
58. Haettner E, Iwase H, Schardt D. Experimental fragmentation studies with ^{12}C therapy beams. Rad Protec Dosim. 2006;122:485.
59. Garcia-Molina R, *et al.* Role of the interaction processes in the depth-dose distribution of proton beams in liquid water. J Phys Conf Ser. 2012;373:012015.
60. Barkas WH. Nuclear research emulsions I: Techniques and theory. Vol. 1. New York: Academic Press. 1963.
61. Kundrát P. A semi-analytical radiobiological model may assist treatment planning in light ion radiotherapy. Phys Med Biol. 2007;52:6813.

62. Hollmark M, *et al.* Influence of multiple scattering and energy loss straggling on the absorbed dose distributions of therapeutic light ion beams: I. Analytical pencil beam model. Phys Med Biol. 2004;49:3247.
63. Inokuti M. Inelastic collisions of fast charged particles with atoms and molecules – the Bethe theory revisited. Rev Mod Phys. 1971; 43:297.
64. Schmidt-Böcking H, *et al.* Dynamics of multiple ionization of atoms and molecules by electron, photon, and ion impact – investigated by the COLTRIMS imaging method. Radiat Phys Chem. 2004; 71:627.
65. Tung CJ, *et al.* Low-energy electron interactions with liquid water and energy depositions in nanometric volumes. Nucl Inst Meth B. 2007;262:231.
66. Surdutovich E, Solov'yov AV. Transport of secondary electrons and reactive species in ion tracks. Eur Phys J D. 2015;69:193.
67. Chandrasekhar S. Stochastic problems in physics and astronomy. Rev Mod Phys. 1943;15:1.
68. Nikjoo H, *et al.* Track structure in radiation biology: Theory and applications. Int J Radiat Biol. 1998;73:355.
69. Waligórski MPR, Hamm RN, Katz R. The radial distribution of dose around the path of a heavy ion in liquid water. Nucl Tracks Radiat Meas. 1986;11:309.
70. Yakubovich AV, Surdutovich E, Solov'yov AV. Atomic and molecular data needs for radiation damage modeling: Multiscale approach. AIP Conf Proc. 2011;1344:230.
71. Yakubovich AV, Surdutovich E, Solov'yov AV. Thermomechanical damage of nucleosome by the shock wave initiated by ion passing through liquid water. Nucl Instr Meth B. 2012;279:135.
72. Komarov FF. Defect and track formation in solids irradiated by superhigh-energy ions. Phys-Usp. 2003;46:1253.
73. Komarov FF. Nano- and microstructuring of solids by swift heavy ions. Phys-Usp. 2017;60:435.
74. Landau LD, Lifshitz EM. Fluid mechanics. 2nd ed. Oxford: Butterworth-Heinemann. 1987.
75. Zel'dovich YB, Raiser YP. Physics of shock waves and high-temperature hydrodynamic phenomena. New York: Academic Press. 1966.
76. Chernyj G. Gas dynamics. Moscow: Nauka. 1994.
77. de Vera P, *et al.* Simulation of the ion-induced shock waves effects on the transport of chemically reactive species in ion tracks. Eur Phys J D. 2018;72:147.

78. de Vera P, Surdutovich E, Solov'yov AV. The role of shock waves on the biodamage induced by ion beam radiation. Cancer Nanotechnol. 2019;10:5.
79. Solov'yov IA, *et al.* MesoBioNano Explorer – A universal program for multiscale computer simulations of complex molecular structure and dynamics. J Comput Chem. 2012;33:2412.
80. Sushko GB, *et al.* Studying chemical reactions in biological systems with MBN Explorer: Implementation of molecular mechanics with dynamical topology. Eur Phys J D. 2016;70:12.
81. Solov'yov IA, Korol AV, Solov'yov AV. Multiscale modeling of complex molecular structure and dynamics with MBN Explorer. Cham: Springer International Publishing. 2017.
82. Solov'yov IA, *et al.* editors. Dynamics of systems on the nanoscale. Cham: Springer International Publishing. 2022.
83. Verkhovtsev A, *et al.* MBN Explorer and MBN Studio Tutorials: Version 5.0. Frankfurt am Main: MesoBioNano Science Publishing. 2024.
84. Schipler A, Iliakis G. DNA double-strand-break complexity levels and their possible contributions to the probability for error-prone processing and repair pathway choice. Nucleic Acids Res. 2013;41:7589–7605.
85. Friis I, *et al.* Modeling the effect of ion-induced shock waves and DNA breakage with the reactive CHARMM force field. J Comput Chem. 2020;41:2429.
86. Bondi A. Van der Waals volumes and radii. J Phys Chem. 1964;68:441.
87. La Verne JA. Radical and molecular yields in the radiolysis of water with carbon ions. Radiat Phys Chem. 1989;34:135.
88. de Vera P, *et al.* Radial doses around energetic ion tracks and the onset of shock waves on the nanoscale. Eur Phys J D. 2017;71:281.
89. Pabis A, Szala-Bilnik J, Swiatla-Wojcik D. Molecular dynamics study of the hydration of the hydroxyl radical at body temperature. Phys Chem Chem Phys. 2011;13:9458.
90. Kreipl M, Friedland W, Paretzke H. Time- and space-resolved Monte Carlo study of water radiolysis for photon, electron and ion irradiation. Radiat Environ Biophys. 2009;48:11.
91. Karamitros M, *et al.* Modeling radiation chemistry in the Geant4 toolkit. Prog Nucl Sci Technol. 2011;2:503.
92. Karamitros M, *et al.* Diffusion-controlled reactions modeling in Geant4-DNA. J Comput Phys. 2014;274:841.
93. Surdutovich E, Verkhovtsev A, Solov'yov AV. Ion-impact-induced multifragmentation of liquid droplets. Eur Phys J D. 2017;71:285.

94. Surdutovich E, Solov'yov AV. Exploration of multifragmentation of Ar clusters with incident protons. Eur Phys J D. 2019;73:241.
95. Abramowitz M, Stegun IA, editors. Handbook of mathematical functions with formulas, graphs, and mathematical tables. New York: Dover. 1972.
96. Surdutovich E, Solov'yov AV. Calculation of survival probabilities for cells exposed to high ion fluences. Eur Phys J D. 2018;72:140.
97. Dang HM, *et al.* Heavy ion induced damage to plasmid DNA: Plateau region vs. spread out Bragg-peak. Eur Phys J D. 2011;63:359.
98. Sushko GB, Solov'yov IA, Solov'yov AV. Molecular dynamics for irradiation driven chemistry: Application to the FEBID process. Eur Phys J D. 2016;70:217.
99. Nickoloff JA, Sharma N, Taylor L. Clustered DNA double-strand breaks: Biological effects and relevance to cancer radiotherapy. Genes. 2020;11:99.
100. Giulietti A, editor. Laser-driven particle acceleration towards radiobiology and medicine. Cham: Springer International Publishing. 2016.
101. Surdutovich E, Solov'yov AV. Cell survival probability in a spread-out Bragg peak for novel treatment planning. Eur Phys J D. 2017; 71:210.
102. von Sonntag C. Free-radical-induced DNA damage and its repair. Berlin: Springer-Verlag. 2006.
103. Hawkins R. A microdosimetric-kinetic model of cell death from exposure to ionizing radiation of any LET, with experimental and clinical applications. Int J Radiat Biol. 1996;69:739.
104. Hawkins R. The relationship between the sensitivity of cells to high-energy photons and the RBE of particle radiation used in radiotherapy. Radiat Res. 2009;172:761.
105. Wéra AC, *et al.* In vitro irradiation station for broad beam radiobiological experiments. Nucl Instrum Meth B. 2011;269:3120.
106. Wéra AC, *et al.* Low-LET proton irradiation of A549 non-small cell lung adenocarcinoma cells: Dose response and RBE determination. Radiat Res. 2013;179:273.
107. Raju MR, *et al.* Radiobiology of α particles: III. Cell inactivation by α-particle traversals of the cell nucleus. Radiat Res. 1991;128:204.
108. Autsavapromporn N, *et al.* The role of gap junction communication and oxidative stress in the propagation of toxic effects among high-dose α-particle-irradiated human cells. Radiat Res. 2011;175:347.
109. Hamada N, *et al.* LET-dependent survival of irradiated normal human fibroblasts and their descendants. Radiat Res. 2006;166:24.

110. Suzuki M, *et al.* Relative biological effectiveness for cell-killing effect on various human cell lines irradiated with heavy-ion medical accelerator in Chiba (HIMAC) carbon-ion beams. Int J Radiat Oncol Biol Phys. 2000;48:241.
111. Tsuboi K, *et al.* Cytotoxic effect of accelerated carbon beams on glioblastoma cell lines with p53 mutation: Clonogenic survival and cell-cycle analysis. Int J Radiat Biol. 1998;74:71.
112. Tsuruoka C, *et al.* LET and ion species dependence for cell killing in normal human skin fibroblasts. Radiol Oncol. 2005;163:494.
113. Weyrather WK, *et al.* RBE for carbon track-segment irradiation in cell lines of differing repair capacity. Int J Rad Biol. 1999;75:1357.
114. Usami N, *et al.* Mammalian cells loaded with platinum-containing molecules are sensitized to fast atomic ions. Int J Radiat Biol. 2008;84:603.
115. Hirayama R, *et al.* Contributions of direct and indirect actions in cell killing by high-LET radiations. Radiat Res. 2009;171:212.
116. Mehnati P, *et al.* Exploration of 'over kill effect' of high-LET Ar- and Fe-ions by evaluating the fraction of non-hit cell and interphase death. J Radiat Res. 2005;46:343.
117. Hawkins RB. A microdosimetric-kinetic model for the effect of non-Poisson distribution of lethal lesions on the variation of RBE with LET. Radiat Res. 2003;160:61.
118. Kase Y, *et al.* Biophysical calculation of cell survival probabilities using amorphous track structure models for heavy-ion irradiation. Phys Med Biol. 2008;53:37.
119. Haume K, *et al.* Gold nanoparticles for cancer radiotherapy: A review. Cancer Nanotechnol. 2016;7:8.
120. Haume K, *et al.* Transport of secondary electrons through coatings of ion-irradiated metallic nanoparticles. Eur Phys J D. 2018;72:116.
121. Verkhovtsev AV, *et al.* Molecular dynamics characterization of radiosensitizing coated gold nanoparticles in aqueous environment. J Phys Chem A. 2022;126:2170.

https://doi.org/10.1142/9781800618183_0010

Chapter 10

Inorganic Nanoparticles and Ionising Radiation

Emilie Brun[*,†,‡] **and Cécile Sicard-Roselli**[*,§]

[*]*Université Paris-Saclay, CNRS,*
Institut de Chimie Physique UMR 8000, 91405 Orsay, France

[†]*School of Geography, Earth and Environmental Sciences,*
University of Birmingham, Edgbaston, Birmingham B15 2TT,
United Kingdom

[‡]*e.c.brun@bham.ac.uk*

[§]*cecile.sicard@universite-paris-saclay.fr*

In 2004, the first evidence of radiosensitisation by a nanomaterial was published. The past 20 years have seen an abundance of literature in this field, from fundamental studies on the interaction between nanomaterials and ionising radiation to successful results of clinical trials. Many different combinations of nano-objects (from metallic spherical nanoparticles to multi-element multi-functional nanoplatforms), irradiation modalities (photons, electrons, or ions at keV and MeV energies), and biological models (biomolecules, cells, and animals) have been tried. Here, we do not aim to write a systematic review but rather to gather useful information from various scientific communities and highlight some specific points. We first identify the therapeutic contexts in which inorganic nanoparticles could interact with ionising radiation. We then focus on the most recent findings about the physicochemical mechanisms underlying nano-radiosensitisation, especially on the role played by the nano–solvent interface. Then, we investigate the fate of the nanoparticles during their journey from the chemist's bench to the patient and how they can be modified by irradiation as well as by their new cellular environment.

Finally, we identify some recent trends in the field, including the latest discoveries on the biological responses that nanomaterials associated with radiation can initiate.

10.1 Introduction

The beginning of the 21st century saw the emergence and then the rapid growth of nanotechnologies with applications in several domains such as chemistry, catalysis, and medicine. In the medical field, nanoparticles (NPs) are considered powerful tools and have been developed for various applications, from gene or drug delivery to regenerative medicine, imaging, and radiotherapeutic treatments, to name a few, grouped under a new name, 'nanomedicine', by the National Institute of Health (NIH) (Bethesda, USA) in the early 2000s. Nanomedicines recently received unprecedented media coverage driven by the worldwide clinical introduction of the Moderna and Pfizer–BioNTech COVID-19 lipid NP mRNA vaccine, but their arrival on the market has been growing for much longer.[1–3] In 2016, 51 nanomedicines were FDA-approved and 77 products were registered in clinical trials.[4] If polymeric and liposomal formulations were the most represented approved nanomaterials, a trend seemed to emerge with a larger variety of inorganic and metallic particles in clinical trials. In 2023, 486 clinical trials were identified,[5] but the percentage of clinical trials involving metallic NPs remained steady at around 12% from 2002 to 2021.

In this chapter, we only focus on inorganic nanomaterials and their interactions with ionising radiation. Only lab-based publications will be considered for the corpus of studies related to numerical simulations in this field, as they are worth a whole review.[6,7] In Section 10.2, we identify the medical contexts where inorganic NPs can be combined with ionising radiation. Indeed, when used internally for labelling NPs or externally for both imaging and therapy, irradiation appears to 'stimulate' NPs. Depending on the radiation type (ions or photons), different mechanisms are considered responsible for the beneficial use of NPs. They are discussed in Section 10.3, with a special focus on the radicals and molecules produced when NPs are irradiated. In Section 10.4, we retrace the journey of a radiosensitising nanomaterial from its synthesis through to its cellular fate, pointing out the modifications that could occur due to the irradiation or the biological environment. Lastly, we identify the new trends in preclinical studies from the literature. We do not aim to write a systematic review, but rather to gather useful information from various scientific communities and highlight points that may have been overlooked. We apologise in advance to the colleagues whose relevant work may not be cited in this review.

This chapter presents a detailed discussion about radiosensitisation. According to the National Cancer Institute at the NIH, radiosensitisation is defined as 'the use of a drug that makes tumour cells more sensitive to radiation therapy'. We acknowledge that different terms now coexist in the literature: nano-amplification, dose-enhancement, potentialisation of ionising radiation, radiotherapy-activation, boosted radiotherapy, etc. By using 'radiosensitisation', we will not presuppose any of the involved mechanisms.

10.2 Inorganic Nanoparticles and Ionising Radiation

Since the first medical X-ray photograph taken by Wilhelm Röntgen of his wife Anna Bertha Ludwig's hand, metals have been known to provide a high contrast when exposed to radiation. As early as the 1950s, dose enhancement at interfaces of materials with higher effective atomic numbers than biological tissues was observed. For gold, Regulla *et al.* reported radiosensitisation by gold films,[8–10] while Herold studied gold microparticles and confirmed dose enhancement by both Fricke dosimetry and cell killing for 100–240 kVp X-rays, but not for Cs-137 γ-rays.[11] However, the real trigger for the craze for nano-radiosensitisation was the study published by Hainfeld in 2004.[12] In that study, 1.9 nm gold nanoparticles (AuNPs) were intravenously delivered to subcutaneous EMT-6 tumours in mice. After 26 Gy irradiation from a 250 kVp source, one-year mice survival was 86%, compared to 20% for radiation alone. These same NPs were also studied as new contrast agents.[13] This illustrates that imaging and radiotherapy have been closely related since the very beginning. Both are based on ionising radiation and will be discussed in the following sections.

10.2.1 *Nanoparticles for radiotherapy*

After 2004 and the pioneering work of Hainfeld, nearly 10 years went by before seeing the first phase I clinical trial involving

NPs associated with external beam radiation (NCT01433068).[14] Surprisingly, it involved hafnium oxide NPs (NBTXR3®), which, at that time, were not among the most commonly described nano-radiosensitisers in preclinical studies.[15] Today, only two nanomaterials are registered in clinical trials for radiotherapy: NBTXR3®, developed by Nanobiotix (Paris, France), and AGuIX® (for Activation and Guidance of Irradiation X) developed by NH TherAGuIX (Lyon, France). What is striking is that these two materials differ in all their characteristics: nature, core material, size, and mode of administration.

10.2.1.1 *NBTXR3®*

NBTXR3®, or Hensify® in its commercial name, is a sterile suspension of 50 nm crystalline NPs of hafnium oxide, presenting a negative charge due to phosphate groups on their surface, with a hydrodynamic diameter distribution of 30–65 nm. Given its size, it is administered *via* an intratumoral injection before radiotherapy. In this respect, international guidelines for intratumoral and intranodal injection of NBTXR3® NPs in head and neck cancers were recently published,[16] allowing standardisation for medical teams worldwide.

As previously stated, it was the first radioenhancer used for treating patients with locally advanced soft tissue sarcoma (NCT01433068)[14] and the first to receive CE mark approval after completing phases II/III (NCT02379845).[17] In 2023, a global licensing, co-development, and commercialisation agreement with Janssen Pharmaceutica NV (Johnson & Johnson) was announced to speed up its development.

In July 2024, eight clinical trials were active, with six in the recruitment phase. Some test NBTXR3® as a single agent for different types of solid cancers: pancreatic (NCT04484909, phase I) and inoperable recurrent non-small cell lung cancer (NCT04505267, phase I). Some others investigate the combination of NPs, radiation, and additional treatments such as chemotherapeutics in oesophageal adenocarcinoma (NCT04615013) or monoclonal antibodies (anti-EGFR cetuximab, NCT04892173). It should be noted that this clinical trial, also named Nano-Ray312, is a randomised phase III study in locally advanced head and neck squamous cell cancers in elderly patients ineligible for platinum-based chemotherapy. The other clinical trials focus on the combination with anti-PD-1 immune checkpoint inhibitors: in advanced cancers (NCT03589339, '1100 study') in head and neck squamous cell carcinoma (NCT04862455), solid tumours with lung and/or liver metastases (NCT05039632), and recurrent or metastatic head and neck squamous cell cancer (NCT04862455). The recent findings regarding its mode of action will be described in Section 10.5.

10.2.1.2 *AGuIX®*

AGuIX® is a 9 kD nano-object of approximately 5 nm in diameter, consisting of about 15 gadolinium atoms in a polysiloxane matrix onto which approximately 10 macrocyclic Gd-chelating ligands (DOTA,[a] DTPA,[b] and DOTAGA[c] in the final version) are covalently grafted. Gd represents 12% in weight.[18] Their size allows for their excretion via the urinary route in a two-step process – rapid elimination of biodegraded and smaller particles and slower elimination of larger particles – without long-term impact on renal function.[19]

The presence of Gd makes AGuIX® an efficient magnetic resonance imaging (MRI) positive contrast agent that enables a precise delineation of where it accumulates for better targeting of the X-ray beams. Recently, Bennet *et al.* published a method to quantify AGuIX uptake in brain metastases (on a per-voxel basis).[20] Quantification has important clinical implications, as it could allow for optimising treatment planning for each patient depending on the amount of Gd present. It is also an essential tool for correlating AGuIX® inter- and intra-tumoral uptake with tumour control efficiency.

In July 2024, five clinical trials were registered with AGuIX®. Most of them take advantage of MR guidance and the numerous pre-clinical studies on glioblastoma or brain metastasis models.[21] NCT03818386 and NCT04899908 are two phase II studies for patients with multiple brain metastases, treated with whole-brain

[a]DOTA is the acronym for dodecane tetraacetic acid.
[b]DTPA stands for diethylenetriaminepentaacetic acid.
[c]DOTAGA is the acronym for tetraazacyclododecane with three acetic acid and one glutaric acid.

therapy and stereotactic brain-directed radiation, respectively, whereas NCT04881032 investigates the impact of concomitant temozolomide treatment in newly diagnosed glioblastoma. Nano-SMART (NCT04789486) entered phase II in July 2024 and concerns patients with locally advanced or unresectable pancreatic ductal adenocarcinoma or non-small cell lung cancer lesions. It should be noted that NANOPRO (NCT04784221), the first clinical trial involving proton irradiation coupled to nano-agents, was announced in June 2024.

In parallel, the number of preclinical studies involving AGuIX® has been on the rise, exploring new radiation combinations or potential modifications of the object. For example, Garcia-Prada *et al.* showed that combining AGuIX® with targeted radionuclide therapy increased the therapeutic efficiency of [^{177}Lu]Lu-DOTA-trastuzumab in ovarian peritoneal carcinomatosis.[22] To our knowledge, it is the first report of a beneficial combination between AGuIX® and a radiopharmaceutical. The question of the functionalisation of AGuIX® for active targeting is still debated,[23] but a recent report demonstrated their successful conjugation to nanobodies targeting Programmed Death Ligand 1 (PD-L1) and Cluster of Differentiation 47 (CD47) receptors, both relevant immunotherapy targets.[24]

So, while AGuIX® is a biodegradable, intravenously injected nano-platform allowing MRI imaging and quantification, for which numerous preclinical publications are available, NBTXR3® is a 50 nm NP requiring an intratumoural deposit, for which most of the information comes from clinical studies.

10.2.2 *Nanoparticles for imaging*

Biomedical imaging has become indispensable in medicine, as it now provides highly detailed structural and morphological information. It is necessary not only for the diagnosis and treatment of diseases but also for interventional procedures. Nevertheless, imaging is often provided by subjecting patients to ionising radiation. For example, computed tomography (CT) scan corresponds to X-ray imaging that was enhanced with rotation around the patient to generate cross-sectional images and thus 3D images.[25] As image quality depends on the attenuation of the different organs (tissues, bones, etc.), the photon energy of the apparatus can vary from tens to a few hundred keV with optimised voltage and current settings. Positron emission tomography (PET) necessitates the injection of radioactive tracers in order to obtain characterisation of metabolic activity or tissues and biochemical functioning of organs.[26] For both CT and PET, the added value of inorganic NPs in order to gain contrast or diagnostic accuracy is now well established.[27,28] Thus, for imaging purposes, NPs can be exposed to not only external ionising radiation but also internal, as they can be doped with or composed of radiolabelled elements. This last point will be further developed in Section 10.2.3.

Several imaging modalities benefitted from the use of high-Z elements as contrast agents,[29] and as a consequence, preclinical studies demonstrated the potential advantages of NPs and, more specifically, inorganic NPs.[30,31] If we focus on ionising radiation, several types of NPs were considered to improve CT scan imaging. Though elements with a Z number between 63 and 73 are expected to be more efficient in enhancing contrast, many other elements were also tested. Among others, lanthanides, hafnium, tantalum, gold, or bismuth, assembled as NPs, were shown to enhance CT images.[32–34] More precisely, gadolinium, ytterbium, and tantalum were shown to have a much higher image contrast than iodine for voltages in the range of 80–140 kVp.[35] Dong *et al.*[36] studied the impact of the size of AuNPs on their CT contrast capacities (5–152 nm and capped with 5 kDa m-PEG). They conclude that there is no impact of the size of the NP on the CT contrast behaviour for several concentrations ranging from 0 to 10 mg/mL. The combination of metals, as illustrated by AuNPs doped with Gd or Fe, also offers the possibility to perform dual imaging, CT and MRI.[37,38] This type of material is also a good candidate for therapy with magnetic-stimulated hyperthermia or radiation enhancement.

The advantages of NPs for imaging are clearly demonstrated in terms of contrast enhancement, but the extent of benefit also relies on their longer circulation duration and ease of functionalisation. Nevertheless, clinical application still faces a few challenges, such as rapid clearance and toxicity. In addition, specific attention should be given to the impact of exposing these NPs to X-ray radiation. First, we should consider the radiation dose of conventional medical examinations. Recalling the definition of 'absorbed dose', one Gray corresponds to 1 $J.kg^{-1}$ of energy deposited. But as different types of radiation produce different biological effects, we should also

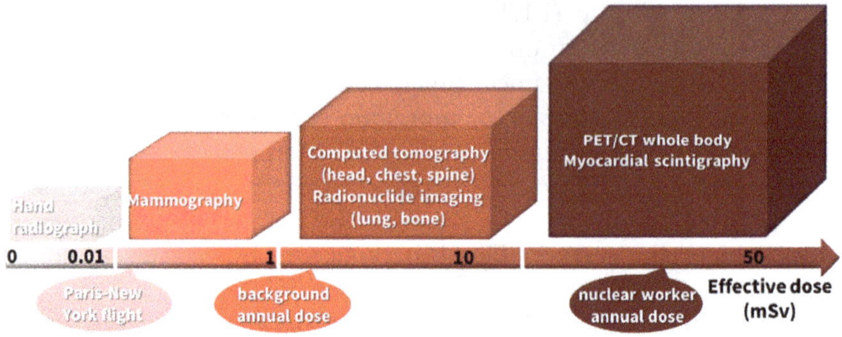

Fig. 10.1. Schematic representation of effective doses received during imaging.

define the 'equivalent dose', expressed in Sievert (Sv), which takes into account the particularity of the beam and the energy deposition in matter, and then the 'effective dose', also expressed in Sv, which takes into account organ sensitivity. For medical imaging, equivalent doses correspond to absorbed doses for X- and γ-rays, i.e., 1 Gy–1 Sv. Depending on the organ, medical imaging can induce a radiation dose of a few mSv.[39,40]

In 2015, McCollough *et al.* reported an effective dose of less than 10 mSv for a CT scan in comparison with background radiation of 3 mSv/year in the United States.[41] From Fig. 10.1, it is clear that the radiation dose due to medical examination is rather low, as it is often lower than natural background radiation. Nevertheless, we should be aware of the high sensitivity of some organs and of the increase that could be caused by repetition over treatment or during image-guided therapy, for example. The presence of NPs can be another issue. The calculation of additional dose due to the NPs is quite difficult, as we should consider the type of NP (composition, size, number of atoms, etc.), the type of radiation, the supplementary energy absorbed, and the physical mechanism.

10.2.3 *Radiolabelled and radioactive nanoparticles*

Some NPs are also considered for non-invasive imaging and are specifically useful for positron emission tomography (PET) and single-photon emission computed tomography (SPECT).[42,43] In both cases, imaging relies on the radioactive decay emissions of the radionuclide. SPECT relies on the detection of emitted gamma photons. For PET, radionuclide decay emits positron particles ($\beta+$) that generate, after interacting with electrons, two gamma photons of 511 keV in opposite directions. Two strategies coexist depending on the type of NPs: either radioactive atoms are incorporated into the NP's structure, or radiolabelled molecules are conjugated to NPs, with the latter taking advantage of existing radiopharmaceuticals.

As mentioned by several authors,[42,43] whether it be for imaging or therapy, it is absolutely necessary to ensure high radiochemical stability of radionuclides to avoid undesirable uptake in healthy tissues. Based on the high surface-to-volume ratio, ease of functionalisation, and surface modification, the role of the NP should be coupled to that of the radionuclide. ^{223}Ra, an alpha emitter, is used for targeted therapy such as metastasis treatment in breast or prostate cancers.[44] To avoid decomplexation, it is often coupled to several types of NPs, such as lanthanum phosphate, nanozeolites, barium sulphate, or lanthanum orthovanadate.[45–47] Gold has also attracted particular attention in this field. AuNPs of different sizes have been conjugated to various radionuclides, mainly to evaluate their biological fate in selected cellular and animal models.[48] For example, preclinical studies show the grafting of AuNPs with dysprosium-166/holmium-166 (β^- emitters). ^{166}Ho emits β^- particles with a maximal energy of 1.85 MeV that result in a long tissue penetration depth of 8.7 mm.[49] The authors also concluded that to prevent radionuclide escape, carriers composed of electron-rich materials should be preferred.

Another strategy to avoid radionuclide decomplexation is to organise radioactive atoms in nanostructures, leading to activated nano-objects.[42] A few radioactive NPs are presented as promising alternatives to brachytherapy, as they could allow for a more effective and uniform distribution of the radiotherapeutic dose inside tumours. The challenge with using such radioactive NPs is that dose delivery has to be done very locally to spare healthy tissues. Therefore, the half-life of the radionuclide, and the nature and energies of the emitted particles – that dictate their penetration range and linear energy transfer – should be carefully considered. For example, if AuNPs are considered, doping them with ^{198}Au is not adequate, as ^{198}Au is a β-emitter with a mean electron energy of 312 keV and a mean spatial penetration from 0.05 to 12 mm,[50] preventing their use for targeting single cells or metastasis. To overcome this difficulty, doping AuNPs with ^{103}Pd seems more appropriate, as it is a low-energy photon emitter (20–30 keV).[51–53]

Importantly, we have to keep in mind that inorganic NPs are in close proximity to various types of radiation (alpha, gamma, β^+, or β^-). As a consequence, they can be 'stimulated', as already mentioned, and may themselves act as harmful providers of deleterious species (as will be detailed in the following section). Therefore, this serves as the rationale in several studies.[54,55] Textbox 10.1 provides a brief summary of the key ideas discussed in this section.

- Both external radiotherapy and brachytherapy can 'stimulate' NPs.
- In the presence of NPs, the effective dose from imaging needs to be evaluated and included in the treatment planning.

Textbox 10.1. Combination of nanoparticles and radiation.

10.3 Nanoparticles Interacting with Radiation

10.3.1 *Radical and molecular products*

10.3.1.1 *Water radiolysis products*

When it comes to radiation, direct and indirect effects are very often mentioned. 'Direct effect' refers to the direct deposition of energy of ionising radiation in a target, whereas 'indirect effect' refers to the attack of the target by radicals arising from the solvent – in biological applications, from water radiolysis. A distinction is often made depending on the linear energy transfer (LET) of the radiation, which is the amount of energy that an ionising particle transfers to the material traversed per unit distance. For low-LET radiation, the contribution of indirect action on cell death is high (~75–85%), and decreases to ~50% under hypoxic conditions.[56] It also decreases with increasing LET. However, it does not reach zero even at very high LETs.[57] This explains why many studies focused on the detection of water radiolysis-induced species in the presence of irradiated NPs.

Water radiolysis leads to radicals ($e-_{aq}$, $^\bullet OH$, $H^\bullet$, and $O_2^{\bullet -}/HO_2^\bullet$) and molecular products (H_2 and H_2O_2), all deriving from excited (H_2O^*) and ionised ($H_2O^{\bullet +}$) water molecules, formed by the initial deposition of energy.[58,59] Their yields of formation are the result of the competition between recombination and diffusion, which strongly depends on how the energy deposition is performed: the initial distribution of ionisation in space, which is reflected by the LET, will decide if more or less radicals will escape the spurs and give more or less molecular products at the microsecond scale. They are usually classified into reducing ($e{-}_{aq}$, $H^\bullet$) and oxidising ($^\bullet OH$, $HO_2^\bullet$, H_2O_2) equivalents. H_2 is considered chemically inert and is often neglected.

The hydroxyl radical, $^\bullet OH$, is the most effective oxidising radical towards biological molecules: it has a high standard potential (E° = 2.34 V/SHE[d] at pH 7). With most substrates, it reacts at nearly diffusion-controlled rates.[60] Even though its biological relevance regarding endogenous oxidative damage is being questioned,[61] it is a key player when irradiation is involved.

Solvated electrons, as well as their conjugate acid ($H^\bullet$, pKa ($H^\bullet/e{-}_{aq}$) = 9.1), are strong reductants ($E^\circ (e{-}_{aq}) = -2.9$ V/SHE and $E^\circ(H^\bullet) = -2.4$ V/SHE). Given the pKa, reactions of $e{-}_{aq}$ with biomolecules are more relevant and so more extensively documented.[62] In oxic conditions, molecular oxygen competes efficiently to scavenge these species and forms superoxide radicals.

Given pKa($HO_2^\bullet/O_2^{\bullet -}$) = 4.8, $O_2^{\bullet -}$ is the predominant species at physiological pH. It has limited reactivity with electron-rich centres due to its negative charge. There are two standard redox potentials for $O_2^{\bullet -}$, showing that it can act as a reducing agent ($E^{\circ\prime}(O_2/O_2^{\bullet -}) = 0.33$ V) or as an oxidizing agent ($E^{\circ\prime}(O_2^{\bullet -}/H_2O_2) = 0.93$ V). Despite this high reduction potential, $O_2^{\bullet -}$ can oxidise very few biological compounds. But it can react with H_2O_2 and iron via the Haber–Weiss reaction, generating $HO^\bullet$.

Hydrogen peroxide is more stable than the radicals cited above and can diffuse over longer distances (its half-life in lymphocytes is about 1 ms).[63] The oxygen atoms of H_2O_2 have an intermediate oxidation number of (−I), explaining both reducing and oxidising properties, with- mono or bi-electronic transfers. Its highest redox potential is as an oxidant ($E^{\circ\prime}(H_2O_2/O_2^{\bullet -}) = 0.93$ V), but it reacts with relatively few biological molecules. Sulphur and seleno compounds are its most favoured targets, with implications in redox signalling.[64] But in the presence of transition metallic centres, H_2O_2 can form $HO^\bullet$ through Fenton chemistry.

[d]SHE: Standard hydrogen electrode.

At this point, it's worth clarifying that the term 'reactive oxygen species' (ROS) is not a synonym for water radiolysis products, as it covers a much larger number of species, both radicals (peroxyl ($RO_2^\bullet$), alkoxyl ($RO^\bullet$), carbonate ($CO_3^{\bullet-}$), etc.) and molecules (organic peroxides (ROOH), hypochlorous acid (HOCl), etc.). This is a generic abbreviation used for a plethora of distinct species and is definitely not a discrete molecular entity. As a community, we need to pay more attention to the specific words we employ to avoid misleading claims. This is especially true when it comes to their quantification.

10.3.1.2 *Radical and molecular species quantification*

Radical quantification has not been extensively documented when NPs are exposed to ionising radiation. Indeed, short lifetimes force very specific short-time resolved equipment. Moreover, each nanomaterial has its own behaviour towards the fluorescent probes or spin traps used,[65] so titration should be handled carefully.[66,67] When it comes to *in cellulo* measurements, there are many assays and commercial kits available, but their use and interpretation are challenging and open to artefacts. We would highly recommend referring to Murphy *et al.*[68] and following their guidelines for best practice. In particular, DCFH-DA,[e] largely used in cells, indiscriminately recognises various oxygen radicals and oxidants and is not competent to specifically detect one of them ($HO^\bullet$, for example).[69] In addition, as the overall amount of ROS generated from primary ionisation events is further propagated via the intracellular activation of endogenous ROS-producing systems (see, for example, Ref. 70), the usual ROS measurements in cellular radiosensitisation studies reflect the global cell response, not the initial amount of ROS and certainly not the extra ROS produced by the nanomaterials. In the following paragraph, we only discuss the quantification of radicals when suspensions of nanomaterials are irradiated.

The first *in vitro* studies focused on the quantification of hydroxyl radicals[71,72] and electrons[73] in the presence of gold or platinum NPs. Considering the high reactivity of both species, indirect detection is more suitable, so fluorescent probes such as coumarins were used. The quantification of both radicals was performed on AuNPs of different

[e]2′,7′-Dichlorodihydrofluorescein diacetate.

sizes and coatings, which showed a very high quantity of radicals produced. These values, when performed on different objects and with different radiation modalities, all agree on the fact that radical formation in irradiated NP suspensions is much higher than what could be expected from theoretical energy absorption by high-Z elements. Chang *et al.* showed that, with EPR detection of BMPO spin-trap adduct, X-irradiated AuNPs overproduce equivalent quantities of hydroxyl radicals and superoxide anion.[74] The NP size was systematically incremented in a few studies,[75–77] with different irradiation types. They show that radical production obtained with the NPs is not proportional to the atom number of gold. The NP surface should be considered, which will be detailed in the following section.

It is important to note that hydroxyl radicals have to be considered together with hydrogen peroxide, as H_2O_2 is formed from the dimerisation of hydroxyl radicals, and at the same time, hydroxyl radicals can be formed from the decomposition of H_2O_2 which can be catalysed by NPs. Indeed, recently, Okasaki *et al.*[78] measured the quantity of $HO^{\bullet}$ when AuDTDTPA NPs were irradiated by X-rays (200kVp) and concluded that it partially arises from H_2O_2 decomposition. For doses below 25 Gy, they didn't detect any hydrogen peroxide, which was interpreted as being due to its catalytic decomposition in the presence of the NPs. Thus, they considered that their NPs have a limited capacity to induce the decomposition of hydrogen peroxide but that both oxidative species are formed when AuDTDTPA NPs are irradiated.

Though these publications all agree on obtaining higher oxidative radical quantities when metallic NPs are irradiated, it is much more confusing with metal oxide ones. Indeed, Guerreiro *et al.*[79] compared 22 metal oxide NPs exposed to clinical megavoltage X-rays. Most of the NPs have similar sizes with a TEM diameter between 40 and 60 nm. Most lanthanide oxides exhibit no effect or, in some cases, even scavenging behaviours, though their radiosensitising capacities are often demonstrated on cellular targets or even cells. Hydroxyl radical, superoxide anion, and singlet oxygen were quantified, and only V_2O_5 ($Z(\mathrm{V}) = 23$) NPs showed a significant increase in $HO^{\bullet}$ production. Even elements of lower atomic numbers, assembled as NPs, were found to overproduce radicals under irradiation. This is especially the case for carbon, in nanodiamonds[80–82] or M-Xenes.[83]

Radiobiologists often consider that therapy and, therefore, biological sample degradation are mostly mediated by oxidative species. Nevertheless, water radiolysis generates reducing species as much as oxidative species.[59] Electrons that can be responsible for important DNA damage should also be considered.[84] Actually, it was shown that AuNP irradiation under X- or γ-rays overproduces equivalent quantities of hydroxyl radicals and solvated electrons. In addition, we should also pay attention to 'non-biological' literature that focuses on electron and H_2 production at the NP surface. Indeed, many publications describing experimental works or numerical simulations were devoted to the overproduction of reducing species at the surfaces of many metal oxide surfaces. McGrady *et al.*, with direct or indirect detections, confirmed the important electron generation in the presence of Fe_2O_3, ZrO_2, or Cr_2O_3 NPs, for example.[85,86] Molecular hydrogen, which is a subject of study for nuclear waste storage, was several times shown to overproduce at the interface of metal oxide particles.[87–89]

In summary, *in vitro* radical and molecular species in the presence of NPs and radiation exceed those from water radiolysis, regardless of the type of inorganic NP that is considered, with very rare exceptions. The different irradiation modalities and inorganic NP types considered suggest that these species are not produced by the same physical phenomenon. Therefore, all these results form a solid basis for deciphering the mechanism(s) involved.

10.3.2 *Mechanism(s)*

When the first publications on the beneficial effects of metallic NPs with irradiation appeared, the proposed mechanism was the higher mass-energy absorption by high-Z elements. Indeed, high-Z element interaction with energetic species was already well described in the literature.[26,50] Particles (photons, ions, and electrons) with a large range of energy can induce electron ejection through different processes, such as Auger or Compton effects. These secondary electrons, whose energy varies with the physical process, interact with the solvent and cause secondary radiolysis with surrounding water molecules. Considering metal oxides, a photocatalysis mechanism is expected to be involved with the structure of the valence and conduction bands. Nevertheless, if we now review the well-documented literature on the subject, we can conclude that there is no consensus about the mechanism of the different NPs considered and that many experimental results contradict these predictions.

First of all, considering the expected higher energy absorption, it is not possible to correlate the number of metal atoms present in the NP with the radical production measured. Indeed, a comparison between gold as a salt and organised as an NP for the same atom number shows that only particle organisation can lead to radical production.[76] Second, low-Z elements were shown to induce additional radical productions *in vitro*.[80,82,83] Third, X- and γ-rays generate similar radical quantities, though photoelectric absorption is expected to be much more efficient for low-energy photons rather than for high-energy ones for high-Z materials such as gold.[73] Moreover, using DNA as a probe, an unexpected shape of dose enhancement versus energy curve was demonstrated, with a minimum occurring at around 40 keV where a maximum would be predicted considering only the ratio between mass energy absorption of gold and water.[75,90] Thus, the numerous discrepancies between prediction and experimental results prompted us to consider that energy absorption is not sufficient to be the only process producing radicals; therefore, we have to consider other pathways.

Whatever the NP type, the solvent at the surfaces of metallic NPs was previously shown to behave differently from the bulk solvent.[91–94] As a consequence, interfacial water is very often cited as a major contributor to catalytic or enhanced reactivity. In the case of AuNPs, many results corroborate the specific reactivity of water[72,73,76] which was also confirmed by theoretical calculations of the orientation of water molecules at the surfaces of gold clusters.[95]

For metal oxide NPs, McGrady[86] demonstrated the increase in H_2 and electron as a function of the surface area. He concluded that this increase is more pronounced with lower water layers and proposed two types of mechanisms for ZrO_2 particles. First, an energy transfer between the oxide and the water at the surface could occur, and second, a direct radiolysis of the adsorbed water. However, to better illustrate the impact of interfacial water, we can rely on results obtained many years ago by the radiation community. Indeed, concerning nuclear waste storage, the production of hydrogen is a crucial issue. In 2002, LaVerne and Tandon measured the generation of H_2 by micron-sized particles of CeO_2, ZrO_2,[96] and UO_2[97] exposed to

γ-rays as well as to 5 MeV He ions. They measured a substantial increase in the yields of H_2 with a decreasing number of adsorbed water layers and confirmed the importance of chemisorbed water molecules. Essehli *et al.* reported similar results for TiO_2 exposed to γ-rays and He ion radiation.[98] Several parameters can affect this enhanced production at the NP surface, such as the LET of the radiation, the presence of oxygen, and the size of the particles.[99] Other reactive species can be formed at the surface of the NP.[100] This was illustrated by Milosavljevic *et al.*[101] with silica particles and electron yield measurements. All these illustrations of the important role of interfacial water should encourage the community to pay more attention to the NP surface and to maintain control over this reactivity even *in vivo*.

Another aspect that is barely considered for NP applications in therapy is Cerenkov light. This phenomenon arises when the speed of a particle exceeds that of light in a medium and thus induces the emission of photons mainly in the UV region.[102,103] Cerenkov luminescence is exhibited by some radioisotopes, as well as by some external beams such as γ-rays. As a consequence, the energetic Cerenkov photons produced can interact first with the biological medium and initiate ROS production or light-induced phenomena. Second, Cerenkov light can activate the NPs either by exciting electrons from the valence band to the conduction band in the case of oxides or through plasmon resonance in the case of metallic particles. AuNPs irradiated with X- or γ-rays were shown to produce equivalent $HO^{\bullet}$ and electrons, though Cerenkov light is only expected for gamma irradiation.[73] This is not in favour of an implication of Cerenkov light in the radical production mechanism for gold nanospheres. Nevertheless, oxide NPs could be more sensitive with a bandgap in the range of the light emitted by the Cerenkov emission. This process could then serve as the basis of many catalytic activities of oxide NPs and also photodynamic reactivity. Pratt *et al.* examined the mechanisms of NP excitation by radionuclides.[104] They proposed Cerenkov light emission by NPs exposed to β radiation; this emission is sometimes used for NP detection and to enhance photodynamic therapy or ROS production.[102,105,106]

Though the mechanism(s) responsible for enhanced molecular degradation by irradiated NPs are not fully understood, it is clear that it is not governed by a unique process, be it physical, chemical, or biological, as will be discussed in the following section. According to the types of NPs and radiation, varied combinations of phenomena can occur, making NP radiosensitisation difficult to predict. Textbox 10.2 provides a brief summary of the key ideas discussed in this section.

- Reliable quantification of radicals produced when NP suspensions are irradiated reveals that mass energy-absorption coefficients are not sufficient to explain this phenomenon.
- Interfacial water plays a significant role in radical production.
- Other phenomena could occur, such as Cerenkov light excitation.

Textbox 10.2. Mechanisms responsible for radical production of irradiated nanoparticles.

10.4 The Fate of Nanoparticles: From Post-Synthesis Microtube to *In Vivo* Destination

10.4.1 *Synthesis of nano-radiosensitisers: Zoology of nano-objects*

Since the first modern synthesis of colloidal gold by Michael Faraday in the 1850s, there has been tremendous progress in the development of new strategies to obtain nanomaterials with highly tuneable physicochemical, optical, and thermal properties. The chemical reduction of metallic salts is the oldest preparation method, but it has remained popular because it can be easily performed in the laboratory. According to the strength of the reductant and its quantity, various sizes can be obtained. Even though this preparation appears easy, it should be noted that the exact mechanism for Turkevich synthesis of AuNPs is still being investigated.[107,108] Moreover, homogeneity in size and shape can be an issue: room temperature and humidity, as well as container cleanliness, could influence the preparation. Therefore, each synthesis should be precisely characterised to increase repeatability and reproducibility. Given the near-infinite variety of available nano-objects, we cannot describe all the synthesis methods; we refer the readers to dedicated reviews.[109–111]

What is striking in the recent literature is the evolution of the tested nanoradiosensitisers. In 2016, 53% of cellular studies concerned spherical gold NPs, but gold in the broad sense represented 62.5% of the publications dealing with cell radiosensitisation, followed by gadolinium (11%) and silver (6%).[112] High-Z

metallic materials were the norm, but it's no longer the case. Low-Z elements, oxides, and diverse combinations are now tested for radiosensitisation. We can cite, among others, iron oxide,[113] titanate,[114] or Mxenes, which are atomically thin layers of titanium carbides ($Ti_3C_2T_x$).[115,116] More researchers are now interested in multi-element nano-objects, mixing elements of low and high Z, such as iron–palladium-decorated carbon nanotubes,[117] gold-nanorod-seeded mesoporous silica NPs,[118] $Au{-}Cu_2 - x$Se NPs,[119] and Au@$AgBiS_2$.[120] With the creativity of chemists knowing no bounds, an increasing number of exotic multimodal nanomaterials are also reported. Mousavi *et al.* created a 'nanocomplex' composed of a superparamagnetic iron oxide (SPIO) core, a gold shell, and an alginate coating that included cisplatin.[121] Ruan *et al.* synthesised Ru-containing organosilica moieties and assembled them onto thiolate-modified gold nanoclusters, with the aim of efficiently releasing activated chemotherapeutic Ru (II) complexes under X-ray irradiation through nanosurface energy transfer.[122] Lyu *et al.* loaded TBP-2 molecules (type-I PDT photosensitiser) into hollow TaOx NPs and decorated the surface with platelet cell membranes.[123] Zhang *et al.* constructed their 'nanosensitiser' through the self-assembly of a CRGDK peptide-conjugated poly(ethylene glycol)-block-poly(glycol lactic acid) (RGD-PEG-b-PLGA) copolymer, HfO_2 NPs, and a hydrogen peroxide activatable prodrug of BRD4 degrader (BPA771).[124] Ayyami *et al.* investigated the potential of methotrexate functionalised, 5-aminolevulinic acid (5-ALA)-modified, chitosan-coated manganese dioxide (MnO_2/CS) NPs.[125] Such achievements aim at acting as crossfires, but we need to be cautious about their exact mechanisms of action.

10.4.2 *Coating*

Intentional coating of nanomaterials aims mainly at increasing their stability in a biological medium, decreasing their toxicity, and optimising their cellular delivery, be it by extending their blood circulation time or by actively targeting some specific cells.[126,127] In the 'nano' community, an almost infinite number of different coatings are available. Among the most popular in radiosensitisation studies are PEG polymers, with different chain lengths, terminations, and chemical modifications such as covalently linked fluorophores or peptides. Dextran is also reported in several studies. When it comes to small molecules, (thio)glucose is mentioned in several papers, as well as folic acid and tiopronin. The nature of the coating material and its thickness can impact the physicochemical characteristics of nanomaterials, and their radiosensitising properties are no exception to the rule.

10.4.2.1 *How does the coating impact in vitro radiosensitisation?*

As early as 2014, Gilles *et al.* investigated how the coating of AuNPs could impact their radiation-induced hydroxyl radical production.[128] They synthesised and thoroughly characterised six different types of 30 nm AuNPs: citrate-capped AuNPs as a reference, two PEGylated AuNPs (molecular weights of 1000 and 4000 g/mol), two thiolated-PEGylated AuNPs (3500 g/mol, with carboxylate or amine terminations), and human serum albumin-coated AuNPs. They showed that, under X-ray irradiation, both the quantity of available $HO^{\bullet}$ and the measured plasmid DNA damage were related to the number of atoms present in the NP coating. This was confirmed by Grellet *et al.* for 2 nm AuNPs bearing sugar moieties and/or thiol-polyethylene glycol-amine irradiated with 220 keV and 6 MV X-rays.[129] Whereas citrate-capped increased the fluorescence signal of 7OH-CCA by 20% compared to water, functionalised AuNPs showed no difference or slightly lower fluorescence. Mai *et al.* reported similar results in a different system.[130] They studied iron oxide nanoparticles (IONPs), coated with four different small molecules (citric acid, sodium phosphate, amino silane, and dopamine), under an alternating magnetic field (AMF). AMF enhances the ROS production of IONPs, which can be followed by the degradation of methylene blue solutions. Uncoated NPs were the most reactive system, degrading about 40% of the dye over 30 min of AMF exposure.

Several hypotheses can be proposed to rationalise these experimental observations. First, if electron emission from the NPs is involved (via photoelectric effect, for example), electrons could be scavenged by the coating before they can react with water to produce radicals through secondary radiolysis. In fact, hole reactivity and charge transfers can also be modified by the presence of the ligands. Second, hydroxyl radicals produced in solution can be scavenged by specific chemical groups present on the NP surface, such as alcohols or thiols. Third, we recently proposed a major role of the NP-water interface.[72,95] Then, ligands grafted at the NP surface could modify this interface, leading to a weaker hydroxyl radical production.[131,132]

Thus, independent experiments testify to the negative impact of coating on radical production when NPs are irradiated. However, *in cellulo* studies reported effective radiosensitisation, even with coated NPs. But when it comes to cellular experiments, the roles played by the coating can be multiple, as it can impact NP uptake and cell machinery.

10.4.2.2 *How does the coating impact in cellulo radiosensitisation?*

The coating can alter NP cellular internalisation, reducing or enhancing the amount of NPs inside the cells. For example, with the incorporation of PEG, lower uptake of GNPs was documented,[133] but it can be counterbalanced by the addition of targeting molecules, such as RGD peptides when the $\alpha v\beta 3$ integrin is overexpressed in the cell model of interest.[134] However, at this point, it is essential to remember that there is no universal relationship between NP cell content and the radiosensitisation factor. Since the work of Butterworth *et al.* in 2010,[135] this has been documented in other systems (cell lines, NPs, and irradiation). For example, although the AuNP content was the same at 6 and 24 h in A549 cells, the reduction of survival fraction after 2 Gy of proton irradiation was only significant after 6 h.[136] More recently, Broekgaarden *et al.* have illustrated this phenomenon.[137] In their study, they compared ultrasmall AuNP (nanoclusters 3 nm) functionalised with two small molecules (glutathione or arginine) in spheroids (U87MG and F98). Despite achieving an 11-fold increase in cancerous cell accumulation, Au-Arg was less efficient than Au-GSH as regards radiosensitisation. The authors suggest that, even though its absolute quantity was less, Au-GSH was more effective in perfusing throughout the tumour compared to Au-Arg, being able to reach inner cells, which could contribute to its higher efficiency. As pointed out in this paper, even though the use of 3D cultures for radiotherapy enhancement is not the norm, we would highly recommend that it be further explored, as 2D cell monolayers cannot reproduce the gradients in oxygen, nutrients, and drugs that develop in 3D. It is all the more important when it comes to nano-objects, as they are less capable of diffusing than chemicals. This could be one of the reasons why the treatment efficacies in 2D cultures can be off by a factor of 10–100 when compared to 3D/*in vivo* models of cancer.

The coating can also impact the cellular machinery. For example, several papers report that dextran could improve cell survival. In Ref. 138, dextran-coated superparamagnetic iron oxide nanoparticles (SPIONs) seem to enhance the clonogenic capacity of unirradiated U87 cells in a concentration-dependent manner, possibly counterbalancing their radiosensitising effect.[138] The coating can also impact the cell cycle distribution. This is of importance because all cell cycle phases do not present the same level of radiosensitivity, with the most radiosensitive being G2/M. For AuNPs, several groups reported an increase in the G2/M cell population after NP exposure prior to irradiation for various cell lines: DU-145,[139] MDA-MB231,[140] SK-OV-3,[141] HepG2,[142] and A549.[143] The NPs used have in common a medium size (4–10 nm diameter) and a coating containing sugar (thioglucose or galactose). To our knowledge, no study has been devoted to examining this phenomenon in detail so far. Other consequences are mentioned in the literature but have not yet been linked to radiosensitisation. For instance, Kepsutlu *et al.* observed that A549 cells undergo major changes in the quantity of cytoplasmic organelles after the uptake of AuNPs with different surface coatings.[144] They used dendritic polyglycerol sulphate (dPGS) and polyethyleneimine (PEI) coatings. Although it was less pronounced with PEI than with dPGS, they found a similar change in cytoplasmic remodelling within the first 6 h of exposure. They also noted that the NP coating influences the localisation of NPs to lipid droplets. This suggests cell physiological changes that could impact the cell's response to ionising radiation.

Recently, the question of the coating's impact on cell radiosensitisation was specifically addressed by Kerr *et al.*[145] AuNPs were functionalised by three coatings, differing only by the PEG linker length between the sulphur anchor (TA) and the macrocycle (DOTAGA): TA-DOTAGA; TA-PEG4-DOTAGA, and TA-PEG11-DOTAGA. Two glioblastoma cell lines (U87MG et U251MG) were exposed for 1 or 24 h to NPs before 225 kVp X-ray irradiation. The analysis included clonogenic survival, DNA double-strand break induction and repair, cell cycle distribution, and cytokines profiles.

These data suggest a potential role of PEG in mediating cytokine expression and give the first evidence of a correlation between radiosensitisation and the extent of the upregulated cytokines that were identified.

However, the question of the coating's impact is even trickier, as NP surfaces are prone to evolve. Klein *et al.* observed that SPION surfaces underwent an ageing process during storage: after three weeks, they had partially lost their catalytic activity, measured through DCF fluorescence in cells.[146] Such effects are rarely documented, but they could probably contribute to the lack of reproducibility experienced in NP-related works. In addition, the stability of the NPs during their action can be questioned.

10.4.3 *Can we expect nanoparticles to be stable during therapy?*

In all studies concerning the effect of NPs and irradiation, the consequences of this combination on a target (radical probes, biomolecules, or cells) are studied. But the reciprocal point of view is rarely adopted. In other words, the consequences of irradiation and/or the biological environment on the NPs themselves are rarely investigated. However, both the core and the coating can be modified.

10.4.3.1 *Core modification*

The reactivity of metallic salts under irradiation is well known since, alongside classical chemical synthesis, radiolytic preparations are available. They allow for control over size, shape, and ligand stabilisation by varying the irradiation dose, dose rate, and beam type and are claimed to be a cleaner alternative.[147] From most of the articles establishing radiolytic protocols, very high doses (tens of kGy) are necessary to induce the formation of NPs. Nevertheless, some examples testify that NP formation can occur for 10 Gy[148] or even lower doses. This radiolytical synthesis also acts as a source for the development of new colorimetric dosimeters to detect therapeutic doses.[149,150] Thus, it is interesting to wonder what happens to the nanomaterials themselves under irradiation.

In fact, some examples of NP reshaping can be found in the literature.[151,152] For example, gold nanocrystals of 17 nm embedded in amorphous SiO_2 were submitted to 4 MeV photon irradiation at 300 K. This resulted in NP dissolution and aggregation to much smaller precipitates of approximately 2 nm.[152] Also, NPs were shown to elongate due to ion impacts in the direction of the beam. Though the mechanism responsible for such a transformation is not well understood, the modification is significant and leads to the formation of new nano-objects.[151]

It is also important to note that not only the impact of the beam can modify the morphology of NPs. For many years, AuNPs were claimed to be inert *in vivo*. This statement was related to the historical usage of gold in medicine and its high resistance to oxidation or corrosion.[153] In 2020, Balfourier *et al.* studied the fate of AuNPs *in cellulo*.[154] They demonstrated the transformation of 4 and 22 nm AuNPs in fibroblasts through a two-step process. First, NPs degrade into gold atoms, which is interpreted as resulting from the action of highly reactive oxygen species produced *in situ*. Second, gold atoms recrystallise in 2.5 nm self-assembled particles. This processing of AuNPs by cells is not an exception, as it was also evidenced for iron oxide-coated NPs.[155] Here, the gradual dissolution of iron oxide nanocrystals occurred first when the objects were incubated in an acidic citrate medium to mimic lysosomal degradation but also when injected into mice. Then, the gold core was also degraded into smaller-sized gold crystals.

10.4.3.2 *Coating modification*

As mentioned above, in the vast majority of studies, nanomaterials are intentionally functionalised. The coating molecules are thus subjected to irradiation but are also the closest molecules to the NPs and the potential reactive species they generate. To our knowledge, the question of the evolution of the coating in such a case has only been addressed once so far.[156] Here, two well-defined polymer-grafted AuNPs were synthesised and studied under gamma irradiation: 4 or 20 kGy did not affect the nano-objects' core. but effects were visible for the polymer corona. In particular, the radius of gyration and apparent molecular weights both showed a clear decrease with irradiation, from 21 to 7.5 kg/mol for PMAA-AuNPs and from 21.85 to 9.8 kg/mol for P(HEMA-MAA)-AuNPs, attesting to shaving and degrafting. This ligand modification is not so surprising for such high doses, but we can expect such an effect even at much lower doses when considering the yield of radicals formed at 1 Gy. Even if it represents an analytical challenge, the community will certainly benefit from such information, not only from a toxicological point of view but also to 'measure' the range of action at the molecular scale of the reactive species produced.

Moreover, the coating can also be modified within the cell, either by being degraded or replaced. As soon as they enter a biological medium (cell culture medium for *in vitro*, blood, or plasma for *in vivo* studies), NPs, in most cases, see their surfaces covered by biomolecules, especially proteins.[157,158] What the cells see is thus not the ideal NP concocted by chemists, meaning the biomolecular corona could have great biological and physiological repercussions, sometimes masking the expected effects of purposely grafted molecules.[159] As the formation and dynamics of the protein corona strongly depend on the surface charge, Feiner-Gracia *et al.* suggested that it could amplify small initial differences in surface chemistry among particles, leading to the coexistence of NPs with very different surfaces.[160] This 'non-intentional' corona can affect the mechanisms of NP uptake by cells,[161] cell recognition,[162] and immune modulation.[163,164] The corona on polymer-coated nanomaterials is known to exchange with other biomolecules that are both internal and external to the cell;[165] however, it was recently proven that the protein corona evolves significantly from the plasma migrating to lysosomal fluids and subsequent cytosolic fluids.[166] This intracellular transformation of the protein corona modulates NP-induced biological processes, including proteostasis, glycolysis, and lipid metabolism. To the best of our knowledge, no such experiment was conducted within the radiosensitisation community, but it raises little doubt that when they reach the cancerous cells, the nanoradiosensitisers, as refined as they are, are different from the chemists' sketch. Textbox 10.3 provides a brief summary of the key ideas discussed in this section.

- Coating reduces the radical production by NPs under irradiation.
- Coating can impact not only the uptake of NPs but also the whole cell machinery.
- Both irradiation and internalisation can modify the NP's core and coating.

Textbox 10.3. Nanoparticle modification after internalisation.

10.5 New Directions

The publications following Hainfeld's pioneering work[12] focused mainly on testing different NPs on isolated biomolecules[75,167,168] or cell monolayers to give a yes or no answer regarding their radiosensitising potential under different irradiation modalities.[169,170] Twenty years later, the literature is abundant and covers a wide range of areas, but, in our opinion, some recent trends are emerging. Our aim is not to be exhaustive but to highlight some new directions.

10.5.1 *More relevant cellular models*

As mentioned in Section 10.4.2.2, 2D monolayers are representative of epithelial and mesothelial cells but not of solid tumours. In the current literature, the use of 3D cultures is beginning to emerge,[137,171] which we can only encourage. Another remarkable trend is the awareness of the role played by the tumoral microenvironment (TME). The TME consists of cancer-associated fibroblasts (CAFs), immune cells including tumour-associated macrophages, vasculatures, and the extracellular matrix. The cell–cell communication within the TME plays a pivotal role in the evolution and progression of cancer. For example, CAFs are known to promote tumour growth, angiogenesis, metastasis, and resistance to chemotherapy and radiotherapy.[172] Recently, Alhussan *et al.*[173] developed a co-culture of pancreatic ductal adenocarcinoma and CAFs that they exposed to AuNPs. Gold uptake levels were not modified between the mono- and co-culture. However, 24h after 2 Gy, the number of γ-H2AX foci was lower in the co-culture compared to the monoculture, suggesting that CAFs can promote DNA repair. Like others,[137] we believe that it is of utmost importance to move towards more integrated cellular models to reduce the gap between *in vitro* and *in vivo* data and, if possible, to provide useful tools for preclinical screening.

10.5.2 *More documented biological effects*

Now that it is clear that a biological step is involved in radiosensitisation, more teams are focusing on a (nano/model of interest) couple to gain deeper insights and decipher the mechanisms. For example, Durand *et al.* used three complementary biological models, based on U87 spheroids (conventional 3D invasion assay, organotypic brain slice cultures, and chronic cranial window), to investigate the impact of irradiation treatments combined with Au@DTDTPA(Gd) NPs on tumour progression.[174] They showed that these NPs altered the adaptive response of glioblastoma cells after irradiation, by reducing their invasiveness and aggressiveness. Even though these mechanisms need to be confirmed *in vivo*, this opens up new research perspectives.

Another important contribution comes from the Nanobiotix company and their studies on the NBTXR3 immunomodulatory capacity. For readers who are not familiar with immunotherapy, we strongly recommend referring to Ref. 175. In 2019, Marill *et al.* proved in colorectal cancerous cells that the association of NBTXR3 with irradiation led to cGAS-STING pathway activation, a cellular response important for the activation of antitumour immunity.[176] This was their rationale for launching phase I/II clinical trials associating NBTXR3 with PD1-inhibitor. Abscopal effects were later reported in mice models.[177,178] More recently, Darmon *et al.* showed that NBTXR3 associated with irradiation may facilitate the antitumour immune response through modification of the immunogenicity of cancer cells, as well as by induction of immunogenic cell death.[179] The authors also observed a significant increase in CD4+, CD8+, and CD68+ cell infiltrates for NBTXR3 activated by radiotherapy, compared to radiotherapy alone. Moreover, they measured significant modifications in T-cell receptor (TCR) repertoire diversity in the radiotherapy + NBTXR3 group, both in treated and distant untreated tumours, compared to radiotherapy alone. This corpus is decisive, as it documents biological effects that were not at all expected even a few years ago. It also invites researchers to include data such as activation of the cGAS/STING pathway and the recruitment of immune cells into the tumour in the development of new *in vivo* projects.

10.5.3 *Multimodality*

Another set of publications explores diverse combinations of treatments in association with NPs and irradiation. These studies included chemotherapeutics, such as docetaxel[180,181] or doxorubicin.[182] In one study, the authors used PEG-encapsulated iron oxide NPs for the intracellular delivery of doxorubicin combined with carbon ion irradiation. Chithrani's team also incorporated pyronaridine (PYD), an approved antimalarial drug, in their current protocol associating AuNPs and radiation.[183] At sub-micromolar concentrations, this molecule acts as a potential inhibitor of a DNA repair enzyme, ERCC1-XPF. Indeed, a significant increase in DNA double-strand breaks was seen with the triple combination as compared to both AuNP/RT and PYD/RT, as well as a significant reduction in the relative cell growth treated with GNP/PYD/RT 24 h after the treatment.

The association of nanomaterials, irradiation, and a third companion, be it an immunomodulator, a DNA repair inhibitor, or a chemotherapeutic, is attracting increasingly more attention. With the clinical trials already in the recruitment phase in the case of NBTXR3, this triple combination is currently a reality for some patients, and we hope for the best results. However, we would like to emphasise that it implies optimising new parameters, such as the different doses and the time intervals between nanomaterials administration, drug administration, and irradiation, and checking for possible negative interactions. Textbox 10.4 provides a brief summary of the key ideas discussed in this section.

- More relevant cellular models are necessary to study radiosensitisation.
- The biological effects are now better documented, including immunomodulation, which requires further investigation.
- Triple combinations of NPs, irradiation, and a third party require more studies to investigate optimal concentrations, best timing, and potential toxicities.

Textbox 10.4. Biological approaches to study the impact of nanoparticles.

10.6 Conclusions

Two decades after the first report of radiosensitisation by NPs, hundreds of publications have addressed this subject with different approaches (simulations or lab-based experiments), different nanomaterials, different models (isolated molecules, cells, or animals), and different irradiation modalities (photons, electrons, or ions), creating a complex landscape. What seems to have reached a consensus is that the pure physical vision, based on mass energy-absorption coefficients, does not account for experimental results and that biological effects play a significant role.

With regard to the physicochemical step, it has still not been elucidated, but the studies seem to agree on the important role played by the surface of the NPs. Here also, cutting-edge techniques used to study the interfaces, often employed in the catalysis community, should give clues to the radiosensitisation mechanism occurring during NP irradiation. A mechanistic understanding is obviously not required for clinical translation – indeed, many other therapeutics do not have a fully elucidated mode of action, but we are convinced that this fundamental research on nanomaterials' behaviour under irradiation is of high interest to a broader community (catalysis, nuclear safety, etc.). This topic is definitely at a crossroads and represents an opportunity for true interdisciplinary collaborations.

In vivo work continues to identify further complexities in understanding radiosensitisation mechanisms. We could hypothesise that there is not a unique universal biological response but probably different cellular responses depending on the nanomaterials and the cells/tissues. Various avenues seem to be opening up, with immunomodulation at the forefront. There will undoubtedly be more studies in the years to come on this subject. Given the relative simplicity of the design of the two nanomaterials in current clinical trials and the surprising recent discoveries of their possible modes of action, we would recommend the utmost caution when interpreting the results obtained with much more complex objects (multi-elements amd multi-functional). This is all the more reasonable given that nanomaterials do evolve during their journey from the chemistry lab to the organism. This topic is currently in its infancy, but it certainly deserves greater attention. The fate and long-term effects of nanomaterials are not necessarily relevant when patients have a short life expectancy, but they will if younger patients can be eligible for such treatments.

References

1. Anselmo AC, Mitragotri S. Nanoparticles in the clinic. Bioeng Transl Med. 2016;1(1):10–29.
2. Anselmo AC, Mitragotri S. Nanoparticles in the clinic: An update. Bioeng Transl Med. 2019;4(3):e10143.

3. Anselmo AC, Mitragotri S. Nanoparticles in the clinic: An update post COVID-19 vaccines. Bioeng Transl Med. 2021;6(3):e10246.
4. Bobo D, Robinson KJ, Islam J, Thurecht KJ, Corrie SR. Nanoparticle-Based Medicines: A Review of FDA-Approved Materials and Clinical Trials to Date. Pharm Res. 2016;33(10):2373–2387.
5. Namiot ED, Sokolov AV, Chubarev VN, Tarasov VV, Schiöth HB. Nanoparticles in Clinical Trials: Analysis of Clinical Trials, FDA Approvals and Use for COVID-19 Vaccines. Int J Mol Sci. 2023;24: 787.
6. Solov'yov AV, Verkhovtsev AV, Mason NJ, Amos RA, Bald I, Baldacchino G, *et al.* Condensed Matter Systems Exposed to Radiation: Multiscale Theory, Simulations, and Experiment. Chem Rev. 2024;124(13):8014–8129.
7. Taheri A, Khandaker MU, Moradi F, Bradley DA. A review of recent advances in the modeling of nanoparticle radiosensitization with the Geant4-DNA toolkit. Rad Phys Chem. 2023;212:111146.
8. Regulla D, Panzer W, Schmid E, Stephan G, Harder D. Detection of elevated RBE in human lymphocytes exposed to secondary electrons released from X-irradiated metal surfaces. Radiat Res. 2001;155(5):744–747.
9. Regulla D, Schmid E, Friedland W, Panzer W, Heinzmann U, Harder D. Enhanced values of the RBE and H ratio for cytogenetic effects induced by secondary electrons from an X-irradiated gold surface. Radiat Res. 2002;158(4):505–515.
10. Regulla DF, Hieber LB, Seidenbusch M. Physical and biological interface dose effects in tissue due to X-ray-induced release of secondary radiation from metallic gold surfaces. Radiat Res. 1998;150(1):92–100.
11. Herold DM, Das IJ, Stobbe CC, Iyer RV, Chapman JD. Gold microspheres: a selective technique for producing biologically effective dose enhancement. Int J Radiat Biol. 2000;76(10):1357–1364.
12. Hainfeld JF, Slatkin DN, Smilowitz HM. The use of gold nanoparticles to enhance radiotherapy in mice. Phys Med Biol. 2004;49(18):N309–N315.
13. Hainfeld JF, Slatkin DN, Focella TM, Smilowitz HM. Gold nanoparticles: a new X-ray contrast agent. Br J Radiol. 2006;79(939): 248–253.
14. Bonvalot S, Le Pechoux C, De Baere T, Kantor G, Buy X, Stoeckle E, *et al.* First-in-Human Study Testing a New Radioenhancer Using Nanoparticles (NBTXR3) Activated by Radiation Therapy in Patients with Locally Advanced Soft Tissue Sarcomas. Clin Cancer Res. 2017;23(4):908–917.

15. Maggiorella L, Barouch G, Devaux C, Pottier A, Deutsch E, Bourhis J, *et al.* Nanoscale radiotherapy with hafnium oxide nanoparticles. Future Oncol. 2012;8(9):1167–1181.
16. Liem X, de Baere T, Vivar OI, Seiwert TY, Shen C, Papai Z, *et al.* International guidelines for intratumoral and intranodal injection of NBTXR3 nanoparticles in head and neck cancers. Head Neck. 2024;46(6):1253–1262.
17. Bonvalot S, Rutkowski PL, Thariat J, Carrere S, Ducassou A, Sunyach MP, *et al.* NBTXR3, a first-in-class radioenhancer hafnium oxide nanoparticle, plus radiotherapy versus radiotherapy alone in patients with locally advanced soft-tissue sarcoma (Act.In.Sarc): a multicentre, phase 2-3, randomised, controlled trial. Lancet Oncol. 2019;20(8):1148–1159.
18. Rocchi P, Labied L, Doussineau T, Julien M, Giroud B, Vulliet E, *et al.* Identification of Molecular Fragments in Equilibrium with Polysiloxane Ultrasmall Nanoparticles. Nanomaterials. 2022;12(5):738.
19. Sancey L, Kotb S, Truillet C, Appaix F, Marais A, Thomas E, *et al.* Long-term in vivo clearance of gadolinium-based AGuIX nanoparticles and their biocompatibility after systemic injection. ACS Nano. 2015;9(3):2477–2488.
20. Bennett S, Verry C, Kaza E, Miao X, Dufort S, Boux F, *et al.* Quantifying gadolinium-based nanoparticle uptake distributions in brain metastases via magnetic resonance imaging. Sci Rep. 2024;14(1):11959.
21. Lux F, Tran VL, Thomas E, Dufort S, Rossetti F, Martini M, *et al.* AGuIX((R)) from bench to bedside-Transfer of an ultrasmall theranostic gadolinium-based nanoparticle to clinical medicine. Br J Radiol. 2019;92(1093):20180365.
22. Garcia-Prada CD, Carmes L, Atis S, Parach A, Bertolet A, Jarlier M, *et al.* Gadolinium-Based Nanoparticles Sensitize Ovarian Peritoneal Carcinomatosis to Targeted Radionuclide Therapy. J Nucl Med. 2023;64(12):1956–1964.
23. Carmès L, Banerjee M, Coliat P, Harlepp S, Pivot X, Tillement O, *et al.* Toward Clinical Transfer of Tumor-Targeted Ultrasmall Inorganic Nanoparticles. Adv Therap. 2023;6(9):2300019.
24. Carmes L, Bort G, Lux F, Seban L, Rocchi P, Muradova Z, *et al.* AGuIX nanoparticle-nanobody bioconjugates to target immune checkpoint receptors. Nanoscale. 2024;16(5):2347–2360.
25. Goldman LW. Principles of CT and CT technology. J Nucl Med Technol. 2007;35(3):115–28; quiz 129–130.

26. Bushberg JT, Seibert JA, Leidholdt EM, Boone JM. The Essential Physics of Medical Imaging. Philadelphia: Wolters Kluwer Health; 2011.
27. Crist RM, Dasa SSK, Liu CH, Clogston JD, Dobrovolskaia MA, Stern ST. Challenges in the development of nanoparticle-based imaging agents: Characterization and biology. WIREs Nanomed Nanobiotech. 2021;13(1):e1665.
28. Khorasani A, Shahbazi-Gahrouei D, Safari A. Recent Metal Nanotheranostics for Cancer Diagnosis and Therapy: A Review. Diagnostics. 2023;13(5):833.
29. Yu SB, Watson AD. Metal-Based X-ray Contrast Media. Chem Rev. 1999;99(9):2353–2378.
30. Kim J, Chhour P, Hsu J, Litt HI, Ferrari VA, Popovtzer R, *et al.* Use of Nanoparticle Contrast Agents for Cell Tracking with Computed Tomography. Bioconjug Chem. 2017;28(6):1581–1597.
31. Smith BR, Gambhir SS. Nanomaterials for In Vivo Imaging. Chem Rev. 2017;117(3):901–986.
32. Aslan N, Ceylan B, Koç MM, Findik F. Metallic nanoparticles as X-Ray computed tomography (CT) contrast agents: A review. J Mol Struct. 2020;1219:128599.
33. Gupta D, Roy I, Gandhi S. Metallic nanoparticles for CT-guided imaging of tumors and their therapeutic applications. OpenNano. 2023;12:100146.
34. Mutreja I, Maalej N, Kaushik A, Kumar D, Raja A. High atomic number nanoparticles to enhance spectral CT imaging aspects. Mat Adv. 2023;4(18):3967–3688.
35. FitzGerald PF, Colborn RE, Edic PM, Lambert JW, Torres AS, Bonitatibus PJ Jr, *et al.* CT Image Contrast of High-Z Elements: Phantom Imaging Studies and Clinical Implications. Radiology. 2016;278(3):723–733.
36. Dong YC, Hajfathalian M, Maidment PSN, Hsu JC, Naha PC, Si-Mohamed S, *et al.* Effect of Gold Nanoparticle Size on Their Properties as Contrast Agents for Computed Tomography. Sci Rep. 2019;9(1):14912.
37. Alric C, Taleb J, Le Duc G, Mandon C, Billotey C, Le Meur-Herland A, *et al.* Gadolinium chelate coated gold nanoparticles as contrast agents for both X-ray computed tomography and magnetic resonance imaging. J Am Chem Soc. 2008;130(18):5908–5915.
38. Feng W, Zhou X, Nie W, Chen L, Qiu K, Zhang Y, *et al.* Au/polypyrrole@Fe_3O_4 nanocomposites for MR/CT dual-modal imaging guided-photothermal therapy: an in vitro study. ACS Appl Mater Interfaces. 2015;7(7):4354–4367.

39. Pantos I, Thalassinou S, Argentos S, Kelekis NL, Panayiotakis G, Efstathopoulos EP. Adult patient radiation doses from non-cardiac CT examinations: a review of published results. Br J Radiol. 2011;84(1000):293–303.
40. Shannoun F, Blettner M, Schmidberger H, Zeeb H. Radiation protection in diagnostic radiology. Dtsch Arztebl Int. 2008;105(3):41–46.
41. McCollough CH, Bushberg JT, Fletcher JG, Eckel LJ. Answers to Common Questions About the Use and Safety of CT Scans. Mayo Clin Proc. 2015;90(10):1380–1392.
42. Pellico J, Gawne PJ, de Rosales RTM. Radiolabelling of nanomaterials for medical imaging and therapy. Chem Soc Rev. 2021;50(5):3355–3423.
43. Pijeira MSO, Viltres H, Kozempel J, Sakmar M, Vlk M, Ilem-Ozdemir D, *et al.* Radiolabeled nanomaterials for biomedical applications: radiopharmacy in the era of nanotechnology. EJNMMI Radiopharm Chem. 2022;7(1):8.
44. van der Zande K, Oyen WJG, Zwart W, Bergman AM. Radium-223 Treatment of Patients with Metastatic Castration Resistant Prostate Cancer: Biomarkers for Stratification and Response Evaluation. Cancers. 2021;13(17):4346.
45. Gaweda W, Pruszynski M, Cedrowska E, Rodak M, Majkowska-Pilip A, Gawel D, *et al.* Trastuzumab Modified Barium Ferrite Magnetic Nanoparticles Labeled with Radium-223: A New Potential Radiobioconjugate for Alpha Radioimmunotherapy. Nanomaterials. 2020;10(10):2067.
46. Toro-González M, Peacock A, Miskowiec A, Cullen DA, Copping R, Mirzadeh S, *et al.* Tailoring the Radionuclide Encapsulation and Surface Chemistry of La(223Ra)VO4 Nanoparticles for Targeted Alpha Therapy. J Nanotheranostics. 2021;2(1):33–50.
47. Trencsenyi G, Csikos C, Kepes Z. Targeted Radium Alpha Therapy in the Era of Nanomedicine: In Vivo Results. Int J Mol Sci. 2024;25(1):664.
48. Silva F, Cabral Campello MP, Paulo A. Radiolabeled Gold Nanoparticles for Imaging and Therapy of Cancer. Materials. 2020;14(1):4.
49. Wang R, Ponsard B, Wolterbeek H, Denkova A. Core-shell structured gold nanoparticles as carrier for (166)Dy/(166)Ho in vivo generator. EJNMMI Radiopharm Chem. 2022;7(1):16.
50. Laprise-Pelletier M, Simao T, Fortin MA. Gold Nanoparticles in Radiotherapy and Recent Progress in Nanobrachytherapy. Adv Healthc Mater. 2018;7(16):e1701460.
51. Fach M, Fliedner FP, Kempen PJ, Melander F, Hansen AE, Bruun LM, *et al.* Effective Intratumoral Retention of [103Pd]AuPd Alloy

Nanoparticles Embedded in Gel-Forming Liquids Paves the Way for New Nanobrachytherapy. Adv Healthc Mater. 2021;10(10):e2002009.

52. Laprise-Pelletier M, Lagueux J, Cote MF, LaGrange T, Fortin MA. Low-Dose Prostate Cancer Brachytherapy with Radioactive Palladium-Gold Nanoparticles. Adv Healthc Mater. 2017;6(4): 1601120.
53. Moeendarbari S, Tekade R, Mulgaonkar A, Christensen P, Ramezani S, Hassan G, *et al.* Theranostic Nanoseeds for Efficacious Internal Radiation Therapy of Unresectable Solid Tumors. Sci Rep. 2016;6:20614.
54. Kiseleva M, Lescot T, Selivanova SV, Fortin MA. Gold-Enhanced Brachytherapy by a Nanoparticle-Releasing Hydrogel and 3D-Printed Subcutaneous Radioactive Implant Approach (Adv. Healthcare Mater. 23/2023). Adv Healthc Mater. 2023;12(23):2370139.
55. Le Goas M, Paquet M, Paquirissamy A, Guglielmi J, Compin C, Thariat J, *et al.* Improving (131)I Radioiodine Therapy By Hybrid Polymer-Grafted Gold Nanoparticles. Int J Nanomed. 2019;14: 7933–7946.
56. Hirayama R, Ito A, Noguchi M, Matsumoto Y, Uzawa A, Kobashi G, *et al.* OH radicals from the indirect actions of X-rays induce cell lethality and mediate the majority of the oxygen enhancement effect. Radiat Res. 2013;180(5):514–523.
57. Hirayama R, Ito A, Tomita M, Tsukada T, Yatagai F, Noguchi M, *et al.* Contributions of direct and indirect actions in cell killing by high-LET radiations. Radiat Res. 2009;171(2):212–218.
58. Baldacchino G, Brun E, Denden I, Bouhadoun S, Roux R, Khodja H, *et al.* Importance of radiolytic reactions during high-LET irradiation modalities: LET effect, role of O2 and radiosensitization by nanoparticles. Cancer Nanotechnol. 2019;10(1):3.
59. Buxton GV. The Radiation Chemistry of Liquid Water: Principles and Applications. Amsterdam: Springer; 2003.
60. Buxton GV, Greenstock CL, Helman WP, Ross AB. Critical-Review of Rate Constants for Reactions of Hydrated Electrons, Hydrogen-Atoms and Hydroxyl Radicals in Aqueous-Solution. J Phys Chem Ref Data. 1988;17(2):513–586.
61. Fleming AM, Burrows CJ. On the irrelevancy of hydroxyl radical to DNA damage from oxidative stress and implications for epigenetics. Chem Soc Rev. 2020;49(18):6524–6528.
62. von Sontag C. Free-Radical-Induced DNA Damage and Its Repair. A Chemical Perspective. Berlin: Springer; 2006.
63. Bienert GP, Schjoerring JK, Jahn TP. Membrane transport of hydrogen peroxide. Biochim Biophys Acta. 2006;1758(8):994–1003.

64. Lennicke C, Cochemé HM. Redox metabolism: ROS as specific molecular regulators of cell signaling and function. Mol Cell. 2021;81(18):3691–3707.
65. Zwiehoff S, Johny J, Behrends C, Landmann A, Mentzel F, Baumer C, *et al.* Enhancement of Proton Therapy Efficiency by Noble Metal Nanoparticles Is Driven by the Number and Chemical Activity of Surface Atoms. Small. 2022;18(9):e2106383.
66. Ducrozet F, Sebastian A, Garcia Villavicencio CJ, Ptasinska S, Sicard-Roselli C. Quantifying hydroxyl radicals generated by a low-temperature plasma using coumarin: methodology and precautions. Phys Chem Chem Phys. 2024;26(11):8651–8657.
67. Leandri V, Gardner JM, Jonsson M. Coumarin as a Quantitative Probe for Hydroxyl Radical Formation in Heterogeneous Photocatalysis. J Phys Chem C. 2019;123(11):6667–6674.
68. Murphy MP, Bayir H, Belousov V, Chang CJ, Davies KJA, Davies MJ, *et al.* Guidelines for measuring reactive oxygen species and oxidative damage in cells and in vivo. Nat Metab. 2022;4(6):651–662.
69. Zhang Y, Dai M, Yuan Z. Methods for the detection of reactive oxygen species. Anal Methods. 2018;10(38):4625–4638.
70. Reisz JA, Bansal N, Qian J, Zhao W, Furdui CM. Effects of ionizing radiation on biological molecules–mechanisms of damage and emerging methods of detection. Antioxid Redox Signal. 2014;21(2):260–292.
71. Cheng NN, Starkewolf Z, Davidson RA, Sharmah A, Lee C, Lien J, *et al.* Chemical enhancement by nanomaterials under X-ray irradiation. J Am Chem Soc. 2012;134(4):1950–1953.
72. Sicard-Roselli C, Brun E, Gilles M, Baldacchino G, Kelsey C, McQuaid H, *et al.* A New Mechanism for Hydroxyl Radical Production in Irradiated Nanoparticle Solutions. Small. 2014;10(16):3338–3346.
73. Gilles M, Brun E, Sicard-Roselli C. Quantification of hydroxyl radicals and solvated electrons produced by irradiated gold nanoparticles suggests a crucial role of interfacial water. J Colloid Interface Sci. 2018;525:31–38.
74. Chang J, Taylor RD, Davidson RA, Sharmah A, Guo T. Electron Paramagnetic Resonance Spectroscopy Investigation of Radical Production by Gold Nanoparticles in Aqueous Solutions Under X-ray Irradiation. J Phys Chem A. 2016;120(18):2815–2823.
75. Brun E, Sanche L, Sicard-Roselli C. Parameters governing gold nanoparticle X-ray radiosensitization of DNA in solution. Colloids Surf B Biointerfaces. 2009;72(1):128–134.

76. Brun E, Tandiana R, Gilles M, Cheref Y, Van-Oanh NT, Clavaguera C, *et al.* Both experimental and molecular dynamics approaches highlight the central role of interfacial water for radical production by irradiated gold nanoparticles. J Phys B: At Mol Op Phys. 2024;57(6):065002.
77. Johny J, van Halteren CER, Cakir FC, Zwiehoff S, Behrends C, Baumer C, *et al.* Surface Chemistry and Specific Surface Area Rule the Efficiency of Gold Nanoparticle Sensitizers in Proton Therapy. Chemistry. 2023;29(50):e202301260.
78. Okazaki Y, Kusumoto T, Roux S, Hirayama R, Fromm M, Bazzi R, *et al.* Increase of OH radical yields due to the decomposition of hydrogen peroxide by gold nanoparticles under X-ray irradiation. RSC Adv. 2024;14(14):9509–9513.
79. Guerreiro A, Chatterton N, Crabb EM, Golding JP. A comparison of the radiosensitisation ability of 22 different element metal oxide nanoparticles using clinical megavoltage X-rays. Cancer Nanotechnol. 2019;10(1):20.
80. Brun E, Girard HA, Arnault JC, Mermoux M, Sicard-Roselli C. Hydrogen plasma treated nanodiamonds lead to an overproduction of hydroxyl radicals and solvated electrons in solution under ionizing radiation. Carbon. 2020;162:510–518.
81. Kurzyp M, Girard HA, Cheref Y, Brun E, Sicard-Roselli C, Saada S, *et al.* Hydroxyl radical production induced by plasma hydrogenated nanodiamonds under X-ray irradiation. Chem Comm. 2017;53(7):1237–1240.
82. Yang C, Zhang B, Lin X, Han Q, Bao H, Liu Y. Universality of hydroxyl radical generation in irradiated nanoenhancer suspensions: Vital role of concentration. AIP Adv. 2024;14(5):055209.
83. Que X, Li S, Hu Y, Zhang Z, Sheng L, Lin T, *et al.* Fluorescence Spectroscopy Study on the Reaction of Hydroxyl Radicals and Hydrated Electrons in MXene Suspension. J Phys Chem C. 2022;126(26):10703–10712.
84. Alizadeh E, Orlando TM, Sanche L. Biomolecular damage induced by ionizing radiation: the direct and indirect effects of low-energy electrons on DNA. Annu Rev Phys Chem. 2015;66:379–398.
85. McGrady J, Kumagai Y, Watanabe M, Kirishima A, Akiyama D, Kitamura A, *et al.* The kinetics and mechanism of H(2)O(2) decomposition at the U(3)O(8) surface in bicarbonate solution. RSC Adv. 2021;11(46):28940–28948.
86. McGrady J, Yamashita S, Kano S, Yang H, Kimura A, Taguchi M, *et al.* Radiolysis of water at the surface of ZrO2 nanoparticles. Rad Phys Chem. 2023;209:110970.

87. Kaddissy JA, Esnouf S, Saffré D, Renault JP. Efficient hydrogen production from irradiated aluminum hydroxides. Int J Hydrogen Energy. 2019;44(7):3737–3743.
88. Klett C, Cui Y, Devineau S, Foy E, Dagnelie R, Renault JP. H2 production through oxide irradiation: Comparison of gamma rays and vacuum ultraviolet excitation. Int J Hydrogen Energy. 2013;38(10):3889–3897.
89. Lousada CM, Jonsson M. Kinetics, Mechanism, and Activation Energy of H2O2 Decomposition on the Surface of ZrO2. J Phys Chem C. 2010;114(25):11202–11208.
90. McMahon SJ, Hyland WB, Brun E, Butterworth KT, Coulter JA, Douki T, *et al.* Energy Dependence of Gold Nanoparticle Radiosensitization in Plasmid DNA. J Phys Chem C. 2011;115(41):20160–20167.
91. Bellarosa L, Garcia-Muelas R, Revilla-Lopez G, Lopez N. Diversity at the Water-Metal Interface: Metal, Water Thickness, and Confinement Effects. ACS Cent Sci. 2016;2(2):109–116.
92. Fenter P, Sturchio NC. Mineral–water interfacial structures revealed by synchrotron X-ray scattering. Prog Surf Sci. 2004;77(5):171–258.
93. Schnur S, Groß A. Properties of metal–water interfaces studied from first principles. New J Phys. 2009;11(12):125003.
94. Thiel PA, Madey TE. The interaction of water with solid surfaces: Fundamental aspects. Surf Sci Rep. 1987;7(6):211–385.
95. Tandiana R, Brun E, Sicard-Roselli C, Domin D, Van-Oanh NT, Clavaguera C. Probing the structural properties of the water solvation shell around gold nanoparticles: A computational study. J Chem Phys. 2021;154(4):044706.
96. LaVerne JA, Tandon L. H_2 Production in the Radiolysis of Water on CeO_2 and ZrO_2. J Phys Chem B. 2002;106(2):380–386.
97. LaVerne JA, Tandon L. H_2 Production in the Radiolysis of Water on UO_2 and Other Oxides. J Phys Chem B. 2003;107(49):13623–13628.
98. Essehli R, Crumière F, Blain G, Vandenborre J, Pottier F, Grambow B, *et al.* H_2 production by γ and He ions water radiolysis, effect of presence TiO2 nanoparticles. Int J Hydrogen Energy. 2011;36(22):14342–14348.
99. Jonsson M. An Overview of Interfacial Radiation Chemistry in Nuclear Technology. Israel J Chem. 2014;54(3):292–301.
100. LaVerne JA. Radiation chemistry of water with ceramic oxides. In: Hatano Y, Katsumura Y, Mozumber A, editors. Charged Particle and Photon Interactions with Matter: Recent Advances, Applications, and Interfaces. Boca Raton: CRC Press; 2010.

101. Milosavljevic BH, Pimblott SM, Meisel D. Yields and Migration Distances of Reducing Equivalents in the Radiolysis of Silica Nanoparticles. J Phys Chem B. 2004;108(22):6996–7001.
102. Boschi F, Spinelli AE. Nanoparticles for Cerenkov and Radioluminescent Light Enhancement for Imaging and Radiotherapy. Nanomaterials. 2020;10(9):1771.
103. Čerenkov PA. Visible Radiation Produced by Electrons Moving in a Medium with Velocities Exceeding that of Light. Phys Rev. 1937;52(4):378–379.
104. Pratt EC, Shaffer TM, Zhang Q, Drain CM, Grimm J. Nanoparticles as multimodal photon transducers of ionizing radiation. Nat Nanotechnol. 2018;13(5):418–426.
105. Teng M, Liang X, Liu H, Li Z, Gao X, Zhang C, *et al.* Cerenkov radiation shining a light for cancer theranostics. Nano Today. 2024;55:102174.
106. Xu P, Lin S, Wang Y, Abdukayum A, Wang Y. Radionuclide-based Cerenkov luminescence in biomedicine: Current research progress and future perspectives. TrAC Trends Ana Chem. 2024;170:117452.
107. Gao YH, Torrente-Murciano L. Mechanistic insights of the reduction of gold salts in the Turkevich protocol. Nanoscale. 2020;12(4):2740–2751.
108. Oliveira AEF, Pereira AC, Resende MAC, Ferreira LF. Gold Nanoparticles: A Didactic Step-by-Step of the Synthesis Using the Turkevich Method, Mechanisms, and Characterizations. Analytica. 2023;4(2):250–263.
109. Abid N, Khan AM, Shujait S, Chaudhary K, Ikram M, Imran M, *et al.* Synthesis of nanomaterials using various top-down and bottom-up approaches, influencing factors, advantages, and disadvantages: A review. Adv Colloid Interface Sci. 2022;300:102597.
110. Cushing BL, Kolesnichenko VL, O'Connor CJ. Recent Advances in the Liquid-Phase Syntheses of Inorganic Nanoparticles. Chem Rev. 2004;104(9):3893–3946.
111. Marcelo GA, Lodeiro C, Capelo JL, Lorenzo J, Oliveira E. Magnetic, fluorescent and hybrid nanoparticles: From synthesis to application in biosystems. Mater Sci Eng C Mater Biol Appl. 2020;106:110104.
112. Brun E, Sicard-Roselli C. Actual questions raised by nanoparticle radiosensitization. Rad Phys Chem. 2016;128:134–142.
113. Ternad I, Penninckx S, Lecomte V, Vangijzegem T, Conrard L, Lucas S, *et al.* Advances in the Mechanistic Understanding of Iron Oxide Nanoparticles' Radiosensitizing Properties. Nanomaterials. 2023;13(1):201.

114. Beaudier P, Vilotte F, Simon M, Muggiolu G, Le Trequesser Q, Deves G, *et al.* Sarcoma cell-specific radiation sensitization by titanate scrolled nanosheets: insights from physicochemical analysis and transcriptomic profiling. Sci Rep. 2024;14(1):3295.
115. Joze-Majidi H, Zabihi E, Arab-Bafrani Z, Mir SM, Reiter RJ. Comparative radiosensitization efficiency assessment of graphene oxide and Ti3C2 MXene as 2D carbon-based nanoparticles against breast cancer cells: characterization, toxicity and mechanisms. 2D Materials. 2023;10(2):025021.
116. Zimmermann M, Gerken LRH, Wee S, Kissling VM, Neuer AL, Tsolaki E, *et al.* X-ray radio-enhancement by Ti(3)C(2)T(x) MXenes in soft tissue sarcoma. Biomater Sci. 2023;11(24):7826–7837.
117. Yang S, Yang Y, Yang Y, Zhao X, Wang Q, Li B, *et al.* Iron-Palladium Decorated Carbon Nanotubes Achieve Radiosensitization via Reactive Oxygen Species Burst. Front Bioeng Biotechnol. 2021;9:683363.
118. Chen MH, Chen MH, Li CY, Tung FI, Chen SY, Liu TY. Using Gold-Nanorod-Filled Mesoporous Silica Nanobeads for Enhanced Radiotherapy of Oral Squamous Carcinoma. Nanomaterials. 2021;11(9):2235.
119. Xu M, Qian Y, Li X, Gu B, He S, Lu X, *et al.* Janus ACSP Nanoparticle for Synergistic Chemodynamic Therapy and Radiosensitization. ACS Appl Mater Interfaces. 2024;16(14):17242–17252.
120. Xiao L, Chen B, Wang W, Tian T, Qian H, Li X, *et al.* Multifunctional Au@AgBiS(2) Nanoparticles as High-Efficiency Radiosensitizers to Induce Pyroptosis for Cancer Radioimmunotherapy. Adv Sci. 2023;10(30):e2302141.
121. Mousavi M, Koosha F, Neshastehriz A. Chemo-radiation therapy of U87-MG glioblastoma cells using SPIO@AuNP-Cisplatin-Alginate nanocomplex. Heliyon. 2023;9(3):e13847.
122. Ruan F, Fang H, Chen F, Xie X, He M, Wang R, *et al.* Leveraging Radiation-triggered Metal Prodrug Activation Through Nanosurface Energy Transfer for Directed Radio-chemo-immunotherapy. Angew Chem Int Ed. 2024;63(10):e202317943.
123. Lyu M, Zhang T, Li Y, Xiang J, Zhu D, Xia L, *et al.* AIEgen-based nanotherapeutic strategy for enhanced FLASH irradiation to prevent tumour recurrence and avoid severe side effects. Chem Eng J. 2023;473:145179.
124. Zhang S, Lai Y, Pan J, Saeed M, Li S, Zhou H, *et al.* PROTAC Prodrug-Integrated Nanosensitizer for Potentiating Radiation Therapy of Cancer. Adv Mater. 2024;36(23):e2314132.

125. Ayyami Y, Ghorbani M, Dastgir M, Malekzadeh R, Mortezazadeh T. Chitosan-modified manganese oxide-conjugated methotrexate nanoparticles delivering 5-aminolevulinic acid as a dual-modal T1-T2* MRI contrast agent in U87MG cell detection. Magn Reson Mater Phy. 2024;37:909–924.
126. Pinals RL, Chio L, Ledesma F, Landry MP. Engineering at the nano-bio interface: harnessing the protein corona towards nanoparticle design and function. Analyst. 2020;145(15):5090–5112.
127. Sanita G, Carrese B, Lamberti A. Nanoparticle Surface Functionalization: How to Improve Biocompatibility and Cellular Internalization. Front Mol Biosci. 2020;7:587012.
128. Gilles M, Brun E, Sicard-Roselli C. Gold nanoparticles functionalization notably decreases radiosensitization through hydroxyl radical production under ionizing radiation. Colloids Surf B Biointerfaces. 2014;123:770–777.
129. Grellet S, Tzelepi K, Roskamp M, Williams P, Sharif A, Slade-Carter R, *et al.* Cancer-selective, single agent chemoradiosensitising gold nanoparticles. Plos One. 2017;12(7):21.
130. Mai T, Hilt JZ. Functionalization of iron oxide nanoparticles with small molecules and the impact on reactive oxygen species generation for potential cancer therapy. Colloids Surf A: Physicochem Eng Asp. 2019;576:9–14.
131. Dickers MD, Verkhovtsev AV, Mason NJ, Solov'yov AV. Atomistic modelling and structural characterisation of coated gold nanoparticles for biomedical applications. Eur Phys J D. 2023;77(7):141.
132. Verkhovtsev AV, Nichols A, Mason NJ, Solov'yov AV. Molecular Dynamics Characterization of Radiosensitizing Coated Gold Nanoparticles in Aqueous Environment. J Phys Chem A. 2022; 126(14):2170–2184.
133. Cruje C, Chithrani BD. Integration of Peptides for Enhanced Uptake of PEGylayed Gold Nanoparticles. J Nanosci Nanotechnol. 2015;15(3):2125–2131.
134. Yang C, Bromma K, Chithrani D. Peptide Mediated In Vivo Tumor Targeting of Nanoparticles through Optimization in Single and Multilayer In Vitro Cell Models. Cancers. 2018;10(3):84.
135. Butterworth KT, Coulter JA, Jain S, Forker J, McMahon SJ, Schettino G, *et al.* Evaluation of cytotoxicity and radiation enhancement using 1.9 nm gold particles: potential application for cancer therapy. Nanotechnology. 2010;21(29):295101.
136. Penninckx S, Heuskin AC, Michiels C, Lucas S. The role of thioredoxin reductase in gold nanoparticle radiosensitization effects. Nanomedicine. 2018;13(22):2917–2937.

137. Broekgaarden M, Bulin AL, Porret E, Musnier B, Chovelon B, Ravelet C, *et al.* Surface functionalization of gold nanoclusters with arginine: a trade-off between microtumor uptake and radiotherapy enhancement. Nanoscale. 2020;12(13):6959–6963.

138. Guerra DB, Oliveira EMN, Sonntag AR, Sbaraine P, Fay AP, Morrone FB, *et al.* Intercomparison of radiosensitization induced by gold and iron oxide nanoparticles in human glioblastoma cells irradiated by 6 MV photons. Sci Rep. 2022;12(1):9602.

139. Roa W, Zhang XJ, Guo LH, Shaw A, Hu XY, Xiong YP, *et al.* Gold nanoparticle sensitize radiotherapy of prostate cancer cells by regulation of the cell cycle. Nanotechnology. 2009;20(37):375101.

140. Wang C, Jiang Y, Li X, Hu L. Thioglucose-bound gold nanoparticles increase the radiosensitivity of a triple-negative breast cancer cell line (MDA-MB-231). Breast Cancer. 2015;22(4):413–420.

141. Geng F, Song K, Xing JZ, Yuan CZ, Yan S, Yang QF, *et al.* Thioglucose bound gold nanoparticles enhance radio-cytotoxic targeting of ovarian cancer. Nanotechnology. 2011;22(28):285101.

142. Zhu CD, Zheng Q, Wang LX, Xu HF, Tong JL, Zhang QA, *et al.* Synthesis of novel galactose functionalized gold nanoparticles and its radiosensitizing mechanism. J Nanobiotechnol. 2015;13:67.

143. Wang C, Li X, Wang Y, Liu Z, Fu L, Hu L. Enhancement of radiation effect and increase of apoptosis in lung cancer cells by thioglucose-bound gold nanoparticles at megavoltage radiation energies. J Nanoparticle Res. 2013;15(5):1–12.

144. Kepsutlu B, Wycisk V, Achazi K, Kapishnikov S, Perez-Berna AJ, Guttmann P, *et al.* Cells Undergo Major Changes in the Quantity of Cytoplasmic Organelles after Uptake of Gold Nanoparticles with Biologically Relevant Surface Coatings. ACS Nano. 2020;14(2): 2248–2264.

145. Kerr BN, Duffy D, McInerney CE, Hutchinson A, Dabaja I, Bazzi R, *et al.* Evaluation of Radiosensitization and Cytokine Modulation by Differentially PEGylated Gold Nanoparticles in Glioblastoma Cells. Int J Mol Sci. 2023;24(12):10032.

146. Klein S, Sommer A, Distel LV, Hazemann JL, Kroner W, Neuhuber W, *et al.* Superparamagnetic iron oxide nanoparticles as novel X-ray enhancer for low-dose radiation therapy. J Phys Chem B. 2014;118(23):6159–6166.

147. Remita H, Lampre I. Synthesis of Metallic Nanostructures Using Ionizing Radiation and Their Applications. Materials. 2024;17(2):364.

148. Akar B, Pushpavanam K, Narayanan E, Rege K, Heys JJ. Mechanistic investigation of radiolysis-induced gold nanoparticle formation

for radiation dose prediction. Biomed Phys Eng Express. 2018;4(6): 065011.

149. Pushpavanam K, Inamdar S, Chang J, Bista T, Sapareto S, Rege K. Detection of Therapeutic Levels of Ionizing Radiation Using Plasmonic Nanosensor Gels. Adv Func Mater. 2017;27(21):1606724.
150. Pushpavanam K, Narayanan E, Chang J, Sapareto S, Rege K. A Colorimetric Plasmonic Nanosensor for Dosimetry of Therapeutic Levels of Ionizing Radiation. ACS Nano. 2015;9(12):11540–11550.
151. Leino AA, Jantunen VE, Mota-Santiago P, Kluth P, Djurabekova F. Insights into nanoparticle shape transformation by energetic ions. Sci Rep. 2023;13(1):6354.
152. Rizza G, Cheverry H, Gacoin T, Lamasson A, Henry S. Ion beam irradiation of embedded nanoparticles: Toward an in situ control of size and spatial distribution. J Appl Phys. 2007;101(1):014321.
153. Balfourier A, Kolosnjaj-Tabi J, Luciani N, Carn F, Gazeau F. Gold-based therapy: From past to present. Proc Natl Acad Sci U S A. 2020;117(37):22639–22648.
154. Balfourier A, Luciani N, Wang G, Lelong G, Ersen O, Khelfa A, *et al.* Unexpected intracellular biodegradation and recrystallization of gold nanoparticles. Proc Natl Acad Sci U S A. 2020;117(1):103–113.
155. Kolosnjaj-Tabi J, Javed Y, Lartigue L, Volatron J, Elgrabli D, Marangon I, *et al.* The One Year Fate of Iron Oxide Coated Gold Nanoparticles in Mice. ACS Nano. 2015;9(8):7925–7939.
156. Le Goas M, Paquirissamy A, Gargouri D, Fadda G, Testard F, Aymes-Chodur C, *et al.* Irradiation Effects on Polymer-Grafted Gold Nanoparticles for Cancer Therapy. ACS Appl Bio Mater. 2019;2(1):144–154.
157. Lynch I, Cedervall T, Lundqvist M, Cabaleiro-Lago C, Linse S, Dawson KA. The nanoparticle-protein complex as a biological entity; a complex fluids and surface science challenge for the 21st century. Adv Colloid Interface Sci. 2007;134-135:167–174.
158. Lynch I, Dawson KA. Protein-nanoparticle interactions. Nano Today. 2008;3(1–2):40–7.
159. Salvati A, Pitek AS, Monopoli MP, Prapainop K, Bombelli FB, Hristov DR, *et al.* Transferrin-functionalized nanoparticles lose their targeting capabilities when a biomolecule corona adsorbs on the surface. Nat Nanotechnol. 2013;8(2):137–143.
160. Feiner-Gracia N, Beck M, Pujals S, Tosi S, Mandal T, Buske C, *et al.* Super-Resolution Microscopy Unveils Dynamic Heterogeneities in Nanoparticle Protein Corona. Small. 2017;13(41):1701631.

161. Francia V, Yang K, Deville S, Reker-Smit C, Nelissen I, Salvati A. Corona Composition Can Affect the Mechanisms Cells Use to Internalize Nanoparticles. ACS Nano. 2019;13(10):11107–11121.
162. Cai R, Ren J, Ji Y, Wang Y, Liu Y, Chen Z, *et al.* Corona of Thorns: The Surface Chemistry-Mediated Protein Corona Perturbs the Recognition and Immune Response of Macrophages. ACS Appl Mater Interfaces. 2020;12(2):1997–2008.
163. Baimanov D, Wu J, Chu R, Cai R, Wang B, Cao M, *et al.* Immunological Responses Induced by Blood Protein Coronas on Two-Dimensional MoS2 Nanosheets. ACS Nano. 2020;14(5):5529–5542.
164. Deng ZJ, Liang M, Monteiro M, Toth I, Minchin RF. Nanoparticle-induced unfolding of fibrinogen promotes Mac-1 receptor activation and inflammation. Nat Nano. 2011;6(1):39–44.
165. Khan AO, Di Maio A, Guggenheim EJ, Chetwynd AJ, Pencross D, Tang S, *et al.* Surface Chemistry-Dependent Evolution of the Nanomaterial Corona on TiO_2 Nanomaterials Following Uptake and Sub-Cellular Localization. Nanomaterials. 2020;10(3):401.
166. Cai R, Ren J, Guo M, Wei T, Liu Y, Xie C, *et al.* Dynamic intracellular exchange of nanomaterials' protein corona perturbs proteostasis and remodels cell metabolism. Proc Natl Acad Sci U S A. 2022;119(23):e2200363119.
167. Brun E, Duchambon P, Blouquit Y, Keller G, Sanche L, Sicard-Roselli C. Gold nanoparticles enhance the X-ray-induced degradation of human centrin 2 protein. Rad Phys Chem. 2009;78(3):177–183.
168. Carter JD, Cheng NN, Qu Y, Suarez GD, Guo T. Nanoscale energy deposition by X-ray absorbing nanostructures. J Phys Chem B. 2007;111(40):11622–11625.
169. Chithrani DB, Jelveh S, Jalali F, van Prooijen M, Allen C, Bristow RG, *et al.* Gold nanoparticles as radiation sensitizers in cancer therapy. Radiat Res. 2010;173(6):719–728.
170. Kong T, Zeng J, Wang XP, Yang XY, Yang J, McQuarrie S, *et al.* Enhancement of radiation cytotoxicity in breast-cancer cells by localized attachment of gold nanoparticles. Small. 2008;4(9):1537–1543.
171. Musielak M, Bos-Liedke A, Piwocka O, Kowalska K, Markiewicz R, Szymkowiak B, *et al.* The Role of Functionalization and Size of Gold Nanoparticles in the Response of MCF-7 Breast Cancer Cells to Ionizing Radiation Comparing 2D and 3D In Vitro Models. Pharmaceutics. 2023;15(3):862.

172. Raskov H, Orhan A, Gaggar S, Gogenur I. Cancer-Associated Fibroblasts and Tumor-Associated Macrophages in Cancer and Cancer Immunotherapy. Front Oncol. 2021;11:668731.
173. Alhussan A, Palmerley N, Smazynski J, Karasinska J, Renouf DJ, Schaeffer DF, *et al.* Potential of Gold Nanoparticle in Current Radiotherapy Using a Co-Culture Model of Cancer Cells and Cancer Associated Fibroblast Cells. Cancers. 2022;14(15):3586.
174. Durand M, Chateau A, Jubreaux J, Devy J, Paquot H, Laurent G, *et al.* Radiosensitization with Gadolinium Chelate-Coated Gold Nanoparticles Prevents Aggressiveness and Invasiveness in Glioblastoma. Int J Nanomed. 2023;18:243–261.
175. Penninckx S, Thariat J, Mirjolet C. Radiation therapy-activated nanoparticle and immunotherapy: The next milestone in oncology? Int Rev Cell Mol Biol. 2023;378:157–200.
176. Marill J, Mohamed Anesary N, Paris S. DNA damage enhancement by radiotherapy-activated hafnium oxide nanoparticles improves cGAS-STING pathway activation in human colorectal cancer cells. Radiother Oncol. 2019;141:262–266.
177. Hu Y, Paris S, Barsoumian H, Abana CO, He K, Wasley M, *et al.* Radiation Therapy Enhanced by NBTXR3 Nanoparticles Overcomes Anti-PD1 Resistance and Evokes Abscopal Effects. Int J Radiat Oncol Biol Phys. 2021;111(3):647–657.
178. Zhang P, Darmon A, Marill J, Mohamed Anesary N, Paris S. Radiotherapy-Activated Hafnium Oxide Nanoparticles Produce Abscopal Effect in a Mouse Colorectal Cancer Model. Int J Nanomed. 2020;15:3843–3850.
179. Darmon A, Zhang P, Marill J, Mohamed Anesary N, Da Silva J, Paris S. Radiotherapy-activated NBTXR3 nanoparticles modulate cancer cell immunogenicity and TCR repertoire. Cancer Cell Int. 2022;22(1):208.
180. Bromma K, Beckham W, Chithrani DB. Utilizing two-dimensional monolayer and three-dimensional spheroids to enhance radiotherapeutic potential by combining gold nanoparticles and docetaxel. Cancer Nanotechnol. 2023;14(1):80.
181. Jackson N, Hill I, Alhussan A, Bromma K, Morgan J, Abousaida B, *et al.* Dual enhancement in the radiosensitivity of prostate cancer through nanoparticles and chemotherapeutics. Cancer Nanotechnol. 2023;14(1):75.

182. Tudor M, Popescu RC, Negoita RD, Gilbert A, Ilisanu MA, Temelie M, *et al.* In vitro hyperspectral biomarkers of human chondrosarcoma cells in nanoparticle-mediated radiosensitization using carbon ions. Sci Rep. 2023;13(1):14878.
183. Jackson N, Alhussan A, Bromma K, Jay D, Donnelly JC, West FG, *et al.* Repurposing Antimalarial Pyronaridine as a DNA Repair Inhibitor to Exploit the Full Potential of Gold-Nanoparticle-Mediated Radiation Response. Pharmaceutics. 2022;14(12):1284.1.

https://doi.org/10.1142/9781800618183_0011

Chapter 11

Radiation-Induced DNA Damage and Repair in a Multiscale Chromatin Landscape: Past Insights, Current Understanding, and Future Complexities

Martin Falk[*,†,§], Myriam Schäfer[†,‡], Jonas Weidner[†], Iva Falk[*], Jiri Toufar[*], Lucie Dobesova[*], Alena Bačíková[*], Olga Kopečná[*], Georg Hildenbrand[†,‡], and Michael Hausmann[†,¶]

[*] *Institute of Biophysics of the Czech Academy of Sciences, Kralovopolska 135, 612 00 Brno, Czech Republic*

[†] *Kirchhoff-Institute for Physics, Heidelberg University, Im Neuenheimer Feld 227, D-69120 Heidelberg, Germany*

[‡] *Faculty of Engineering, University of Applied Sciences, Würzburger Straße 45, D-63743 Aschaffenburg, Germany*

[§] *falk@ibp.cz*

[¶] *hausmann@kip.uni-heidelberg.de*

This chapter delves into the intricate relationship between ionising radiation (IR), chromatin architecture, and the cellular response to DNA damage. With the advent of hadron radiotherapy and advancements in space research, understanding how the physical properties of IR, particularly high linear energy transfer heavy ions, affect biological systems has become increasingly critical. The text emphasises the role of chromatin architecture in the induction and repair of radiation-induced

DNA damage, specifically double-strand breaks (DSBs). It explores how the chromatin environment at both local and global levels influences the damage response, which ultimately dictates the repair mechanism selected at each damage site. The use of advanced microscopy techniques, including super-resolution microscopy, is highlighted as indispensable for studying these processes at the nanoscale. This chapter presents key findings that challenge existing paradigms in radiobiology and offers new insights into DNA repair mechanisms, chromosomal aberrations, and cellular radioresistance in the context of IR.

11.1 Introduction

With the development of hadron radiotherapy and space research,[1–3] the importance of our understanding of the relationship between the physical properties and the biological effects of various types of ionising radiation (IR), including heavy ions with high linear energy transfer (LET),[4,5] is increasing. In past years, it has become apparent that chromatin architecture and dynamics play a crucial role both in DNA radiation damage induction and repair,[6–10] which resulted in the development of new experimental and theoretical tools, including mathematical procedures for the complex analysis of chromatin architecture at the nanoscale[11] and software packages for multiscale modelling of radiation-induced processes in the cell.[4,12–14]

The cell nucleus is a complex biological system in which simultaneous reactions and functions take place to keep the cell as an individualised, specialised system running well.[15] The cell nucleus contains a chromatin network[16,17] interspersed with interchromatin channels, where there is enough 'free' space for floating RNAs, proteins, enzymes, ATPs, ions, water molecules, etc., which get to the places where they are needed by super- and supradiffusion. This form of trafficking seems to work sort of automatically and control the system perfectly.[18] After exposure to IR, leading to DNA damage ranging from single base modifications to double-strand breaks (DSBs), the entire system – the cell nucleus – responds to this stress and initiates repair processes to restore its fully functional state.[15,16,19]

However, chromatin is variably condensed and separated into volumes of chromosomal territories and subchromosomal domains with specific architectures and functions.[20] DNA damage due to radiation thus occurs in locally different chromatin environments.[7,10,21]

Hence, the DNA repair of individual DNA lesions is controlled by two basic components: the local environment and processes at the sites of individual DNA lesions[9] and the global, systemic response of the chromatin network.[15] In the following text, we first address the question of how the architecture of chromatin and the physical characteristics of IR influence the features of DSB lesions. Next, we address the questions of how the multiscale properties of DSBs and the local chromatin architecture at a given DSB site determine the local component of the chromatin response, what drives the systemic chromatin response, and how these two components of the chromatin response are subsequently reflected in the selection and propagation of a given repair mechanism at each individual DSB site. We limit our focus to DNA DSBs because they represent the most severe type of DNA damage effectively induced by IR,[22] and the efficiency of DSB repair (i.e., its capacity, kinetics, robustness, and fidelity) therefore plays a critical role in the cellular response to irradiation and contributes significantly to cellular radioresistance.[23]

In parallel, we point out the indispensable role of microscopy and the recent super-resolution microscopy (or 'nanoscopy')[11,24–30] in contemporary radiobiological research. Specifically, we illustrate the necessity of the unique feature of correlated nano-microscopy – the possibility of direct observation of radiation-induced cellular events at a multidimensional level – for current research into DNA radiation damage and repair. We present some important results, which we achieved using confocal microscopy, later combined with single-molecule localisation microscopy (SMLM),[11,25,26,31–35] which may eventually change some paradigms regarding radiation damage and

Table 11.1. Types of ionising radiation (IR) related to the discussed results and their basic parameters. The 'geometry' column presents the angles at which the cell monolayer was irradiated, and Z denotes the proton number.

IR	Z	Geometry [°]	Energy [MeV/n]	LET [keV/μm]	Photons/Particles per nucleus
^{60}Co (γ)	–	90	1.33 + 1.17	0.3	Many
p^{+}	1	90	30	1.90	Many
^{11}B	5	10	7.5	148.3	3.3
^{15}N	7	10	13.1	181.4	2.1
^{20}Ne	10	10	33.9	170.9	2.7

DNA repair, as well as the development of chromosomal aberrations in cells exposed to different types of IR. As a key method of this chapter, the SMLM principle is briefly presented; for a description of other essential experimental methods and procedures, references are made to the cited papers. The types of IR to which the results presented in the following relate and the basic characteristics of these radiations are summarised in Table 11.1.[36]

11.2 Chromatin Architecture: Structure, Function, and Significance

Before proceeding to explore the relationship between multiscale chromatin architecture and DSB induction and repair following irradiation, let us briefly review some basic features of this architecture. The DNA is arranged in a non-random manner already at its linear sequence.[37] For example, we can observe large functional clusters – chromosomal bands – representing either euchromatin (Eu) or heterochromatin (Hc). The formation of the DNA complex together with histones and non-histone proteins – called chromatin – then allows DNA to form higher-order spatial (3D) structures that hold regulatory and functional significance.[38–40]

As a result, the chromatin in the cell nucleus is intricately organised in specific domains to facilitate a balance between accessibility and compaction, allowing the cell to manage fundamental processes,[17,18] such as gene expression, DNA replication, and, most importantly in the context of this chapter, DNA repair. With some simplification, Eu is generally less condensed and transcriptionally active, whereas Hc is more compact and mostly genetically inactive.[41] Heterochromatin is further classified into constitutive and facultative heterochromatin, where the constitutive heterochromatin is often found around centromeres and telomeres and remains highly condensed and typically inactive throughout the cell cycle. Facultative heterochromatin, on the other hand, can switch between an active and an inactive state depending on developmental or environmental signals. This correlated structural and functional switching represents an illustrative example pointing to the fact that chromatin architecture is highly dynamic and that this dynamism is essential for its regulatory role.

Chromosomal territories (reviewed in Ref. (42)) represent another level of chromatin organisation. Each chromosome occupies a distinct, non-overlapping space within the nucleus, known as a chromosomal territory. These territories are not random; rather, they reflect the functional state and activity level of the DNA sequence they contain.[38–40] Gene-rich, transcriptionally active regions often localise towards the nuclear interior, whereas gene-poor regions tend to be positioned near the nuclear periphery.[38] This spatial organisation within territories influences how chromosomes interact with each other and participate in fundamental nuclear processes. This principle of organisation holds true also for the whole territories.

It is worth mentioning that there are more types of chromatin domains.[43,44] Less well-known but structurally and functionally even more homogeneous domains than Eu and Hc with respect to the level of gene transcription and chromatin condensation are the strongly decondensed RIDGE (i.e., region of increased gene expression) domains, which contain clusters of highly expressed genes in a particular cell type, and their condensed counterparts, the anti-RIDGEs, which contain silenced genes.[37] We can also mention the lamina-associated domains (LADs) and nucleolus-associated domains (NADs).[44] LADs are regions of chromatin that interact with the nuclear lamina and are usually associated with gene repression, while NADs are linked to the nucleolus and often contain ribosomal DNA and other repressive chromatin regions. These specialised domains play a crucial role in maintaining nuclear architecture and point to the functional importance of chromatin organisation within the cell nucleus.

Maintaining this highly structured chromatin organisation involves various molecular mechanisms, including histone modifications, DNA methylation, and ATP-dependent chromatin remodelling complexes.[45] Histone modifications, such as acetylation and methylation, act as signals that either promote chromatin relaxation (euchromatin formation) or compaction (heterochromatin formation), influencing accessibility for transcriptional and other protein machineries. An example of an epigenetic marker of heterochromatin is the trimethylation of lysine 9 on histone H3 (H3K9me3).[46] DNA methylation,[47] typically associated with gene silencing, contributes to the stability of heterochromatin regions. Chromatin remodelers, such as the SWI/SNF and INO80 complexes, use ATP to alter the position of nucleosomes, allowing or restricting access to specific regions of DNA, which is essential for all processes on the DNA molecule, including DNA repair.[48,49]

The organisation of chromatin can be thus broadly categorised into distinct domains and territories, each with specific functions and structural features. The spatial organisation of chromatin reflects and, at the same time, directly impacts all fundamental nuclear processes. For instance, during gene expression, euchromatin regions provide accessibility for transcription factors and RNA polymerase, ensuring efficient transcription of active genes. Ongoing transcription can subsequently open the chromatin domain further. Similarly, chromatin architecture is crucial in DNA repair, especially for DSBs.[8,9,50] Euchromatin, which is less compact, allows repair proteins to quickly access and repair breaks, while heterochromatin presents a selective barrier and often requires chromatin rearrangement to facilitate the repair process.[51] Thus, we can assume that repair complexes that form at a particular DSB site under conditions of specific accessibility of the surrounding chromatin to different proteins and molecules can subsequently reorganise the structure of the damaged chromatin domain in a way that favours a particular repair mechanism.[9] Thus, understanding how chromatin architecture is affected by irradiation and how it changes over time following irradiation is essential to gain new insights into DNA repair mechanisms and their selection at individual DSB sites.

11.3 Ionising Radiation-Induced Foci (IRIFs) as Microscale Markers of Radiation Damage and Repair

Several minutes after irradiation, chromatin around the interrupted DNA at DSB sites undergoes phosphorylation of histone H2AX at serine 139.[52] This event is one of the processes that initiate DSB repair. Phosphorylated histones H2AX (further referred to as γH2AX) spread over a large DNA region (of about 2 Mbp in humans),[52] forming thus microscopically visible 'foci' in the cell nucleus (Fig. 11.1).[52]

γH2AX foci act as a multilateral DNA damage signal functioning in a positive feedback loop and initiating DNA repair, cell cycle

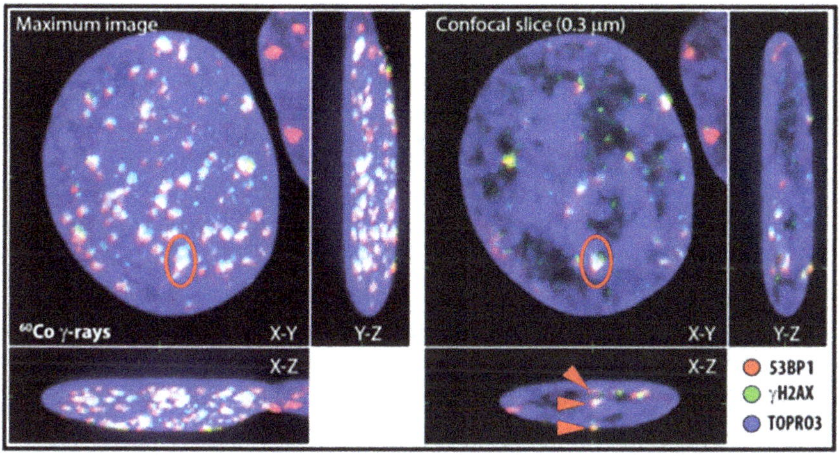

Fig. 11.1. Detection of IRIFs (γH2AX in green, 53BP1 in red) using immunofluorescence confocal microscopy. Left: a 2D image of the nucleus (stained with TOPRO3, blue) of a human dermal fibroblast irradiated with 4 Gy of gamma rays, presented as a maximum intensity projection from 30 optical sections, each 0.25 μm thick. Right: single confocal sections of the same nucleus in the x–y, x–z, and y–z planes. These 0.25 μm thick sections intersect the IRIF cluster indicated by the red ellipse. Red arrows in the x–z section highlight IRIFs that are not distinguishable in the 2D maximal image (left). The confocal sections also reveal the localisation of IRIFs within distinct chromatin domains: euchromatin (dark or light coloured regions) and heterochromatin (intensely blue areas). The differential colocalisation of γH2AX with 53BP1, as well as other morphological characteristics of IRIFs, can be analysed at the microscopic level using appropriate software described in this chapter. The horizontal dimension of the images is approximately 10 μm.

arrest, and eventually cell death.[53,54] At the same time, γH2AX foci form platforms helping assemble huge repair protein complexes at DSB sites.[53,54] Thus, the individual repair proteins in the time-specific sequence and amount gradually colocalise with γH2AX foci and dissociate from them again if their activity is no longer needed. If the DSB is successfully repaired, the γH2AX foci also dissolve, although it is uncertain whether γH2AX can persist at DSB sites where the DSB has been successfully rejoined but the chromatin architecture has not returned to its original state.[55] In cooperation with other proteins, γH2AX sets up a chromatin structure suitable for further repair.[56,57] Complexes of γH2AX associated with a plethora of repair proteins, hereinafter called ionising radiation-induced foci (IRIFs), or DNA repair foci in short, are therefore critical structures in the context of DNA damage and repair, as well as radiobiology in general. The discovery of IRIFs and methods of their detection opened a new era in radiobiology with previously unimaginable possibilities for research.[8,10,58–60]

Since the number of IRIFs corresponds more or less to the number of DSBs for sparsely ionising X-rays and γ-rays,[8,61] one can estimate the absorbed dose of IR by simply counting the IRIFs in the nucleus. IRIFs, therefore, have found an irreplaceable place also in biodosimetric applications.[62–66] IRIF quantification is particularly useful in situations of radiation mass incidents where normal physical dosimetric measurements are not available and in personalised radiotherapy, when we need to identify patients hypersensitive to radiation prior to radiotherapy or monitor radiation damage to specific tissues, such as tumours.[23]

In addition to the ability of counting the numbers of IRIFs (DSBs),[61] microscopy allows for observing their distribution patterns in the nucleus as they occur following irradiation with different types of IR (Figs. 11.1 and 11.2).[36]

By combining GFP technology to label repair proteins and fluorescence *in situ* hybridisation (FISH) to label DNA, IRIFs can be visualised in structurally and functionally distinct chromatin domains.[7,21] For instance, we used this approach to quantify IRIFs in the territories of individual chromosomes and in regions of increased gene expression (RIDGEs) and their structural and functional counterparts (anti-RIDGEs).[7]

Microscopy thus shed new light on the long-standing dilemma about the sensitivity of heterochromatin and euchromatin to DSB induction by radiation (discussed later)[7] and about the kinetics and persistence of DSBs in these chromatin domains.[36,58,67–72]

Microscopy of IRIFs can also be used to analyse mutual interactions (colocalisation) between γH2AX and specific repair proteins in the frame of these structures. Hence, one can explore the formation and development during the post-irradiation period (repair) of specific IRIF types in different chromatin environments.[7,59] An example could be the formation of RAD51 foci (marking DSB sites repaired by homologous recombination) in heterochromatin and euchromatin. Importantly, IRIFs can be analysed in spatially (3D) fixed[71] or even living cells.[51]

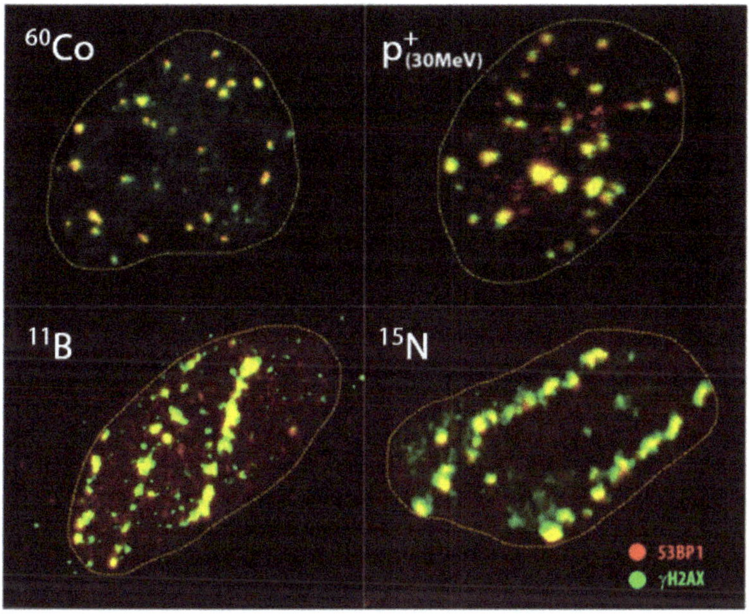

Fig. 11.2. Microscopic structure and nuclear distribution of IRIFs (γH2AX in green, 53BP1 in red) induced by various types of ionising radiation with different energies and linear energy transfer (LET) (refer to Table 11.1). The images represent maximum intensity projections created by overlaying 30 optical confocal sections, each 0.25 μm thick. Dashed lines denote the boundaries of cell nuclei, which have been excluded to enhance the visibility of the IRIF structures. The lower panels illustrate the multiplicity and clustering of IRIFs in cells exposed to high-LET accelerated ions. For 30 MeV protons, the observed pattern resembles that of gamma radiation (top left) rather than high-LET ions (bottom right), consistent with their low LET; however, several instances of multiple or clustered IRIFs are also visible. The differential colocalisation of γH2AX with 53BP1, as well as other morphological characteristics of IRIFs, can be analysed at the microscopic level using appropriate software described in this chapter. The horizontal dimension of the images measures approximately 10 μm.

Finally, microscopy enables us to study DNA damage and DNA repair in individual cells. This is critical for research for numerous reasons, as follows. First, individual cells in the population occur in different phases of the cell cycle or different states in general, characterised by preferred repair mechanisms and different radiosensitivity. The averaging of results for all cells, as it occurs, for example, in experiments performed on isolated cell material, often leads to masking some important phenomena or even misinterpretation of the results. Second, microscopy enables us to study DNA damage and DNA repair under maximum physiological conditions. Third, live cell microscopy then allows continuous monitoring of the development of individual IRIFs over time. Finally, using appropriate software,[11,61] the geometric and other parameters of IRIFs and surrounding chromatin can be studied in relation to the chromatin architecture at the DSB site, the ongoing repair mechanism,[11,59,61] etc.

11.4 DNA Double-Strand Breaks Induced by Various Types of Ionising Radiation in Distinct Chromatin Environments

The most effective inducer of DSBs is IR, especially its types with high LET.[36] While endogenous cellular processes,[73] radiomimetic chemicals,[74–76] and sparsely IR[6] generate predominantly simple, spatially isolated DSBs, high-LET radiations give rise to multiple, frequently complex DSB lesions or even clusters of this damage forming a damage 'tunnel' along the particle's track, as predicted theoretically[5,77–80] and observed experimentally (Fig. 11.2).[36,81–83] A DSB is defined as 'multiple' when two or more DSBs occur within one to two helical turns of DNA (i.e., <3–4 nm) and as 'complex' when it appears at the damaged site together with some other DNA lesion type, such as base alterations.[36]

Since both complex and multiple DSBs and their clusters have been recognised as difficult to repair[36] and therefore determining the biological effectiveness of radiation (RBE),[14,36,79,84–89] high-LET IR is biologically more dangerous than low-LET IR and personifies the double-edged sword in relation to human health even more clearly. Applied in modern radiotherapy, high-LET IR is one of the most effective tools we currently have for tumour elimination.[90–93] On the other hand, highly ionising energetic particles pose significant hazards to treated patients[94] and, for instance, for space flight crews. Such hazards critically complicate the application of hadron therapy and current flight planning to Mars.[1,2,94–97]

We do not yet have accurate data for the different particles and irradiation conditions that could be included in radiotherapy planning and estimating the risk of space travel. Although we are aware of the worse reparability of DSB lesions induced by high-LET radiation, we do not know whether and possibly how the cell response to high-LET particles mechanistically differs from that to low-LET photons and how it depends on specific properties of various particles. A crucial factor with respect to the biological effectiveness of different types of IR is therefore the architecture of the chromatin at the site of radiation damage, or, more precisely, the way in which a given type of IR interacts with a given type of chromatin domain.[9] Hence, we need to investigate more deeply how the structure of DSB (clusters) and surrounding affected chromatin domain influence the selection of a specific repair pathway at a particular DSB site, the overall biochemical activity at that site, and ultimately the efficiency of the repair process.[9]

Microscopic studies have suggested that after exposure to sparse IR (e.g., gamma rays or X-rays), chromatin structure affects susceptibility to DSB formation.[7,21] Due to its condensation and composition, heterochromatin is better protected against damage from this type of radiation than euchromatin, as it contains less water, from which reactive radicals are formed by radiolysis, and more proteins, which protect DNA from these radicals.[7,21] On the other hand, in the case of densely ionising particles, exposure to heterochromatin probably leads to more severe damage because the architecture of (hetero)chromatin cannot provide any protection from the direct effect of energetic particles but instead provides more potential DNA targets per unit volume than euchromatin.[9,10] The enormous importance of the interaction between IR and chromatin stems from the discovery that the character of the generated DSBs (clusters and tracks) differs not only for the same ions differing significantly in their LET and their energy (E) but also for different ions with similar values of LET and E.[36] Different ions, even with similar LET and E, due to their distinct properties, create a differently wide core track and emit delta electrons with varying range and distribution, thus causing specific DNA damage at the micro level.[36] Different microdosimetric patterns of DNA damage are subsequently manifested by the formation of DSB clusters of varying multiplicity or complexity, which are repaired by the cell with quite different efficiency.[36]

While these paradigm-shifting results on the relationship between radiation type, DSB induction, and DNA repair were obtained by high-resolution immunofluorescence confocal microscopy,[36] the real breakthrough has come only with the breaking or circumventing, respectively, of the Abbe limit and the development of super-resolution microscopy (nanoscopy).[11,25,26,26,27,30,33] This progress opened the door to the hitherto unknown nanocosmos of the cell.[60] New fundamental questions have emerged: what are the nanostructured parameters of DNA damage and DNA repair objects under different irradiation conditions and how do these nano-parameters relate to earlier observations at the microscopic level?[11] In other words, we must determine the functional relationship between the structural aspects of chromatin radiation damage and repair at the microscopic and nanoscopic levels. Correlated microscopic research at the microscopic and nanoscopic (molecular) levels thus takes on a fundamental role in DNA damage and repair research, as well as in radiobiology in general.

11.5 Mechanisms of DNA Double-Strand Break Repair

Because DSBs are also induced by the vital processes of the cell itself – namely energy metabolism, DNA replication,[98] and antibody production[99] – in addition to IR and radiomimetics,[74,75,100] it is no exaggeration to say that life would not have been possible without DNA repair, at least in its current form. Up to date, individual biochemical processes – serial steps of DSB repair – have been described in a relatively detailed way and are related to several repair pathways. In humans, two main pathways were identified: non-homologous end joining (NHEJ)[101–104] and homologous recombination (HR).[105,106] NHEJ and HR differ in mechanism, kinetics, fidelity, robustness, and probably other parameters. This suggests that they are not only mutually supportive but are specialised for different DSB substrates and/or function in different environments. In addition, less well-characterised alternative (or backup) repair pathways also exist, which use short homologies for recombination and are always mutagenic.[107] However, the hierarchy of these repair pathways and how they are selected at each DSB site remain poorly understood.[9] This is a crucial question, as these repair pathways differ significantly in their speed, accuracy, robustness, and overall repair result with respect to cell survival and the risk of mutagenesis. Finding effective ways of modifying the activity of particular repair processes can be thus an important and highly versatile therapeutic approach applicable, for example, in cancer radiotherapy,[108–112] treatment of various non-malignant (e.g., neurodegenerative) diseases,[113,114] and fighting ageing.[115,116]

The protein composition of IRIF complexes depends on the chosen repair mechanism and evolves over time. Therefore, the colocalisation of individual repair proteins and γH2AX within IRIFs also varies,[117] so it can be assumed that these differences are also reflected in the 3D architecture (topology/geometry) of IRIF.[9] In addition, it can be predicted that the IRIF architecture is also influenced by the chromatin architecture at the DSB site.[9,59]

Theoretically, cells in the G2 phase may follow two repair strategies, each having its own logical rationale. In the first case, the cell initiates repair primarily as NHEJ, which, due to its speed, provides a chance for a timely repair process, especially in organisms with large genomes. Although NHEJ suffers from some inaccuracies (especially small deletions or insertions at breakpoints), it can largely prevent the occurrence of chromosomal aberrations, which would be much more dangerous or even fatal for the cell. Given that human coding sequences make up only about 2% of DNA, the risk of mutation in important genes due to NHEJ can be considered relatively small and more than offset by its benefits. Thus, it appears that only certain DSBs activate high-precision but slow HR. However, the opposite hierarchy of repair pathways would also make good sense. Given the evolutionary age of HR[118] and its dominant role in stem cells, and in line with its significantly higher accuracy compared to NHEJ, it does not sound illogical if the repair always starts as HR and switches to NHEJ only if it were inhibited for some reason.[119] However, whether a cell initiates NHEJ primarily or only after HR failure, the question remains as to which factors determine the final activation of a particular repair mechanism at each individual DSB site.[9]

In fact, the question of repair mechanism selection does not only concern the G2 cells. Penny Jeggo and her group revealed that while the core NHEJ proteins are sufficient for a successful repair in euchromatin, the same process in heterochromatin cannot proceed without additional proteins and is slower.[120–124] Accordingly, DNA repair has biphasic kinetics not only in the G2 phase of the cell cycle, where NHEJ and HR occur in parallel, but also in the G1 phase, where, until recently, NHEJ was assumed to be the only active repair pathway.

However, the situation in the G1 phase is likely to be even more complex. For example, complex and multiple DSBs that can hardly be accurately repaired by NHEJ occur even in G1. Since these dangerous DSBs can also affect critical genes, we can assume that there must be a repair process that is able to repair these DSBs even in G1. It has been discovered only recently that HR can also occur in G1, at least at DSBs at intensively transcribed genomic loci.[125–128] The trick here is that the homologous template for repair is provided by nascent RNA strands instead of missing sister chromatids.[125–128] This interesting discovery explains how a cell protects important (genetically active) genome loci from mutations also in G1 while, at the same time, illustrating our attempts at understanding the relationships between repair pathways. To further complicate the situation, alternative repair pathways do not need sister chromatids to proceed and can thus operate throughout the cell cycle.[107]

11.6 Influence of Local Chromatin Environment on DNA Double-Strand Break Repair

Understanding the relationship among the physical characteristics of IR, chromatin architecture, and the character of DSB damage described earlier in this chapter is the first step necessary to understand cellular response to irradiation. However, it appears that chromatin architecture also plays a pivotal role in the DNA repair of generated DSBs. Various studies suggest, for instance, that resection-based repair, but not necessarily HR, is primarily focused on repairing DSBs that are difficult to repair, such as multiple or complex DSBs,[129] DSBs found in complex heterochromatin architectures,[122,130–132] and those found in DNA loci with a need for highly precise repair, such as intensively transcribed genes.[108,133–137] However, heterochromatin is a very dangerous substrate for HR because it allows illegal recombination between repetitive sequences,[138,139] which are present in large amounts. A solution to this paradox is provided by the precise regulation of the individual HR steps in the context of the chromatin architecture. The damaged heterochromatin domain must first decondense, thereby displacing the damaged chromatin outside the domain with repetitive sequences.[6,140] Only then can the SSB proteins protecting the resected single-stranded DNA ends be replaced with RAD51 recombinase, which is required for continued HR. After accomplishing the DNA repair, the epigenetic status of chromatin must be restored to its original state.[6] The need for these reorganisation steps may explain a slower course of DSB repair in heterochromatin. In addition, epigenetic changes in chromatin and its extensive decondensation in the initial phase of repair also pose the risk of epimutations[51] and chromosomal translocations,[6] which is discussed later in this chapter. The obtained results thus point to the crucial role of the as-yet-unexplored aspects of DSB repair, namely its organisation in space and time and its regulation through the structure of chromatin.

Therefore, the heterochromatin architecture plays a contradictory role in cells exposed to sparse gamma radiation: it protects DNA from the induction of DSBs but complicates their repair if they have already occured.[7,8] In the case of the exposure to accelerated ions, the whole chromatin domain may be disintegrated due to extensive chromatin fragmentation, which again influences DSB repair in a specific way, though we still do not know if increasing complexity/multiplicity of DSB clusters leads to a preference for a particular repair mechanism and whether this choice also depends on the structure of surrounding chromatin.

Therefore, to understand the cellular response to irradiation, in addition to the relationship between chromatin architecture and radiation damage to DNA, we must also discern the relationship between the local chromatin architecture at the sites of individual DSBs, the formation of repair complexes at these sites, and the initiation/progression of specific repair pathways (mechanisms). Clarifying the relationships between the chromatin architecture and repair processes could, for instance, shed new light also on our understanding of the hyper-[141,142] or hyposensitivity[143] of cells to low and very low doses of IR.

We present for discussion the possibility that the structure of DSB (clusters) and the corresponding chromatin domain dictate how the repair complexes are assembled at individual DSB sites in the initial phase of the DNA damage response. We propose that this could be a topology of individual early repair complexes that further supports the assembly of repair complexes in a way that points towards a particular repair mechanism. The architecture of a damaged chromatin domain can thus unify and reflect effects of various repair mechanism-(co)determining factors; it can therefore be considered an integral signal for the binding of specific chromatin structure remodelers which subsequently modify the affected domain in the characteristic manner required to trigger the most appropriate repair pathway.[9] The structure of DSB (clusters) and the affected chromatin domain (see the following), later in combination with the composition and topology of the basics of only forming repair complexes, may thus significantly participate in the decision-making for the activation of the most suitable repair pathway and propagation at a particular DSB site. It is not necessary to emphasise that the structure of DSBs (or DSB clusters), the architecture of the affected chromatin domain, and the chosen repair mechanism all influence the success of DNA repair.

11.7 Pan-Nuclear Chromatin Texture and Its Role in DNA Double-Strand Break Repair and Misrepair

DSB repair is not always successful and may result in the formation of various chromosomal aberrations. Defective DSB repair, specifically the formation of chromosomal aberrations, is a process that well illustrates that not only the local architecture of the damaged chromatin domain but also the overall organisation of chromatin in the cell nucleus plays a crucial role in the response of cells to irradiation. We already know from the previous text that densely ionising particles form clusters of DSBs in the DNA molecule that are difficult to repair. These (IRIF) clusters, which are formed as a result of highly concentrated energy deposition and extensive local chromatin fragmentation, can be termed primary clusters.[10,51] It is not hard to imagine that these clusters with many mutually close free DNA ends represent an ideal substrate for the easy formation of complex chromosomal translocations and other aberrations. In contrast, sparsely IR deposits its energy much more homogeneously over the nucleus, so that it does not produce clusters of DSBs but usually single DSBs.

However, complex chromosomal translocations can be rarely observed even in cells irradiated with sparsely IR. So, how do these translocations arise?

Our measurement of total IRIF mobility in gamma-irradiated cell nuclei showed that while most IRIFs exhibit only limited movement, the movement of some of them is much higher.[6] These mobile IRIFs were later identified with DBSs localised in heterochromatin, and their increased mobility reflected the decondensation of the damaged domain. It was found that the movement of heterochromatic DSBs leads away from the damaged heterochromatin domain to sites of lower chromatin density. In these spatially restricted subdomains of the cell nucleus, two or, rarely, more DSBs can then interact and form secondary DSB clusters. Therefore, secondary DSB clusters, like primary ones, represent potential seeds for complex chromosomal translocations; however, unlike primary clusters, they only form as a result of repair processes.[10,51] A paradigm-shifting finding from the above observations[6,8] is that the movement of free DNA ends is driven by the architecture of chromatin (chromatin topology) in the cell nucleus, so that the probability of a given chromosomal translocation also depends not only on the distance between the involved chromosomal loci but also on the chromatin topology.

11.8 Advances in Super-Resolution Microscopy: Overcoming the Abbe Diffraction Limit

The results above, along with extensive literature, demonstrate that the microscopic architecture of chromatin and IRIFs is fundamentally involved in the induction and repair of DSBs following irradiation,[6–10,144,145] providing insight into the spatial and temporal organisation of DNA repair processes.[8] However, while standard confocal microscopy makes it possible to monitor the formation and evolution of IRIFs in terms of their protein composition and micro-architecture, as well as to analyse the architecture of the surrounding chromatin, its resolution is insufficient for a detailed molecular-level description of these structures and processes.[36]

Until recently, the only alternative was electron microscopy, which, despite its exceptionally high resolution, requires relatively destructive specimen fixation and ultrathin sectioning, making it challenging to visualise spatial structures in an undamaged state and often introducing experimental artefacts.[146,147] Consequently, the resolution range between optical microscopy (200 nm) and electron microscopy (nanometre scale) has long remained a gap, concealing the meso- and nanoscale properties of chromatin and repair complexes.[10]

Hence, the development of pioneering super-resolution optical microscopy techniques,[25,31,32,36,59,148] which have broken the Abbe limit in various ways and improved resolutions down to 10 nm (Fig. 11.3), represents a major breakthrough in radiobiology.

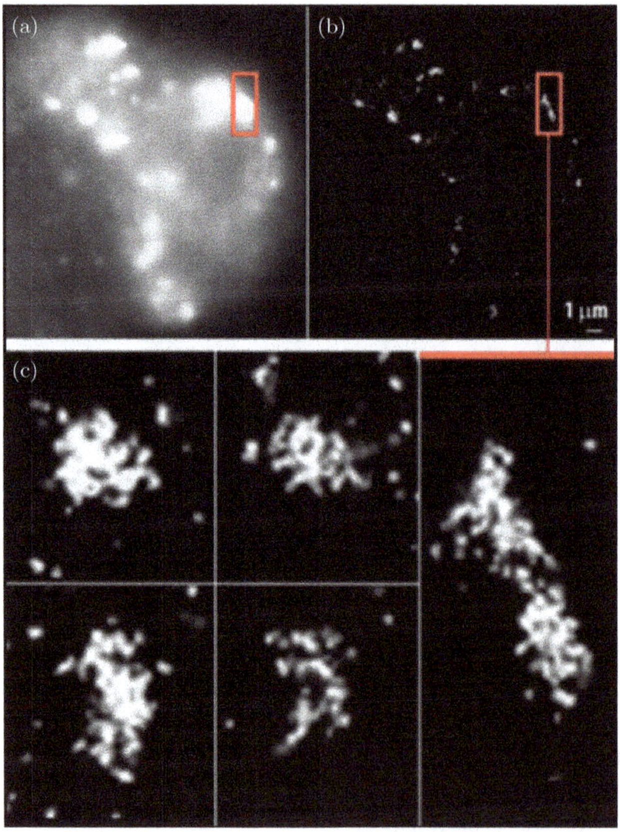

Fig. 11.3. Visualisation of γH2AX foci using (a) standard immunofluorescence microscopy and (b) single-molecule localisation microscopy (SMLM). Image (b) was generated based on a coordinate matrix derived from 4,000 SMLM images acquired over a short period. (c) Detailed nanostructure of the selected γH2AX foci (highlighted with the red rectangle). The topology, geometry, and mutual distribution of molecular γH2AX clusters can be analysed at the nanoscopic level using appropriate mathematical and software procedures described in this chapter.

Although in biological practice this resolution still falls short of that achieved by electron microscopy, it is sufficient to visualise individual molecules while retaining most of the benefits of standard optical microscopy. Since a detailed description of individual super-resolution techniques is beyond the scope of this work, we refer readers to previously published articles.[27,30,31,149–158] Here, we focus only on explaining the principles of the SMLM method, which we used in combination with confocal microscopy[59] to address key radiobiological questions.

SMLM is a type of super-resolution optical localisation microscopy method that offers an impressive resolution of around 10 nm (defined by the minimal measurable distance between two labelling molecules) while using the same sample preparation and labelling methods as standard confocal microscopy.[25,32,35,59] The only requirement is that the fluorochromes must be capable of 'blinking' after initial illumination, as described in the following. This compatibility allows the same samples to be analysed using both confocal microscopy and SMLM, providing correlated results for micro-, meso-, and nano-level preparations,[32,158] which is crucial for a multiscale approach.[4] This approach can provide new insights into DNA damage induction and DNA repair mechanisms. The combined use of confocal microscopy and SMLM is also advantageous methodologically, as it aids in interpreting SMLM results, which can be challenging due to their novel nature.

SMLM achieves super-resolution by separating the signals of individual fluorochromes over time, effectively circumventing the Abbe limit.[25,29,32,35,59] Initially, a standard widefield image of the sample, such as a cell nucleus with immunofluorescently labelled histones (γH2AX) and other repair proteins involved in IRIF formation, is captured. This provides an overview of IRIF properties, such as their number and distribution. In this state, all fluorochrome molecules emit light simultaneously, causing signal overlap and limiting resolution. However, by applying a strong laser pulse ($kW \cdot cm^{-2}$), fluorochromes are temporarily switched to an off-state, randomly returning to an on-state, or 'flashing'. This allows only a few fluorochromes to emit light at a time, enabling spatial and temporal isolation of their signals. Without signal interference, the intensity profile of each can be accurately fitted with a Gaussian function to precisely determine its position. By capturing a time series of thousands of image frames with short intervals (100 ms), a matrix of all the signals' coordinates and their parameters, including measurement errors, is generated. This matrix can then be analysed mathematically[159] without the need for problematic and potentially error-generating image processing. Alternatively, a pointillistic image can be constructed from both raw and processed data, visually highlighting the phenomena observed (Fig. 11.3).[25,32,59,159] Although pointillistic images may appear less intuitive, they are closer to reality than our vision, as they lack the noise from signal interference present in standard images.

SMLM allows the direct measurement of mutual spatial relationships between signals from individual investigated molecules and the determination of the number of these molecules; therefore, we can also use SMLM to visualise differently defined protein clusters and their properties and quantify their molecular composition. The available results suggest that IRIFs, as observed at the microscopic scale, are composed of molecular γH2AX clusters and molecular clusters of repair proteins at the nano-scale.[25,29,32,35,59] Whether these molecular clusters are organised and functional nanostructure subunits within a microscopic IRIF, single DSBs clustered in a form of a microscopic IRIF, or artefacts of immunofluorescence labelling at such a high resolution remains the subject of further research. Because different repair proteins form molecular clusters that are unequal in size and characterised by different topologies and colocalise with γH2AX molecular clusters in a protein-specific manner, the latter possibility does not seem likely.[25,29,32,35,59] The application of SMLM for the study of DNA damage caused by densely ionising radiation is presented in Refs. (25,29,32,35,59).

11.9 Nanoscale Organisation of IRIFs and the Surrounding Chromatin Landscape

In the previous sections, we have already revealed some aspects of the involvement of the architecture of IRIFs and damaged chromatin in DNA repair. Here, using SMLM and newly applied mathematical procedures to study the nanoscale topology (e.g., by persistent homology, persistent imaging, or principal component analysis) and geometry (e.g., by DBScan or Ripley pairwise distance frequency distributions) of objects,[11] we attempt to analyse the relationship between chromatin architecture and IRIFs even more deeply, down to the level of topology and geometry of individual molecules (see also Chapter 10). We are still analysing data that can potentially answer the question of whether the IRIF nanoarchitecture reflects the chromatin nanoarchitecture at DSB sites and whether, potentially, the chromatin nanoarchitecture at the DSB sites can direct DNA repair to a particular mechanism in this way (via oblivion of the IRIF architecture). Indeed, the preliminary results show that γH2AX foci in heterochromatin exhibit different structural features and greater homogeneity compared to those in euchromatin.[35,59] Additionally, we observed that the structure, protein composition, and the ratio

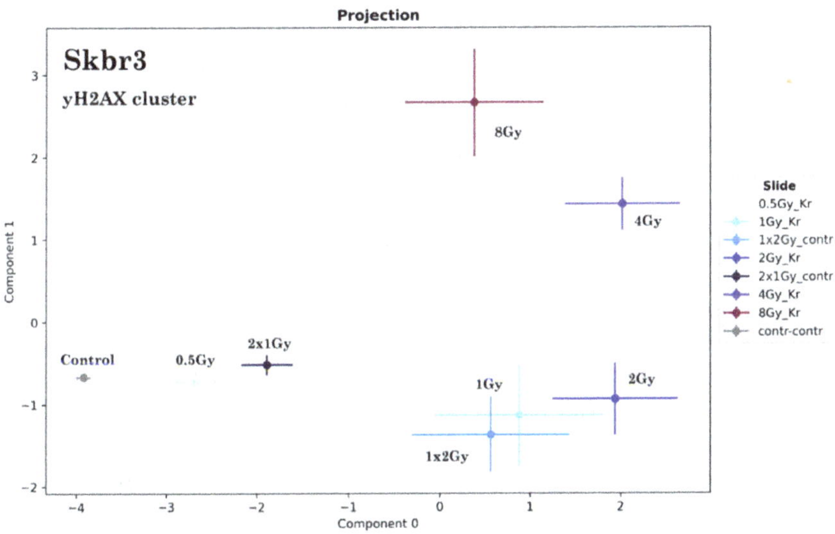

Fig. 11.4. PCA of the persistent homology data of the γH2AX labelling loci within the clusters formed around DSBs 30 min after the application of radiation of different doses to SkBr3 breast cancer cells. Mean values of component 1 vs. component 0 are shown. The error bars represent the standard deviation. The data indicate that, up to 2 Gy, the topologies of the γH2AX clusters are very similar. The irradiation scheme of 1 Gy – 30 min – 1 Gy (= 2x 1 Gy) shows that, after repair having already started after the first dose fraction, the topology is transferred back to the situation of low dose. For heavy damages at 4 Gy and 8 Gy, the topological organisation showed higher changes, indicating that other repair mechanisms such as NHEJ and HR may be involved. For methodological details, see the explanation in Fig. 11.5.

of repair proteins within and outside IRIFs vary among different cell types, such as normal human dermal fibroblasts (NHDFs) and radioresistant U87 glioblastoma cells.[32] These differences appear to correlate with DSB repair efficiency and post-irradiation cell survival.

For instance, we found that U87 cells require fewer 53BP1 proteins than NHDFs to form IRIFs.[32] In addition, although both cell types appeared to exhibit similar overall DNA repair kinetics, U87 cells were more efficient at repairing clustered DSB lesions, consistent with their high radioresistance. Thus, the various results above provide a picture, albeit one that is still very incomplete, in which the chromatin architecture at the DSB site influences the architecture of the just-forming IRIFs. The architecture of IRIFs is likely to be influenced by a number of other factors affecting the choice of repair mechanism (see also the following). Thus, the IRIF architecture may represent a complex 'repair signal' (Fig. 11.4) that integrates individual repair signals and allows the cell to rapidly interpret these different signals and select the most appropriate repair mechanism at a given DSB site.

11.10 Systemic Chromatin Network Response to Radiation-Induced DNA Damage

While molecular and systems biology have advanced our understanding of cellular functions and the responses of cells to various physiological and stress signals, the mechanisms underlying the organisation of interactions among molecules, molecular complexes, and networks in a highly complex chemical environment remain insufficiently understood.[160] Cells and their organelles function as self-organised systems within distinct, enclosed spaces. Consequently, the distribution and behaviour of water, ions, macromolecules, and larger molecular complexes are influenced by fundamental physical and chemical principles, such as thermodynamics and electrodynamics.[161,162]

The formation and organisation of molecules and their complexes at the meso- and nanoscale establish their functional roles and influence their interactions with other molecular entities, such as chromatin interacting with DNA repair proteins. This creates a dynamic system where alterations in the structure of molecular complexes can modify their interactions and functions within the cellular environment, and *vice versa*.[15,16] These insights highlight the importance of studying the multiscale architecture of cellular organelles and structures, along with their networks.

The chromatin enclosed within the cell nucleus is one of the most illustrative examples of a biological self-organising network. Chromatin – the complex of DNA and histone proteins within the cell nucleus – is folded into intricate structures that significantly influence genetic programs and cellular functions, including DNA replication, transcription, and DNA repair.[163,164] Importantly, it has been demonstrated that the disruption of chromatin's 3D structure can lead to various diseases, including cancer[165] and diabetes,[166] as these folding patterns are essential for proper gene regulation and cellular function. Moreover, many studies have demonstrated that the chromatin landscape undergoes changes throughout development and differentiation.[163,164]

However, the results, including those presented in this chapter, also suggest that the chromatin network can be, as a system, involved in the regulation of faster responses, e.g., the repair of DSB lesions following cell irradiation.[15,16] The mathematical and software approaches we developed to analyse chromatin network (re)organisation after DNA DSB induction (by radiation) and during repair procedures are described in Ref. (11).

The first experiments to test the hypothesis that chromatin architecture influences (epi)genetic regulatory pathways and *vice versa* were performed on irradiated NHDFs and various cancer cell lines using SMLM and the mathematical analyses mentioned above. They revealed a very interesting phenomenon: the spatial organisation of chromatin is altered following irradiation in a dose-dependent manner (Fig. 11.5).

Based on calculations involving persistent homologies and persistent images, topological variations can be represented in an n-dimensional vector space. Using principal component analysis (PCA), this n-dimensional space can be reduced to a two-dimensional latent space that represents the major variations in chromatin, neglecting 'biological noise' (small fluctuations represented by higher components of PCA).[11] If the topology of the chromatin network had been expressed in this way before irradiation of cells and then at different times after irradiation, an interesting behaviour was observed.

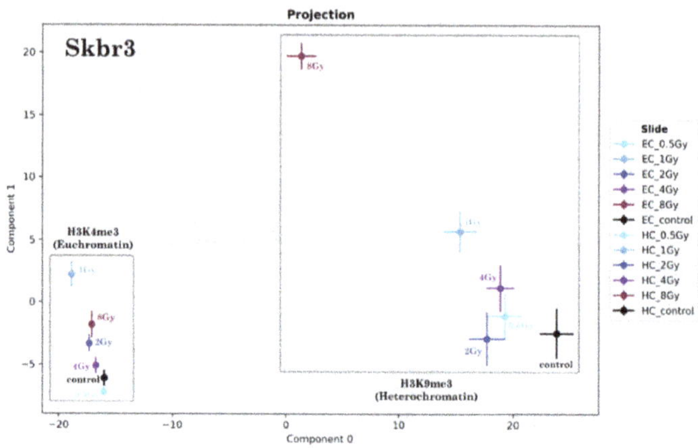

Fig. 11.5. PCA of the persistent homology data of the H3K9me3 loci (heterochromatin) and H3K4me3 loci (euchromatin) of SkBr3 breast cancer cells after the application of radiation of different doses. The data indicate that heavy damage (8 Gy) induces the highest changes in heterochromatin organisation while the euchromatin organisation seems to be less affected. Mean values of component 1 vs. component 0 are shown. The error bars represent the standard deviation. **In detail:** For topological analysis, persistent homology analysis,[11,159] persistent imaging, and PCA were applied, as described in detail in Ref. (11). Significant structures of a point pattern are obtained by the following process: each point (= component = dimension 0) registered in the SMLM coordinate matrix is surrounded by a circle with an increasing radius. Each component is represented by a bar that ends at the radius value where two circles meet. In this way, two components merge into one. During this process, enlarged components can form a closed loop (in its simplest form, a triangle) with free space inside not covered by the increasing circles. The area of this free space is considered a hole, and another bar starts (bars of dimension 1). When the hole is completely covered by the increasing circles, this bar also ends. With connecting components and thus reducing their number, the number of holes is first increasing and then decreasing because more and more holes are closed. For each cell, the results of the barcodes are transferred into a diagram of 'bar lifetime' vs. 'bar birth', which is then transferred into a pixel scan (persistent image) in which the pixel intensity correlates with the number of points in this diagram. These persistent images are obtained for each cell of an experiment. Afterwards the persistent images are subjected to PCA. This means that the values of the first, second, third, fourth, and so on up to the nth pixel represent an n-dimensional space. The dimension with the largest variances then forms PCA component 0 in the final diagram. PCA component 1 is a dimension perpendicular to component 0 and shows the second-largest variance of values. In this way, the results of a complex point pattern can be reduced to a diagram of two dimensions (2D latent space). This means that only two principal components describe the main features and their changes in the labelled molecules, while small variations ('biological noise') are neglected.

In normal healthy cells (NHDFs), the positions in space at different post-irradiation times varied both from the position of non-irradiated cells and from one another, but they formed a kind of closed cycle. Closure of the cycle usually occurred in cells analysed 24 h after irradiation, when the position in space returned to near the value observed in non-irradiated cells.

Surprisingly, in tumour cells, a wandering of the positions of the topologies in mathematical space was also observed, similar to normal (non-tumour) cells, but no cycle closure occurred (Schäfer *et al.*, manuscript in preparation). These results suggest that, in normal cells, the overall topology of the chromatin network changes significantly very early (5 min) after irradiation, then further changes during DNA repair, and finally its topology (functional state) returns to the initial state. In the case of cancer cells, the topological changes in the chromatin network resemble those of normal cells, but the chromatin network does not have the ability to return to its original topological state, at least not within the time interval observed (24 h). Therefore, this may lead to epigenetically determined genomic instability in cancer cells, and disruption of the regulation of other processes cannot be excluded. Overall, current research emphasises that the spatial organisation of chromatin is not merely a structural feature but a dynamic and essential element of regulation.

11.11 Conclusions and Perspectives

Microscopic analyses of γH2AX nuclear foci formed at DSB sites with the participation of a number of other repair proteins have contributed significantly to our current understanding of radiation damage and DNA repair processes. In addition to the biochemical processes taking place at DSB sites, microscopy also revealed important relationships between chromatin architecture at DSB sites, DSB repair mechanisms, and mechanisms of chromosomal translocations following exposure of cells to various types of IR. Studying the organisation of DNA repair processes in space and time would be impossible without the aid of microscopy. Circumventing the Abbe diffraction limit and developing various methods of super-resolution microscopy that push structural analysis of IRIFs and repair processes down to the molecular level is a milestone not only in radiobiology; it will undoubtedly be extremely exciting to see how nanoscopic studies broaden and potentially change our understanding of NHEJ, HR, alternative reparation pathways, and their mutual regulation.

Radiation damage to DNA and subsequent DNA repair of this damage occur in different chromatin environments. The nature and extent of DNA damage depend on the type of IR and the architecture of the damaged chromatin domain. Chromatin responds to radiation damage to DNA (here, induction of DSBs) both locally at DSB sites and as a whole network, i.e., system. The chromatin response is also dependent on the cell type and is probably important in terms of regulation of DSB repair mechanisms (pathways).

References

1. Boice JD. Space: The Final Frontier – Research Relevant to Mars. Health Phys. 2017;112(4):392–397.
2. Boice JD. The Million Person Study relevance to space exploration and mars. Int J Radiat Biol. 2022;98(4):551–559.
3. Zhang J, Guo J, Dobynde MI, Wang Y, Wimmer-Schweingruber RF. From the Top of Martian Olympus to Deep Craters and Beneath: Mars Radiation Environment Under Different Atmospheric and Regolith Depths. J Geophys Res Planets. 2022;127(3):e2021JE007157.
4. Solov'yov AV, Verkhovtsev AV, Mason NJ, Amos RA, Bald I, Baldacchino G, *et al.* Condensed Matter Systems Exposed to Radiation: Multiscale Theory, Simulations, and Experiment. Chem Rev. 2024;124(13):8014–8029.
5. Solov'yov AV, editor. Nanoscale Insights into Ion-Beam Cancer Therapy. Cham: Springer International Publishing; 2017.
6. Falk M, Lukasova E, Gabrielova B, Ondrej V, Kozubek S. Chromatin dynamics during DSB repair. Biochim Biophys Acta. 2007;1773(10): 1534–1545.
7. Falk M, Lukášová E, Kozubek S. Chromatin structure influences the sensitivity of DNA to gamma-radiation. Biochim Biophys Acta. 2008;1783(12):2398–2414.
8. Falk M, Lukasova E, Kozubek S. Higher-order chromatin structure in DSB induction, repair, and misrepair. Mutat Res. 2010;704(1–3): 88–100.
9. Falk M, Hausmann M. A Paradigm Revolution or Just Better Resolution — Will Newly Emerging Superresolution Techniques Identify

Chromatin Architecture as a Key Factor in Radiation-Induced DNA Damage and Repair Regulation? Cancers (Basel). 2021;13(1):18.

10. Falk M, Hausmann M, Lukasova E, Biswas A, Hildenbrand G, Davidkova M, *et al.* Determining Omics Spatiotemporal Dimensions Using Exciting New Nanoscopy Techniques to Assess Complex Cell Responses to DNA Damage: Part – Structuromics. Crit Rev Eukaryot Gene Expr. 2014;24(3):225–247.
11. Weidner J, Neitzel C, Gote M, Deck J, Küntzelmann K, Pilarczyk G, *et al.* Advanced image-free analysis of the nano-organization of chromatin and other biomolecules by Single Molecule Localization Microscopy (SMLM). Comput Struct Biotechnol J. 2023;21:2018–2034.
12. Surdutovich E, Solov'yov AV. Multiscale approach to the physics of radiation damage with ions. Eur Phys J D. 2014;68(11):353.
13. Verkhovtsev A, Surdutovich E, Solov'yov AV. Phenomenon-based evaluation of relative biological effectiveness of ion beams by means of the multiscale approach. Cancer Nanotechnol. 2019;10(1):4.
14. Surdutovich E, Solov'yov AV. Multiscale modeling for cancer radiotherapies. Cancer Nanotechnol. 2019;10(1):6.
15. Erenpreisa J, Giuliani A, Yoshikawa K, Falk M, Hildenbrand G, Salmina K, *et al.* Spatial-Temporal Genome Regulation in Stress-Response and Cell-Fate Change. Int J Mol Sci. 2023;24(3):2658.
16. Erenpreisa J, Krigerts J, Salmina K, Gerashchenko BI, Freivalds T, Kurg R, *et al.* Heterochromatin Networks: Topology, Dynamics, and Function (a Working Hypothesis). Cells. 2021;10(7):1582.
17. Brandani GB, Gu C, Gopi S, Takada S. Multiscale Bayesian simulations reveal functional chromatin condensation of gene loci. PNAS Nexus. 2024;3(6):226.
18. Martin RM, Cardoso MC. Chromatin condensation modulates access and binding of nuclear proteins. FASEB J. 2010;24(4):1066–1072.
19. Matzke AJM, Lin WD, Matzke M. Evidence That Ion-Based Signaling Initiating at the Cell Surface Can Potentially Influence Chromatin Dynamics and Chromatin-Bound Proteins in the Nucleus. Front Plant Sci. 2019;10:1267.
20. Nozaki T, Imai R, Tanbo M, Nagashima R, Tamura S, Tani T, *et al.* Dynamic Organization of Chromatin Domains Revealed by Super-Resolution Live-Cell Imaging. Mol Cell. 2017;67(2):282–293.e7.
21. Falk M, Lukasova E, Gabrielova B, Ondrej V, Kozubek S. Local changes of higher-order chromatin structure during DSB-repair. J Phys Conf Ser. 2008;101:012018.
22. Penninckx S, Pariset E, Cekanaviciute E, Costes SV. Quantification of radiation-induced DNA double strand break repair foci to evaluate

and predict biological responses to ionizing radiation. NAR Cancer. 2021;3(4):zcab046.

23. Falk M, Horakova Z, Svobodova M, Masarik M, Kopecna O, Gumulec J, *et al.* γH2AX/53BP1 foci as a potential pre-treatment marker of HNSCC tumors radiosensitivity – preliminary methodological study and discussion. Eur Phys J D. 2017;71(10):241.
24. Cremer C, Ketteler A, Lemmer P, Kaufmann R, Weiland Y, Mueller P, *et al.* Far-Field Fluorescence Microscopy of Cellular Structures at Molecular Optical Resolution. In: Diaspro A, editor. Nanoscopy and Multidimensional Optical Fluorescence Microscopy. Boca Raton: Chapman and Hall/CRC; 2010. 3-1-3–35.
25. Depes D, Lee JH, Bobkova E, Jezkova L, Falkova I, Bestvater F, *et al.* Single-molecule localization microscopy as a promising tool for γH2AX/53BP1 foci exploration. Eur Phys J D. 2018;72(9):158.
26. Hausmann M, Neitzel C, Bobkova E, Nagel D, Hofmann A, Chramko T, *et al.* Single Molecule Localization Microscopy Analyses of DNA-Repair Foci and Clusters Detected Along Particle Damage Tracks. Front Phys. 2020;8:578662.
27. Kaufmann R, Lemmer P, Gunkel M, Weiland Y, Müller P, Hausmann M, *et al.* SPDM: single molecule superresolution of cellular nanostructures. In: Enderlein J, Gryczynski ZK, Erdmann R, editors. Proc SPIE 7185, Single Molecule Spectroscopy and Imaging II, 71850J. Bellingham: SPIE; 2009.
28. Khater IM, Nabi IR, Hamarneh G. A Review of Super-Resolution Single-Molecule Localization Microscopy Cluster Analysis and Quantification Methods. Patterns (N Y). 2020;1(3):100038.
29. Lemmer P, Gunkel M, Baddeley D, Kaufmann R, Urich A, Weiland Y, *et al.* SPDM: Light microscopy with single-molecule resolution at the nanoscale. Appl Phys B. 2008;93(1):1–12.
30. Sydor AM, Czymmek KJ, Puchner EM, Mennella V. Super-Resolution Microscopy: From Single Molecules to Supramolecular Assemblies. Trends Cell Biol. 2015;25(12):730–748.
31. Hausmann M, Ilić N, Pilarczyk G, Lee JH, Logeswaran A, Borroni AP, *et al.* Challenges for super-resolution localization microscopy and biomolecular fluorescent nano-probing in cancer research. Int J Mol Sci. 2017;18(10):2066.
32. Bobkova E, Depes D, Lee JH, Jezkova L, Falkova I, Pagacova E, *et al.* Recruitment of 53BP1 Proteins for DNA Repair and Persistence of Repair Clusters Differ for Cell Types as Detected by Single Molecule Localization Microscopy. Int J Mol Sci. 2018;19(12):3713.

33. Hausmann M, Falk M, Neitzel C, Hofmann A, Biswas A, Gier T, *et al.* Elucidation of the Clustered Nano-Architecture of Radiation-Induced DNA Damage Sites and Surrounding Chromatin in Cancer Cells: A Single Molecule Localization Microscopy Approach. Int J Mol Sci. 2021;22(7):3636.
34. Falk M, Hausmann M. Advances in research of DNA damage and repair in cells exposed to various types of ionizing radiation in the era of super-resolution optical microscopy. Cas Lek Cesk. 2020;159(7–8):286–297.
35. Bach M, Savini C, Krufczik M, Cremer C, Rösl F, Hausmann M. Super-Resolution Localization Microscopy of γ-H2AX and Heterochromatin after Folate Deficiency. Int J Mol Sci. 2017;18(8):1726.
36. Jezkova L, Zadneprianetc M, Kulikova E, Smirnova E, Bulanova T, Depes D, *et al.* Particles with similar LET values generate DNA breaks of different complexity and reparability: A high-resolution microscopy analysis of γh2AX/53BP1 foci. Nanoscale. 2018;10(3):1162–1179.
37. Caron H, van Schaik B, van der Mee M, Baas F, Riggins G, van Sluis P, *et al.* The human transcriptome map: clustering of highly expressed genes in chromosomal domains. Science. 2001;291(5507): 1289–1292.
38. Kozubek S, Lukásová E, Jirsová P, Koutná I, Kozubek M, Ganová A, *et al.* 3D Structure of the human genome: order in randomness. Chromosoma. 2002;111(5):321–331.
39. Lukásová E, Kozubek S, Kozubek M, Falk M, Amrichová J. The 3D structure of human chromosomes in cell nuclei. Chromosome Res Int J Mol Supramol Evol Asp Chromosome Biol. 2002;10(7):535–548.
40. Falk M, Lukásová E, Kozubek S, Kozubek M. Topography of genetic elements of X-chromosome relative to the cell nucleus and to the chromosome X territory determined for human lymphocytes. Gene. 2002;292(1–2):13–24.
41. Allshire RC, Madhani HD. Ten principles of heterochromatin formation and function. Nat Rev Mol Cell Biol. 2018;19(4):229–244.
42. Cremer T, Cremer M. Chromosome Territories. Cold Spring Harb Perspect Biol. 2010;2(3):a003889.
43. Dixon JR, Gorkin DU, Ren B. Chromatin Domains: The Unit of Chromosome Organization. Mol Cell. 2016;62(5):668–680.
44. Pontvianne F, Liu C. Chromatin domains in space and their functional implications. Curr Opin Plant Biol. 2020;54:1–10.
45. Huang T, editor. Epigenetics – Regulation and New Perspectives. London: IntechOpen; 2023.

46. Padeken J, Methot SP, Gasser SM. Establishment of H3K9-methylated heterochromatin and its functions in tissue differentiation and maintenance. Nat Rev Mol Cell Biol. 2022;23(9):623–640.
47. Dhar GA, Saha S, Mitra P, Nag Chaudhuri R. DNA methylation and regulation of gene expression: Guardian of our health. Nucleus. 2021;64(3):259–270.
48. Gerhold CB, Gasser SM. INO80 and SWR complexes: relating structure to function in chromatin remodeling. Trends Cell Biol. 2014;24(11):619–631.
49. Reyes AA, Marcum RD, He Y. Structure and Function of Chromatin Remodelers. J Mol Biol. 2021;433(14):166929.
50. Arnould C, Rocher V, Saur F, Bader AS, Muzzopappa F, Collins S, *et al.* Chromatin compartmentalization regulates the response to DNA damage. Nature. 2023;623(7985):183–192.
51. Falk M, Lukášová E, Štefančíková L, Baranová E, Falková I, Ježková L, *et al.* Heterochromatinization associated with cell differentiation as a model to study DNA double strand break induction and repair in the context of higher-order chromatin structure. Appl Radiat Isot. 2014;83:177–185.
52. Rogakou EP, Pilch DR, Orr AH, Ivanova VS, Bonner WM. DNA double-stranded breaks induce histone H2AX phosphorylation on serine 139. J Biol Chem. 1998;273(10):5858–5868.
53. Podhorecka M, Skladanowski A, Bozko P. H2AX Phosphorylation: Its Role in DNA Damage Response and Cancer Therapy. J Nucleic Acids. 2010;2010:920161.
54. Prabhu KS, Kuttikrishnan S, Ahmad N, Habeeba U, Mariyam Z, Suleman M, *et al.* H2AX: A key player in DNA damage response and a promising target for cancer therapy. Biomed Pharmacother. 2024;175:116663.
55. Ježková L, Falk M, Falková I, Davídková M, Bačíková A, Štefančíková L, *et al.* Function of chromatin structure and dynamics in DNA damage, repair and misrepair: γ-rays and protons in action. Appl Radiat Isot. 2014;83:128–136.
56. Scully R, Xie A. Double strand break repair functions of histone H2AX. Mutat Res. 2013;750(1-2):5–14.
57. Osley MA, Tsukuda T, Nickoloff JA. ATP-dependent chromatin remodeling factors and DNA damage repair. Mutat Res. 2007; 618(1-2):65–80.
58. Falk M, Hausmann M, Lukasova E, Biswas A, Hildenbrand G, Davidkova M, *et al.* Determining Omics Spatiotemporal Dimensions Using Exciting New Nanoscopy Techniques to Assess Complex Cell

Responses to DNA Damage: PART A-Radiomics. Crit Rev Eukaryot Gene Expr. 2014;24(3):205–223.
59. Hausmann M, Wagner E, Lee JH, Schrock G, Schaufler W, Krufczik M, *et al.* Super-resolution localization microscopy of radiation-induced histone H2AX-phosphorylation in relation to H3K9-trimethylation in HeLa cells. Nanoscale. 2018;10(9): 4320–4331.
60. Hausmann M, Neitzel C, Hahn H, Winter R, Falkova I, Heermann DW, *et al.* Space and Time in the Universe of the Cell Nucleus after Ionizing Radiation Attacks: A Comparison of Cancer and Non-Cancer Cell Response. In: The 1st International Electronic Conference on Cancers: Exploiting Cancer Vulnerability by Targeting the DNA Damage Response. Basel: MDPI; 2021:5.
61. Vicar T, Gumulec J, Kolar R, Kopecna O, Pagacova E, Falkova I, *et al.* DeepFoci: Deep learning-based algorithm for fast automatic analysis of DNA double-strand break ionizing radiation-induced foci. Comput Struct Biotechnol J. 2021;19:6465–6480.
62. Barnard S, Ainsbury EA, Al-hafidh J, Hadjidekova V, Hristova R, Lindholm C, *et al.* The first gamma-H2AX biodosimetry intercomparison exercise of the developing European biodosimetry network RENEB. Radiat Prot Dosimetry. 2015;164(3):265–270.
63. Moquet J, Barnard S, Staynova A, Lindholm C, Monteiro Gil O, Martins V, *et al.* The second gamma-H2AX assay inter-comparison exercise carried out in the framework of the European biodosimetry network (RENEB). Int J Radiat Biol. 2017;93(1):58–64.
64. Moquet J, Barnard S, Rothkamm K. Gamma-H2AX biodosimetry for use in large scale radiation incidents: comparison of a rapid '96 well lyse/fix' protocol with a routine method. PeerJ. 2014;2:e282.
65. Raavi V, Perumal V, Paul SFD. Potential application of γ-H2AX as a biodosimetry tool for radiation triage. Mutat Res Rev Mutat Res. 2021;787:108350.
66. Zahnreich S, Ebersberger A, Kaina B, Schmidberger H. Biodosimetry Based on γ-H2AX Quantification and Cytogenetics after Partial- and Total-Body Irradiation during Fractionated Radiotherapy. Radiat Res. 2015;183(4):432.
67. Hofer M, Falk M, Komůrková D, Falková I, Bačíková A, Klejdus B, *et al.* Two New Faces of Amifostine: Protector from DNA Damage in Normal Cells and Inhibitor of DNA Repair in Cancer Cells. J Med Chem. 2016;59(7):3003–3017.
68. Pagáčová E, Štefančíková L, Schmidt-Kaler F, Hildenbrand G, Vičar T, Depeš D, *et al.* Challenges and Contradictions of Metal Nano-Particle Applications for Radio-Sensitivity Enhancement in Cancer Therapy. Int J Mol Sci. 2019;20(3):588.

69. Sevcik J, Falk M, Kleiblova P, Lhota F, Stefancikova L, Janatova M, *et al.* The BRCA1 alternative splicing variant δ14-15 with an in-frame deletion of part of the regulatory serine-containing domain (SCD) impairs the DNA repair capacity in MCF-7 cells. Cell Signal. 2012;24(5):1023–1030.
70. Sevcik J, Falk M, Macurek L, Kleiblova P, Lhota F, Hojny J, *et al.* Expression of human BRCA1δ17-19 alternative splicing variant with a truncated BRCT domain in MCF-7 cells results in impaired assembly of DNA repair complexes and aberrant DNA damage response. Cell Signal. 2013;25(5):1186–1193.
71. Dobešová L, Gier T, Kopečná O, Pagáčová E, Vičar T, Bestvater F, *et al.* Incorporation of Low Concentrations of Gold Nanoparticles: Complex Effects on Radiation Response and Fate of Cancer Cells. Pharmaceutics. 2022;14(1):166.
72. Štefanciková L, Lacombe S, Salado D, Porcel E, Pagáčová E, Tillement O, *et al.* Effect of gadolinium-based nanoparticles on nuclear DNA damage and repair in glioblastoma tumor cells. J Nanobiotechnology. 2016;14(1):63.
73. Vilenchik MM, Knudson AG. Endogenous DNA double-strand breaks: Production, fidelity of repair, and induction of cancer. Proc Natl Acad Sci U S A. 2003;100(22):12871–12876.
74. Rittich B, Španová A, Falk M, Beneš MJ, Hrubý M. Cleavage of double stranded plasmid DNA by lanthanide complexes. J Chromatogr B Analyt Technol Biomed Life Sci. 2004;800(1-2):169–173.
75. Inturi S, Tewari-Singh N, Agarwal C, White CW, Agarwal R. Activation of DNA damage repair pathways in response to nitrogen mustard-induced DNA damage and toxicity in skin keratinocytes. Mutat Res. 2014;763-764:53–63.
76. Jan YH, Heck DE, Laskin DL, Laskin JD. DNA damage signaling in the cellular responses to mustard vesicants. Toxicol Lett. 2020;326:78–82.
77. de Vera P, Surdutovich E, Solov'yov AV. The role of shock waves on the biodamage induced by ion beam radiation. Cancer Nanotechnol. 2019;10(1):5.
78. Surdutovich E, Solov'yov AV. Shock wave initiated by an ion passing through liquid water. Phys Rev E Stat Nonlin Soft Matter Phys. 2010;82(5 Pt 1):051915.
79. Friis I, Verkhovtsev AV, Solov'yov IA, Solov'yov AV. Lethal DNA damage caused by ion-induced shock waves in cells. Phys Rev E. 2021;104(5-1):054408.

80. Verkhovtsev A, Surdutovich E, Solov'yov AV. Multiscale approach predictions for biological outcomes in ion-beam cancer therapy. Sci Rep. 2016;6:27654.
81. Anderson RM, Stevens DL, Sumption ND, Townsend KMS, Goodhead DT, Hill MA. Effect of linear energy transfer (LET) on the complexity of alpha-particle-induced chromosome aberrations in human CD34+ cells. Radiat Res. 2007;167(5):541–50.
82. Lorat Y, Brunner CU, Schanz S, Jakob B, Taucher-Scholz G, Rübe CE. Nanoscale analysis of clustered DNA damage after high-LET irradiation by quantitative electron microscopy – The heavy burden to repair. DNA Repair (Amst). 2015;28:93–106.
83. Lorat Y, Reindl J, Isermann A, Rübe C, Friedl AA, Rübe CE. Focused Ion Microbeam Irradiation Induces Clustering of DNA Double-Strand Breaks in Heterochromatin Visualized by Nanoscale-Resolution Electron Microscopy. Int J Mol Sci. 2021;22(14):7638.
84. Tracy BL, Stevens DL, Goodhead DT, Hill MA. Variation in RBE for Survival of V79-4 Cells as a Function of Alpha-Particle (Helium Ion) Energy. Radiat Res. 2015;184(1):33.
85. Mavragani I, Nikitaki Z, Souli M, Aziz A, Nowsheen S, Aziz K, *et al.* Complex DNA Damage: A Route to Radiation-Induced Genomic Instability and Carcinogenesis. Cancers (Basel). 2017;9(12):91.
86. Dokic I, Mairani A, Niklas M, Zimmermann F, Chaudhri N, Krunic D, *et al.* Next Generation Multi-Scale Biophysical Characterization of High Precision Cancer Particle Radiotherapy Using Clinical Proton, Helium-, Carbon- and Oxygen Ion Beams. Oncotarget. 2016;7(35):56676–56689.
87. Asaithamby A, Chen DJ. Mechanism of Cluster DNA Damage Repair in Response to High-Atomic Number and Energy Particles Radiation. Mutat Res. 2011;711(1-2):87–99.
88. Barendsen GW. RBE—LET Relationships for Different Types of Lethal Radiation Damage in Mammalian Cells: Comparison with DNA DSB and an Interpretation of Differences in Radiosensitivity. Int J Radiat Biol. 1994;66(5):433–436.
89. Sørensen BS, Overgaard J, Bassler N. In Vitro RBE-LET Dependence for Multiple Particle Types. Acta Oncol. 2011;50(6):757–762.
90. Helm A, Fournier C. High-LET Charged Particles: Radiobiology and Application for New Approaches in Radiotherapy. Strahlenther Onkol. 2023;199(12):1225–1241.
91. Durante M, Orecchia R, Loeffler JS. Charged-Particle Therapy in Cancer: Clinical Uses and Future Perspectives. Nat Rev Clin Oncol. 2017;14(8):483–495.

92. Obolensky OI, Surdutovich E, Pshenichnov I, Mishustin I, Solov'yov AV, Greiner W. Ion Beam Cancer Therapy: Fundamental Aspects of the Problem. Nucl Instrum Methods Phys Res B. 2008;266(8): 1623–1628.
93. Pompos A, Durante M, Choy H. Heavy Ions in Cancer Therapy. JAMA Oncol. 2016;2(12):1539.
94. Durante M, Cucinotta FA. Heavy Ion Carcinogenesis and Human Space Exploration. Nat Rev Cancer. 2008;8(6):465–472.
95. Kamsali N, Chakravarty SC, Basuvaraj PK. Space Radiation Hazards and Future Mars Missions. ArXiv preprint. 2019. arXiv: 1908.08362.
96. Kamsali N, Chakravarty SC, Basuvaraj PK. Investigation of HZE Particle Fluxes as a Space Radiation Hazard for Future Mars Missions. Heliyon. 2019;5(12):e02972.
97. Curtis SB. Risk from Relativistic Heavy Ions on Manned Space Missions. Radioprotection. 1993;28(2):179–181.
98. Aparicio T, Baer R, Gautier J. DNA Double-Strand Break Repair Pathway Choice and Cancer. DNA Repair (Amst). 2014;19:169–175.
99. Soulas-Sprauel P, Rivera-Munoz P, Malivert L, Le Guyader G, Abramowski V, Revy P, *et al.* V(D)J and Immunoglobulin Class Switch Recombination: A Paradigm to Study the Regulation of DNA End-Joining. Oncogene. 2007;26(56):7780–7791.
100. Woods D, Turchi JJ. Chemotherapy Induced DNA Damage Response: Convergence of Drugs and Pathways. Cancer Biol Ther. 2013;14(5):379–389.
101. Shibata A, Jeggo P, Löbrich M. The Pendulum of the Ku-Ku Clock. DNA Repair (Amst). 2018;71:164–171.
102. Chang HHY, Pannunzio NR, Adachi N, Lieber MR. Non-Homologous DNA End Joining and Alternative Pathways to Double-Strand Break Repair. Nat Rev Mol Cell Biol. 2017;18(8):495–506.
103. Davis AJ, Chen DJ. DNA Double Strand Break Repair via Non-Homologous End-Joining. Transl Cancer Res. 2013;2(3):130–143.
104. Yang K, Guo R, Xu D. Non-Homologous End Joining: Advances and Frontiers. Acta Biochim Biophys Sin (Shanghai). 2016;48(7):632–640.
105. Bordelet H, Dubrana K. Keep Moving and Stay in a Good Shape to Find Your Homologous Recombination Partner. Curr Genet. 2019;65(1):29–39.
106. Wright WD, Shah SS, Heyer WD. Homologous Recombination and the Repair of DNA Double-Strand Breaks. J Biol Chem. 2018;293(27):10524–10535.

107. Iliakis G, Murmann T, Soni A. Alternative End-Jointing Repair Pathways Are the Ultimate Backup for Abrogated Classical Non-Homologous End-Jointing and Homologous Recombination Repair: Implications for the Formation of Chromosome Translocations. Mutat Res. 2015;793:166–75.
108. Jachimowicz RD, Goergens J, Reinhardt HC. DNA Double-Strand Break Repair Pathway Choice – From Basic Biology to Clinical Exploitation. Cell Cycle. 2019;18(13):1423–1434.
109. Lim YC, Roberts TL, Day BW, Stringer BW, Kozlov S, Fazry S, *et al.* Increased Sensitivity to Ionizing Radiation by Targeting the Homologous Recombination Pathway in Glioma Initiating Cells. Mol Oncol. 2014;8(8):1603–1615.
110. Stadler J, Richly H. Regulation of DNA Repair Mechanisms: How the Chromatin Environment Regulates the DNA Damage Response. Int J Mol Sci. 2017;18(8):1715.
111. Zhao Y, Chen S. Targeting DNA Double-Strand Break (DSB) Repair to Counteract Tumor Radio-resistance. Curr Drug Targets. 2019; 20(9):891–902.
112. Toulany M. Targeting DNA Double-Strand Break Repair Pathways to Improve Radiotherapy Response. Genes (Basel). 2019;10(1):25.
113. Merlo D, Mollinari C, Racaniello M, Garaci E, Cardinale A. DNA Double Strand Breaks: A Common Theme in Neurodegenerative Diseases. Curr Alzheimer Res. 2016;13(11):1208–1218.
114. Madabhushi R, Pan L, Tsai LH. DNA Damage and Its Links to Neurodegeneration. Neuron. 2014;83(2):266–282.
115. Gorbunova V, Seluanov A. Making ends meet in old age: DSB repair and aging. Mech Ageing Dev. 2005;126(6-7):621–628.
116. Mah LJ, El-Osta A, Karagiannis TC. γH2AX as a molecular marker of aging and disease. Epigenetics. 2010;5(2):129–136.
117. Bekker-Jensen S, Mailand N. Assembly and function of DNA double-strand break repair foci in mammalian cells. DNA Repair (Amst). 2010;9(12):1219–1228.
118. Takeda S, Nakamura K, Taniguchi Y, Paull TT. Ctp1/CtIP and the MRN Complex Collaborate in the Initial Steps of Homologous Recombination. Mol Cell. 2007;28(3):351–352.
119. Kousholt AN, Menzel T, Sørensen CS. Pathways for genome integrity in G2 phase of the cell cycle. Biomolecules. 2012;2(4):579–607.
120. Fortuny A, Polo SE. The response to DNA damage in heterochromatin domains. Chromosoma. 2018;127(3):291–300.
121. Baldock RA, Day M, Wilkinson OJ, Cloney R, Jeggo PA, Oliver AW, *et al.* ATM Localization and Heterochromatin Repair Depend

on Direct Interaction of the 53BP1-BRCT 2 Domain with γH2AX. Cell Rep. 2015;13(10):2081–2089.

122. Goodarzi AA, Jeggo P, Lobrich M. The influence of heterochromatin on DNA double strand break repair: Getting the strong, silent type to relax. DNA Repair (Amst). 2010;9(12):1273–1282.

123. Goodarzi AA, Noon AT, Deckbar D, Ziv Y, Shiloh Y, Löbrich M, *et al.* ATM Signaling Facilitates Repair of DNA Double-Strand Breaks Associated with Heterochromatin. Mol Cell. 2008;31(2): 167–177.

124. Goodarzi AA, Noon AT, Jeggo PA. The impact of heterochromatin on DSB repair. Biochem Soc Trans. 2009;37(Pt 3):569–576.

125. Jalan M, Olsen KS, Powell SN. Emerging Roles of RAD52 in Genome Maintenance. Cancers (Basel). 2019;11(7):1038.

126. Wei L, Nakajima S, Böhm S, Bernstein KA, Shen Z, Tsang M, *et al.* DNA damage during the G0/G1 phase triggers RNA-templated, Cockayne syndrome B-dependent homologous recombination. Proc Natl Acad Sci U S A. 2015;112(27):E3495–E3504.

127. Ouyang J, Lan L, Zou L. Regulation of DNA break repair by transcription and RNA. Sci China Life Sci. 2017;60(10):1081–1086.

128. McDevitt S, Rusanov T, Kent T, Chandramouly G, Pomerantz RT. How RNA transcripts coordinate DNA recombination and repair. Nat Commun. 2018;9:1091.

129. Mladenova V, Mladenov E, Stuschke M, Iliakis G. DNA Damage Clustering after Ionizing Radiation and Consequences in the Processing of Chromatin Breaks. Molecules. 2022;27(5):1540.

130. Chiolo I, Minoda A, Colmenares SU, Polyzos A, Costes SV, Karpen GH. Double-Strand Breaks in Heterochromatin Move Outside of a Dynamic HP1a Domain to Complete Recombinational Repair. Cell. 2011;144(5):732–744.

131. Tsouroula K, Furst A, Rogier M, Heyer V, Maglott-Roth A, Ferrand A, *et al.* Temporal and Spatial Uncoupling of DNA Double Strand Break Repair Pathways within Mammalian Heterochromatin. Mol Cell. 2016;63(2):293–305.

132. Janssen A, Breuer GA, Brinkman EK, van der Meulen AI, Borden SV, van Steensel B, *et al.* A single double-strand break system reveals repair dynamics and mechanisms in heterochromatin and euchromatin. Genes Dev. 2016;30(14):1645–1657.

133. Clouaire T, Legube G. DNA double strand break repair pathway choice: a chromatin based decision? Nucleus. 2015;6(2):107–113.

134. Kakarougkas A, Jeggo PA. DNA DSB repair pathway choice: an orchestrated handover mechanism. Br J Radiol. 2014;87(1035): 20130685.

135. Shibata A, Conrad S, Birraux J, Geuting V, Barton O, Ismail A, *et al.* Factors determining DNA double-strand break repair pathway choice in G2 phase: DSB repair pathway choice in G2 phase. EMBO J. 2011;30(6):1079–1092.
136. Marnef A, Cohen S, Legube G. Transcription-Coupled DNA Double-Strand Break Repair: Active Genes Need Special Care. J Mol Biol. 2017;429(9):1277–1288.
137. Aymard F, Bugler B, Schmidt CK, Guillou E, Caron P, Briois S, *et al.* Transcriptionally active chromatin recruits homologous recombination at DNA double-strand breaks. Nat Struct Mol Biol. 2014; 21(4):366–374.
138. George CM, Alani E. Multiple cellular mechanisms prevent chromosomal rearrangements involving repetitive DNA. Crit Rev Biochem Mol Biol. 2012;47(3):297–313.
139. Amaral N, Ryu T, Li X, Chiolo I. Nuclear Dynamics of Heterochromatin Repair. Trends Genet. 2017;33(2):86–100.
140. Ryu T, Spatola B, Delabaere L, Bowlin K, Hopp H, Kunitake R, *et al.* Heterochromatic breaks move to the nuclear periphery to continue recombinational repair. Nat Cell Biol. 2015;17(11):1401–1411.
141. Wykes SM, Piasentin E, Joiner MC, Wilson GD, Marples B. Low-Dose Hyper-radiosensitivity is not Caused by a Failure to Recognize DNA Double-Strand Breaks. Radiat Res. 2006;165(5):516–524.
142. Rothkamm K, Löbrich M. Evidence for a lack of DNA double-strand break repair in human cells exposed to very low x-ray doses. Proc Natl Acad Sci U S A. 2003;100(9):5057–5062.
143. Lau YS, Chew MT, Alqahtani A, Jones B, Hill MA, Nisbet A, *et al.* Low Dose Ionising Radiation-Induced Hormesis: Therapeutic Implications to Human Health. Appl Sci. 2021;11(19):8909.
144. Jakob B, Dubiak-Szepietowska M, Janiel E, Schmidt A, Durante M, Taucher-Scholz G. Differential Repair Protein Recruitment at Sites of Clustered and Isolated DNA Double-Strand Breaks Produced by High-Energy Heavy Ions. Sci Rep. 2020;10(1):1443.
145. Jakob B, Splinter J, Taucher-Scholz G. Positional stability of damaged chromatin domains along radiation tracks in mammalian cells. Radiat Res. 2009;171(4):405–418.
146. Ayache J, Beaunier L, Boumendil J, Ehret G, Laub D. Artifacts in Transmission Electron Microscopy. In: Sample Preparation Handbook for Transmission Electron Microscopy. New York: Springer; 2010:125–170.
147. Klumperman J. Arts and artifacts of electron microscopy. Trends Cell Biol. 1999;9(6):247.

148. Varga D, Majoros H, Ujfaludi Z, Erdélyi M, Pankotai T. Quantification of DNA damage induced repair focus formation via super-resolution dSTORM localization microscopy. Nanoscale. 2019;11(30):14226–1436.
149. Schermelleh L, Ferrand A, Huser T, Eggeling C, Sauer M, Biehlmaier O, *et al.* Super-resolution microscopy demystified. Nat Cell Biol. 2019;21(1):72–84.
150. Dedecker P, Flors C, Hotta J, Uji-i H, Hofkens J. 3D Nanoscopy: Bringing Biological Nanostructures into Sharp Focus. Angew Chem Int Ed Engl. 2007;46(44):8330–8332.
151. Egner A, Hell SW. Fluorescence microscopy with super-resolved optical sections. Trends Cell Biol. 2005;15(4):207–215.
152. Garini Y, Vermolen BJ, Young IT. From micro to nano: recent advances in high-resolution microscopy. Curr Opin Biotechnol. 2005;16(1):3–12.
153. Gould TJ, Hess ST. Chapter 12 Nanoscale Biological Fluorescence Imaging. In: Methods in Cell Biology. Amsterdam: Elsevier; 2008:329–358.
154. Galbraith CG, Galbraith JA. Super-resolution microscopy at a glance. J Cell Sci. 2011;124(10):1607–1611.
155. Baddeley D, Bewersdorf J. Biological Insight from Super-Resolution Microscopy: What We Can Learn from Localization-Based Images. Annu Rev Biochem. 2018;87(1):965–989.
156. Thorley JA, Pike J, Rappoport JZ. Super-resolution Microscopy. In: Fluorescence Microscopy. Amsterdam: Elsevier; 2014:199–212.
157. Vangindertael J, Camacho R, Sempels W, Mizuno H, Dedecker P, Janssen KPF. An introduction to optical super-resolution microscopy for the adventurous biologist. Methods Appl Fluoresc. 2018;6(2):022003.
158. Friedrich T, Ilicic K, Greubel C, Girst S, Reindl J, Sammer M, *et al.* DNA damage interactions on both nanometer and micrometer scale determine overall cellular damage. Sci Rep. 2018;8(1):16063.
159. Hofmann A, Krufczik M, Heermann D, Hausmann M. Using Persistent Homology as a New Approach for Super-Resolution Localization Microscopy Data Analysis and Classification of γH2AX Foci/Clusters. Int J Mol Sci. 2018;19(8):2263.
160. Bizzarri M, Naimark O, Nieto-Villar J, Fedeli V, Giuliani A. Complexity in Biological Organization: Deconstruction (and Subsequent Restating) of Key Concepts. Entropy (Basel). 2020;22(8):885.
161. Hausmann M, Hildenbrand G, Pilarczyk G. Networks and Islands of Genome Nano-architecture and Their Potential Relevance for Radiation Biology: (A Hypothesis and Experimental Verification Hints).

In: Kloc M, Kubiak JZ, editors. Nuclear, Chromosomal, and Genomic Architecture in Biology and Medicine. Cham: Springer International Publishing; 2022:3–34.

162. Erenpreisa J, Giuliani A. Resolution of Complex Issues in Genome Regulation and Cancer Requires Non-Linear and Network-Based Thermodynamics. Int J Mol Sci. 2019;21(1):240.
163. Boltsis I, Grosveld F, Giraud G, Kolovos P. Chromatin Conformation in Development and Disease. Front Cell Dev Biol. 2021;9:723859.
164. Luo C, Dong J, Zhang Y, Lam E. Decoding the role of chromatin architecture in development: coming closer to the end of the tunnel. Front Plant Sci. 2014;5:374.
165. Wang M, Sunkel BD, Ray WC, Stanton BZ. Chromatin structure in cancer. BMC Mol Cell Biol. 2022;23(1):35.
166. Schones DE, Leung A, Natarajan R. Chromatin Modifications Associated With Diabetes and Obesity. Arterioscler Thromb Vasc Biol. 2015;35(7):1557–1561.

https://doi.org/10.1142/9781800618183_0012

Chapter 12

Foundations of Astrochemistry I: Observations and Ion- and Electron-Induced Processes in Ices

Nigel J. Mason,[*,†,**] **Sergio Ioppolo,**[‡] **Heidy M. Quitián-Lara,**[*,§] **Felipe Fantuzzi,**[†,¶,††] **Duncan V. Mifsud,**[†] **and Jana Bocková**[||]

[*] *Centre for Astrophysics and Planetary Science, School of Engineering, Mathematics and Physics, University of Kent, Canterbury CT2 7NH, United Kingdom*

[†] *HUN-REN Institute for Nuclear Research (Atomki), Debrecen H-4026, Hungary*

[‡] *Centre for Interstellar Catalysis, Department of Physics and Astronomy, Aarhus University, DK 8000 Aarhus, Denmark*

[§] *Max Planck Institute for Extraterrestrial Physics, 85748 Garching, Germany*

[¶] *Supramolecular, Interfacial and Synthetic Chemistry, School of Natural Sciences, University of Kent, Canterbury CT2 7NH, United Kingdom*

[||] *Institut de Chimie de Nice (ICN), CNRS UMR 7272, Université Côte d'Azur, 06108, Nice, France*

[**] *n.j.mason@kent.ac.uk*

[††] *f.fantuzzi@kent.ac.uk*

Astrochemistry is one of the newest but also most rapidly growing interdisciplinary fields of modern astronomy and space science. It is now recognised that space is a rich chemical environment in which more than 320 different molecular species are synthesised as part of star and planet formation, with such chemistry playing a crucial role in these formation processes. Furthermore, the evidence of this 'universal' chemistry is essential for unravelling the mechanisms that led to the origins of life on Earth and for assessing the likelihood of life existing elsewhere, both within our own solar system and among those thousands of exoplanets we can now detect and study. Built upon the foundations of atomic and molecular physics, astrochemistry relies on spectroscopy to observe the Universe and on collisional and radiation-induced processes to understand the evolution pathways of molecular synthesis, planet formation, and planetary atmospheres. This chapter provides an overview of the critical role atomic and molecular processes play in astrochemistry, emphasising the significance of both gas and condensed phases. The latter is particularly vital for the study of dust grains in the interstellar medium, where they frequently serve as sites for molecular synthesis. We also discuss the essential contribution of experimental research and the considerable challenges involved in replicating space environments. As we enter a new era of astronomy and space science, a comprehensive understanding of chemical processes, alongside physical phenomena, is indispensable for addressing some of the most profound questions in modern science – questions that demand an even deeper insight into atomic and molecular physics.

12.1 Introduction

One of the largest users of atomic and molecular data is the astronomy community. Most astronomical data are collected remotely and presented in spectral formats, creating a demand for extensive spectroscopic data repositories that cover a broad spectral range, from radio to X-ray regions.[1,2] Much of this data need to be derived computationally using state-of-the-art electronic and structural models for atoms, molecules,[3] and ions, as the ionised state is the most prevalent form of matter.[4]

Within our own solar system, astronomical observations can be complemented by fly-by and *in situ* missions, which enhance spectroscopic observational data with mass spectrometric studies. These studies, focusing on, e.g., magnetospheric plasmas, cometary tails, and planetary atmospheres, rely heavily on a large amount of data concerning electron and ion collisions.[5]

Over the past 50 years, it has become evident that the Universe is molecular in nature, with molecular systems ranging from the simplest diatomics to complex fullerenes abundantly present in the interstellar medium (ISM), where stars and planets form.[6] This discovery implies a rich and dynamic chemistry in these regions.

The study of chemistry within the ISM, along with its role in star and planet formation and its implications for the synthesis and stability of molecular systems necessary for creating the prebiotic compounds and structures necessary for the emergence of life, has given rise to a new field in astronomy, that of *astrochemistry*. Today, astrochemistry encompasses both the solar system and the ISM,[7] investigating chemical processes in space environments vastly different from the chemistry on Earth, which mostly occurs at standard temperatures and pressures (STP). Astrochemical processes involve radiation-induced reactions and low-temperature quantum regimes.[8,9] Astrochemistry is rooted in atomic and molecular physics (AMP), involving the study of collisions and radiation-induced processes both in the gaseous and solid phases,[10] the latter reflecting the presence of molecular 'ices' in the ISM and on planetary and cometary surfaces.

In this chapter, we review the current state of our understanding of the 'molecular universe' and its foundations in AMP. We discuss current challenges, topics for further research, and the critical need for the compilation and storage of large-scale databases such as VAMDC[11] and the Leiden database.[12] The use of artificial intelligence (AI) and machine learning to extend existing spectral and collisional databases is also examined. Finally, we conclude with a forward-looking perspective, addressing potential threats to foundational research in this field.

12.2 Observations and the Need for Laboratory Studies

The discovery of molecules in the ISM has been a pivotal milestone in consolidating modern astrochemistry.[13,14] The identification of the first molecules in the ISM dates back to the early 1940s.[15] The CH^+, CH, and CN[16–18] systems marked the beginning of the 'zoo' of chemical species that make up the space between the stars. However, it took over two decades for radio telescopes to revolutionise and dominate the field, enabling the detection of new molecular species, a process that continues to expand our knowledge to this day.[19,20]

The era of the molecular universe emerged prominently thanks to the continuous discoveries of various chemical species, enabled by radio telescopes operating in the centimetre, millimetre, and submillimetre ranges. Notable facilities contributing to these discoveries include the NRAO 12m and the Robert C. Byrd Green Bank Telescope (GBT) 100m of the National Radio Astronomy Observatory (NRAO),[21] the IRAM 30m of the Institut de radioastronomie millimétrique,[22] and, more recently, the Yebes 40m radio telescope of the Yebes Observatory.[23] Large-scale programmes, such as the Astrochemical Surveys at IRAM 30m (ASAI), have been particularly instrumental in studying the chemistry of solar-type protostars and expanding our knowledge of interstellar molecular complexity.[24–26] By 1 December 2024, observations conducted using these and other instruments collectively led to the identification of 327 gas-phase molecules in the ISM. A comprehensive catalogue of interstellar molecules can be found in The Cologne Database for Molecular Spectroscopy (CDMS).[a] The first 241 species were documented by McGuire and coworkers[27] as of 1 June 2021. Between that date and 1 December 2024, 86 additional molecules were identified (see Tables 12.1 and 12.2). Of the 327 molecules catalogued, 170 are classified as simple species, consisting of molecules with up to five atoms in their structure,[14] while 157 are classified as complex organic molecules (COMs). This latter group includes seven members of the polycyclic aromatic hydrocarbon (PAH) family: 1- and 2-cyanonaphthalene (1-$C_{10}H_7CN$, 2-$C_{10}H_7CN$),[28] indene (C_9H_8),[29] 1-cyanocyclopentadiene (c-C_9H_7CN),[30] 1- and 5-cyanoacenaphthylene (1-$C_{12}H_7CN$, 5-$C_{12}H_7CN$),[31] and

Table 12.1. Simple molecules identified in the ISM, from 1 June 2021 to 1 December 2024.

2 Atoms	3 Atoms	4 Atoms	5 Atoms
PO^+	MgC_2	HNCN	HCSCN
SiP	HSO	H_2NC	HCCCO
FeC	CaC_2	$HCCS^+$	NaCCCN
NaS		HCNS	MgC_3N^+
MgS		$HOCS^+$	HC_3N^+
CaO		HNSO	NC_3S
			HC_3S

Source: Adapted from the CDMS database.

[a]https://cdms.astro.uni-koeln.de/classic/molecules.

Table 12.2. COMs identified in the ISM, from 1 June 2021 to 1 December 2024.

6 Atoms	7 Atoms	8 Atoms	9 Atoms
C_5O	H_2C_5	MgC_6H	HOCHCHCHO
C_5H^+	MgC_5N	$C_2H_3NH_2$	$CH_2{=}CH\text{-}CH{=}NH$
$HCCNCH^+$	CH_2CCCN	*Z*-$(CHOH)_2$	HC_7N^+
c-C_5H	NC_4NH^+	HCCCHCCC	$HCCCH_2CCH$
HCCCCS	MgC_5N^+	C_7N^-	
HMgCCCN	$NCCONH_2$	CH_3CHCO	
MgC_4H^+	HC_5N^+	MgC_6H^+	
H_2CCCH^+	HNC_5	*Z*-NCCHCHCN	
H_2CCCN	$CH_2(CN)_2$		
HOCOOH			
H_2CNCN			
NCCHCS			
HCSCCH			

10 Atoms	11 Atoms	12 Atoms	13 Atoms
o-C_6H_4	$NH_2CH_2CH_2OH$	$C_2H_5CONH_2$	1-*c*-C_5H_5CCH
H_2CCCHC_3N	CH_2CCHC_4H	CH_3C_7N	2-*c*-C_5H_5CCH
$C_2H_5NH_2$	$C_{10}H^-$	*i*-C_3H_7OH	*c*-$C_5H_4CCH_2$
C_2H_5NCO	$C_{10}H$	*n*-C_3H_7OH	$CH_3OCH_2CH_2OH$
HC_7NH^+	*E*-1-C_4H_5CN	$(CH_3)_2C{=}CH_2$	
t-$CH_3CHCHCN$			
c-$CH_3CHCHCN$			
$CH_2C(CH_3)CN$			
g-CH_2CHCH_2CN			
c-CH_2CHCH_2CN			
$NH_2C(O)CH_2OH$			

14 Atoms	PAH
C_6H_5CCH	*c*-C_9H_7CN
	1-$C_{12}H_7CN$
	5-$C_{12}H_7CN$
	$C_{16}H_9CN$

Source: Adapted from the CDMS database.

1-cyanopyrene ($C_{16}H_9CN$),[32] as well as three fullerene species: C_{60}, ${C_{60}}^+$, and C_{70}.[33]

Reports of newly detected molecules in the ISM are published with increasing regularity. For the purposes of this chapter, we focus

Table 12.3. Number of molecules identified per observatory. The left section of the table lists facilities with the highest number of molecules reported up to 1 June 2021, while the right section highlights facilities involved in the identification of new interstellar molecules between 1 June 2021 and 1 December 2024. Please note that some detections are counted twice in instances where distinct facilities collaborated on the same detection.

Facility	Number of detections	Facility	Number of detections
IRAM 30m	64	Yebes 40m	72
NRAO 12m	33	IRAM 30m	34
GBT 100m	28	GBT 100m	8
NRAO/ARO 12m	27	APEX	4
Yebes 40m	19	ALMA	3
Nobeyama 45m	15	ARO	2
NRAO 43m	13	ACA	1
ALMA	8		
Bell 7m	8		

Note: Information until 1 June 2021 taken from McGuire et al. 2022.[27]
ACA: Atacama Compact Array.
APEX: Atacama Pathfinder Experiment.
ARO: Arizona Radio Observatory.

on the state of knowledge as of 1 December 2024, reflecting the most up-to-date scenario available at the time of writing.

As of 1 June 2021, approximately 80% of the molecules identified in the ISM were discovered using just 9 out of a total of 46 facilities. However, from that date until December 2024, the majority of new molecular identifications have been achieved primarily through the collaboration between the Yebes and IRAM observatories. Using the Yebes 40m and IRAM 30m radio telescopes, these two facilities have been responsible for nearly 90% of the species identified during this period (see Table 12.3).

This highlights that significant advancements in acquisition instruments used in various single-dish radio telescopes, along with the commissioning of modern interferometers such as the very large array (VLA) and the Atacama Large Millimeter/submillimeter Array (ALMA), have provided superior resolution and sensitivity. These improvements have enabled the detection of a larger number of chemical species and have refined observational limits, allowing for more precise determination of the regions where gas-phase molecules are present.[23,34–36]

On the other hand, the use of space telescopes, operating in the infrared range, particularly in the mid-infrared, provides a window into the chemistry of the condensed phase.[37,38] Astrophysical dust particles and ices serve as catalytic media for the formation of complex species in the ISM,[39,40] playing a crucial role in the chemistry of obscured bodies where low-temperature processes dominate.[41,42] Observations with the Spitzer Space Telescope have revealed molecules of the fullerene family[33,43,44] and underscored the widespread presence of PAHs in diverse objects and regions of space,[45,46] including extragalactic environments.[47,48] Some of these species have been conclusively identified through high-resolution observations using facilities such as the GBT 100m and Yebes 40m radio telescopes.[28,29,49]

The combined efforts of various radio and space telescopes, including the state-of-the-art James Webb Space Telescope (JWST), launched on 25 December 2021, have significantly transformed our understanding of the ISM.[50] JWST can penetrate dense clouds of interstellar dust and observe deep space and point objects such as exoplanets. Its unparalleled sensitivity in the infrared allows for high-resolution detection of molecules in environments that were previously beyond the reach of astronomical observation. Some of the most remarkable discoveries made with JWST include the detection of prebiotic molecules, such as ammonium (NH_4^+), in dark molecular clouds,[51] as well as the identification of formaldehyde (H_2CO), formic acid (HCOOH), and COMs such as ethanol (C_2H_5OH), acetaldehyde (CH_3CHO), methyl formate ($HCOOCH_3$), and acetic acid (CH_3COOH) in ices associated with protostellar objects.[52] Additionally, molecules such as methyl cyanide (CH_3CN) and ethyl cyanide (CH_3CH_2CN) have been tentatively identified in these environments,[53] although further investigation is required to confirm their presence.

By 1 December 2024, JWST has facilitated the identification of 13 condensed-phase molecular species, marking a significant breakthrough in detection limits within ice astrochemistry.[b] The high sensitivity and spatial resolution of JWST provide critical data, enabling the study of distinct chemical states in transitional environments,

[b]https://www.astrochymist.org/astrochymist_ism_ice.html.

such as those surrounding protostars.[54] It is widely believed that the synthesis of complex molecular species occurs via catalytic processes on the surface of ices.[55,56] However, despite the extensive catalogue of species identified in the gas phase, the extent to which these species originate from the desorption of ices remains uncertain. The mechanisms underlying this transition, as well as the connection between gas-phase and condensed-phase chemistries, are not yet fully understood.[57,58]

This gap is addressed through computational models[58–60] and experimental studies conducted in specialised laboratories.[61–64] These laboratories simulate the extreme physicochemical conditions of interstellar environments, enabling the exploration of various aspects of astrochemical processes, including formation routes,[65–69] dissociation mechanisms,[70–73] structural rearrangements,[74–77] and the stability of different species under radiation.[78–80]

12.2.1 *Importance of laboratory experiments*

Specialised laboratories around the world play a crucial role in advancing our understanding of chemical reactions in space. While almost 90% of all known interstellar molecules have been identified through their gas-phase rotational spectra, studying the chemical content and processes occurring on the surface of interstellar grains remains a significant challenge. To address this, laboratory experiments recreate the extreme conditions of the ISM, such as low temperatures and intense radiation, enabling the study of molecular formation and evolution on grain surfaces. Among the techniques utilised in these laboratories is vibrational spectroscopy, which, although less effective at identifying discrete molecules compared to rotational spectroscopy, is instrumental in characterising the types of chemical bonds and functional groups present in icy grain mantles. These experimental efforts provide critical data that complement astronomical observations, offering a more comprehensive understanding of interstellar chemistry.[81]

In condensed-phase astrochemistry laboratories, ultrahigh-vacuum chambers and cryogenic coolers are utilised to replicate the extremely low temperatures of interstellar space. These conditions allow for the simulation of ice mantle formation on dust grains, akin to those found in molecular clouds, exoplanets, and icy surfaces such as Jupiter's moons. Experiments have demonstrated that irradiation of these ices with photons, electrons, and/or charged particles can drive the formation of complex molecules, such as methanol (CH_3OH) and formaldehyde (H_2CO), from simple precursors such as water and carbon monoxide.[82,83] This experimental approach also enables the investigation of the chemical enrichment of diverse celestial bodies, including icy moons, through the study of organic and inorganic molecules.

Sulphur, one of the most abundant elements in the Universe and a key player in prebiotic chemistry,[84,85] exemplifies the significance of such studies. Found on the surfaces of moons such as Io and Europa[86] and in the atmospheres of exoplanets identified by the JWST,[87] sulphur undergoes complex chemical processes. Recent experiments at the Ice Chamber for Astrophysics-Astrochemistry (ICA) at the Institute for Nuclear Research (Atomki) in Debrecen, Hungary, revealed the potential formation of SO_2 through S^+ ion bombardment of a CO_2 surface at 20 K and the synthesis of H_2SO_4 on water ices. Such findings are particularly relevant for understanding the icy surfaces of Europa, Ganymede, and Callisto, which are continuously bombarded by magnetospheric ions from Jupiter. This bombardment penetrates the ice sheets, inducing chemical reactivity and driving the formation of new molecular species.[64]

Another crucial aspect that underlines the importance of laboratory use in astrochemistry is the creation of experimental databases, which enable direct comparison with astronomical observations. These databases are indispensable, particularly when analysing data from instruments like JWST, which often involve mixed and complex ices. In such cases, substances such as water, methane (CH_4), and methanol (CH_3OH) can obscure the intensity of spectral bands associated with other molecules or functional groups.[88–90] Resources such as the Leiden Ice Database for Astrochemistry (LIDA[c]), which offers access to infrared spectra of numerous astrophysical ice analogues at various temperatures and under simulated chemical conditions, are therefore of significant importance.[91]

In experimental astrochemistry, spectroscopy and spectrometry are fundamental tools for identifying molecules in samples,

[c]https://icedb.strw.leidenuniv.nl/.

investigating their reactivity and transformations, and studying their properties. Techniques such as infrared, terahertz, and Raman spectroscopy, as well as mass spectrometry, are particularly effective for analysing irradiated ice samples. These methods provide detailed insights into the chemical composition of samples and the reactions occurring within them.[63,85,92] Furthermore, they play a pivotal role in the development and implementation of future laboratories aimed at studying *in situ* samples from solar system objects.[93]

The combination of laboratory and observational data has facilitated the development of detailed models describing the chemistry of the ISM.[60,94,95] These models integrate reaction rates and products derived from laboratory experiments and computational calculations with observations of molecular abundances in interstellar clouds. Comparing model predictions with observations enables the identification of dominant chemical processes in different regions of space and informs the refinement of these models.

This interdisciplinary approach, which combines astronomical observations, laboratory experimentation, and computational modelling, has been essential in advancing our understanding of the formation and evolution of both simple and complex molecules in space. This synergy not only sheds light on the processes shaping the chemistry of the universe but also opens pathways for exploring the potential for life on other planets. With ongoing advancements in observational and experimental technologies, further discoveries about the universe's rich chemical diversity and its capacity to support life are anticipated.

12.3 Ion-Induced Processes in Astrochemistry

Radiation-induced chemistry is a key driver of molecular complexity and cycling in the extraterrestrial environments of primary interest to astrochemistry, such as the nano- and microscale ices adsorbed on interstellar dust grains. Indeed, the radiative processing of interstellar icy grain mantles is believed to be among the main sources of molecules (including complex organic molecules relevant to life's origins) in the cosmos,[96–100] which are then released from the ices primarily as a result of thermal desorption (e.g., during the birth of a nearby young stellar object) or non-thermal desorption (e.g., chemical or reactive desorption or photon-induced desorption) processes.[101–106] Although a large proportion of the radiation chemistry leading to these important molecules is anticipated to be driven by low-energy (25 eV or less) electrons,[107–109] it is also possible for it to be triggered as a result of ion impact processes.

In the context of astrophysics, ions may be supplied to ices from a number of sources, including galactic cosmic rays, stellar winds, giant planetary magnetospheric plasmas, and the decay of radioactive isotopes in comets, asteroids, or moons. Each source is associated with different ion energies and ion compositions; for example, galactic cosmic rays are overwhelmingly composed of protons along with a significantly smaller abundance of α-particles and the fully ionised atoms of heavier elements (termed HZE ions), whose energy distribution typically peaks at a few hundred MeV.[110] Conversely, giant planetary magnetospheric plasmas may be enriched in more exotic ions with lower energy distributions. The classic example is the Jovian magnetosphere, whose ion component is sourced largely from volcanic activity on the moon Io, which emits copious amounts of sulphurous ejecta, primarily in the form of SO_2. A portion of these SO_2 molecules is subsequently dissociated and ionised within the Jovian magnetosphere, yielding sulphur and oxygen ions, along with electrons.[111,112] The energy distribution of these ions varies considerably depending on the distance from Jupiter, although thermal ions with energies of about 1 keV, together with higher energy ions ranging up to a few MeV, have been measured at the distance of Europa's orbit.[111–117]

12.3.1 *Energy loss mechanisms and molecule destruction*

The irradiation of low-temperature films analogous to interstellar icy grain mantles may therefore lead to the radiolytic destruction and formation of new molecules. However, before considering the chemical effects of the irradiation of such ices by energetic ions, it is important to first discuss the energy loss mechanisms of an ion traversing a solid medium (in this case, an ice in an extraterrestrial environment), which eventually leads to this chemistry. A natural starting point for such a discussion is the concept of stopping (also referred to as stopping power or linear energy transfer), i.e., the average linear rate of energy loss of an ion traversing a medium. This parameter is usually symbolised as $-dE/dx$ and is expressed in units of eV/nm (or some multiple thereof, e.g., MeV/cm). Overall, there are two types of stopping that may contribute to the energy loss of a projectile ion: (1) electronic (or inelastic) stopping during collisions between the ion and the bound electrons of the medium through which it is transiting, and (2) nuclear (or elastic) stopping during collisions with atomic nuclei.

A mathematical description of stopping for ions travelling through a solid medium was first provided by Bohr using a semi-classical method. In this method, the stopping of an ion of atomic number z and Coulombic charge e moving at a velocity V past a bound electron of mass m in a medium with an electron density of n is given as

$$-\frac{dE}{dx} = \frac{4\pi k^2 z^2 e^4 n}{mV^2} \ln \frac{mV^2}{hf}, \tag{12.1}$$

where $k = 1/4\pi\epsilon_0$ is the Coulombic force proportionality constant and hf is interpreted as an average energy associated with the electronic quantum states of the medium. A full derivation of this formula may be found in the work of Turner.[118]

An alternative description of stopping, which relies on a quantum mechanical treatment, was provided by Bethe,[118] who derived the formula

$$-\frac{dE}{dx} = \frac{4\pi k^2 z^2 e^4 n}{mc^2 \beta^2} \left(\ln \frac{2mc^2 \beta^2}{I(1-\beta^2)} - \beta^2 \right), \tag{12.2}$$

where c is the speed of light in a vacuum, $\beta = V/c$ (i.e., the speed of the ion relative to that of light in a vacuum), and I is the mean excitation energy of the medium. It is important to note that this latter term is explicitly defined in terms of properties of the atoms in the target medium, whereas the parameter hf in the semi-classical derivation of Bohr only has a vague meaning.

Empirical formulae are available to estimate I with reasonable accuracy for a given element with an atomic number of z. In the case of hydrogen (i.e., $z = 1$), $I = 19$ eV. For heavier elements, $I = 11.2 + 11.7z$ eV if $2 \leq z \leq 13$, and $I = 52.8 + 8.71z$ eV if $z > 13$.[118] Of course, ices in extraterrestrial environments are molecular in nature rather than atomic, and thus an approximation for the mean excitation energy I of a molecular target medium may be calculated by adding the individual contributions of each of its constituent elements. Thus, the mean excitation energy of a molecular target I_{mol} is given as

$$I_{\mathrm{mol}} = \exp\left\{\frac{1}{n}\sum_{i} N_{\mathrm{i}} z_{\mathrm{i}} \ln I_{\mathrm{i}}\right\}, \tag{12.3}$$

where, once again, n is the total electron density in the molecular solid and N_{i}, z_{i}, and I_{i}, respectively, refer to the atomic density, atomic number, and mean excitation energy of each individual element in the molecular solid.

Although the stopping of a projectile ion was previously described as the average linear rate of energy loss as the ion traverses a medium, it is important to note that the stopping of an ion is not itself linear when expressed as a function of path length. Indeed, a peak in energy loss occurs just before the ion comes to rest within the target medium; this is referred to as the Bragg peak (Fig. 12.1).[118] This peak arises due to the fact that the energy loss experienced by the projectile ion is inversely proportional to the square of its velocity and is a characteristic feature of the irradiation of solid materials by ions. Notably, a Bragg peak (or some equivalent thereof) is not associated with irradiation processes involving high-energy photons such as ultraviolet photons or X-rays.

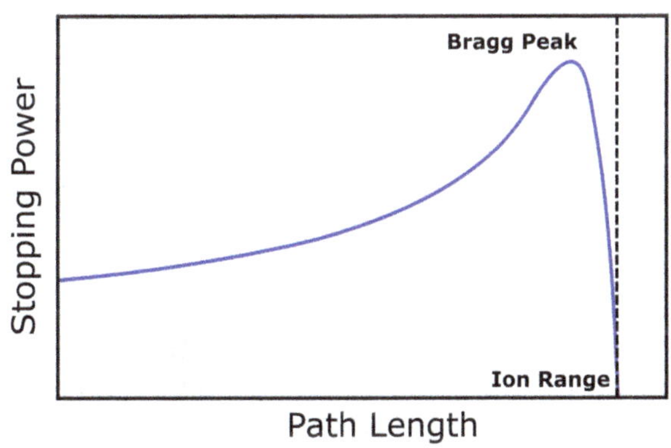

Fig. 12.1. A generic diagram exhibiting the stopping (or stopping power) of an ion traversing through a medium as a function of its path length through the medium. The range of the ion (i.e., the distance it travels before it comes to rest) and the Bragg peak (i.e., the maximum stopping) are highlighted.

In the context of interstellar astrochemistry, it is often the case that galactic cosmic rays have sufficiently high kinetic energies, and interstellar icy grain mantles are sufficiently thin that the former penetrates through the latter completely. In such a scenario, it is said that the range of the projectile ion exceeds the thickness of the target medium. The ion irradiation of the target molecular medium will result in the destruction of the constituent molecules of the medium due to the transfer of energy from the former to the latter. The extent to which molecules in the target medium undergo radiolytic destruction will depend on a number of parameters (e.g., the ion's stopping, its range within the medium, the mean excitation energy of the medium, the ionisation energy of the individual molecules, and bond dissociation energies), and as such it is desirable to quantify this molecular destruction.

There are two commonly cited terms by which radiolytic destruction of a molecular target is quantified: (1) G-values (symbolised as G and expressed in units of molecules per 100 eV) and (2) destruction cross-sections (symbolised as σ_d and typically expressed in units of cm^2).[119–126] The G-value of an irradiation process is defined as the number of target molecules destroyed for each 100 eV of energy absorbed by the medium, while the destruction cross-section is a measure of the probability that a molecule in the target medium will be destroyed as a result of the interaction between that molecule and the projectile ion. It should be noted that, when destruction cross-sections are defined, they are assumed to be constant for a given irradiation process. However, such an assumption is correct only if the overall energy loss of the projectile ion as it traverses through the target medium is negligible compared to its initial kinetic energy, such as in the case of galactic cosmic rays having kinetic energies of at least a few MeV interacting with nanometre- or micrometre-thick interstellar icy grain mantles.

Of course, it is also possible that the thickness of an ice in an astrophysical setting exceeds the range of the projectile ion. This is frequently the case on the surfaces of outer moons of the solar system or Kuiper Belt objects, where ices may be several metres or kilometres thick. In such instances, it is not appropriate to define a classical destruction cross-section for a molecular component of the ice undergoing radiolytic destruction since the assumption that the kinetic energy of the projectile ion remains constant does not hold. Instead, as the ion slows down and eventually comes to rest within the ice, it expends most of its energy just before it comes to rest at the Bragg peak, thus resulting in a non-constant destruction cross-section. Alternative formulae that consider an amended destruction cross-section that is integrated over the energy range of the projectile ion have been derived for such scenarios.[127–130]

12.3.2 *Ion irradiation of ices: Physical and chemical effects*

In this section, a number of laboratory studies are discussed briefly to exemplify the physical and chemical effects of ion irradiation on astrophysical ice analogues. In particular, the following three topics are discussed: (1) compaction, amorphisation, and sputtering of ices, (2) the ion-impact-induced fragmentation of selected molecular ices (i.e., H_2O, CO_2, and CH_3OH), and (3) the implantation of reactive ions into astrophysical ice analogues.

12.3.2.1 *Ice compaction, amorphisation, and sputtering*

The irradiation of ices with keV or MeV ions mimicking galactic cosmic rays or stellar winds is not only capable of inducing chemical changes through the destruction and formation of molecular bonds but is also able to induce physical changes to the structure of the ice. One of the most widely studied processes is that of compaction, i.e., the removal of structural pores from a solid ice. Ion-induced ice compaction has been extensively analysed with H_2O ice due to the fact that this process can be readily studied *in situ* using standard mid-infrared spectroscopic techniques.

This is because the mid-infrared spectrum of low-temperature amorphous H_2O ice presents two small absorption features at around 3700 cm^{-1} that are attributable to the stretching modes of the so-called dangling O–H bonds within the micropores of the ice structure.[131] These absorption features are, of course, only visible if the ice is indeed a porous structure. Several experimental studies have repeatedly demonstrated that the exposure of a porous amorphous H_2O ice to an incident ion beam results in these absorption features diminishing greatly or disappearing entirely, which has been interpreted to be a result of the compaction of the ice during ion beam irradiation.[132–136]

Indeed, the work of Palumbo demonstrated this for amorphous H_2O ice exposed to 200 keV H^+ ions at 15 K.[132] An interesting result of that study was that the rate of compaction (as measured through the change in the intensity of the dangling O–H bonds) was more rapid in the case of pure H_2O ices than in H_2O ices mixed with CO, CO_2, or CH_4.[132] The results of this study were later corroborated by the work of Raut *et al.*, who made use of gas adsorption studies (in addition to the more traditionally used mid-infrared absorption spectroscopy) to demonstrate that the gas adsorption ability of H_2O ice is reduced as a result of its irradiation by ions.[133] Indeed, that study concluded that ion irradiation reduces the accessibility of pores within the H_2O ice, thus decreasing the ability of foreign gas molecules to adsorb within the porous ice.

If crystalline ices are subjected to ion irradiation, then yet another potential change to their physical structure that may be induced is amorphisation. This has been studied for a number of ices, including H_2O, CH_3OH, and CO_2,[137–140] as well as more refractory substances such as forsterite and diamond.[141,142] The amorphisation of crystalline structures as a result of their interaction with ionising radiation has interesting implications for the state of physical matter in extraterrestrial icy material in the ISM and outer solar system since it is conceivable that various equilibria may be set up between radiation-induced amorphisation and thermally induced crystallisation in environments characterised by different radiation doses and thermal gradients.

Finally, it is worth briefly discussing the sputtering of icy material that may be induced by ion irradiation. This process, which contributes to the depletion of solid material, may contribute to the establishment of transient or tenuous exospheres on outer solar system bodies. For instance, several components of the exosphere of Jupiter's moon Europa are believed to originate from the sputtering of surface ices and other materials caused by magnetospheric ion impacts. A notable example is the observation of atomic sodium in Europa's exosphere, which has been attributed to the ion sputtering of sodium-enriched surface components, such as sodium encapsulated within H_2O ice or within sulphate and chloride minerals.[143–148]

Laboratory studies and models have sought to understand trends in the ion-induced sputtering of ices by making use of various targets, projectile ions, and by varying experimental parameters, such as temperature and phase of the target ice. Although a detailed review of such studies would go well beyond the scope of this discussion, an example is provided to illustrate recent experimental advances. A recent study by Ivlev *et al.* considered the irradiation of pure CO ices using He^+ ions of different energies between 0.3 and 2 MeV in order to study, among other things, the energy dependence of the sputtering yield.[149] That study found that the commonly predicted dependence of sputtering yield on the square of the electronic stopping power of the projectile ion tends to break down at lower ion energies (i.e., those energies below the peak of the stopping power curve) and is actually significantly underestimated by this dependence relation.[149] Moreover, the charge state of the projectile He ion (i.e., He^+ vs. He^{2+}) was found to have a negligible effect on the overall CO sputtering yield. The study concluded by proposing that electronic excitation of CO molecules was likely a more dominant route towards ice sputtering (as well as CO radiolysis) compared to ionisation.[149]

12.3.2.2 *Fragmentation of H_2O, CO_2, and CH_3OH*

Interstellar icy grain mantles are overwhelmingly dominated by the presence of H_2O,[150] although other chemically simple molecules (such as CO_2 and CH_3OH) are also relatively abundant.[96,151] A summary of the major reactions occurring during the irradiation of H_2O, CO_2, and CH_3OH ices is provided in Fig. 12.2. Accordingly, there have been a number of studies that have sought to characterise the ion-induced radiation chemistry of these molecular solids under laboratory conditions that mimic the dense ISM. Such studies have demonstrated, for example, that the irradiation of solid H_2O ice results in a number of products, most prominently H_2O_2, H_2, and O_2.[152–158]

The most interesting of these products is arguably H_2O_2 due to its potent oxidising ability. If produced on the H_2O-rich surfaces of outer solar system moons, which are hypothesised as potential habitats for life (e.g., Europa, Ganymede, and Enceladus), it could play a critical role in facilitating putative prebiotic chemistry by inducing redox

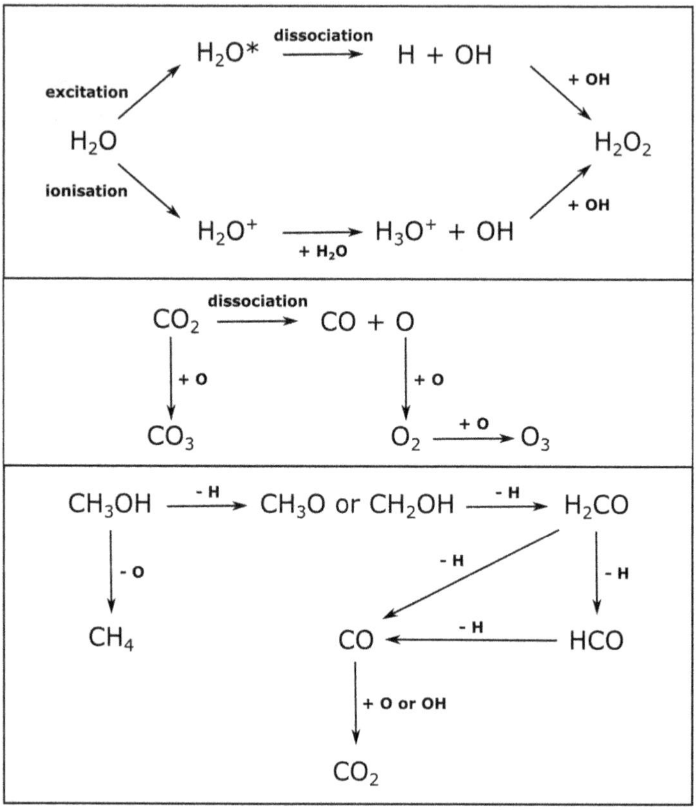

Fig. 12.2. Summary of the major reactions occurring during the irradiation of H_2O (top panel), CO_2 (middle panel), and CH_3OH (bottom panel) ices.

disequilibrium. Studies performed by Gomis *et al.*[152] on the irradiation of H_2O ice at 16 and 77 K using 200 keV H^+, 200 keV He^+, and 400 keV Ar^{2+} ions were able to confirm the formation of H_2O_2 after each irradiation and provided cross-sections for each irradiation process. Interestingly, irradiations performed at lower temperatures yielded more H_2O_2, and there was also a positive correlation between the yield of this product and the size of the projectile ion. The temperature trends revealed by these studies were also observed in the works of Loeffler *et al.*[153] and Hudson and Moore,[154] who considered the irradiation of H_2O ice using 50 and 100 keV Ar^+ ions and 100 and 800 keV H^+ ions. Later studies by Mifsud *et al.* have also suggested that the phase of the irradiated H_2O sample may play a role in determining the amount of H_2O_2 produced, with amorphous phases yielding more product compared to more ordered or crystalline phases.[158]

Loeffler *et al.* provided a detailed overview of the reaction network that leads to the formation of H_2O_2 during the irradiation of H_2O ice.[153] Irradiation may result in the ionisation of H_2O to H_2O^+ or its electronic excitation to H_2O^*. The ionised H_2O^+ species is thought to react with a neutral H_2O molecule to yield the H_3O^+ ion and the OH radical, while the electronically excited H_2O^* species may subsequently dissociate to the H and OH radicals. The formation of H_2O_2 may then proceed as a result of the combination of two OH radicals.[153] In structurally disordered H_2O ices, this process may be assisted by the existence of L-type Bjerrum defects, wherein two oxygen atoms from two separate H_2O molecule are in proximity with one another due to the rotation of one of the H_2O molecules about its oxygen atom in the solid lattice.[158,159]

The irradiation of solid CO_2 at cryogenic temperatures has also been considered in a number of studies.[160–162] Previous studies have shown that this irradiation results in the production of a number of molecules, among them being CO, CO_3, and O_3. Indeed, the work of Mejía *et al.* demonstrated the formation of these molecules and quantified the relevant cross-sections during the processing of pure CO_2 by a series of swift heavy ions having energies of a few tens or a few hundreds of MeV.[160] The mechanism by which CO, CO_3, and O_3 form during the irradiation of CO_2 ice was elucidated by Bennett *et al.*, who determined that this reaction network is initiated by the radiation-induced dissociation of CO_2 to CO and an oxygen atom.[161] The oxygen atoms produced by this dissociation process are suprathermal, that is, they are not in thermal equilibrium with their surrounding environment and typically have energies of a few eV.[161,163–165] This energy may be used to overcome activation energy barriers and therefore allows for the oxygenation of other molecules or molecular fragments within the ice. The addition of such a suprathermal oxygen atom to CO_2, for example, yields CO_3. The combination of two oxygen atoms leads to the formation of O_2, which, upon the addition of another oxygen atom, yields O_3.

Finally, the ion irradiation of CH_3OH ice is briefly discussed. Early studies on this topic, such as those of Moore *et al.* and Palumbo *et al.*,[166,167] observed the formation of a large number of products as a result of the irradiation of CH_3OH ice using 3–30 keV He^+ ions and 1 MeV protons, respectively. These molecules included simple inorganic and organic species, such as H_2O, CO, CO_2, CH_4, H_2CO, C_2H_5OH, and CH_3COCH_3. Later studies on the irradiation of CH_3OH ice using various radiation sources confirmed the formation of these products during radiolytic processing, along with a number of radical species such as HCO, CH_3O, and CH_2OH, as well as various complex organic molecules, including CH_3CHO, CH_3OCHO, and $CHOCH_2OH$.[168–176]

A detailed description of the mechanistic reaction pathway leading to the formation of all of these molecules would go beyond the scope of the present brief discussion, which will therefore be limited to the most frequently reported radiolytic products: CO, CO_2, and CH_4. Mechanistic pathways leading to the formation of these products were investigated and quantified by Bennett *et al.*, whose study identified several unimolecular radiolytic dissociation routes for the CH_3OH molecule.[168] These include dissociation into: (1) CH_3O and H, (2) CH_2OH and H, (3) CH_4 and O, and (4) H_2CO and H_2. Further dehydrogenation of the H_2CO molecule could then proceed either via the sequential removal of a hydrogen atom or, alternatively, via the one-step elimination of H_2, thereby yielding CO. Finally, the formation of CO_2 could either occur as a result of the addition of a suprathermal oxygen atom to CO or via a non-thermal reaction between CO molecules and any OH radicals in the ice matrix.[168,177]

A recent study on the ion irradiation of CH_3OH using several H^+ ion beams having kinetic energies in the range of 200–1000 keV has demonstrated that the rate of radiolytic-induced decay of the target CH_3OH ice does not significantly vary among the investigated ion beam energies, nor does the evolution of the abundance of CH_4 within the irradiated ice.[173] On moving to a 6 MeV S^{2+} projectile ion, it was noted that the target CH_3OH ice was radiolytically destroyed at a faster rate and, indeed, the abundance of the CH_4 radiolytic product peaked at an earlier dose compared to the H^+ ion irradiation experiments. Several potential explanations were proposed for the faster CH_3OH radiolysis rate observed with 6 MeV S^{2+} ions compared to the 200–1000 keV H^+ ions. These include differences in the relative contributions of nuclear and electronic stopping, as well as the potential for enhanced ice sputtering induced by the larger projectile ions.[173] It should also be noted that, as was the case with H_2O, there is also evidence for the phase of CH_3OH playing a role in the extent of its radiolytic dissociation, with crystalline CH_3OH being more radio-resistant compared to the amorphous structure.[174]

12.3.2.3 *Reactive ion implantation*

If the range of a projectile ion does not exceed the thickness of the ice through which it is travelling, then the ion will come to rest within the ice. This ion implantation process may allow for the formation of new product molecules that incorporate the implanted ion and are of particular relevance to outer solar system bodies, such as the icy Galilean moons of Jupiter. Laboratory astrochemistry studies have primarily focused on the implantation of four key ions: deuterium (D^+), carbon (C^+), nitrogen (N^+), and sulphur (S^+). Each of these ions has been extensively employed to simulate processes relevant to interstellar and planetary environments. The use of each of these in astrochemical implantation studies is briefly reviewed here.

Deuterium ions have been implanted into pure O_2 and CO_2 ices.[178–180] In the case of the former, a number of products incorporating the implanted deuterium ion were identified, including D_2, DO_2, D_2O, and D_2O_2. These products highlight the potential existence of a hydrogenation reaction mechanism towards the formation of interstellar H_2O, starting from O_2 ice adsorbed on dust grain surfaces. Such a reaction mechanism would be complementary to experimental results that have shown H_2O formation as a result of the addition of thermalised hydrogen atoms to O_2 ice.[181–183] The implantation of deuterium ions into CO_2 ice also results in the incorporation of the projectile ions into newly formed molecules. Detected species resulting from this process include D_2O, D_2CO_3, D_2CO, CD_3OD, and DCOOD, showcasing the chemical complexity that can arise under such conditions.[179]

When it comes to the implantation of carbon ions in astrophysical ice analogues, the vast majority of experimental studies have focused on H_2O ice as a target ice.[184–188] These studies have repeatedly shown that a major product of this implantation process is CO_2, with formation efficiencies as high as 0.57 molecules per implanted ion, although this value is dependent on the energy of the projectile ion and the temperature at which the implantation process takes place. CO has also been detected as a major product of this implantation process.[188] An interesting study considered the chemistry induced by the implantation of 30 keV $^{13}C^{2+}$ ions into H_2O ice mixed with both N_2 and O_2.[189] The results revealed the formation of $^{13}CO_2$ and ^{13}CO, alongside molecules produced from the irradiative processing of mixed N_2 and O_2 ices (e.g., O_3 and nitrogen oxides). However, no radiation products containing both carbon and nitrogen atoms were detected.

The implantation of nitrogen ions into astrophysical ice analogues is arguably the least well investigated of the ion implantation processes discussed herein. Previous work by Strazzulla *et al.* demonstrated that the implantation of 15–30 keV N^+ ions into H_2O ice at 16 and 77 K did not produce any nitrogen-bearing product molecules that were discernible in acquired infrared spectra of the irradiated ice.[184] However, if the target ice was changed to a mixed ice of H_2O:CH_4, then infrared absorption features attributable to =CN functional groups are readily observable[190] and are indeed very similar to absorption features attributed to nitriles observed after the irradiation of, for instance, H_2O:NH_3:CH_4 mixed ices.[191]

Lastly, the implantation of sulphur ions into astrophysical ice analogues has been considered largely to address the possible formation of sulphur-bearing molecules on the surfaces of the icy Galilean moons of Jupiter, which orbit within the sulphur-enriched giant Jovian magnetosphere.[192] Previous studies have found that the implantation of sulphur ions into H_2O ices results in the formation of hydrated H_2SO_4, with negligible formation of any SO_2.[193,194] Other target ices have also been considered, with implantations into pure O_2 ice having been shown to yield large quantities of SO_2,[195] while the implantation of sulphur ions into CO ice has also been shown to yield SO_2 albeit at lower formation efficiencies.[196]

An interesting result was recently found in the case of the implantation of sulphur ions into CO_2 ice, which could have important implications for the radiation-induced chemistry occurring on the surfaces of the Galilean moons.[197] The implantation of sulphur ions into CO_2 ice has been shown to result in the formation of SO_2 if performed at very low temperatures (10–20 K)[195,196,198] but not at higher temperatures that are more representative of the surfaces of Galilean moons (70–80 K).[195,198] As such, the implantation of sulphur ions into CO_2 may be an important contributor to sulphur cycling in low-temperature environments within the solar system, such as in Kuiper Belt, but not so in higher-temperature regions, such as on the surfaces of the Galilean moons.

12.4 Electron-Induced Processes in Astrochemistry

In any ionising-induced process, secondary electrons are produced in copious amounts. A single high-energy cosmic ray interacting with an ice-covered surface may liberate hundreds, or even thousands, of secondary electrons (Fig. 12.3). This 'avalanche' of electrons, in turn, produces both physical and chemical alterations in the irradiated target. Accordingly, the final products of the irradiation event may be more representative of processes induced by the secondary electrons than those of the primary ionising particles. Understanding electron-driven processes and electron-induced chemistry is,

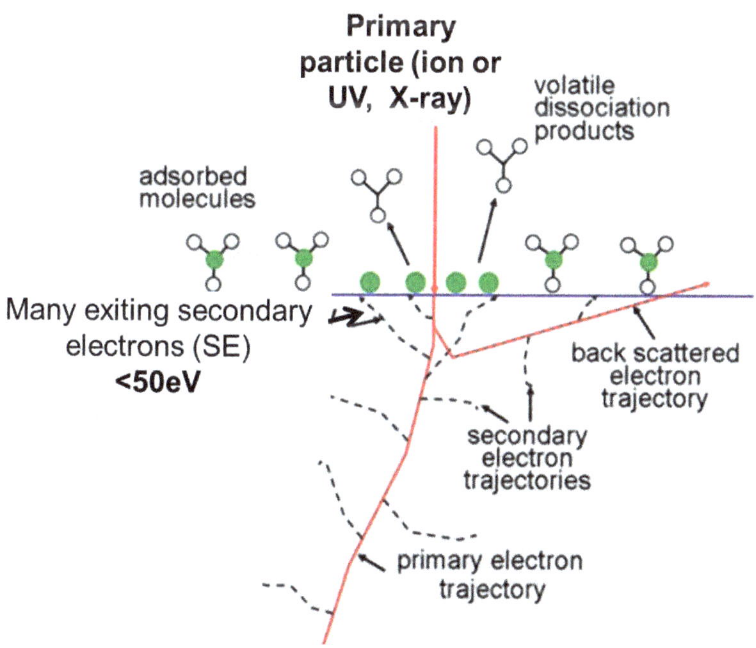

Fig. 12.3. Secondary electrons are generated abundantly during irradiation processes, often serving as the primary agents driving the physical and chemical transformations observed.

therefore, vital for developing a comprehensive knowledge of both astrophysics and astrochemistry.[107,109,199] Studies employing multiscale approaches, such as those detailed in a recent review by Solov'yov *et al.* (2024),[200] have been instrumental in elucidating these processes, offering valuable frameworks for understanding complex systems exposed to irradiation.

12.4.1 *Electron reactions in the early universe*

Electron collisions were among the first collision processes to occur after the Big Bang and played a crucial role in the development of the Universe. The first elements were formed during a process known as 'Big Bang nucleosynthesis', with protons and neutrons colliding to produce hydrogen and helium. Initially, the hydrogen and helium in the early Universe were ionised; however, as the Universe cooled, electrons were captured by these ions through a process known as electron recombination.[201] Accurate electron–ion recombination data are therefore crucial for modelling the early Universe. Extensive evaluations of electron–proton recombination cross-sections have been conducted, with both 'complete' and semi-analytical results reported and reviewed for each final state of the resulting hydrogen atom, H(n), where n denotes the principal quantum number of the atom's energy level. These studies cover an energy range from 0.1 to 10.4 eV.[201] Similarly, comprehensive data sets have been compiled for electron–He^{2+}/He^{+} interactions, including

$$He^{2+} + e^{-} \rightarrow He^{+} + h\nu, \tag{12.4}$$

$$He^{+} + e^{-} \rightarrow He + h\nu. \tag{12.5}$$

A small amount of lithium is also created in the early stages of the Universe via $p + {}^{4}He \rightarrow {}^{5}Li + h\nu$, which requires electron interactions with Li cations as an initial step.

The earliest molecules formed in the Universe were simple combinations of the lightest elements, formed in their cationic forms. HeH^{+} is believed to be the very first molecule to form, predating other primordial species such as D_2^{+}, H_3^{+}, H_2D^{+}, D_2H^{+}, D_3^{+}, HeD^{+}, LiH^{+}, and LiD^{+}. HeH^{+} was only recently observed in interstellar space, particularly within the planetary nebula NGC 7027.[202] At the equilibrium geometry, HeH^{+} exhibits a covalent He^{+}–H bond, which arises from electron charge transfer from helium to H^+.[203] These primordial ions are believed to have been formed through 'barrierless' ion–molecule collisions, with reaction rates often increasing at lower temperatures and occasionally showing quantum tunnelling phenomena. Notably, strong isotope effects are observed in these early molecular reactions.[204] Once again, electron recombination reactions were dominant, such as

$$H_3^+ + e^- \rightarrow H + H + H, \tag{12.6}$$

$$H_3^+ + e^- \rightarrow H + H_2, \tag{12.7}$$

$$HeH^+ + e^- \rightarrow He + H. \tag{12.8}$$

These dissociative recombination reactions have been studied extensively using ion storage rings and are strongly dependent on the internal (ro-vibrational) energy of the target ion.

Electron collisions also led to the appearance of the first anions in the Universe. Electron attachment to atomic hydrogen, $H + e^- \rightarrow H^- + h\nu$, provides a key pathway for the production of molecular hydrogen via $H^- + H \rightarrow H_2 + e^-$, a process known as associative detachment. This mechanism is considered the most probable route for the initial formation of molecular hydrogen in the early Universe, preceding the emergence of the first stars and the subsequent production of dust, which is now believed to be the dominant surface catalyst for molecular hydrogen formation.

12.4.2 *Electron collisions in molecular clouds*

Molecular clouds are regions in space that consist of both gases and dust, where stars and planets are formed. Due to the dust, they may appear as 'empty' regions in space, obscuring stars behind them. Observations have shown that these clouds harbour an extensive variety of molecules, with over 320 species identified to date (see Section 12.2), and they serve as the birthplace for numerous molecular species. Photon- and ion-induced processes generate substantial fluxes of electrons within these clouds, which play a crucial role in driving chemical synthesis. The initial average energy of these secondary electrons is approximately 20 eV, though energies can reach up to 1 keV.

12.4.2.1 *Collisions in the gas phase*

Once again, since the ISM is a weakly ionised plasma, electron recombination processes are important and may lead to reactive atomic (H) and molecular species in the gas phase. However, electron collisions also play a vital role in 'cooling' the ISM through inelastic collisions. For example when electrons ionise H_2, they lose 15.4 eV, and when they excite electronic states of H_2, the minimum energy lost is about 10 eV. Vibrationally inelastic collisions, e.g., $e^- + H_2\ (\nu = 0) \rightarrow e^- + H_2\ (\nu = 1)$, occur with a loss of 0.555 eV, and rotationally inelastic collisions, e.g., $e^- + H_2\ (J = 0) \rightarrow e^- + H_2\ (J = 2)$, occur with a loss of 60 meV. Successive electron collisions lead to a process known as 'electron cooling', which reduces the mean 'electron temperature'. For electron energies below 60 meV, further cooling may occur through rotationally inelastic collisions with polar molecules, such as CO and H_2O, the two most abundant gas-phase molecules in the ISM after molecular hydrogen.

12.4.2.2 *The condensed phase*

The ISM and molecular clouds contain ample amounts of 'dust' created as part of the stellar cycle and the death of stars. These grains accrete to form the primordial dust, which is the building block of planets. These dust grains are composed of iron and magnesium silicates plus some graphitic components.

The cold surface of the dust grains allows atomic hydrogen to recombine to form the bulk of the molecular hydrogen in the Universe. Indeed, there is no gas-phase mechanism that can synthesise the observed molecular hydrogen. However, these dust grains also accrete other molecules made in the gas phase of the ISM to form an icy mantle.

It is now recognised that there is a rich electron-driven chemistry within these ice mantles and that understanding electron interactions with dust grains is essential if we are to understand both physical and chemical processes in molecular clouds and how these lead to star and planet formation. A 'free' electron interacting with the grain may lead to electron capture, electron-induced chemistry, or electron-stimulated desorption (ESD).

Electron capture. There is no gradient in electrochemical potential between the free electrons in the gas phase and those on the grains; as a result, the grain becomes charged. Positive ions in the gas phase are subsequently attracted to the grain surfaces, following the electrons to maintain electroneutrality within the molecular cloud plasma. These positive ions then recombine with the mobile surface electrons on the grain surface.[205]

However, electric fields can also form spontaneously on the grain surface due to the phenomenon known as the spontelectric effect.[206] This effect can generate electric fields as strong as 3.7×10^7 V m^{-1} through the spontaneous orientation of molecular dipoles. As the dipolar film grows, the grain surface becomes less negatively charged, with the spontelectric effect effectively offsetting its electrochemical potential from that of the surrounding plasma. Consequently, the electrochemical potential of the grain surface no longer equals that of the gas phase.

This imbalance drives a net flow of electrons towards the grain, leading to the deposition of more positive ions on the dust grains. On the surface, electrons and ions recombine, establishing a new equilibrium between cosmic ray ionisation and recombination, driven by the requirement to equalise the electrochemical potential of electrons in the gas phase and on the grain surface. Interestingly, this grain surface recombination mechanism is 40 times more efficient than gas-phase recombination.

Electron-induced surface chemistry. Electron collisions play a key role in molecular synthesis on dust grains, as they serve as a pathway for the dissociation of adsorbed molecules. Various electron collision mechanisms, including electron impact excitation, ionisation, and dissociative attachment, can induce molecular dissociation, thereby facilitating chemical transformations on grain surfaces. The most important processes are briefly described as follows.

Electron-induced neutral dissociation (ND) involves interactions such as $e^- + \mathrm{AB} \rightarrow e^- + \mathrm{A} + \mathrm{B}$, where the fragments are neutral atoms, or $e^- + \mathrm{ABC} \rightarrow e^- + \mathrm{AB} + \mathrm{C}$ (or A + BC), where the fragments can be molecules. These dissociation products may exist in electronically excited states (A*, B*, AB*, and BC*) or in rovibrational states

(AB(J, ν), BC(J, ν)), and they may also carry considerable kinetic energy. The process requires the parent molecule to be excited to a dissociative electronic state (AB* or ABC*), resulting in an onset energy for ND, which is typically > 4 eV.

Electron impact dissociative ionisation (DI), in turn, is related to processes such as $e^- + \mathrm{AB} \rightarrow e^- + \mathrm{A}^+ + \mathrm{B}$ or $e^- + \mathrm{ABC} \rightarrow e^- + \mathrm{AB}^+ + \mathrm{C}$ (or $\mathrm{A}^+ + \mathrm{BC}$). In this process, one of the fragments is ionised, while the others may remain neutral. Once again, the fragments can exist in electronically or rovibrationally excited states, and they may also be ejected with significant kinetic energy. Since this process requires both electronic excitation and ionisation, the energy threshold for this process lies above the ionisation energy of the neutral parent and is generally greater than 20 eV.

Dissociative electron attachment (DEA) is associated with the following mechanisms: $e^- + \mathrm{AB} \rightarrow \mathrm{A}^- + \mathrm{B}$ or $e^- + \mathrm{ABC} \rightarrow \mathrm{AB}^- + \mathrm{C}$. DEA results in the production of anionic species, but it is highly selective, occurring only over a narrow energy range.[207] This selective behaviour arises because DEA is a resonant process, where the incoming electron temporarily forms a transient negative ion (TNI): $e^- + \mathrm{AB} \rightarrow \mathrm{AB}^-$ or $e^- + \mathrm{ABC} \rightarrow \mathrm{ABC}^-$. The TNI is inherently unstable and may decay through either (1) autodetachment, e.g., $\mathrm{ABC}^- \rightarrow \mathrm{ABC}(J,\nu) + e^-$ or $\mathrm{AB}(J,\nu) + \mathrm{C} + e^-$; or (2) dissociation, e.g., $\mathrm{AB} \rightarrow \mathrm{A}^- + \mathrm{B}$ or $\mathrm{ABC}^- \rightarrow \mathrm{AB}^- + \mathrm{C}$. DEA processes can occur at very low electron energies, including energies as low as 0 eV. This is because the potential energy curve of the transient negative ion lies below that of the corresponding neutral state (see Fig. 12.4), allowing the dissociation channel to remain open even at zero energy.

DEA is also bond-selective, allowing for the rupture of specific molecular bonds in the target molecule by tuning the energy of the incoming electron. This selectivity enables the controlled production of specific dissociation products, facilitating reaction pathways that are localised to the fragmentation site. Such site-specific chemistry may play a crucial role in molecular synthesis on dust grain surfaces, where electron energies are typically low. For certain 'electrophilic' molecular species, such as those containing halogens, low-energy DEA processes exhibit exceptionally large cross-sections, ranging from hundreds to thousands of Å^2. These cross-sections significantly exceed those for ND and DI, highlighting the efficiency of DEA under these conditions.

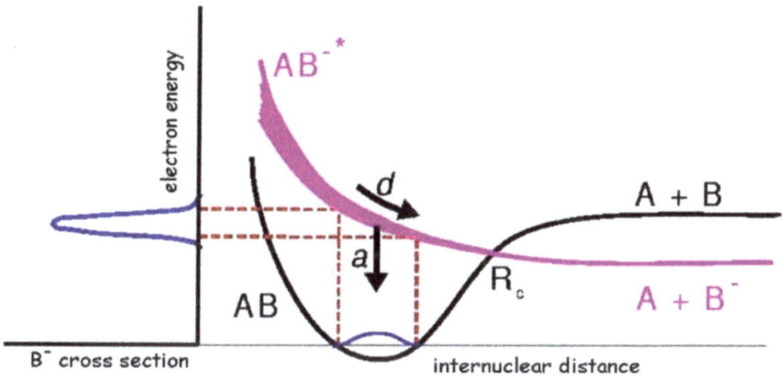

Fig. 12.4. Illustrating the DEA process through potential energy curves of the neutral and temporary negative anion.

Dipolar dissociation (DD), also referred to as ion-pair formation (IPF), represents a hybrid process between DEA and DI.[208] This process involves the production of an ion pair, such as $e^- + \text{AB} \rightarrow \text{A}^- + \text{B}^+$ or $e^- + \text{ABC} \rightarrow e^- + \text{A}^- + \text{BC}^+$. While the cross-section for DD is relatively low (typically $< 10^{-20}$ cm^2), it becomes a crucial channel for anion formation in molecular targets where the DEA pathway is weak or entirely prohibited, as observed in most hydrocarbons.

Ion-pair (IP) states are super-excited states of the neutral molecule, situated near the ionisation continuum. As a result, IPF exhibits a higher threshold energy compared to ND, DI, and DEA, making it a distinct and energetically demanding process for ion-pair generation.

Figure 12.5 provides a schematic representation of the energetic regions associated with each of these electron-induced processes.[108] It is important to note that molecules can undergo dissociation across a broad energy range via electron collisions, with the peak of the secondary electron energy distribution typically lying below the ionisation potential of most molecules.

The dissociation fragments generated by electron collisions are then able to initiate complex surface chemistry. Indeed, electron irradiation of a simple molecule like methanol has been shown to yield 19 stable products.[208] A one-to-one correspondence has been observed between the yield of synthesised products and the incident electron energy, with resonant profiles compatible with the DEA yield of anions from nascent surface species. This demonstrates that DEA fragments directly participate in the chemical synthesis. At higher electron energies, the yield of the final product tends to follow the DI cross-section.[209,210]

The open question in electron-induced chemistry of astronomical ices lies in determining the extent of chemical complexity that can be achieved. Provided that the necessary reactants are generated, complex synthesis becomes feasible. Accordingly, in multi-component ices composed of the most common ISM or planetary/lunar surface species (e.g., CO, H_2O, CO_2, NH_3, CH_4, and CH_3OH), electron irradiation has been shown to yield complex hydrocarbons, such as acetylenediol (HOCCOH) and glyoxal (HCOCHO),[211] and vinylamine ($C_2H_3NH_2$).[212] These molecules are, by themselves, potential precursors for even more complex molecules, including prebiotic amino acids. Recent evidence highlighting the importance of electron-induced tertiary reactions suggests that such surface-mediated synthesis, initiated by high-energy primary interactions, is non-negligible.[213] These experiments have led to electrons being

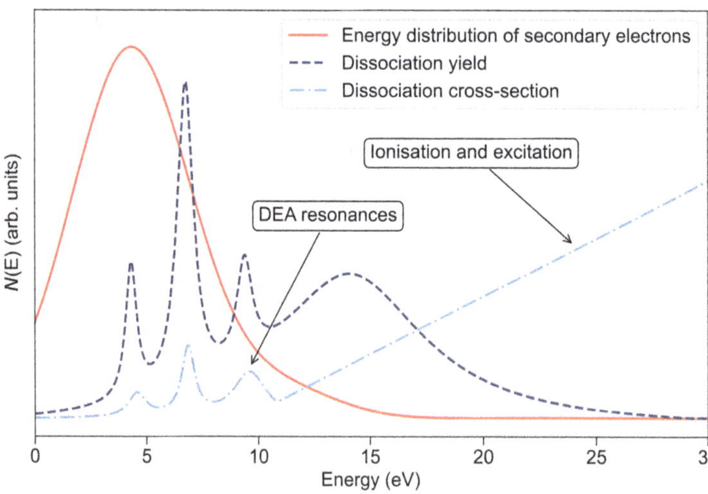

Fig. 12.5. Schematic diagram showing the expected secondary electron energy distribution from cosmic ray impact on a dust grain compared with typical product yields induced by dissociative ionisation, dissociative excitation, and the resonant-driven process of dissociative electron attachment (DEA).
Source: Adapted with permission from Boyer *et al.*[108] Copyright 2016 Elsevier Science & Technology Journals.

described as catalytic agents in chemical reactions, with catalytic activity exemplified in the production of CO_2 through electron interactions with formic acid.[214]

However, under continued electron irradiation, such complex species may also be fragmented and reduced back to their basic building blocks. Secondary or tertiary electrons may, therefore, also determine a 'lifetime' for such species in an astrochemical environment, and the generation of such species from primary irradiation (UV, X-rays, and cosmic rays) may have defined those lifetimes reported in reported experiments.[215]

Electron-stimulated desorption. Despite the advancements brought by the JWST and ALMA, the majority of the more than 320 molecular species detected in the ISM are identified in the gas phase, even though many of these molecules are likely synthesised in the condensed phase. This raises the question: what are the routes by which molecules formed on the surface of an ice-covered dust grain are transported into the gas phase? Ice grains are subject to cycles through regions with varying temperatures and radiation fluxes, facilitating the release of surface-adsorbed molecules via several pathways. These include thermal desorption driven by heating, release due to shock waves, sputtering by ions, photodesorption caused by UV irradiation, or ESD.

ESD can produce both neutral and ionised species, including anions and cations, with neutrals often generated in their ground, excited, or even metastable states. Typically, the ESD cross-sections for ion production are lower than those for gas-phase production, and their effectiveness varies depending on the adsorbate. The ratio of ESD cross-sections for ions to those for neutrals ranges from 10^{-1} to 10^{-5} or even lower, highlighting the predominance of neutral species in ESD processes.

Desorption induced by fast charged particles is often rationalised by assuming that the desorption yield is proportional to a specific power of the stopping power for the particle–target combination under consideration. Many experimental studies have sought to explore the relationship between desorption yields and stopping power for molecular ices, though the majority have focused on ions as projectiles, with fewer investigations involving electrons, often at keV energies. Recent work has shown that ESD at lower energies exhibits greater complexity, challenging the simplistic stopping power framework.

Since ESD is a consequence of electronic excitation of the surface–adsorbate bonding, it provides fundamental information about the bonding geometry and surface dynamics of adsorbed species. The concept of a *desorption-relevant depth* has been introduced to characterise how energy or particle transport from the bulk to the surface influences the desorption process. This concept highlights how layers beneath the surface can contribute to desorption and may account for differences in ESD behaviours in amorphous and crystalline phases.[216]

Recent advancements in ESD characterisation have focused on two metrics: (i) yield per incident particle, which assesses ESD relative to the fluence of the incoming beam, and (ii) yield per deposited energy, which enables comparisons across different projectiles. For instance, a 10.2 eV photon and a 10.2 eV electron may exhibit different desorption yields, with the electron typically producing a higher yield than the corresponding photon.

Although ESD has been a subject of study for over 50 years, its application as an astrophysical method has only been explored in depth over the past two decades.[217–219] Its significance remains a topic of debate; however, comparisons between laboratory ESD studies and observed chemical abundances in regions of the ISM and protoplanetary discs have the potential to provide key information about the morphology of dust grain mantles – insights that cannot be easily obtained through spectroscopic methods alone.

12.4.3 *Electron processes in planetary atmospheres*

12.4.3.1 *The aurorae*

The 'capture' of solar wind protons and electrons by the planetary magnetosphere may lead to their penetration deeper into the planetary atmosphere, where they produce an avalanche of secondary electrons that collide with the molecules comprising that atmosphere. The most dramatic example of such electron interactions with terrestrial molecules is the bright-coloured curtains often seen in northern and southern polar regions, where the electron density is highest. These are the aurorae.

In the terrestrial atmosphere, the aurorae are characterised by different colours. Electron excitation of atomic oxygen, i.e., e_{E} + $\mathrm{O}(^3\mathrm{P}) \rightarrow e_{\mathrm{E}-\Delta\mathrm{E}} + \mathrm{O}(^1\mathrm{D})$, followed by the decay $\mathrm{O}(^1\mathrm{D}) \rightarrow \mathrm{O}(^3\mathrm{P})$, produces transitions at 630 and 634 nm, creating the red aurorae. Similarly, $e_{\mathrm{E}} + \mathrm{O}(^3\mathrm{P}) \rightarrow e_{\mathrm{E}-4.17\ \mathrm{eV}} + \mathrm{O}(^1\mathrm{S})$, followed by the decay $\mathrm{O}(^1\mathrm{S}) \rightarrow \mathrm{O}(^3\mathrm{P})$, gives the strong yellow-to-green light at 557.7 nm, which is the most commonly observed aurorae. Dinitrogen molecules produce a crimson or purple light at 427.8 nm, often forming the lower fringes of auroral curtains. This emission arises from the molecular cation N_2^+ in the B $^2\Sigma_u^+$ electronic state transitioning to the first vibrational level of the X $^2\Sigma_g^+$ electronic ground state of N_2^+.

Aurorae are not restricted to the visible spectrum but extend into the X-ray and UV regions. Excitation of dinitrogen molecules into the $\mathrm{a}^1\Pi_g$ electronic state, followed by its decay (the so-called Lyman–Birge–Hopfield band), produces some of the brightest auroral emissions, although these occur in the UV range. Additionally, auroral emissions are observed in the infrared spectrum, driven by electron interactions with nitric oxide (NO).[220]

Aurorae and related dayglow emissions have also been seen on other planets with large magnetospheres, such as Jupiter, and may represent some of the first aeronomic phenomena detectable in exoplanetary atmospheres. Jupiter's aurorae are much more extensive than those on Earth and are primarily caused by emissions from H_3^+ and hydrocarbon cations.[221] Similarly, aurorae have been observed on Saturn, appearing in the UV spectrum due to hydrogen Lyman-α and H_2 Lyman- and Werner-band transitions, as well as in the infrared spectrum due to H_3^+ emissions.

Interestingly, these large gas giants (Jupiter and Saturn) may also host 'internal' aurorae, driven by the ejection of material from their moons. This material, upon ionisation by solar radiation, penetrates the planetary atmosphere and induces localised auroral activity. For example, Io's volcanic activity generates a persistent UV and X-ray aurora on Jupiter, while Enceladus, with its water plumes, injects electrons and ions along magnetic field lines into Saturn's northern pole, giving rise to the observed aurorae.

Several planetary moons may have their own aurorae. For example, the Jovian moon Ganymede, which is embedded in Jupiter's large magnetosphere, has its own aurorae arising from its intrinsic magnetic field. This magnetic field is believed to be induced by a subsurface saltwater ocean beneath its ice-covered surface.[222] Aurorae have also been observed on planets without magnetospheres, such as Mars. On Mars, the solar wind strikes the planet's atmosphere directly, allowing high-energy electrons to penetrate deeply and excite CO_2 molecules, resulting in UV auroral emissions.

Thus, understanding planetary aurorae necessitates extensive knowledge of electron interactions with a wide range of molecular systems, from simple hydrogen to complex hydrocarbon species.[223] Investigations into the fluorescence induced by electron impact and the dissociation of molecular species are critical, yet few experimental setups currently exist to examine these significant astronomical phenomena.[224]

12.4.3.2 *Ionospheric processes*

The study of planetary aurorae is intrinsically linked to the physical and chemical structure of the planetary ionosphere. Detailed models of the Jovian ionospheres have been developed to interpret mission data[225] and are being developed to support the forthcoming ESA JUICE mission.[226] The Saturn Thermosphere-Ionosphere Model (STIM) was developed to support and interpret Cassini data.[227] Such models contain extensive electron scattering cross-sections over the necessary wide energy range.

In the terrestrial atmosphere, the E-region exhibits a high concentration of anions, primarily attributed to the formation of O^- ions via DEA to molecular oxygen: $e^- + O_2 \rightarrow O + O^-$. Until recently, this was the only known example of anion formation in any planetary atmosphere. However, in 2004, the Cassini–Huygens mission reached the Saturnian system, and the Huygens probe explored Titan, Saturn's largest moon. Titan is unique among solar system moons in having a dense atmosphere, composed of 98.4% nitrogen and 1.4% methane, with a surface pressure exceeding that of Earth at 1.5 bars.

The Electron Spectrometer (ELS), one of the sensors making up the Cassini Plasma Spectrometer (CAPS), detected significant quantities of negative ions in Titan's upper atmosphere, with masses reaching up to 10,000 amu/q. Additionally, ELS identified distinct mass peaks at 22 ± 4 and 44 ± 8 amu/q, and evidence for a third peak at 82 ± 14 amu/q (see Fig. 12.6).[228] The formation of these anions is thought to result from DEA to a set of molecules known as

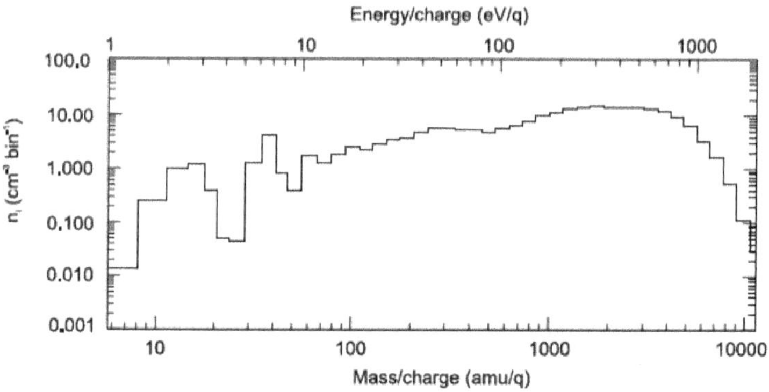

Fig. 12.6. The anion spectrum of Titan's atmosphere recorded by the Cassini–Huygens mission at an altitude of 953 km.[228]

tholins – a class of heteropolymer molecules produced by solar UV irradiation of methane and nitrogen. The simplest building blocks of tholins include HCN and CH_3N, whose anions can transfer their negative charge to more acidic molecules such as HC_3N, HC_5N, or C_4H_2.

The three low-mass peaks observed by the ELS are attributed to CN^-, C_3N^-, and C_5N^-,[229] which likely act as intermediates in the formation of larger negative ions. These larger anions, in turn, serve as precursors to the aerosols observed at Titan's lower altitudes, shedding light on the complex chemistry of its atmosphere.

Detailed chemical and physical models of Titan's ionosphere have been developed to interpret the data gathered by the Cassini–Huygens mission, focusing on the complex interactions of electrons with nitriles and tholin compounds.[230–232] These models underscore the need for additional laboratory data to better understand these electron-driven processes. The discovery of tholins on Pluto and its moon Charon by the New Horizons mission further emphasises the importance of this research.[233]

Indeed, Pluto's atmospheric interactions with the solar wind also warrant more detailed modelling. Neutral atoms and molecules escaping from Pluto's exosphere not only deposit material onto Charon but also extend for millions of kilometres into interplanetary space. These neutrals become ionised either through interactions with solar EUV photons or collisions with solar wind electrons, creating a comet-like interaction between the solar wind and Pluto. Such phenomena highlight the need for comprehensive studies of electron interactions in these unique environments.

12.4.3.3 *Electron collisions with cometary systems*

One of the unexpected yet fascinating discoveries made by the ESA Rosetta mission to Comet 67P/Churyumov–Gerasimenko was the revelation that electrons, rather than photons as previously assumed, are the primary agents responsible for the dissociation of water and carbon dioxide erupting from the comet's surface as it approached the Sun.[234–236] These observations provide the first evidence for discrete electron collisions in an astronomical object. By analysing the emission spectra recorded by the OSIRIS camera and the ALICE spectrometer, it was found that the observed OH and [O I] 630 nm emissions in the inner coma could not be matched to any photon-induced model. Instead, the findings suggest electron dissociative excitation of H_2O and CO_2 as the source. Electron impact on water was found to produce OH at rates 100 times higher than photo-dissociation, while electron impact on CO_2 yielded 40 times more O I than photo-dissociation. These electrons are generated when UV photons from the Sun ionise water molecules in the comet's coma. The resulting photoelectrons then dissociate water and CO_2, producing the excited oxygen and OH emissions detected by ALICE and OSIRIS.

Similarly, the emissions detected in the NH filter of the OSIRIS camera are most likely attributable to electron dissociative excitation, producing OH^+ ions from H_2O molecules in excited states. Data from the CN filter further suggest emissions arising from CO_2^+ ions generated by electron impact dissociative excitation, as well as fluorescent emission from CN radicals, most likely originating from HCN molecules. However, the lack of experimental cross-sections significantly hampers the interpretation of these observations. Absolute cross-sections for the production of excited species from key cometary molecules, such as CO_2, H_2O, and HCN, via electron impact are urgently required. Unfortunately, few experimental systems currently exist to generate such crucial data.[237]

Electron collisions may also account for physical phenomena observed in comets. Notably, as Comet 67P/Churyumov–Gerasimenko approached the Sun, the intensity of emissions in its inner coma was observed to decrease, even as gas production rates increased. The most plausible explanation is that the rapid rise in gas production outpaced the increase in ionising solar radiation. This led to collisions between electrons and neutral water molecules, effectively reducing electron temperatures in the inner coma below the activation threshold required for dissociative impact excitation reactions.[238]

12.5 Concluding Perspectives

The exploration of the molecular universe has undergone a remarkable transformation over the past several decades, driven by advancements in observational astronomy, laboratory experimentation, and theoretical modelling. The identification of over 320 gas-phase molecules in the ISM (as of December 2024) underscores the complexity and richness of cosmic chemistry. Radio telescopes such as the Yebes 40m and IRAM 30m have been instrumental in these discoveries, revealing not only simple molecules but also complex organic compounds, including PAHs and fullerenes. The advent of space-based observatories, notably the JWST, has further expanded our ability to detect and characterise molecular species in both the gas and condensed phases, offering unprecedented insights into the chemistry of ices and dust grains in various astrophysical environments. Gaining a greater understanding of astrochemistry may also unravel two of the greatest scientific unknowns: how did life emerge on Earth? and can life be sustained elsewhere in the Universe?

Laboratory studies have played a pivotal role in interpreting astrochemical observations, allowing the replication and investigation of the extreme conditions of space within controlled settings. Experiments simulating the low temperatures, high radiation fields, and unique chemical compositions of interstellar ices have been crucial in understanding the formation and destruction pathways of molecules detected in the ISM. Techniques such as infrared spectroscopy and mass spectrometry have provided detailed information on the physical and chemical processes occurring in these ices, bridging the gap between observational data and theoretical models. The development of comprehensive databases, like LIDA, has been essential in cataloguing experimental results and facilitating comparative studies.

The interaction of ions and electrons with cosmic ices has emerged as a fundamental aspect of astrochemistry, influencing both the physical state and chemical evolution of interstellar and planetary ices. Ion-induced processes, including radiolytic fragmentation, compaction, amorphisation, and sputtering, have been shown to significantly alter ice compositions and structures. These processes contribute to the synthesis of new molecules and the modification of existing ones, impacting the chemical inventory of regions where stars and planets form. Similarly, electron-induced processes have been recognised for their efficiency in driving chemical reactions at the molecular level. Secondary electrons, generated abundantly during ionising events, initiate a cascade of reactions leading to the formation of complex organic molecules and even prebiotic compounds. These mechanisms are not only critical in the ISM but also in planetary atmospheres and cometary environments, where they can lead to phenomena such as aurorae and influence atmospheric chemistry.

In this chapter, we have highlighted the crucial role of AMP in the study and development of astrochemistry. AMP is the foundation upon which astrochemistry, and thus our understanding of the Universe, is based. It is therefore imperative to ensure sustained support for AMP and the infrastructure required for such foundational research. Conducting this 'foundation science' requires access to large-scale facilities, such as accelerators and synchrotrons. However, the integration of research programmes at these facilities with the astronomy and space research communities often remains suboptimal, with only a limited number of beamlines and end-stations specifically designed to simulate space environments, as discussed in Section 11.2. Moreover, the past two decades have seen a marked decline in fundamental studies of electron and ion impacts on atoms and molecules, largely due to the retirement of key researchers and the diversification of AMP groups into fields such as cold atom physics and other applied sciences.

With the decline in experimental data and the growing demand for atomic and molecular data, there has been an increased reliance on theoretical approaches to address the gaps. The accuracy of theoretical calculations and their capacity to produce extensive data sets have become crucial. For example, several semi-empirical methods for deriving electron scattering cross-sections have been developed to meet the needs of the plasma community.[239–241] While these methods may provide a guide to total cross-sections for certain processes, they often lack the detail required for understanding fragmentation pathways and are typically reliable only at higher impact energies. Recently, machine learning and artificial intelligence methodologies have been explored as potential tools for generating large data sets.[242–244] Although these approaches provide a means to produce significant amounts of data, their accuracy depends heavily on carefully curated training data sets, often derived from experiments. Furthermore, many subtle quantum effects in atomic and molecular collisions, particularly at low energies, may not be captured by these methods. As a result, such data must be treated with caution and thoroughly validated by AMP experts.

The vast amounts of AMP data expected to be required in the coming decades present significant challenges for their assembly, archiving, and accessibility, highlighting ongoing debates about funding and database support. AMP databases have long been essential to astronomical research, with resources like the U.S. National Institute of Standards and Technology (NIST) spectral database widely used to identify atomic and ionic species during the late 20th century. However, NIST remains an exception in providing sustained funding and infrastructure, as most databases were historically compiled and maintained by individuals, leading to infrequent updates and limited accessibility. In the 21st century, initiatives such as the European Open Science Cloud (EOSC) and the adoption of Findable, Accessible, Interoperable, and Reusable (FAIR) principles have introduced opportunities to harmonise AMP data access, with major databases such as VAMDC, LIDA, Europlanet's VESPA, and Belgrade's BEAM and MOLD registering thousands of downloads annually.[245] Despite these advances, significant issues remain, including the lack of consistent monitoring for data usage and citations, as users often select data without adhering to established standards. Developing consistent and reliable datasets for models and simulations requires closer collaboration between users and providers, while funding is urgently needed to support, maintain, and upgrade existing databases. This includes sensitivity analyses of astrochemical models and observational data reductions, which depend on input data with clearly defined uncertainties for both experimental and theoretical contributions.

Looking forward, it is essential to maintain and strengthen the synergy between observational astronomy, laboratory astrochemistry, and theoretical modelling. Investments in next-generation telescopes, advanced laboratory infrastructure, and computational resources are necessary to tackle the outstanding questions in astrochemistry. Furthermore, fostering interdisciplinary collaborations will be key to integrating knowledge across different domains, from the microphysics of molecular interactions to the macroscopic dynamics of star and planet formation. Continuing to build on the foundations laid in recent years will position the field to unravel the mysteries of the molecular universe, shedding light on the origins of chemical complexity and the potential for life beyond Earth.

Acknowledgements

The authors gratefully acknowledge support from the Europlanet 2024 RI, funded by the European Union's Horizon 2020 Research and Innovation Programme (grant agreement no. 871149). Additional support was received from the Research, Development, and Innovation Fund of Hungary (grant no. K128621). This chapter is based on the work from the COST Action CA20129 – Multiscale Irradiation and Chemistry Driven Processes and Related Technologies (MultIChem), supported by the European Cooperation in Science and Technology (COST). S.I. acknowledges the Danish National Research Foundation for funding through the Centre of Excellence 'InterCat' (grant agreement no. DNRF150). F.F. and N.J.M. extend their gratitude for partial financial support from the European Commission via the RADON project (grant agreement no. 872494) under the H2020-MSCA-RISE-2019 call. H.M.Q.L. acknowledges additional support from the University of Kent and the Max Planck Institute for Extraterrestrial Physics (MPE). J.B. is supported by a postdoctoral fellowship from the National Centre for Space Studies (CNES).

References

1. Tennyson J. Astronomical spectroscopy: An introduction to the atomic and molecular physics of astronomical spectroscopy. Singapore: World Scientific. 2019.

2. Heiter U, *et al.* Astron Astrophys. 2021;645:A106.
3. Barone V, *et al.* Nat Rev Methods Primers. 2021;1:38.
4. Parks GK. Physics of space plasmas: An introduction. Boca Raton: CRC Press. 2019.
5. Marinković BP, *et al.* Atoms. 2017;5:46.
6. Tielens A. Rev Mod Phys. 2013;85:1021–1081.
7. Makishima A. Introductory astrochemistry: From inorganic to life-related materials. Amsterdam: Elsevier. 2024. p. 1–399.
8. Williams DA, Cecchi-Pestellini C. Astrochemistry: Chemistry in interstellar and circumstellar space. Cambridge: Royal Society of Chemistry. 2023.
9. Shaw AM. Astrochemistry: The physical chemistry of the universe. Hoboken: John Wiley & Sons. 2021.
10. Cuppen HM, Linnartz H, Ioppolo S. Annu Rev Astron Astrophys. 2024;62:243–286.
11. Albert D, *et al.* Atoms. 2020;8:76.
12. Rocha W, *et al.* Astron Astrophys. 2022;668:A63.
13. Lada CJ. CO twenty-five years millimeter-wave spectroscopy. Dordrecht: Springer Netherlands. 1997. p. 387–396.
14. Herbst E, van Dishoeck EF. Annu Rev Astron Astrophys. 2009;47: 427–480.
15. Dunham JT. Publ Astron Soc Pacific. 1937;49:26.
16. Adams WS. Astrophys J. 1941;93:11.
17. Swings P, Rosenfeld L. Astrophys J. 1937;86:483.
18. McKellar A. Publ Astron Soc Pacific. 1940;52:187.
19. Wlodarczak G. J Mol Struct. 1995;347:131–142.
20. Orchiston W. Phys Today. 2022;75:54–55.
21. Kellermann KI, Bouton EN, Brandt SS. 2020. p. 583–614.
22. Carter M, *et al.* Astron Astrophys. 2012;538:A89.
23. Tercero F, *et al.* Astron Astrophys. 2021;645:A37.
24. Lefloch B, *et al.* Mon Not R Astron Soc. 2018;477:4792–4809.
25. Vastel C, *et al.* Mon Not R Astron Soc. 2018;478:5514–5532.
26. Quitián-Lara HM, *et al.* Mon Not R Astron Soc. 2023;527:10294–10308.
27. McGuire BA. Astrophys J Suppl Ser. 2022;259:30.
28. McGuire BA, *et al.* Science. 2021;371:1265–1269.
29. Cernicharo J, *et al.* Astron Astrophys. 2021;649:L15.
30. McCarthy MC, *et al.* Nat Astron. 2020;5:176–180.
31. Cernicharo J, *et al.* Astron Astrophys. 2024;690:L13.
32. Wenzel G, *et al.* Science. 2024;386:810–813.
33. Cami J, *et al.* Science. 2010;329:1180–1182.
34. Warnick KF, *et al.* Proc IEEE. 2016;104:607–622.

35. Tanarro I, *et al.* Astron Astrophys. 2018;609:A15.
36. Uzawa Y, *et al.* Radio Sci. 2021;56:e2020RS007157.
37. Leisawitz DT, *et al.* In: Breckinridge JB, Stahl HP, Barto AA, editors. UV/Optical/IR sp. telesc. instruments innov. technol. concepts IX. Bellingham: SPIE. 2019. p. 25.
38. Ceccarelli C. Faraday Discuss. 2023;245:11–51.
39. van Dishoeck EF. Faraday Discuss. 2014;168:9–47.
40. Materese CK, Gerakines PA, Hudson RL. Acc Chem Res. 2021;54:280–290.
41. Herbst E. Int Rev Phys Chem. 2017;36:287–331.
42. Sandford SA, *et al.* Chem Rev. 2020;120:4616–4659.
43. Kroto H. Angew Chem Int Ed. 1997;36:1578–1593.
44. Sellgren K, *et al.* Astrophys J. 2010;722:L54–L57.
45. Tielens A. EAS Publ Ser. 2011;46:3–10.
46. Peeters E, *et al.* Acc Chem Res. 2021;54:1921–1933.
47. Kaneda H, Onaka T, Sakon I. Astrophys J. 2005;632:L83–L86.
48. Li A. Nat Astron. 2020;4:339–351.
49. Sita ML, *et al.* Astrophys J Lett. 2022;938:L12.
50. McElwain MW, *et al.* Publ Astron Soc Pacific. 2023;135:058001.
51. McClure MK, *et al.* Nat Astron. 2023;7:431–443.
52. Rocha WRM, *et al.* Astron Astrophys. 2024;683:A124.
53. Nazari P, *et al.* Astron Astrophys. 2024;686:A71.
54. van Dishoeck EF, *et al.* Faraday Discuss. 2023;245:52–79.
55. Burke DJ, Brown WA. Phys Chem Chem Phys. 2010;12:5947.
56. Inostroza N, *et al.* Astron Astrophys. 2019;629:A28.
57. Bergantini A, Zhu C, Kaiser RI. Astrophys J. 2018;862:140.
58. Zamirri L, *et al.* ACS Earth Space Chem. 2019;3:1499–1523.
59. Barone V, Biczysko M, Puzzarini C. Acc Chem Res. 2015;48:1413–1422.
60. Puzzarini C. Front Astron Space Sci. 2022;8:811342.
61. Linnartz H, Ioppolo S, Fedoseev G. Int Rev Phys Chem. 2015;34:205–237.
62. Santoro G, *et al.* Rev Sci Instrum. 2020;91:124101.
63. Mason NJ, *et al.* Front Astron Space Sci. 2021;8:739046.
64. Mifsud DV, *et al.* Geophys Res Lett. 2022;49:e2022GL100698.
65. Kaiser RI. Chem Rev. 2002;102:1309–1358.
66. Abplanalp MJ, Kaiser RI. Phys Chem Chem Phys. 2019;21:16949–16980.
67. Qasim D, *et al.* Nat Astron. 2020;4:781–785.
68. Gutiérrez-Quintanilla A, *et al.* Mon Not R Astron Soc. 2021;506:3734–3750.
69. Barrera NF, *et al.* Mon Not R Astron Soc. 2023;524:3741–3748.

70. Fantuzzi F, *et al.* Mon Not R Astron Soc. 2011;417:2631–2641.
71. Jones AP, Nuth JA. Astron Astrophys. 2011;530:A44.
72. Monfredini T, *et al.* Mon Not R Astron Soc. 2019;488:451–469.
73. Bouallagui A, *et al.* Astrophys J. 2023;956:22.
74. Monfredini T, *et al.* Astrophys J. 2016;821:4.
75. Quitián-Lara HM, *et al.* Mon Not R Astron Soc. 2020;499:6066–6083.
76. Thakur AK, *et al.* J Chem Phys. 2022;156:154704.
77. Chakraborty S, *et al.* Astron Astrophys. 2024;681:A39.
78. Quitián-Lara HM, *et al.* Astrophys J. 2018;854:61.
79. Fourré I, *et al.* Astron Astrophys. 2020;639:A16.
80. Ramachandran R, *et al.* J Chem Sci. 2023;135:77.
81. Ziurys LM. Annu Rev Phys Chem. 2024;75:307–327.
82. Potapov A, *et al.* Astrophys J. 2021;920:111.
83. Turner AM, *et al.* Astrophys J. 2021;916:74.
84. Bacchus-Montabonel M-C. Phys Chem Chem Phys. 2019;21:1428–1434.
85. Mifsud DV, *et al.* Front Astron Space Sci. 2021;8:757619.
86. de Kleer K, *et al.* Science. 2024;384:682–687.
87. Powell D, *et al.* Nature. 2024;626:979–983.
88. Potapov A, *et al.* J Phys Chem A. 2022;126:1627–1639.
89. Rosa CA, *et al.* Life. 2023;13:2208.
90. Slavicinska K, *et al.* Astron Astrophys. 2023;677:A13.
91. Rocha WRM, *et al.* Astron Astrophys. 2022;668:A63.
92. Öberg KI. Chem Rev. 2016;116:9631–9663.
93. Elsaesser A, *et al.* npj Microgravity. 2023;9:43.
94. Frankland VL, *et al.* Rev Sci Instrum. 2015;86:055103.
95. Wiesenfeld L, Oomens J, Cheung ASC. Phys Chem Chem Phys. 2018;20:5341–5343.
96. Öberg KI. Chem Rev. 2016;116:9631–9663.
97. Wang J, *et al.* Sci Adv. 2024;10:eadl3236.
98. Muñoz Caro GM, *et al.* ACS Earth Space Chem. 2019;3:2138–2157.
99. Nuevo M, *et al.* Proc Int Astron Union. 2019;15:123–126.
100. Hudson RL, *et al.* Astrobiology. 2008;8:771–779.
101. Vasyunin A, Herbst E. Astrophys J. 2013;769:34.
102. Fredon A, Radchenko AK, Cuppen HM. Acc Chem Res. 2021;54:745–753.
103. Caro GM, *et al.* Astron Astrophys. 2010;522:A108.
104. Thrower JD, *et al.* Astrophys J. 2008;673:1233.
105. Collings MP, *et al.* Mon Not R Astron Soc. 2004;354:1133–1140.
106. Minissale M, *et al.* ACS Earth Space Chem. 2022;6:597–630.
107. Mason NJ, *et al.* Faraday Discuss. 2014;168:235–247.
108. Boyer MC, *et al.* Surf Sci. 2016;652:26–32.

109. Wu QT, *et al.* ACS Earth Space Chem. 2023;8:79–88.
110. Boezio M, Munini R, Picozza P. Prog Part Nucl Phys. 2020;112: 103765.
111. Bagenal F, Dols V. J Geophys Res Space Phys. 2020;125: e2019JA027485.
112. Clark G, *et al.* J Geophys Res Space Phys. 2020;125:e2020JA028052.
113. Addison P, Liuzzo L, Simon S. J Geophys Res Space Phys. 2022;127: e2021JA030136.
114. Bagenal F, Delamere PA. J Geophys Res Space Phys. 2011;116:A5.
115. Kim TK, *et al.* J Geophys Res Space Phys. 2020;125:e2019JA027696.
116. Cooper JF, *et al.* Icarus. 2001;149:133–159.
117. Mauk B, *et al.* J Geophys Res Space Phys. 2004;109:A9.
118. Turner JE. Atoms, radiation, and radiation protection. Hoboken: John Wiley & Sons. 2008.
119. De Barros A, *et al.* Astron Astrophys. 2011;531:A160.
120. Andrade DP, *et al.* Mon Not R Astron Soc. 2013;430:787–796.
121. de Barros A, *et al.* Astrophys J. 2015;810:156.
122. Pereira R, *et al.* Mon Not R Astron Soc. 2020;495:40–57.
123. Ada Bibang PC, *et al.* Low Temp Phys. 2019;45:590–597.
124. Loeffler M, *et al.* Astron Astrophys. 2005;435:587–594.
125. Johnson R. J Geophys Res Planets. 1991;96:17553–17557.
126. Shingledecker CN, *et al.* Astrophys J. 2019;876:140.
127. Souza-Corrêa J, da Costa C, Da Silveira E. Astrobiology. 2019;19: 1123–1138.
128. da Costa C, Souza-Corrêa J, da Silveira E. Mon Not R Astron Soc. 2021;502:2105–2119.
129. Maté B, *et al.* Astrophys J. 2015;806:151.
130. Pilling S, *et al.* Eur Phys J D. 2014;68:58.
131. Rowland B, Fisher M, Devlin JP. J Chem Phys. 1991;95:1378–1384.
132. Palumbo M. Astron Astrophys. 2006;453:903–909.
133. Raut U, *et al.* J Chem Phys. 2007;126:244511.
134. Raut U, *et al.* Astrophys J. 2008;687:1070.
135. Dartois E, *et al.* Astron Astrophys. 2013;557:A97.
136. Mejía C, *et al.* Icarus. 2015;250:222–229.
137. Hudson RL, Moore MH. Radiat Phys Chem. 1995;45:779–789.
138. Dartois E, *et al.* Astron Astrophys. 2015;576:A125.
139. Strazzulla G, *et al.* Europhys Lett. 1992;18:517.
140. Famá M, *et al.* Icarus. 2010;207:314–319.
141. Brucato JR, *et al.* Astron Astrophys. 2004;413:395–401.
142. Brunetto R, Baratta GA, Strazzulla G. In: J Phys Conf Ser. 2005;6:120.
143. Johnson RE. Icarus. 2000;143:429–433.

144. Johnson R, *et al.* Icarus. 2002;156:136–142.
145. Leblanc F, Johnson R, Brown M. Icarus. 2002;159:132–144.
146. Leblanc F, *et al.* Icarus. 2005;178:367–385.
147. Cipriani F, *et al.* Geophys Res Lett. 2008;35:L19201.
148. Johnson R, *et al.* Europa. 2009;21:507–528.
149. Ivlev AV, *et al.* Astrophys J. 2023;944:181.
150. Van Dishoeck EF, Herbst E, Neufeld DA. Chem Rev. 2013;113: 9043–9085.
151. Van Dishoeck EF. Faraday Discuss. 2014;168:9–47.
152. Gomis O, Leto G, Strazzulla G. Astron Astrophys. 2004;420:405–410.
153. Loeffler M, *et al.* Icarus. 2006;180:265–273.
154. Moore M, Hudson R. Icarus. 2000;145:282–288.
155. Zheng W, Jewitt D, Kaiser RI. Astrophys J. 2006;639:534.
156. Zheng W, Jewitt D, Kaiser RI. Astrophys J. 2006;648:753.
157. Zheng W, Jewitt D, Kaiser RI. Chem Phys Lett. 2007;435:289–294.
158. Mifsud DV, *et al.* Eur Phys J D. 2022;76:87.
159. Bjerrum N. Science. 1952;115:385–390.
160. Mejía C, *et al.* Nucl Instrum Methods Phys Res B. 2015;365:477–481.
161. Bennett CJ, Jamieson CS, Kaiser RI. Phys Chem Chem Phys. 2010;12:4032–4050.
162. Mifsud DV, *et al.* J Mol Spectrosc. 2022;385:111599.
163. McConkey J, *et al.* Phys Rep. 2008;466:1–103.
164. Cosby P. J Chem Phys. 1993;98:7804–7818.
165. Cosby P. J Chem Phys. 1993;98:9560–9569.
166. Moore M, Ferrante R, Nuth JA III. Planet Space Sci. 1996;44: 927–935.
167. Palumbo M, Castorina A, Strazzulla G. Astron Astrophys. 1999;342: 551–562.
168. Bennett CJ, *et al.* Astrophys J. 2007;660:1588.
169. De Barros A, *et al.* Mon Not R Astron Soc. 2011;418:1363–1374.
170. Torres-Díaz D, *et al.* ChemPhysChem. 2023;24:e202200912.
171. Boamah MD, *et al.* Faraday Discuss. 2014;168:249–266.
172. Brunetto R, *et al.* Icarus. 2005;175:226–232.
173. Herczku P, *et al.* Rev Sci Instrum. 2021;92:084501.
174. Mifsud DV, *et al.* Phys Chem Chem Phys. 2022;24:10974–10984.
175. Schmidt F, Swiderek P, Bredehöft JH. ACS Earth Space Chem. 2021;5:391–408.
176. Saenko EV, Feldman VI. Phys Chem Chem Phys. 2016;18:32503–32513.
177. Ioppolo S, *et al.* Mon Not R Astron Soc. 2011;413:2281–2287.
178. Bennett CJ, Ennis CP, Kaiser RI. Astrophys J. 2014;782:63.
179. Bennett CJ, Ennis CP, Kaiser RI. Astrophys J. 2014;794:57.

180. Sivaraman B, *et al.* J Chem Phys. 2013;139:074706.
181. Ioppolo S, *et al.* Astrophys J. 2008;686:1474.
182. Ioppolo S, *et al.* Phys Chem Chem Phys. 2010;12:12065–12076.
183. Cuppen H, *et al.* Phys Chem Chem Phys. 2010;12:12077–12088.
184. Strazzulla G, *et al.* Icarus. 2003;164:163–169.
185. Dawes A, *et al.* Phys Chem Chem Phys. 2007;9:2886–2893.
186. Hunniford C, *et al.* In: J Phys Conf Ser. 2009;163:012078.
187. Lv X, *et al.* Astron Astrophys. 2012;546:A81.
188. Muntean EA, *et al.* Mon Not R Astron Soc. 2016;462:3361–3367.
189. Boduch P, *et al.* Astron Astrophys. 2012;544:A30.
190. Strazzulla G. Planet Space Sci. 1999;47:1371–1376.
191. Strazzulla G, Palumbo ME. Planet Space Sci. 1998;46:1339–1348.
192. Mifsud DV, *et al.* Space Sci Rev. 2021;217:14.
193. Strazzulla G, *et al.* Icarus. 2007;192:623–628.
194. Ding J, *et al.* Icarus. 2013;226:860–864.
195. Mifsud DV, *et al.* Icarus. 2024;411:115926.
196. Lv X, *et al.* Mon Not R Astron Soc. 2014;438:922–929.
197. Strazzulla G, Palumbo ME, Boduch P, Rothard H. Earth Moon Planets. 2023;127:2.
198. Mifsud DV, *et al.* Geophys Res Lett. 2022;49:e2022GL100698.
199. Arumainayagam CR, *et al.* Chem Soc Rev. 2019;48:2293–2314.
200. Solov'yov AV, *et al.* Chem Rev. 2024;124:8014–8129.
201. Larsson M, Orel AE. Dissociative recombination of molecular ions. Cambridge: Cambridge University Press. 2008.
202. Güsten R, *et al.* Nature. 2019;568:357–359.
203. Araujo L, *et al.* Phys Chem Chem Phys. 2025;27:707–716.
204. Janev RK, *et al.* 2012.
205. Rosu-Finsen A, *et al.* Phys Chem Chem Phys. 2016;18:5159–5171.
206. Cassidy A, McCoustra MR, Field D. Acc Chem Res. 2023;56:1909–1919.
207. Fabrikant II, *et al.* Adv At Mol Opt Phys. 2017;66:545–657.
208. Chakraborty D, Paul A. J Phys B. 2023.
209. Bredehöft JH, *et al.* ACS Earth Space Chem. 2017;1:50–59.
210. Schmidt F, *et al.* Eur Phys J D. 2021;75:302.
211. Wang J, *et al.* Astrophys J. 2024;967:79.
212. Zhang C, *et al.* Astrophys J. 2023;952:132.
213. Bilgilisoy E, *et al.* Beilstein J Nanotechnol. 2023;14:1178–1199.
214. Davis D, *et al.* J Chem Phys. 2018;149:064305.
215. Pernet A, *et al.* Astron Astrophys. 2013;552:A100.
216. Dupuy R, *et al.* J Appl Phys. 2020;128:174502.
217. Almeida GC, *et al.* J Phys Chem C. 2012;116:25388–25394.
218. Shulenberger KE, *et al.* ACS Earth Space Chem. 2019;3:800–810.

219. Schmidt F, *et al.* Atoms. 2022;10:25.
220. Campbell L, Brunger M. Geophys Res Lett. 2007;34:L22102.
221. Gérard J-C, *et al.* Icarus. 2023;389:115261.
222. de Kleer K, *et al.* Planet Sci J. 2023;4:37.
223. Campbell L, Brunger M. Int Rev Phys Chem. 2016;35:297–351.
224. Blaško J, *et al.* Eur Phys J D. 2023;77:22.
225. Al Saati S, *et al.* J Geophys Res Space Phys. 2022;127: e2022JA030586.
226. Plainaki C, *et al.* Astrophys J. 2022;940:186.
227. Galand M, *et al.* J Geophys Res Space Phys. 2011;116:A9.
228. Coates A, *et al.* Geophys Res Lett. 2007;34:L22.
229. Horvath G, *et al.* Eur Phys J Appl Phys. 2010;49:13105.
230. Vuitton V, *et al.* Planet Space Sci. 2009;57:1558–1572.
231. Vuitton V, *et al.* Titan after Cassini-Huygens. 2024.
232. Mukundan V, Bhardwaj A. Astrophys J. 2018;856:168.
233. Gladstone GR, Young LA. Annu Rev Earth Planet Sci. 2019;47: 119–140.
234. Feldman PD, *et al.* Astron Astrophys. 2015;583:A8.
235. Feldman PD, *et al.* Astrophys J Lett. 2016;825:L8.
236. Feldman PD, *et al.* Astron J. 2017;155:9.
237. Bodewits D, *et al.* Astrophys J. 2019;885:167.
238. Bodewits D, *et al.* Astron J. 2016;152:130.
239. Joshipura K, Mason N. Atomic-molecular ionization by electron scattering: Theory and applications. Cambridge: Cambridge University Press. 2019.
240. Tomer H, *et al.* Int J Mass Spectr. 2021;470:116708.
241. Chakraborty I, Sinha N, Antony B. Radiat Phys Chem. 2024;216: 111421.
242. Harris A, Nepomuceno J. J Phys B. 2024;57:025201.
243. Stokes PW, *et al.* J Chem Phys. 2021;155:084305.
244. Isik N, Eskicioglu OC. Open Phys. 2023;21:20220261.
245. Marinković BP, *et al.* Atoms. 2019;7:11.

https://doi.org/10.1142/9781800618183_0013

Chapter 13

Foundations of Astrochemistry II: Photon- and Radical-Induced Processes in Ices, Computational Approaches, and Biomolecular Chirality

Nigel J. Mason,[*,†,**] **Sergio Ioppolo,**[‡] **Heidy M. Quitián-Lara,**[*,§] **Felipe Fantuzzi,**[†,¶,††] **Duncan V. Mifsud,**[†] **and Jana Bocková**[||]

[*] *Centre for Astrophysics and Planetary Science, School of Engineering, Mathematics and Physics, University of Kent, Canterbury CT2 7NH, United Kingdom*

[†] *HUN-REN Institute for Nuclear Research (Atomki), Debrecen H-4026, Hungary*

[‡] *Centre for Interstellar Catalysis, Department of Physics and Astronomy, Aarhus University, DK 8000 Aarhus, Denmark*

[§] *Max-Planck Institute for Extraterrestrial Physics, 85748 Garching, Germany*

[¶] *Supramolecular, Interfacial and Synthetic Chemistry, School of Natural Sciences, University of Kent, Canterbury CT2 7NH, United Kingdom*

[||] *Institut de Chimie de Nice (ICN), CNRS UMR 7272, Université Côte d'Azur, 06108, Nice, France*

[**] *n.j.mason@kent.ac.uk*

[††] *f.fantuzzi@kent.ac.uk*

In this chapter, key atomic and molecular processes shaping modern astrochemistry are examined, with a focus on the role of spectroscopy in interpreting astronomical data. Photon-induced processes – absorption, ionisation, and dissociation – are explored for their impact on the chemical evolution of interstellar environments. The chapter discusses experimental investigations utilising synchrotron radiation facilities to study the chemistry of interstellar ices, dust grains, and molecular clouds. Surface reactions on dust grains within icy mantles, including Langmuir–Hinshelwood and non-diffusive mechanisms, are highlighted as pathways for the formation of fundamental and complex organic molecules. Computational astrochemistry is introduced as a cornerstone of the field, encompassing wavefunction-based methods, density functional theory, and multiscale modelling, with applications that address complex astrochemical phenomena. Chirality is also explored, highlighting mechanisms such as circularly polarised light and spin-polarised electrons as potential drivers of biological homochirality. Emphasis is placed on interdisciplinary efforts and robust data repositories to advance astrochemistry and unravel the chemical complexity of the Universe and the origins of life.

13.1 Introduction

Most astronomical data are obtained remotely and typically provided in a spectral format, necessitating extensive spectroscopic data repositories spanning a wide spectral range from the radio to X-ray regions.[1,2] Spectroscopic measurements are therefore fundamental to observational astronomy and to studies of photon-induced processes, such as absorption, ionisation, and dissociation, which are central to advancing our understanding of astrochemistry and planetary sciences. Given the vast amounts of data required, a significant portion must be derived computationally using advanced atomic and molecular structure models for atoms, molecules,[3] and ions, as ionised states are the most prevalent in the universe.[4]

In this chapter, the second reviewing the atomic and molecular physics underpinning modern astrochemistry, photonic processes are examined in both gaseous and condensed phases. The role of photon-induced processing of astronomical ices in the interstellar medium (ISM) and on planetary bodies is emphasised, with photodissociation shown to generate radical species that drive the synthesis of numerous molecular compounds. Recent pioneering experiments in radical chemistry are reviewed, shedding light on these processes. Alongside experimental approaches, computational methods are discussed, covering both the fundamentals of techniques used in computational astrochemistry and recent findings that reveal reaction mechanisms, energetics, and the behaviour of complex astrochemical environments. Additionally, the potential influence of light polarisation on chemical processing is explored, highlighting its possible role in the origin of the chiral nature of life.

As in the complementary chapter, the importance of further studies on atomic and molecular processes is emphasised as essential for interpreting the wealth of data collected by astronomical and space missions. These missions address some of the most fundamental scientific questions, including the evolution of the universe from the Big Bang to the present, the origins of life on Earth, and the possibility of life elsewhere in the Universe.

13.2 Photon-Induced Processes

Space is pervaded by light; for example, photons generated by the stars permeate their own solar system irradiating planets. In the ISM, dust and molecules are irradiated by light from new stars as they are 'born', while throughout all space there is the Cosmic Microwave Background (CMB) radiation, the cooled remnant of the first light that could ever travel freely throughout the Universe. In space, there are several interstellar photo-radiation sources that can interact with ice material in prestellar clouds, hot cores, hot corinos, and the edge of protoplanetary discs, such as the interstellar radiation field (ISRF), the protostar black-body radiation, and the secondary ultraviolet (UV) radiation generated by the interaction of cosmic rays and H_2 molecules in the gas phase following the reactions

$$H_2 + CR \rightarrow H_2^+ + e^*, \tag{13.1}$$

$$H_2 + e^* \rightarrow H_2^* + e^-, \tag{13.2}$$

$$H_2^* \rightarrow H_2 + h\nu, \tag{13.3}$$

where e^* refers to secondary electrons with a typical energy of 19 eV and H_2^* denotes electronically excited H_2.[5–7]

Depending on its energy, a photon interacting with an ice material can induce various processes: ionisation (e.g., gamma-rays, X-rays, and vacuum ultraviolet (VUV) photons), electronic excitation (e.g., UV and visible (vis) photons), vibrational excitation (e.g., infrared (IR) and terahertz (THz) photons), or rotational excitation (e.g., microwave and radio-wave photons). High-energy gamma-ray (γ-ray) photons, X-rays, and VUV photons are classified as ionising radiation, while lower-energy UV, visible light, near/mid/far infrared (NIR/MIR/FIR), THz, microwaves, and radio waves are classified as non-ionising radiation.

The boundary between ionising and non-ionising radiation in the UV region cannot be sharply defined, as different molecules and atoms ionise at different energies. Furthermore, charged radiation species, such as cosmic rays, interact with the electrons in the absorbing medium through Coulomb forces. These interactions occur continuously along the particle's trajectory, gradually depleting its kinetic energy. In contrast, uncharged radiation species, such as γ-ray photons and X-rays, undergo discrete single-photon interactions. In these cases, the photon's energy is completely absorbed in a process known as photoabsorption, which can result in the ejection of an electron from an atom or molecule (photoionisation) or the dissociation of a molecule into reactive fragments (photodissociation).

Additionally, a single photon can, through one or more steps, induce photodesorption – the release of an atom or molecule from a solid surface, such as an ISM dust grain. These various 'photon-induced processes' are fundamental to astrochemistry, playing a dominant role in the chemical evolution of interstellar environments. They were also among the earliest processes studied in laboratory settings and incorporated into astronomical models.

13.2.1 *Experimental studies*

Laboratory experiments can replicate photo-processes in the ISM, on planetary surfaces, and in planetary atmospheres by serving as mimics of the different photon sources found in space. Most of the earliest experiments mimicked 'starlight' using lamps, particularly microwave-discharge hydrogen flow lamps, that provided strong sources of the hydrogen Lyman-α radiation. This radiation is a dominant photon source in space and is capable of inducing photoexcitation, photodissociation, and photoionisation of many molecular species. However, such lamps often lack the ability to provide broadband, continuously tuneable radiation.

Since the 1980s, synchrotron laboratories have emerged as the most versatile and tunable sources of electromagnetic radiation, spanning a broad range from IR to X-rays, and in some cases extending into the γ-ray region. This broad tunability offers unparalleled advantages for astrochemistry studies, enabling the investigation of a wide variety of photon-induced processes. Experiments utilising synchrotron radiation have been, and continue to be, conducted at facilities worldwide, including the Aarhus Synchrotron Radiation Facility (ASTRID2) in Denmark,[8] the Source Optimisée de Lumière d'Énergie Intermédiaire du LURE (SOLEIL) in France,[9] Indus in India,[10] the National Synchrotron Radiation Research Center (NSRRC) in Taiwan,[11] and the Laboratório Nacional de Luz Síncrotron (LNLS/Sirius) in Brazil,[12] where dedicated beamlines and experimental chambers have been developed to support cutting-edge astrochemistry research.[13–17]

Experimental setups include those designed to study photoabsorption, the electronic state spectroscopy of molecules in both gaseous and condensed phases, and photoionisation. These studies often utilise analytical techniques such as photoelectron spectroscopy (PES) and mass spectrometry (MS). While these processes represent some of the most fundamental atomic and molecular interactions, their detailed characterisation is crucial for understanding and quantifying astrochemical processes. The collection and organisation of such data into comprehensive databases are essential for identifying molecular species in astronomical observations and integrating them into astrochemical models.

The role of higher-energy photons (with energies exceeding that of Lyman-α at 121.567 nm or 10.2 eV) is increasingly recognised as critical in astrochemistry. 'Soft' X-rays, in particular, are essential for the chemical evolution of the ISM and other astrophysical environments close to young, luminous stars. These photons can penetrate deeply into molecular clouds and protostellar discs, initiating chemistry in regions inaccessible to UV stellar photons.

When soft X-rays interact with astrophysical ices, they release secondary electrons within and on the ice surfaces, triggering a cascade of chemical reactions.[18] Synchrotron facilities, with their ability to produce broadband, tunable X-ray sources, are invaluable for studying these processes. Moreover, synchrotrons can generate circularly polarised light, enabling investigations into chirality – an intrinsic property of biochemistry and a fundamental feature of life (see Section 13.5).

In space, high-energy γ-rays are generated by some of the Universe's most extreme and energetic phenomena and objects, such as neutron stars, pulsars, supernova explosions, and the regions around black holes. γ-ray bursts represent some of the most energetic events encountered in space, releasing more energy in 10 s than the Sun will emit over its entire 10-billion-year expected lifespan.

Gamma rays are also routinely used to determine the elemental composition on other planets. For instance, the Mercury Surface, Space Environment, Geochemistry, and Ranging (MESSENGER) Gamma-Ray Spectrometer (GRS) has been used to map Mercury's surface composition. γ-rays, in this context, are produced when cosmic rays irradiate the surface, causing chemical elements in the soils and rocks to emit uniquely identifiable signatures. These γ-ray signatures allow for the creation of surface maps that highlight geologically significant elements, such as hydrogen, magnesium, silicon, oxygen, iron, titanium, sodium, and calcium.

Traditionally, γ-ray sources are derived from the decay of radioactive materials, with Cobalt-60 (^{60}Co) being one of the most widely used. In astrochemistry, numerous experiments have investigated the effects of gamma radiation, including molecular dissociation and chemical synthesis in astrochemical ices.[19,20]

13.2.2 *Photoabsorption*

Photoabsorption studies provide spectroscopic data, characterising molecular excitation processes, including rotational, vibrational, and electronic excitation, and revealing spectral features that may be used to identify chemical species. Microwave spectra are routinely used to detect molecules in the ISM, with over 320 species identified to date. IR spectra are used to probe the composition of ices in the ISM, protoplanetary discs, and related environments, while THz spectra are needed for James Webb Space Telescope (JWST) observations. In contrast, UV–vis studies are more commonly applied to solar system research. For example, most NASA space exploration missions include UV–vis spectrometers to measure reflectance spectra of planetary, lunar, cometary, and asteroidal surfaces. The recorded ice spectra are then compared with laboratory-derived data, enabling the identification of abiotic ozone on Jovian and Saturnian moons and the tentative identification of hydrazine (N_2H_4) on Rhea.[21,22]

It is important to note that spectra obtained in the condensed (ice) phase often differ significantly from those in the gas phase.

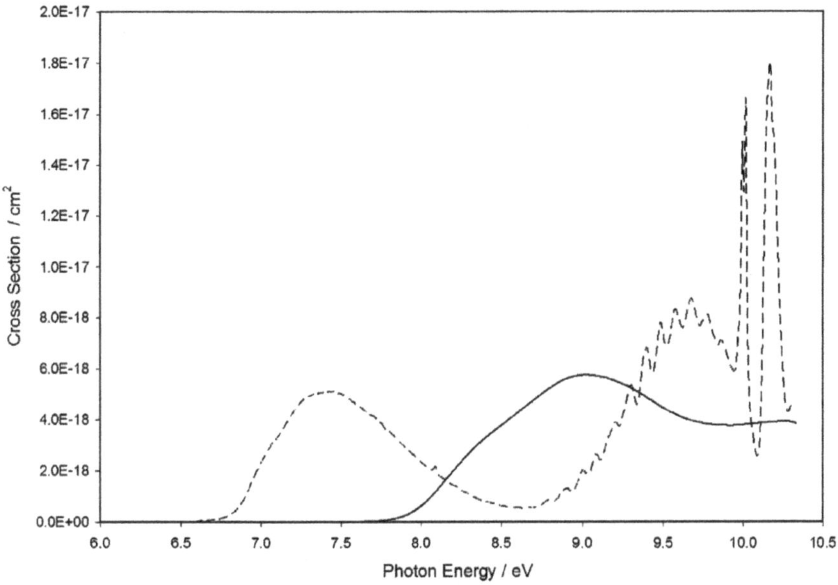

Fig. 13.1. Illustration of the UV photoabsorption spectra of gaseous (dashed line) and condensed (solid line) water.[23]

In the condensed phase, there are shifts in electronic state excitation energies, as well as the eventual suppression of Rydberg states seen in the gas phase. For instance, water – the most common ice species – exhibits a 1 eV blue shift in its electronic excitation energies, which has significant consequences for photodissociation models (see Fig. 13.1).[23] UV–vis studies of ice phases can also provide information on the morphology of ice mantles and solid-state effects, such as the excitation of plasmons and excitons.[24]

13.2.3 *Photon-induced processing and synthesis*

Experiments simulating the energetic processing of interstellar ices to form prebiotic molecules through photochemistry (involving electronic excitation) and radiation chemistry (involving both electronic excitation and ionisation) are routinely conducted in various laboratories, ranging from small-scale setups to large-scale research facilities. A comprehensive review of the literature is beyond the scope of this discussion; however, readers are directed to relevant references for further details, and one illustrative example is provided here.[25–29]

Photo-processing of methanol ices has been identified as a crucial step in the formation of many complex organic molecules (COMs), such as methyl formate ($HCOOCH_3$), dimethyl ether (CH_3OCH_3), glycolaldehyde ($HOCH_2CHO$), acetic acid (CH_3COOH), and methoxymethanol (CH_3OCH_2OH), all of which have been detected in star-forming regions. Interestingly, most of the products resulting from UV photon irradiation of methanol ices are also observed as products following low-energy ($\leq$20 eV) electron irradiation of condensed methanol, suggesting that photochemistry studies of astrochemically relevant ices often overlap with radiation chemistry processes.

In their review on photochemistry versus radiation chemistry of interstellar ices, Arumainayagam *et al.* (2019)[30] highlighted that the ionisation threshold in the condensed phase is lower than in the gas phase. This underscores the importance of conducting photoabsorption spectroscopy studies of molecules in the solid phase, complemented by density functional theory (DFT) calculations, to gain a comprehensive understanding of how photons interact with solids.[31–36]

13.2.4 *Photodesorption*

Photodesorption is, by far, the most extensively studied non-thermal desorption mechanism. Upon absorption of a photon, a (sub)surface molecule is excited with sufficient energy to either overcome its binding energy or transfer this energy to a neighbouring surface molecule that is subsequently ejected. Most studies have focused on the impact of vacuum UV photons, typically in the 120–170 nm range, which are commonly generated by microwave-discharge hydrogen flow lamps.[27] Studies utilising monochromatic light at synchrotron facilities, such as SOLEIL in France and ASTRID2 in Denmark, have demonstrated that the photodesorption rate correlates directly with the rovibronic absorption profile of the A–X electronic band system (see Fig. 13.2).[37] These findings confirm that the photodesorption process is predominantly governed by electronic excitation.[38–41]

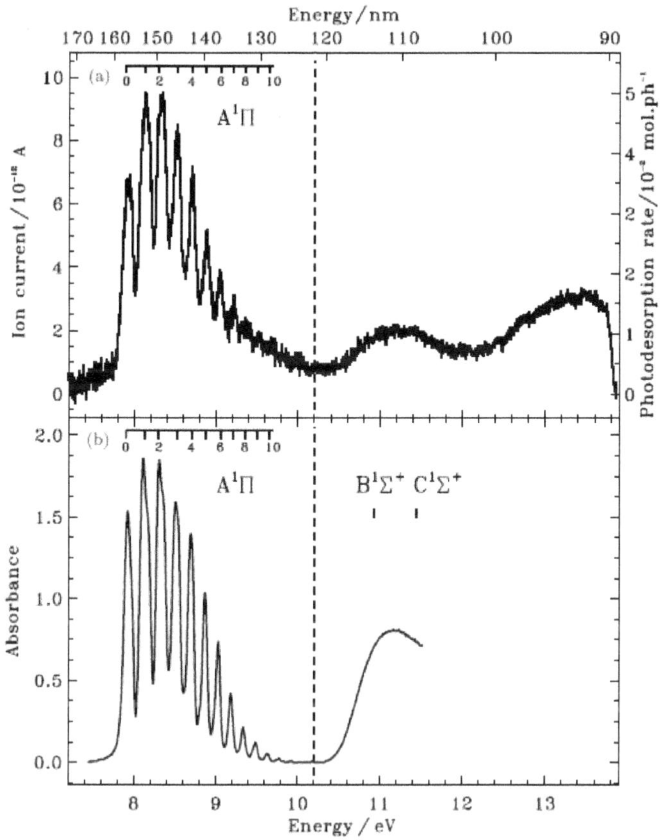

Fig. 13.2. (a) Photodesorption spectrum of CO ice at 18 K. (b) Absorption spectrum of CO ice at 10 K.[37]

13.3 Atom/Radical Processes

The ISM is a complex environment where gas, dust, and ice coexist at extremely low temperatures and densities. Despite these harsh conditions, a rich and diverse chemistry takes place, contributing to the formation of a wide variety of molecules, from simple diatomics to COMs. Surface reactions on dust grains, particularly within the icy mantles that accumulate on these grains, play a crucial role in driving molecular synthesis.[42]

Among these surface processes, hydrogen atom addition and atom abstraction reactions are essential in furthering chemical evolution within molecular clouds, star-forming regions, and protoplanetary discs. This section explores the mechanisms and significance of these reactions, discussing how they shape the chemical landscape of the ISM and contribute to the formation of key molecules such as water, methanol, and COMs.

13.3.1 *Surface reactions on interstellar ice grains*

Interstellar dust grains typically consist of silicate or carbonaceous cores enveloped by icy mantles, which form through the accretion of gas-phase molecules. These grains serve as catalysts for surface reactions that are inefficient or unlikely to occur in the gas phase, given the extremely low temperatures and densities characteristic of the denser regions of the ISM, where stars and planets are born. The icy mantles primarily consist of water (H_2O), carbon monoxide (CO), carbon dioxide (CO_2), ammonia (NH_3), and methanol (CH_3OH), among other species. These mantles provide a surface for atoms and molecules to interact, facilitating the formation of both simple and complex molecules through a series of non-energetic atom addition/abstraction surface reactions.

Atom addition reactions occur when an atom interacts with an existing species on the surface of the ice and bonds covalently to it to form a new species. These reactions are especially significant at the low temperatures (10–20 K) found in dense molecular clouds, where thermal energy is insufficient to overcome activation barriers in the gas phase. On the other hand, atom abstraction reactions involve the removal of an atom from a molecule, often initiated by an atomic or molecular species that impinges on the icy surface. These abstraction reactions are often followed by subsequent addition reactions, driving the growth of complex molecules on grain surfaces.

One of the most fundamental atom addition processes in the ISM is hydrogenation. Given the abundance of hydrogen atoms, they play a central role in surface chemistry on interstellar grains, serving as key reactants in molecular synthesis.

In the past, three primary mechanisms were recognised as the main pathways for molecule formation on dust grains in the absence of energetic triggers: the Langmuir–Hinshelwood, Eley–Rideal, and hot-atom mechanisms.[43] More recently, a non-diffusive mechanism

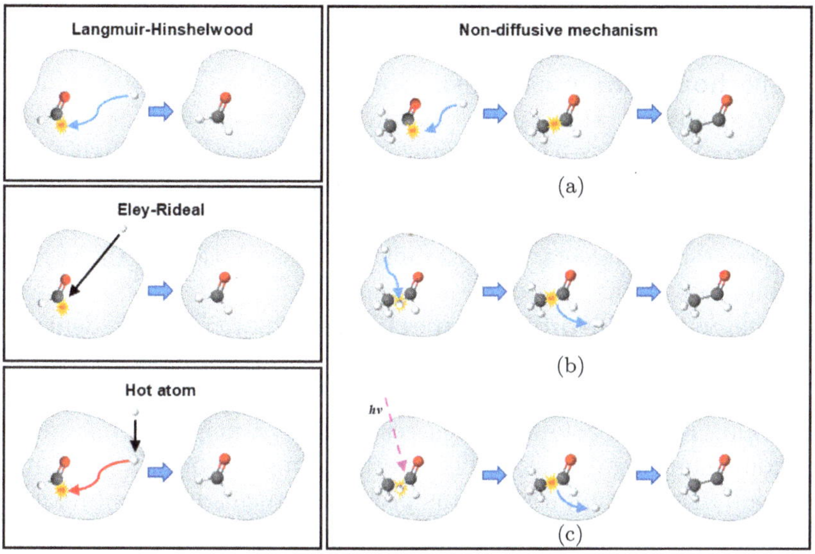

Fig. 13.3. Schematic diagram of different surface reaction mechanisms. Left panels from top to bottom: the Langmuir–Hinshelwood mechanism, in which one or both reactants diffuse over the surface to meet; the Eley–Rideal mechanism, which occurs when a species from the gas phase directly hits a reactant on the surface; and the hot-atom mechanism which combines Langmuir–Hinshelwood and Eley–Rideal with the incoming species being non-thermalised to the surface temperature. Right panel: the non-diffusive mechanism which occurs when two radicals are in the vicinity of one another due to a previous reaction, which can be induced by (a) atom-addition reaction, (b) atom-abstraction reaction, or (c) photodissociation event, after which they can react together without the need for the radicals to move.

has been introduced as a fourth pathway.[44,45] These four mechanisms are summarised and illustrated in Fig. 13.3.

The **Langmuir–Hinshelwood mechanism**, proposed by I. Langmuir in 1921 and further developed by C. Hinshelwood in 1926, relies on the diffusion of reactants across the ice surface. This mechanism is particularly important for reactions involving highly mobile species such as H or H_2, which migrate on the surface until they meet and react.

The **Eley–Rideal mechanism**, introduced in 1938 by D. D. Eley and E. K. Rideal, involves a gas-phase reactant directly interacting with a molecule adsorbed on the dust grain surface, without itself adsorbing. This process is rare unless the surface is densely covered with molecules and radicals.

The **hot-atom mechanism** combines aspects of the previous two mechanisms. Here, a 'hot' atom – one with significant kinetic or internal energy – travels across the surface without being fully thermalised, enabling it to react with another species. This mechanism becomes significant when the gas is warmer than the surface or when energetic triggers, such as molecule dissociation on the surface, produce hot atoms *in situ*.

Finally, the **non-diffusive mechanism**, more recently identified, involves a two-step process. In this scenario, a stable species becomes reactive due to a reaction or energetic event (e.g., UV irradiation or cosmic ray impact) and then reacts with a neighbouring species without requiring diffusion. This mechanism often occurs when atom addition or abstraction reactions produce radicals in close proximity or when dissociation events create reactive intermediates near each other.

These mechanisms collectively provide a comprehensive framework for understanding the diverse pathways of molecular formation on interstellar dust grains.

13.3.2 *Formation of simple interstellar ice molecules*

Molecular hydrogen (H_2) is the simplest molecule in space, and it is formed primarily on the surface of dust grains in the ISM. The formation of H_2 on the surfaces of bare silicate dust grains, such as forsterite (Mg_2SiO_4), and carbonaceous materials, including polycyclic aromatic hydrocarbons (PAHs) and graphite, has been extensively investigated through both modelling and experimental studies.[46–50] However, dust grains are not cold enough to allow for H_2 molecules to form an ice layer on top of them. Therefore, the first ice layer on dust grains is mostly made of water molecules, the most ubiquitous and abundant ice species in the cosmos,[51,52] and occurs in the transition between diffuse and translucent clouds.[53] It has been shown that water can be formed through different gas- and solid-phase routes that are dominant under specific conditions in the ISM.[52,54]

The idea that water forms on the surface of ice grains in the coldest regions of the ISM dates back to 1949, when van der Hulst proposed that gas-phase species such as H, H_2, and O_2 could accrete onto the surface of dust grains to produce H_2O and H_2O_2.[55] A more detailed description of the potential surface reactions leading to water formation in the cold ISM was provided by Tielens and Hagen (1982),[56] who identified three main formation routes: the hydrogenation of atomic oxygen (O), molecular oxygen (O_2), and ozone (O_3). According to Cuppen *et al.* (2007),[57] the dominant reaction under diffuse and translucent conditions is the formation of H_2O through the hydrogenation of atomic oxygen on interstellar dust surfaces. In contrast, molecular species such as O_2 and O_3 become more significant in denser environments, such as cold dense cores.

Laboratory evidence for the efficient formation of water through the hydrogenation of atomic oxygen on bare grain analogues is well documented in the literature.[58,59] In the late 2000s, three independent research groups conducted laboratory studies on the hydrogenation of molecular oxygen, further elucidating this formation pathway.[60–62] Experimental and theoretical work has since demonstrated that the reaction network proposed by Tielens and Hagen (1982)[56] was incomplete, with additional processes identified and characterised.[63–66]

The reaction H + O_3 was subsequently studied under cold cloud conditions by Mokrane *et al.* (2009)[67] and Romanzin *et al.* (2011)[68] and on warm silicate surfaces by He *et al.* (2014),[69] confirming water formation under various space-analogue conditions. Amorphous solid water (ASW), formed on dust grains at temperatures as low as 10 K, is the most abundant form of ice in the Universe. ASW plays a crucial catalytic role in interstellar chemistry, providing a surface for reactions and a matrix for molecule formation while also dissipating excess reaction energy.[70–74] In addition to liberating water, the water-rich ice layer on interstellar dust grains enables the formation of various simple inorganic and complex organic molecules containing H, O, N, and C, such as CO_2, NH_3, methane (CH_4), and NH_2CH_2COOH.[75]

With abundances ranging between 20 and 30% that of H_2O ice, CO_2 is one of the most abundant solid-phase species in the ISM.[76] In dense molecular clouds, CO_2 is formed in the solid phase primarily through the CO + OH surface reaction route, which includes the subsequent HOCO + H reaction in water-rich ice.[77–79] The HOCO radical can evolve into CO_2 through H-abstraction reactions, but these compete with addition reactions that lead to the formation of carboxylic acids (R-COOH).[77,78,80–82]

Moreover, CO_2 can also be formed along CO + O and HCO + O reaction routes in a CO-rich ice. Although the HCO + O reaction is experimentally challenging to investigate in the solid phase due to competing CO_2 formation routes, the CO + O reaction has been extensively studied by different groups.[83–86] These studies have shown that O atoms preferentially react with O to form O_2 and with O_2 to form O_3, as the CO + O surface reaction has a gas-phase barrier of ~2000 K.[87] Additionally, the onset of tunnelling occurs at temperatures too low for the reaction to significantly contribute to CO_2 formation under interstellar conditions.[88] Thus, the CO + O pathway to CO_2 formation becomes efficient only when energetic processing is involved, leaving CO + OH as the dominant surface formation route to CO_2 in the ISM.[89]

Ammonia (NH_3) ice was first unambiguously detected in 24 low-mass young stellar objects (YSOs) by the *c2d Spitzer Spectroscopic Survey of Ices* team, with abundances ranging from ~2% to 15% relative to H_2O and an average abundance of $5.5\pm2.0\%$.[90] Observations combined with laboratory studies suggest that ammonia forms via hydrogenation of atomic nitrogen (N) alongside water ice formation in a relatively low-density molecular phase of star formation.[90]

The sequential hydrogenation of N atoms embedded in a solid N_2 matrix was first investigated mass spectrometrically by Hiraoka *et al.* (1995).[91] Subsequent work by Hidaka *et al.* (2011) spectroscopically confirmed the formation of ammonia within a solid N_2 matrix under cryogenic conditions.[92] In 2015, Fedoseev *et al.*[93,94] explored the surface formation of NH_3 through sequential hydrogenation of N atoms at 15 K using interstellar-relevant ice analogues. They demonstrated that the reactions proceed via a Langmuir–Hinshelwood mechanism, which is rapid and likely barrierless, refining and constraining the earlier findings of different research groups.

With abundances of 2–10% relative to solid water, methane (CH_4) ice is considered an important player in prebiotic chemistry.[95] Observational surveys of CH_4 ice towards low- and high-mass YSOs suggest that much of the CH_4 is formed through the hydrogenation of atomic C on dust grains within water-rich ices.[96] The solid-state formation of methane involves four sequential atomic hydrogenation steps of carbon in its ^3P ground state, a process that has been experimentally confirmed.[96,97]

However, recent findings indicate that H_2 plays a more critical role in the formation of methane than previously assumed. Although the reaction $C + H_2 \rightarrow CH_2$ is not barrierless for all binding sites on water ice, it can still occur under ISM conditions. Additionally, the reaction $CH + H_2 \rightarrow CH_3$ proceeds without a barrier, while reactions such as $CH_2 + H_2 \rightarrow CH_3 + H$ and $CH_3 + H_2 \rightarrow CH_4 + H$ can only occur through tunnelling mechanisms.[98]

13.3.3 *Formation of complex organic interstellar ice molecules*

Qasim *et al.* (2018)[99] demonstrated that the formation pathway of methane is linked to the formation of methanol (CH_3OH) ice in a water-rich layer by the sequential surface reactions $CH_4 + OH \rightarrow CH_3 + H_2O$ and $CH_3 + OH \rightarrow CH_3OH$. The $CH_4 + OH$ channel can explain the observation of CH_3OH at low visual extinctions, that is, prior to the heavy CO freeze-out stage of dense cold cores, where CH_3OH is efficiently formed through the CO + H reaction route.[99] Another CH_3OH formation pathway involving CH_4 molecules was tested by Bergner *et al.* (2017).[100] They provided experimental constraints on the insertion of oxygen atoms in their first electronically excited state, $O(^1D)$, into methane to form methanol in astrophysical ice analogues.

Beyond CH_3OH, there is a strong evidence for other larger COMs that are relevant to life on Earth to be formed in water-rich ice grains. For instance, the combined laboratory and modelling work by Ioppolo *et al.* (2021)[75] showed that glycine ($C_2H_5NO_2$) can form within such an ice layer through a non-diffusive reaction mechanism. This process involves the recombination of radicals and molecules formed in close proximity, with only hydrogen diffusing at 10 K within the ice.

Particularly, glycine forms through the non-diffusive reaction of NH_2CH_2, a precursor of methylamine (NH_2CH_3), with the HOCO radical. The production of NH_2CH_2 radicals has a substantial contribution from the reaction of CH, a precursor of CH_4, with NH_3. As with CO_2, the production of HOCO is dominated by non-diffusive reactions of CO and OH.[101] Following the same non-diffusive mechanism, in principle, once formed, other functional groups can be added to the glycine backbone by hydrogenation abstraction reactions, resulting in the formation of other amino acids, such as alanine ($C_3H_7NO_2$) and serine ($C_3H_7NO_3$) in dark clouds in space.[102]

The catastrophic condensation of CO molecules, which takes place once the density of interstellar clouds exceeds about 10^5 molecules cm^{-1}, leads to the formation of a second layer of ice on top of water-rich ice grains. Unlike water ice, solid CO forms a non-polar environment, lacking an OH-bonding network. Despite this, it has been shown that hydrogenation-initiated reactions can drive a complex surface chemistry even at 10 K. Since the early 2000s, studies have demonstrated that formaldehyde (H_2CO) and methanol form efficiently at 10–15 K through the sequential hydrogenation of CO ice.[103–107] However, Fedoseev *et al.* (2015c)[108] experimentally showed that methanol is not the only end product of the surface reaction network initiated by CO hydrogenation.

Subsequent laboratory and modelling studies revealed that precursors of ribose (a simple sugar) and xylitol (a sugar alcohol), such as glycolaldehyde ($HOCH_2CHO$), ethylene glycol ($HOCH_2CH_2$ OH), glyceraldehyde ($HOCH_2CH(OH)CHO$), and glycerol ($HOCH_2$ $CH(OH)CH_2OH$), as well as other COMs such as methyl formate ($HC(O)OCH_3$) and methoxymethanol (CH_3OCH_2OH), are efficiently formed through non-diffusive radical–radical recombination reactions at 10 K.[108–113]

In conclusion, atom addition and abstraction non-energetic surface reactions are responsible for the formation of many of the simple and most abundant molecules in interstellar ices, including H_2O, CO_2, NH_3, CH_4, H_2CO, and CH_3OH. These processes also contribute to the synthesis of a variety of O-rich COMs (e.g., methyl formate isomers and simple sugars) in CO-dominated layers, as well as and C- and N-rich COMs (e.g., amino acids) in water-rich ices. Therefore, the synergistic and sequential combination of energetic triggers (e.g., ions, photons, and electrons) and non-energetic triggers (e.g., atoms and radicals) must be further investigated in laboratory settings to fully characterise the formation of species in the solid phase under space-like conditions.

13.4 Computational Astrochemistry

Astrochemistry can be conceptualised as a triadic discipline underpinned by three main pillars: observations, experiments, and theory.[114] Within this framework, computational astrochemistry constitutes an indispensable component, providing a theoretical backbone to interpret observational data and guide experimental efforts. Its importance arises from the unique ability to study complex astrochemical phenomena that are often inaccessible to direct experimental or observational approaches. For instance, computational methods facilitate the prediction of high-resolution spectroscopic data, enabling the attribution of molecular transitions critical for identifying interstellar species. Additionally, they allow for the detailed exploration of reaction mechanisms underlying the formation and destruction of molecules under astrophysical conditions, often revealing pathways that remain elusive in laboratory settings. Computational astrochemistry is also instrumental in predicting potential molecular candidates for interstellar detection, modelling the physicochemical growth of molecular ices on dust grains, and simulating large-scale processes, such as chemical evolution within dense molecular clouds or protostellar environments.

The subsequent discussion in this section introduces the methodological landscape of computational astrochemistry. We begin with an overview of theoretical approaches, including *ab initio* and DFT-based quantum-mechanical methods for high-accuracy electronic structure calculations, classical mechanics techniques for modelling larger molecular systems, and stochastic methods such as Monte Carlo and molecular dynamics simulations for studying mesoscale and macroscale phenomena. This technical foundation sets the stage for applications discussed in the following sections, encompassing the generation of spectral data, exploration of potential energy surfaces and reaction dynamics, and multiscale modelling of surface astrochemistry. Together, these topics illustrate the critical role of computational tools in elucidating the fundamental chemistry of the universe.

13.4.1 *A brief journey through computational approaches*

The theoretical description of atomic and molecular electronic structure is a vast and intricate field within quantum chemistry and molecular physics. This section provides only a succinct overview, introducing key concepts and methods frequently employed in computational astrochemistry. Readers seeking a more rigorous, in-depth introduction to electronic structure theory may consult dedicated texts, such as the textbooks by Szabó and Ostlund[115] and Jensen.[116] Additionally, a recent review published by one of us focuses on computational methods for both ground and excited states in organometallic systems, which may serve as a valuable resource for interested readers.[117]

The fundamental starting point in most quantum-chemical treatments is the *electronic Schrödinger equation*, which, within the Born–Oppenheimer approximation, governs the behaviour of the electrons in a fixed nuclear framework. The solution of this equation is complicated by the *Pauli principle*, which demands that the total electronic wavefunction must be antisymmetric with respect to the exchange of any two electrons. A convenient way to enforce antisymmetry is to represent the many-electron wavefunction as a single *Slater determinant*:

$$\Psi(\mathbf{x}_1, \mathbf{x}_2, \ldots, \mathbf{x}_N) = \frac{1}{\sqrt{N!}} \begin{vmatrix} \phi_1(\mathbf{x}_1) & \phi_2(\mathbf{x}_1) & \cdots & \phi_N(\mathbf{x}_1) \\ \phi_1(\mathbf{x}_2) & \phi_2(\mathbf{x}_2) & \cdots & \phi_N(\mathbf{x}_2) \\ \vdots & \vdots & \ddots & \vdots \\ \phi_1(\mathbf{x}_N) & \phi_2(\mathbf{x}_N) & \cdots & \phi_N(\mathbf{x}_N) \end{vmatrix}, \tag{13.4}$$

where $\phi_i(\mathbf{x})$ are *spin-orbitals*, one-electron functions that depend on both spatial and spin coordinates $\mathbf{x}$, and N represents the number of electrons in the system. To solve practical problems, these spin-orbitals are expressed as linear combinations of *basis functions*, typically atom-centred functions such as Gaussian-type orbitals. The set of all chosen basis functions is termed the *basis set*, and its choice greatly influences the accuracy and computational cost of the calculations.

A cornerstone approach for obtaining a first approximation to the electronic wavefunction is the *Hartree–Fock (HF) theory*. In HF theory, one assumes that the exact many-electron wavefunction can be approximated by a single Slater determinant built from a set of optimised, variationally determined spin-orbitals. The optimal spin-orbitals are obtained by solving the *Roothaan–Hall equations*:

$$\mathbf{FC} = \mathbf{SC}\epsilon, \tag{13.5}$$

where **F** is the Fock matrix, **S** is the overlap matrix of the chosen basis set, **C** is the matrix of molecular orbital coefficients, and $\boldsymbol{\epsilon}$ is the diagonal matrix of orbital energies. The HF energy can be written as

$$E_{\mathrm{HF}} = \sum_{i}^{\mathrm{occ}} \langle i|h|i\rangle + \frac{1}{2}\sum_{i,j}^{\mathrm{occ}} (2\langle ij|ij\rangle - \langle ij|ji\rangle), \tag{13.6}$$

where h is the one-electron Hamiltonian (kinetic plus nuclear attraction), and 'occ' refers to the set of occupied molecular orbitals. The term $\langle ij|ij\rangle$ corresponds to the two-electron Coulomb interaction, describing the classical repulsion between electrons, while $\langle ij|ji\rangle$ represents the exchange interaction, arising from the antisymmetric nature of the wavefunction.

While HF theory properly accounts for the antisymmetry of the wavefunction, it treats electron–electron interactions only at an average (mean-field) level. This neglect of *electron correlation* often results in significant quantitative errors, especially for systems where many electronic configurations are energetically important.

To recover the missing correlation energy not included in HF, one can employ *post-HF* methods. Among these, *Møller–Plesset (MP)* perturbation theory treats correlation as a perturbation around the HF reference. The second-order MP2 energy correction is given by

$$E_{\mathrm{corr}}^{\mathrm{MP2}} = \sum_{i<j}^{\mathrm{occ}} \sum_{a<b}^{\mathrm{virt}} \frac{\langle ij||ab\rangle^2}{\epsilon_i + \epsilon_j - \epsilon_a - \epsilon_b}, \tag{13.7}$$

where $\langle ij||ab\rangle$ are antisymmetrised two-electron integrals, the summation over a and b running through the virtual molecular orbitals, and ϵ are the orbital energies. MP2 is cost-effective and improves upon HF, but it can fail in systems with strong *static* (or *non-dynamical*) *correlation*, such as those involving bond dissociation.

Another approach is *configuration interaction (CI)*, which expands the wavefunction as a linear combination of all possible Slater determinants that can be formed from a given set of orbitals. A *full-CI* solution, although exact within the chosen basis, is computationally prohibitive for anything but the smallest systems. Truncated expansions such as CISD (CI with single and double excitations) improve over HF but lose desirable properties such as size-extensivity, meaning that the energy does not scale correctly with system size, leading to errors in larger systems.

A more systematically improvable route is offered by *coupled cluster (CC)* theory. In CC, the wavefunction is expressed in an exponential form:

$$|\Psi_{\text{CC}}\rangle = e^{\hat{T}}|\Phi_{\text{HF}}\rangle, \tag{13.8}$$

where $\hat{T}$ is the cluster operator, typically truncated at singles and doubles (CCSD), plus perturbative triples (CCSD(T)). CCSD(T) is often regarded as a gold standard for small- and medium-sized systems, providing accurate energetics and properties. However, the computational cost grows rapidly, limiting its direct application to larger molecules.

For systems with pronounced multireference character, where a single Slater determinant is an inadequate representation, *multiconfigurational (MC)* methods are essential. The *complete active space self-consistent field (CASSCF)* approach, for example, defines an active space of selected orbitals and includes all possible electronic configurations within this space:

$$|\Psi_{\text{CASSCF}}\rangle = \sum_{I} c_I|\Phi_I\rangle. \tag{13.9}$$

CASSCF optimises orbitals (Φ_I) and configuration coefficients (c_I) simultaneously, capturing the essential non-dynamical correlation. However, it does not recover the *dynamical correlation* that arises from the subtle instantaneous electron–electron interactions.

To incorporate dynamical correlation on top of a multireference reference function, perturbative corrections such as the complete active space with second-order perturbation theory (*CASPT2*)[118] or the N-electron valence state perturbation theory (*NEVPT2*)[119–121] are employed. These methods combine a carefully chosen multiconfigurational reference with a perturbative treatment of excitations outside the active space:

$$E_{\text{CASPT2}} = E_{\text{CASSCF}} + E^{(2)}_{\text{corr}}, \tag{13.10}$$

where $E^{(2)}_{\text{corr}}$ includes dynamical correlation effects. Although highly accurate, CASPT2 and NEVPT2 calculations can become prohibitively expensive for larger systems. Techniques such as *domain-based local pair natural orbitals (DLPNO)*[122] can approximate the correlation treatment and make these calculations more tractable.

An alternative framework, which is often more computationally efficient, is provided by *DFT*.[123] In DFT, the complex many-body electronic structure problem is reformulated in terms of the electron density $\rho(\mathbf{r})$, rather than the many-electron wavefunction. According to the Hohenberg–Kohn theorems, the ground-state energy of an N-electron system is a unique functional of the density, i.e.,

$$E[\rho] = T_s[\rho] + V_{\text{ext}}[\rho] + J[\rho] + E_{\text{xc}}[\rho], \tag{13.11}$$

where $T_s[\rho]$ is the kinetic energy of a fictitious non-interacting system with the same density $\rho(\mathbf{r})$, $V_{\text{ext}}[\rho]$ is the external potential energy (primarily due to the nuclear framework), and $J[\rho]$ is the classical Coulomb energy. The term $E_{\text{xc}}[\rho]$ is the exchange-correlation functional, encompassing all the many-body effects, including exchange and electron correlation contributions that go beyond the mean-field approximation.

The practical implementation of DFT is carried out through the Kohn–Sham (KS) formalism, which introduces a system of non-interacting electrons moving in an effective potential chosen to yield the exact ground-state density of the interacting system. The KS equations have a similar form to the HF equations, but the central challenge in DFT is to develop accurate approximations for $E_{\text{xc}}[\rho]$. Although the exact form of $E_{\text{xc}}[\rho]$ remains unknown, numerous approximate functionals have been devised. These functionals are often categorised along what is commonly referred to as *Jacob's ladder*, a metaphorical hierarchy connecting the simplest approximations to increasingly sophisticated ones, thereby aiming to approach the 'heaven' of chemical accuracy.

At the lowest rung of Jacob's ladder lies the *local density approximation (LDA)*, where $E_{\text{xc}}[\rho]$ depends solely on the local density at each point in space. Although LDA often yields surprisingly reasonable results for bulk solids and certain other systems, it can be qualitatively inaccurate for molecular geometries and energetics due to its overly local description of electron correlation.

The next rung comprises *generalised gradient approximations (GGAs)*, which incorporate the gradient of the density $\nabla\rho(\mathbf{r})$ to improve upon LDA. GGAs, such as PBE and BLYP, often provide better atomisation energies, improved equilibrium geometries, and more reliable descriptions of molecular properties relevant in astrochemistry. However, they may still under- or overestimate certain reaction barriers and fail to provide the highly accurate energetics required for some delicate processes.

Climbing further, *meta-GGAs* introduce additional ingredients such as the kinetic energy density or the Laplacian of the density. Meta-GGAs (e.g., SCAN) often achieve a more balanced description of both molecular and condensed-phase systems, making them appealing for complex astrochemical environments, such as grain surfaces and molecular clusters.

The subsequent rung involves *hybrid* functionals, which mix a portion of exact exchange from a wavefunction-based calculation (often from HF theory) with a GGA or meta-GGA correlation component. Prominent hybrids include B3LYP and PBE0. By incorporating exact exchange, hybrids often correct some of the self-interaction errors and significantly improve the prediction of molecular properties, excitation energies, and reaction barriers. However, they are more computationally expensive than pure DFT approaches, which can be challenging for very large systems.

Finally, the top rung of Jacob's ladder (at least in commonly used functionals to date) includes *double-hybrid* functionals.[124] These methods blend the concepts of hybrid DFT with second-order perturbation theory, effectively incorporating both explicit exchange and a perturbative correlation treatment. Double hybrids can yield accuracies approaching or even surpassing post-HF methods for certain properties. However, their computational costs and scaling with system size are higher than conventional hybrids, imposing practical limits on their applicability to large systems.

Beyond the scales where quantum chemical methods remain tractable, classical and stochastic approaches provide invaluable frameworks for exploring nano- and mesoscale phenomena. These methods are particularly relevant for understanding complex astrochemical processes, such as ice formation on dust grains, cluster aggregation, and the evolution of solid-state environments where fully *ab initio* treatments become prohibitively expensive.

At the simplest level, *molecular mechanics (MM)* methods describe atomic interactions using classical force fields rather than explicitly solving the electronic Schrödinger equation. Here, the total energy is expressed as a sum of simple functional forms, typically including bond stretches, angle bends, torsions, and non-bonded interactions such as van der Waals and electrostatic terms:

$$E_{\mathrm{MM}} = \sum_{\text{bonds}} \tfrac{1}{2}k_b(r - r_0)^2 + \sum_{\text{angles}} \tfrac{1}{2}k_\theta(\theta - \theta_0)^2 + \sum_{\text{torsions}} V_n[1 + \cos(n\varphi - \gamma)] + \sum_{i<j} \left(\frac{A_{ij}}{r_{ij}^{12}} - \frac{B_{ij}}{r_{ij}^{6}} + \frac{q_i q_j}{4\pi\epsilon_0 r_{ij}} \right). \tag{13.12}$$

These simple expressions allow the rapid computation of energetics and geometries for thousands to millions of atoms, enabling coarse but fast models of large astrophysical systems, such as icy mantles and dust grain aggregates.

To capture dynamic behaviour and temperature effects, *molecular dynamics (MD)* simulations integrate the classical equations of motion:

$$m_i \frac{d^2\mathbf{r}_i}{dt^2} = -\nabla_{\mathbf{r}_i} E_{\mathrm{MM}}(\{\mathbf{r}_j\}), \tag{13.13}$$

where m_i and $\mathbf{r}_i$ denote the mass and position of atom i, respectively. MD provides time-resolved trajectories of atomic motion, offering insights into processes such as molecular diffusion on surfaces, solvation dynamics in icy environments, and the thermal evolution of molecular assemblies under astrophysical conditions.

In scenarios where bond breaking and formation are integral to the studied dynamical process, two possibilities emerge. *Born–Oppenheimer MD*, employing either *ab initio* or semi-empirical methods, is well suited for small-scale systems and has been applied in astrochemistry to investigate, e.g., potential pathways for the synthesis of interstellar glycine ($C_2H_5NO_2$),[125] fragmentation pathways of saturated interstellar molecules,[126] and the formation of N-heterocyclic compounds.[127] In the latter case, the *ab initio nanoreactor* method, developed by the Martínez group,[128] was utilised. For larger systems or extended timescales, *reactive force fields* offer a more sophisticated framework than traditional fixed-bond force fields, which cannot describe bond breakage or bond formation processes. Approaches such as the reactive force field (ReaxFF)[129] or the reactive CHARMM (rCHARMM)[130] implementation in the MBN Explorer software[131] dynamically adjust bonding environments according to instantaneous atomic configurations, thereby enabling simulations of chemical reactions without prior specification of reaction pathways. This capability is especially relevant for astrochemical contexts where complex reaction networks, radical formation, or catalytic processes on dust grain surfaces play an essential role. Moreover, a recent development in MBN Explorer allows for the incorporation of irradiation-driven phenomena[132] through the *irradiation-driven molecular dynamics (IDMD)* method.[133,134] IDMD extends conventional reactive force field simulations by explicitly accounting for the interaction of energetic particles (e.g., electrons, ions) with the simulated system. As a result, it enables the examination of radiation-induced chemical reactions, bond scission, and structural reorganisation processes on larger spatial and temporal scales.

When phenomena occur on timescales or length scales further beyond the reach of classical MD, more coarse-grained or stochastic methods[135] become necessary. For example, *kinetic Monte Carlo (KMC)*[136,137] simulations and other stochastic approaches consider reaction events as probabilistic steps governed by predefined rate constants, thereby bypassing the need to simulate every atomic vibration. In KMC, the system evolves through a sequence of discrete events, such as adsorption, desorption, and reactive encounters, according to a probabilistic algorithm:

$$P(\text{event}) \propto k_{\text{event}} e^{-E_a/k_B T}, \tag{13.14}$$

where k_{event} is a rate constant and E_a is the activation energy of the process. Such methods enable the study of slow processes such as grain mantle growth and the progressive formation of COMs on interstellar dust particles over astronomical timescales.

MM, MD, and stochastic approaches provide a *multiscale modelling* framework, which can therefore bridge the gap between fully quantum chemical treatments and large-scale astrophysical phenomena.[132] In practice, one often uses a combination of these techniques: for instance, parameterising force fields or reaction rate constants with DFT or post-HF calculations, and then employing classical and stochastic methods to extend these insights to longer timescales and larger system sizes.

In the following sections, we present distinct examples illustrating how computational astrochemistry provides crucial findings and valuable information to the field, ranging from the generation of accurate spectral data for molecules and interstellar ices to the modelling of ice growth processes.

13.4.2 *Generation of spectral data I: Rotational and vibrational features*

The advent of modern facilities for radioastronomy and infrared spectroscopy, such as JWST and ground-based observatories like those discussed in the previous chapter, has revolutionised the detection of interstellar molecules and significantly improved the quality of astrochemical spectral data. The continuous recording of spectral signatures from these advanced instruments necessitates detailed interpretation and identification. As more state-of-the-art facilities are developed and become operational, the volume of observational data is poised to grow exponentially, emphasising the need for robust, accurate, and comprehensive theoretical data sets to enable effective interpretation and contextualisation of these groundbreaking findings.

Computational astrochemistry has, indeed, a unique role to play in this context. While laboratory experiments and observational campaigns produce invaluable data, they cannot match the sheer throughput attainable by high-performance computing. The generation of potential energy surfaces (PESs), which serve as the foundation for accurate anharmonic vibrational frequency calculations and precise rotational constants, can be designed as parallel tasks and executed on multiple computing nodes simultaneously. With current software and computational approaches, it is now routine to compute, e.g., anharmonic fundamental vibrational frequencies for moderate-sized molecules to within a few percent of experimental values, using only a modest high-performance computing cluster. Such results can be obtained in a matter of days, providing a rapid and cost-effective means of populating spectral databases.

The required software tools for constructing PESs, performing the required high-level calculations, and automating the subsequent analyses of results are already widely available. The integration of automated workflows now allows for the near-continuous generation of predictive spectra. Moreover, recent advances in computational chemistry have made it possible to automatically generate trial molecular structures by combining predefined fragments, potentially leading to the automated prediction of spectroscopic data for entire classes of astrochemically relevant molecules. Although some molecular targets may present difficulties, the vast majority can be successfully processed, and the problematic cases can be refined by hand. This synergy between automation and targeted human intervention sets the stage for an enormous increase in theoretical spectral data, potentially keeping pace with the rapid inflow of observational measurements.

Seminal works in this field highlight the potential of these computational strategies. For instance, the research conducted by Puzzarini and collaborators[138] has demonstrated that quantum-chemical computations, combined with advanced nuclear-motion treatments, can yield rotational constants and vibrational features of benchmark accuracy for key astrochemical molecules. Studies on molecules such as oxirane (C_2H_4O)[139] and protonated oxirane ($C_2H_4OH^+$),[140] as well as various prebiotic candidates, including aminoacetonirile ($C_2H_4N_2$)[141] and 3-aminoisoxazole ($C_3H_4N_2O$),[142] provide a clear view of how rotational spectra can be predicted to guide laboratory assignments and assist in the interpretation of astronomical line surveys. These efforts have also encompassed COMs and molecular ions of potential astrochemical significance,[143] where precise rotational data are essential for interpreting the dense spectral line forests captured by ALMA and other advanced radio observatories.

For infrared spectroscopy, the situation is equally promising. Large databases such as the NASA Ames PAHdb[144] provide a wealth of IR spectral data for PAHs, which are believed to be ubiquitous in space.[145–147] Refined IR simulations help disentangle overlapping spectral features and identify structural motifs, ionisation states, and substituents. Beyond PAHs, dedicated efforts to compute and curate IR spectra for a broad range of molecular species, including inorganic compounds, prebiotic precursors, and radicals, have supported the identification of previously unassigned interstellar absorption and emission bands.

As we move forward, the combination of enhanced computational efficiency, automated molecular generation, and the increasing availability of high-quality *ab initio* methods will drive theoretical spectroscopy to meet, and possibly exceed, the challenges posed by the torrent of new observational data. In turn, these theoretical spectra will be essential for extracting meaningful astrophysical insights from the wealth of information delivered by modern and future astronomical observatories.

13.4.3 *Generation of spectral data II: Electronic features*

While rotational and vibrational spectroscopies have traditionally been the main pillars for detecting and analysing molecules in space, electronic spectroscopy is emerging as an essential complementary tool. The ability to probe electronic transitions opens the door to exploring aspects of astrochemical environments that may not be fully accessible through other means. This is especially significant for interstellar ices, where molecules – ranging from simple species such as H_2O, NH_3, and CO, to more complex organics – become trapped on dust grain surfaces under extremely cold, tenuous conditions. Electronic spectroscopy thus stands poised to complement IR-based methods, potentially enhancing our understanding of ice composition and chemistry in both the ISM and solar system objects.

A recent study by Woon (2024)[148] developed a protocol utilising time-dependent density functional theory (TD-DFT) to predict electronic excitation spectra of astrophysical ices. The approach began with careful benchmarking against high-level multireference configuration interaction (MRCI) calculations for a training set of key astromolecules (CO, H_2O, HCN, NH_3, CH_3OH, and OH). Then, the work was extended from isolated molecules to cluster models representing amorphous ices.

Constructing such cluster models is a non-trivial task. Typically, a series of finite clusters (10–30 molecules) is employed to approximate the local environments within an amorphous ice. These cluster models are generated by starting from random or carefully chosen initial arrangements of the constituent molecules and then optimising their geometries, ensuring that the final cluster structures correspond to local minima on the PES, capturing some of the complexity and heterogeneity intrinsic to amorphous solids. After computing multiple clusters of varying sizes and configurations, the individual predicted spectra are summed, averaging out peculiarities of any single configuration and yielding a more representative electronic spectrum for the amorphous ice as a whole.

When applied to amorphous water and ammonia ices, these carefully constructed cluster models and optimised TD-DFT protocols yielded spectra that qualitatively reproduce key experimental features, often differing by only a few tenths of an electronvolt to the experimental spectra. While not perfectly matching every detail, these calculations mark a substantial step forward, confirming that TD-DFT-based electronic spectroscopy can serve as a valuable guide for laboratory investigations and the interpretation of astronomical data, potentially identifying radical intermediates or tracing the influence of energetic processing. In the future, improved functionals, larger and more diverse cluster sets, and integrated approaches combining vibrational and electronic analyses may further refine our understanding, ultimately providing a more coherent and comprehensive picture of the chemistry and physics dictating the properties and evolution of interstellar ices.

13.4.4 *Minimum energy landscape and reaction mechanisms*

The search for new interstellar molecules is intimately connected to identifying low-energy molecular structures for given stoichiometries involving the most common elements in the Universe. Historically, the principle that the most stable isomer of a given molecular formula is the most likely to be abundant and thus detectable in the ISM has served as a guiding heuristic. This concept, known as the *minimum energy principle (MEP)*, was proposed by Lattelais *et al.* (2009).[149] Several detections support this principle. For example, in the C_3H_4 stoichiometry, methyl acetylene (CH_3CCH), the most stable isomer, is indeed observed in space.[150] Similarly, in the CH_4N_2O system, urea ($CO(NH_2)_2$), which holds the global minimum position on its corresponding PES,[151] has also been detected.[152]

However, deviations from this simple rule have emerged. Higher-energy isomers such as propanal (C_2H_5CHO) and propylene oxide (CH_3CHCH_2O), which lie 5.4 and 30.9 kcal/mol above acetone ($(CH_3)_2CO$), the global minimum with the C_3H_6O stoichiometry, have been observed in the ISM.[153] Likewise, the C_2H_5NO and C_4H_3N systems saw the detections of N-methylformamide

(CH_3NHCHO) in 2017[154] and propargyl cyanide ($HCCCH_2CN$) in 2020, even though acetamide (CH_3CONH_2),[155] detected earlier in 2006,[156] and methylcyanoacetylene (CH_3C_3N),[157] detected in 1984,[158] are the corresponding global minima. Another illustrative case is N-cyanomethanimine (H_2CNCN), a higher-energy isomer of $C_2H_2N_2$ observed with abundances comparable to its more stable counterpart.[159] Also, recent work by Noriega *et al.* (2024)[160] on $C_3H_8O_2$ isomers exemplifies the complexity of isomeric landscapes. Their comprehensive exploration of the PES revealed that 2-methoxyethanol, currently the only $C_3H_8O_2$ species detected in the ISM, is far from the global minimum. But perhaps the most striking example is the case of the $C_2H_4O_2$ isomers. Despite being the most thermodynamically stable isomer, acetic acid (CH_3COOH) is consistently at least 10 times less abundant than its isomer, methyl formate ($HCOOCH_3$), in virtually all astronomical sources where it has been detected.[149]

These cases clearly show that while the global minimum on the PES is often a good first candidate for molecular detection, it is not the sole determinant. Ellinger *et al.* (2020)[161] proposed a set of factors that influence detectability: (i) more rigid molecular structures yield simpler rotational spectra, aiding line assignments; (ii) a dipole moment of at least 2 D enhances the strength of observed transitions; (iii) the detected molecule should ideally be the most stable or at least reside within a $\sim$30 kcal/mol window above the global minimum; and (iv) species that are only weakly bound to grain surfaces are more easily released into the gas phase, increasing their chances of detection. More recently, Hrodmarsson *et al.* (2024)[162] pointed to VUV photostability as a key factor, proposing that the non-detection of the most stable $C_3H_6N_2$ isomer, 2-aminopropionitrile,[163] may be due to its susceptibility to photodestruction.

Beyond neutral species, molecular ions also play significant roles in interstellar chemistry. Interestingly, ionisation can lead to structural rearrangements, sometimes resulting in highly stable or entirely new cationic geometries without neutral analogues. For instance, the $C_2H_4O_2$ system exhibits intriguing cationic and dicationic structures that can deviate substantially from their neutral minima. Investigations have shown that the most stable $C_2H_4O_2^+$ species are enolic, stabilising rare tautomers not favoured in the neutral manifold. The doubly charged $C_2H_4O_2^{2+}$ species introduce even more unusual bonding motifs, with its global minimum featuring terminal H_2O^+ and $C{\equiv}O^+$ groups.[164,165]

Another example of differences between the neutral PES and those of the corresponding cations and anions is provided by the C_4H_nN species (with n ranging from 3 to 5), as investigated by Perlin, Wolff and Oliveira.[166] For C_4H_3N, the global minima of the cationic and anionic states correspond to $CH_2C_3NH^+$ and $CH_2CHC_2N^-$, respectively, rather than CH_3C_3N, the global minimum for the neutral, as stated above.[157] In the case of C_4H_4N, the anionic global minimum adopts a five-membered ring structure, while the neutral and cationic global minima are distinct open-chain structures. For C_4H_5N, both the neutral and cationic global minima feature a five-membered ring, whereas the anionic global minimum exhibits a branched open-chain configuration.

Benzene (C_6H_6) and PAHs are also susceptible to drastic loss of structural integrity in ionised states. For C_6H_6, the characteristic six-membered planar carbon ring is disrupted in its doubly charged analogue, which features a pentagonal-pyramidal C_5H_5 ring with an additional CH motif oriented perpendicular to the pentagonal base.[167,168] Methylation of benzene leads to toluene (C_7H_8), the simplest alkyl-functionalised benzene derivative. Monfredini *et al.* (2016)[169] performed a computational investigation of ${C_7H_n}^{2+}$ species, formed from X-ray photoionisation and photodissociation of C_7H_8, and found that ${C_7H_2}^{2+}$ and ${C_7H_3}^{2+}$ have polyyne-like forms, a low-lying ${C_7H_4}^{2+}$ minimum contains a planar tetracoordinate carbon, ${C_7H_6}^{2+}$ features dual cyclopropenyl groups, and the ${C_7H_8}^{2+}$ dication forms a meta-protonated benzyl cation, confirming a previous attribution by by Roithova *et al.* (2006).[170]

Chlorobenzene (C_6H_5Cl) is the simplest chlorinated benzene ring and belongs to the organohalogen class. Organohalogens became focus of astrochemical research following the detection of two isotopologues of CH_3Cl in the low-mass protostar IRAS 16293–2422 and the coma of comet 67P/Churyumov–Gerasimenko.[171] Interestingly, ionisation of C_6H_5Cl can yield highly exotic species. Dication and trication species have been observed in proton-impact experiments, with the most stable structures featuring unusual C–Cl double bonds ($r_{\mathrm{C-Cl}}$ = 1.542–1.621 Å).[172]

Complete hydrogenation of benzene produces cyclohexane (C_6H_{12}). Ionisation of cyclohexane leads preferentially to C–C bond rupture and molecular reorientation to branched, open-chain isomers.[17] Notably, the most stable cyclic ionic structure is a three-membered ring. Its surprisingly large stability is attributed to the formation of an endocyclic C–C one-electron bond, which reduces steric repulsion through bond elongation. Moreover, this isomer has three methyl groups connected to the two endocyclic carbon atoms involved in the one-electron bond, providing additional stabilisation.[173]

Turning to PAHs, naphthalene ($C_{12}H_8$) remains fused-ring-based at low charge states, but high charge states (e.g., 4+) favour open-chain isomers.[147] Increasing charge reduces the dominance of aromatic structures, increasing isomerisation and promoting forms such as cyclopropenyl rings and open chains. Biphenyl ($C_{12}H_{10}$), when ionised, similarly rearranges. Quitián-Lara *et al.* (2020)[174] showed that ${C_{12}H_{10}}^+$ favours an acenaphthene-like structure that is 20 kcal/mol more stable than the biphenyl form. The most stable ${C_{12}H_{10}}^{2+}$ dication also prefers an acenaphthene backbone, as steric repulsion destabilises biphenyl-like geometries. Such acenaphthene cores relate to cyanoacenaphthylene ($C_{12}H_7CN$), whose two isomers were recently detected in TMC-1.[175]

In addition to examining the minimum energy landscape of PESs, understanding mechanistic reaction pathways and kinetic factors is essential for unravelling the formation and destruction of astrochemically relevant molecules. Distinct research groups, including those led by Balucani,[176] Kaiser,[177] Ugliengo,[178] Rimola,[179] and Geppert,[180] to name a few, have been thoroughly investigating these processes. Their efforts encompass bottom-up reactions, such as CO hydrogenation on interstellar ices, as well as top-down mechanisms involving the fragmentation of PAHs, alongside complex gas-phase and surface-mediated pathways.[181] These theoretical approaches, ranging from DFT to post-HF methods, are vital for interpreting experimental results, guiding observations, and bridging the gap between laboratory conditions and the true complexity of the ISM. The synergy of advanced quantum-chemical calculations and kinetic modelling provides insights into the origins and abundances of COMs, whether formed via surface-catalysed reactions on icy dust grains or through intricate radical–radical couplings in the gas phase. Moreover, these theoretical frameworks aid in interpreting data from the latest observational facilities and sample return missions,[182] thereby helping us understand the molecular heritage and complexity that define our cosmic surroundings.

13.4.5 *Surface astrochemistry and multiscale modelling*

The chemistry that unfolds on dust-grain surfaces is fundamental to astrochemistry, enabling the formation of saturated, hydrogen-rich molecules and more complex organic species that gas-phase reactions alone struggle to produce.[183] Detailed, quantum-level insights from DFT and related methods now guide the construction of accurate input parameters – binding energies, reaction barriers, and branching ratios – for large-scale models. Such data have revealed, for example, that radicals such as HCO can form and recombine on cold grain surfaces without the need for thermal activation or external irradiation, directly yielding glyoxal (HC(O)CHO)[108] and, upon successive hydrogenation, glycoaldehyde ($HC(O)CH_2OH$) and ethylene glycol ($H_2C(OH)CH_2OH$).[183] These findings challenge previous assumptions that UV photolysis or elevated temperatures are prerequisites for COM synthesis, demonstrating that purely local, low-temperature radical–radical interactions can suffice under certain conditions.

This radiation-free route to complexity is not the whole story, however. Radiation – whether from UV photons, cosmic rays, or X-rays – continues to be a key factor in interstellar environments. Photochemistry and radiative processing can dissociate stable molecules, generate new radicals, and broaden the chemical network, while non-thermal excitation from irradiation can stir species into different parts of the mantle. Balancing these contrasting routes requires a concerted modelling effort. Key parameters, such as molecular binding energies to heterogeneous grain surfaces, are still poorly constrained, and even small deviations can drastically alter predicted abundances and chemical routes. Moreover, exothermic surface reactions introduce non-thermal diffusion processes, reshaping ice morphologies and altering the distribution of reactive intermediates. KMC simulations, fed by DFT-derived energy landscapes, can capture how subtle changes in local structure and energetics influence radical pairing and product formation. MD techniques provide atomistic-level insight into how energy released by reactions or radiation-induced fragmentation can lead to rearrangements, desorption events, or further reactivity.

Yet, no single simulation method suffices on its own. Multiscale modelling emerges as a powerful strategy by integrating distinct computational frameworks.[132] Rate-equation models help track bulk chemical evolution over astronomical timescales, KMC simulations refine how microscopic detail affects ice composition, and MD provides a window into immediate atomistic rearrangements following photodissociation or radiolysis. Coupling these layers ensures that the microscopic accuracy gleaned from DFT and MD feeds into KMC and rate-equation models, connecting reaction pathways and radiation-driven chemistry to global ice properties and molecular distributions. A small shift in a binding energy measured or estimated from quantum calculations can, for example, transform an initially porous ice morphology into a more compact structure, influencing whether radicals remain trapped long enough to form prebiotic species or become reprocessed by radiation.

As facilities like JWST deliver new spectroscopic data and broaden the known census of interstellar molecules, these multiscale astrochemical models will be crucial. They will bridge laboratory scale constraints, computational predictions, and observations, helping to pinpoint the most influential reactions and conditions. Multiscale models can unravel the interplay of factors shaping the diverse chemical complexity found in interstellar ices.

13.5 Chirality and Biomolecular Processes

13.5.1 *Manifestations of chirality across disciplines*

Chirality – the property of an object not being superimposable on its mirror-image – is omnipresent in nature. In chemistry, chirality manifests on the structural level, with point chirality being the most common form. It is characterised by a central atom (often carbon) having four non-identical substituents around a tetrahedral geometry (Fig. 13.4). Molecules, however, can also exhibit other types of chirality, such as axial, helical, propeller, inherent, or planar.[184,185] In biology, chirality spans multiple scales: from the molecular level, such as the right-handed helical structure of DNA; through biological

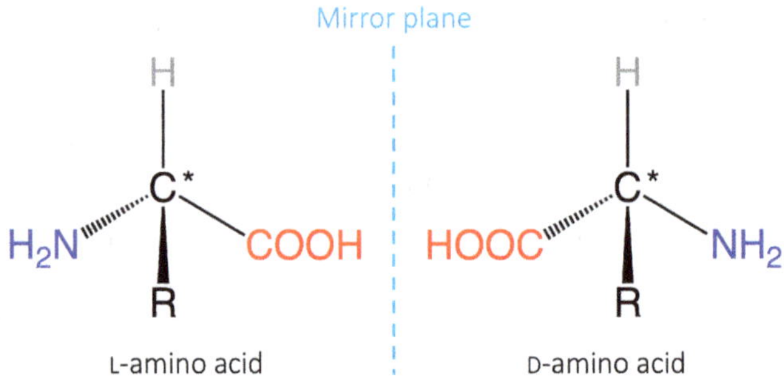

Fig. 13.4. Point chirality of amino acids. A pair of enantiomers of a chiral α-hydrogen-amino acid that are non-superimposable mirror images. Left-handed (L-) amino acids build proteins as opposed to their non-biological right-handed (D-) counterparts.

nanostructures, such as the intricate arrangements of butterfly wing scales;[186] to the cellular level, exemplified by cortical structures in ciliates; tissues, such as the left–right asymmetry observed in *Drosophila*; and organs, such as hands.[187] In physics, chirality manifests in parity violation in weak interactions,[188] in spin polarisation of elementary particles,[189] or in circular polarisation of electromagnetic radiation.[190]

One of the biggest yet unresolved questions in the field of chirality is how life's homochirality originated. While life's key biopolymers, proteins, are built from homochiral left-handed L-amino acids, the backbones of RNA and DNA are solely based on the right-handed D-enantiomers of sugars bound to phosphate. There have been multiple detections of enantioenriched chiral organics in carbonaceous chondrites (organic carbon-rich meteorites),[191] with amino acids standing out due to their numerous systematic L-enrichments. These enrichments mirror the homochirality of amino acids in Earth's proteins, suggesting a potential extraterrestrial origin for the foundations of biological homochirality.

13.5.2 *Mechanisms of symmetry breaking in chiral molecules*

While several experiments on interstellar ice analogues irradiated by cosmic ray proxies or UV photons provide strong evidence for rich synthesis of organics in the ISM,[192–194] the underlying mechanisms for the enantioenrichment of chiral species remain under discussion. Symmetry breaking requires the presence of a chiral force, and several scenarios have been outlined to date, which could explain the generation of an enantiomeric imbalance in chiral prebiotics, the amplification of which might have led to the evolution of homochiral life.

Given the focus on astrochemistry, we provide here an overview of the physicochemical mechanisms that could trigger symmetry breaking in chiral organics synthesised in the interstellar environments in space and that could, possibly, explain their enantioenrichments detected in carbonaceous chondrites.[191] A recent comprehensive review on symmetry-breaking processes can be found in the work of Sallembien *et al.*[195]

Circularly polarised light (CPL) is a true chiral force capable of inducing enantioenrichment. The enantioselective interaction of CPL with enantiomers of chiral molecules stems in different absorption cross-sections for the absorption of left- and right-handed CPL (Fig. 13.5(a)). This phenomenon, known as circular dichroism (CD), is quantified as $\Delta\varepsilon = \varepsilon_L - \varepsilon_R$, where ε_L and ε_R are the extinction coefficients for left- and right-handed CPL, respectively. The magnitude of the CD signal is proportional to the scalar product of the electric and magnetic transition dipole moments associated with the electronic transition induced by the absorption of CPL.

The enantioselective response to CPL is quantitatively described by the anisotropy factor g, defined as the ratio of the differential

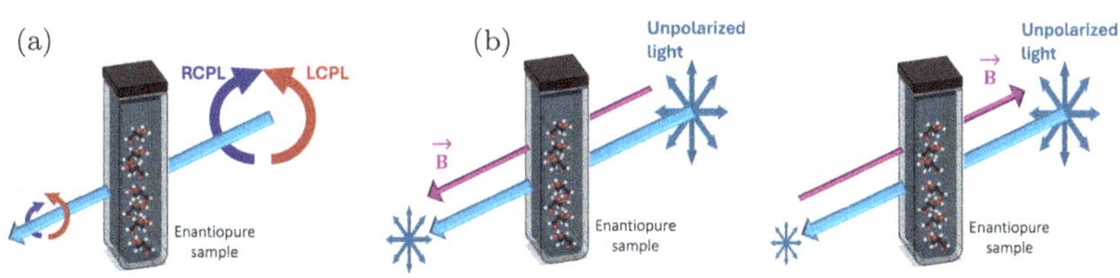

Fig. 13.5. Principle of (a) circular dichroism and (b) magneto-chiral dichroism. In (a), the enantiopure sample preferentially absorbs right-handed circularly polarised light (RCPL) as opposed to left-handed circularly polarised light (LCPL). In (b), the absorbance of unpolarised light by the enantiopure sample depends on the direction of the magnetic field (parallel or anti-parallel) with respect to the direction of light propagation.

absorption of left- and right-handed CPL to the absorption of unpolarised light:

$$g = \frac{\Delta\varepsilon}{\varepsilon}, \tag{13.15}$$

where $\varepsilon = (\varepsilon_L + \varepsilon_R)/2$ is the average extinction coefficient.

Three mechanisms have been identified as capable of inducing chiral imbalance upon interaction with CPL: (1) asymmetric photolysis, (2) asymmetric photoresolution, and (3) asymmetric photosynthesis.[196] Several experimental studies have addressed the latter two, i.e., the interconversion of enantiomers of a racemic mixture in the excited state following absorption of UV CPL,[197,198] and the UV CPL-induced enantioselective synthesis of chiral molecules starting from achiral precursors condensing on substrates cooled down to several tens of Kelvin, mimicking interstellar ice conditions.[199,200] Yet, the most studied of the three processes, yielding the highest enantiomeric enrichments to date, is asymmetric photolysis. This process induces a chiral imbalance in a racemic mixture due to the preferential photodegradation of the more absorbing enantiomer at a given wavelength and helicity of UV CPL.[201–206]

The enantiomeric excess (ee) is defined as

$$ee_L = \frac{(L - D)}{(L + D)}, \tag{13.16}$$

and its lower limit achievable in asymmetric photolytic reactions was first derived by Kagan *et al.* (1974)[202] and later simplified by Meinert *et al.* (2012)[207] as

$$|\%ee| \geq (1 - (1 - \xi)^{|g|/2}) \times 100\%, \tag{13.17}$$

where ξ is the extent of reaction and g is the anisotropy factor, as defined above. The value of g is typically close to zero for most molecules, reflecting the subtle enantioselectivity of CPL interactions.

The anisotropy factor g, and hence the efficiency of CPL-induced asymmetric reactions, depends not only on the wavelength of CPL but also on the immediate environment of the molecule.[208–210] This is because the environment can influence molecular structure, with different conformations of the same molecule potentially exhibiting

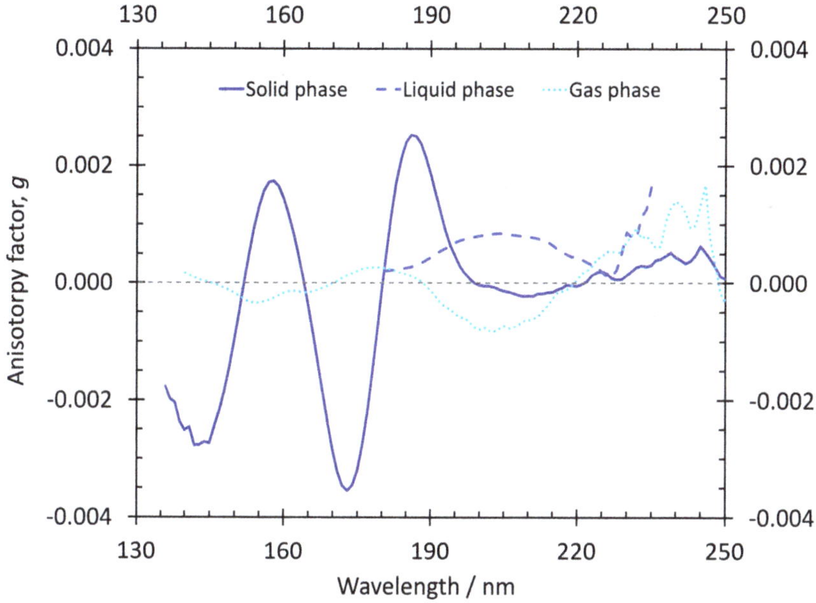

Fig. 13.6. The effect of molecular environment on the chiroptical properties of L-isovaline. A comparison of the anisotropy spectra of L-isovaline in the form of a solid amorphous film (dark blue, solid line), in aqueous solution (blue, dashed line), and in the gas phase (light blue, dotted line) reveals significant differences in the sign, position, and magnitude of the anisotropy bands. It should be noted that amino acids are present in the solid and liquid phases as zwitterions, whereas they are present as neutrals in the gas phase.
Source: Reproduced from Bocková *et al.* (2023)[210] (http://creativecommons.org/licenses/by/4.0/, enantiomeric excess values were converted to anisotropy values).

different circular dichroism and absorption spectra.[208] A textbook example is the striking variation in the intensity and band positions of the anisotropy spectra of the amino acid isovaline ($C_5H_{11}NO_2$) in the solid, liquid, and gas phases (Fig. 13.6).[210] Therefore, a thorough investigation of the chiroptical spectra of a large set of prebiotically relevant chiral organics across a broad wavelength range, under conditions closely resembling astrophysical environments, offers the greatest potential for unravelling the role of stellar CPL in symmetry breaking in space.

Interestingly, unpolarised light propagating parallel or antiparallel to a static magnetic field acts also as a chiral force, an effect known as magneto-chiral dichroism (MChD, Fig. 13.5(b)).

The enantioenrichment resulting from MChD is opposite for optical antipodes, and its sign reverses when the direction of the magnetic field is reversed relative to the direction of light propagation.[211]

Compounds exhibiting strong CD and magnetic circular dichroism (MCD) – the differential absorption of left- and right-handed CPL induced by a strong magnetic field aligned with or opposite to the direction of light propagation – are also expected to exhibit MChD.[212] MChD was first experimentally confirmed in a chiral europium(III) complex by Rikken and Raupach (1998),[213] taking advantage of the strong magnetic dipole character and MCD response of the $5D_0 \rightarrow 7F_{1,2}$ luminiscent transitions of Eu(III).

To achieve MChD in purely organic compounds, high orbital angular momentum of the aromatic π-conjugated system of porphyrins was exploited.[214] Later on, theoretical calculations also predicted MChD in the amino acid L-alanine, albeit with intensities at the edge of experimental capabilities.[215]

In general, MChD requires very strong magnetic fields, and hence areas in the vicinity of neutron stars with magnetic fields of 10^8–10^{12} T have been suggested as potential sites for MChD-induced symmetry breaking.[212] Nevertheless, experimental evidence of MChD-induced enantioenrichment in prebiotically relevant chiral organics remains absent in the literature to date.

13.5.3 *Spin-polarised electrons*

Interactions with spin-polarised particles have been identified as another potential mechanism capable of inducing chiral imbalance in a racemic mixture. In particular, nuclear-decay- or cosmic-ray-derived leptons (electrons, muons, and neutrinos) exhibit a predefined helicity due to parity violation in weak interactions. This means that their spin angular momentum is polarised anti-parallel (left-handed) or parallel (right-handed) to their kinetic momentum (Fig. 13.7).[195,216] Among these leptons, electrons have captured the most attention due to their omnipresence in ordinary matter.

In nature, left-handed spin-polarised electrons are emitted during β^- (beta-minus) decay of a radioactive nucleus mediated by the weak interaction. This phenomenon was first experimentally demonstrated in the famous Wu experiment, which revealed asymmetry in

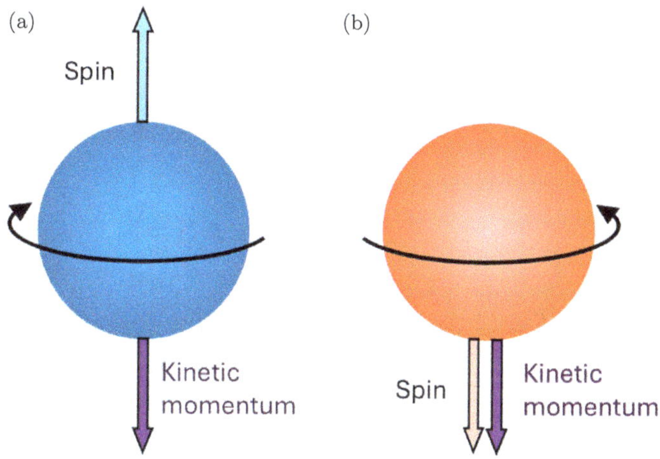

Fig. 13.7. (a) Left- and (b) right-handed spin-polarised electrons with spin angular momentum and kinetic momentum aligned anti-parallel and parallel, respectively.

the angular distribution of electrons emitted during the β^- decay of polarised ^{60}Co nuclei.[217]

Despite considerable experimental efforts, neither the direct effect of high-energy spin-polarised electrons in enantioselective destruction, nor stereoselective degradation or synthesis via circularly polarised '*Bremsstrahlen*' photons – resulting from irradiation of matter with high-energy spin-polarised electrons[218,219] – have proven efficient mechanisms for inducing enantiomeric enrichment.[220–224]

Importantly, much more promise has been shown by low-energy (<15 eV) spin-polarised electrons.[216] Such electrons can be readily released by UV photon-induced valence-level photoionisation,[225–229] ionising radiation interacting with a magnetic substrate,[230–232] or via the spin filtering of chiral molecules adsorbed on gold surfaces irradiated by circularly or linearly polarised light, the so-called chiral-induced spin selectivity (CISS) effect (Fig. 13.8).[233–239]

Despite their energies being below ionisation thresholds, low-energy electrons can dissociate molecules via a mechanism called dissociative electron attachment (DEA, see previous chapter). In DEA, the incident electron is captured in a virtual molecular orbital, leading to a transient negative ion, which can then dissociate to a negatively charged fragment and a neutral molecule.[241]

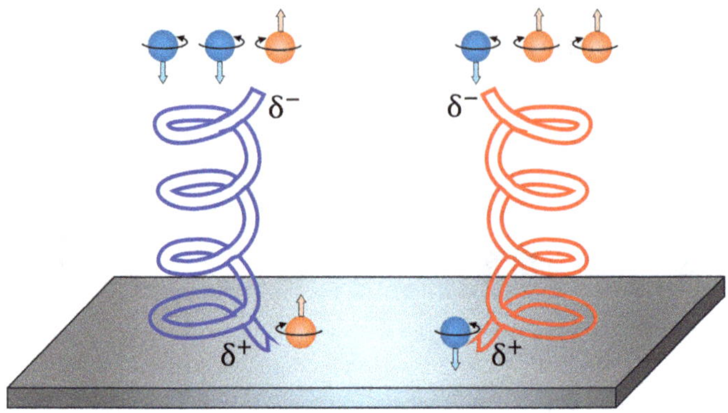

Fig. 13.8. The chiral-induced spin selectivity (CISS) effect. Enantiomers of chiral molecules act as spin filters for electron transport, initiated by their adsorption on the substrate. The charge polarisation is accompanied by a transient spin polarisation at the electric poles.[240]

Several experimental works have demonstrated enantioselective interaction of low-energy spin-polarised electrons with chiral molecules. Notably, Rosenberg *et al.* (2008)[242] observed about 10% enhancement in the CO bond cleavage rate of *R*- or *S*-2-butanol by spin-polarised (10–15% spin polarisation) secondary electrons produced by X-ray irradiation of a magnetic Permalloy ($Fe_{0.2}Ni_{0.8}$). The effect depended on both the chirality of the molecule and the degree of electron spin polarisation. A predicted *ee* value of 25% was estimated for the enantioselective degradation of the original racemic sample after 98.6% of the starting material was destroyed. Such an *ee* is about 100 times greater than the maximum *ee* predicted for the enantioselective destruction of the same compound in an aqueous solution by monochromatic circularly polarised light[243] at the same extent of reaction, highlighting the significantly higher enantioselective efficiency of spin-polarised electrons as opposed to chiral photons.

In a later study, natural spin filtering (60% spin polarisation) by a DNA layer self-assembled on a gold substrate irradiated by X-rays resulted in a 16% difference in the quantum yield of Cl dissociation from *R*- or *S*-epichlorohydrin adsorbed onto the DNA layer.[244] Furthermore, the CISS effect has been shown to cause different adsorption rates of the enantiomers of chiral molecules on a magnetic substrate depending on the direction of magnetisation.[245]

These and many other examples[195] provide a substantial support for future investigations into whether low-energy spin-polarised electron interactions in astrophysical settings could have been critically involved in inducing symmetry breaking in space. Strong magnetic fields produced around neutron stars could align magnetic domains of the interstellar dust particles enriched in iron or nickel compounds, thereby producing spin-polarised electrons upon irradiation by synchrotron light or electrons. While low-energy spin-polarised electrons could induce chiral-selective interactions with the adsorbed molecules, magnetic surfaces could also serve as enantioselective filters for accreting enantiomers of chiral compounds via the CISS effect.[216]

Finally, parity non-conserving neutral current weak interactions give rise to parity-violating forces, theoretically predicted to induce an energy shift between enantiomers of chiral molecules on the order of femtojoule to picojoule per mole.[246] Such energy differences could possibly imbalance the equilibrium composition of chiral mixtures. Despite numerous attempts,[247–253] such small energy differences have never been unequivocally confirmed experimentally. Moreover, they were found to be five to seven orders of magnitude smaller than the minimum energy difference required to consistently break the chiral symmetry and direct the Soai reaction[254] – the only yet known asymmetric autocatalytic reaction – towards a desired enantioenrichment.[255] Thus, despite being fundamentally interesting, the parity-violating energy difference (PVED) has been questioned regarding its relevance for explaining the origins of biological homochirality. Nevertheless, a complete understanding of amplification mechanisms is necessary before ultimately drawing conclusions about PVED.

In summary, several physicochemical processes have been outlined as potential triggers of symmetry breaking in chiral prebiotics. Among those discussed here, circularly polarised light and spin-polarised low-energy electrons appear to be more compelling inducers of chiral imbalance possibly involved in the origins of biological homochirality, as opposed to MChD or PVED. The former requires very strong magnetic fields, while the latter induces only a tiny energy difference between optical antipodes, which has not yet been verified even experimentally. While the asymmetric effect of circularly polarised light has been relatively well understood, the much more recent discovery of the CISS effect has, in the past few years, prompted significant research towards disentangling its physical underpinnings and mechanistic features.[239] Strengthening our foundational understanding of this effect and its extent may not only shed light on its potential involvement in the origin of biological homochirality but is also likely to foster further applications in fields such as spintronics, quantum information, enantioseparation, chemical reaction control, and beyond.

13.6 Final Remarks

The comprehensive exploration of atomic and molecular processes in astrochemistry has profoundly enhanced our understanding of the Universe and the potential origins of life. Spectroscopic measurements, both experimental and computational, are indispensable for interpreting astronomical data and investigating photon-induced processes such as absorption, ionisation, and dissociation. The extensive spectral range – from radio waves to X-rays – demands vast data repositories and the development of sophisticated models to generate the necessary computational data.

Photon-induced processes play a crucial role in the chemical evolution of interstellar environments. High-energy photons, including γ-rays and X-rays, are instrumental in initiating reactions that lead to the formation of complex molecules. Laboratory simulations using synchrotron facilities have been pivotal in replicating these processes under controlled conditions. These studies have not only deepened our understanding of fundamental interactions such as photoabsorption and photodissociation but have also provided invaluable data for refining astronomical models and expanding spectroscopic databases.

Computational approaches are equally vital in advancing astrochemical research, complementing experimental and observational efforts. Wavefunction-based methods and DFT allow for accurate modelling of complex molecular structures and interactions, while multiscale modelling bridges the gap between molecular-level processes and macroscopic phenomena. These techniques provide information about the formation pathways of molecules under interstellar conditions, from simple diatomics to biologically relevant COMs. Additionally, computational simulations are essential for interpreting astronomical observations, enabling predictions of molecular spectra, reaction rates, and surface processes on dust grains. Such models are indispensable for constructing robust databases and understanding the intricate chemistry of the ISM, planetary atmospheres, and cometary systems.

The study of chirality and biomolecular processes adds a profound dimension to our understanding of life's potential cosmic origins. The enantioenrichment of chiral molecules, such as amino acids detected in carbonaceous chondrites, suggests extraterrestrial influences on biological homochirality. Mechanisms involving circularly polarised light and spin-polarised low-energy electrons emerge as the most plausible inducers of chiral imbalance in prebiotic molecules. While magneto-chiral dichroism and parity-violating energy differences present intriguing theoretical possibilities, their practical impact appears limited due to the requirement of extreme conditions or minimal energy differentials.

Future research should continue to focus on replicating and studying these processes under laboratory conditions that closely mimic interstellar environments. The utilisation of advanced facilities such as synchrotrons and the development of sophisticated computational models will be instrumental in this endeavour. Additionally, expanding our understanding of the chiroptical properties of prebiotically relevant molecules across various phases and conditions can provide deeper insights into the mechanisms underlying homochirality.

Ultimately, the interdisciplinary nature of astrochemistry – integrating astronomy, physics, chemistry, and biology – is crucial for addressing some of the most profound questions about the Universe. The launch of JWST, coupled with ground-based telescope arrays such as ALMA and, in the future, the Square Kilometre Array (SKA), has inaugurated a new era in astronomy where the chemical complexity of the Universe is being unveiled. Advancing this field relies heavily on the collective efforts of the scientific community to build and maintain extensive databases of atomic and molecular data. Our understanding of the Universe is fundamentally rooted in spectroscopic studies, yet the availability of such data faces challenges due to limited experimental facilities and the vast amounts of data required. Therefore, sustained collaboration between atomic and molecular physics research groups and the broader astronomy community is essential to support observational programmes and provide foundational insights into interstellar chemistry, planetary formation, and life's potential beyond Earth.

Acknowledgements

The authors gratefully acknowledge support from the Europlanet 2024 RI, funded by the European Union's Horizon 2020 Research and Innovation Programme (grant agreement no. 871149). Additional support was received from the Research, Development, and Innovation Fund of Hungary (grant no. K128621). This chapter is based on the work from the COST Action CA20129 – Multiscale Irradiation and Chemistry Driven Processes and Related Technologies (MultIChem), supported by the European Cooperation in Science and Technology (COST). S.I. acknowledges the Danish National Research Foundation for funding through the Centre of Excellence 'InterCat' (grant agreement no. DNRF150). F.F. and N.J.M. extend their gratitude for partial financial support from the European Commission via the RADON project (grant agreement no. 872494) under the H2020-MSCA-RISE-2019 call. H.M.Q.L. acknowledges additional support from the University of Kent and the Max Planck Institute for Extraterrestrial Physics (MPE). J.B. is supported by a postdoctoral fellowship from the National Centre for Space Studies (CNES).

References

1. Tennyson J. Astronomical spectroscopy: An introduction to the atomic and molecular physics of astronomical spectroscopy. Singapore: World Scientific. 2019.
2. Heiter U, *et al.* Astron Astrophys. 2021;645:A106.
3. Barone V, *et al.* Nat Rev Methods Primers. 2021;1:38.
4. Parks GK. Physics of space plasmas: an introduction. Boca Raton: CRC Press. 2019.
5. Mathis JS, Mezger PG, Panagia N. Astron Astrophys. 1983;128:212–229.
6. Prasad SS, Tarafdar SP. Astrophys J. 1983;267:603–609.
7. Hunt LK, *et al.* Astron Astrophys. 2019;621:A51.
8. Bianchi M, *et al.* Eur Phys J Plus. 2023;138:132.
9. Susini J, *et al.* Eur Phys J Plus. 2024;139:80.
10. Nandedkar R, Singh G. 2003.

11. Liang KS, *et al.* Synchrotron Radiat News. 2009;22:13–20.
12. Liu L, Neuenschwander RT, Rodrigues ARD. Philos Trans R Soc A Math Phys Eng Sci. 2019;377:20180235.
13. Śmiałek MA, *et al.* Eur Phys J D. 2017;71:129.
14. Gerlach M, *et al.* J Chem Phys. 2023;158:134303.
15. Shastri A, Das AK, Rajasekhar B. J Quant Spectrosc Radiat Transf. 2020;242:106782.
16. Pavithraa S, *et al.* Astrophys Space Sci. 2017;362:126.
17. Quitián-Lara HM, *et al.* Astrophys J. 2018;854:61.
18. Pilling S, Andrade DP. X-Ray Spectroscopy. 2012.
19. Shingledecker CN, *et al.* Proc Int Astron Union. 2019;15:454–455.
20. Pavlov AA, *et al.* Astrobiology. 2024;24:698–709.
21. Ramachandran R, *et al.* Icarus. 2024;410:115896.
22. Elowitz M, *et al.* Sci Adv. 2021;7:eaba5749.
23. Mason NJ, *et al.* Faraday Discuss. 2006;133:311–329.
24. Dawes A, *et al.* J Chem Phys. 2007;126:244711.
25. Gerakines PA, Schutte WA, Ehrenfreund P. Astron Astrophys. 1996;312:289–305.
26. Muñoz Caro GM, Schutte WA. Astron Astrophys. 2003;412:121–132.
27. Öberg KI, *et al.* Astron Astrophys. 2009;504:891–913.
28. Paardekooper DM, Bossa JB, Linnartz H. Astron Astrophys. 2016;592:A67.
29. Ciaravella A, *et al.* Astrophys J. 2019;879:21.
30. Arumainayagam CR, *et al.* Chem Soc Rev. 2019;48:2293–2314.
31. Mason NJ, *et al.* Faraday Discuss. 2006;133:311.
32. Cruz-Diaz GA, *et al.* Astron Astrophys. 2014;562:A119.
33. Cruz-Diaz GA, *et al.* Astron Astrophys. 2014;562:A120.
34. Ioppolo S, *et al.* Astron Astrophys. 2020;641:A154.
35. Ioppolo S, *et al.* Astron Astrophys. 2021;646:A172.
36. Zhang J, *et al.* Mon Not R Astron Soc. 2024;533:826–840.
37. Fayolle EC, *et al.* Astrophys J. 2011;739:L36.
38. Fayolle EC, *et al.* Astron Astrophys. 2013;556:A122.
39. Fillion J, *et al.* Faraday Discuss. 2014;168:533.
40. Féraud G, *et al.* ACS Earth Space Chem. 2019;3:1135.
41. Del Fré S, *et al.* Phys Rev Lett. 2023;131:238001.
42. Cuppen HM, Linnartz H, Ioppolo S. Annu Rev Astron Astrophys. 2024;62:243–286.
43. Cuppen HM, Karssemeijer LJ, Lamberts T. Chem Rev. 2013;113: 8840–8871.
44. Jin M, Garrod RT. Astrophys J Suppl Ser. 2020;249:26.
45. Garrod RT, *et al.* Astrophys J Suppl Ser. 2022;259:1.
46. Hornekar L, *et al.* Science. 2003;302:1943–1946.

47. Thrower JD, *et al.* Astrophys J. 2012;752:3.
48. Mennella V, *et al.* Astrophys J Lett. 2021;908:L18.
49. Jaganathan R, *et al.* Astron Astrophys. 2022;663:A136.
50. Simonsen FDS, *et al.* Astrophys J. 2024;972:42.
51. Klinger J. J Phys Chem. 1983;87:4209–4214.
52. Van Dishoeck EF, Herbst E, Neufeld DA. Chem Rev. 2013;113: 9043–9085.
53. Whittet DCB, *et al.* Astrophys J. 2001;547:872–884.
54. Wakelam V, Herbst E, Cuppen H. Astrochemistry: Synthesis and modelling. In: Smith I, Cockell C, Leach S, editors. Astrochemistry and astrobiology. London: Springer. 2013.
55. van de Hulst HC. The solid particles in interstellar space. Utrecht: Drukkerij Schotanus & Jens. 1949.
56. Tielens AGGM, Hagen W. Astron Astrophys. 1982;114:245–260.
57. Cuppen HM, Herbst E. Astrophys J. 2007;668:294–309.
58. Dulieu F, *et al.* Astron Astrophys. 2010;512:A30.
59. Jing D, *et al.* Astrophys J Lett. 2011;741:L9.
60. Miyauchi N, *et al.* Chem Phys Lett. 2008;456:27–30.
61. Ioppolo S, *et al.* Astrophys J. 2008;686:1474–1479.
62. Matar E, *et al.* Astron Astrophys. 2008;492:L17–L20.
63. Ioppolo S, *et al.* Phys Chem Chem Phys. 2010;12:12065.
64. Cuppen HM, *et al.* Phys Chem Chem Phys. 2010;12:12077.
65. Lamberts T, *et al.* Phys Chem Chem Phys. 2013;15:8287.
66. Lamberts T, *et al.* Phys Chem Chem Phys. 2016;18:33021.
67. Mokrane H, *et al.* Astrophys J Lett. 2009;705:L195–L198.
68. Romanzin C, *et al.* J Chem Phys. 2011;134:084504.
69. He J, Vidali G. Astrophys J. 2014;788:50.
70. Fraser HJ, *et al.* Mon Not R Astron Soc. 2001;327:1165–1172.
71. Noble JA, *et al.* J Phys Chem Lett. 2014;5:826.
72. Noble JA, *et al.* J Phys Chem C. 2020;124:20864.
73. Coussan S, *et al.* J Phys Chem A. 2022;126:2262.
74. Cuppen HM, *et al.* J Phys Chem A. 2022;126:8859–8870.
75. Ioppolo S, *et al.* Nat Astron. 2021;5:197–205.
76. Boogert AA, Gerakines PA, Whittet DC. Annu Rev Astron Astrophys. 2015;53:541.
77. Oba Y, *et al.* Astrophys J Lett. 2010;712:L174–L178.
78. Ioppolo S, *et al.* Mon Not R Astron Soc. 2011;413:2281–2287.
79. Noble JA, *et al.* Astrophys J. 2011;735:121.
80. Oba Y, *et al.* Astrophys J. 2010;722:1598–1606.
81. Ioppolo S, *et al.* Mon Not R Astron Soc. 2011;410:1089–1095.
82. Ishibashi A, *et al.* arXiv e-prints. arXiv:2410.01373 (2024).
83. Roser JE, *et al.* Astrophys J Lett. 2001;555:L61–L64.

84. Madzunkov S, *et al.* Phys Rev A. 2006;73:020901.
85. Raut U, Baragiola RA. Astrophys J Lett. 2011;737:L14.
86. Ioppolo S, *et al.* Rev Sci Instrum. 2013;84:073112.
87. Talbi D, Chandler GS, Rohl AL. Chem Phys. 2006;320:214–228.
88. Goumans TPM, Andersson S. Mon Not R Astron Soc. 2010;406: 2213–2217.
89. Loeffler MJ, *et al.* Astron Astrophys. 2005;435:587–594.
90. Bottinelli S, *et al.* Astrophys J. 2010;718:1100–1117.
91. Hiraoka K, *et al.* Astrophys J. 1995;443:363.
92. Hidaka H, *et al.* Phys Chem Chem Phys. 2011;13:15798.
93. Fedoseev G, *et al.* Mon Not R Astron Soc. 2015;446:439.
94. Fedoseev G, Ioppolo S, Linnartz H. Mon Not R Astron Soc. 2015;446:449.
95. Kobayashi K, *et al.* Astrobiology. 2017;17:786–812.
96. Qasim D, *et al.* Nat Astron. 2020;4:781–785.
97. Qasim D, *et al.* Rev Sci Instrum. 2020;91:054501.
98. Lamberts T, *et al.* Astrophys J. 2022;928:48.
99. Qasim D, *et al.* Astron Astrophys. 2018;612:A83.
100. Bergner JB, Öberg KI, Rajappan M. Astrophys J. 2017;845:29.
101. Garrod RT, Pauly T. Astrophys J. 2011;735:15.
102. Oba Y, *et al.* Chem Phys Lett. 2015;634:53–59.
103. Hiraoka K, *et al.* Astrophys J. 2002;577:265.
104. Watanabe N, Kouchi A. Astrophys J Lett. 2002;571:L173–L176.
105. Watanabe N, *et al.* Astrophys J. 2004;616:638–642.
106. Hidaka H, *et al.* Astrophys J. 2004;614:1124.
107. Fuchs GW, *et al.* Astron Astrophys. 2009;505:629–639.
108. Fedoseev G, *et al.* Mon Not R Astron Soc. 2015;448:1288–1297.
109. Chuang KJ, *et al.* Mon Not R Astron Soc. 2016;455:1702.
110. Chuang K-J, *et al.* Mon Not R Astron Soc. 2017;467:2552.
111. Fedoseev G, *et al.* Astrophys J. 2017;842:52.
112. Simons MAJ, Lamberts T, Cuppen HM. Astron Astrophys. 2020;634:A52.
113. He J, *et al.* Astron Astrophys. 2022;659:A65.
114. Fortenberry RC. J Phys Chem A. 2024;128:1555–1565.
115. Szabó A, Ostlund NS. Modern Quantum Chemistry. Dover Publications, Mineola. 1989.
116. Jensen F. Introduction To Computational Chemistry. 3rd ed. John Wiley & Sons, Chichester. 2017.
117. Souza CP, Fantuzzi F. In: Bakewell C, Costa N, Musgrave R, Owen G, editors. Organomet Chem. Vol. 45. 2024. pp. 271–316.
118. Andersson K, Malmqvist P-Å, Roos BO. J Chem Phys. 1992;96: 1218–1226.

119. Angeli C, *et al.* J Chem Phys. 2001;114:10252–10264.
120. Angeli C, Cimiraglia R, Malrieu J-P. Chem Phys Lett. 2001;350: 297–305.
121. Angeli C, Cimiraglia R, Malrieu J-P. J Chem Phys. 2002;117:9138–9153.
122. Riplinger C, *et al.* J Chem Phys. 2016;144:024109.
123. Mardirossian N, Head-Gordon M. Mol Phys. 2017;115:2315–2372.
124. Goerigk L, Grimme S. WIREs Comput Mol Sci. 2014;4:576–600.
125. Rimola A, *et al.* Int J Mol Sci. 2022;23:4252.
126. Londoño-Restrepo J, *et al.* Chem Sci. 2025;16:3051–3065.
127. Johansen SL, *et al.* ACS Earth Space Chem. 2024;8:1771–1783.
128. Wang L-P, *et al.* Nat Chem. 2014;6:1044–1048.
129. Senftle TP, *et al.* NPJ Comput Mater. 2016;2:15011.
130. Sushko GB, *et al.* Eur Phys J D. 2016;70:12.
131. Solov'yov IA, *et al.* J Comput Chem. 2012;33:2412–2439.
132. Solov'yov AV, *et al.* Chem Rev. 2024;124:8014–8129.
133. Sushko GB, Solov'yov IA, Solov'yov AV. Eur Phys J D. 2016;70:217.
134. Verkhovtsev AV, Solov'yov IA, Solov'yov AV. Eur Phys J D. 2021;75:213.
135. Solov'yov IA, *et al.* J Comput Chem. 2022;43:1442–1458.
136. Voter AF. In: Radiation Effects In Solids. Springer Netherlands, Dordrecht. 2007. pp. 1–23.
137. Andersen M, Panosetti C, Reuter K. Front Chem. 2019;7.
138. Biczysko M, Bloino J, Puzzarini C. WIREs Comput Mol Sci. 2018;8.
139. Puzzarini C, *et al.* Astrophys J. 2014;785:107.
140. Puzzarini C, *et al.* Astrophys J. 2014;792:118.
141. Jiang N, *et al.* Phys Chem Chem Phys. 2023;25:4754–4763.
142. Melli A, *et al.* Molecules. 2022;27:3278.
143. Cerqueira HBA, *et al.* J Phys Chem A. 2020;124:6845–6855.
144. Bauschlicher CW, *et al.* Astrophys J Suppl Ser. 2018;234:32.
145. Tielens A. Annu Rev Astron Astrophys. 2008;46:289–337.
146. Monfredini T, *et al.* Mon Not R Astron Soc. 2019;488:451–469.
147. Santos JC, *et al.* Mon Not R Astron Soc. 2022;512:4669–4682.
148. Woon DE. Mol Phys. 2024;122.
149. Lattelais M, *et al.* Astrophys J. 2009;696:L133–L136.
150. Kakkar R, Padhi BS. Int J Quantum Chem. 1998;58:389–398.
151. Fourré I, *et al.* Astron Astrophys. 2016;589:A18.
152. Belloche A, *et al.* Astron Astrophys. 2019;628:A10.
153. Elango M, *et al.* J Phys Chem A. 2010;114:9864–9874.
154. Belloche A, *et al.* Astron Astrophys. 2017;601:A49.
155. Simmie JM. J Phys Chem A. 2022;126:924–939.
156. Hollis JM, *et al.* Astrophys J. 2006;643:L25–L28.

157. Custer T, *et al.* J Phys Chem A. 2016;120:5928–5938.
158. Broten NW, *et al.* Astrophys J. 1984;276:L25.
159. San Andrés D, *et al.* Astrophys J. 2024;967:39.
160. Noriega L, *et al.* J Phys Chem A. 2024;128:9964–9971.
161. Ellinger Y, *et al.* Astron Astrophys. 2020;633:A49.
162. Hrodmarsson HR, *et al.* Astrophys J. 2024;964:26.
163. Shoji M, *et al.* Astrobiology. 2022;22:1129–1142.
164. Fantuzzi F, *et al.* Mon Not R Astron Soc. 2011;417:2631–2641.
165. Fantuzzi F, *et al.* Int J Quantum Chem. 2012;112:3303–3311.
166. Perlin AL, Wolff W, Oliveira RR. J Phys Chem A. 2023;127: 2481–2488.
167. Fantuzzi F, de Sousa DWO, Nascimento MAC. Comput Theor Chem. 2017;1116:225–233.
168. Wolff W, *et al.* J Phys Chem A. 2020;124:9261–9271.
169. Monfredini T, *et al.* Astrophys J. 2016;821:4.
170. Roithová J, *et al.* J Phys Chem A. 2006;110:2970–2978.
171. Fayolle EC, *et al.* Nat Astron. 2017;1:703–708.
172. Fantuzzi F, *et al.* J Am Chem Soc. 2018;140:4288–4292.
173. Fantuzzi F, *et al.* Phys Chem Chem Phys. 2019;21:24984–24992.
174. Quitián-Lara HM, *et al.* Mon Not R Astron Soc. 2020;499:6066–6083.
175. Cernicharo J, *et al.* Astron Astrophys. 2024;690:L13.
176. Balucani N, *et al.* Faraday Discuss. 2023;245:327–351.
177. Kaiser RI, *et al.* Nat Commun. 2022;13:786.
178. Rimola A, Sodupe M, Ugliengo P. Phys Chem Chem Phys. 2010;12:5285.
179. Enrique-Romero J, *et al.* Mon Not R Astron Soc. 2020;493: 2523–2527.
180. Richardson V, *et al.* Front Astron Space Sci. 2021;8.
181. Campisi D, Perrero J, Balucani N. 2024.
182. Lauretta DS, *et al.* Space Sci Rev. 2017;212:925–984.
183. Cuppen HM, *et al.* Proc Int Astron Union. 2017;13:293–304.
184. Furlan F, *et al.* Nat Photon. 2024.
185. Brandt JR, Salerno F, Fuchter MJ. Nat Rev Chem. 2017;1:0045.
186. Saranathan V, *et al.* Proc Natl Acad Sci USA. 2010;107:11676–11681.
187. Inaki M, Liu J, Matsuno K. Philos Trans R Soc B Biol Sci. 2016;371:20150403.
188. Quack M, Seyfang G, Wichmann G. Chem Sci. 2022;13:10598–10643.
189. Phys Chem Chem Phys. 2020;22:21570–21582.
190. Cameron RP, *et al.* Philos Trans R Soc A. 2017;375.
191. Glavin DP, *et al.* Chem Rev. 2020;120:4660–4689.
192. Muñoz Caro GM, *et al.* Nature. 2002;416:403–406.
193. Schutte WA. Adv Space Res. 2002;30:1409–1417.

194. Abplanalp MJ, Kaiser RI. Phys Chem Chem Phys. 2019;21:16949–16980.
195. Sallembien Q, *et al.* Chem Soc Rev. 2022;51:3436–3476.
196. Inoue Y, Ramamurthy V. Chiral Photochemistry. New York. 2004.
197. Stevenson KL, Verdieck JF. J Am Chem Soc. 1968;90:2974–2975.
198. Feringa BL, *et al.* Chem Rev. 2000;100:1789–1816.
199. De Marcellus P, *et al.* Astrophys J Lett. 2011;727:L27.
200. Modica P, *et al.* Astrophys J. 2014;788:79.
201. Flores JJ, Bonner WA, Massey GA. J Am Chem Soc. 1977;99:3622–3625.
202. Balavoine G, Moradpour A, Kagan HB. J Am Chem Soc. 1974;96:5152–5158.
203. Meierhenrich UJ, *et al.* Chem Biodiversity. 2010;7:1651–1659.
204. Meinert C, *et al.* Angew Chem Int Ed. 2014;53:210–214.
205. Nishino H, *et al.* Org Lett. 2001;3:921–924.
206. Nishino H, *et al.* J Am Chem Soc. 2002;124:11618–11627.
207. Meinert C, *et al.* Angew Chem Int Ed. 2012;51:4484–4487.
208. Bredehöft JH, *et al.* Chirality. 2014;26:373–378.
209. Bocková J, *et al.* Chirality. 2022;34:245–252.
210. Bocková J, *et al.* Nat Commun. 2023;14:3381.
211. Rikken GLJA, Raupach E. Nature. 1997;390:493–494.
212. Ishii K, Hattori S, Kitagawa Y. Photochem Photobiol Sci. 2020;19:8–19.
213. Rikken GL, Raupach E. Phys Rev E. 1998;58:5081–5084.
214. Kitagawa Y, Segawa H, Ishii K. Angew Chem Int Ed. 2011;50:9133–9136.
215. Cukras J, *et al.* Phys Chem Chem Phys. 2016;18:13267–13279.
216. Rosenberg RA. Symmetry. 2019;11:528.
217. Wu CS, *et al.* Phys Rev. 1957;105:1413–1415.
218. Vester F, Ulbricht TLV, Krauch H. Sci Nat. 1959;46:68.
219. Ulbricht TLV. Q Rev Chem Soc. 1959;13:48–60.
220. Bonner WA. Orig Life Evol Biosph. 1991;21:59–111.
221. Bonner WA. Orig Life. 1984;14:383–390.
222. Bonner WA. Chirality. 2000;12:114–126.
223. Burkov VI, *et al.* Orig Life Evol Biosph. 2008;38:155–163.
224. Gusev GA, *et al.* Orig Life Evol Biosph. 2008;38:509–515.
225. Xu Y, Greig D, Seddon E. Phys Rev B. 1997;55:11442.
226. Eib W, Alvarado SF. Phys Rev Lett. 1976;37:444.
227. Pierce DT, Meier F. Phys Rev B. 1976;13:5484.
228. Heinzmann U, Dil JH. J Condens Matter Phys. 2012;24:173001.
229. Heinzmann U. Physica Scripta. 1987;T17:77–88.
230. Koike K, Kirschner J. J Phys D Appl Phys. 1992;25:1139.

231. Penn DR, Apell SP, Girvin SM. Phys Rev B. 1985;8:15.
232. Solleder B, *et al.* Phys Rev B. 2007;76:075115.
233. Carmeli I, *et al.* J Chem Phys. 2003;118:10372–10375.
234. Naaman R, Vager Z. Acc Chem Res. 2003;36:291–299.
235. Naaman R, Vager Z. Phys Chem Chem Phys. 2006;8:2217–2224.
236. Ray SG, *et al.* Phys Rev Lett. 2006;96:036101.
237. Göhler B, *et al.* Science. 2011;331:894–897.
238. Naaman R, Paltiel Y, Waldeck DH. Chimia. 2018;72:394.
239. Bloom BP, *et al.* Chem Rev. 2024;124:1950–1991.
240. Naaman R, Paltiel Y, Waldeck DH. Nat Rev Chem. 2019;3:250–260.
241. Boudaïffa B, *et al.* Science. 2000;287:1658–1660.
242. Rosenberg RA, Abu Haija M, Ryan PJ. Phys Rev Lett. 2008;101:1–4.
243. Meinert C, *et al.* J Photochem Photobiol A. 2016;331:130–137.
244. Rosenberg RA, Mishra D, Naaman R. Angew Chem Int Ed. 2015;54:7295–7298.
245. Banerjee-Ghosh K, *et al.* Science. 2018;360:1331–1334.
246. Quack M. Angew Chem Int Ed. 2002;41:4618–4630.
247. Crassous J, *et al.* Org Biomol Chem. 2005;3:2218–2224.
248. Fiechter MR, *et al.* J Phys Chem Lett. 2022;13:10011–10017.
249. Cournol A, *et al.* Quantum Electron. 2019;49:288–292.
250. Darquié B, *et al.* Chirality. 2010;22:870–884.
251. Fábri CF, Horný L, Quack M. ChemPhysChem. 2015;16:3584–3589.
252. Albert S, *et al.* J Phys Chem Lett. 2016;7:3847–3853.
253. Albert S, *et al.* Phys Chem Chem Phys. 2017;19:11738–11743.
254. Soai K, *et al.* Nature. 1995;378:767–768.
255. Blackmond DG. Chem Rev. 2020;120:4831–4847.

https://doi.org/10.1142/9781800618183_bmatter

Index

G

H

I

K

L

M

N

www.ingramcontent.com/pod-product-compliance
Lightning Source LLC
Chambersburg PA
CBHW070941220226
39790CB00004B/4